T0226575

Developments in Soil Science – Volume 32

Chemical Bioavailability In Terrestrial Environments

Developments in Soil Science
Series Editors: A.E. Hartemink and A.B. McBratney

Developments in Soil Science – Volume 32

Chemical Bioavailability In Terrestrial Environments

Editor-in-Chief

Ravendra Naidu

Associate Editors

**N.S. Bolan, M. Megharaj, A.L. Juhasz,
S.K. Gupta, B.E. Clothier and R. Schulin**

ELSEVIER

Amsterdam • Boston • Heidelberg • London • New York • Oxford • Paris
San Diego • San Francisco • Singapore • Sydney • Tokyo

Elsevier
Radarweg 29, PO Box 211, 1000 AE Amsterdam, The Netherlands
Linacre House, Jordan Hill, Oxford OX2 8DP, UK

First edition 2008

Library of Congress Cataloging in Publication Data
A catalog record for this book is available from the Library of Congress

British Library Cataloguing in Publication Data
A catalogue record for this book is available from the British Library

ISBN: 978-0-444-52169-9
ISSN: 0166-2481

For information on all Elsevier publications
visit our website at books.elsevier.com

Transferred to Digital Printing, 2010
Printed and bound in the United Kingdom

Working together to grow
libraries in developing countries

www.elsevier.com | www.bookaid.org | www.sabre.org

ELSEVIER   BOOK AID
            International   Sabre Foundation

CONTENTS

LIST OF CONTRIBUTORS

Adriano, D. Savannah River Ecology Laboratory, PO Drawer E, Aiken, SC 29802, USA.

Anastasio, C. Department of Land, Air, and Water Resources, University of California, Davis, CA 95616, USA.

Anderson, C.W.N. Soil and Earth Science Group, Massey University, Palmerston North, New Zealand.

Bolan, N.S. Cooperative Research Centre for Contamination Assessment and Remediation of the Environment (CRC CARE), Centre for Environmental Risk Assessment and Remediation of the Environment, Mawson Lakes, South Australia 5095, Australia; Soil and Earth Sciences, Institute of Natural Resources, Massey University, Private Bag 11222, Palmerston North, New Zealand.

Clothier, B.E. Environment and Risk Management Group, HortResearch, PB 11-030, Palmerston North, New Zealand.

Crowe, S. Great Lakes Institute for Environmental Research, University of Windsor, Windsor, Ontario, Canada.

Curtin, D. Crop and Food, Christchurch, New Zealand.

de la Luz Mora, M. Departamento de Ciencas Químicas, Instituto de Agroindustria, Universidad de La Frontera, Temuco, Chile.

Deurer, M. Institute of Soil Science, University of Hannover, Hannover, Germany.

Diels, L. Flemish Institute for Technological Research (Vito), Department of Environmental Technology, Boeretang 200, B-2400 Mol, Belgium.

Fowle, D. Great Lakes Institute for Environmental Research, University of Windsor, Windsor, Ontario, Canada.

Fuentes, B. Programa de Doctorado en Ciencas de Recursos Naturales, Universidad de La Frontera, Temuco, Chile.

Gangaiya, P. Wollongong City Council, 41 Burelli Street, Locked Bag 8821, Wollongong, NSW 2500, Australia.

Geets, J. Limburgs Universitair Centrum Environmental Biology, Universitaire Campus, D-3590 Diepenbeek, Belgium; LabMET, Ghent University, Coupure Links 653, B-9000 Ghent, Belgium.

Gerson, A.R. Applied Centre for Structural and Synchrotron Studies, University of South Australia, Mawson Lakes, South Australia 5095, Australia.

Granel, Th. ENITA de Clermont Ferrand, Clermont Ferrand, France.

Green, S.R. Environment and Risk Management Group, HortResearch, PB 11-030, Palmerston North, New Zealand.

Guo, B. Department of Mechanical and Aeronautical Engineering, University of California, Davis, CA 95616, USA.

Gupta, S.K. Agroscope FAL Recknholz, Swiss Federal Research Station of Agroecology and Agriculture, Recknholzstrasse 191, 8046 Zürich, Switzerland.

Haemmann, M. Agroscope FAL Recknholz, Swiss Federal Research Station of Agroecology and Agriculture, Recknholzstrasse 191, 8046 Zürich, Switzerland.
Hedley, M.J. Soil and Earth Sciences, Institute of Natural Resources, Massey University, Private Bag 11222, Palmerston North, New Zealand.
Hooda, P.S. School of Earth Sciences and Geography, Kingston University London, KT1 2EE, UK.
Juhasz, A.L. Cooperative Research Centre for Contamination Assessment and Remediation of the Environment (CRC CARE) and Centre for Environmental Risk Assessment and Remediation (CERAR), University of South Australia, Mawson Lakes, South Australia 5095, Australia.
Kalis, E.J.J. Department of Soil Quality, Wageningen University, P.O. Box 8005, 6700 EC Wageningen, The Netherlands.
Keatinge, Z. School of Biology, Institute for Research on Environment and Sustainability, Devonshire Building, University of Newcastle-upon-Tyne, Newcastle-upon-Tyne, NE1 7RU, UK.
Kennedy, I. Department of Mechanical and Aeronautical Engineering, University of California, Davis, CA 95616, USA.
Kirkham, M.B. Department of Agronomy, Kansas State University, Manhattan, KS 66506-5501, USA.
Ko, B.G. Soil and Earth Science Group, Massey University, Palmerston North, New Zealand.
Krishnamurti, G.S.R. 313-855 West 16th A Street, North Vancouver, Vancouver, BC V7P 1R2, Canada.
Liphadzi, M.S. Water Linked Ecosystems, Water Research Commission, Gezina, Pretoria, South Africa.
Lombi, E. CSIRO Land and Water, Centre for Environmental Contaminants Research, PMB2 Glen Osmond, South Australia 506, UDS.
Lyons, B. Centre for Environment, Fisheries and Aquacultural Science (CEFAS), Lowestoft, Suffolk, NR33 0HT, UK.
Ma, Y. Croucher Institute for Environmental Sciences Department of Biology, Hong Kong Baptist University, Hong Kong, PR China.
Madrid, F. Instituto de Recursos Naturales y Agrobiología de Sevilla, Consejo Superior de Investigaciones Científicas, Apartado de Correos 1052.
Mahendra, N. Department of Chemistry, School of Pure and Applied Sciences, University of the South Pacific, Suva, Fiji.
Mahimairaja, S. Department of Environmental Science, Tamil Nadu Agricultural University, Coimbatore, India.
Marcus, M.A. Advanced Light Source, Lawrence Berkeley National Laboratory, Berkeley, CA 94720, USA.
Martin, R.R. Department of Chemistry, University of Western Ontario, Chemistry Building, 1151 Richmond Street, London, Ontario, N6A 5B7, Canada.
Megharaj, M. Centre for Environmental Risk Assessment and Remediation of the Environment South Australia, Cooperative Research Centre for

Contamination Assessment and Remediation of the Environment (CRC CARE), University of South Australia, Mawson Lakes, South Australia 5095, Australia.

Mills, T.M. Environment and Risk Management Group, HortResearch, PB 11-030, Palmerston North, 5301 New Zealand.

Naftel, S.J. Department of Chemistry, University of Western Ontario Chemistry Building, 1151 Richmond Street, London, Ontario, N6A 5B7, Canada.

Naidu, R. Cooperative Research Centre for Contamination Assessment and Remediation of the Environment (CRC CARE) and Centre for Environmental Risk Assessment and Remediation of the Environment (CERAR), Mawson Lakes Boulevard, Mawson Lakes, South Australia 5095, Australia.

Nelson, A.J. Department of Anthropology, University of Western Ontario, Chemistry Building, 1151 Richmond Street, London, Ontario, N6A 5B7, Canada.

Nico, P.S. Earth Sciences Division, Lawrence Berkeley National Laboratory, Berkeley, CA 94720, USA.

Owens, G. Cooperative Research Centre for Contamination Assessment and Remediation of the Environment (CRC CARE), Mawson Lakes, South Australia 5095, Australia.

Paktunc, D. Department of Geology, University of Kansas, 1475 Jayhawk Building, Lawrence, KS 66045, USA.

Pollard, S.J.T. Cooperative Research Centre for Contamination Assessment and Remediation of the Environment (CRC CARE), Mawson Lakes, South Australia 5095, Australia.

Pruszinski, A.W. Environment Protection Authority, 77 Grenfell Street, Adelaide, South Australia 5000, Australia.

Roberts, J.A. CANMET, Mining and Mineral Sciences Laboratories, 555 Booth Street, Ottawa, Canada.

Robinson, B.H. Environment and Risk Management Group, HortResearch, PB 11-030, Palmerston North, 5301 New Zealand.

Roembke, J. ECT Oekotoxikologie GmbH, Boettgerstrasse 2-14, D-65439 Floersheim am. Main, Germany.

Rowarth, J. University of Melbourne, Melbourne, Australia.

Saggar, S. Landcare Research, Private Bag 11052, Palmerston North, New Zealand.

Schulin, R. ETH Zurich Institute of Terrestrial Ecosystems, Universitaetstr. 16, CH-8092 Zurich, Switzerland.

Scotter, D.R. Environment and Risk Management Group, HortResearch, PB 11-030, Palmerston North, 5301 New Zealand; Institute of Natural Resources, Massey University, Palmerston North, New Zealand.

Semple, K.T. Department of Environmental Science, Lancaster University, Lancaster, LA1 4YQ, UK.

Sethunathan, N. Flat no. 103, Ushodaya Apartments, Sri Venkateswara Officers Colony, Ramakrishnapuram, Secunderbad 5000056, India.

Singleton, I. School of Biology, King George VI Building, University of Newcastle-upon-Tyne, Newcastle-upon-Tyne, NE1 7RU, UK; School of Biology, Institute for Research on Environment and Sustainability, Devonshire Building, University of Newcastle-upon-Tyne, NE1 7RU, UK.

Singh, N. Division of Agricultural Chemicals, Indian Agricultural Research Institute, New Delhi-110012, India.

Singh, J. Soil and Earth Sciences, Institute of Natural Resources, Massey University, Private Bag 11222, Palmerston North, New Zealand.

Speir, T.W. Institute of Environmental Science and Research, PO Box 50348, Porirua, New Zealand.

Taghavi, S. Brookhaven National Laboratory (BNL), Biology Department, Building 463, Upton, NY 11973-5000, USA.

Temminghoff, E.J.M. Soil Quality, Environmental Sciences Group, Department of Soil Quality, Wageningen University, PO Box 8005, 6700 EC Wageningen, The Netherlands.

Thayalakumaran, T. Environment and Risk Management Group, HortResearch, PB 11-030, Palmerston North, 5301 New Zealand; Institute of Natural Resources, Massey University, Palmerston North, New Zealand.

van der Lelie, D. Brookhaven National Laboratory (BNL), Biology Department, Building 463, Upton, NY 11973-5000, USA.

van der Velde, M. Institute for Biodiversity and Ecosystem Dynamics Physical Geography, University of Amsterdam, The Netherlands.

van der Zee, S.E.A.T.M. Soil Physics, Ecohydrology, and Groundwater Group Wageningen University, PO Box 47, 6700 AA Wageningen, The Netherlands.

Vangronsveld, J. Limburgs Universitair Centrum, Environmental Biology, Universitaire Campus, D-3590 Diepenbeek, Belgium.

Van Riemsdijk, W.H. Department of Soil Quality, Wageningen University, P.O. Box 8005, 6700 EC Wageningen, The Netherlands.

van Straalen, N.M. Vrije Universiteit, Institute of Ecological Science, De Boelelaan 1087, 1081 HV Amsterdam, The Netherlands.

Vogeler, I. Environment and Risk Management Group, HortResearch, PB 11-030, Palmerston North, 5301 New Zealand.

Weisener, C.G. Great Lakes Institute for Environmental Research, University of Windsor, Windsor, Ontario, Canada.

Weng, L. Department of Soil Quality, Wageningen University, P.O. Box 8005, 6700 EC Wageningen, The Netherlands.

Wenger, K. TECAN Schweiz AG, Seestrasse 103, CH-8708 Männedorf, Switzerland; Cornell University, Wing Hall, Ithaca, NY 14853, USA; Institute of Terrestrial Ecology, ETH Zurich, Grabenstrasse 3, CH-8952 Schlieren, Switzerland.

Werner, M.L. Department of Land, Air, and Water Resources, University of California, Davis, CA 95616, USA.

Wilson, S.C. School of Rural Science and Agriculture, University of New England, Armidale, NSW 2351, Australia.

Wong, M.H. Croucher Institute for Environmental Sciences, Department of Biology, Hong Kong Baptist University, Hong Kong, PR China.

Zaman, M. Summit-Quinphos NZ Ltd, PO Box 24-020, Royal Oak, New Zealand.

Zhang, H. Environmental Science Department, Lancaster University, Lancaster, LA1 4YQ, UK.

PREFACE

Remediation of contaminated sites using traditional technology is estimated to cost in excess of $US 1 trillion. These technologies operate within legislation that considers total contaminant content rather than the fraction that is free or bioavailable. Where contaminants are tightly bound by the soil and not bioavailable, as evidenced by bioavailability bioassays and other toxicity assay, exhaustive clean-up of soils may not be necessary as the contaminants may not pose a risk to end users. This approach to remediation is termed "risk-based land management" (RBLM) and is considered attractive as it may save millions of dollars in remediation costs. Despite the importance of bioavailability, there is still considerable controversy regarding its definition and also what constitutes the bioavailable fraction. This book brings together much of the current knowledge on bioavailability, discussing the fundamental principles governing RBLM and the application of chemical bioavailability towards using such an approach for managing contaminated land.

The book is divided into six sections including

- New concepts and definitions
- The role of chemical speciation in bioavailability
- Bioavailability and ecotoxicity of contaminants
- Bioavailability of nutrients and agrichemicals
- Tools for assessing bioavailability
- The role of bioavailability in risk assessment and remediation.

The initial chapters of the book focus on definitions of bioavailability in relation to environmental, ecological and human receptors including key concepts for each receptor. Bioavailability, as affected by speciation and its impact on ecotoxicity, is then discussed for both organic and inorganic contaminants. The following chapters investigate bioavailability implications for nutrients and chemicals in agricultural settings, while the final sections focus on tools for assessing bioavailability and the role of bioavailability in risk assessment, with special focus on RBLM.

This book is intended as a text for postgraduate students, as well as remediators and risk-assessment experts to understand the application of various conventional and innovative tools for assessing bioavailability and risks posed by chemicals at contaminated sites. It is also intended for regulatory authorities and environmental planners who wish to learn more about metal bioavailability, risk assessments and site remediation.

R. Naidu, N.S. Bolan, M. Megharaj, A.L. Juhasz,
S.K. Gupta, B.E. Clothier and R. Schulin

Developments in Soil Science, volume 32
Ravendra Naidu (Editor)
© 2008 Elsevier B.V. All rights reserved

Chapter 1

CHEMICAL BIOAVAILABILITY IN TERRESTRIAL ENVIRONMENTS

R. Naidu, N.S. Bolan, M. Megharaj, A.L. Juhasz, S.K. Gupta, B.E. Clothier
and R. Schulin

1.1 Introduction

Bioavailability refers to how much of a chemical is available to a living biota. However, the bioavailability of a chemical defines the relationship between the concentration of the chemical in the terrestrial environment and the level of the chemical that actually enters the receptor causing either positive or negative effect on the organism. Bioavailability is species-specific because the dose which reaches an organism's target organs or tissues, and results in a biological response, may vary among receptors (Stokes et al., 2005).

Bioavailability is also often referred to as the extent to which a chemical can desorb, dissolve, or otherwise dissociate from the environmental medium in which it occurs to become accessible (i.e. bioaccessibility) for absorption (Alexander and Alexander, 2000). Chemical bioavailability is now considered an important consideration in the environment because the availability of chemicals may be mitigated once the chemical comes in contact with the soil and sediment. For this reason, both fertility status of soils as well as risk assessment of contaminated sites requires quantification of chemical bioavailability much like any other parameter in a risk calculation (Hrudey et al., 1996). Where contaminated site assessment is concerned, bioavailability addresses the fundamental issue of exposure of the contaminant to a receptor. Biota may be negatively impacted following exposure to contaminants. However, exposure is not dictated by the total concentration of the contaminant in the environmental media, but rather by the fraction of the total concentration of the contaminant that is biologically available. Exposure is affected by the interaction of the non-sequestered fraction of the contaminant with an organism through its life history, which takes into account route and duration of exposure. Hence exposure can only occur following the release (e.g. desorption) of the contaminant from the soil particle and if the contaminant is then transported to the receptor (see Chapter 2).

The amount of a chemical in the terrestrial environment that is bioavailable depends on a variety of factors including the properties of both the chemical and the environmental media (Beyerle-Pfnur and Lay, 1990; Traina and Laperchie,

1999; Naidu et al., 2003). Soil plays a very significant role in reducing the potential bioavailability of contaminants in the environment. However, bioavailability is impacted not only by soil type, but also contact time (i.e. ageing) (Hatzinger and Alexander, 1995; Vig et al., 2002), which reduces the concentration of chemical available for uptake by organisms or which causes toxicity. There are a number of physico-chemical and biological processes occurring in soil systems which have a significant impact on the bioavailability of chemicals (see e.g. Naidu et al., 2003). The primary soil factors controlling the potential bioavailability of all contaminants are soil pH, cation or anion exchange capacity (depending upon available charged sites on soil surfaces), clay content, and soil organic matter. Chemical reactions, either abiotic or biologically mediated, can result in changes in chemical structure, which in turn may alter chemical solubility. Changes in the geochemical equilibrium, which can result due to oxidation–reduction reactions, complexation, or sorption, can result in the precipitation of a chemical due to a change in the chemical's structure and solubility.

For heavy metals, physico-chemical factors influencing dissolution and precipitation into the mineral phase ultimately determine their aqueous concentrations. Precipitation is a commonly occurring process and controls the solubility of many elements including aluminium, iron, manganese, magnesium, and calcium. Modern analytical techniques such as synchrotron have shown that precipitation of these metals occurs even when the ionic product of the bulk solution does not exceed the solubility products (see Chapter 7). The sorption of a chemical onto soil plays an important role in determining its bioaccessibility (see Chapter 2) and subsequent bioavailability. Metals in the solid or sorbed phase are usually unavailable and the uptake of metals by an organism usually occurs when metals are soluble and remain in soil solution (pore water). Thus, processes influencing the equilibrium between soluble and solid or bound phases for heavy metal impact their overall bioavailability.

Organic chemicals exist in soils in free or bound form; "free" indicates that the compound can be extracted from soil without altering chemical structures, while "bound" (i.e. recalcitrant) chemicals are those resistant to extraction (Bollag and Huang, 1998; see Chapter 10). The sorption/desorption process often results in sequestering organic compounds within the soil particles, resulting in an overall decrease in bioavailability of the sorbed compound (Kelsey and Alexander, 1997). Unlike heavy metals, most organic chemicals undergo degradation, which is an important factor controlling bioavailability. Long-term sequesteration can be altered by changes in the physical–chemical–biological environment of soil systems (Chung and Alexander, 1999). Changes in redox state have been shown to effect both sorption/desorption and degradation reactions of organic chemicals. Oxidation of soil humic material can be responsible for releasing organic contaminant, which were previously bound to the soil organic fraction. Since contaminants immobilized in soil/sediment systems may be released back to the aqueous environment, one of the key issues concerns the long-term stability of non-bioavailable fraction.

While there are standard procedures for assessing chemical bioavailability of nutrients in soil (e.g. Olsen P test) to manage sustainable production and sustainable food crop quality, there are no standardized approaches for evaluating contaminant bioavailability (Naidu et al., 2003). The presence of multiple chemicals further complicates the evaluation of bioavailability as the evaluation of bioavailability for a single contaminant may not reflect the true risk posed by contaminants to the biota. The dynamic nature of the interactions between soil and most contaminants (especially organic contaminants) further complicates the efforts in establishing standard procedures for assessing their bioavailability, although it is not impossible to develop assays that indicate chemical bioavailability in soils.

Since bioavailability is influenced by factors such as the properties of the media (e.g. soil), chemical form, and duration of exposure (Reible et al., 1999; Naidu et al., 2003; Krishnamurti et al., 2004), an evaluation of bioavailability and its incorporation into ecological risk assessment must consider site-specific factors particularly when conducting contaminated site assessments. Consideration of bioavailability during risk assessment is important because the bioavailability of a contaminant is inversely related to risk-based clean-up levels. This means that the lower a contaminant's bioavailability, the higher the risk-based clean-up levels will be. Considering bioavailability in risk assessment can provide a better determination of risk (e.g. less uncertainty), which in turn may result in a significantly lower risk-management cost than would have occurred without its consideration. As the shift to risk-based clean-up continues, it is critical to establish "How clean is clean?" or "When is a site clean enough?". Although the concepts of bioavailability are well established, incorporating these principles into risk assessments and the regulatory framework has proven problematic. The requirement for "zero tolerance" of a specific contaminant in soil is incompatible with the existing literature, which has shown that the toxicity of many contaminants decreases with time. Current regulatory decisions often assume that the amount of chemical detected equals the amount which could cause an effect to a receptor. Ignoring bioavailability has often resulted in conservative estimates of transfer of contaminants from soils to the biota. The question then becomes "How to incorporate bioavailability when setting soil limits for regulatory purposes?" or alternatively, "What concentration of a chemical poses a significant risk?". This decision process has significant implications when establishing endpoints for remediation goals for soil since such endpoints will ultimately determine the removal/treatment volume and cost of remediation. In an excellent report titled "How clean is clean", Tomson and Tan (1997) reported that there is a point in soil remediation when it is acceptable and even preferable to leave contamination remaining in the soil given that contaminants bound tightly to soil may never desorb. Thus, bioavailability of chemicals in soils can be manipulated using soil amendments to either manage remediation of contaminated sites or enhance nutrient availability. For instance, in strongly weathered and oxidic soils a very large amount of phosphate fertilizer may remain unavailable to plants unless the soil properties are manipulated to enhance P release (Bolan et al., 2003a). Likewise, metal contaminated soils are managed for risk by chemical immobilization of metals using phosphate compounds (Berti and Cunningham, 1997; Bolan et al., 2003b). Increased bioavailability of aged phenanthrene in soils have also been reported by competitive displacement with pyrene (White et al., 1999) suggesting that considerable caution must be emphasized while developing strategies to manage contaminated soils.

Soil scientists appear to have concentrated on bioavailability as being a process mediated by chemical exchange mechanisms between the soil solution and the matrix. Biologists, on the other hand, have sought to understand bioavailability from the perspective of the roots and biota themselves. A convergence of approaches has occurred as a result of improved measurement devices and monitoring techniques for observing transport, exchange, and uptake processes in the rootzone, and this has been complemented by the rapid development of quantitative modelling tools that can describe the complexity of biophysical mechanisms that operate in the soil.

This measurement-modelling dualism attempts to decipher the mechanisms, and genetic variation in the controls on bioavailability, transport and plant uptake of both metals, plus organic and inorganic contaminants in soil. Whereas the theory describing water flow and chemical transport through unsaturated soil, in the presence of distributed uptake of both by roots, has long been defined, it has been the phenomenal growth in computing power for numerically solving these coupled, non-linear, partial differential equations that has led to readily available packages for quantitatively predicting solute movement and uptake in the rootzone. Comprehensive mechanistic models can now provide the biophysics researcher with detailed predictions of chemical status, solute mobility, bioavailability, and plant uptake under realistic boundary conditions. This surge in model availability has been complemented by the development of new measurement devices that are capable of measuring rootzone processes on spatial and temporal scales unthought of just a decade ago. Nonetheless, problems of model verification and process parameterization remain. Recent work not only highlights the predictive power of mechanistic models, but it serves to identify remaining difficulties in unravelling the plant-controlled and soil-based connections that determine the bioavailability of chemicals in the rootzone.

Nuttle (2000) posed the question "... can ecosystem managers rely on mechanistic simulation models to guide their decisions?" He warned that they "... are vulnerable as long as managers regard the mechanistic numerical models as the primary tool for synthesising scientific information [but they] can reduce this vulnerability by relying on other components of the whole-system approach. Observation, experiment and modelling together are the essential components of the whole system approach". Perspicacious experimentation using novel devices, in tandem with comprehensive modelling will lead to new understanding of what controls the bioavailability, uptake, transport, and fate of chemicals in soil.

1.2 Conclusion

Since chemical bioavailability was first introduced as central to the assessment of contaminated sites and also for predicting fertilizer requirements of soils, much research has been conducted worldwide to investigate key factors influencing bioavailability. In this book, an attempt is made to present an overview of the methods used to estimate chemical bioavailability along with the implications of bioavailability to ecological and human health risk assessment. The role of bioavailability in risk-based land management is also reviewed to provide context to current shift towards this cost-effective means of managing contaminated sites. Although chemical bioavailability has been the central feature of scientific research and often the subject of much discussion amongst regulators and industries, much more needs to be done for it to be accepted in the risk-assessment framework.

References

Alexander, R.R., Alexander, M., 2000. Bioavailability of genotoxic compounds in soils. Environ. Sci. Technol. 34, 1589–1593.

Berti, W.R., Cunningham, S.D., 1997. In-place inactivation of Pb in Pb contaminated soils. Environ. Sci. Technol. 31, 1359–1364.

Beyerle-Pfnur, R., Lay, J.P., 1990. Adsorption and desorption of 3,4-dichloroaniline on soil. Chemosphere 21, 1087–1094.

Bolan, N.S., Adriano, D.C., Curtin, D., 2003a. Soil acidification and liming interactions with nutrient and heavy metal transformation and bioavailability. Adv. Agron. 78, 216–272.

Bolan, N.S., Adriano, D.C., Naidu, R., 2003b. Role of phosphorus in (im)mobilization and bioavailability of heavy metals in the soil-plant system. Rev. Environ. Contam. Toxicol. 177, 1–44.

Bollag, J.M., Dec, J., Huang, P.M., 1998. Formation mechanisms of complex organic structures in soil habitats. Adv. Agron. 63, 237–266.

Chung, N., Alexander, M., 1999. Effect of concentration on sequestration and bioavailability of two polycyclic aromatic hydrocarbons. Environ. Sci. Technol. 33, 3605–3608.

Hatzinger, P.B., Alexander, M., 1995. Effect of aging of chemicals in soil on their biodegradability and extractability. Environ. Sci. Technol. 29, 537–545.

Hrudey, S.E., Chen, W., Rousseaux, C.G., 1996. Bioavailability in Environmental Risk Assessment. Lewis Publishers, Boca Raton, FL.

Kelsey, J.W., Alexander, M., 1997. Declining bioavailability and inappropriate estimation of risk of persistent compounds. Environ. Toxicol. Chem. 16, 582–585.

Krishnamurti, G.S.R., Megharaj, M., Naidu, R., 2004. Bioavailability of cadmium-organic complexes to soil alga – an exception to free ion model. J. Agric. Food Chem. 52, 3894–3899.

Naidu, R., Rogers, S., Gupta, V.V.S.R., Kookana, R.S., Bolan, N.S., Adriano, D., 2003. Bioavailability of metals in the soil-plant environment and its potential role in risk assessment: An overview. In: R. Naidu, S. Rogers, V.V.S.R. Gupta, R.S. Kookana, N.S. Bolan and D. Adriano (Eds.), Bioavailability and its Potential Role in Risk Assessment. Science Publishers, New York, pp. 21–59.

Nuttle, W.K., 2000. Ecosystem managers can learn from past successes. Eos 81(25), 278, 284, American Geophysical Union, Washington, DC.

Reible, D.D., Chaney, R.L., Hughes, J.B., 1999. Bioavailability. In: W.C. Anderson, R.C. Loehr and B.P. Smith (Eds.), Environmental Availability in Soils. American Academy of Environmental Engineers, Washington, DC.

Stokes, J.D., Paton, G.I., Semple, K.T., 2005. Behaviour and assessment of bioavailability of organic contaminants in soil: relevance for risk assessment and remediation. Soil Use Manage. 21, 475–486.

Tomson, M., Kan, A., 1997. How clean is clean enough if contaminants are irreversibly adsorbed? South and Southwest hazardous substance Research Centre Research Brief No. 16, 2pp.

Traina, S.J., Laperchie, V., 1999. Contaminant bioavailability in soils, sediments, and aquatic environment. Proc. Natl. Acad. Sci. 96, 3365–3371.

Vig, K., Megharaj, M., Sethunathan, N., Naidu, R., 2002. Bioavailability and toxicity of cadmium to microorganisms and their activities in soil: a review. Adv. Environ. Res. 8, 121–135.

White, J.C., Hunter, M., Pignatello, J.J., Alexander, M., 1999. Increase in bioavailability of aged phenanthrene in soils by competitive displacement with pyrene. Environ. Toxicol. Chem. 18, 1728–1732.

A: Bioavailability – new concepts and definitions

Considerable controversy exists in the literature related to 'what constitutes the bioavailable fraction'; including the definition itself and the methods used for its measurements. For instance; microbiologists often regard the concentration that can induce a change either in morphology or physiology of an organism as the bioavailable fraction; plant scientists regard the plant available fraction as the bioavailable fraction. Consequently, terms such as 'bioavailable', phytoavailable' and 'available' are used by researchers, regulators and the environment management industry. Thus there is no single adequate definition of bioavailability yet! Moreover, chemists and plant scientists have often used a single chemical extraction as an index of bioavailability, assuming that bioavailability is a static phenomenon, while it may change with time. This section presents an overview of current thinking on bioavailability; its definition, cutting edge research in speciation and advancement in tools for assessing chemical bioavailability in the terrestrial environment.

Developments in Soil Science, volume 32
Ravendra Naidu (Editor)

Chapter 2

CONTAMINANT CHEMISTRY IN SOILS: KEY CONCEPTS AND BIOAVAILABILITY

R. Naidu and N.S. Bolan

2.1 Introduction

Contamination of soils and water from improper disposal of hazardous industrial and municipal wastes has long been recognized as an environmental issue of public concern, regulatory activity, and scientific investigation. Such disposals have resulted in the occurrence of in excess of 80,000 potentially contaminated sites in Australia (Natusch, 1997) and over 3 million such sites in the Asia region (Naidu et al., 2003). In the United States, there are estimated to be about 400,000 waste disposal sites where soil and groundwater contamination is deemed to be of sufficient extent and magnitude that some type of remedial action is warranted to protect public health or to minimize adverse environmental and ecological impacts (Rao et al., 1996; USEPA, 2004). Contaminants can have a range of long- and short-term impacts on human and animal health through various exposure pathways including ingestion of contaminated soil, inhalation of vapours, gases or dust, skin contact, ingestion through food (plants and animals) and drinking contaminated surface or groundwater, and for this reason restricted use or remediation of contaminated sites is often recommended (USEPA, 1994, http://www.epa.gov/safewater/mcl.html).

However, in many countries including Australia, remediation is not a compulsory act established by their relevant environment protection authorities since the clean-up process is governed by both the risk that it poses to human health (depending on land use) and the potential commercial value of the site. This is in keeping with the position paper written by the Australian and New Zealand Environment and Conservation Council (1992) that states

> Governments should not intervene to direct that remedial action be taken in the case of contaminated sites where the existing land poses no threat to human health or the environment. Where it is intended to put land in a more sensitive use for which the present level of contamination poses unacceptable risks, the requisite standard of clean up should be achieved in a manner determined by the owner or developer.

Remediation of contaminated sites to the extent practicable is expected to cost approximately $5–8 billion in Australia (Powell, 1992; CMPS&F – Environment Australia, 1997) alone, while it may exceed 1 trillion dollars in USA (Rao et al., 1996; USEPA, 2004). Despite much effort towards clean up of contaminated sites, it is now recognized by scientists and environmental managers that the required regulatory levels may not be achieved at most sites, even after several decades of clean-up efforts using the existing technologies. Remediation is a complex process requiring site assessment, risk characterization and removal of contaminants or minimizing their entry to food chain from the contaminated matrix. While the remediation endpoints required by regulators, the scope of monitoring pro-grammes, and the assignment of legal/financial liability for remediation efforts may vary greatly from country to country, the public perception of the relative risks of soil and groundwater contamination (in comparison to other hazards) can influence the regulatory policy, and levels of allowed contamination as well as the expectations of the required clean up of contaminated sites.

The decision to remediate is often based on the total contaminant loading in soil and groundwater. This is still the case in many countries including Australia. However, there is increasing evidence that as contaminants persist in the environment (see Chapters 3–5, 21 and 27), they become increasingly less available for uptake by organisms, for exerting toxic effects, and for biodegrada-tion and bioremediation by microorganisms. These changes in the impact of contaminants to ecosystem have been attributed to a decline in the labile fraction of contaminant available for causing such adverse impacts. The fraction of contaminant that is mobile and available for uptake by plants, animals and human is often defined as the bioavailable fraction. However, the definition of bioavailability varies both amongst the scientists and the origin of their discipline. Readers interested in the definition of bioavailability are referred to Chapter 3.

The amount of a contaminant in soil that is bioavailable depends on a variety of factors including the properties of both the contaminant and the soil environment (Adriano et al., 2004). Bioavailability defines the relationship between the concentration of a contaminant in soil and the amount of the contaminant that actually enters biological receptors, and thus may cause harm (Reichenberg and Mayer, 2006). Consideration of bioavailability during risk assessment is important given that the mobility and the subsequent availability of a contaminant is inversely related to risk-based clean-up levels (see Chapter 4). This means that the risk-based clean-up level of a contaminant increases with its decreasing bioavailability. The risks contaminants pose to the environment can be determined adequately with minimum uncertainty by giving consideration to bioavailability in hazard characterization and remedia-tion goals based on bioavailability. Such an approach is likely to result in significant decreases in risk-management costs. Similar risk-based strategy linking contaminant bioavailability was used by Naidu and his co-researchers (Naidu et al., 2003a, 2003b; see Chapter 4) to assess and manage an arsenic-contaminated site in Australia. The total cost of implementing this strategy was reduced significantly compared to excavation and transport of contaminated soils to prescribed landfill. In this chapter, following a brief overview of the nature and sources of contaminants, we outline the various biogeochemical processes controlling the fate and dynamics of contaminants in the soil environment and its implications to contaminant bioavailability. Readers are directed to Chapter 3 for a detailed review of the definition of bioavailability and techniques for assessing soil bioavailability.

2.2 Nature and sources of contaminants

The main activities contributing to contamination include industrial, mining, agricultural, commercial activities as well as transport and services (Table 2.1). Uncontrolled mining, manufacturing and disposal of wastes inevitably cause environmental pollution. Military land and land used for recreational shooting are also important sites of contamination. The contaminants associated with such activities are listed in Table 2.1. The legacy of contamination at many of these sites appears to have resulted because of lax regulatory measures prior to the establishment of legislation protecting the environment.

As shown in Table 2.1, the nature of contaminants present in the environment can range from toxic heavy metal(loid)s to persistent organic pollutants. The interactions of contaminants with soil components depend on both the soil properties and the intrinsic properties of the contaminants. Metal(loid)s can exist as either cations (heavy metals such as Cd, Cu, Zn and Pb) or anions (metalloids such as Cr, As) in the soil environment, which significantly affects their sorption, mobility and solubility in soils. For instance, the sorption of cationic metals increases with increasing pH while that of anionic metals (e.g., Cr(VI)) decreases with increasing pH (Adriano et al., 2004). However, for organics (such as DDT), lipophilicity and persistence alter their bioavailability, as well as ionic potential in the case of organic contaminants with ionizable functional groups (such as atrazine) (Aislabie and Lloydjones, 1995).

2.2.1 Metal(loid)s

Both soil solid properties and soil solution composition determine the dynamic equilibrium between metal(loid)s in solution and the soil solid phase. The soil solution composition may also be influenced in the field through continuous depletion of solutes by plant uptake, surface run-off and leaching. The soil solution concentration of metal(loid)s is influenced by the nature of both organic (citrate, oxalate, fulvic, dissolved organic carbon (DOC)) and inorganic ($H_2PO_4^-$, NO_3^-,

Table 2.1. *Industries, land uses and associated chemicals.*

Industry	Type of chemical	Associated chemicals
Agricultural/ horticultural activities		See fertilizer, insecticides, fungicides, herbicides under chemicals manufacture and use
Airports	Hydrocarbons metals	Aviation fuels particularly aluminium, magnesium, chromium
Asbestos production and disposal		Asbestos
Battery manufacture and recycling	Metals	Lead, manganese, zinc, cadmium, nickel, cobalt, mercury, silver, antimony
	Acids	Sulphuric acid
Breweries/distilleries	Alcohol	Ethanol, methanol, esters
Chemicals manufacture and use	Acid/alkali	Mercury (chlor/alkali), sulphuric, hydrochloric and nitric acids, sodium and calcium hydroxides
	Adhesives/resins	Polyvinyl acetate, phenols, formaldehyde, acrylates, phthalates
	Dyes	Chromium, titanium, cobalt, sulphur and nitrogen organic compounds, sulphates, solvents
	Explosives	Acetone, nitric acid, ammonium nitrate, pentachlorophenol, ammonia, sulphuric acid, nitroglycerine, calcium cyanamide, lead, ethylene glycol, methanol, copper, aluminium, *bis*(2-ethylhexyl) adipate, dibutyl phthalate, sodium hydroxide, mercury, silver
	Fertilizer	Calcium phosphate, calcium sulphate, nitrates, ammonium sulphate, carbonates, potassium, copper, magnesium, molybdenum, boron, cadmium
	Flocculants	Aluminium
	Foam production	Urethane, formaldehyde, styrene
	Fungicides	Carbamates, copper sulphate, copper chloride, sulphur, chromium
	Herbicides	Ammonium thiocyanate, carbamates, organochlorines, organophosphates, arsenic, mercury
	Paints	
	Heavy metals	Arsenic, barium, cadmium, chromium, cobalt, lead, manganese, mercury, selenium, zinc
	General	Titanium dioxide
	Solvent	Toluene, oils natural (e.g., pine oil) or synthetic
	Pesticides	Arsenic, lead, organochlorines,
	Active ingredients	organophosphates, sodium, tetraborate,
	Solvents	carbamates, sulphur, synthetic pyrethroids, xylene, kerosene, methyl isobutyl ketone, amyl acetate, chlorinated solvents

Table 2.1. (*Continued*)

Industry	Type of chemical	Associated chemicals
	Pharmacy	Dextrose, starch
	General Solvents	Acetone, cyclohexane, methylene chloride, ethyl acetate, butyl acetate, methanol, ethanol, isopropanol, butanol, pyridine methyl ethyl ketone, methyl isobutyl ketone, tetrahydrofuran
	Photography	Hydroquinone, pheidom, sodium carbonate, sodium sulphite, potassium bromide, monomethyl paraaminophenol sulphates, ferricyanide, chromium, silver, thiocyanate, ammonium compounds, sulphur compounds, phosphate, phenylene diamine, ethyl alcohol, thiosulphates, formaldehyde
	Plastics	Sulphates, carbonates, cadmium, solvents, acrylates, phthalates, styrene
	Rubber	Carbon black
	Soap/detergent General	Potassium compounds, phosphates, ammonia, alcohols, esters, sodium hydroxide, surfactants (sodium lauryl sulphate), silicate compounds
	Acids	Sulphuric acid and stearic acid
	Oils	Palm, coconut, pine, tea tree
	Solvents	Ammonia
	General	e.g., BTEX (benzene, toluene, ethylbenzene, xylene)
	Hydrocarbons Chlorinated organics	e.g., trichloroethane, carbon tetrachloride, methylene chloride
Defence works		See explosives under chemicals manufacture and use, foundries, engine works, service stations
Drum reconditioning		See chemicals manufacture and use
Dry cleaning		Trichlorethylene and ethane Carbon tetrachloride Perchlorethylene
Electrical		PCBs (transformers and capacitors), solvents, tin, lead, copper
Engine works	Hydrocarbons Metals Solvents Acids/alkalis Refrigerants Antifreeze	Ethylene glycol, nitrates, phosphates, silicates
Foundries	Metals	Particularly aluminium, manganese, iron, copper, nickel, chromium, zinc, cadmium and lead and oxides, chlorides, fluorides and sulphates of these metals

Table 2.1. (*Continued*)

Industry	Type of chemical	Associated chemicals
Gas works	Acids	Phenolics and amines Coke/graphite dust
	Inorganics	Ammonia, cyanide, nitrate, sulphide, thiocyanate
	Metals	Aluminium, antimony, arsenic, barium, cadmium, chromium, copper, iron, lead, manganese, mercury, nickel, selenium, silver, vanadium, zinc
	Semivolatiles	Benzene, ethylbenzene, toluene, total xylenes, coal tar, phenolics and PAHs
Iron and steel works		Metals and oxides of iron, nickel, copper, chromium, magnesium and manganese, and graphite
Landfill sites		Methane, hydrogen sulphides, heavy metals, complex acids
Marinas		Engine works, electroplating under metal treatment
Metal treatments	Antifouling paints	Copper, tributyltin (TBT)
	Electroplating Metals	Nickel, chromium, zinc, aluminium, copper, lead, cadmium, tin
	Acids	Sulphuric, hydrochloric, nitric, phosphoric
	General	Sodium hydroxide, 1,1,1-trichloroethane, tetrachloroethylene, toluene, ethylene glycol, cyanide compounds
	Liquid carburizing baths	Sodium, cyanide, barium, chloride, potassium chloride, sodium chloride, sodium carbonate, sodium cyanate
Mining and extractive industries		Arsenic, mercury and cyanides and also refer to explosives under chemicals, manufacture and use
Power stations		Asbestos, PCBs, fly ash, metals
Printing shops		Acids, alkalis, solvents, chromium (see photography under chemicals, manufacture and use)
Railway yards		Hydrocarbons, arsenic, phenolics (creosote), heavy metals, nitrates and ammonia
Scrap yards		Hydrocarbons, metals, solvents
Service stations and fuel storage facilities		Aliphatic hydrocarbons BTEX (i.e., benzene, toluene, ethylbenzene, xylene) PAHs (e.g., benzo(a) pyrene) Phenols Lead

Table 2.1. (*Continued*)

Industry	Type of chemical	Associated chemicals
Sheep and cattle dips		Arsenic, organochlorines and organophosphates, carbamates, and synthetic pyrethroids
Smelting and refining		Metals and the fluorides, chlorides and oxides of copper, tin, solver, gold, selenium, lead, aluminium
Tanning and associated trades	Metals	Chromium, manganese, aluminium
	General	Ammonium sulphate, ammonia, ammonium nitrate, phenolics (creosote), formaldehyde, tannic acid
Wood preservation	Metals	Chromium, copper, arsenic
	General	Naphthalene, ammonia, pentachlorophenol, dibenzofuran, anthracene, biphenyl, ammonium sulphate, quinoline, boron, creosote, organochlorine pesticides

Cl^- and SO_4^{2-}) ligand ions through their influence on metal complexation/ precipitation processes (Shuman, 1986; Homann and Zasoski, 1987; Naidu et al., 1994; Harter and Naidu, 1995; Naidu and Harter, 1997; Bolan et al., 1999a). Metals can interact with inorganic soil constituents, e.g., phosphates, carbonates, sulphates, hydroxides, sulphides, to form either precipitates or positively charged complexes (Bolan et al., 2003a). Both complexation and precipitation reactions are pH-dependent (see Section 2.4). Therefore, although metals can form complexes with a net negative charge, under most environmentally relevant scenarios (pH = 4–8.5), these elements either precipitate or exist as cationic species.

2.2.2 Organic contaminants

The fate and behaviour of organic contaminants depend on a variety of processes including sorption–desorption, volatilization, chemical and biological degradation, plant uptake, surface run-off and leaching. Sorption–desorption and degradation are perhaps the two most important processes as the bulk of the chemicals is either sorbed by organic and inorganic soil constituents, and chemically or microbially transformed/degraded. The degradation process is not always a detoxification process. This is because in some cases the transformation or degradation process leads to intermediate products that are more mobile, more persistent or more toxic to non-target organisms (e.g., conversion of DDT into DDE; Boul, 1995). The relative importance of these processes is determined by the chemical nature of the compound (Aislabie et al., 1995). For ionizable organics, sorption–desorption reactions are governed by all factors that govern the surface chemical properties of soils and the dissociation of charged organics. Thus, sorption of ionizable organic compounds with acidic characteristics increases with increasing pH while the converse would be true for basic organic compounds (Naidu et al., 1998).

The sorption of non-ionic organic compounds (NOCs) to soil is primarily related to their hydrophobicity and the amount of soil organic matter (La Grega et al., 1994; Lee et al., 1990; Boivin et al., 2005), with the exception of the more polar, nitro-substituted organic contaminants (i.e., the explosives). Differences in the sorption of several NOCs in diverse soil- and sediment-water systems have been minimized by normalization of their sorption to organic matter or more specifically organic carbon (OC) with OC-normalized distribution or partition coefficients, referred to as K_{oc} values (e.g., Lyman et al., 1990; Gertsl, 1990; Wauchope et al., 2002). The greater the affinity of a contaminant for organic matter, the larger the K_{oc}, and the soil with higher amounts of organic matter has a higher propensity to sorb NOCs. The hydrophobicity of organic compounds, thus the K_{oc}, increases with the size of the compound and with increasing chlorine content, in the case of chlorinated organics (Wauchope et al., 2002). Therefore, sorption by soils of polycyclic aromatic hydrocarbons (PAHs) increases with the number of aromatic rings. For compounds like PCBs, sorption increases with increasing chlorination. Increasing compound hydrophobicity also reflects increasing lipophilicity, which will result in a greater propensity to bioaccumulate in the lipid fraction of biota (Krauss and Wilcke, 2005; Muller et al., 2001).

Although organic matter is the primary sorption domain in soils, all contaminants have some affinity to be associated with any reactive surface through weak physical forces (Schwarzenbach et al., 1993; Baskaran et al., 1996). In addition, the nitro-substituted NOCs have been shown to have specific interactions with clay surfaces that are impacted by the inorganic cations present and clay charge density, and less so by the amount of organic matter present (Weissmahr et al., 1998, 1999).

2.3 Contaminant interactions in soil

Following input into the soil environment contaminants interact with soil solid phase via a series of sorption–desorption (all ionic and non-ionic solutes), precipitation–dissolution (polar and ionic compounds) reactions including physical migration into subatomic pores and diffusion into solid phase (Fig. 2.1).

Once the chemical substance is released into soil solution (1) it crosses the solution–solid phase boundary (2) following which it is transported into a liquid-filled soil macropores (3). During transport, the chemical substance interacts with colloid surface via either columbic interaction (non-specific

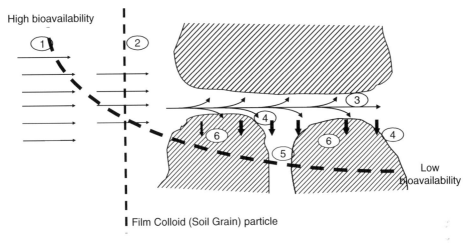

Figure 2.1. *Fate of chemicals in soil environment (modified from Sparks, 1998).*

sorption) or covalent bonding (specific sorption) (4). This leads to the binding of the chemical to the surface of soil colloids (4). Once bound to the solid surface, the chemical substance redistributes between soil solution and solid mineral and organic phases. This redistribution of the solute as quantified by its distribution constant, K_d, is a function of both specific and non-specific sorption (Eq. (2.1)), with the soil solution concentration being governed by environmental factors (including contaminant loading), all of which influencing the above processes.

$$K_d = f(\text{non-specific})(\text{specific binding}) \qquad (2.1)$$

With ageing, the initially surface-sorbed chemicals are slowly redistributed to interior of soil aggregates including the subatomic pores (5, Fischer et al., 1996) with a small proportion diffusing into the solid particles (6, Barrow et al., 1989) from which desorption becomes slow and often difficult and hence contaminants become inaccessible to microorganisms (Karickhoff, 1980). Thus the sorption of a chemical onto soil plays an important role in determining its mobility and bioavailability. Ageing of contaminated soils results in sequestering compounds within the soil particles, resulting in an overall decrease in bioavailability of the sorbed compound (Manilal and Alexander, 1991). There is evidence that sequestration of certain organic chemicals with time results in significant decline of their bioavailability. These sequestered organic compound are known as bound or recalcitrant compounds. For instance, ethylene dibromide, a soil fumigant with relatively high water solubility, volatility and biodegradability was reported to persist up to 19 years after its last application (Steinberg et al., 1987, 2003).

Similarly, DDT has been shown to persist in soils for a long period (Boul, 1995). While inorganic chemicals such as heavy metal(loid)s often exist in bound form (unless present in concentrations that saturate the binding sites), organic chemicals exist in soils in both free or bound form; "free" indicates that the compound can be extracted from soil without altering chemical structures, whereas "bound" chemicals are those resistant to extraction.

Enzymes and certain minerals can influence oxidation of organic compounds to free radicals, which then results in coupling the contaminant to the organic fraction of soils. For example, enzymes found in soil fungi (e.g., laccases) have been shown to oxidize phenolic compounds and initiate the polymerization process (Hatzinger and Alexander, 1995). This results in the oxidative coupling of the organic contaminant to organic fraction, resulting in the incorporation of these material (covalently) into humic material (Farenhorst, 2006). Studies have shown that certain pesticides become covalently bound to the humic fraction of soils and the bioavailability of these humic-bound fragments were practically negligible (Kelsey and Alexander, 1997; Tao et al., 2004).

Contaminants present in soil solution (porewater) are the most bioavailable fraction and it is this fraction to which organisms, plants and other species are directly in contact with, thereby having direct access for uptake. It is this fraction that is mobile, thereby resulting in the leaching through soil to contaminate groundwater. Often the soil solution fraction is defined by soil scientists as the "intensity factor'. In long-term contaminated soils, chemicals present in soil solution reach equilibrium with the most labile fraction (i.e., chemicals non-specifically bound (exchangeable), Eq. (2.2) – 'capacity factor'). As chemicals present in porewater deplete, they are replenished through sequential release from other pools (specific and non-specifically sorbed fraction, see Fig. 2.2) with the rate of replenishment (buffering) being governed by the partition coefficient (k_d) controlling the equilibrium between the different chemical-binding substrates (Fig. 2.2 and Eq. (2.2)).

Metal(loid)s in the solid or sorbed phase are usually unavailable and the uptake by an organism usually occurs when these substances become soluble and reach the solution phase or porewater. Thus, processes influencing the equilibrium between soluble and solid or bound phases for heavy metal(loid)s impact their overall bioavailability. For example, orthophosphate can complex metals such as Pb, Cd and Zn to form precipitates which are insoluble, relatively stable and reduce the overall toxicity of the metal contaminants (Bolan et al., 2003a).

Unless conditions support precipitation, metal(loid)s and ionizable organics generally sorb to charged surface such as layer silicate minerals and organic matter (Naidu et al., 1996). In contrast, the NOCs are sorbed generally through

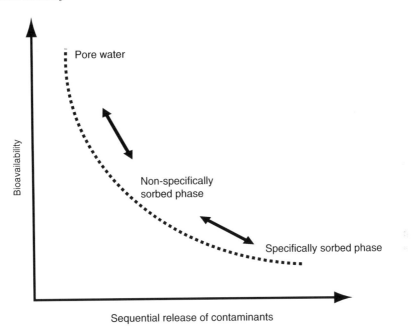

Bioavailability

Pore water

Non-specifically
sorbed phase

Specifically sorbed phase

Sequential release of contaminants

Figure 2.2. *Chemicals are sequentially released as the porewater concentration depletes.*

partitioning into organic matter (Chiou, 1989). In general, low water solubility of NOCs and high organic matter content in soils manifest increased sorption.

Chemicals that persist in the environment can exhibit a wide range of toxicities across different soil types. Bioavailability is impacted not only by soil type, but also contact time between soil and contaminant (i.e., ageing), which reduces the level of contaminant available for degradation/bioremediation (Feng et al., 2000; Pignatello, 1989; Karickhoff, 1980). There are important physical–chemical–biological processes occurring in soil systems which have a significant impact on the bioavailability of organic and inorganic (metal) contaminants. Thus the bioavailable fraction varies with time, the nature of soil types, organisms and the environmental factors. Conceptually, the bioavailable fraction may be defined as

$$\text{Bioavailability } (F_b, t) = f(\text{soil properties})(\text{environmental conditions})$$
$$\cdot (\text{plant characteristics}) \cdot (\text{microbes}) \cdot dT \qquad (2.2)$$

$$F_b, t = F_b, 0 \; K_{t1} \qquad (2.3)$$

$$F = \text{fraction bioavailable at time } 't' \qquad (2.4)$$

$$K_1 = a \, [\text{clay}] + b \, [\text{organic matter}] + \text{etc.} \qquad (2.5)$$

Given the dependence of bioavailability on plant type, microbial genre, the definitions currently used to define bioavailability may be limiting (see Chapter 3). Soil and solution factors influencing chemical bioavailability in the soil environment are briefly discussed in the following sections.

2.4 Key soil properties influencing chemical bioavailability in soils

Soil plays a significant role in controlling the potential bioavailability of contaminants in the environment. The primary soil factors influencing the potential bioavailability of contaminants are soil pH, cation and anion exchange capacities (available charged sites on soil surfaces), texture (clay content), soil type and soil organic matter.

2.4.1 Soil pH

Soil pH is one of the key parameters influencing the sorption of both inorganic and ionizable organic contaminants given that it controls virtually all aspects of contaminant and biogeochemical processes in soils. These processes include solubility, precipitation, speciation and sorption as well as microbial activity. In most variable, charge soils such as the strongly weathered tropical soils and less weathered Andisols, increasing soil pH results in an increase in the number of negatively charged soil sites with a concomitant decrease in the positively charged sites (Naidu et al., 1994, 1996; Bolan et al., 1999b; Fig. 2.3).

In contrast to variable charge soils, the effect of pH on surface charge characteristics in permanent charged soils such as those from temperate Mediterranean region is less marked unless the soil is high in organic matter content. Nevertheless, changing the soil pH directly impacts the sorption and removal from the porewater of metal(loid) (Fig. 2.4) or organic solutes (Fig. 2.5) (Bohn et al., 1985; Naidu et al., 1996).

For organic bases such as atrazine, the fraction of contaminant existing as an anion decreases with increasing pH. The anion has a lower affinity for the soil relative to the neutral species (Fig. 2.6).

In addition to its effect on surface chemical properties, soil pH also controls the speciation of both ionizable organic contaminants such as 2,4-D, sulphonyl urea, PCP and metal(loid)s. For metals, the net charge of the metal complexes and their precipitation/dissolution reactions are directly impacted by soil pH. For organic bases such as atrazine, the fraction of contaminant existing as a cation decreases with increasing pH (see Fig. 2.6) while for organic acids such as

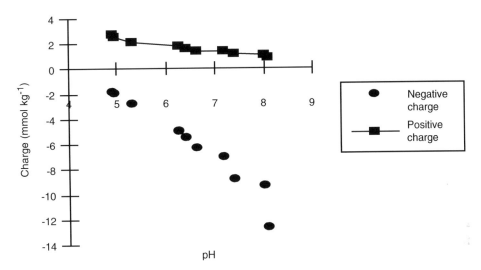

Figure 2.3. *Effect of soil pH on surface charge characteristics of an Oxisol (R. Naidu, unpublished).*

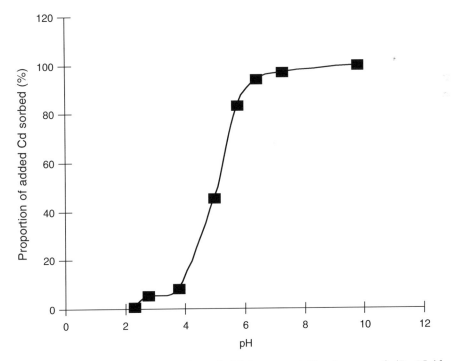

Figure 2.4. *Effect of pH on sorption of Cd by a variable charge soil (R. Naidu, unpublished).*

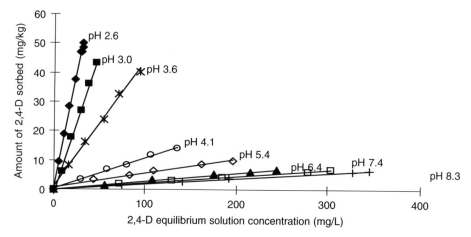

Figure 2.5. *Effect of pH on sorption of 2,4-D by a variable charge soil (R. Naidu, unpublished).*

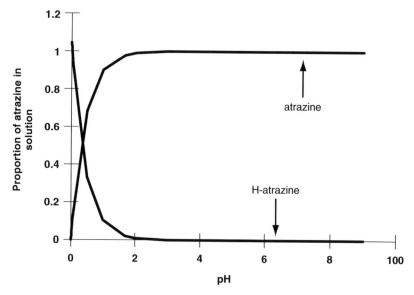

Figure 2.6. *Effect of soil pH on speciation of atrazine (R. Naidu, unpublished).*

PCP, the fraction of contaminant existing as an anion increases with increasing pH. The anion has a lower affinity for the soil relative to the neutral species.

The impact of pH on the behaviour and bioavailability of non-ionic organic contaminants is less marked and is generally achieved through its influence on organic matter (see Chapters 3 and 10, this volume) and on microbial activity. Thus, for both metal(loid)s and ionizable organics soil pH plays a major role in controlling contaminant bioavailability.

2.4.2 Organic matter content

Soil organic matter is a complex polymeric mixture arising from microbial and chemical degradation processes, the exact structure of which has not been fully elucidated (Khan, 1978). Organic matter has a high affinity to bind organic compounds as well as some metals in soils thereby, reducing their availability. Organic contaminants preferentially partition to the organic domain of organic matter relative to the polar aqueous phase (Chiou, 1989), while the organic acid functional groups typically present in organic matter have a high affinity to attract metal cations. For non-polar or neutral organic contaminants at equilibrium, sorption is positively correlated to the amount of organic matter, usually reported as the fraction of organic carbon (f_{oc}), and inversely proportional to aqueous solubility (Guerin and Boyd, 1993). Studies involving PAHs have attributed the decrease in microbial mineralization of PAHs to their association with the soil organic matrix (Hatzinger and Alexander, 1995; Michelcic and Luthy, 1991; Gomez-Lahoz and Ortega-Calvo, 2005). The greater the hydrophobicity or lipophilicity of an organic contaminant, the greater the potential for its sorption to organic matter (Christman and Pfaender, 2006). The latter has led to the use of the OC-normalized partition coefficients (K_{oc}) for estimating contaminant sorption with the soil-specific distribution coefficient, K_d ($K_{oc} = K_d/f_{oc}$, where f_{oc} is fraction of organic carbon). Another indirect effect of soil organic matter is its role on limiting contaminant mass-transfer. The rate of mass-transfer of an organic contaminant from soil particles to the surrounding porewater is inversely proportional to the contaminant's soil-water distribution coefficient (Pignatello, 2000). Therefore, with increasing organic matter content, retention of an organic contaminant increases and rates of release decrease, thereby, decreasing overall contaminant bioavailability (Stokes et al., 2005). For instance, Guerin and Boyd (1993) observed that naphthalene sorbed to soils of high OC contents was less available to *Pseudomonas putida* strain 17484 than that sorbed to low OC soils. These investigators suggested that a large proportion of naphthalene was present in a non-labile phase in a high OC soil, and, therefore, naphthalene was less accessible to potential microbial degraders, compared with that in a low OC soil.

2.4.3 Ion exchange capacities of soils

As discussed above, pH influences sorption through its effect on surface charge as quantified by cation and anion exchange capacities of soils. The anion exchange capacity (AEC) which is generally associated with oxidic minerals is a measure of available positively charged surface sites which decreases with increasing soil pH (see Fig. 2.3 and Naidu et al., 1990). The cation exchange capacity (CEC) is a measure of negatively charged sites and is generally associated with layer silicate minerals (such as 2:1 and 1:1 silicates) and organic matter.

Both AEC and CEC vary with the clay mineral content and type, organic matter and soil pH. CEC is greater for 2:1 clays such as montmorillonite (60–100 cmol/kg) compared to 1:1 clays such as kaolin (2–16 cmol/kg), while the CEC of organic matter ranges from 200 to 400 cmol/kg. CEC in soil ranges from values as low as 1.0 cmol/kg for extremely coarse-textured sandy soil to as much as 60 cmol/kg for fine-textured soil, containing large amounts of 2:1 clays and organic matter (Bohn et al., 1985). High clay soils will have a high affinity to sorb cationic species of both organic and inorganic solutes due to high CEC, and to sorb non-ionic organic contaminants due to large surface area, thus making contaminants less bioavailable relative to sandy soils. In addition to charged sites available in clays, siloxane oxygen-reactive sites present in clays can interact specifically with contaminants such as the nitro-substituted explosives. As the number of positively charged sites (i.e., AEC) on the majority of soil types is very small, and in environmentally relevant pH ranges is usually negligible, it is not generally considered an important parameter in assessing contaminant availability at most sites in the temperate region although in the tropics and less weathered soils such as Andepts it could be an important parameter because the later soils tend to carry higher positive charges.

In addition to soil pH there are a number of other environmental variables that influence ion exchange characteristics of soils. Amongst these is the presence of inorganic and organic ligands that bind specifically to soil colloid surface. Specific sorption of anions onto variable charge components has often been shown to increase the net surface charge (i.e., CEC) of soils and consequently increase the capacity of soils to bind cationic metals and positively charged organic contaminants (Table 2.1). This process has been described by a number of researchers as anion-induced cation sorption (Ryden and Syers, 1976; Bolland et al., 1977; Wann and Uehara, 1978; Shuman, 1986; Table 2.2).

Shuman (1986) observed that the specific sorption of anions, such as $H_2PO_4^-$ and SO_4^{2-}, increased the sorption of Zn^{2+} by variable charge components. Bolland et al. (1977) and Kuo and McNeal (1984) reported that $H_2PO_4^-$ sorption increases the sorption of Zn^{2+} and Cd^{2+} by hydrous ferric oxide. Anion-induced cation sorption depends on the variable charge components of the soils. Naidu et al. (1994) and Bolan et al. (1999, 2003b) have demonstrated $H_2PO_4^-$- and SO_4^{2-}-induced Cd sorption in strongly weathered Oxisols from Australia and less weathered but net positively charged Andept from New Zealand. Metals can also form precipitates with inorganic soil constituents, such as carbonate and

Table 2.2. *Increase in surface charges due to specific adsorption of anions and cations (Bolan et al., 1999).*

Soil constituents	Solute	pH	Charge added ($mol\,mol^{-1}$anion)	Reference
Iron hydrous oxide	Phosphate	6.5	1.25	Bolan et al. (1985)
Allophane	Phosphate	5.1	0.5	Rajan et al. (1974)
Soil	Phosphate	6.5	0.65	
	Sulphate	6.5	0.26	
Soil	Phosphate	5.0	0.38	Sawhney (1974)
		6.5	0.47	
		7.5	0.77	
Soil	Phosphate	7.0	0.35–0.7	Schalscha et al. (1974)
Soils	Phosphate	5.8	0.52	Naidu et al. (1990)
Aluminium oxide	Sulphate	5.0	1.06	Rajan (1978)
Soils	Sulphate	5.6	0.25	Curtin and Syers (1990)
Soil	Calcium	5.8	0.35–0.58	Bolan et al. (1993)
Soil	Calcium	5.8	0.52	Ryden and Syers (1975)

phosphate minerals under certain soil conditions (Bolan et al., 2003a; Naidu et al., 1996). Carbonate- and phosphate-metal complexes have varying degrees of solubility and reactivity depending on the metal, its oxidation state, the ligand to which it is bound and pH. The above studies suggest that two different mechanisms may operate depending on the nature of the soil surface (Harter and Naidu, 1995). Changes in surface chemical properties of soils as influenced by pH and ligand ions such as P have been used by many environmental managers to reduce metal bioavailability in soils (Basta et al., 2001). It has long been reported that raising the soil pH using lime amendments reduces plant uptake of heavy metal such as Cd due to enhanced binding and hence reduced bioavailability of the metal (Bolan et al., 2003c). However, the effect of pH and ligand ions on heavy metal binding by soils may vary depending on the nature of soils. For example, Turpeinen et al. (2000) investigating the mobility and bioavailability of Pb in soils reported that liming that increased the soil pH did not reduce the solubility, mobility or bioavailability of Pb in soils.

2.4.4 Soil type

The nature of soil types varies considerably depending on the geographical location. Thus Alfisols, Entisols, Inceptisols, Ultisols, Vertisols and Oxisols are all commonly found in tropical and subtropical regions receiving more than 500 mm mean annual rainfall. Landscapes throughout the tropics and subtropics are, however, dominated by Oxisols and Ultisols occupying extensive areas of potentially highly productive soils. The mineral fractions of these soils consist primarily of low-activity clays having variable surface charge that differs from high-activity clays in the origin of that charge. Low-activity clays are dominated by iron (Fe) and aluminium (Al) oxyhydroxides and 1:1 layer silicates (kaolin) (Naidu et al., 1996). In contrast to soils from the tropical region, temperate soils such as that from the Mediterranean region consist of less weathered soils such as the Alfisols, Mollisols, Vertisols which receive less than 500 mm mean annual rainfall. The mineral fraction of these soils consists primarily of high-activity clays such as 2:1 layer silicate minerals (smectites, illites, etc.) having permanent surface charge. The organic matter content of such soils also vary significantly and may range from low <1% to high (>2%) in the surface soils. The temperate region may also consist of less weathered soils that are young and dominated by poorly ordered silicate minerals such as allophonic and imogolites. Such soils that are commonly found in Japan and New Zealand are usually dominated with variable charge minerals (Naidu et al., 1996; Bolan et al., 1999b). Given the widely different surface charge and chemical properties of the soils, their ability to sorb contaminants varies considerably (Fig. 2.7). Consequently, contaminant bioavailability varies significantly with soil type due to differences in soil properties. For instance, Chung and Alexander (1998, 1999) demonstrated that the bioavailability of phenanthrene and atrazine, whether freshly added or aged, as measured by biodegradability varied greatly amongst 16 soils which varied widely in their physical and chemical properties including texture, porosity, percentage OC, clay content and mineralogy, CEC and surface area. Similarly, the effect of soil types on sorption of Cd was investigated by Naidu et al. (1996). These investigators demonstrate that soils from the temperate region consisting of 2:1 layer silicate minerals generally sorbed the highest amount of Cd, while the least was recorded for an Oxisol with a pH of 5. There was a marked increase in the amount of Cd sorbed with increasing soil pH confirming many previous reports on the effect of pH on metal sorption.

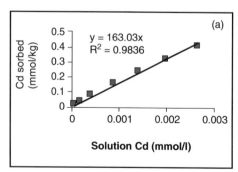 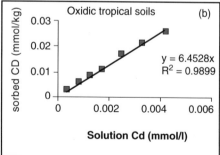

Figure 2.7. *Sorption of Cd varies with soil type. (a) Smectitic soil and (b) oxidic soils.*

Numerous investigators (Feng et al., 2000) have investigated the effect of soil type on sorption and bioavailability of contaminants by relating K_d value to either uptake of contaminants by plants or organisms. In these studies they found that soils with large K_d values have lower contaminant bioavailability compared to those with low K_d values. In their studies Feng et al. (2000) found that the K_d increased with ageing of soils but the rate of increase varied with soil type which they attributed to varying soil chemical properties. In a paper entitled "more unified approach to assessing contaminant bioavailability", Naidu and Bolan (2004) recognized that soil consists of a wide range of mineral and organic constituents and that the metal-binding capacity of soils is a function of the partition coefficient (K_d) describing the thermodynamics of the interaction between metals and each mineral and organic fraction as shown below:

$$K_d = f(K_{d1},\ K_{d2},\ \ldots,\ K_{dx}) \tag{2.6}$$

Naidu (2004) found that the K_d for Cd varied significantly amongst the different soil types and when the Cd-treated soils were planted with spinach, there was an exponential decline in the amount of plant tissue Cd with increasing K_d (Fig. 2.8). This suggests that Cd bioavailability (indicated by plant Cd content) decreases with increasing K_d. However, the relationship between K_d and bioavailability (metal uptake) may not be applicable to soils saturated with contaminants given the mass action effect under these conditions may be the major driving force controlling contaminant bioavailability. Further research on

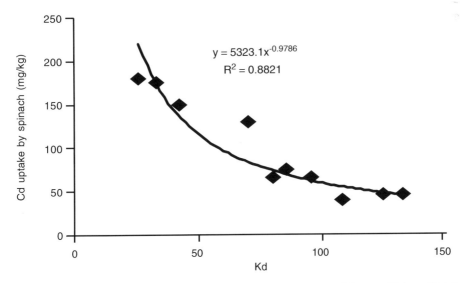

Figure 2.8. *Relationship between Cd uptake and cadmium partition coefficient (K_d) in 10 soils.*

the role of partition coefficient on metal and organic bioavailability is currently being researched by Naidu and his co-researchers.

In the case of pesticides, while the mobility of pesticide is affected only by its K_d value, the bioavailability as measured by the rate of degradation (half life – $t_{1/2}$) is affected by both the K_d value and the microbial activity of soil (Fig. 2.9). For instance, Roberts et al. (1999) noticed that the leaching of pesticides decreased with increasing K_d value (Fig. 2.10). However, Bolan and Baskaran (1996) noticed that both the K_d value of 2,4-D herbicide and microbial activity as measured by respiration increased with soil organic matter content (Fig. 2.11). However, the $t_{1/2}$ value increased with an initial increase in K_d value, but it decreased when the K_d value was greater than 8W dm^3/kg. The initial increase

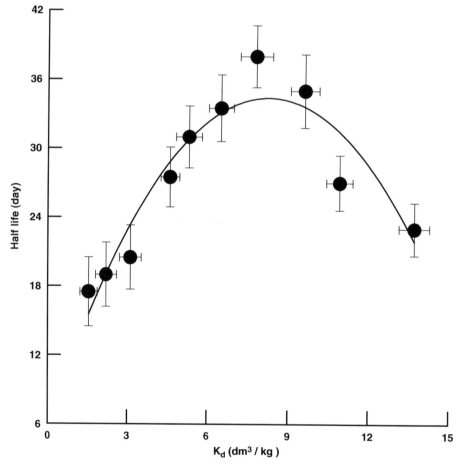

Figure 2.9. *The relationships between* K_d *values and half life* $(t_{1/2})$ *of 2,4-D herbicide (Bolan and Baskaran, 1996).*

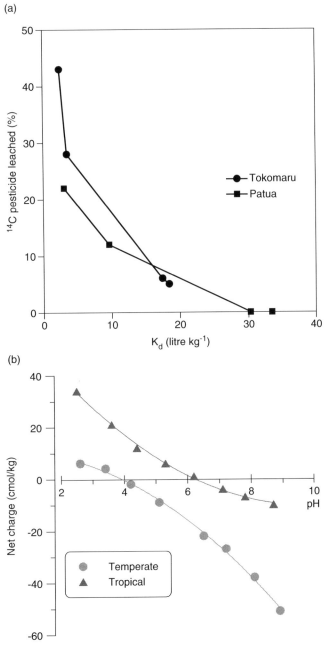

Figure 2.10. *(a) Relationship between distribution coefficient (k_d) and the percentage of pesticide leached in (•) Tokomaru and (▲) Patua soils. (b) Net surface charge of temperate and tropical soils as affected by pH.*

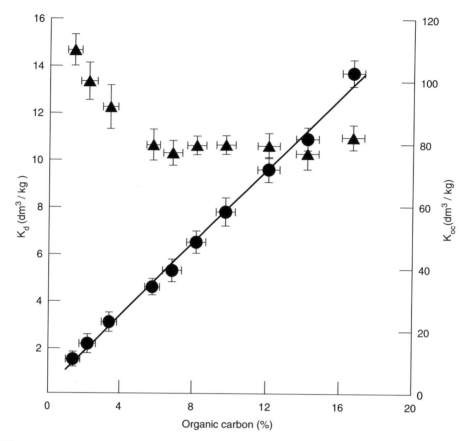

Figure 2.11. *The relationship between soil organic carbon (OC), K_d (dm³/kg) and K_{oc} (dm³/kg) values for 2,4-D.*

in $t_{1/2}$ value (decrease in the rate of degradation) was probably caused by the adsorption of the pesticides on to the soil organic matter. This could result in its inaccessibility to microbial attack (Smith et al., 1992) and a low concentration of the pesticide in solution. Processes which lead to a decrease in the concentration of the pesticide in soil solution are likely to decrease the rate of degradation of the pesticides by the microorganisms (Alexander, 1999). Adsorbed pesticides become "bound" or "recalcitrant" which resist degradation and become more persistent in soils (Ogram et al., 1985; Fewson, 1988). The increase in the rate of degradation of 2,4-D in soils with high K_d values ($>8W$ dm³/kg) was intriguing. The soils with high K_d values tended to have high microbial activity and the decrease in the concentration of pesticides in soil solution with increasing adsorption is compensated by the increased microbial activity thereby increasing the rate of degradation.

Research on the mobility and bioavailability of contaminants in soils varying in organic and mineral contents is still in infancy. Further work is needed on the fate and behaviour of contaminants in different soil types and in particular how it relates to bioavailability.

2.4.5 Time dependence

As discussed above, contaminants may undergo a time-dependent sequestration in soil that results in a decline in bioavailability. This process known as ageing, is attributed to the sequestration of contaminants into solid phase (Alexander, 1995; Stewart et al., 2003) either in mineral nanopores (Hatzinger and Alexander, 1995; Pignatello and Xing, 1996; Hatzinger and Alexander, 1997; Nam and Alexander, 1998) or partitioning to specific and non-specific sites of organic matter (Pignatello and Xing, 1996; Hatzinger and Alexander, 1997). For instance, Naidu and his co-researchers (see for example, Hamon et al., 1998) investigating the long-term changes in Cd bioavailability in soils reported that despite an increase in the total Cd content due to fertilization, a significant proportion of added Cd was found to exist in a non-bioavailable pool in the soil. Similar decline in the long-term availability of Zn was reported by Fischer et al. (1996). Although the mechanisms responsible for the decline in Zn availability was not conclusively established, numerous investigators have postulated migration of Zn into micropores or the movement of Zn into the solid matrix through soil diffusion mechanisms (Barrow, 1998). Similar to metals, studies of organic contaminants fate have shown that microbial mineralization decreases with increasing contaminant residence time (Boucard et al., 2005). In particular, it has been reported that the time-dependent decline in bioavailability of organics is not reflected in the decline of the concentration of the compounds determined by vigorous extraction with organic solvents (see Chapter 10).

A clear link exists between biodegradation of contaminants, toxicity and bioavailability. The phenomenon of ageing (sequestration) in general results in the apparent decline of bioavailability associated with persistence and increased resistance of the contaminant to extraction with solvents. For example, laboratory studies involving the possible bioremediation of long-term car-baryl-contaminated soil have confirmed the reduced bioavailability of carbaryl in that soil. Inoculation of the long-term contaminated soil with carbaryl-degrading bacteria showed approximately 45% of the carbaryl was apparently unavailable to the microorganisms. The amount of carbaryl degraded in this soil was related to the water extractable carbaryl content. This indicates that the reduced bioavailability of contaminant in aged soils is a major constraint in the bioremediation. A similar effect of ageing on metal bioavailability and metal impact on soil biota has also been observed in our laboratory. Also, a similar effect of ageing on arsenic bioavailability has been demonstrated in our laboratory (R. Naidu and J. Smith, unpublished).

2.5 Accessibility to contaminants bound to soil determines biodegradation of contaminants

Although bioremediation is considered to be one of the potential remediation options, the lack of contaminant bioavailability in long-term contaminated soils makes it difficult to be successful remediation option. As discussed above, as contaminated soils age, contaminants gradually disperse and migrate into subatomic pores (Ogram et al., 1985). While diffusion into subatomic pores is spontaneous (though kinetically slow), the reverse process is difficult and requires significant input of energy. For example, several reports suggest the use of surfactants and co-solvents to increase the solubility and consequent bioavailability of organic contaminants to enhance bioremediation (Müller et al., 2007). However, the use of surfactants can cause soil damage besides being expensive. Kantachote et al. (2000) have recently examined the ability of sodium (Na) to increase DDT transformation as Na is known to cause clay dispersion and increase DOC content, thereby increasing the DDT bioavailability. This work clearly demonstrated the increased solubility of DDT as shown by an increase in soil solution DDT concentration which is further supported by an increase in bacterial growth in the Na-amended long-term DDT-contaminated soil.

2.6 Concluding remarks

In this chapter, we have presented an overview of the transformation and fate of contaminants in soils as controlled by both the properties of the soil and the nature and type of contaminants. Soil properties, including pH, organic matter and clay contents, surface charge, soil solution composition, and contaminant physicochemical properties, including ionic and oxidation state, molecular structure, polarity, aqueous solubility, and lipophilicity, are responsible for controlling the behaviour of contaminants. These properties in turn affect the fraction of contaminants that may be bioavailable in a given soil. This is further impacted by the duration of contact between contaminants and soils with the bioavailability declining with ageing.

Since the bioavailable fraction is the underlying basis for toxicity, biodegradation, and mobility of contaminants, bioavailability plays vital role in the remediation of contaminated soils. Thus bioavailability of soils can be manipulated to manage contaminated soils. To quantify bioavailability, a large number of techniques have been employed ranging from organic and aqueous-based solvent extractions to the use of bioassay tests involving micro, meso, and macro biota. These tests could be used in the assessment and remediation of contaminates sites. Remediation can be achieved either by reducing the bioavailability of contaminants using soil amendments thereby minimising their chance to reach food chain or my enhancing their bioavailability and subsequent removal through bioremediation and phytoremediation techniques.

References

Adriano, D., Wenzel, W.W., Vangronsveld, J., Bolan, N.S., 2004. Role of assisted natural remediation in environmental cleanup. Geoderma 122(2–4), 121–142.

Aislabie, J., Lloydjones, G., 1995. A review of bacterial-degradation of pesticides. Aust. J. Soil Res. 33, 925–942.

Alexander, M., 1995. How toxic are toxic chemicals in soil. Environ. Sci. Technol. 29, 2713–2717.

Alexander, M., 1999. Biodegradation and Bioremediation, 2nd Ed., Elsevier, San Diego, CA, p. 453.

Australian and New Zealand Environment and Conservation Council., 1992. National Health and Medical Research Council. Australian and New Zealand Guidelines for the assessment and management of contaminated sites. ANZECC, NHMRC, Canberra, p. 57.

Barrow, N.J., 1998. Effects of time and temperature on the sorption of cadmium, zinc, cobalt, and nickel by a soil. Aust. J. Soil Res. 36(6), 941–950.

Barrow, N.J., Gerth, J., Brümmer, G.W., 1989. Reaction kinetics of the adsorption and desorption of nickel, zinc and cadmium by goethite. II. Modelling the extent and rate of reaction. J. Soil Sci. 40, 437–450.

Baskaran, S., Bolan, N.S., Rahman, A., Tillman, R.W., 1996. Adsorption of ionic and non-ionic pesticides by allophanic and non-allophanic soils of New Zealand. N. Z. J. Agric. Res. 39, 297–310.

Basta, N.T., Gradwohl, R., Snethen, K.L., Schroder, J.L., 2001. Chemical immobilisation of lead, zinc and cadmium in smelter-contaminated soils using biosolids and rock phosphate. J. Environ. Qual. 30, 1222–1230.

Bohn, H., McNeal, B., O'Connor, G., 1985. Soil Chemistry. , 2nd Ed., Wiley, New York.

Boivin, A., Cherrier, R., Schiavon, M., 2005. A comparison of five pesticides adsorption and desorption processes in thirteen contrasting field soils. Chemosphere 61, 668–676.

Bolan, N.S., Adriano, D.C., Naidu, R., 2003a. Role of phosphorus in (im)mobilization and bioavailability of heavy metals in the soil-plant system. Rev. Environ. Contam. Toxicol. 177, 1–44.

Bolan, N.S., Adriano, D.C., Mani, P., Duraisamy, A., Arulmozhiselvan, S., 2003b. Immobilization and phytoavailability of cadmium in variable charge soils: I. Effect of phosphate addition. Plant Soil 250, 83–94.

Bolan, N.S., Adriano, D.C., Mani, P., Duraisamy, A., Arulmozhiselvan, S., 2003c. Immobilization and phytoavailability of cadmium in variable charge soils: II. Effect of lime addition. Plant Soil 250, 187–198.

Bolan, N.S., Baskaran, S., 1996. Biodegradation of 2,4-D herbicide as affected by its adsorption-desorption behaviour and microbial activity of soils. Aust. J. Soil Res. 34, 1041–1053.

Bolan, N.S., Naidu, R., Tillman, R.W., Khan, A., Syers, J.K., 1999a. Effect of anion sorption on cadmium sorption by soils. Aust. J. Soil Res. 37, 445–460.

Bolan, N.S., Naidu, R., Syers, J.K., Tillman, R.W., 1999b. Surface charge and solute interactions in soils. Adv. Agron. 67, 88–141.

Bolland, M.D.A., Posner, A.M., Quirk, J.P., 1977. Zinc adsorption by geothite in the absence and presence of phosphate. Aust. J. Soil Res. 15, 279–286.

Boucard, T.K., Bardgett, R.D., Jones, K.C., Semple, K.T., 2005. Influence of plants on the chemical extractability and biodegradability of 2,4-dichlorophenol in soil. Environ. Pollut. 133, 53–62.

Boul, H.L., 1995. DDT residues in the environment – a review with a New Zealand perspective. N. Z. J. Agric. Res. 38(2), 257–277.

Chiou, C.T., 1989. Theoretical considerations for the partition uptake of nonionic organic compounds by soil organic matter. In: B.L. Sawhney, K. Brown (Eds.), Reactions and Movement of Organic Chemicals in Soil. Soil Science Society of America, Madison, WI, pp. 1–29.

Christman, R.F., Pfaender, F.K., 2006. Molecular implications of hydrophobic organic partitioning theory. Acta Hydrochim. Hydrobiol. 34, 367–374.

Chung, N., Alexander, M., 1998. Differences in sequestration and bioavailability of organic compounds aged in dissimilar soils. Environ. Sci. Technol. 32, 855–860.

Chung, N., Alexander, M., 1999. Effect of concentration on sequestration and bioavailability of two polycyclic aromatic hydrocarbons. Environ. Sci. Technol. 33, 3605–3608.

CMPS&F – Environment Australia. 1997. Appropriate technologies for the treatment of scheduled wastes. Review Report No. 4, November.

Curtin, D., Syers, J.K., 1990. Extractability and adsorption of sulphate in soils. J. Soil Sci. 41, 305–312.

Farenhorst, A., 2006. Importance of soil organic matter fractions in soil-landscape and regional assessments of pesticide sorption and leaching in soil. Soil Sci. Soc. Am. J. 70, 1005–1012.

Feng, Y., Park, J.H., Voice, T.C., Boyd, S.A., 2000. Bioavailability of soil-sorbed biphenyl to bacteria. Environ. Sci. Technol. 34, 1977–1984.

Fewson, C.A., 1988. Biodegradation of xenobiotic and other persistent compounds: the causes of recalcitrance. Trends Biotechnol. 6, 148–153.

Fischer, S.F., Poschlod, P., Beinlich, B., 1996. Experimental studies on the dispersal of plants and animals on sheep in calcareous grasslands. J. Appl. Ecol. 33, 1206–1222.

Gertsl, Z., 1990. Estimation of organic chemical sorption by soils. J. Contam. Hydrol. 6, 357–375.

Gomez-Lahoz, C., Ortega-Calvo, J.J., 2005. Effect of slow desorption on the kinetics of biodegradation of polycyclic aromatic hydrocarbons. Environ. Sci. Technol. 39, 8776–8783.

Guerin, W.F., Boyd, S.A., 1993. Sorption and degradation of pesticides and organic chemicals in soil. In: D.M. Linn (Ed.). Soil Science Society of America, Madison, WI, pp. 197–208.

Hamon, R., McLaughlin, E., Naidu, R., Correll, R., 1998. Long-term changes in cadmium bioavailability in soil. Environ. Sci. Technol. 32(23), 3699–3703.

Harter, R.D., Naidu, R., 1995. Role of metal-organic complexation in metal sorption by soils. Adv. Agron. 55, 219–263.

Hatzinger, P.B., Alexander, M., 1995. Effect of aging of chemicals in soil on their biodegradability and extractability. Environ. Sci. Technol. 29, 537–545.

Hatzinger, P.B., Alexander, M.A., 1997. Biodegradation of organic compounds sequestered in organic solids or in nanopores within silica particles. Environ. Toxicol. Chem. 16, 2215–2221.

Homann, P.S., Zasoski, R.J., 1987. Solution composition effects on cadmium sorption by forest soil profiles. J. Environ. Qual. 16(4), 429–433.

Kantachote, D., Singleton, I., Naidu, R., Williams, B.D., McClure, N., Megharaj, M., 2004. Bioremediation of DDT contaminated soil: enhancement by seaweed addition. J. Chem. Technol. Biotechnol. 79, 632–638.

Karickhoff, S.W., 1980. Sorption kinetics of hydrophobic pollutants in natural sediments. In: R.A. Baker (Ed.), Contaminants and Sediments. Vol. 2, Ann Arbor Scientific Publishers, Ann Arbor, MI, pp. 193–205.

Kelsey, J.W., Alexander, M., 1997. Declining bioavailability and inappropriate estimation of risk of persistent compounds. Environ. Toxicol. Chem. 16, 582–585.

Khan, S.U., 1978. The interaction of organic matter with pesticides. In: M. Schnitzer, S.U. Khan (Eds.), Soil Organic Matter. Elsevier, Amsterdam, pp. 137–171.

Krauss, M., Wilcke, W., 2005. Persistent organic pollutants in soil density fractions: distribution and sorption strength. Chemosphere 59, 1507–1515.

Kuo, S., McNeal, B.L., 1984. Effects of pH and phosphate on cadmium sorption by a hydrous ferric oxide. Soil Sci. Soc. Am. J. 48, 1040–1044.

La Grega, M.D., Buckingham, P.L., Evans, J.C., 1994. Hazardous Waste Management. McGraw-Hill, New York, p. 859.

Lee, L.S., Rao, P.S.C., Nkedi-Kizza, P., Delfino, J.J., 1990. Influence of solvent and sorbent characteristics on distribution of pentachlorophenol in octanol-water and soil-water systems. Environ. Sci. Technol. 24, 654–661.

Lyman, W.J., Reehl, W.F., Rosenblatt, D.H., 1990. Handbook of Chemical Property Estimation Methods: Environmental Behavior of Organic Compounds. American Chemical Society, Washington, DC.

Manilal, V.B., Alexander, M., 1991. Factors affecting the microbial degradation of phenanthrene in soil. Appl. Microbiol. Biotechnol. 35, 401–405.

Michelcic, J.R., Luthy, R.G., 1991. Sorption and microbial degradation of naphthalene in soil-water suspensions under denitrification conditions. Environ. Sci. Technol. 25(1), 169–177.

Müller, K., Magesan, G.N., Bolan, N.S., 2007. A critical review of the influence of wastewater irrigation on organic chemical transformation and transport in soil. Agric. Ecosyst. Environ. 120, 93–116.

Muller, S., Wilcke, W., Kanchanakool, N., Zech, W., 2001. Polycyclic aromatic hydrocarbons (PAH) and polychlorinated biphenyls (PCB) in density fractions of urban soils in Bangkok, Thailand. Soil Sci. 166, 672–680.

Naidu, R., Bolan, N.S., 2004. Risk based land management. Abstract Fertilizer and Lime Research Conference, Palmerston North, New Zealand, March.

Naidu, R., Bolan, N.S., Adriano, D.C. (Eds.), 2003a. Bioavailability, toxicity and risk relationships in ecosystems: the path ahead. In: Bioavailability and its Potential Role in Risk Assessment. Science Publishers, New York, pp. 331–339.

Naidu, R., Kookana, R.S., Baskaran, S., 1998. Fate and behavior of pesticides in tropical soils. Invited keynote address presented at the pesticide risk assessment workshop in Jakarta, Indonesia, February 14–19.

Naidu, R., Kookana, R.S., Oliver, D.P., Rogers, S., McLaughlin, M.J., 1996. Contaminants in the Soil Environment in the Australasia Pacific Region. Kluwer Academic Publishers, Netherlands, p. 717.

Naidu, R., Rogers, V.V., Gupta, S.R., Kookana, R.S., Bolan, N.S., Adriano, D.C. (Eds.), 2003b. Bioavailability of metals in the soil-plant environment and its potential role in risk assessment: an overview. In: Bioavailability and its Potential Role in Risk Assessment. Science Publishers, pp. 21–59.

Naidu, R., Harter, R.D., 1997. Effect of different organic ligands on cadmium sorption by and extractability from soils. Soil Sci. Soc. Am. J. 62, 644–650.

Naidu, R., Bolan, N.S., Kookana, R.S., Tiller, K.G., 1994. Ionic strength and pH effects on sorption of cadmium and charge in soils. J. Soil Sci. 45, 419–429.

Naidu, R., Syers, J.K., Tillman, R.W., Kirkman, J.H., 1990. Effect of liming and added phosphate on charge characteristics of acid soils. Eur. J. Soil Sci. 41(1), 157–164.

Nam, K., Alexander, M., 1998. Role of nanoporosity and hydrophobicity in sequestration and bioavailability. Tests Model Solids Environ. Sci. Technol. 32, 71–74.

Natusch, J., 1997. Application and development of contaminated site remediation technologies in Australia. ANZAC Fellowship Report to Department of Internal Affairs, Wellington, New Zealand and Department of Foreign Affairs and Trade, Canberra, Australia.

Ogram, A.V., Jessup, R.E., Ou, L.T., Rao, P.S.C., 1985. Effects of sorption on biological degradation rates of (2,4-dichlorophenoxy) acetic acid in soils. Appl. Environ. Microbiol. 49, 582–587.

Pignatello, J.J., 1989. Sorption dynamics of organic compounds in soils and sediments. In: B.L. Sawhney, K. Brown (Eds.), Reactions and Movement of Organic Chemicals in Soil. SSSA Spec. Publ. 22. SSSA, Madison, WI, pp. 45–80.

Pignatello, J.J., 2000. The measurement and interpretation of sorption and desorption rates for organic compounds in soil media. Adv. Agron. 69, 1–73.

Pignatello, J.J., Xing, B., 1996. Mechanisms of slow sorption of organic chemicals to natural particles. Environ. Sci. Technol. 30, 1–11.

Powell, I., 1992. Contaminated land – who pays? NSW State Pollution Commission Impact Statement Document, NEPC.

Rajan, S.S.S., 1978. Sulfate adsorption on hydrous alumina ligands displaced and changes in surface charge. Soil Sci. Soc. Am. J. 42, 39–44.

Rajan, S.S.S., Perrott, K.W., Saunders, W.M.H., 1974. Identification of phosphate-reactive sites of hydrous alumna from proton consumption during phosphate adsorption at constant pH values. J. Soil Sci. 25, 438–447.

Rao, D.P., Navalgund, R.R., Krishnamurthy, Y.V.N., 1996. Cadastral applications using IRS-1C data – some case studies. Curr. Sci. 70(7), 624–628.

Reichenberg, F., Mayer, P., 2006. Two complementary sides of bioavailability. Accessibility and chemical activity of organic contaminants in sediments and soils. Environ. Toxicol. Chem. 25(5), 1239–1245.

Ryden, J.C., Syers, J.K., 1976. Calcium retention in response to phosphate adsorption by two soils. Soil Sci. Soc. Am. J. 40, 845–846.

Schwarzenbach, R.P., Gschwend, P.M., Imboden, D.M., 1993. Environmental Organic Chemistry. Wiley, New York.

Shuman, L.M., 1986. Effect of ionic strength and anions on zinc adsorption by two soils. Soil Sci. Soc. Am. J. 50, 1438–1442.

Smith, S.C., Ainsworth, C.C., Traina, S.J., Hicks, R.J., 1992. Effect of sorption on the biodegradation of quinoline. Soil Sci. Soc. Am. J. 56, 737–746.

Sparks, D.L., 1998. Kinetics of soil chemical phenomena: future directions. In. P.M. Huang, D.L. Sparks and S.A. Boyd (Eds.), Future Prospects for Soil Chemistry. SSSA Spec. Publ. 55, Soil Science Society of America, Madison, WI, pp. 81–101.

Steinberg, C.E.W., Paul, A., Pflugmacher, S., Meinelt, T., Klöcking, R., Wiegand, C., 2003. Pure humic substances have the potential to act as xenobiotic chemicals — a review. Fresenius Environ. Bull. 12, 391–401.

Steinberg, S.M., Pignatello, J.J., Sawhney, B.L., 1987. Persistence of 1,2-dibromoethane in soils: entrapment in intraparticle micropores. Environ. Sci. Technol. 21, 1201–1208.

Stewart, M.A., Jardine, P.M., Brandt, C.C., Barnett, M.O., Fendorf, S.E., McKay, L.D., Mehlhorn, T.L., Paul, K., 2003. Effects of contaminant concentration, aging, and soil properties on the bioaccessibility of Cr(III) and Cr(VI) in soil. Soil Sediment Contam. 12, 1–21.

Stokes, J.D., Paton, G.I., Semple, K.T., 2005. Behaviour and assessment of bioavailability of organic contaminants in soil: relevance for risk assessment and remediation. Soil Use Manage. 21, 475–486.

Tao, S., Guo, L.Q., Wang, X., Liu, W.X., Ju, T.Z., Dawson, R., Cao, J., Xu, F.L., Li, B.G., 2004. Use of sequential ASE extraction to evaluate the bioavailability of DDT and its metabolites to wheat roots in soils with various organic carbon contents. Sci. Total Environ. 320, 1–9.

Turpeinen, R., Salminen, J., Kairesalo, T., 2000. Mobility and bioavailability of lead in contaminated boreal forest soil. Environ. Sci. Technol. 34(24), 5152–5156.

USEPA., 1994. www.epa.gov/safewater/mcl.html.

USEPA. 2004. http://www.epa.gov/superfund/news/30years.htm.

Wann, S.S., Uehara, G., 1978. Surface charge manipulation of constant surface potential soil colloids. I: relation to sorbed phosphorus. Soil Sci. Soc. Am. J. 42, 565–570.

Wauchope, R.D., Yeh, S., Linders, J.B.H.J., Klaskowski, R., Tanaka, K., Rubin, B., Katayama, A., Kördel, W., Gerstl, Z., Lane, M., Unsworth, J.B., 2002. Review. Pesticide soil sorption parameters: theory, measurement, uses, limitations and reliability. Pest Manag. Sci. 58, 419–445.

Weissmahr, K.W., Haderlein, S.B., Schwarzenbach, R.P., 1998. Complex formation of soil minerals with nitroaromatic explosives and other B acceptors. Environ. Sci. Technol. 27, 316–326.

Weissmahr, K.W., Hildenbrand, M., Schwarzenbach, R.P., Haderlein, S.B., 1999. Laboratory and field scale evaluation of geochemical controls on groundwater transport of nitroaromatic ammunition residues. Environ. Sci. Technol. 33, 2593–2600.

Developments in Soil Science, volume 32
Ravendra Naidu (Editor)

Chapter 3

BIOAVAILABILITY: DEFINITION, ASSESSMENT AND IMPLICATIONS FOR RISK ASSESSMENT

R. Naidu, K.T. Semple, M. Megharaj, A.L. Juhasz, N.S. Bolan, S.K. Gupta,
B.E. Clothier and R. Schulin

3.1 Introduction

Bioavailability is often used as the key indicator of potential risk that chemicals pose to environment and human health. It is now regarded as a priority research area for both remediation and risk assessment as it is an important yet poorly quantified regulatory factor. From a regulatory perspective, the potential for bioavailability to influence decision-making is greatest where:

- the contaminant is (or is likely to be) the risk driver at a site;
- the default assumptions made during risk assessment that affect the final clean-up goal are inappropriate;
- significant change to remedial goals is likely, for example, because substantial quantities of contaminated soil or sediment are involved;
- conditions present at the site are unlikely to change substantially over time; and
- regulatory and public acceptance is high.

However, the term 'bioavailability' is defined inadequately, used ambiguously in the literature and the concept on which it is based is unclear (Naidu et al., 2003; Ehlers and Luthy, 2003; Semple et al., 2004). Given this uncertainty in the definition, there is considerable indecision regarding:

- How bioavailability is measured?
- How bioavailability varies between receptor organisms?
- How bioavailability data are interpreted?
- How this data will be used or incorporated into a regulatory setting?

The uncertainty arises because of

- the complexity of the issues and a limited understanding of bioavailability and
- the difficulty of measuring relevant environmental and human health indicators of bioavailability.

These uncertainties in risk, liability and technology application necessitate the need for better definitions of bioavailability. In this chapter, we present an overview of the current understanding of bioavailability, definitions and their limitations, methods for the assessment of bioavailability and bioavailability implications to risk assessment.

3.2 Definition of bioavailability

Considerable ambiguity exists within scientific and regulatory communities over what is actually meant by 'bioavailability'. The term 'bioavailability' was originally developed in pharmacology/toxicology; where it relates to the systemic availability of a xenobiotic after intravenous or oral dosing. Since the term bioavailability was first introduced, its definition and the methods used to assess it in the terrestrial environment has varied considerably depending on environmental discipline and receptor organism. In terms of environmental risk assessment, contaminant bioavailability has significant impact on a wide variety of ecological receptors including bacteria, fungi, algae, terrestrial invertebrates, vertebrates and humans. Some definitions that have been used to describe contaminant bioavailability for these receptors include:

- A chemical element is bioavailable if it is present as, or can be transformed readily to, the free ion species; if it can move to plant roots on a time scale that is relevant to plant growth and development and if once absorbed by the root it affects the life cycle of the plant (bioavailability in terms of plant uptake) (Sposito, 1989);
- The degree to which a chemical in a potential source is free for uptake (movement into an organism) (general bioavailability term) (Newman and Jagoe, 1992; Benson and Albert, 1992);
- A physiologically driven desorption process (general bioavailability term) (Peijnenburg et al., 1997);
- The degree of absorption across the gastrointestinal tract as determined by the characteristics of the ingested source (bioavailability in terms of higher organisms) (Beresford et al., 2000);
- The accessibility of a contaminant to microorganisms from the standpoint of their metabolism, their ability to grow in these chemicals, to change their physiology and perhaps modulation of genetic response (bioavailability to microorganisms) (Sayler et al., 1998);
- The fraction of an administered dose that reaches the central (blood) compartment from the gastrointestinal tract (bioavailability in terms of higher organisms) (Ruby et al., 1999);
- The extent to which a contaminant is available for biological conversion which is a function of the biological system, physicochemical properties of the contaminant and environmental factors (bioavailability in terms of biodegradation) (Juhasz et al., 2000);

- The accessibility of a chemical for assimilation and possible toxicity (bioavailability to microorganisms) (Alexander, 2000);
- The fraction of extractable metals that correlates with the total metal uptake by plants (bioavailability in terms of plant uptake) (Naidu et al., 2004); and
- A measure of the physicochemical access that a toxicant has to the biological process of an organisms (general bioavailability term) (USEPA, 1997).

Other related terms include:

- *Absolute bioavailability* – the fraction or percentage of a compound which is ingested, inhaled or applied to the skin that actually is absorbed and reaches systemic circulation (Battelle and Exponent, 2000).
- *Relative bioavailability* – refers to the comparative bioavailability of different forms of a chemical or for different exposure media containing the chemical and is expressed as a fractional relative absorption factor (Ruby et al., 1999). In the context of environmental risk assessment, relative bioavailability is the ratio of the absorbed fraction from the exposure medium in the risk assessment (e.g., soil) to the absorbed fraction from the dosing medium used in the critical toxicity study (Battelle and Exponent, 2000).
- *Bioaccessibility* – the fraction of a contaminant that is soluble and which is available for uptake into the circulatory system (Ruby et al., 1996, 1999; Kramer and Ryan, 2000; Moore, 2003). In light of the differing viewpoints on the definition of bioavailability, the Committee on Bioavailability of Contaminants in Soils and Sediments (NRC, 2002; Ehlers and Luthy, 2003) coined the term 'bioavailability processes' to encapsulate the mechanisms involved in the dissolution, transport and absorption of environmental contaminants by a receptor organism. Figure 3.1 provides a visual representation of the term bioavailability processes.

Phase A of Figure 3.1 represents the physical and biological processes involved in the release of the contaminant from the associated soil or sediment (see Chapter 2). Once solubilised, the contaminant is transported to the membrane of the organisms via diffusion, advection, etc. processes (Phase B). Contaminants may also be transported to the site of uptake bound to a solid phase as in the case of dermal contact of soil, incidental ingestion of soil or exposure of terrestrial invertebrates to soil (Phase C). Phases A–C may depict processes that occur in the external environment as well as those occurring internal to organisms such as in the gut lumen. Phase D in bioavailability processes entails movement of the contaminant across the membrane, resulting in absorption of the contaminant in the organism. Whilst Phase E represents the response of the organisms to the 'administered dose', it does not truly encompass a process influencing contaminant bioavailability, more so it represents the toxic action or metabolic effect of a chemical on the receptor organism (measured end point).

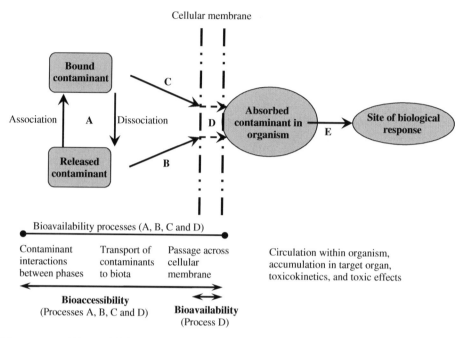

Figure 3.1. *Bioavailability processes in soil or sediment, including release of a solid-bound contaminant (A) and subsequent transport (B), direct contact of a bound contaminant (C), uptake by passage through a membrane (D) and incorporation into a living system (E). Process E is not considered bioavailability processes per se because soil and sediment are no longer a factor (Semple et al., 2004).*

Utilising etymological routes, Semple et al. (2006) proposed unambiguous definitions that apply directly to the broad spectrum of organisms exposed to soil-borne contaminants, e.g., soil micro- and meso-fauna, plants and higher animals. Thus, bioavailability was defined as "that which is freely available to cross an organism's (cellular) membrane from the medium the organism inhabits at a given point in time" (processes B and D in Fig. 3.1). Whereas, bioaccessibility was defined as "that which is available to cross an organisms' (cellular) membrane from the environment it inhabits, if the organism had access to it; however, it may be either physically removed from the organism or only bioavailable after a period of time" (processes A then B and D, Fig. 3.1, with temporal constraints). The bioavailable fraction, therefore, is that compound which is interacting with biota at a given moment in time, whereas bioaccessibility "encompasses what is actually bioavailable now *plus* what is *potentially* bioavailable" (e.g., isolated compound that may become available with time or through spatial rearrangements). Caussy (2003) conceptually defined bioavailability as the integral sum of a four-component process consisting ingestion, bioaccessibility, adsorption and first-pass effect. Here, the author defines "first pass" as the process that occurs in the liver immediately following adsorption. Clearly Caussy's definition focuses on processes A–E in the concept defined by NRC (2002).

3.3 Methods for the assessment of contaminant bioavailability

There are no standardised approaches for evaluating bioavailability. Since bioavailability will be influenced by factors such as the properties of the media (e.g., soil and sediment), chemical form and duration of exposure, an evaluation of bioavailability and its incorporation into risk assessment must consider site-specific factors. Thus assessment of bioavailability has resulted in the development of numerous assays, including chemical extraction techniques, and tests employing microorganisms, invertebrates, plants and higher organisms.

3.3.1 Chemical tests

Chemical extractions are commonly used to estimate the fraction of the contaminant in soil or sediment that is either taken up by plants, is mobile or impacts on soil microbiota. A variety of extraction techniques are available for extracting contaminants from soil (see Chapters 10, 14, 18–22). The specific extraction method used should be selected based on considerations of: the type of contaminant (e.g., metals or organics), media properties and objective of the assessment.

3.3.2 Toxicity tests

Bioavailability is also being estimated using a number of endpoints including a toxic response, by the inhibition of a metabolic function, by changes in microbial population structure, by mortality or malformations or by the accumulation of chemical species in organs or the bloodstream. Toxicity tests or bioassays are routinely used to make site-specific determinations of a contaminated media's toxicity, and indirectly the bioavailability of the contaminant. These tests can be used to show that the fraction of the contaminant that is bioavailable is within an acceptable or tolerable impact level. Typically, toxicity testing would be conducted to provide supplemental qualitative information as to contaminant bioavailability, and can contribute to a weight-of-evidence used for decision-making.

3.3.3 Estimating or measuring bioaccumulation directly from the environmental media

Bioavailability of a contaminant is also evaluated by determining the bioaccumulation of the contaminant. Direct measures of bioaccumulation can be obtained by analysing contaminant levels in tissues of organisms. For example, earthworms collected from contaminated soil could be analysed to determine how much contaminant has accumulated in the earthworm tissue.

Bioaccumulation may also be estimated as:

$$BAF = Cb/Cm$$

where, Cb is the concentration of contaminant in biota tissue, Cm the concentration of contaminant in media (e.g., soil or sediment) and BAF the bioaccumulation factor (biota concentration/media concentration).

In addition, bioavailability may be assessed by the dissolution of contaminants after extraction of particular mineral phases.

3.3.4 Estimating bioavailability using soil properties

Qualitative estimates of bioavailability can often be made based on information on the chemical and physical properties of the contaminant and media (e.g., soil or sediment). For example, the bioavailability of many contaminants in soil is inversely related to the organic content of the soil. The bioavailability of inorganic contaminants, such as lead, is influenced by soil pH and the form of the contaminant. With the appropriate information, a qualitative estimate of bioavailability can be determined.

Readers are directed to Chapters 14, 18–23 for a detailed discussion on chemical and biological methods used for the assessment of contaminant bioavailability in soils.

3.3.5 Extractants for assessing bioremediation potential

Recently, a number of methodologies have been developed for assessing the bioavailability of soil-borne contaminants for bioremediation potential. The majority of work has focussed on the assessment of non-exhaustive extraction methods using mild extractants. Low molecular weight primary alcohols (methanol, ethanol, propanol and butanol) have been used to assess PAH bioavailability with a number of researchers observing correlations between extracted (phenanthrene and pyrene) and biodegraded PAHs in PAH spiked soils (Liste and Alexander, 2002).

Reid et al. (2000) proposed the use of an aqueous-based extraction technique, utilising hydroxypropyl-β-cyclodextrin (HPCD), for estimating the PAH bioavailable fraction in soil. HPCD was chosen because of its high solubility, the prevalence of hydroxyl functional groups on the exterior of the torus and a hydrophobic organic cavity, making it possible to form an inclusion complex with PAHs (Reid et al., 2000; Stokes et al., 2005). This non-exhaustive extraction method was tested against the PAH catabolic activity of a *Pseudomonas* sp. in phenanthrene spiked soils. The bioavailable phenanthrene fraction correlated well with phenanthrene biodegradation ($R^2 = 0.964$), however, the butanol extraction method overestimated phenanthrene bioavailability by 60%. Since this initial study, a number of papers from Semple's research group and collaborators have shown that the HPCD extraction directly predicts the fraction available for microbial degradation in soil spiked and genuinely contaminated with individual chemicals, simple and complex mixtures of organic contaminants (Doick et al., 2005; Stokes et al., 2005; Allan et al., 2006; Doick et al., 2006; Semple et al., 2006). This extraction method offers a rapid, robust and reproducible direct determination of the fraction available for microbial degradation in soil (Doick et al., 2005). Chapter 23 provides an overview of bioavailability research and its application for predicting the efficacy of bioremediation.

3.3.6 Assessment of contaminant bioaccessibility

The physiologically based extraction test (PBET) and gastrointestinal test are *in vitro* screening-level assays used for predicting the bioaccessibility of contaminants from a soil matrix. Whilst the PBET method has been applied to both organic and inorganic contaminants, it is more commonly recognised as an assay for assessing metal/metalloid bioaccessibility. The *in vitro* methods simulate the leaching of metals from soil in a fluid matrix, which replicates the gastrointestinal tract of a 2–3-year-old infant (Ruby et al., 1996). A gastrointestinal tract representing an infant of this age is used because it is believed that 2–3-year-old infants are at the greatest risk to metal exposure due to soil ingestion.

Parameters including gastric and small intestine pH, soil mass and fluid volume, stomach mixing and stomach emptying rates and small intestine transit time can be varied to simulate different physiological states. The PBET procedure involves adding metal-contaminated soil to gastric solution buffered with glycine (pH 1.5) at a soil:solution ratio of 1:100 while the IVG methods utilises pepsin as the buffer (pH 1.8) and a soil:solution ratio of 1:150. The solution is incubated in a water bath at 37 °C, mixed with hydrated argon and maintained at the appropriate pH by the addition of HCl (where necessary). Samples are collected during the first hour (stomach residence time) after which the gastric solution is titrated to pH 7 by the addition of a dialysis bag containing $NaHCO_3$. After reaching neutral pH, the dialysis bag is removed and bile salts and pancreatin are added to simulate conditions in the small intestines. Incubation is continued and samples removed for a further 3 h. The concentration of metals in solution taken during stomach or small intestines incubations are then determined. Although both methods include an intestinal absorption phase following the gastric phase, extending the *in vitro* methods beyond the gastric phase has not been shown to improve the estimation of metal bioaccessibility. This suggests that metal bioaccessibility is controlled by dissolution kinetics within the gastric phase.

3.3.7 Microbial biosensors

Over the past 10–15 years, with the advent of more sophisticated techniques, microbial biosensors have developed from laboratory-based research to environmental application. A biosensor is a device that detects, transmits and records information relating to physiological or biochemical change (D'Souza and Jaykus, 2006). The use of microorganisms in such devices is advantageous because of their ubiquitous nature, their resistance to a variety of metals and the wide range of compounds they can metabolise. In addition, microorganisms are able to adapt to adverse conditions, are able to develop over time cata-bolic systems for new compounds and are amenable for genetic manipula-tion.

In terms of environmental biosensors, biological recognition may be developed using biocatalytic (enzymes), bioaffinity (antibodies and nucleic acids) or microorganism-based elements. Several reviews extensively describe the development and application of these recognition systems for use in environmental monitoring and analysis (Rogers and Mascini, 1998; Köhler et al., 2000; D'Souza and Jaykus, 2006; Paton et al., 2005). A number of reporter genes or regulatory proteins have been used in recombinant sensing to produce a detectable reaction in response to the analyte. Responses may be readily detectable, as in the case of green fluorescent proteins (GFP) and biolumines-cence or their activities easily measured (β-galactosidase).

Responses produced by the biological recognition systems may be detected using a variety of signal transducers including electrochemical (potentiometric striping analysis), optical-electronic (light addressable potentiometric, surface plasmon resonance), optical (absorbance, luminescence, fluorescence) or acoustic (quartz crystal microbalance, surface acoustic wave, surface transverse wave) methods.

3.3.8 Animal trials

A number of *in vivo* studies utilising a range of animal models (monkey, pig, dog, rabbit, rat) have been developed for predicting human exposure to metals in soil (Table 3.1). *In vivo* bioavailability assessment involves animal dosing experiments where the test animal is dosed with a soluble form of the contaminant (administered via gavage or intravenous injection) or contaminated material (administered orally). Contaminant absorption (or bioavailability) is

Table 3.1. *Assessment of Arsenic bioavailability in contaminated soil.*

Bioavailability assay	Source of material	As bioavailability (%)	Reference
In vitro (PBET/IVG)	Railway corridors	5–43	Juhasz et al. (2003)
In vivo (rabbits)	Smelter soils	22–31	Freeman et al. (1993)
In vivo (monkeys)	Smelter soils	~14	Freeman et al. (1995)
In vivo (dogs)	Mine sites	~8	Groen et al. (1994)
In vitro and *in vivo* (PBET, pigs)	Mine sites	0–52	Casteel et al. (1997)
In vitro and *in vivo* (PBET/IVG, pigs)	Mine sites	3–43	Rodriguez et al. (1999)
In vitro and *in vivo* (PBET/IVG, pigs)	Railway corridors, mine sites, dip sites	7–75	Juhasz et al. (2007a,b)

Note: As bioavailability was site-specific – bioavailability was dependent on As form, encapsulating matrix and soil properties.

determined by monitoring the concentration of the contaminant in the blood and/or urine over time. The absolute bioavailability of an oral dose can be calculated by dividing the area under the blood/urine time curve (AUC) by the AUC for an intravenous treatment ×100 following dose normalisation. The relative bioavailability can be determined by comparing the AUC of the soil dose to the AUC of the reference dose following dose normalisation.

3.4 Bioavailability implications to risk assessment

Currently, most risk assessment models assume that the target chemical is 100% available. However, from studies reported in the scientific literature and discussed in this chapter, this clearly not the case. As discussed above bioavailability is the underlying basis for risk assessment and remediation where contaminants pose risk to environment and human health (Naidu et al., 2004 – PN conference proceeding). The United Kingdom's Contaminated Land Regulations under Part IIA of the Environmental Protection Act of 1990 defines contaminated land as "land that appears to the local authority to be in such a condition, by reason of substances in, on, or under the land, that significant harm is being caused, or there is significant possibility that harm is being caused" (DETR, 2000). Thus, just the presence of substances of concern is not sufficient; harmful interaction with a receptor must be a possibility. Because toxic effects require that an organism takes up the contaminant, the extent to which substances are bound to soil particles or are available to cause harm needs to be considered (Semple et al., 2004). In Australia, for both ecological and human health risk assessment, bioavailability is usually reflected in default

Table 3.2. *Examples where Arsenic bioavailability adjustments have been included in remediation targets for contaminated sites in the USA (NFESC, 2000).*

Site	Contaminant	Test	Bioavailability (%)	Clean-up target ($\mu g\,g^{-1}$)	Regulator agency
National Zn Co.,	Pb	*In vivo*, rat	40	925[a] (500)[b]	Oklahoma DEQ
Bartlesville, OK	Cd	*In vivo*, rat	33	100[a] (30)[b]	
	As	*In vitro*, PBET	25	60[a] (20)[b]	
Anaconda, MT	As (soil)	*In vivo*, monkey	18	250[b]	USEPA Region VIII
	As (dust)	*In vivo*, monkey	26		
Crego Park, MI	As	*In vitro*, PBET	10	68[a] (6.8)[b]	Michigan DEQ
Union Pacific Railroad Yard, Sacramento, CA	As	*In vivo*, pig	<0.1	No clean up (up to 1,800 $\mu g\,g^{-1}$ As in slag)[a]	Cal-EPA

[a]Clean-up target after site-specific bioavailability assessment.
[b]Clean-up target prior to bioavailability adjustment.

values or site-specific data that are inserted into exposure equations. Generally a highly conservative approach is considered in that the default values are considered 100% bioavailable although a multitude of processes can affect bioavailability (see Fig. 3.1). Such an approach usually leads to remediation of sites where contaminants may not be bioavailable and hence may not pose risk to human and environmental health. Given this and the uncertainty in the decision-making process, there is now a move towards the estimation of bioavailability using animals as surrogates for humans. In the USA, bioavailability assessment has been used to adjust clean-up targets for some inorganic contaminants, such as arsenic and lead (Table 3.2). The inclusion of bioavailability data from *in vivo* (rat, monkey and swine) and *in vitro* assays (PBET) led to a lowering of clean-up targets at a number of sites.

Whilst bioavailability adjustments are scientifically justified in terms of human health risk assessment, bioavailability adjustments are impractical for ecological risk assessment due to the variability in bioavailability responses for different ecological receptors. Naturally a conservative approach will be adopted in these situations in order to protect ecological health.

To minimise uncertainty in the decision-making process, there is an urgent need to conduct further work that addresses bioavailability assessment, its scientific validity, tools and predictive models.

References

Alexander, M., 2000. Aging, bioavailability, and overestimation of risk from environmental pollutants. Environ. Sci. Technol. 20, 4259–4265.

Allan, I.J., Semple, K.T., Arya, R., Reid, B.J., 2006. Prediction of mono- and polycyclic aromatic hydrocarbon degradation in spiked soils using cyclodextrin extraction. Environ. Pollut. 44, 562–571.

Battelle and Exponent, 2000. Bioavailability of contaminants in soils and sediments: Processes, tools, and applications. Committee on Bioavailability of Contaminants in Soils and Sediments, National Research Council, 432 pp.

Benson, W.H., Alberts, J.J., 1992. Bioavailability: physical, chemical and biological interactions, Session 3, Chapter 2. Synopsis of Discussion Session on the Bioavailability of Inorganic Contaminants, Thirteenth, Pellston Workshop, Pellston, Michigan, August 17–22, Lewis Publishers.

Beresford, N.A., Mayes, R.W., Cooke, A.I., Barnett, C.L., Howard, B.J., Lamb, C.S., Naylor, P.L., 2000. The importance of source-dependent bioavailability in determining the transfer of ingested radionuclides in ruminant-derived food products. Environ. Sci. Technol. 34, 4455–4462.

Casteel, S.W., Brown, L.D., Dunsmore, M.E., Weis, C.P., Henningsen, G.M., Hoffman, E., Brattin, W.J., Hammon, T.L., 1997. Relative bioavailability of arsenic in mining wastes. Document Control No. 4500-88-AORH. US Environmental Protection Agency, Region 8, Denver, CO.

Caussy, D., 2003. Case studies of the impact of understanding the bioavailability: arsenic. Ecotoxicol. Environ. Saf. 56, 164–173.

DETR Circular 2/2000, 2000. Contaminated land: implementation of Part IIA of the Environmental Protection Act 1990. DETR, London, ISBN: 0 11 753544 3.

Doick, K.J., Clasper, P.J., Urmann, K., Semple, K.T., 2006. Further validation of the HPCD-technique for the evaluation of PAH microbial availability in soil. Environ. Pollut. 144, 345–354.

Doick, K.J., Dew, N.M., Semple, K.T., 2005. Linking catabolism to cyclodextrin extractability: determination of the microbial availability of PAHs in soil. Environ. Sci. Technol. 39, 8858–8864.

D'Souza, D.H., Jaykus, L., 2006. Molecular approaches for the detection of foodborne viral pathogens. In: J. Maurer (Ed.), PCR Methods in Foods. Springer, New York, NY.

Ehlers, L.J., Luthy, R.G., 2003. Contaminant bioavailability in soil and sediment. Environ. Sci. Technol. 37, 295A–302A.

Freeman, G.B., Johnson, J.D., Killinger, J.M., Liao, S.C., Davis, A.O., Ruby, M.V., Chaney, R.L., Lovre, S.C., Bergstrom, P.D., 1993. Bioavailability of arsenic in soil impacted by smelter activities following oral administration in rabbits. Fund. Appl. Toxicol. 21, 83–88.

Freeman, G.B., Schoof, R.A., Ruby, M.V., Davis, A.O., Dill, J.A., Liao, S.C., Lapin, C.A., Bergstrom, P.D., 1995. Bioavailability of arsenic in soil and house dust impacted by smelter activities following oral administration in cynomolgus monkeys. Fund. Appl. Toxicol. 28, 215–222.

Groen, K., Vaessen, H., Kliest, J.J.G., de Boer, J.L.M., Ooik, T.V., Timmerman, A., Vlug, F.F., 1994. Bioavailability of inorganic arsenic from bog ore-containing soil in the dog. Environ. Health Perspect. 102, 182–184.

Juhasz, A.L., Naidu, R., 2000. Bioremediation of high molecular weight polycyclic aromatic hydrocarbons: a review of the microbial degradation of benzo[a]pyrene. Inter. Biodeter. Biodegrad. 45, 57–88.

Juhasz, A.L., Smith, E., Naidu, R., 2003. Estimation of human bioavailability of arsenic in contaminated soil. In: A. Langley, M. Gilbet and B. Kennedy (Eds.), Proceedings of the 5th National Workshop on Assessment of Site Contamination. National Environmental Protection Council Service Corporation 2003. Adelaide, South Australia, pp. 183–194.

Juhasz, A.L., Smith, E., Weber, J., Rees, M., Rofe, A., Kuchel, T., Sansom, L., Naidu, R., 2007a. In vitro assessment of arsenic bioaccessibility in contaminated (anthropogenic and geogenic) soils. Chemosphere 69, 69–78.

Juhasz, A.L., Smith, E., Weber, J., Rees, M., Rofe, A., Kuchel, T., Sansom, L., Naidu, R., 2007b. Comparison of in vivo and in vitro methodologies for the assessment of arsenic bioavailability in contaminated soils. Chemosphere 69, 961–966.

Kohler, S., Belkin, S., Schmid, R.D., 2000. Reporter gene bioassays in environmental analysis. Fresenius, J. Anal. Chem. 366, 769–779.

Kramer, B.K., Ryan, P.B., 2000. Soxhlet and microwave extraction in determining the bioaccessibility of pesticides from soil and model solids. Proceedings of the 2000 Conference on Hazardous Waste Research. Denver, Colorado, US, pp. 196–210.

Liste, H.H., Alexander, M., 2000. Plant-promoted pyrene degradation in soil. Chemosphere 40, 7–10.

Moore, M.R., 2003. Risk assessment in environmental contamination and environmental health. In: R. Naidu, S. Rogers, V.V.S.R. Gupta, R.S. Kookana, N.S. Bolan, D. Adriano (Eds.), Bioavailability and its Potential Role in Risk Assessment. Science Publishers, New York.

Naidu, R., Rogers, S., Gupta, V.V.S.R., Kookana, R.S., Bolan, N.S., Adriano, D., 2003. Bioavailability of metals in the soil-plant environment and its potential role in risk assessment: An overview. In: R. Naidu, S. Rogers, V.V.S.R. Gupta, R.S. Kookana, N.S. Bolan and D. Adriano (Eds.), Bioavailability and its Potential Role in Risk Assessment. Science Publishers, New York, pp. 21–59.

National Research Council (2002). Bioavailability of Contaminants in Soils and Sediments: Processes, Tools and Applications. National Academies Press, Washington, DC.

Newman, M.C., Jagoe, C.H., 1992. Bioavailability: physical, chemical and biological interactions, Session 3, Chapter 1. Ligands and Bioavailability of metals in aquatic environments, Thirteenth, Pellston Workshop, Pellston, Michigan, August 17–22, Lewis Publishers.

NFESC, 2000. Guide for incorporating bioavailability adjustments into human health and ecological risk assessments at US Navy and Marine Corps facilities, Part 1: Overview of metals bioavailability, User's Guide UG-2041-ENV.

Paton, G.I., Weitz, H., Killham, K., Semple, K.T., 2005. Biological tools for the assessment of contaminated land. Soil Use Manag. 21, 487–499.

Peijnenburg, W.J.G.M., Posthuma, L., Eijsackers, H.J.P., Allen, H.E., 1997. A conceptual framework for implementation of bioavailability of metals for environmental management purposes. Ecotoxicol. Environ. Saf. 37, 163–172.

Reid, B.J., Stokes, J.D., Jones, K.C., Semple, K.T., 2000. Nonexhaustive cyclodextrin-based extraction technique for the evaluation of PAH bioavailability. Environ. Sci. Technol. 34, 3174–3179.

Rodriguez, R.R., Basta, N.T., Casteel, S.W., Pace, L.W., 1999. An in vitro gastrointestinal method to estimate bioavailable arsenic in contaminated soils and solid media. Environ. Sci. Technol. 33, 642–649.

Rogers, K.R., Mascini, M., 1998. Biosensors for field analytical monitoring. Field Anal. Chem. Technol. 2, 317–331.

Ruby, M.V., Davis, A., Schoof, R., Eberle, S., Sellstone, C.M., 1996. Estimation of lead and arsenic bioavailability using a physiologically based extraction test. Environ. Sci. Technol. 30, 422–430.

Ruby, M.V., Schoof, W.B., Brattin, W., Goldade, M., Post, G., Harnois, M., Mosby, D.E., Casteel, S.W., Berti, W., Carpenter, M., Edwards, D., Cragin, D., Chappel, W., 1999. Advances in evaluating the oral bioavailability of inorganics in soil for use in Human Health Risk Assessment. Environ. Sci. Technol. 33, 3697–3705.

Semple, K.T., Dew, N.M., Doick, K.J., Rhodes, A., 2006. Does mineralisation predict the microbial availability of organic chemicals in soil?. Environ. Pollut. 140, 164–172.

Semple, K.T., Doick, J.K., Jones, K.C., Bureauel, P., Craven, A., Harms, H., 2004. Defining bioavailability and bioaccessibility of contaminated soil and sediments is complicated. Environ. Sci. Technol. 2, 229A–231A.

Sposito, G., 1989. The Chemistry of Soils. Oxford University Press, New York, 277pp.

Stokes, J.D., Wilkinson, A., Reid, B.J., Jones, K.C., Semple, K.T., 2005. Validation of the use of an aqueous hydroxypropyl-β-cyclodextrin (HPCD) extraction as an indicator of PAH bioavailability in contaminated soils. Environ. Toxicol. Chem. 24, 1325–1330.

USEPA, 1997. Engineering bulletin: Technology alternatives for the remediation of soils contaminated with As, Cd, Cr, Hg, and Pb. O. o. E. a. R. Response. Washington, DC, United States Environmental Protection Agency, pp. 1–20.

Developments in Soil Science, volume 32
Ravendra Naidu (Editor)

Chapter 4

BIOAVAILABILITY: THE UNDERLYING BASIS FOR RISK-BASED LAND MANAGEMENT

R. Naidu, S.J.T. Pollard, N.S. Bolan, G. Owens and A.W. Pruszinski

4.1 Introduction

An inevitable legacy of industrialisation has been the introduction of contaminants into soils and groundwaters in a number of countries throughout the world. In particular, in the Australasia-Pacific region where legislation guiding disposal of industrial and municipal wastes was only introduced within the past 15 years. This resulted in the continued disposal of waste material onto land and water bodies up until the mid-1980s. These activities still persist in many of these countries due to a lack of adherence to regulation and, in some cases, poor enforcement.

Waste disposal has resulted in some 3,000,000 contaminated sites across the Australasia-Pacific region with many suspected of contamination. Of these, nearly 80,000 contaminated sites have been identified in Australia (Natusch, 1997, see Chapter 2). Sites range in scale from localised point sources of contamination (e.g. a single leaking underground storage tank or old landfill sites) to large industrial or defence mega-sites. On a regional basis, examples of contamination include areas of land surrounding smelters and the overuse of fertilisers that have widespread impact on groundwater.

The scale of the problem is difficult to assess, as the definition of "contaminated land" or "site contamination" has often been subjective or poorly defined. Only limited efforts have been made to develop an inventory of contaminated sites in the Australasian-Pacific region. Uncertainties about the nature and extent of contamination can be a major constraint to sustainable development in cities and rural areas, thereby increasing pressure on the use of limited uncontaminated land. Moreover, the techniques available for *in-situ* or *ex-situ* remediation can be prohibitively expensive, resulting in poor adoption in developing countries.

Policy makers and regulators have grappled with the decision criteria for remediation over decades. A central theme has been the distinction between hazard and risk. The presence of chemical substances in soils and groundwater is of concern, but harm to the environment, human health and/or ecosystems require exposure.

Two policy shifts have been notable over the last 25 years – firstly, recognition that it is the management of the risks posed by the chemical substances that should drive remedial action; and secondly, in line with this, that not all chemical substances require risk management to the same degree.

The assessment and management of site contamination is dependent upon the proposed land use, community and governmental expectations, and the resources available for the task. Social, financial and environmental factors must all be taken into account in decision-making. Together, these concepts form the basis of risk-based land management (RBLM), which represents a mature, sustainable approach to the challenges of contamination.

4.2 Risk-based land management

The protection and management of soils and groundwater has slowly climbed the environmental agenda. In common with many international jurisdictions, and in order to protect the users of land and water, environmental regulations have gradually evolved in the region. While legislation differs amongst the regional Asia-Pacific countries, the underlying principle of environmental regulation is to ensure land and water is fit for use and does not pose an unacceptable risk to human health or the environment. Within "modern regulation" is the specific intention to avoid prescriptive remedies and unnecessary cost and to allow solutions that are appropriate for future use.

Flexibility is key and decision-makers now prefer to secure a greater evidence base for their decisions on risk management, than was historically the case. Thus, where sites do not pose an unacceptable risk to end users, remedial options may include risk management rather than total clean-up of the site.

"Risk-based land management" in Europe and Australia fundamentally embodies the axiom that, for chemical risk, "the dose makes the poison".[1] That is, that there has to be exposure before harm from the exposure can be realised. This underlying principle draws out the distinction between hazard (a substance or situation with the potential to cause harm) and risk (a combination of the likelihood and consequences of a hazard being realised). This is not semantics. The distinction cautions against an overemphasis on the presence of contamination in soils or groundwater without reference to the concept of exposure. Without someone being exposed to site contamination, the likelihood of exposure is zero and harm cannot occur and thus the risk cannot be realised.

---

[1]"All substances are poisons; there is none which is not a poison. The right dose differentiates a poison ..." Paracelsus (1493–1541).

This is a powerful concept for decision-makers – it allows them to:

- consider the relationship between the source, exposure pathway and receptor (the thing/person we value);
- evaluate whether these components are, or are likely to be linked; and
- determine the strategies for reducing exposure, say through source removal (dig and dump), pathway management (soil capping, vapour extraction, hydroseeding) or receptor control (land use planning, restriction on groundwater use).

Whilst critics retain reservations about such strategies in that they may not permanently remove the hazard (source), the risk-based approach does offer a more scientific, toxicologically robust and proportionate approach in keeping with effective environmental protection. Its implementation however is not without difficulty.

One issue with not removing the chemical substance is the perpetual management of the hazard. Questions of ongoing liability and transfer of information remain. US studies have shown that after three sales of a property, the information relating to site contamination is lost.

Many regulators, professional advisors and decision-makers have routinely adopted a "brightline", standards-based approach to regulation and decision-making, where apparent clarity is assured by reference to a definitive concentration above which the matrix is considered contaminated. The concept of risk management brings with it a demanding responsibility to evidence decisions with sound science. Often the science can be contentious and may mistakenly appear to work against the tenets of human health protection, especially if improperly communicated.

Communities have become sophisticated contributors to scientific knowledge and retain an acute sense of natural justice. Risk-based arguments for contamination management may not always secure favour when there is a strong focus on hazards in a climate of outrage. Often risk is defined as the sum of Hazard and Outrage.

$$Risk = Hazard + Outrage \qquad (4.1)$$

The Site Contamination NEPC (1999) states that outrage is a factor that includes the personal inequities, emotions or concerns that the hazard, situation or the responsible party may evoke. Outrage may also be determined by people's perceptions of the organisation and people who are working on solving the problem, i.e. trust and credibility issues can also drive outrage and perceptions of risks. Outrage may also determine what people see as important technical issues which the consultant or regulator may not have included, e.g. noise problems, smell, appearances affecting property values, etc. These may be seen by the community to be personal inequities.

4.3 Bioavailability and risk management

The concept of bioavailability deals with the availability of contaminants beyond the immediate soil matrix and thus with their availability for onward exposure (see Chapters 2 and 3). For many chemical substances, bioavailability decreases with aging (see for example metals, Fig. 4.1 and Chapters 2–3), such that the toxic effects of substance may eventually decrease to concentrations below concern for human and environmental health. Thus, through this approach, it may not be a requirement to secure the full clean-up and removal of all contamination, where this is not necessary, depending on the available exposure pathways. In practice, recognition of this reduced likelihood of exposure and adoption of this approach has already avoided enormous costs to the Australian community.

   In the case of some of the largest contaminated sites in Australia, such as those in the Homebush Bay area (the Sydney Olympic site) and where large industrial operations have taken place (such as steel making and gas manufacture), adopting risk-based clean-up strategies have resulted in remediation cost savings of tens of millions of dollars for each site. However, the risk-based approach to remediation is an area that remains controversial.

   Regulators throughout the world are quite rightly demanding evidence that contaminants left behind in the soil do not pose a risk under changing

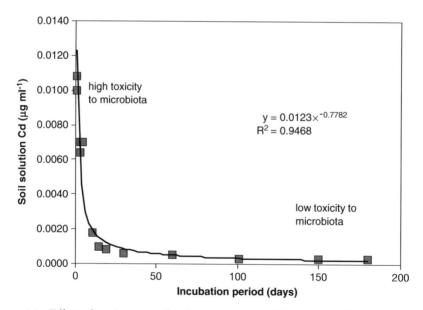

Figure 4.1. *Effect of ageing on soil solution cadmium (Vig et al., 2002).*

conditions of exposure through:

- the introduction of new receptors (change of land use) for example, or of new pathways for exposure (new basements being built); or
- the introduction of a new source of contamination (solubilising solvents, for example).

Under these scenarios, there are changes to the likelihood of exposure and these may dramatically affect the risk of harm being realised.

4.4 Risk management – policy

As a policy response to the problems of site contamination then, RBLM is seen to be far from straightforward. Even though there is ongoing remediation of contamination and hundreds of millions of dollars are being spent each year on satisfying clean-up criteria, there remains considerable uncertainty regarding:

- the basis for the clean-up criteria;
- the magnitude of the resulting and residual risk; and
- the application of potentially more cost-effective management and remediation methods which reduce residual contamination levels and risk.

Many people adopt a conservative approach of using NEPM HIL or EIL as clean-up criteria. Such an approach misuses the intention of the published values. Uncertainty arises because of:

- the complexity of the issues and a limited understanding of the factors involved; and
- the difficulty of measuring relevant environmental and human health indicators.

These uncertainties are further exacerbated by:

- the lack of data relating to ecological systems and human health for the various chemical substances under the conditions that apply in soils and water; and
- the implications and risks associated with residual contamination and how this should be managed in perpetuity.

These uncertainties in risk, liability and technology application, together with increasing incidences of toxicity reported in the Asia-Pacific region and elsewhere, necessitate the need for focussed research to develop approaches (including technology) for the management and remediation of contaminated environments. In this chapter, we further discuss the sources of contaminants, their interactions in soils (see Chapter 2), potential risks to the environment and human health and cost-effective strategies including RBLM that could be adopted to minimise environmental impact of such pollutants.

4.5 Sources and definition of contamination

Environmental contaminants may have either anthropogenic or geogenic origin. Typically, the introduction to the environment of substances liable to cause a hazard to human health, harm to natural resources and ecological systems, damage to structures and amenities or interference with any legitimate use of the environment is usually associated with anthropogenic sources. Geogenic sources are typically natural sources of substance (or energy) such as the natural weathering process of mineral deposits (for example, acid sulphate soils).

In risk terms, of course, such distinctions are immaterial. The Site Contamination NEPM provides a definition for contamination as "the condition of land or water where any chemical substance or waste has been added at above background level and represents, or potentially represents, an adverse health or environmental impact". The NEPM goes on to define a chemical substance as "any organic or inorganic substance, whether liquid, solid or gaseous".

The chemical substance is thus categorised by the risk it poses. In terms of the potential consequences of exposure, a simple distinction may be made between a chemical substance which may have no observed harmful effects, and contamination, which is likely to cause toxicity and toxic or hazardous materials.

Indeed, contaminants are categorised in a variety of different ways including:

- the medium to which they are released, i.e. air, water or land;
- according to their physical characteristics, i.e. gaseous, liquid or solid;
- type of hazard they pose, i.e. corrosive, reactive, biological or radioactive; or
- according to their origin, i.e. naturally occurring or municipal and industrial/manufactured waste.

The most common way of classifying contaminants is according to the source which has two major advantages:

- it takes into account the processes that generate the contaminants, and thus permits identification of areas where waste generation can be minimised or recycled;
- it permits the tracking of the source such that proper and safe management and/or disposal of the contaminants can be implemented.

On the basis of source of origin, contaminants are therefore commonly grouped into two major classes:

- agricultural; and
- industrial and municipal.

4.6 Contaminant interaction

As introduced above, the extent to which contaminants are bound to the soil matrix may have a marked influence on their availability for specific exposure pathways. Contaminants are present in the environment either as free species, bound to mineral and organic colloidal surfaces or soluble species with the extent of binding dependent on the duration of contact (see Chapter 2).

If the environment is soil, then binding generally includes both mineral and organic constituents of the soil as shown in Figure 4.2.

4.7 Key contaminants

Key contaminants of importance are:

- *Heavy metals*: Research to date has concentrated mainly on the factors that affect the bioavailability and plant uptake of heavy metals.
- *Persistent organic compounds(or pollutants) and mixed organic/inorganic contaminant systems*: There is a need for new remedial methods that are both cost-effective and less energy intensive than the current remedial methods available that can still reduce risk efficiently.
- *Radionuclides*: The long-term persistence of radioactive substances is of special concern. There is therefore a need to assess the impact and identify the most appropriate options for remediation and future utilisation of sites contaminated with radioactive substances over much longer timescales than any other class of contaminant.
- *Non-aqueous phase liquids (NAPLs) in groundwater*: The challenge with this contaminant is to develop methods capable of achieving and defining acceptable "end-point" concentrations of residual NAPL.

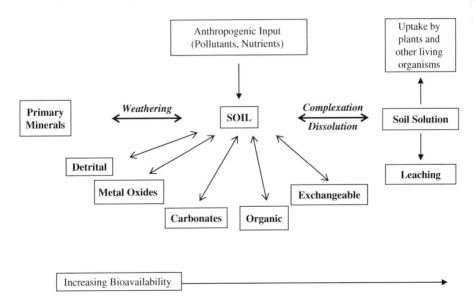

Figure 4.2. *Interactions of contaminants with soil and soil solution.*

- *Dissolved contaminants in groundwater*: one remediation strategy for sustainable long-term solutions to groundwater contaminant problems uses reactive barriers.
- *Contaminants that require containment*: Such contaminants require the successful integration of containment with continued land use, management and other risk reduction techniques.
- *Contaminants that naturally degrade*: The factors that influence the dynamics of potentially degradable compounds (e.g. organics, nutrients, etc.) in soil and water environments, and the critical factors that enhance natural attenuation need to be examined in greater detail and optimised for each site.

Contaminants present in the soil matrix are generally the most readily bioavailable and as this pool is exhausted, replenishment occurs from those pools that contain labile forms of contaminants.

For metals, the most labile fraction is often that associated with the ion-exchange sites of the soil while for organics the labile contaminant fraction is more often associated with the soil organic matter pool. The replenishment capacity of contaminants in soil solution is generally dictated by the reaction capacity (size of the sorption sites) and the extent of saturation with respect to the particular contaminant. However, soils with a large reaction capacity may not necessarily have higher replenishment capacity, because under saturation of any particular contaminant, a lower rate of replenishment may result. For this reason, it is important to initially assess the reaction capacity of the soils by estimating their capacity to retain additional contaminants and this factor must be included in any estimation of risk that contaminants pose to environmental and human health.

4.8 Contaminant interaction varies with soil type

The capacity for soils to bind contaminants may vary with soil type depending on the nature of mineral and organic fractions constituting the soils (see Chapter 2). In general, temperate soils, such as Xeralfs and Vertisols, containing expanding minerals often retain much larger amounts of metals relative to oxidic tropical soils with pH values below 6 (Fig. 4.3). This suggests that metal bioavailability, in strongly weathered tropical soils with low binding capacity, would be much higher than that recorded in temperate less weathered soils such as Xeralfs, Vertisols, etc.

Detailed sorption studies conducted by Naidu et al. (1998) shows increasing metal sorption in the general order: Oxisol < Inceptisol < Ultisol < Vertisol/Mollisol.

As observed with metals, pesticide sorption, as measured by partition coefficient (K_d) has been reported to vary widely with different soil types and organic matter content (Brusseau and Rao, 1989; Johnson and Sims, 1993).

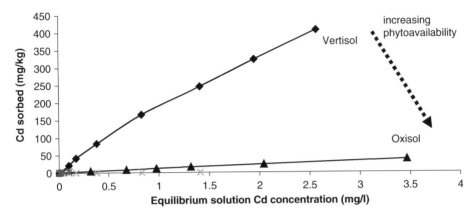

Figure 4.3. *The effect of soil type on Cd sorption.*

For example, Bolan and Baskaran (1996) observed that the K_d values for 2,4-D herbicide increased with an increase in the organic carbon content of the soils (see Fig. 2.11). Since organic carbon (OC) is considered to be the principal soil component involved in pesticide sorption, the K_d values are generally normalised to the organic carbon content of the soils $[K_{oc} = (K_d) (100)/(\% \ OC)]$.

It has often been observed that although the K_d values of organic pesticides increase with organic matter content of soils, the K_{oc} values varied less between soils. Generally, temperate soils, that are higher in organic matter content, show higher pesticide retention compared to tropical soils.

This variation in contaminant binding capacity with soil type has major implications for bioavailability and the consequent potential risks that contaminants pose to the environment and human health and is discussed further in the following sections after a brief definition of risk assessment.

4.9 Historical approach to remediation

The South Australian EPA define[2] remediation as "the treatment, containment, removal or management of chemical substances or wastes so that they no longer represent an actual or potential risk to human health or the environment, taking into account the current and intended use of the site".

Historically, the most commonly used method of remediation has been to excavate contaminated soils and dispose of them to landfill. In Australia this is termed dig and dump. This approach consumes valuable landfill space and may perpetuate the contamination problem, albeit in a different location. It does not

---

[2]Environmental management of on-site remediation; March 2006.

directly address groundwater contamination. However, source removal is considered to be one of the most effective tools in groundwater remediation.[3]

It is highly desirable to apply remedial approaches that reduce the risk of contamination while allowing the soil to remain on site, and allow *in situ* soil and groundwater treatment.

The Site Contamination NEPM provides a hierarchy for clean-up and/or management. On-site treatment of the contamination so that it is destroyed or the associated risk is reduced to an acceptable level is the most preferred option.

There is however, a general lack of knowledge worldwide on how remedial technologies and management options can be applied to realise cost-effective alternative strategies that achieve sufficient risk reduction to make these *in-situ* remediation applications feasible.

It is now recognised, however, that drastic or overly conservative risk control; for instance cleaning up sites to background concentrations or to concentrations suitable for the most sensitive land use is generally neither technically, nor economically feasible.

Ferguson and Kasamas (1999) reported the number of contaminated sites in the Netherlands growing from 350 in 1981 to 300,000 (following surveys that identify these) with the estimated clean-up cost of these sites at the time being €13 billion, all on account of a policy of "multi-functionality" that allowed for sites being cleaned up for a range of end-uses.

A similar number of contaminated sites exist in other industrialised countries with clean-up costs exceeding billions of dollars. These massive clean-up costs, together with the technological challenge of attempting to reduce chemical substances to background levels led remediators and regulatory bodies to reassess legislation and question whether there is need for total removal of the chemical substances from the soil (Ferguson et al., 1998a, b).

Recognition of the practical and economic infeasibility of multi-functionality led to uptake of the risk-based approach to land management whereby sites are prepared suitable for use. The risk-management paradigm is seeing broader adoption among policy makers and the focus is on risk reduction and management rather than hazard elimination. However, there still appears to be concern among some regulators with respect to implementation of the risk-based approach. Less enthusiastic advocates cite a lack of consistency in the selection of remedial objectives and varying degrees of competency among professional advisors as reasons for this.

---

[3]Dennis Paustenbach, "Lessons Learned Over the Past 30 Years Regarding the Use of Risk Assessment to Deal with Contaminated Sites – Reflections on Experiences at more than 50 Sites in the U.S. and Elsewhere", 2006.

Whilst the "standards" approach offers clarity and simplicity for decision-makers, it is widely recognised that risk assessment requires well-developed thinking to be applied to the site-specificities of site contamination. This controversy is generally related to the scarcity of scientific understanding and consequently acceptance of these risk-based regimes under the regulatory arrangements. However much of the information presented is at the inception phase and there is not a lot of experienced practitioners or demonstrated successes that are suitable for the promotion of such concepts.

Even where regulatory authorities have accepted risk-based approaches, the uncertain foundations for these regimes have left significant exposure in public debate and potential liability. The public is unclear about the basis for community and environmental protection, which in key instances has led to heated public debate and further government intervention. In response, regulators have been considering afresh how best to engage communities in decisions on site contamination (Fischer, 1993; Petts et al., 2003).

Specifically, there continues to be considerable uncertainty regarding:

- the basis for the clean-up criteria;
- the magnitude of the resulting risk;
- the application of new cost-effective management and remediation methods which target residual contamination and risk; and
- regulatory and community acceptance of the management strategies.

Addressing these requires a better understanding of the legal, economic and societal factors that drive the need for remediation and affect the acceptance of management and remedial strategies. These uncertainties in risk and technology application can create significant exposure to liability.

4.10 Risk assessment

4.10.1 Definition

Many definitions of the concept of risk assessment exist amongst the scientific and regulatory community. Such definitions may range from informal qualitative (empirical or operational) to formal quantitative (mathematical) (pragmatic or functional) definitions. However, there are still ongoing debates about the measurability or predictability of risk. This is further complicated by perceived risks that makes risk communication an extremely important component of the risk assessment process. While definitions of risk assessment vary slightly a well-accepted one is that of the United States National Academy of Science (1983), which recommends that:

> risk assessment ... mean(s) the characterisation of the potential adverse health effects of human exposures to environmental hazards. Risk

assessments include several elements: description of the potential adverse health effects based on an evaluation of results of epidemiological, clinical, toxicological, and environmental research [hazard identification]; extrapolation from those results to predict the type and estimate the extent of health effects in humans under given conditions of exposure [dose response assessment]; judgments on the number and characteristics of persons exposed at various intensities and durations [exposure assessment]; and summary judgments on the existence and overall magnitude of the public-health problem [risk characterisation]. Risk assessment also includes characterisation of the uncertainties inherent in the process of inferring risk.

In Australia, the NEPM provides the definition of risk management as "the decision-making process involving considerations of political, social, economic, environmental and engineering factors associated with site contamination together with risk-related information to identify, develop, analyse and compare the range of options for site management and select the appropriate response to a potential health or environmental hazard".

The risk assessment of site contamination involves an assessment of potential risk posed by chemical substances to environmental and human health. Generally in most countries, site contamination risk assessment considers the following end points:

(i)   human health;
(ii)   ecological risk; and
(iii) groundwater and surface water resources.

An early categorisation of the different components used for assessing risks posed by contaminants are presented in Figure 4.4.

4.10.2 Move towards risk-based land management

More recent conceptualisations of the process of RBLM reflect the risk analysis–options appraisal–risk management paradigm (Fig. 4.5). Here, problem definition and conceptual model development precede tiered risk assessment (qualitative first; quantitative if supportable), which then informs the identification and appraisal of options for risk management (accept risk, reduce it, transfer it), being finally followed by implementation of risk management with the monitoring of residual risk. The Environment Agency's (England and Wales) "Model Procedures" (for managing land contamination), for example, adopt this approach which is viewed as consistent with UK government approaches to environmental risk management in general.

The model procedures map the stages of risk analysis, options appraisal and risk management (Fig. 4.5) across to the more conventional process of land management (Fig. 4.4). Importantly, the procedures offer a formal role for the

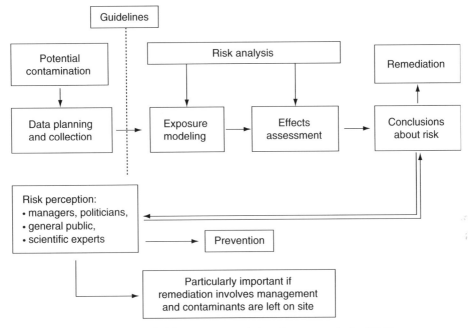

Figure 4.4. *Components of risk assessment (after Ferguson et al., 1998).*

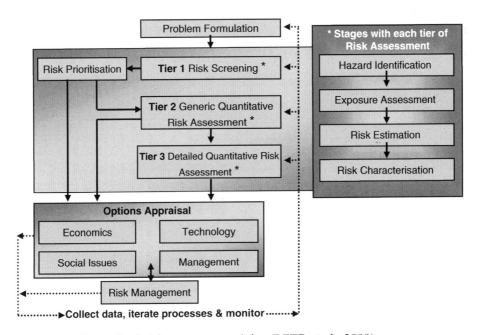

Figure 4.5. *Generalised risk management (after DETR et al., 2000).*

consideration of technological capability, social issues, economic factors and management practicalities alongside risk significance as factors that influence the selection of risk management options (Pollard et al., 2000; 2004).

Under this approach, issues of economics and practicality play a more formal role. Here, many long-term contaminated sites may not require intensive remediation given the positive role that nature plays with time. RBLM stresses the importance of integrative sustainable solutions for restoring the usability and economic value of land (Kasamas et al., 2000). Such solutions require adoption of multicriteria decisions tools that can handle hard and soft data (Linkov et al., 2005). According to Kasamas et al. (2000) these solutions can be characterised by:

(i)   risk reduction;
(ii)   environment protection; and
(iii) reduction of aftercare requirements.

The RBLM process was evolved during the Contaminated Land Rehabilitation Network for Environmental Technologies (http://www.clarinet.at).

The process was applied in Australia as early as 1996 to demonstrate to residents and regulatory body that presence of contaminants at a site does not necessarily represent unacceptable risk (R Naidu, unpublished, 1996). As discussed above the underlying principle of Australian environmental regulation is to seek to ensure that land and water is fit for use, and does not pose a risk to human health or the environment. It is a specific intention of the regulations to avoid prescriptive remedies and unnecessary cost, and to allow solutions that are appropriate for future use (ANZECC, 1992). This approach was used to demonstrate that contaminants present in the backyard of residential properties were not bioavailable and did not pose risk to soil, water and human health.

The RBLM strategy requires that all of the key elements outlined in Table 4.1 are addressed in order to ensure the sustainability of the solution for a site contamination problem.

4.10.3 Selecting optimal risk-management strategies

Selection of optimal risk-management strategies requires consideration of core objectives such as risk management, technical practicability, feasibility, cost effectiveness of the strategy and wider environment, social and economic impacts. Deciding on which management options are most appropriate for a particular site needs to be considered in a holistic manner (Bardos and Vik, 2001a, b, c). Key factors that need to be considered during the decision-making process are listed in Table 4.2.

The six factors listed in Table 4.2 are the key elements that need consideration during the application of risk-based management strategies. Arriving at an

Table 4.1. *Key elements of risk-based land management (after Bardos et al., 1998).*

Elements	Requirements
Risk reduction	Achieved by removing the source and either breaking or eliminating the path and/or removing the receptors.
Land use	Need to consider land use requirements and tailor risk management accordingly.
Spatial planning	Spatial planning requirements include the proposed and any future land use requirements including land use impacts on subsoil especially in relation to surface and groundwater.
Management	Choice of certain solutions may depend on other factors such as funding mechanisms, communication with stakeholders and the general public. Decision-making process must include all of these elements.

Table 4.2. *Key factors in decision-making involving risk-based land management.*

Key factors	Components considered in risk-based land management
Drivers and goals for remediation	To protect human health and environment To enable redevelopment To limit future potential liabilities.
Risk management	Identify source and nature of contaminants Identify receptor (e.g. humans, ground water, ecosystems, etc.) Identify pathway (route to receptor).
Sustainable development	Development that meets the needs of the present without comprising the ability of future generations to meet their own needs. Underpinning sustainable development are: 1. Economic growth 2. Environmental protection; and 3. Social progress.
Stakeholder satisfaction	Important to seek consensus between the different stakeholders to achieve sustainable development.
Cost effectiveness	Costs and benefits attributed between different strategies for managing site contamination must be considered.
Technical suitability and feasibility	Suitable technology is one that meets the technical and environmental criteria for dealing with a particular remediation problem.

optimal risk-management solution for site contamination, however, involves three main phases of the decision-making process. These include problem identification, development of problem solving alternatives and selection of the best alternative.

In the discussion below, we consider how this process works using a practical example for an arsenic (As) contaminated site.

4.11 Case study: As contaminated soil – application of risk-based land management

4.11.1 Background

The site, in Australia, had previously been shown to be extensively contaminated with As. These studies found that As exceeded both the ecological (20 mg kg^{-1}) and health investigation levels (100 mg kg^{-1}) (NEPC, 1999) and was appreciably water soluble indicating that large amounts of As were potentially mobile at this site.

The historical source of the As appears to be the ubiquitous use of As-based herbicides to limit plant growth on site. Exposure pathway analyses showed that the highly mobile As posed risk to both groundwater and residents living in the area. Risk assessment determined that the soils, unmanaged, would continue to act as an ongoing source for groundwater impact. The site was identified for industrial development.

Options for managing contaminated soil included *in-situ* clean-up, excavation and transport to landfill sites or application of an *ex-situ* RBLM strategy. Both *in-situ* clean-up and excavation and transport to landfill were found to be prohibitively expensive. An added complication with the transport to landfill site was the regulatory requirement for the *ex-situ* stabilisation of the As prior to transport to the landfill as the total As content exceeded the concentration allowed at the landfill.

A risk-based management strategy was therefore considered the most cost-effective option, provided that the strategy was acceptable to stakeholders and the public. Key elements considered during the development of this strategy included: source characterisation and identification of pathways and receptors. In implementing this strategy, the community and regulatory bodies were consulted. The effectiveness of a risk reduction strategy where As remained at the site such that it would not pose risk or hinder planned future land use requirements was discussed.

Risk reduction included consideration of basic As chemistry that was used to reduce the risk posed by both mobile and immobile As present at the site.

4.11.2 Basic As chemistry and remediation strategies

It is commonly accepted that in most soils As primarily exits either as arsenite (As(III)) or arsenate (As(V)) and their relative distribution is determined by the prevailing pH and redox potential of the soil. However, the presence of colloidal material can also significantly influence the bioavailability of As.

The pH not only controls the charge of the As oxy-anion but also the charge of a potential sorption surface. The surfaces of Mn, Al and Fe oxides/hydroxides are positively charged below pH 3, 5 and 8, respectively (Sadiq, 1997) and hence adsorption on soil colloids is more pronounced at low pH, since at high pH most colloids have a net negative charge which favors As solubility.

The fundamental strategy upon which risk reduction was based was to increase the reaction capacity of the soil for As. This may be achieved by either:

1. changing the physical properties of the soil so that As is more tightly bound and therefore less bio-available;
2. chemically immobilizing As either by sorption onto a mineral surface or by precipitation as a discrete insoluble molecule; or
3. diluting the contaminated soil with uncontaminated soil thus increasing the number of As binding sites.

However, the last option is unacceptable to most environmental regulators.

With this basic knowledge in mind the project concentrated mainly on strategy (2) and bench-scale experiments assessed if new mineral phases could be generated *in situ* to immobilse As. This should theoretically be possible by amending the soil with a ferrous salt and allowing the Fe to oxidise and precipitate as described below.

$$4FeSO_4 + O_{2(g)} + 6H_2O \Leftrightarrow 4FeOOH_{(s)} + 4SO_4{}^{2-} + 8H^+$$

This reaction requires oxygen to be available to the soil and also generates considerable amounts of acid, which may be counter productive to As immobilisation in poorly buffered soils. The increased acidity could additionally be neutralised by the amendment with lime, for example as $CaCO_3$, as described below.

$$2FeSO_4 + \tfrac{1}{2}O_2(g) + 2CaCO_3 + 5H_2O \Leftrightarrow 2FeOOH_{(s)} + 2CaSO_4 \cdot 2H_2O + 2CO_{2(g)}$$

The redox conditions of the soil will also influence the speciation of As and an example of two possible redox couples is given below.

$$Fe_2O_3\ (s) + 4H^+ + AsO_3{}^{3-} \Leftrightarrow 2Fe^{2+} + AsO_4{}^{3-} + 2H_2O \qquad E_o = 0.21\ V$$

$$MnO_2\ (s) + 2H^+ + AsO_3{}^{3-} \Leftrightarrow Mn^{2+} + AsO_4{}^{3-} + H_2O \qquad E_o = 0.67\ V$$

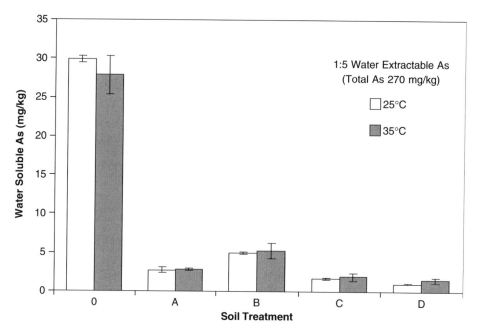

Figure 4.6. *Variation of water extractable As (1:5) for a subsurface contaminated soil with soil treatment and incubation temperature. The treatments were (0) control soil; (A) Fe; (B) Fe + lime; (C) Fe + Mn and (D) Fe + Mn + Al.*

Under alkaline conditions, such as those expected at the site, these reactions would also favour the formation of the oxides. The formation of Mn(IV) is also advantageous since Mn oxides can oxidise As(III) to the less mobile As(V) and may also catalyse the adsorption of As onto Fe and Al oxides (Oscarson et al., 1983). Following initial detailed laboratory studies, a mixture of Fe/Mn/gypsum was used as the stabilizing chemical. As shown in Figure 4.6, application of the mixed chemical led to significant decline in mobile As. Subsequent studies involving aging of the treated soil showed complete elimination of risk posed by As.

This strategy provided significant saving (25%) to the client.

4.12 Conclusion

In many cases contaminants will naturally degrade to concentrations below concern without external influence. This is especially true of low molecular weight organic contaminants that provide good substrates for microbial activity. Larger and complex organic contaminants and heavy metals are generally more persistent in the environment and will generally require intervention to ensure low risk to human and environmental health. In all cases a risk assessment, based on the land use, must be undertaken to determine the best strategy for managing the contaminants present on or below the site.

The level of intervention required in any remediation strategy needs to be reviewed on a site-by-site basis. As such a practitioner should carefully consider the natural reaction capacity of the soil environment and use it to its full potential. Hence full consideration of *in-situ* methods or *ex-situ* methods that promote the low bioavailability of contaminants and hence low risk in the most cost-effective manner need to be considered.

Remediation of site contamination to background concentrations should rarely be necessary and can result in overly expensive programs. Risk to human and environmental health can be minimised by remediation to levels necessary to meet regulatory agreed safe endpoints.

References

Australian and New Zealand Environment and Conservation Council 1992. National Health and Medical Research Council: Australian and New Zealand Guidelines for the Assessment and Management of Contaminated Sites. ANZECC, NHMRC, Canberra.

Bardos, R.P., Vik, E., 2001a. Decision making for the remediation of contaminated sites. Presented at the ICS-UNIDO Expert Group Meeting on "Remediation of Polluted Sites in CEE Countries, Current Status and Perspectives".

Bardos, R.P., Vik, E., 2001b. A framework for selecting remediation technologies for contaminated Sites. Proceedings of the CLARINET Final Conference, Vienna, 21–22 June.

Bardos, R.P., Vik, E., 2001c. Summary of CLARINET's key findings on risk management solutions and decision support in Europe. Proceedings of the CLARINET Final Conference, Vienna, 21–22 June.

Bolan, N.S., Baskaran, S., 1996. Biodegradation of 2,4-D herbicide as affected by its adsorption-desorption behaviour and microbial activity of soils. Aust. J. Soil Res. 34, 1041–1053.

Brusseau, M.L., Rao, P.S.C., 1989. The influence of sorbate-organic matter interactions on sorption nonequilibrium. Chemosphere 18, 1691–1706.

Department of the Environment, Transport and the Regions (DETR), Environment Agency and Institute for Environment and Health 2000. Guidelines for Environmental Risk Assessment and Management: Revised departmental guidance. The Stationery Office, London.

Ferguson, C., Darmendrail, D., Freier, K., Jensen, B.K., Jensen, J., Kasamas, H., Urselai, A., Vegter, J. (Eds.), 1998a. Risk Assessment for Contaminated Sites in Europe Volume 1: Scientific Basis. LQM Press, Nottingham.

Ferguson, C., Darmendrail, D., Freier, K., Jensen, B.K., Jensen, J., Kasamas, H., Urselai, A., Vegter, J., 1998b. Risk Assessment for Contaminated Sites in Europe Volume II–Policy, Land Quality Press, Nottingham.

Ferguson, C., Kasamas, H., 1999. Risk Assessment for Contaminated Sites in Europe, Volume 2: Policy Frameworks, LQM Press, Nottingham.

Fischer, F., 1993. The greening of risk assessment towards a participatory approach. In: D. Smith (Ed.), Business and the Environment: Implications of the New Environmentalism. Paul Chapman, London, pp. 98–115.

Johnson, R.M., Sims, J.T., 1993. Influence of surface and subsoil properties on herbicide sorption by Atlantic and coastal plain soils. Soil Sci. 155, 339–348.

Linkov, I., Satterstrom, K., Kiker, G., Ferguson, E., Bridges, T., 2005. Multi-criteria decision analysis and adaptive management: a review and framework for application to Superfund sites. In: G.P. Macey, J. Cannon (Eds.), Reclaiming the Land: Rethinking Superfund Institutions, Methods and Practices. Springer, New York.

Naidu, R., 1996. Health risk assessment of arsenic contaminated site, unpublished.

Naidu, R., Sumner, M.E., Harter, R.D., 1998. Sorption of heavy metals in strongly weathered soils: an overview. J. Environ. Geochem. Health 20, 5–9.

Natusch, J., 1997. Application and development of contaminated site remediation technologies in Australia. ANZAC Fellowship Report to Department of Internal Affairs, Wellington, New Zealand and Department of Foreign Affairs and Trade, Canberra, Australia.

National Academy of Sciences 1983. Risk Assessment in the Federal Government: Managing the Process. National Academy Press, Washington, DC.

NEPC, 1999. Impact Statement, in Assessment of Site Contamination. Draft National Environment Protection Measure and Impact Statement. National Environment Protection Council, Adelaide, 29 March.

Oscarson, D.W., Huang, P.M., Liaw, W.K., Hammer, U.T., 1983. Kinetics of oxidation or arsenite by various manganese dioxides. Soil Sci. Society Am. J. 47, 644–648.

Petts, J., Pollard, S.J.T., Gray, A.J., Orr, P., Homan, J., Delbridge, P., 2003. Involving lay audiences in environmental risk assessments. In: ConSoil 2003. Proceedings of the 8th International FZK/TNO Conference on Contaminated Land (Ghent, 12–16 May 2003), Theme D, pp. 2969–2978.

Pollard, S.J.T., Brookes, A., Earl, N., Lowe, J., Kearney, T., Nathanail, C.P., 2004. Integrating decision tools for the sustainable management of land contamination. Sci. Tot. Environ. 325(1–3), 15–28.

Pollard, S.J.T., Fisher, J., Twigger-Ross, C., Brookes, A., 2000. Approaches to decision support in the context of sustainable development. Evaluation of Demonstrated and Emerging Technologies for the Treatment of Contaminated Land (Phase III), Special Session, Decision Support Tools, NATO/CCMS Pilot Study.

Sadiq, M., 1997. Arsenic chemistry in soils: an overview of thermodynamic predictions and field observations. Water Air Soil Pollut. 93, 117–136.

Vig, K., Megharaj, M., Sethunathan, N., Naidu, R., 2002. Bioavailability and toxicity of cadmium to microorganisms and their activities in soil: a review. Adv. Environ. Res. 8, 121–135.

Developments in Soil Science, volume 32
Ravendra Naidu (Editor)
© 2008 Elsevier B.V. All rights reserved

Chapter 5

BIOAVAILABILITY OF SORBED PESTICIDES TO BACTERIA: AN OVERVIEW

N. Singh, N. Sethunathan, M. Megharaj and R. Naidu

5.1 Introduction

The bioavailability of organic pollutants governs their ecotoxicology and degradation in contaminated soils (see Chapters 2 and 11–13). This is apparent in soils with high organic matter in which easily biodegradable organic compounds persist for long periods even under conditions favourable for biodegradation. The lack of biodegradation in such soils has been attributed to the low bioavailability of the contaminants. Environmental microbiologist defines bioavailability as the contaminant fraction which represents the accessibility of a chemical to a living organism for assimilation, degradation and ecotoxicology expression (see Chapter 3 for definitions of bioavailability). Consequently, the bioavailability of contaminants varies with soil type, nature of contaminants and environmental factors (see Chapters 2 and 4). Factors such as sorbent type, the residence time, desorption rate and nature of microorganisms influence bioavailability and consequently biodegradation and toxicity of sorbed compounds. In this chapter, we present an overview of pesticide bioavailability in soils in particular the factors that underpin biodegradation.

5.2 Sorption influences bioavailability

The influence of sorption on the biodegradation of organic compounds has been recognized as an important (see Chapter 3), albeit poorly understood phenomenon, in bioremediation. Currently, two schools of thoughts exist regarding the bioavailability and biodegradation of organic contaminants (see Chapter 23). These include

(a) Contaminants must be present in available form for the microbes to biodegrade and that the release of sorbed contaminant into aqueous phase is a prerequisite for its bioavailability (Weber and Coble, 1968; Ogram et al., 1985; Steinberg et al., 1987; Rijnaarts et al., 1990; Estrella et al., 1993; Harms and Zehnder, 1994; Shelton and Doherty, 1997).

(b) Contaminants may biodegrade even from sorbed phase in the presence of enzymes (Singh et al., 2003).

Generally, weakly sorbed and easily desorbed compounds are more readily available for biodegradation than strongly sorbed and weakly desorbed compounds. Therefore, soil organic matter (SOM) plays an important role in the bioavailability of sorbed compounds (Singh and Sethunathan, 1992; Sreedharan, 1995 (see Chapters 10 and 23). Thus, pesticides sorbed to soils low in organic matter content are easily degraded by microorganisms as they are not strongly held by SOM or mineral fraction. On the contrary, pesticides, sorbed to high SOM soils, are not desorbed easily and thereby resist degradation in sorbed state.

Presence of bacteria in the system containing sorbed pesticides may serve to extract the nondesorbable fraction, much as an organic solvent. Living bacterial cultures continuously produce a wide array of soluble organic materials (biosurfactants) and it is possible that some of them may facilitate desorption of organic molecule. Biosurfactants have been shown to promote the solubilization of hydrophobic compounds and also increase their rate of release into solution (Zhang and Miller, 1994; Zhang et al., 1997; Garcia et al., 2001; Mata-Sandoval et al., 2000, 2001, 2002). However, in such cases the bioavailability is limited by the desorption rate.

The bioavailability of sorbed compounds may be affected by the nature of microorganisms themselves. Literature suggests that microbial ability to directly utilize soil-sorbed compounds is a species-specific characteristic (Calvillo and Alexander, 1996; Guerin and Boyd, 1997; Tang et al., 1998). Probably, these organisms associate more intimately and/or extensively with sorbent surface and bioavailability of sorbed compound may be the result of direct partitioning of the contaminant to cell membrane or via extracellular enzymes.

Higher degradation of sorbed compounds at rates slightly greater than the rate of compound's abiotic (uninoculated) desorption from sorbent is not a convincing evidence to suggest that sorbed compounds are bioavailable to microorganisms without being desorbed into solution (Harms and Zehnder, 1995; McGhee et al., 1999). These reports did not account for the:

(a) longer incubation periods (week) which may provide sufficient time for desorption; and

(b) shift in equilibrium once the compound available in solution phase is degraded.

In a recent study, Singh et al. (2003) have shown that fenamiphos, an organophosphorus insecticide and nematicide, intercalated on cetyltrimethy-lammonium-modified montmorillonite (CTMA-Mt) was bioavailable to *Brevibacterium* sp. for degradation at sorption sites, without being desorbed into solution phase. These researchers used a combination of X-ray diffraction (XRD) and infra-red (IR) analysis to demonstrate the hydrolysis of fenamiphos in the interlayers at sorption sites by extracellular enzymes produced by bacteria. In the following sections, we summarize some of the important findings on the bioavailability of sorbed pesticides including the effect of surfactants/biosurfactants, species-specific interaction, desorption-limited degradation, etc.

However, most of the work on bioavailability is limited to nonionic compounds (NOCs) and limited reports on bioavailability of pesticides. Therefore, to get a better understanding of the concept, reports dealing both with pesticides and NOC have been included in this chapter.

5.3 Desorption-limited degradation

Biodegradation is by far the best and probably the only environmentally sound means of organic contaminants detoxification in soils and waters. However, despite environment conditions favouring degradation, contaminant degradation is often limited by its nonavailability in aqueous phase (Ogram et al., 1985; Rijnaarts et al., 1990; Mueller et al., 1992; Jacobsen et al., 2001). Numerous studies suggest that only pesticides present in aqueous phase are instantaneously available for degradation and that the sorbed phase must first desorb into aqueous phase for any significant degradation. Sorption appears to be one of the crucial factors limiting the intracellular degradation of pesticides.

Numerous authors have conceptualized model assuming (Ogram et al., 1985) that sorbed 2,4-D is completely protected from degradation and that both sorbed and solution bacteria were capable of degrading 2,4-D in solution. This could explain the degradation of sorbed 2,4-D by a bacterial species. However, these researchers were unable to differentiate between whether bacteria were unable to reach sorbed 2,4-D, probably located in nanopores within the soil organic matrix or bacteria were unable to dislodge sorbed 2,4-D. A variety of other models using diffusion-limited sorption–desorption and biodegradation (Scow and Hutson, 1992; Scow and Alexander, 1992; Shelton and Doherty, 1997) confirmed that availability of pesticide in solution phase is a prerequisite of its bioavailability to microorganisms.

Carbofuran sorbed to a low organic carbon soil exhibiting low sorption and high desorption was readily degraded by a mixed enrichment culture obtained from carbofuran-treated *Azolla* plots (Singh and Sethunathan, 1992). The same enrichment culture was unable to degrade carbofuran sorbed in high organic matter soil with more extensive sorption and low desorption. Likewise, β-HCH (βhexachlorocyclohexane) sorbed to low organic matter content soil was readily degraded by HCH-degrading *Sphingomonas paucimobilis* (Sreedharan, 1995), but it was unable to degrade sorbed β-HCH in high organic matter containing soil. These findings demonstrate that biodegradation is to a large extent limited by the soil processes that influence contaminant biavailability. Thus in low organic matter containing soils with limited sorption sites and also in freshly contaminated sites biodegradation will be greatest in the presence of bioavailable contaminant. Once the free phase is exhausted degradation is likely to be restricted by the rate of contaminant transfer.

5.4 Role of surfactant molecules

Surfactants are surface-active compounds which at concentrations exceeding their critical micellar concentration (CMC) greatly enhance the apparent water solubility of hydrophobic compounds. Such chemicals are often used to assist biodegradation of contaminants that either have low water solubility or are strongly sorbed to colloid particles. Chemicals with surfactant properties enhance biodegradation through its effect on contaminant desorption from strongly bound sites. Strong sorption may be the case when contaminants are adsorbed to natural organic matter (Pignatello and Xing, 1996; Sreedharan et al., 2001) or adsorbed to the walls of submicron-sized pores inside soil particles and such sites are physically inaccessible by microorganisms. Under such conditions, it has been claimed that the use of surfactants, homogenization or dispersal of contaminants may increase the degradation rates (Rijnaarts et al., 1990; Ahmad et al., 2004).

However, considerable caution needs to be exercised with the use of surfactants given the observations that such chemicals can either stimulate or inhibit contaminant biodegradation. Inhibitory effects have been attributed to the toxicity of surfactant to microorganisms. Garcia et al. (2001) showed that Brij 35 (a nonionic surfactant) doubled the initial desorption of sorbed dibenzofuran from Teflon. Surprisingly, surfactant slowed the biodegradation of desorbed dibenzofuran by *Sphingomonas* sp. Low bioavailability of sorbed dibenzofuran in presence of surfactant was due to (i) accumulation of desorbed compound in surfactant micelles leading to reduced truly water-soluble dibenzofuran and (ii) suppression of contact between bacteria and Teflon.

5.5 Species-specific interactions and bioavailability

Bacteria greatly differ in their ability to degrade sorbed contaminants that are partitioned in solid organic phase. There have been reports which suggest that not all bacteria which can degrade contaminant in aqueous phase have the ability to degrade contaminant in sorbed state. This may be attributed to:

(i)   variation in attachment contaminant to sorbent (Holm et al., 1992; Guerin and Boyd, 1992; Neu, 1996);
(ii)   production of surfactants (Herman et al., 1997; Noordman et al., 1998); and
(iii)   metabolic pathways used to mineralize contaminant (Tang et al., 1998).

Conventional approach to enrich the contaminant-degrading ability of a microorganism is to provide contaminant to microorganism as a sole source of carbon in aqueous solution. However, knowing the fact that sorption plays an important role in bioavailability, an enrichment strategy where contaminant is provided to microorganisms in sorbed state may provide better know-how of the processes affecting bioavailability of sorbed compounds. In a recent study by

Tang et al. (1998), a bacterium was isolated by enrichment technique by providing phenentherene sorbed on polyacrylic porous resin. This bacterium has the ability to degrade sorbed phenentherene faster than another bacterium, which was isolated using conventional enrichment technique. This study clearly demonstrated the importance of enrichment techniques and conditions which simulate natural environment that can help in isolating more meaningful bacteria (Friedrich et al., 2000; Grosser et al., 2000).

Recent research (Tang et al., 1998; Chander et al., 2005) on bioavailability of sorbed NOCs provides evidence which confirms that sorbed contaminant was bioavailable. But, suggested possible mechanism of degradation in such degradation studies is microbially facilitated desorption of contaminant via surfactant production or development of steep concentration gradient between solid phase and interfacial contaminant. This shows that contaminant is degraded after desorption and it is the species-specific characteristic of some bacteria to desorb and degrade the sorbed contaminant.

None of these studies provides any direct evidence of contaminant's bioavailability to microorganisms without being desorbed into aqueous phase. Recently, our research team have demonstrated convincingly that a soil bacterium, *Brevibacterium* sp., having an exceptionally high ability to hydrolyse fenamiphos [ethyl-3-methyl-4-(methylthio) phenyl-1-(1-methylethyl) phosphoramidate] in aqueous solution (Megharaj et al., 2003) was able to degrade fenamiphos intercalated into cationic-surfactant-modified montmorillonite clay (Fig. 5.1; Singh et al., 2003).

Fenamiphos was immobilized via sorption by cetyltrimethylammonium-exchanged montmorillonite (CTMA-Mt). XRD, IR and negative electrophoretic mobility indicated that fenamiphos was intercalated in bacterially inaccessible interlayer of clay lattice. Within 24 h, *Brevibacterium* sp. hydrolysed 82% of the fenamiphos sorbed in CTMA-Mt complex with concomitant accumulation of fenamiphos phenol, a hydrolysis product, in stoichiometric amounts. During corresponding period, only 3% of the sorbed fenamiphos was chemically degraded in abiotic (uninoculated) control. In 24 h, a total of 4.6% of sorbed fenamiphos was released into aqueous phase, even though the aqueous solubility of fenamiphos is $>400 \, \mu g/ml$.

Desorption of fenamiphos was also studied (Fig. 5.2) in the presence of activated charcoal, having very high affinity for organic compounds, which acts as a sink for pesticide and should provide a gradient force for desorption of fenamiphos from adsorption sites in CTMA-Mt clay. Even in the presence of activated charcoal, only 6% of sorbed fenamiphos was desorbed from CTMA-Mt clay. This indicates that even under the influence of strong desorption gradient forces only a fraction of sorbed fenamiphos was released into the aqueous phase. Thus, we can neglect this possibility that *Brevibacterium* sp. may have

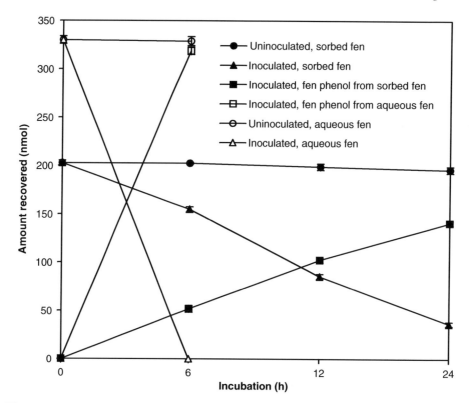

Figure 5.1. *Quantities of fenamiphos recovered from uninoculated aqueous solution (circle); inoculated aqueous solution (triangle); uninoculated suspension of CTMA-Mt clay complex with 0.2 wt% sorbed fenamiphos (solid circle) and inoculated suspension of CTMA-Mt clay complex with 0.2 wt% sorbed fenamiphos (solid triangle). Quantities of the hydrolysis product fenamiphos phenol recovered from the inoculated aqueous solution (square) and inoculated suspension of CTMA-Mt clay complex with 0.2 wt% of sorbed fenamiphos (solid square). Measurement errors are within the dimensions of the symbols. Tests for fenamiphos in aqueous solutions were carried out only for 6 h as trends were obvious (Singh et al., 2003).*

created a strong desorption gradient for fenamiphos and fenamiphos was degraded in aqueous phase.

Earlier studies on bioavailability of NOC have suggested the possibility of surfactant production by bacteria to dislodge sorbed contaminant. Even synthetic surfactants have enhanced contaminants desorption. The *Brevibacter-ium* sp. does not appear to produce any surfactant as there was no measurable change in the surface tension of growth medium after 4 days of inoculation with bacterium. Also, XRD patterns of bacteria-treated CTMA–fenamiphos complex

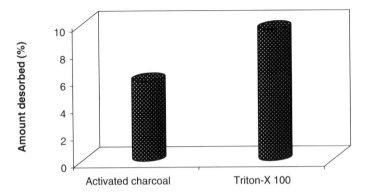

Figure 5.2. *Desorption of sorbed fenamiphos from CTMA-Mt clay complex using activated charcoal and Triton-X 100, a nonionic surfactant after 24 h.*

did not show noticeable changes in the lattice structure, indicating that bacteria do not produce any compound which can disturb the arrangement of clay lattice to dislodge the sorbed fenamiphos. Even Triton-X 100, a synthetic nonionic surfactant at 10,000 µg/ml levels could desorb only 10% of fenamiphos from CTMA-Mt–fenamiphos complex. Both these findings suggest that fenamiphos was not easily released from the sorption sites and surfactant was not produced by the bacteria to facilitate its desorption before degrading it.

Brevibacterium sp. does not have access to sorbed fenamiphos intercalated into CTMA-Mt complex as interlayer space of CTMA-Mt clay (1.8 nm) is >200 times smaller than the diameter of bacterium (400 nm). However, scanning electron micrograph showed that bacterial cells were attached to the surface of organoclay complex. Viable bacterial cell number in aqueous solution decreased from 2.0×1012 to 0.4×107 after 24 h incubation of bacterial suspension with CTMA-Mt clay. Reports indicate that attached bacterial cells are more active in degrading sorbed compounds either through direct partitioning to the cell membrane or via degradation by extracellular enzymes. As bacterial cells do not have direct access to the intercalated fenamiphos, we can neglect the first possibility that degradation occurs through direct partitioning of sorbed contaminant to the cell membrane. However, probably bacteria degrade the intercalated fenamiphos through the production of extracellular enzymes. But, results indicated that fenamiphos-degrading enzyme was extracellular in nature and successfully degraded fenamiphos in aqueous solution. However, enzyme added in the free state was more effective in degrading fenamiphos in free state compared to fenamiphos intercalated on CTMA-Mt complex. This may be the result of inactivation of enzyme. Results indicate that both CTMA bromide and unmodified montmorillonite do not affect the degrading activity of the enzyme. However, hydrolytic activity of the enzyme was suppressed by 50% after equilibration with CTMA-Mt clay, probably because of the sorption of enzyme by CTMA-Mt clay complex. Enzyme protein value in the supernatant decreased from 97 to 26 µg/ml following 2 h equilibration with CTMA-Mt clay. CTMA-Mt clay complex with sorbed enzyme easily hydrolysed fenamiphos in solution. Therefore, enzyme sorbed on organoclay retains its activity.

These results prove that fenamiphos intercalated on CTMA-Mt clay complex is bioavailable and is degraded at adsorption site without being desorbed. This assumption is supported by the facts that:

(i)   degradation time allowed is very short (24 h);
(ii)   even in the presence of a strong gradient for desorption, only a little amount of sorbed fenamiphos was released into solution;
(iii) hydrolytic enzyme was extracellular in nature;
(iv)   no surfactant was produced by the bacterium; and
(v)   accumulation of hydrolytic product, fenamiphos phenol, on solid phase.

5.6 Conclusion

Sorption has a limiting effect on the degradation rate of pesticides, which can be explained by the fact that it is the amount of pesticide available in the aqueous phase of the soil that can be microbially degraded. Even though this factor plays a major role, degradation of sorbed chemicals still take place. It may be facilitated by microorganisms by production of surfactants or by development of concentration gradient between solid phase and interfacial contaminant or by extracellular enzymes released by the specific microorganisms, enabling the degradation of contaminant at the sorption site itself.

References

Ahmad, R., Kookana, R.S., Megharaj, M., Alston, A.M., 2004. Aging reduces the bioavailability of even a weakly sorbed pesticide (carbaryl) in soil. Environ. Toxicol. Chem. 23, 2084–2089.

Calvillo, Y.M., Alexander, M., 1996. Mechanism of microbial utilization of biphenyl sorbed to polyacrylic beads. Appl. Microbial. Biotechnol. 45, 383–390.

Chander, Y., Kumar, K., Goyal, S.M., Gupta, S.C., 2005. Antibacterial activity of soil bound antibiotics. J. Environ. Qual. 34, 1952–1957.

Estrella, M.R., Brusseau, M.L., Maier, R.S., Pepper, I.L., Wierenga, P.J., Miller, R.M., 1993. Appl. Environ. Microbiol. 59, 4266–4273.

Friedrich, M., Grosser, R.J., Kern, E.A., Inskeep, W.P., Ward, D.M., 2000. Effect of model sorptive phases on phenanthrene biodegradation: molecular analysis of enrichment and isolation suggests selection based on bioavailability. Appl. Environ. Microbiol. 66, 2703–2710.

Garcia, J.M., Wick, L.Y., Harms, H., 2001. Influence of the nonionic surfactant Brij 35 on the bioavailability of solid and sorbed dibenzofuran. Environ. Sci. Technol. 35, 2033–2039.

Grosser, R.J., Friedrich, M., Ward, D.M., Inskeep, W.P., 2000. Effect of model sorptive phases on phenanthrene biodegradation: different enrichment conditions influence bioavailability and selection of phenanthrene-degrading isolated. Appl. Environ. Microbiol. 66, 2695–2702.

Guerin, W.F., Boyd, S.A., 1992. Bioavailability of naphthalene associated with natural and synthetic sorbents. Water Res. 51, 1504–1512.

Harms, H., Zehnder, A.J.B., 1994. Influence of substrate diffusion on degradation of dibenzofuran and 3-chlorodibenzofuran by attached and suspended bacteria. Appl. Environ. Microbiol. 60, 2736–2745.

Harms, H., Zehnder, A.J.B., 1995. Bioavailability of sorbed 3-chlorodibenzofuran. Appl. Environ. Microbiol. 61, 27–33.

Herman, D.C., Lemhard, R.J., Miller, R.M., 1997. Formation and removal of hydrocarbon residuals in porous media: effect of attached bacteria and biosurfactants. Environ. Sci. Technol. 31, 1290–1294.

Holm, P.E., Nielson, P.H., Albrechtsen, H.J., Cristensen, T.H., 1992. Importance of unattached bacteria and bacteria attached to sediment in determining potentials for degradation of xenobiotic organic contaminants in an aerobic aquifer. Appl. Environ. Microbiol. 58, 3020–3026.

Jacobsen, C.S., Shapir, N., Jensen, L.O., Juhler, R.K., Streibig, J.C., Mandelbaum, R.T., Helweg, A., 2001. Bioavailability of triazine herbicides in a sandy soil profile. Biol. Fert. Soil 33, 501–506.

Mata-Sandoval, J.C., Karns, J., Torrents, A., 2000. Effect of rhamnolipids produced by *Pseudomonas aeruginosa* UG2 on the solubilization of pesticides. Environ. Sci. Technol. 34, 4923–4930.

Mata-Sandoval, J.C., Karns, J., Torrents, A., 2001. Influence of rhamnolipids and Triton X-100 on the biodegradation of three pesticides in aqueous phase and soil slurries. J. Agric. Food Chem. 49, 3296–3203.

Mata-Sandoval, J.C., Karns, J., Torrents, A., 2002. Influence of rhamnolipids and triton X-100 on the desorption of pesticides from soils. Environ. Sci. Technol. 36, 4669–4675.

McGhee, I., Sannino, F., Gianfreda, L., Burns, R.G., 1999. Bioavailability of 2,4-D sorbed to a chlorite like complex. Chemosphere 39, 285–291.

Megharaj, M., Singh, N., Kookana, R.S., Naidu, R., Sethunathan, N., 2003. Hydrolysis of fenamiphos and its oxidation products by a soil bacterium in pure culture, soil and water. Appl. Microbiol. Biotechnol. 61, 252–256.

Mueller, T.G., Moorman, T.B., Snipes, C.E., 1992. Effect of concentration, sorption and microbial biomass on the degradation of herbicide fluometron in surface and subsurface soils. J. Agric. Food Chem. 40, 2517–2522.

Neu, T.R., 1996. Significance of bacterial surface active compounds in interaction of bacteria with interfaces. Microbiol. Rev. 60, 151–166.

Noordman, W.H., Ji, W., Brusseau, M.L., Janssen, D.B., 1998. Effects of rhamnolipid biosurfactants on removal of phenanthrene from soil. Environ. Sci. Technol. 32, 1806–1812.

Ogram, A.V., Jessup, R.E., Ou, L.T., Rao, P.S.C., 1985. Effects of sorption on biological degradation rates of (2,4-dichlorophenoxy) acetic acid in soils. Appl. Environ. Microbiol. 49, 582–587.

Pignatello, J.J., Xing, B., 1996. Mechanisms of slow sorption of organic chemicals to natural particles. Environ. Sci. Technol. 30, 1–11.

Rijnaarts, H.H., Bachmann, M.A., Jumelet, J.C., Zehnder, A.J.B., 1990. Effect of desorption and intraparticle mass transfer on the aerobic mineralization pf a-hexachlorocyclohexane in a contaminated calcareous soil. Environ. Sci. Technol. 24, 1349–1354.

Scow, K.M., Alexander, M., 1992. Effect of diffusion on the kinetics of biodegradation: experimental results with synthetic aggregates. Soil Sci. Soc. Am. J. 56, 128–134.

Scow, K.M., Hutson, J., 1992. Effect of diffusion and sorption on biodegradation: theoretical considerations. Soil Sci. Soc. Am. J. 56, 119–127.

Shelton, D.R., Doherty, M.A., 1997. A model describing pesticide bioavailability and biodegradation in soil. Soil Sci. Soc. Am. J. 61, 1078–1084.

Singh, N., Megharaj, M., Gates, W.P., Churchmann, G.J., Anderson, J., Kookana, R.S., Naidu, R., Chen, Z., Slade, P.G., Sethunathan, N., 2003. Bioavailability of an organophosphorus pesticide, fenamiphos intercalated in an organo-clay complex. J. Agric. Food Chem. 51, 2653–2658.

Singh, N., Sethunathan, N., 1992. Degradation of soil-sorbed carbofuran by an enrichment culture from carbofuran-treated *Azolla* plots. J. Agric. Food Chem. 40, 1062–1065.

Sreedharan, B., 1995. Interaction of isomers of hexachlorocyclohexane with soils. Ph.D. Thesis, Utkal University, Bhubaneswar, India, p. 187.

Sreedharan, B., Singh, N., Sethunathan, N., 2001. Degradation of soil-sorbed isomers of hexachlorocyclohexane in soils under flooded conditions. In: J. Vijgen, S. Stobiecki, S. Pruszynski and W. Sliwnski (Eds.), 6th HCH and Pesticide Forum Book. Plant Protection Institute, Miczurine, Poznan, Poland, pp. 229–235.

Steinberg, S.M., Pignatello, J.J., Sawhney, B.L., 1987. Persistence of 1,2-dibromoethane in soils: entrapment in intraparticle micropores. Environ. Sci. Technol. 21, 1201–1208.

Tang, W.C., White, J.C., Alexander, M., 1998. Utilization of sorbed compounds by microorganisms specially isolated for that purpose. Appl. Microbial. Biotechnol. 49, 117–121.

Weber, J.B., Coble, H.D., 1968. Microbial decomposition of diquat adsorbed on montmorillionite and kaolinite clays. J. Agric. Food Chem. 16, 475–478.

Zhang, Y., Maier, W.J., Miller, R.M., 1997. Effect of rhamnolipids on the dissolution, bioavailability and biodegradation of phenentherene. Environ. Sci. Technol. 31, 2211–2217.

Zhang, Y., Miller, R.M., 1994. Effect of *Pseudomonas* rhamnolipis biosurfactant on cell hydrophobicity and biodegradation of octadecane. Appl. Environ. Microbiol. 60, 2101–2106.

Developments in Soil Science, volume 32
Ravendra Naidu (Editor)

83

Chapter 6

MECHANISTIC APPROACH FOR BIOAVAILABILITY OF CHEMICALS IN SOIL

Sjoerd E.A.T.M. van der Zee and Erwin J.M. Temminghoff

6.1 Introduction

Both soil fertility and contaminants research requires an understanding of chemical bioavailability. In soil fertility research, much emphasis has always been given to the pragmatic approach as our understanding of the soil and rhizosphere system had to grow before alternative routes were feasible (see Chapter 18 for nutrient bioavailability). Also in soil contaminants research, the need to address soil quality and issue regulations on soil pollution motivated a realistic approach. Such an approach was commonly based on the assessment of contaminant concentrations, without consideration of the hazards that were associated with these concentrations. However, soon after satisfying the need to develop standards to distinguish uncontaminated and contaminated soils, the debate to relate standards with effects intensified again.

The current understanding in most European Union (EU) countries is that bioavailability of nutrients and contaminants differs as a function of soil type (see Chapters 2 and 3) and of receptors (vegetation, crops, soil biota, animals, and humans). For many soil types and receptors there is also some evidence of how bioavailability depends on them, although this evidence is still sketchy and often based on correlative studies with rather basic parameters such as concentration in soil and the presence of major sorbing constituents (clay and organic matter content) and on pH. Furthermore, integration of understanding is not benefiting from the use of standardised (e.g. Organization for Economic Cooperation and Development (OECD)) materials to assess effects on receptors.

This chapter presents an overview of the mechanistic approach to understand bioavailability using some recent examples to illustrate the complexity of the soil system and the opportunities that may arise with such an approach. The intention is not to replace the more empirical approaches to bioavailability, as alternatives have their own merits, but rather to emphasise the added value if approaches are used in combination, and when mechanistic studies provide background understanding for empirical ones.

6.2 Soil and rhizosphere system

The system that we are interested in comprises of an abiotic and a biotic component. The abiotic component involves the solid phase made up from minerals and organic matter and the water and air filled pore space. This part might be considered as the supplier of chemicals that are distributed in the solid, liquid, and gaseous phases (a distribution process that is comprehensively called speciation; see Chapters 8–10). At the receptor side, are vegetation, crops, and soil biota (e.g. springtails, nematodes, earthworms) and somewhat further away animals (e.g. mice, birds) and humans via the food chain. The receptors found in the soil system form a food web, with its own characteristics and which is not passive with regard to the supply by the abiotic part. For instance, it is well known that plants may exude acids that favour uptake of certain nutrients and pollutants, whereas mycorrhizal fungi (that are important for most crops) are known to affect bioavailability of metals and phosphorus.

A schematic diagram of the soil system is presented in Figure 6.1 which shows that all distinguishing compartments may transfer chemicals to adjacent compartments. Furthermore, different receptors are shown, with the intention to draw attention to differences in response for each organism. However, the figure does not show that under field environment a broad range of chemicals may be involved, and these chemicals may compete with each other. Such a competition may occur both with respect to sorption by soil as well with the uptake by organisms. It has been broadly recognised that exposure to either single or multiple contaminants affects the response to the contaminants. Furthermore, ecological aspects may control the response of the biosphere to the supply of nutrients and contaminants,

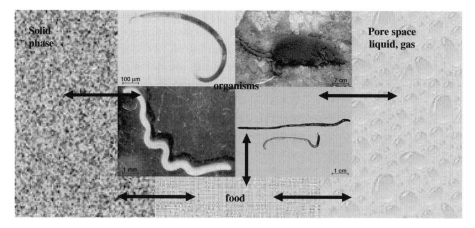

Figure 6.1. *Schematic of the soil/rhizosphere system, with arrows indicating flows of chemicals.*

for instance involving the food web. Also, spatiotemporal variability has been recognised as a profoundly complicating factor in assessing bioavailability.

Without intending to suggest completeness, a number of these complications will be explored, to identify gaps in understanding, to suggest a useful role for mechanistic bioavailability research, and to motivate multidisciplinary approaches.

6.3 Rhizosphere, a competitive environment

Currently, it may be regarded as common knowledge that the bioavailability of metals increases with increasing acidity (*de*creasing pH). However, the toxicity of metals may increase with decreasing acidity (*in*creasing pH) if plants or microorganisms are grown on nutrient solutions (Lexmond, 1980, 1981; Plette et al., 1996). Such findings suggest that changes in soil pH may impact both contaminant bioavailability (see Chapter 2) and their toxicity with subsequent effect on biota. Indeed, this has been demonstrated by numerous researchers (Plette et al., 1999; Temminghoff et al., 1995; Weng et al., 2003). This has major implications to land use. For instance, if major land use changes are the result of the EU's agricultural policies, the anticipation of their indirect effects is probable to be partly due to soil acidification. Irrespective of soil type, metal binding usually decreases with decreasing pH (Fig. 6.2) for both a sandy and a clayey soil, where sorption is shown as a function of free Cd-concentration.

A similar effect of pH is shown in Figure 6.3 for a soil bacterium that is exposed to a solution containing Cd. In both cases (soil and bacterium), the sorption may be quantified using a three component Freundlich equation (Temminghoff et al., 1995).

Such a parameterisation may then be combined for soil and bacterium, to assess how the bacterium would react to metal exposure in the soil–solution–bacterium system, a system that is experimentally involved as it is almost impossible to separate soil and bacterium in batch systems. Metal sorption by bacteria depends on total metal sorbed by soil, for different pH-values and calcium concentrations (Temminghoff et al., 1995) (Fig. 6.4).

Two major differences can be observed for the sandy and the clayey soil:

(i)   for the sandy soil, sorption *increases* as pH decreases, whereas for the clayey soil the opposite occurs; and

(ii) the effect of calcium concentration is much smaller for the clayey than for the sandy soil.

Both differences illustrate the effect of competition (i: for soil vs. bacterium sorption sites; ii: between Ca and Cd).

The effect of pH on the bioaccumulation of nickel (Ni) by plants is also reversed when using a nutrient solution or a soil as a growing medium (Weng et al., 2003).

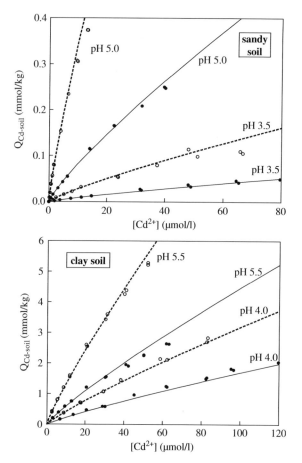

Figure 6.2. *Sorption of Cd by sandy and by clayey soil, as a function of free Cd, for varied pH-value (from Temminghoff et al., 1995).*

This paradox can be understood if the pH effect on the bioaccumulation, on the chemical speciation in the soil solution, and on binding to the soil of Ni is all taken into account (Weng et al., 2003, 2004). It was shown that bioaccumulation by the oat plants, binding to the soil solid phase, and binding to the dissolved organic matter all increase with increasing pH. However, the magnitude of the effect is the least for bioaccumulation for the oat plants as a function of pH, causing the apparent paradox.

A study that considered Cu accumulation by the springtail (Collembolan) *Folsomia candida* Willem, however, failed to observe a large pH effect on Cu accumulation, despite the strong pH-dependence of Cu-sorption by soil (Bruus Pedersen et al., 1997). This observation was surprising enough to warrant a closer inspection because it has often been suggested that heavy metal

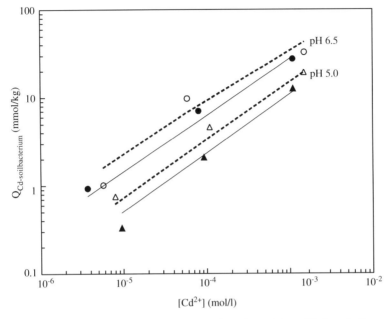

Figure 6.3. *Sorption of Cd by bacterium as a function of free Cd in solution.*

bioavailability and hence their adverse effect increases with decreasing pH (Van der Zee et al., 2004).

It appeared that Cu sorption by soil was indeed pH-dependent, but so was Cu-sorption by yeast (Fig. 6.5), which was supplied to the bioassay as a food source for collembolans (a common practice). Thus, both yeast and soil competed for free Cu. To further elucidate the phenomena, the Freundlich type of equation was fitted to both yeast and soil data for four pH-values, which gave a good description of the observations.

Combination of the soil and yeast equations provided a means to describe the speciation of Cu in the soil–soil solution–yeast system, and indeed the pH-effect appeared to be small due to competition effects between yeast and soil. Furthermore, it was possible to relate calculated Cu sorbed by yeast in the full system with accumulation by collembolans (Fig. 6.5). We observe that the data for different pH-values follow a smooth, though not monotonously increasing or decreasing, line, where the part at large exposure of Cu may be related with toxicity effects. The studies revealed that competition effects may be expected between pollutants, between the solid soil phase and organic surfaces (microbiota, roots), and between the solid soil phase and food for organisms, and that addition of food in ecotoxicological bioassays may involve a significant bias in the observations.

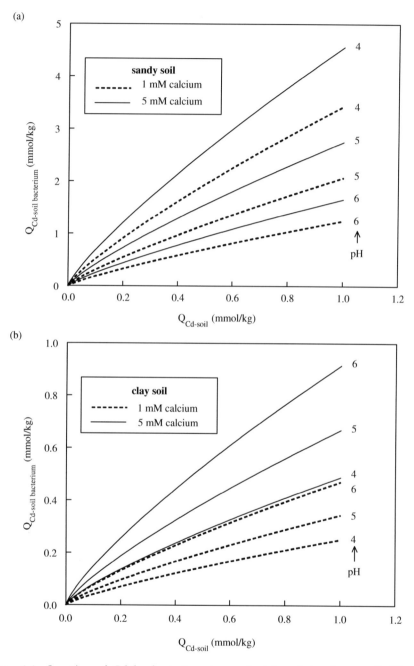

Figure 6.4. *Sorption of Cd by bacterium in sandy (a) and in clayey soil (b), as a function of Cd sorbed by soil, for varied pH-value (from Temminghoff et al., 1995).*

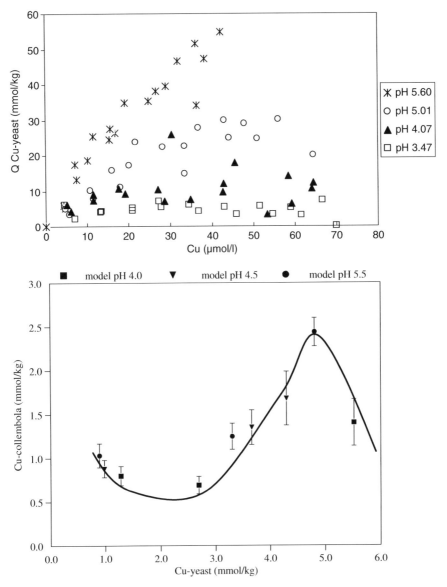

Figure 6.5. *Cu sorption by yeast as a function of total dissolved Cu (left), and Cu accumulated by colembolans as a function of Cu adsorbed by yeast (right) (from Van der Zee et al., 2004). Line of right figure is fitted by the 'eye'.*

6.4 Food web complications

Exposure may occur dermally and orally for soil organisms. The latter route of exposure may be complicated by competition for contaminants, but the food web may be involved also because the food is affected itself by the contaminants. In a recent study (Wang et al., 2004), the effect of Cd on the mulberry plant (*Morus alba* L.) and the silkworm (*Bombyx moril* L.) feeding on mulberry leaves was considered. This work focussed on the toxic effects of Cd on the mulberry plant and on the silkworm. For a soil Cd-concentration of 75 mg/kg, mulberry root weights were significantly reduced, whereas at twice that concentration, the plant exhibited marginal growth. Leaf Cd-contents increase linearly with soil Cd, and indicators of leaves as a source of food (water and chlorophyll contents) adversely reacted on Cd-exposure. Nutritive value (proteins and soluble sugar concentration) decreased as Cd increased, and superoxide dismutase and peroxidase concentrations revealed significant stress for mulberry.

The silkworms were given mulberry leaves that were exposed to soil Cd or to Cd by spraying on the leaves. Adverse effects were observed in both cases, but remarkably, direct toxic effects (exogenous applied Cd) were observed only for levels that did not occur in mulberry exposed to soil Cd. At levels where endogenously exposed mulberry caused problems, they were not observed for the exogenously exposed silkworms. Consequently, the silkworms were not affected adversely by Cd-concentrations. Rather, it was concluded that the mulberry leaf quality as food was adversely affected by Cd, causing it to become of poor nutritive value and causing malnutrition for the silkworms. At least for the considered soil, mulberry and silkworm, it appears that the food chain protects the silkworm against toxic effects by Cd.

6.5 Spatiotemporal variability of exposure

One of the main concerns that have been expressed in ecotoxicological research, is whether or not bioassays that are conducted in the laboratory have implications to field situations, i.e., do adverse effects occur at similar exposure levels under laboratory and field conditions. The concern is due to differences in both situations, such as the presence of different contaminants, with either synergistic or antagonistic effects, the spatial variability, and the presence of more than one stress factor in field situations.

For a field site (former breaking-up yard) that was contaminated with Cu, Pb, and Zn, a very detailed soil sampling was conducted to provide maps of the contaminants (Marinussen et al., 1997). At this site, few earthworms were found in the most contaminated part (Fig. 6.6A–D, bottom right), which suggests either avoidance behaviour for or toxicity in the most contaminated part.

Twice a large number of earthworms *Dendrabaena venata* were introduced at several places of the site and sampled again after varied exposure times, along with native earthworms species. For all sampled worms, the exact location was registered, to enable comparison with the contamination levels at those

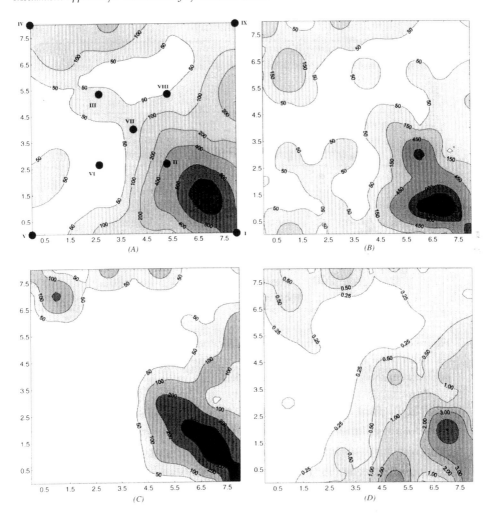

Figure 6.6. *Contour maps, with distances (m) at the axes, for total Cu (A), Pb (B), Zn (C), and easily extractable Cu (D) with values in (mg/kg) (from Marinussen et al., 1997).*

locations, as toxicokinetic studies suggested a rapid response to changing exposure (due to earthworm mobility) of earthworm tissue metal concentrations. Figure 6.7 shows the Cu-concentration factors (ratio of tissue over soil concentrations) of the field experiment together with data from other experiments. The agreement between the different data reveals that properly conducted field and laboratory experiments are well compatible. At the same time, the field experiment was costly and labour intensive, and required a density of sampling locations that is prohibitive for routine assays. The mechanistic approach to deal with spatiotemporal variability is therefore

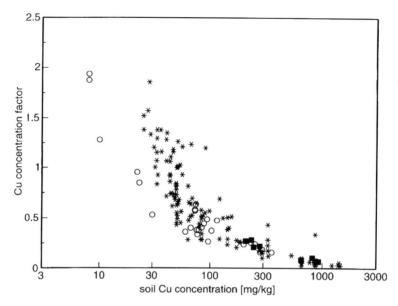

Figure 6.7. *Cu bioconcentration factors BCF (Cu-tissue/Cu-soil) for earthworms collected at a field (*), and two different laboratory studies (■, ○), as a function of Cu-soil (from Marinussen et al., 1997).*

worthwhile, but perhaps only well possible to acquire detailed understanding of specific gaps in understanding.

A significant problem in the interpretation of experimental field data is the complexity of the spatiotemporally variable soil and rhizosphere system. In a recent study (Ettema et al., 1998), this was implicitly demonstrated for a field site in the USA. At this field site, detailed sampling of nematode species and different times revealed a significant spatial variability. Nematode abundancies have been recognised as an indicator of soil quality, but the abundancies for different nematode species were shown to be quite differently distributed as is shown in Figure 6.8. Moreover, these patterns changed quite profoundly as a function of time. This suggests that the interpretation of these field data is difficult because it makes a difference at what moment in the year this 'indicator' is measured. Moreover, the patterns for nematode species differs from the patterns of other indicators, such as microbial respiration, and nutrient status, which are also spatiotemporally variable. Apparently, different parameters are indicative of different soil quality attributes, but it is not immediately clear which these are, and it is not ascertained that they hold for different field sites.

A methodological concern is the use of geostatistical tools to generate the various maps of this chapter. For instance, nematodes are miniscule organisms

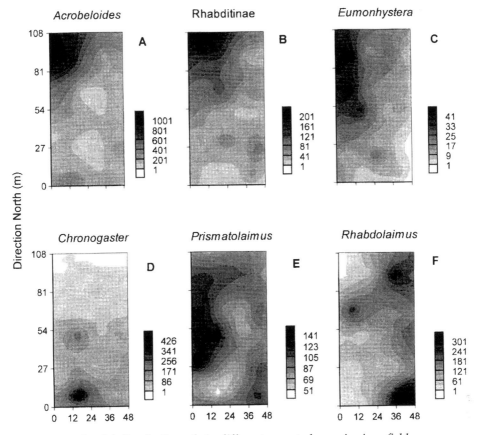

Figure 6.8. *Spatial distribution of six different nematode species in a field.*

and orders of magnitude smaller that the correlation scales underlying Figure 6.8.

Correlation scales of other parameters, as nitrate, organic carbon, respiration, also differ for each parameter. Whether these correlation scales have any mechanistic meaning, as suggested by the maps, is unclear. Nematodes may be subject to the correlation scale related with the niches, where they survive, e.g. during cold or dry seasons, but this scale is probably much smaller than the one used in the figure. Nematodes may also react on the correlation scale of an underlying parameter such as moisture content or nutrient status, but apparently these parameters differ from different nematode species and identification requires advanced soil ecological understanding. In conclusion, the interpretation of detailed, labour and cost intensive, high quality obserrvations may still be problematic with our current understanding of the system.

6.6 Conceptual simplifications

A mechanistic approach of the concept of bioavailability is discussed as complementing more empirical approaches in the following sections. In such a mechanistic approach, the soil system is simplified, and a major shortcoming of this approach is where the simplifications are carried too far. To recognise which detail is needed, it serves to explicitly recognise, that bioavailability depends on the (soil or groundwater) system at hand, and on the receptor (organism) under consideration. Hence, the commonly made (implicit) assumption that a bioavailable fraction can be distinguished and measured, has to be recognised as being inappropriate except for the most simplified bioassays (where all other environmental and habitat factors are being kept constant).

For the soil system, it is important to remember that it consists of different phases and that the solid phase consists of different constituents. For the liquid phase, this is also the case, e.g. due to dissolved organic matter, for instance. In general, the different phases and indeed their constituents, each form competitors for the interaction of the chemicals of interest. Moreover, the biosphere that involves among others fungal, microbial, and root surfaces may also be regarded as 'interaction competitors' for chemicals. This multicompetitive speciation process which includes interactions between different chemicals is one of the main factors that complicate the soil system (Weng et al., 2001, 2002). For instance, release of chemicals may be due to dissolution and desorption from abiotic pools but also to exudation by organic pools, and such releases may profoundly affect the chemical speciation in the system.

For the organisms, it is well known that the exposure routes may be diverse and complex. Thus, two examples were given where exposure via food significantly affected the effect of contamination. In the collembolan assay, effects were related by consumption of Cu-contaminated yeast and an artefact of the bioassay, whereas for silkworm, the primary effect of Cd-pollution of soil may be malnutrition because of poor nutrition on mulberry leaves rather than direct Cd-toxic effects.

As the soil system is spatiotemporally variable, this is also the case for bioavailability. This does not render laboratory assessments invalid, but it does complicate the extrapolation of understanding from laboratory to field. Often, in the field we have to deal with more than one pollutant, and indeed other non-pollution stressors, where the intensity of all stress factors differ as a function of space and time. Such spatiotemporal variability is the essence of ecosystems, rather than a side effect that calls for simplification. Major concepts and ways to deal with variability are yet lacking. This is due to gaps in process and system understanding as well as due to different characteristic scales of abiotic and biotic variability and of receptors. The latter, for instance, raises questions on whether the scales found for different soil attributes as respiration, nematode, and densities are relevant or artefacts of the sampling design.

As has been said earlier, the mechanistic approach does not replace other approaches. However, it is valuable to understand the results of other approaches, to fill gaps in understanding, and to ask critical questions.

Acknowledgement

This work was partly funded by SOWA – Integrated Soil and Water Protection, funded by the 5th Framework Programme of the European Union.

References

Bruus Pedersen, M., Temminghoff, E.J.M., Marinussen, M.P.J.C., Elmegaard, N., van Gestel, C.A.M., 1997. Copper accumulation and fitness of *Folsomia candida* willem in a copper contaminated sandy soil as affected by pH and soil moisture. Appl. Soil Ecol. 6, 135–146.

Ettema, C.H., Coleman, D.C., Vellidis, G., Lawrance, R., Rathbun, S.L., 1998. Spatiotemporal distributions of bacterivorous nematodes and soil resources in a restored riparian wetland. Ecology 79, 2721–2734.

Lexmond, Th.M., 1980. The effect of soil pH on copper toxicity to forage maize grown under field conditions. Neth. J. Agric. Sci. 28, 164–184.

Marinussen, M.P.J.C., van der Zee, S.E.A.T.M., de Haan, F.A.M., 1997. Cu accumulation in the earthworm *Dendrabaena venata* in a heavy metal (Cu, Pb, Zn) contaminated site compared with Cu accumulation in laboratory experiments. Environ. Pollut. 79(2), 227–233.

Plette, A.C.C., van Riemsdijk, W.H., Benedetti, M.F., 1996. Competitive binding of protons, calcium, cadmium and zinc to isolated cell walls of gram-positive soil bacterium, Environ. Sci. Technol. 30, 1902–1910.

Plette, A.C.C., Nederlof, M.M., Temminghoff, E.J.M., van Riemsdijk, W.H., 1999. Bioavailability of heavy metals in terrestrial and aquatic systems: a quantitative approach. Environ. Toxicol. Chem. 18, 1882–1890.

Temminghoff, E.J.M., van der Zee, S.E.A.T.M., de Haan, F.A.M., 1995. Speciation and calcium competition effects on cadmium sorption by soil at various pHs. Eur. J. Soil Sci. 46, 649–655.

Van der Zee, S.E.A.T.M., Temminghoff, E.J.M., Marinussen, M.P.J.C., 2004. Competition effects for copper between soil, soil solution, and yeast in a bioassay for *Folsomia candida* Willem, Environ. Toxicol. Chem., 23(7), 1743–1750.

Wang, K.R., Gong, H., Wang, Y., van der Zee, S.E.A.T.M., 2004. Toxic effects of cadmium on *Morus alba* L. & *Bombyx moril* L. Plant and Soil. 261(1–2), 171–180.

Weng, L., Lexmond, Th.M., Wolthoorn, A., Temminghoff, E.J.M., van Riemsdijk, W.H., 2003. Phytotoxity and bioavailability of nickel: chemical speciation and bioaccumulation. Environ. Toxicol. Chem. 22, 2180–2187.

Weng, L., Temminghoff, E.J.M., van Riemsdijk, W.H., 2001. Contribution of individual sorbents to the control of heavy metal activity in sandy soil. Environ. Sci. Technol. 35, 4436–4443.

Weng, L., Temminghoff, E.J.M., Lofts, S., Tipping, E., van Riemsdijk, W.H., 2002. Complexation with dissolved organic matter and solubility control of heavy metals in a sandy soil. Environ. Sci. Technol. 36, 4804–4810.

Weng, L., Wolthoorn, A., Lexmond, Th.M., Temminghoff, E.J.M., van Riemsdijk, W.H., 2004. Understanding the effects of soil characteristics on phytotoxity and bioavailability of nickel using speciation models. Environ. Sci. Technol. 38, 156–162.

B: The role of chemical speciation in bioavailability

The characterisation of contaminant speciation, as well as the definition of speciation has varied significantly during the past 40 years. Earlier, speciation included either association of contaminants to various mineral and organic phases present in the solid matrix, or quantification of contaminants associated with organic and inorganic phases. While the former has been the subject of extensive study, and numerous speciation techniques have been developed, recognition for the need for solution phase speciation arose only during the past 20 years. Ure and Davidson (1995) define speciation as either (a) the process of identifying and quantifying the different, defined species, forms or phases present in a material, or (b) the description of the amounts and kinds of these species, forms or phases present. They suggest that whichever approach is taken, the species, forms or phases are either defined (i) functionally, (ii) operationally, or (iii) as chemical compounds or oxidation states. The functionally defined speciation is more closely related to solution-phase study and is reflected by the plant-available species. Of late, this form of speciation has also been related to microbiota bioavailable form, or the intensity factor. The operational definition reflects the association of contaminants with the various phases present in soil-solid phase. Although solid phase speciation is an area of much interest, the overview only focuses on solution phase speciation, including both functionally defined speciation and various oxidation states. This section presents an overview of techniques currently used for metal and organic contaminant speciation, and discusses how these various species relate to contaminant bioavailability and toxicity.

Developments in Soil Science, volume 32
Ravendra Naidu (Editor)

Chapter 7

FRONTIERS IN ASSESSING THE ROLE OF CHEMICAL SPECIATION AND NATURAL ATTENUATION ON THE BIOAVAILABILITY OF CONTAMINANTS IN THE TERRESTRIAL ENVIRONMENT

A.R. Gerson, C. Anastasio, S. Crowe, D. Fowle, B. Guo, I. Kennedy, E. Lombi,
P.S. Nico, M.A. Marcus, R.R. Martin, S.J. Naftel, A.J. Nelson, D. Paktunc,
J.A. Roberts, C.G. Weisener and M.L. Werner

7.1 Introduction

It has long been recognised that contaminants interact with the solid soil phase via a series of sorption–desorption (all chemicals) and precipitation–dissolution (polar and ionic compounds) reactions including physical migration into micropores and diffusion into the solid phase (see Chapter 2). Until recently the extent of such interactions and the binding with different solid pools was established via a series of chemical extractions (see Chapter 20) and a combination of instrumental techniques. However, none of these techniques provide a true indication of the speciation of contaminants on or in solid minerals and organics. The last decade has witnessed the emergence of tools that provide an assessment of the speciation and attenuation of chemicals at molecular level.

In this chapter we provide an overview of the current state-of-the-art for assessing speciation and attenuation of contaminants in the terrestrial environment. Given that speciation and attenuation influence chemical bioavailability, we devote part of this chapter to the application of isotopic dilution techniques to bioavailability assessment. We have not attempted to be exhaustive but rather to highlight a number of studies in sufficient detail so that the reader will be provided with an insight as to whether such approaches are applicable to their field of endeavour and what the realistic outcomes might be.

Isotopic dilution techniques have been developed to measure the pool of an element that is present in an exchangeable form, within a specific timeframe, in the soil system. As such these techniques do not chemically speciate an element but provide information regarding the "chemical reactivity" of an element in the soil. Isotopic dilution techniques were originally developed to assess nutrient availability, in particular phosphorus, in soil. More recently, these techniques have been used increasingly to investigate various aspects of the biogeochemical behaviour of contaminants in soils. In fact, even if these techniques do not provide a direct measurement of bioavailability they are useful in quantifying the pool of metals in soil that is in direct equilibrium with the solution phase and can therefore buffer the free ion activity of metals in solution. The use of isotopic dilution techniques for the assessment of rhizosphere processes, redox sensitive species and natural attenuation is described in Section 7.2.

The metals found in and/or associated with past cultures can provide insights into the level of available technologies, trade routes, travel and migration in ancient civilizations. They can be of primary importance if written records are absent. Trace metals in human remains are, in principle, rich sources of information indicative of environmental exposure, diet and/or disease. Bone, hair and teeth (Lorentz, 2002; Martin et al., 2005; Attramadal and Jonsen, 1976) are all potential repositories of metals and consequently detailed analysis of these materials is of interest to archaeologists and anthropologists. Naturally any results will be of interest in modern medicine as well since the location and mechanism of metal sequestration in tissues can help to elucidate their role in human physiology where they may serve as essential elements and/or toxins. Section 7.3 reports the findings of a study, conducted using synchrotron microprobe X-ray fluorescence (XRF), of trace elements in 1000-year-old Peruvian mummy teeth.

X-ray absorption near edge spectroscopy (XANES) falls into the general analytical category of X-ray absorption fine structure (XAFS) spectroscopy, as does extended XAFS (EXAFS). XAFS measurements are frequently carried out by measuring, as a function of incident X-ray energy, the absorption of X-rays near and above the energy required to excite a core-level electron for a specific element. XAFS spectra may also be measured via other processes that also reflect the X-ray absorption characteristics. These processes include fluorescence emission and the emission of photoelectrons from the sample. The method of measurement of XAFS determines the depth of analysis of the measurement, which can vary from a few nanometers to more than 100 μm. Thus, the bulk or surface of a sample may be investigated. These measurements, for environmental studies, can only practically be carried out using a synchrotron radiation source. A significant strength of XAFS is it enables structural and electronic information to be derived from samples unsuitable for crystallographic analysis, for instance liquids, glasses, amorphous solids, metals, multicomponent systems, surfaces etc.

The absorption edge is defined as the energy required to excite an electron from its ground state to the first allowed unfilled orbital. Up to about 50 eV above the absorption edge, the outgoing photoelectron is strongly affected by its environment and may be captured into unoccupied bound or resonant states or multiply scattered. This part of the XAFS spectra is termed the XANES region and contains information about the local electronic structure, valence and coordination geometry for a given element. Section 7.4 describes the application of synchrotron XANES and synchrotron micro-XRF to the analysis of Cr-containing species within particulate matter.

EXAFS occurs at higher incident X-ray energies than XANES. In this region the electron is nearly free and only weakly scattered by neighbouring atoms. The photoelectron can be backscattered from the surrounding atoms to interact either constructively or destructively with the outgoing photoelectron wave. Whether the interference is constructive or destructive depends on the energy of the photoelectron and the local atomic geometry. Hence by changing the energy of the photoelectron by varying the energy of the incident X-ray radiation an interference pattern can be obtained. By analysis of this interference pattern, the structure number, type and distances to neighbouring atoms may be determined. Excellent introductory synchrotron XAFS material is presented by Newville (2004) with more detailed texts available by Koningsberger and Prins (1988), Stern and Heald (1983) and Als-Nielsen and Marrow (2001).

Bulk concentration levels alone are not adequate to assess the toxicity of elements. Elevated levels of an element in soil do not necessarily mean that they are harmful to the environment and public. Instead, toxicity and health effects are determined by bioavailability and exposure criteria. Accordingly, comprehensive mineralogical characterisation is required to determine the form, nature and distribution of the elements in soil. Synchrotron-based XAFS is a unique structural probe to obtain information on the oxidation state and chemical coordination of contaminants. An example is provided in Section 7.5 that outlines the application of XANES and EXAFS to the analysis of As in soils affected by a base-metal smelter.

Freshwater lacustrine sediments contain a variety of Mn minerals and amorphous coatings. The manganese phases in the sedimentary reservoir are comprised of a mixture of lithogenic material, authigenic material precipitated from overlying waters and diagenetic material formed in the sediment after deposition. Lithogenic material has a highly variable composition that is dependent on the source catchment. For example, in lateritic soils, where intense weathering causes the development of extensive horizons rich in metal oxyhydroxides, manganese exists largely as the minerals asbolane, lithiophorite, or in Fe-rich horizons, substituted into goethite (Manceau et al., 1987). Manganese oxides can also exist as thin veneer coatings on sediment grains effectively isolating the underlying mineral substrate and suppressing its reactivity (Fuller and Harvey, 2000).

Thus, although Mn and its respective solid-phase reservoirs are ubiquitous in these settings the complexity of their interactions requires an integrative geochemical-mineralogical approach to ascertain their influence on trace element geochemistry. It is often necessary, in order to fully understand a natural system, to apply a range of complementary techniques. Section 7.6 outlines a study aimed at examining the forms of Mn and determining the effect on Mn speciation of bacterial action in a lacustrine environment. The complementary techniques of scanning electron microscopy and synchrotron XAFS and XRF have been applied.

7.2 Isotopic dilution techniques

7.2.1 The isotopic dilution principle

Isotopic dilution techniques are based on the assumption that when a small amount of a stable or radioactive isotope (an amount that does not significantly change the overall equilibrium of the system) is introduced into a soil suspension it will exchange with the other isotopes of the same elements (Hamon et al., 2002). Therefore, a quantity of an isotope added to a system will, within a few days of its addition, distribute itself between the solution- and exchangeable-phase such that

$$\frac{a^*_{sol}}{a^*_{exch}} = \frac{M_{sol}}{M_{exch}} \tag{7.1}$$

In the case of a radioisotope, sol and exch represent the activity (a^*) of the isotope, or the concentration of the metal (M), in the solution-phase and exchangeably adsorbed on the soil solid-phase, respectively. The sum of a^*_{sol} and a^*_{exch} will be equal to the total radioactivity introduced (A); similarly the sum of M_{sol} and M_{exch} will be equal to the labile metal pool (M_E). Sampling and analysis of the solution-phase enables determination of the isotope distribution between the solution- and solid-phases, and the labile pool. In this case the labile pool is traditionally called an E-value (M_E) and can be calculated by rearranging Eq. (7.1).

$$M_E \equiv (M_{sol} + M_{exch}) = \frac{M_{sol}}{a^*_{sol}} \times A \tag{7.2}$$

A method that uses plants to assess the labile, or in this case the biologically accessible pool of metals in the soil was developed by Larsen several decades ago (Larsen, 1952) and has recently been extended to other soil organisms such as snails (Scheifler et al., 2003) and earthworms (Scott-Fordsmand et al., 2004). In this case the labile metal pool is defined as L-value (M_L) and Eq. (7.2) can be rewritten as

$$M_L = \frac{M_{shoot} - M_{seed}}{a^*_{shoot}} \times A \qquad (7.3)$$

where $M_{shoot} - M_{seed}$ is the concentration of the metal in the shoot minus the contribution from the seed and a^*_{shoot} the activity of the isotope in the shoot.

The isotopic dilution principle is valid only under the following conditions:

(i)   All metal species measured in solution (M_{sol}) are isotopically exchangeable, i.e. there is no interference from non-exchangeable metals associated with colloids (Hamon and McLaughlin, 2002; Lombi et al., 2003);

(ii) The isotope has physically mixed with the entire labile metal pool (along enough equilibration time is allowed); and

(iii) The isotope has not accessed the non-labile metal pool through processes such as (co)precipitation (equilibrium of the system has not been significantly changed by the addition of the isotope).

The investigation of metal/metalloids lability in soil has been hindered for some elements due to the short half-life of radioisotopes (i.e. 12.4 hour for ^{64}Cu) and/or because of safety issues. However, rapid and sensitive analytical techniques, such as ICP-MS, have led to the use of stable metal isotopes. In this case, some of the assumptions reported previously need to be carefully tested.

7.2.2 Assessment of rhizosphere processes

Rhizosphere processes, in particular metal mobilisation, can be investigated using isotopic dilution techniques. If a plant's roots do not modify the surrounding soil, in terms of metal mobility, the specific activity (i.e. the ratio between the radioactivity of an isotope and the metal concentration in solution) of an element in the rhizosphere soil and in the bulk soils will be identical. Consequently, the specific activity in soil solution and in the plant (once corrected for the seed metal contribution) will be also identical and so will the E- and L-values (when measured over the same timeframe). However, if rhizosphere processes mobilise metals from a pool that was not in equilibrium with the soil solution (e.g. metals occluded in minerals) then the specific activity in the solution will decrease and, assuming that the plants take up metal primarily from the rhizosphere soil, the L-value will be larger than the E-value of the bulk soil.

Small differences between E- and L-values were reported by Smolders et al. (1999) for Cd, Tiller et al. (1972) for Zn, Hamon et al. (1997) for Cd and Zn, and Echevarria et al. (1998) for Ni. These results were all obtained by growing plants, such as wheat and clover, which do not accumulate metals. However, comparisons between E- and L-values have been recently employed to assess whether metal hyperaccumulating plants are able to mobilise metal in the rhizosphere (one of the possible mechanisms responsible for hyperaccumlation). So far the results obtained by Hutchinson et al. (2000), Gérard et al. (2000) and Hamon and McLaughlin (1999) who investigated the well known Cd and Zn hyperaccumulator *Thlaspi caerulescens* indicated that these metals are not mobilised in the rhizosphere of this hyperaccumulator.

7.2.3 Behaviour of redox sensitive species

Measurement of isotopic exchangeability of redox sensitive species has been challenging due to the possibility that the oxidation state of the introduced isotope equilibrates rapidly with the oxidation states of the element in exchangeable pools in the system. If this is the case, there is a need to combine the isotopic dilution technique with a solution speciation procedure to assess the oxidation state of stable and radioactive isotopes in solution. To our knowledge the only example of such study was conducted by Hamon et al. (2004) who assessed the labile pools of As(III) and As(V) in two soils under various redox conditions and microbial activity. Using this technique the authors were able to differentiate various mechanisms responsible for As mobility in the soil. In particular, they could differentiate and quantify changes in solution concentration of As due to changes in solid-solution partitioning (caused by pH and redox reactions) from changes in the size of the exchangeable pool of both As oxidation states.

7.2.4 Assessment of remediation strategies

A modified *E*-value technique was employed by Hamon et al. (2002) and Lombi et al. (2003, 2004) to investigate the reversibility of metal/metalloid fixation in contaminated soils treated with various amendments such as lime, phosphate and Fe-rich materials. This technique consisted of a modified isotopic dilution method coupled with a step-wise acidification treatment. This technique allows assessment of whether a reduction in soluble metals is due to simple changes in metal partitioning (which are often caused by changes in pH) or to a reduction in the exchangeable pool in the soil. Furthermore, this procedure enables differentiation between three classes of attenuation mechanisms that are hypothesised to increase in their resilience to environmental change as follows: reversible sorption < irreversible "fixation" at constant pH < irreversible "fixation" across a range of pH. This modified isotopic dilution technique has regulatory merit in that it can simultaneously provide an assessment as to the stability/reversibility of remediation treatments for metals and also can be used to measure potential mobilisation of metals from fixed to available pools (Hamon et al., 2002).

7.2.5 Assessment of natural attenuation

Ageing is defined as the slow reactions that occur following the rapid partitioning of added soluble heavy metals (such as Cu, Zn and Cd) between solution and solid phases in soil, which can take years to attain equilibrium (see Chapter 2). These slow reactions are attributed to micropore diffusion, occlusion in solid phases by co-precipitation and co-flocculation, and cavity entrapment. Although the slow reactions play a key role in metal bioavailability in the field soils, their rates, mechanisms and controlling factors have not been comprehensively investigated.

Evidence of ageing processes is provided by studies of metal extractability and lability. It has been frequently observed that easily extractable pools revert with time (<1 year) to more strongly bound forms. Also, biological evidence is provided by studies examining the efficiency and residual value of micronutrient (e.g. Cu and Zn) fertilisers. However, at high rates of metal addition to soil (i.e. toxicity experiments using plants, invertebrates or microorganisms), biological evidence is limited and inconclusive. This is probably due to confounding factors associated with spiking soils with soluble metal salts at high rates (pH and salt effects) and due to biological adaptation/tolerance with time.

It is reasonable to assume that key factors influencing ageing reactions are soil properties, loading levels of metals, temperature, drying and wetting cycles, and microbial activity. Hamon et al. (1998) measured the rate of fixation of Cd in agricultural soils where this metal was added with phosphate fertilisers. Using a radioisotopic technique they developed a model that estimated Cd fixation in the order of 1–1.5% of the total added Cd per year. Using a similar technique Young et al. (2001) studied the fixation of Cd and Zn in 23 metal spiked soils over a period of 811 days. They observed that the rate of fixation increased with soil pH. Recently, Ma et al. (2006a, 2006b) used stable and radioactive isotopes of Cu to assess natural attenuation in 19 European soils. Cu was added at two effective concentrations shown to inhibit plant growth by 10% and 90%. The results showed that the lability of Cu added to soils rapidly decreased after addition, especially in the soils with pH > 6.0, followed by a slow decrease in Cu lability. The attenuation of Cu lability was modelled on the basis of three processes: precipitation/nucleation of Cu on soil surfaces, Cu occlusion within organic matter, and diffusion of Cu into micropores. The lability (E-values) of Cu equilibrated in the soil for 1 and 12 months are reported in Figure 7.1 as a function of soil pH. This figure clearly illustrates that the mechanisms and the extent of Cu natural attenuation in soil is pH dependent. At pH > 6.5 Cu exchangeability was low even after 1 month of equilibration time and this is most likely due to precipitation reactions. In contrast, at soil pH < 6.5

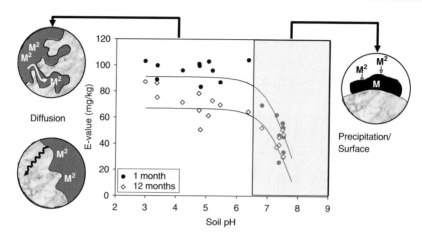

Figure 7.1. *Isotopic exchangeability of Cu added to 19 European soils as a function of time and soil pH (drawn from data taken by Ma et al., 2006b).*

diffusion reactions such as solid-phase and micropore diffusions are likely to dominate.

7.3 Microbeam synchrotron X-ray fluorescence of human teeth

7.3.1 The cementum of human teeth

The research reported here is part of an interdisciplinary study that includes among its objectives the development of advanced analytical techniques to examine samples of special anthropological interest. Teeth were chosen because they represent the most robust of the suite of hair, bones and teeth and are the least likely to undergo diagenesis (post-burial changes in chemical and physical properties). Figure 7.2 shows the anatomy of a human tooth. It is generally assumed that the best record of trace element exposure is found in the tooth crown. This is the hardest material in the human body and hence is generally the best preserved. In routine analysis the outer 100 μm are removed both to avoid metals that may have diffused inward from the tooth surface after death and to remove the permeable dentin layer from the bottom of the crown. The remaining material is then subjected to analysis by such techniques as ICP-MS. The results are indicative of metal exposure only during the period of secondary tooth formation, which is essentially complete within two years of secondary tooth eruption.

The cementum is a thin calcified tissue that serves two major functions. It seals the tubules of the dentin and acts as an attachment for the periodontal

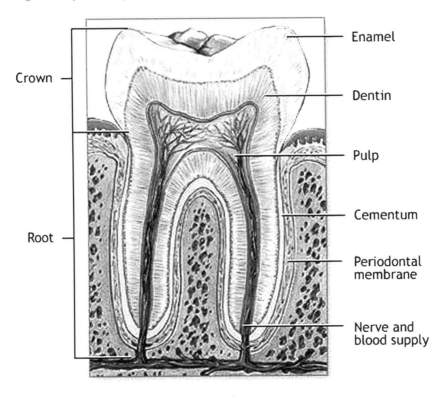

Figure 7.2. *The major components of a tooth.*

fibres thus helping to maintain the position of the teeth in the jaw. Cementum is deposited throughout the individual's life and displays a well-defined series of layers under polarised light resulting from changes in orientation of the associated collagen fibres in the cementum (Stutz, 2002). There is evidence that these layers are deposited annually in mammals (Lieberman, 1994) and consequently the trace element composition may preserve a chronological record of metal exposure throughout any individual's life. The resulting mineralised region on the root surface varies in thickness from the root tip to the gum line (thickest at the root tip) and naturally varies with age. Finally the cementum consists of two different types: cellular cementum, which contains remnants of cementoblasts, cells specialised to mineralise cementum and acellular cementum, which contains no cellular material but is also mineralised, presumably by a protein-mediated mechanism. The entire cementum system is between 20 and 600 μm wide with layers having a thickness between 5 and 20 μm. The resulting complex is not unlike the annual growth rings in temperate zone trees. It will be convenient in this study to refer to the cementum layers

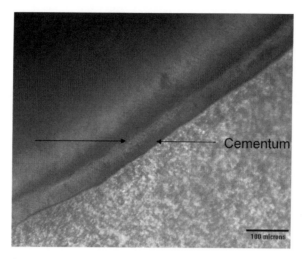

Figure 7.3. *Tooth embedded in resin, showing the cementum region.*

as the cementum ring system. Figure 7.3 shows the cementum region in a tooth that had been exposed for a long period to humic acid before lateral sectioning.

The work reported here was carried out as part of an interdisciplinary study of teeth collected at sites in Northern Peru (Fig. 7.4) and are approximately 500 years old. The objectives were to determine if the trace metal component of the cementum differed from that of other regions of the tooth and if the metal concentration varied within the cementum (and might therefore provide a temporal record).

7.3.2 Trace metal analysis by synchrotron microprobe

The teeth (supplied by the Department of Anthropology, University of Western Ontario) were embedded in epoxy resin. Cross sections through the root were cut using a low speed saw, the resulting sections were polished with successively smaller silicon carbide grit and were finished using successively 3, 1 and 0.24 μm diamond grit with Varsol as the lubricant. The samples were finally sonicated in methanol and were then used for synchrotron radiation analysis without further preparation.

Microprobe synchrotron beamlines operate in an identical manner to bulk XAFS or XRF lines in terms of the type of spectroscopic data collected. The beam is reduced in size by demagnification of a larger beam. It should be pointed out that although it is convenient to think of the beam as illuminating an area on the

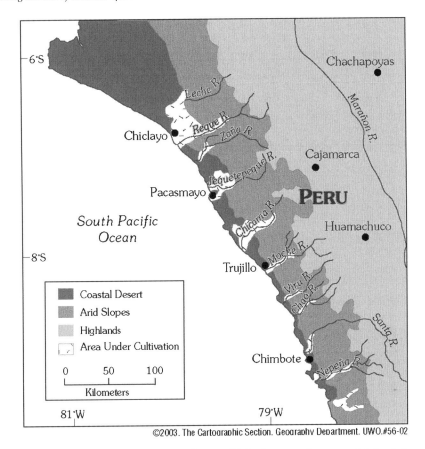

©2003. The Cartographic Section. Geography Department. UWO.#56-02

Figure 7.4. *The burial site in Peru from which those teeth were obtained is called Tucume. Tucume is located near the modern city of Chiclayo.*

sample, it does in fact always illuminate a volume of sample. The exact dimensions of this volume depend on the angle of orientation of the sample to the beam and the penetration depth of the X-rays, which depends on the type of material being probed and the energy of the incident and fluorescent X-rays. On a microprobe beamline the sample is mounted on a moveable stage that makes it possible to scan over an area of the sample.

This new capability is used to create 2-dimensional "maps" of the elements present in a sample. This is done by adjusting the incident X-ray energy to a value above the absorption edges of all the elements of interest in the sample. When a given location on the sample is illuminated by a beam of this energy, all the elements in the illuminated volume fluoresce with characteristic fluorescence X-ray energies. These X-rays can be sorted and counted using a solid state X-ray detector and the identities of the elements in the illuminated volume determined. The intensity of the fluorescence X-rays for a given element is a function of several factors, including the amount of that element in the illuminated volume. Although "concentrations" obtained this way are not quantitative unless special strict precautions are taken, they can be used to determine qualitative or semi-quantitative differences in the elemental distributions within a sample. For an introduction to synchrotron XRF and microprobe measurements see Manceau et al. (2002) and other chapters within the same volume of *Reviews in Mineralogy and Geochemistry*.

The cementum region is uniquely well suited to synchrotron radiation analysis using an intense synchrotron X-ray micro-beam (Martin et al., 2004). The metals of interest are in low concentration, the areas under study are small and the samples (in this case often from remains obtained from ancient Peruvian burial sites) should not be subjected to destructive analytical techniques.

Analysis was carried out at the Pacific Northwest Consortium – Collaborative Access Team (PNC-CAT) located at the Advanced Photon Source (APS) Argonne National Laboratory using beamline 20ID focused to a $5 \times 5\,\mu m$ spot. X-ray fluorescence (XRF) was collected for Sr K_α, Pb L_α, Zn K_α, Ca K_α and Br K_α.

Figure 7.5 shows the XRF maps collected for Ca, Zn and Br from a large area of the ancient tooth (Martin et al., 2004) including the central pulp chamber, the dentin and the narrow cementum region. Since the minerals in teeth are almost exclusively calcium based (the dominant mineral is bioapatite) it is not surprising that that tooth is defined by the calcium X-ray fluorescence image. While the detailed changes close to the pulp region are due to intrusion of blood and other body fluids and may be ignored, the band of enriched Zn is striking

Figure 7.5. *X-ray fluorescence maps of a tooth showing top Ca, middle Zn and bottom Br. The Zn enriched cementum region is particularly apparent.*

and may be used to define the cementum region. This Zn enrichment (5 to 10 times that elsewhere in the tooth) was observed in all the teeth studied as part of this research. The Zn distribution shows that the cementum does display a trace metal distribution unique to this tissue.

Br is also enriched in the cementum of this tooth (Fig. 7.5). The apparent width of the Br region is an experimental artefact. The exciting X-ray is at 45° to the tooth and hence excites fluorescence X-rays from a point below the tooth surface before it reaches the plane along which the tooth was sectioned. Accordingly XRF is detected early in the map and the intensity rises till the exciting beam reaches the upper tooth surface. The resulting broadening of the fluorescence is a function of the excited X-ray escape depth, greatest for Br and least for Ca. We frequently observe comparable enrichment in other teeth (or in teeth in general). Bromine is taken to be indicative of a marine diet. Taken together these results show enrichment of both Zn and Br in the cementum of this tooth.

Figure 7.6 shows the Zn XRF image from the cementum region of a second ancient tooth. Not only is the cementum region clearly delineated by this image but the image also shows variations in the Zn concentration throughout the cementum region. Internal variations in this form are indicative of endogenous metal deposition. Figure 7.7 shows the cementum in this tooth to contain

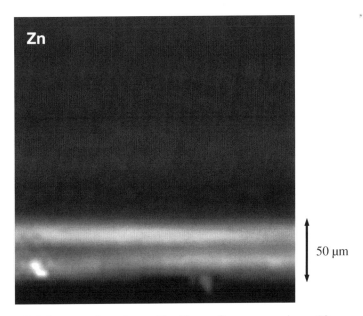

Figure 7.6. *Brightest region shows Zn X-ray fluorescence in a 50 μm wide band corresponding to the cementum. Note two regions of enrichment.*

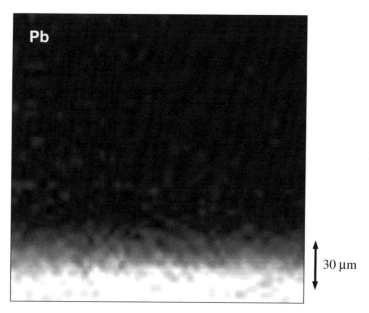

Figure 7.7. *Pb enrichment in a 30 μm band.*

appreciable Pb. The latter may be the result of lead release during smelting of Pb ores (Abbott and Wolfe, 2003).

7.4 XANES and XRF studies of Cr speciation in particulate matter

7.4.1 Atmospheric particles and Cr

The US Environmental Protection Agency estimates that 50,000 deaths a year are caused by inhaled particulate matter, i.e. tiny solid and liquid particles suspended in air (http://www.epa.gov/ttn/naaqs/). Inhalation of particulate matter (PM) is strongly associated with premature cardiopulmonary deaths and is known to aggravate existing respiratory and cardiovascular conditions, especially in vulnerable populations such as children and the elderly (Chalupa et al., 2004; Pope et al., 1995; Schlesinger, 2000). Currently PM is classified and regulated based on its size (aerodynamic diameter). Particles with an average size of less than 10 μm are called PM_{10}, while those smaller than 2.5 μm are $PM_{2.5}$. $PM_{2.5}$, also known as fine particles, are of special regulatory interest because they are able to penetrate deep into the lungs when inhaled (http://www.epa.gov/ttn/naaqs/).

While the adverse health effects of PM are well known, little is understood about how the chemical composition of the particles, as opposed to only their size, alters their toxicity. Presumably the reactivity and inherent toxicity of the material forming a particle could have a dramatic effect on the type and/or extent of damage it causes once inhaled. One reason for this lack of understanding is that the composition of ambient aerosol particles is extremely complex, and it is difficult to identify and classify the components. The second problem is that given the tiny dimensions of individual particles, there are few techniques capable of quantitative analysis of single particles, whereas bulk particulate analyses provide no information on the single particle level.

Of the metals found in particulate matter, Cr is of particular interest for two reasons: abundance and toxicity. While the highest exposures to Cr(VI)-containing particulate matter are likely to occur to workers during welding, chrome plating, spray painting and chrome pigment production (Cohen et al., 1993), a study in Los Angeles showed that ambient fine particles can also contain significant amounts of Cr and that the amount of Cr (relative to the total mass of metals) is greater in the smallest particles (Hughes et al., 1998). In that study, Cr accounted for up to $\approx 10\%$ of the transition metal mass in ultrafine particles (diameters $< 0.097\,\mu m$). Cr is also among the top five most abundant metals in particle emissions from diesel fuel combustion (Wang et al., 2003). Secondly, Cr has several oxidation states under environmental conditions, including Cr(III) and Cr(VI). Cr(III) has very low toxicity and solubility and is actually an essential nutrient in small doses. In contrast, Cr(VI) is found in the form of the soluble oxyanion chromate (CrO_4^{2-}) and is highly toxic and carcinogenic (Cieslak-Golonka, 1996). Therefore, it seems likely that the toxicity of Cr-bearing particles will depend on the oxidation state of the Cr.

7.4.2 Bulk XANES analysis

XANES is an ideal technique for the analysis of chemical speciation of Cr in atmospheric particles as XANES spectra of Cr(III) and Cr(VI) are very distinct, with the Cr(VI) compounds having a characteristic pre-edge feature (Fig. 7.8). The height of this pre-edge feature, in a normalised spectrum, is related to the percentage of Cr that is present as Cr(VI) in the sample (Patterson et al., 1997; Peterson et al., 1997). In addition to clearly identifying the Cr(VI) content of a sample, the XANES spectra can be used to distinguish between a variety of oxidation states and chemical forms of Cr, as can be seen by the standard spectra in Figure 7.8. The spectra of Cr metal and chromium carbide, Cr_3C_2, spectra are distinct from the rest by the shift to lower energy of their edge positions. Furthermore, even within Cr(III) compounds, $Cr(OH)_3$, Cr_2O_3 and Cr_2FeO_4 (chromite or Cr–Fe spinel) can be distinguished based on their XANES spectra.

We had two goals in this work, to determine how the amount of Cr(VI) present in the particles generated from a laboratory flame depended upon flame

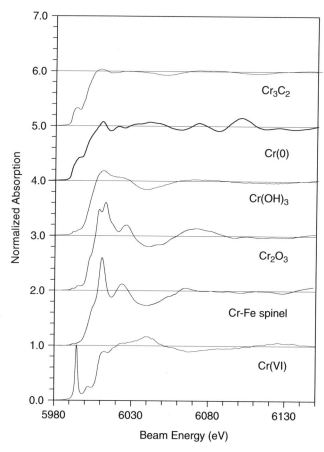

Figure 7.8. *Standard XANES spectra of different Cr-containing compounds. (Spectra are offset by integer values for clarity.)*

conditions and to examine how simulated atmospheric ageing affected levels of Cr(VI) in the particles. Particles were produced in a laboratory combustion chamber by introducing chromium hexacarbonyl $(Cr(CO)_6)$ vapour into a flame and collecting the resulting particles. Particles were collected on Teflon filters and stored under $N_{2(g)}$ in the dark until analysis. To understand how the form and oxidation state of Cr might change during particle ageing in the ambient atmosphere, a section of each filter was removed and aged in a simulated atmospheric chamber where it was exposed to simulated sunlight, humidified air (75% RH) and ozone (≈ 1 ppmv). We used bulk XANES collected at Sector 13, GSECARS, of the Advanced Photon Source (Argonne National Laboratory, Chicago, USA) to analyse particles for Cr(VI) (Werner et al., 2006, 2007).

There were two main results from this study. Firstly, the temperature of the flame had a dramatic effect on the amount of Cr(VI) found in the particles, with hotter (100% H_2) flame producing more Cr(VI) ($\approx 43\%$ Cr(VI)) than cooler (33% H_2 in air) flame ($\approx 5\%$ Cr(VI)). Secondly, it was observed that the kinetics of Cr transformation under simulated atmospheric conditions were relatively slow and tended toward reduction of Cr(VI) to Cr(III). Estimated half-lives from these experiments were in the range 50–100 h which is within the range 13–200 h reported previously by Seigneur and Constantinou (1995) and Grouse et al. (1998). The differences in the estimated half-lives in the previous work depended mostly on whether or not the model systems included organics, with the longer half-lives being observed in the absence of organics (Seigneur and Constantinou, 1995; Grouse et al., 1998).

Interestingly, when Fe was introduced into the flame, as iron pentacarbonyl (Fe(CO)$_5$), along with the Cr (in a 10 to 1, Fe to Cr ratio), the form of Cr in the particles changed dramatically. There was significantly less Cr(VI) present in the particles for a given flame condition. For example, in a hot, 100% H_2 flame, Cr(VI) represented only 12% of the Cr in the presence of Fe, in contrast to accounting for 43% of Cr in the sample without Fe. The dominant form of the Cr(III) in the samples also changed from a mixture of Cr_2O_3 and $Cr(OH)_3$ in the absence of Fe to a mixed Cr/Fe phase with a spinel-like structure in the presence of Fe. This phase is represented in our studies by the Cr_2FeO_4, Cr–Fe spinel, standard (Fig. 7.8) but it should be pointed out that spinel structures containing various amounts of Fe and Cr are possible and could potentially give almost identical XANES spectra.

7.4.3 X-ray fluorescence mapping

The next phase of our work was to test whether our laboratory-based results were applicable to ambient aerosol samples. Ambient samples of PM$_{2.5}$ were collected on Teflon filters from three different locations in California's Central Valley: urban (Sacramento), suburban (Davis) and rural (near Placerville, CA). Because of the low concentrations of Cr in these samples and the spatial variability of particles on the filters, bulk XANES was no longer an appropriate analytical technique. For this reason we used the X-ray microprobe beamlines 10.3.2 at the Advanced Light Source (Berkeley, CA) and 13-BM, GSECARS at the Advanced Photon Source (Argonne, IL) to analyse our samples.

An example of the results from this micro-XRF mapping is shown in Figure 7.9. Figure 7.9a shows the spatial distribution of Ca within the sample, while Figure 7.9b shows the distribution of Cr and Figure 7.9c the distribution of Mn. The repeating pattern of circles seen in the Ca map is an artefact of the sample collection technique and is due to the fact that the collection filter

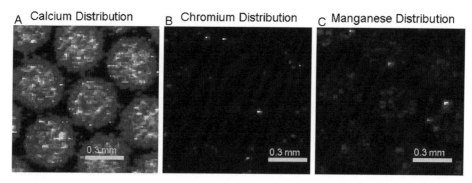

Figure 7.9. *XRF map showing calcium, chromium, and manganese distribution in an ambient aerosol sample from Sacramento, California. Warmer colours indicate higher concentrations and cooler colours indicate lower concentrations.*

sits on a support screen containing a series of circular holes through which the air flows. Therefore, the particles collected on the filter are concentrated in the areas above these holes. In contrast to the universal abundance of Ca-containing particles, the Cr map shows that there is only a small number of Cr-containing particles in the sample as represented by the bright spots on Figure 7.9a.

7.4.4 Micro-XANES analysis

After the Cr-containing areas have been identified by XRF mapping, the sample can be moved so that the focused beam is centered upon one of these spots. At this point, a XANES spectra of the spot can be taken in the usual way, by scanning the energy of the incident X-rays and recording the intensity of the resulting fluorescence. The XANES data obtained in this manner are usually of slightly lower quality than their bulk XANES counterparts. Several factors contribute to this, including the fact that the absolute quantity of the Cr fluorescing is smaller and therefore the fluorescence signal is also smaller. Also, given the small size of the beam, and the rapidity with which its intensity varies with distance from the center, the data are susceptible to any motion of the sample relative to the beam from either motion of the beam or from motion of the sample. The latter is frequently seen in samples that are not securely mounted or that warp as they are heated by the beam (as may be observed for kapton mounted samples at the microprobe end-station at GSECARS at the Advanced Photon Source). Successive scans of the same particle frequently showed beam induced reduction of Cr(VI) to Cr(III). Therefore, we were usually only able to use one scan of each particle, which limited the quality of the data obtainable. Once the data were collected and processed, the spectra of each particle was reproduced by a linear combination of standard spectra. Several example spectra of Cr particles are shown in Figure 7.10. Some of the ambient samples analysed were clearly dominated by single phases (Fig. 7.10a and b) while other particles were mixtures of different phases (Fig. 7.10c and d).

We acquired XANES spectra from nine to twenty-eight particles from each sampling location. The overall distribution of Cr species at each site (Table 7.1) shows several interesting points. The first is that the Cr/Fe-mixed phase spinel first observed in the laboratory samples is a dominant phase at all the sampling locations. This is perhaps not surprising given the ubiquitous nature of Fe and the chemical similarity between Cr and Fe. However, it is gratifying to observe such a clear link between the laboratory experiments and the ambient aerosol samples. The second observation from the data is that the majority of the Cr is found as Cr(III) with minor contributions from other species including Cr(0), Cr-carbide and Cr(VI). This too is consistent with the laboratory experiments and previous work showing an overall tendency for Cr reduction in the atmosphere. (Note that the only Cr-carbide standard used was Cr_3C_2 and no attempt is made to distinguish between different types of carbides and in addition the spectra provided in Fig. 7.8 is over adsorbed.)

Cr(VI)-containing particles were only identified in the sample collected on a Friday at the Sacramento site (there were five Cr(VI)-containing particles out of 22 particles analysed from this sample). The sample taken at the same location in Sacramento on Saturday did not show any particles containing Cr(VI). Admittedly, the total number of Cr particles identified in the Saturday sample was small ($N = 6$). As in the example shown in Figure 7.10d, in particles with Cr(VI) these species accounted for only a small fraction, 8.5–17%, of the total Cr. The remainder of the Cr in the Cr(VI)-containing particles consisted of Cr(III)-containing species including, $Cr(OH)_3$, Cr_2O_3 and Cr–Fe spinel, with $Cr(OH)_3$ being the most common other constituent. This is perhaps to be expected, because if Cr(VI) were being reduced in the particle the first Cr(III)-containing species formed would be $Cr(OH)_3$. The sampling site in Sacramento is not only the most urban, but also ≈ 13 km downwind from a Cr plating facility known to emit Cr(VI) although this plant is not the only possible source of the Cr(VI) in this sample.

Table 7.1. *Average chromium speciation in particles from different sampling locations.*

Location	Percentage composition						Number of particle spots
	Cr/Fe spinel	$Cr(OH)_3$	Cr_2O_3	Cr_2C_3	Cr(0)	Cr(VI)	
Davis	56	16	14	10	5	0	18
Sacramento	41	33	12	6	5	3	28
Placerville	33	43	16	7	1	0	9

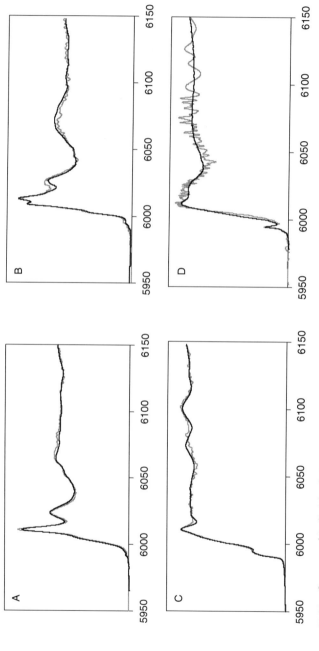

Figure 7.10. *Spectra of individual Cr particles. Original data shown in grey line and reconstructed spectra are shown in solid line. A, 100% Cr-Fe spinel; B, 100% Cr_2O_3; C, 13% Cr-Fe spinel, 49% Cr(0), 38% Cr_3C_2; D, 88% Cr(OH)$_3$, 12% Cr(VI).*

7.5 XANES and EXAFS analyses of As in soils

7.5.1 Contaminated soil near a base-metal smelter

The objective of this study was to characterise the form and nature of As in soil samples by XAFS. This molecular-scale characterisation information can be used in the assessment of the stability of As carriers in soil and in the development of a conceptual model for contaminant release mechanisms including exposure pathways such as ingestion, inhalation, dermal absorption and plant uptake. Furthermore, the data is essential in evaluating health risks and determining remediation options.

A molecular-scale characterisation study was undertaken in support of an environmental impact baseline study examining human health effects due to exposure to contaminants such as As, Ni, Co and Cu in soils impacted by a base-metal smelter. A small subset of samples representing near surface soil samples were collected in the vicinity of the smelter. The samples represent three different sites including residential properties and public sites, and three different soil horizons to a depth of 20 cm (totalling nine samples in all). As concentrations in the soil samples ranged from less than 20–270 ppm.

7.5.2 XAFS measurement and analysis methodology

The samples analysed by XAFS are listed in Table 7.2 along with their location, depth and bulk As concentrations. The XAFS measurements were carried out at the Stanford Synchrotron Radiation Laboratory (SSRL) as part of a broader study on the speciation of As in solid metallurgical wastes and contaminated soil. The soil samples were loaded into Teflon holders with Mylar windows. Experiments were performed at the wiggler beamline 4-1. The energy in the storage ring was 2.99 GeV and the current was around 60 mA. Si (220) double-crystal monochromator with 2 mm vertical slits were used. The monochromator

Table 7.2. *Samples and number of scans performed for EXAFS analysis.*

Site	Sample number	Depth (cm)	As (ppm)	XANES	EXAFS
1	A-1	0–5	200	4	NA
	A-2	5–10	254	4	6
	A-3	10–20	270	6	12
2	B-1	0–5	220	4	12
	B-2	5–10	160	4	NA
	B-3	10–20	150	4	NA
3	C-1	–	–	4	NA
	C-2	–	–	4	NA
	C-3	–	–	4	9

Note: –, Unkown; NA, not analysed.

was detuned to 50% to get rid of the harmonics. The samples were placed at 45°
angle to the incident beam. As foil was placed between the second and third
ionisation chambers for energy calibration at the inflection point of 11,867 eV.
XAFS spectra were collected in the fluorescence mode using a 13-element Ge
detector at room temperature. Aluminium filters were placed between the
sample and the Ge detector.

XANES spectra were obtained by scanning the monochromator at 0.2 eV
steps over the edge region and the EXAFS spectra at 1.8–6.2 eV steps over the
EXAFS region. The XANES analysis involved scanning of each sample at least 4
times and EXAFS up to 12 times to increase signal-to-noise ratio (Table 7.2). Data
reduction and analysis were made by the use of EXAFSPAK (George and
Pickering, 1993), LSFitXAFS (Paktunc, 2004) and IFFEFIT software (Newville,
2001; Ravel and Newville, 2005).

Individual scans were calibrated for edge shift and averaged. Following this,
background subtraction was performed by fitting a polynomial equation in the
pre-edge region to 11840 eV. EXAFS spectra were fitted by a spline function in
the region above 11885 eV. Fourier transformation of the EXAFS function was
performed from 3 to 13 Å^{-1}.

7.5.3 XANES and EXAFS interpretation

XANES spectra of the soil samples indicate that As occurs only as As(V)
(Fig. 7.11) regardless of depth. The k^3-weighted EXAFS spectra of the samples
show more than one frequency (distance) and resemble spectra from As-
adsorbed goethite model compound and natural samples (Paktunc et al., 2003,
2004). Fourier transformation of the k^3-weighted frequencies produces radial
structure functions which provide quantitative estimates of interatomic
distances between the As atom and its nearest neighbours and compositional
information about neighbour atoms. The distances shown on the abscissa in
Figure 7.12 are less than the true interatomic distances as they are not corrected
for phase shift. The major peak on all the spectra centered at approximately 1.3 Å
corresponds to the As–O distance in tetrahedral coordination of O atoms around
an As atom. The second prominent peak occurring at approximately 2.7 Å
corresponds to scattering from Fe atoms. These peaks were isolated and back-
transformed to k-space to form individual filtered EXAFS oscillations. Using
phase and amplitude functions derived from model compounds, the filtered
spectra were fit by non-linear least squares methods to estimate coordination
numbers, distances and Debye-Waller parameters. As–O interatomic distances
of 1.69 ± 0.01 Å and coordination numbers ranging from 4.6 ± 1.7 to 5.5 ± 1.7
indicate tetrahedral coordination (Table 7.3). A shell of 4 O atoms or OH

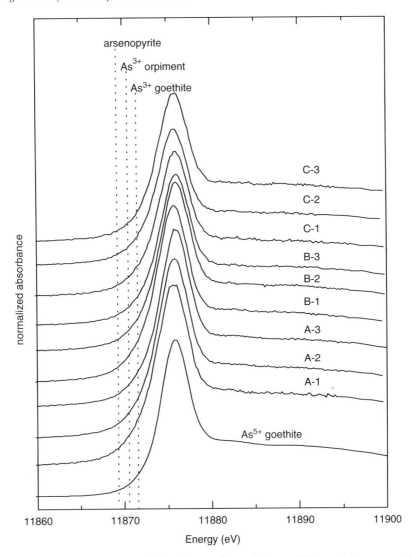

Figure 7.11. *XANES spectra of the soil samples and the goethite model compound with* *As^{5+}. White line energy positions of As(I) in arsenopyrite and As^{3+} in orpiment and* *As^{3+} in goethite are shown by vertical dotted lines as references.*

moieties surrounding an As atom has a radial distance of 1.69 Å. Radial As–Fe distances vary from 3.25 ± 0.02 to 3.30 ± 0.02 Å. Along with the coordination numbers varying between 1.2 and 2.9, the radial distances suggest inner-sphere corner-sharing polyhedra or bidentate-binuclear arrangement. In this

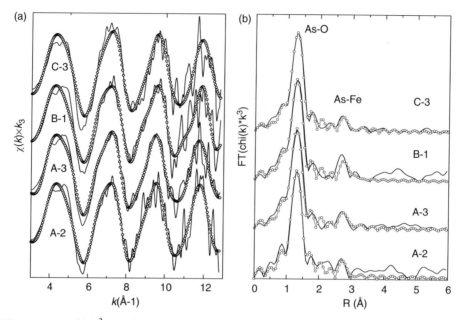

Figure 7.12. *(a)* k^3*-weighted EXAFS spectra of the soil samples. (b) Fourier-transformed EXAFS spectra. Radial distance not correct for phase shift. Fitting was performed with a k interval of 3–13 Å. Experimental curves are shown by solid lines and the fitted spectra are by circles.*

Table 7.3. *Local coordination environment based on EXAFS data analysis within* k *interval of 3–13* $Å^{-1}$.

Sample number	Shell	CN	R(Å)	σ^2 (Å^2)	E_0 (eV)
A-2	As–O	4.55	1.69	0.0008	−5.27
	As–Fe	1.17	3.26	0.0042[a]	–
A-3	As–O	5.48	1.69	0.0026	−4.51
	As–Fe	2.88	3.30	0.0096[a]	–
B-1	As–O	5.49	1.69	0.0025	−5.14
	As–Fe	1.49	3.25	0.0040[a]	–
C-3	As–O	5.08	1.69	0.0022	−4.77
	As–Fe	2.46	3.30	0.0105[a]	–

Note: CN, Coordination number; R, interatomic distance; σ^2, Debye-Waller parameter; E_0, energy offset; –, value fixed to the 1st shell E_0.
[a]Fixed value.

arrangement, two adjacent Fe–O octahedra share an O atom each with an As–O tetrahedron.

The inner-sphere bidentate-binuclear adsorption of As on Fe oxyhydroxide surfaces is the least labile form of attachment because desorption of arsenate would require simultaneous removal of the two bonds joining the As atom to the adjacent Fe atoms through O. Along with the occurrence of As in pentavalent form, these findings suggest that the As in smelter-affected soil would be in a less bioavailable and less toxic form in comparison to its trivalent form.

7.6 Complementary spectroscopic analyses of lacustrine sediments

7.6.1 Lacustrine sediments: The role of manganese

As Mn oxides have a large surface area due in part to their small particle size and malleable crystal structure, they tend to act as potent scavengers for trace metals and contaminants in both terrestrial and aquatic environments (Nelson and Lion, 2003; O'Reilly and Hochella, 2003; Tebo et al., 2004). Manganese oxides (Mn(IV)) are formed by the oxidation of soluble or relatively mobile Mn(II), which occur in soils, lake sediments and groundwaters at concentrations between 1 and 10^{-3} mM (Kabata-Pendias, 2001). Under most conditions the oxidation of Mn(II) is thermodynamically favoured, but is generally very slow in natural water (Morgan, 2000). Microbes, however, in both field and laboratory-based studies have been shown to speed up oxidation of Mn(II) by several orders of magnitude (Tebo et al., 2004; Morgan, 2000; Morgan, 2005).

XAFS has been used to investigate eutrophic lake waters and their sediments (Friedl et al., 1997). The authors found that H^+-birnessite was the predominant Mn-oxide forming in the water column via the oxidation of Mn(II). Since the mineral birnessite is a potent sorbent of cations due to the large number of vacant sites between alternating layers of octahedral Mn(IV) its formation was clearly influencing trace element chemistry in the suboxic zone of the lake. With settling and burial in the lakes anoxic sediments, birnessite is rapidly reduced resulting in the release and porewater accumulation of Mn(II), trace elements and the precipitation of mixed cationic carbonate and phosphate minerals such as $(Ca, Mn)CO_3$ and $(Fe, Mn)(PO_4)_2 \cdot 8H_2O$ (Friedl et al., 1997).

Marcus et al. (2004) used a combination of synchrotron-based X-ray microdiffraction, as well as X-ray fluorescence and micro-XANES to evaluate the speciation and spatial relationships for Mn, Fe, As and Zn in a fast-growing (0.02 mm/yr), shallow-marine, ferromanganese nodule. It was observed that the nodule exhibited alternating Fe-rich and Mn-rich layers reflecting redox variations associated with the water chemistry. The Mn in this case was in the form of birnessite. Despite these initial investigations there has been little detailed spectroscopic evidence on the sequestration and subsequent lability of trace metals in Mn/Fe oxide rich sediments in equatorial regions (one example of a related study is that by Takahashi et al., 2007).

We have begun a series of studies in tropical lake settings, that receive significant inputs of lateritic soils via erosion and changes in land use, to understand the impact on trace metal mobility in these lakes. Once soil enters the lake and is eventually integrated as sediment, organic carbon degradation leads to a series of microbially catalysed reactions that consume oxidants in the order of decreasing free energy yield (i.e. O_2, NO_3^-, Mn(IV), Fe(III) and SO_4^{2-}). This reaction series can be observed in sediments in the form of a vertical zonation of the reactants and products. However, the observed sequence is complicated by the availability of reactive organic carbon, the relative abundance and mineralogy of the various oxidants and the specific microbial community. The reduction of Fe and Mn oxyhydroxides below the sediment water interface can cause fluxes of various metals in the sediment porewater. These metals then diffuse upward and/or downward through the sediment according to aqueous concentration gradients in the porewater. Thus, the potential exists for heavy metal fluxes from the sediment to the water column as a result of Fe and Mn diagenesis and the production of a concentration gradient. However, in the presence of an oxygenated water column, upwardly diffusing Fe(II) and Mn(II) is frequently oxidised in the sediment by downward diffusing O_2. In other words, the presence of an oxygenated zone in the surface sediment will often preclude heavy metal fluxes from the sediment to the water column.

In contrast, in sub-oxic waters the potential for metal release from the sediment is much higher. Therefore to quantify the role of dissimilatory metal reducing bacteria (DMRB) on the mobility of trace metals in Fe-rich lacustrine sediment (LS) we have performed a series of microbial incubation experiments. Based on prior selective extraction experiments it was shown that there is a Mn-rich phase in our LS samples that potentially serves as an important reservoir of trace elements in our Fe-rich sediment. To confirm this hypothesis we utilise a combination of scanning electron microscopy (SEM) and synchrotron-based XAFS and XRF to first investigate the lability of trace metals within this lake sediment and secondly to test the influence that DMRB may have on the Fe and more specifically Mn oxide components in sub-oxic lacustrine environments.

7.6.2 SEM and synchrotron XAFS and XRF measurements

Sediments used in this study were collected from Lake Matano (Sulawesi Island, Indonesia; 121°20′E, 2°28′S) using a gravity-coring device. The sediments of Lake Matano are very Fe-rich due to extensive weathering of the lateritic soils that dominate the catchments. The LS sample selected in this study represents material from the upper 10 cm of a core collected below the sediment–water interface. The LS sample was freeze-dried and homogenised by mixing (sample LS). Sediment-microbial assemblages were prepared for electron microscopy by utilising a rapid room temperature chemical fixation technique (Nation, 1983; Vandevivere and Beveye, 1992). Individual sediment grains were stub-mounted, gold-sputter coated for 90 s, then imaged using a LEO 1550 field emission scanning electron microscope (FESEM) operating at 15 kV. This fixation technique was used on the samples to ensure their integrity and preserve cellular structures associated with the sediments analysed.

The Cr K (5.9892 keV) and Ni K (8.3328 keV) edge XAFS measurements and micro-XRF maps were performed at the National Synchrotron Light Source (NSLS), Brookhaven National Laboratory (Upton, NY). All the bulk XAFS data were collected on beamline 23-B, which is equipped with fixed-exit-position double-crystal monochromator. Beamline X26-A was used for the acquisition for all the XRF maps.

7.6.3 Scanning electron microscopy

The coupling of observations provided by FESEM and synchrotron-based techniques to soil investigations can provide a wealth of information. Valuable information is provided by FESEM techniques with respect to the spatial associations and morphological changes of the lateritic soil sample in the presence of the DMRB. There are clear indications that the coatings associated with some grains proximal to the cells are being dissolved leaving behind a pronounced heterogeneous material. An example of this is illustrated in Figure 7.13 of LS exposed to facultative anaerobes for 17 days (sample LSM). The composite LSM grain on the left in Figure 7.13a shows a combination of primary and secondary mineral reaction products. Some signs of secondary mineral precipitation (e.g. siderite) on the cells and exopolysaccharides (EPS) are also observed. The micrograph in Figure 7.13b shows a significant fraction of the LSM sample is coated with biologically produced EPS.

There is significant contention about the role of EPS in the weathering of sediments and soils. Firstly, EPS has been linked to its ability to promote dissolution via proton gradients at the surface of the minerals and via

 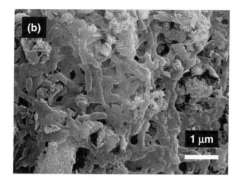

Figure 7.13. *SEM micrograph of the lake sediment incubated for 17 days with a DMRB (sample LSM). (a) Cells are preserved although they show signs of desiccation due to the low temperature field preservation technique. (b) The micrograph shows extensive EPS coverage intimately associated with the mineral grains from the sediment. Evidence of secondary mineral precipitates is also observed proximal and on the top of the EPS.*

adsorption of released metal cations. This reaction could lead to the formation of secondary precipitates that we are observing. Secondly, EPS blocks surface sites and therefore prevents mass transfer of cations to the bulk fluid which in turn decreases the rate of weathering. Thirdly the existence of these biological polymers leads to increased sediment stabilisation that may maintain porosity longer during compaction. One important observation, particularly in the context of nanophase structural information that can be obtained from the acquisition of high-resolution microscopies, is the nature of the porosity of these films (Fig. 7.14). This figure shows the utility of high-resolution FESEM imagery by essentially showing that although the sample is quite heterogeneous dissolution is proceeding quite uniformly on the mineral surfaces.

7.6.4 Synchrotron XAFS

Synchrotron XAFS was performed on the homogenised lake sediments (LS) sample and compared to a range of Cr standards representing different structural and chemical oxidation states.

The Cr in the LS sample (freeze dried and homogenised) was in the reduced form of Cr(III). The radial distribution function in Figure 7.15 shows a comparison of the Cr standards and the LS sample characterised by similar bonding lengths. The measured Cr bond lengths for the Cr standards agree favourably with the theoretical values for chromite (i.e. Cr–O 1.977 Å; Cr–Cr 2.982 Å; Cr–Fe 3.474 Å, measured radial distribution function shown as A in

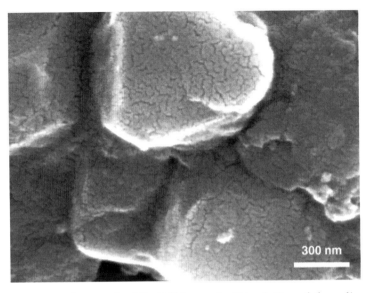

Figure 7.14. *SEM micrograph of the LSM sample. Dissolution of the sediment grains appears to be proceeding quite uniformly on the mineral surfaces (i.e. along edges and terraces). The porosity of the EPS encapsulating the individual grains is also evident.*

Fig. 7.15) and Pb chromate (i.e. Cr–O 1.64 Å, measured radial distribution function shown as C in Fig. 7.15). The chromate standard was examined to rule out the possibility of Cr(VI) species being present in the sediments examined. The modeled radial distribution function plots for the chromite standard exhibits a Cr–O bond distance of 1.97 Å and a Cr–Cr bond distance of 2.96 Å. The homogenised LS sample showed a slightly shorter bond distances for Cr–O of 1.95 Å and a Cr–Cr bond distance of 2.94 Å (B in Fig. 7.15). The LS sample also shows an additional Cr–Fe bond at 3.44 Å. The observed shorter bond distances in the LS sample, compared to the reference compounds, may arise from inherent heterogeneity of Cr(III) phases in the homogenised core sediment. It may be that several phases of Cr are present (i.e. reduced chromite or more oxidised forms may be present in the sample).

The Ni $k^2\chi$ data along with the radial distribution function plots for the Ni in goethite standard, LS (freeze dried and homogenised), LSC (control, 17 day incubation; no microbial culture) and the LSM (LS exposed to facultative anaerobes for 17 days) is shown in Figure 7.16. The k^2-weighted EXAFS spectra in Figure 7.16a provide a clue to the structural changes that may be occurring within the experiment. In this particular case the Ni phase in both the LS sample (starting material) and the LSC are structurally similar suggesting that Ni does

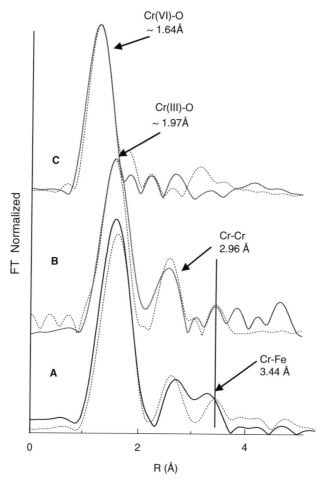

Figure 7.15. *Structurally the chromite(III) standard (A) and the LS sample (B) contain similar features with little overlap with the chromate standard (C). Both the standard and the LS sample contain a strong Cr–O bond and Cr–Cr bond. The LS sample exhibits slightly shorter bond distances for the Cr–O bond distance at 1.95 Å vs. 1.97 Å for the measured chromite standard) and a Cr–Cr bond distance of 2.94 Å. The experimental data is shown by solid lines and the simulated radial distribution functions by dotted lines.*

not undergo abiotic alteration during the course of the 17-day experiment. In contrast the LS sample exposed to the consortia of facultative anaerobes over the 17-day period shows a dramatic structural reorientation to a phase resembling more the Ni-substituted goethite standard (Fig. 7.16a). Based on this observation the contributing Ni-bearing phase associated with Mn may be preferentially removed by the bacteria under these experimental conditions.

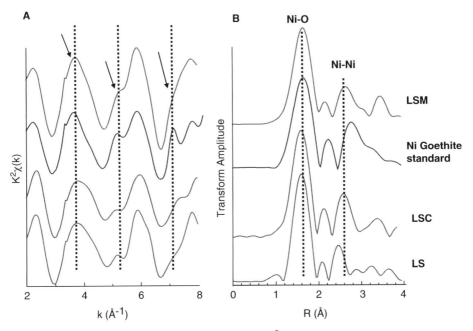

Figure 7.16. *(a) The solid line represents the raw $k^2\chi$ data for the Ni K-edge. Structural differences for Ni are evident between the LS (starting material) and the LSC (abiotic 17 days incubation) compared to the LSM (microbial leach 17 day exposure) and are highlighted by the arrows. The bound Ni associated with the LSM sample has similar structural characteristics to the Ni-goethite standard. (b) The radial distribution functions of the LSM (microbial leach) sample also show a close match with the Ni-goethite standard.*

Nickel substitution within manganese oxide typically exhibits bond lengths and nearest neighbour distances similar to 1.88 Å (Ni–O), 2.04 Å (Ni–Ni) and 3.06 Å (Ni–Ni) (Manceau et al., 2003). The Ni–O polyhedra are identical in both the LS (starting material) and the LS control (LSC), at 1.94 Å. This is shorter than one would expect if all the Ni present in the lake sediment was directly associated with Fe-bearing phases (i.e. goethite, hematite). In Fe oxides (i.e. goethite) the Ni–O bond distance from the first shell should be close to 1.98 Å based on theoretical calculations (Carvalho-E-Silva et al., 2003). The shorter bond distances observed in the control and starting material may be a result of Ni associated with mixed Mn and Fe oxides present in the sample. This assumption is supported by further evidence based on the measured Ni–O bond determined for the LSM sample which showed not only structural changes to the incorporated Ni signature over the 17-day period but also an increase in the average bond distance to 1.97 Å resembling more closely the synthetic Ni–goethite standard.

This observation suggests that Mn oxides in the form of coatings may be removed via the mobilisation of Ni due to bacterial metabolic activity, resulting in Ni-bearing goethite as the predominant phase after Mn reduction.

7.6.5 Synchrotron XRF

To corroborate the interpretation of the XAFS data sets, micro beam XRF mapping was performed on the LS and LSM samples. The micro-XRF maps collected from the untreated LS sample shows the distribution of Cr and Ni within the Fe and Mn bearing phases (Fig. 7.17). The distribution of Cr and Ni is quite heterogenous with Ni being primarily associated with the Fe-bearing and to some extent the Mn-bearing phases. Ni also occurs as discrete-isolated

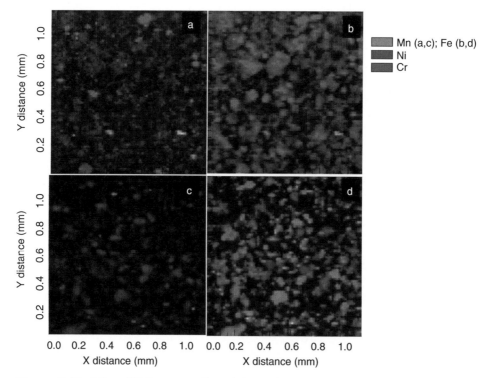

Figure 7.17. *Micro-XRF maps collected from the LS and LSM samples shows the deportment of Mn, Cr, Ni and Fe. (a) The distribution of metals in the LS sample, (b) the distribution of Cr and Ni in the Fe bearing phases associated with the LS sample, (c) the distribution of Ni and Cr in the LSM sample and (d) after microbial treatment a Ni is enriched in several areas of the map and shows stronger correlation with the remaining Mn-bearing phases (r = 0.43) than Fe phases (r = 30).*

moieties. Cr on the other hand does not correlate strongly with the Fe bearing phases but does show some affiliation with the Mn bearing phases.

After exposure to the microbial consortia the LSM sample showed alteration of some principal metal carrying phases. Based on several map surveys there was a remarkable reduction in the proportion of Mn bearing phases suggesting that the Mn oxide was being preferentially removed by the microbial consortia. Localised Cr appears to have no physical correlation with either Fe or Mn bearing phases in this sample and in some areas shows enrichment of Cr suggesting possible remobilisation of Cr perhaps as isolated chromites. The Ni distribution shows both enriched areas and a compelling correlation with the remaining Fe-bearing phases.

7.7 Summary

This chapter has sought to provide the reader a series of case studies that exemplify cutting edge techniques and methodologies for assessing the role of chemical speciation and natural attenuation on the bioavailability of contaminants in the terrestrial environment. We provide here a summary of the findings of the techniques and case studies presented to reinforce the strength of the techniques described.

Isotopic dilution techniques represent a flexible and powerful tool to assess changes in metal exchangeability as a result of a number of processes such as natural attenuation, rhizosphere reactions, remediation processes, etc. However, it must be remembered that this information is operationally defined mainly as a function of the equilibration time. Nonetheless, a combination of isotopic dilution techniques together with spectroscopic methods, such as those described in this chapter, represent an exciting possibility to obtain quantitative and mechanistic information regarding important processes controlling contaminant's bioavailability in soil.

The strength of synchrotron-based spectroscopic techniques has been described. For instance through the application of XANES it has been demonstrated that Cr is present in ambient particulate matter in a wide variety of forms that probably represent an equally broad set of sources. The majority of the Cr analysed within the aerosol particulate case study presented in Section 7.3 was present in relatively unreactive phases such as Cr(III) oxy(hydro)oxides, Cr(0) and Cr-carbides, but Cr(VI) is also present in some ambient particles. This finding was consistent with complementary laboratory ageing experiments, which suggest that ambient particulate Cr(VI) has a lifetime of days.

Analysis of Peruvian mummy teeth by using synchrotron micro-XRF mapping has provided an insight into the environmental exposure of individuals to trace metals that would be difficult to obtain by other means. It is apparent from this study that the trace metal concentration in the cementum ring system differs from that of the rest of the tooth. Metals such as Br and Pb appear in the cementum of teeth from different individuals, indicative of different lifetime exposure. In addition the two distinct Zn bands suggest an endogenous origin for the Zn.

It is clear that micro-focused XAFS is an ideal tool for the analysis of ambient particulate matter. The specific advantages of micro-XAFS include spatial resolution of a few microns, element specific analysis, oxidation state determination, elemental correlations and detailed structural information, even in non-crystalline samples. These are all important factors that must be determined for a more complete understanding of particulate matter composition and its impacts on health. A better understanding of the complex chemical composition of ambient particulate matter may lead to improvements in the effectiveness of particulate matter regulations.

Soil samples affected by a smelter were analysed by XAFS spectroscopy to determine the speciation and local coordination environment of As in the samples. The level of understanding achieved through the determination of local coordination and oxidation state provided by these studies would be extremely difficult to acquire by other means. This study stemmed from the need to assess the stability of As carriers in soil for the development of a conceptual model for human health and environmental risks associated with the soil containing up to 270 ppm As.

The idea of remobilisation of trace metals due to microbial activity is supported by XAFS measurements along with the spatial associations provided by the SEM. This spectroscopic evidence suggests that trace element mobility can be directly associated with dissimilatory metal reducing bacteria. By combining both electron microscopy and synchrotron-based techniques with other available selective extraction procedures one can investigate the partitioning of trace metals in lacustrine sediment and similarly iron rich lateritic soil profiles. In particular, the ability to estimate concentrations of metals associated with carbonate Fe (particularly siderite and ankerite) and magnetite Fe materials in weathered sediment profiles contributes to our increased understanding with respect to bioavailability. The application of the spectroscopic procedures to laterite environments provides insight into the dominant modes of occurrence of different trace metals associated with Fe and Mn minerals in weathered laterites and how they will behave in the presence of biologically mediated redox environments.

Acknowledgements

P.S. Nico, M.L. Werner, M.A. Marcus, C. Anastasio: X-ray absorption spectroscopy was performed at GeoSoilEnviroCARS (Sector 13), Advanced Photon

Source (APS), Argonne National Laboratory and at beamline 10.3.2 of the Advanced Light Source, Lawrence Berkeley National Laboratory. GeoSoil-EnviroCARS is supported by the National Science Foundation-Earth Sciences (EAR-0217473), Department of Energy-Geosciences (DE-FG02-94ER14466) and the State of Illinois. Use of the APS was supported by the U.S. Department of Energy, Basic Energy Sciences, Office of Science, under Contract No. W-31-109-Eng-38. Partial support was also provided by the U.S. Department of Energy under contract number DE-AC02-05CH11231. The Advanced Light Source is supported by the Director, Office of Science, Office of Basic Energy Sciences, of the U.S. Department of Energy under Contract No. DE-AC02-05CH11231.

C.G. Weisener: Portions of this work were performed at Beamline X26A, and 23B National Synchrotron Light Source (NSLS), Brookhaven National Laboratory. X26A is supported by the Department of Energy (DOE)-Geosciences (DE-FG02-92ER14244 to the University of Chicago-CARS) and DOE-Office of Biological and Environmental Research, Environmental Remediation Sciences Division (DE-FC09-96-SR18546 to the University of Georgia). Use of the NSLS was supported by DOE under Contract No. DE-AC02-98CH10886. This work was supported by grant number 5 P42 ES04699 from the National Institute of Environmental Health Sciences (NIEHS) of the NIH. The contents of this chapter are solely the responsibility of the authors and do not necessarily represent the official views of the NIEHS, NIH.

R.R. Martin: The study of trace metal in the cementum of human teeth was partially supported by a UWO, ADF grant. PNC-CAT facilities at the Advanced Photon Source, and research at these facilities, are supported by the US DOE Office of Science Grant No. DEFG03-97ER45628, the University of Washington. A major facilities access grant from National Science and Engineering Research Council of Canada, Simon Fraser University and the Advanced Photon Source. Use of the Advanced Photon Source is also supported by the U. S. Department of Energy, Office of Science, Office of Basic Energy Sciences, under Contract No. W-31-109-Eng-38. Archaeological material was exported from Peru under the Dirección General de Patrimonio Arqueologico, Oficio No. 514-2001-INC/DGPA-D. Archaeological work in Peru has been supported by the Social Sciences and Humanities Research Council of Canada and by the University of Western Ontario Social Sciences Alumni Travel Grant.

D. Paktunc: As XAFS work was carried out at the Stanford Synchrotron Radiation Laboratory (SSRL), a national user facility operated by Stanford University on behalf of the U.S. Department of Energy, Office of Basic Energy Sciences. Dr. Andrea Foster's help with data collection and interpretation over many years of collaborative work at SSRL is greatly appreciated.

References

Abbott, M.B., Wolfe, A.P., 2003. Intensive pre-Incan metallurgy recorded by lake sediments from the Bolivian Andes. Science 301, 1893–1895.

Als-Nielsen, J., McMorrow, D., 2001. Elements of Modern X-ray Physics. John Wiley & Sons, USA.

Attramadal, A., Jonsen, J., 1976. The content of lead, cadmium, zinc and copper in deciduous and permanent teeth. Acta Odontol. Scand. 34, 127–131.

Carvalho-E-Silva, M.L., Ramos, A.Y., Tolentino, H.C., Enzweiler, J., Netto, S.M., Martins-Alves, M., 2003. Incorporation of Ni into natural goethite: an investigation by X-ray absorption spectroscopy. Am. Mineral. 88, 876–882.

Chalupa, D.C., Morrow, P.E., Oberdorster, G., Utell, M.J., Frampton, M.W., 2004. Ultrafine particle deposition in subjects with asthma. Environ. Health Persp. 112, 879–882.

Cieslak-Golonka, M., 1996. Toxic and mutagenic effects of chromium(VI): a review. Polyhedron 15, 3667–3689.

Cohen, M.D., Kargacin, B., Klein, C.B., Costa, M., 1993. Mechanism of chromium carcinogenicity and toxicity. Crit. Rev. Toxicol. 23, 255–281.

Echevarria, G., Morel, J.L., Fardeau, J.C., Leclerc-Cessac, E., 1998. Assessment of phytoavailability of nickel in soils. J. Environ. Qual. 27, 1064–1070.

Friedl, G., Wehrli, B., Manceau, A., 1997. Solid phases in the cycling of manganese in eutrophic lakes: new insights from EXAFS spectroscopy. Geochim. Cosmochim. Acta 61, 3277.

Fuller, C.C., Harvey, J.W., 2000. Reactive uptake of trace metals in the hyporheic zone of a mining-contaminated stream, Pinal Creek, Arizona. Environ. Sci. Technol. 34, 1150–1155.

George, G.N., Pickering, I.J., 1993. EXAFSPAK: a suite of computer programs for analysis of X-ray absorption spectra. Stanford Synchrotron Radiation Laboratory.

Gérard, E., Echevarria, G., Sterckeman, T., Morel, J.L., 2000. Cadmium availability to three plant species varying in cadmium accumulation pattern. J. Environ. Qual. 29, 1117–1123.

Grouse, P.M., Gutknect, W.F., Hodson, L., Wilson, B.M., 1998. The fate of hexavalent chromium in the atmosphere. Research Triangle Institute, North Carolina, CARB Contract No. A6-096-32.

Hamon, R.E., Lombi, E., Fortunati, P., Nolan, A.L., McLaughlin, M.J., 2004. Coupling speciation and isotope dilution techniques to study arsenic mobilization in the environment. Environ. Sci. Technol. 38, 1794–1798.

Hamon, R.E., McLaughlin, M.J., 1999. Use of the hyperaccumulator Thlaspi caerulescens for bioavailable contaminant stripping. In: W.W. Wenzel, D. Adriano, B. Alloway, H. Doner, C. Keleer, N. Lepp, M. Mench, R. Naidu, G. Pierzynski (Eds.), Fifth International Conference on the Biogeochemistry of Trace Metals, Vienna, Austria, July 11–15, ISTEB, pp. 908–909.

Hamon, R.E., McLaughlin, M.J., 2002. Interferences in the determination of isotopically exchangeable P in soils and a method to minimise them. Aust. J. Soil Res. 40, 1383–1397.

Hamon, R.E., McLaughlin, M.J., Cozens, G., 2002. Mechanisms of attenuation of metal availability in in situ remediation treatments. Environ. Sci. Techol. 36, 3991–3996.

Hamon, R.E., McLaughlin, M.J., Naidu, R., Correll, R., 1998. Long-term changes in cadmium bioavailability in soil. Environ. Sci. Technol. 32, 3699–3703.

Hamon, R.E., Wundke, J., McLaughlin, M.J., Naidu, R., 1997. Availability of zinc and cadmium to different plant species. Aust. J. Soil Res. 35, 1267–1277.

Hughes, L.S., Cass, G.R., Gone, J., Ames, M., Olmez, I., 1998. Physical and chemical characterization of atmospheric ultrafine particles in the Los Angeles area. Environ. Sci. Technol. 32, 1153–1161.

Hutchinson, J.J., Young, S.D., McGrath, S.P., West, H.W., Black, C.R., Baker, A.J., 2000. Determining uptake of 'non-labile' soil cadmium by Thlaspi caerulescens using isotopic dilution techniques. New Phytol. 146, 453–460.

Kabata-Pendias, A., 2001. Trace Elements in Soils and Plants, 3rd edition. CRC Press, Boca Raton, p. 432.

Koningsberger, D.C., Prins, R., 1988. X-ray Absorption: Principles, Applications, Techniques of EXAFS, SEXAFS, and XANES in Chemical Analysis 92. New York, Wiley.

Larsen, S., 1952. The use of P32 in studies on the uptake of phosphorus by plants. Plant Soil 4, 1–10.

Lieberman, D.E., 1994. The biological basis for seasonal increments in dental cementum and their applications to archaeological research. J. Archaeol. Sci. 21, 525–539.

Lombi, E., Hamon, R.E., McGrath, S.P., McLaughlin, M.J., 2003. Lability of Cd, Cu, and Zn in polluted soils treated with lime, beringite, and red mud and identification of a non-labile colloidal fraction of metals using isotopic techniques. Environ. Sci. Technol. 37, 979–984.

Lombi, E., Hamon, R.E., Wieshammer, G., McLaughlin, M.J., McGrath, S.P., 2004. Assessment of the use of industrial by-products to remediate a copper/arsenic contaminated soil. J. Environ. Qual. 33, 902–910.

Lorentz, W., 2002. Archaeological contributions of skeletal lead analysis. Accounts Chem. Res. 35, 669–675.

Ma, Y.B., Lombi, E., Nolan, A., McLaughlin, M.J., 2006a. Short-term natural attenuation of copper in soils: effects of time, temperature and soil characteristics. Environ. Toxicol. Chem. 25, 652–658.

Ma, Y.B., Lombi, E., Oliver, I.W., Nolan, A.L., McLaughlin, M.J., 2006b. Long-term aging of copper added to soils. Environ. Sci. Technol. 40, 6310–6317.

Manceau, A., Llorca, S., Calas, G., 1987. Crystal chemistry of cobalt and nickel in lithiophorite and asbolane from New Caledonia. Geochim. Cosmochim. Acta 51, 105–113.

Manceau, A., Marcus, M.A., Tamura, N., 2002. Quantitative speciation of heavy metals in soils and sediments by synchrotron X-ray techniques. In: P. Fenter and N.C. Sturchio (Eds.), Applications of Synchrotron Radiation in Low-Temperature Geochemistry and Environmental Science. Reviews in Mineralogy and Geochemistry, Vol. 49. Mineralogical Society of America, Washington, DC, pp. 341–428.

Manceau, A., Tamura, N., Celestre, R.S., Macdowell, A.A., Geoffroy, N., Sposito, G., Padmore, H.A., 2003. Molecular-scale speciation of Zn and Ni in soil ferromanganese nodules from loess soils of the Mississippi Basin. Environ. Sci. Technol. 37(1), 75–80.

Marcus, M.A., Manceau, A., Kersten, M., 2004. Mn, Fe, Zn and As speciation in a fast-growing ferromanganese marine nodule. Geochim. Cosmochim. Acta 68, 3125–3136.

Martin, R.R., Naftel, S.J., Feilen, A.B., Narvaez, A., 2004. Synchrotron X-ray fluorescence and trace metals in the cementum rings of human teeth. J. Environ. Monit. 6, 1–5.

Martin, R.R., Kempson, I.M., Naftel, S.J., Skinner, W.M., 2005. Preliminary synchrotron analysis of lead in hair from a lead smelter worker. Chemosphere 80, 1385–1390.

Morgan, J.J., 2000. Manganese in natural waters and earths crust: its bioavailability to organisms. Metal ions Biol. Syst. 37, 1–34.

Morgan, J.J., 2005. Kinetics of reaction between O-2 and Mn(II) species in aqueous solutions. Geochim. Cosmochim. Acta 69, 35–48.

Nation, J.L., 1983. A new method using hexamethyldislazane for preparation of soft tissues for scanning electron microscopy. Stain Technol. 58(6), 347–351.

Nelson, Y.M., Lion, L.W., 2003. Formation of biogenic manganese oxides and their influence on scavenging of toxic trace metals. In: H.M. Selim, W.L. Kingerly (Eds.), Geochemical and Hydrological Reactivity of Heavy Metals in Soils. CRC Press, Boca Raton, pp. 169–186.

Newville, M.J., 2001. IFEFFIT: interactive XAFS analysis and FEFF fitting. J. Synchrotron Radiat. 8, 322–324.

Newville, M., 2004. Fundamental of XAFS. Consortium for Advanced Radiation Sources, University of Chicago, Illinois, USA. Available at http://cars9.uchicago.edu/xafs/xas_fun/xas_fundamentals.pdf#search='fundamental%20of%20XAFS'.

O'Reilly, S.E., Hochella, M.F., 2003. Lead sorption efficiencies of natural and synthetic Mn and Fe-oxides. Geochim. Cosmochim. Acta 67, 4471–4487.

Paktunc, D., 2004. A computer program for analysing complex bulk XAFS spectra and for performing significance tests. J. Synchrotron Radiat. 11, 295–298.

Paktunc, D., Foster, A., Heald, S., Laflamme, G., 2004. Speciation and characterization of arsenic in gold ores and cyanidation tailings using X-ray absorption spectroscopy. Geochim. Cosmochim. Acta 68, 969–983.

Paktunc, D., Foster, A., Laflamme, G., 2003. Speciation and characterization of arsenic in Ketza River Mine tailings using X-ray absorption spectroscopy. Environ. Sci. Technol. 37–10, 2067–2074.

Patterson, R.R., Fendorf, S., Fendorf, M., 1997. Reduction of hexavalent chromium by amorphous iron sulfide. Environ. Sci. Technol. 31, 2039–2044.

Peterson, M.L., Brown, G.E., Parks, G.A., Stein, C.L., 1997. Differential redox and sorption of Cr (III/VI) on natural silicate and oxide minerals: EXAFS and XANES results. Geochim. Cosmochim. Acta 61, 3399–3412.

Pope, C.A., Dockery, D.W., Schwartz, J., 1995. Review of epidemiological evidence of health effects of particulate air pollution. Inhal. Toxicol. 7, 1–18.

Ravel, B., Newville, M., 2005. ATHENA, ARTEMIS, HEPHAESTUS: data analysis for X-ray absorption spectroscopy using IFEFFIT. J. Synchrotron Radiat. 12, 537–541.

Scheifler, R., Schwartz, C., Echevarria, G., de Vaufleury, A., Badot, P.M., Morel, J.L., 2003. "Nonavailable" soil cadmium is bioavailable to snails: evidence from isotopic dilution experiments. Environ. Sci. Technol. 37, 81–86.

Schlesinger, R.B., 2000. Properties of ambient PM responsible for human health effects: coherence between epidemiology and toxicology. Inhal. Toxicol. 12, 23–25.

Scott-Fordsmand, J.J., Stevens, D., McLaughlin, M.J., 2004. Do earthworms mobilize fixed zinc from ingested soil? Environ. Sci. Technol. 38, 3036–3039.

Seigneur, C., Constantinou, E., 1995. Chemical kinetic mechanism for atmospheric chromium. Environ. Sci. Technol. 29, 222–231.

Smolders, E., Brans, K., Foldi, A., Merkx, R., 1999. Cadmium fixation in soils measured by isotopic dilution. Soil Sci. Soc. Am. J. 63, 78–85.

Stern, A., Heald, S.M., 1983. Principles and applications of EXAFS, Chapter 10. In: E.E. Koch (Ed.), Handbook of Synchrotron Radiation, North-Holland, Amsterdam, pp. 995–1014.

Stutz, A.J., 2002. Polarizing microscopy identification of chemical digenesis in archaeological vementum. J. Archaeol. Sci. 29, 1327–1347.

Takahashi, Y., Manceau, A., Geoffroy, N., Marcus, M.A., Usui, A., 2007. Chemical and structural control of the partitioning of Co, Ce and Pb in marine ferromanganese oxides. Geochim. Cosmochim. Acta 71, 984–1008.

Tebo, B.M., Bargar, J.R., Clement, B.G., Dick, G.J., Murray, K.J., Panker, D., Verity, R., Webb, S.M., 2004. Biogenic manganese oxides: properties and mechanisms of formation. Annu. Rev. Earth Planet. Sci. 32, 287–328.

Tiller, K.G., Honeysett, J.L., deVries, M.P.C., 1972. Soil zinc and its uptake by plants. I. Isotopic exchange equilibria and the application of tracer techniques. Aust. J. Soil Res. 10, 151–164.

Vandevivere, P., Baveye, P., 1992. Improved preservation of bacterial exopolymers for scanning electron microscopy. J. Microsc. 167, 323–330.

Wang, Y.F., Huang, K.L., Li, C.T., Mi, H.H., Luo, J.H., Tsai, P.J., 2003. Emissions of fuel metals content from a diesel vehicle engine. Atmos. Environ. 37, 4637–4643.

Werner, M., Nico, P., Bing, G., Kennedy, I., Anastasio, C., 2006. Laboratory study of simulated atmospheric transformations of chromium in ultrafine combustion aerosol particles. Aerosol Sci. Technol. 40, 545–556.

Werner, M.L., Nico, P.S., Marcus, M.A., Anastasio, C., 2007. Use of micro-XANES to speciate chromium in airborne fine particles in the Sacramento Valley. Environ. Sci. Technol. 41, 4919–4924.

Young, S.D., Tye, A., Crout, N.M.J., 2001. Rates of metal ion fixation in soils determined by isotopic dilution. Proceedings of the 6th International Conference on the biogeochemistry of trace elements, Guelph, Canada, p. 105.

Developments in Soil Science, volume 32
Ravendra Naidu (Editor)

Chapter 8

PROCESS-BASED APPROACH IN THE STUDY OF BIOAVAILABILITY OF IONS IN SOILS

E.J.M. Temminghoff, L. Weng, E.J.J. Kalis and W.H. Van Riemsdijk

8.1 Introduction

In the decision-making process concerning the protection of the soil environment, it is necessary to make accurate assessments of the bioavailability of native or added chemicals (see Chapters 2–4 and 7). Such an assessment is the underlying basis for the evaluation of both the ecological risk posed by contaminants and by the excessive use of fertilizers (see Chapter 11–13).

It has long been recognized that the bioavailability of elements is not a simple function of their total concentration in the media (see Chapters 2 and 3; Naidu et al., 2003). Factors influencing the bioavailability can be both biotic, e.g. the biological species, the growth stage of the organism and physiological characteristics; and abiotic, e.g. the type of environmental medium like soil type, soil structure, pH and presence of other ions. A dominant factor through which the abiotic factors affect the bioavailability is the influence on the chemical speciation of the elements (see Chapters 3, 7 and 9). Chemical speciation refers to the distribution of one element among its various chemical forms. The energy level of one molecule of a particular species, e.g. one $CdCl^+$, is the same independent of its environment. The distribution of the element over its species at (partial) equilibrium is found by minimization of the free energy of the system. The main groups of chemical species in soil, sediment and water are various precipitates, e.g. particular mineral phases, adsorbed to reactive surfaces like metal(hydr)oxides, complexed with organic and inorganic ligands in the solution phase and the 'free ion'. The concentration or activity of the free ion is a convenient choice as a reference (energy) level, since in principle all other species that contain this element can be expressed in terms of the concentration of this reference species (in combination with the concentration of other elements that are present in the species).

In soil, ions are often taken up by organisms via the soil solution and chemical speciation affects the bioavailability, in the first place, via influencing the partitioning of the elements between the solid phase and the dissolved phase (see Chapter 2). In the solution phase, elements can be present in the form of the free (hydrated) ion and as complexes with inorganic ligands or dissolved organic matter (DOM). The bioavailability of an element can be expressed in terms of the concentration of the free ion and the concentration of other ions that are involved in the direct uptake (see Chapter 3 for other definitions of bioavailability). Since speciation influences the concentration of all species, it is likely to affect the concentration of the free ion and therefore the bioavailability is at least a function of the concentration of the free ion. For metal ions, in many studies strong correlations have been obtained between the free metal ion concentration and metal uptake. For example, Slaveykova and Wilkinson (2002) found a strong relationship between the free lead concentration and lead uptake by algae, rather than the total or labile lead concentrations. Based on such observations, the free-ion-activity model (FIAM) has been proposed. The FIAM states that the free ion activity (or concentration) is the determining factor in the uptake (Campbell, 1995).

The observation that the free ion activity is not always the single determining factor for the bioavailability of metal ions has led to the preference of the biotic ligand model (BLM) over the FIAM (Pagenkopf, 1983; Paquin et al., 2002). The BLM considers the binding of the metal ions to the reactive sites on the membrane or cell wall of the organisms (biotic ligands) as the first step of the uptake. It is clear from the above that when more ions are involved in the expression that regulates the binding to the reactive biotic ligand that the concentration of these ions should be included in the model, similarly like the formulation of other chemical processes. The amount of the biotic ligand that is bound to a metal ion is regarded as the key parameter in determining its bioavailability. The BLM model is a useful concept to integrate knowledge from the discipline of chemistry, physiology and toxicology (Paquin et al., 2002). The competition effects of other ions on the uptake of the ion under concern can be qualitatively and sometimes also quantitatively explained with the BLM approach. However, it should be realized that within the BLM framework one still needs to find the proper model expression that determines the extent of binding to the metal to the reactive biotic ligands. This model expression is linked to the uptake mechanism and is expected to be at least a function of the free metal ion activity, but it can also be a function of the proton activity (pH), the calcium activity, etc. Even when it is known which ions affect the uptake of the metal ion of interest, one still has to find the proper chemical expression relating the various concentrations or activities.

Both the FIAM and BLM were initially developed for metal ions in aqueous media. A terrestrial BLM is currently being developed for regulatory use in Europe (Antunes et al., 2006). In soils, plants take up ions from the soil solution in the vicinity of the root (rhizosphere). The depletion of the chemicals in this small volume of soil solution creates a concentration gradient. The continuous diffusion of the soluble species of the ions will eventually decrease the concentration of the ions in the solution close to the solid particles in the soil. The disturbance of the equilibrium due to the decrease of the concentration of ions in the solution will lead to release of the chemicals from the solid phase. The dissolved ions will be replenished through the dissolution and/or desorption of chemicals initially present in the solid phase. The mobile fraction of the ion includes both the soluble complexes and the free ion.

The BLM is normally considered as a steady-state model, in which the processes that control the supply flux of the free ions to the organisms is much faster than the uptake rate. The rapid and reversible adsorption to the biotic surfaces is coupled with an irreversible and rate-limiting internalization of the ions bound to the carrier (Van Leeuwen, 1999; Hassler et al., 2004). Under this condition, ions in the solid and solution phase in the soil can be considered to be in equilibrium and the concept of the BLM is applicable. However, if the supply flux of the free ion to the biotic surface is the rate-limiting step, the uptake is controlled by the supply and the BLM concept is no longer directly applicable. More careful analysis of the supply flux of the ions in a soil system shows that a slow flux can be due to slow kinetics of several processes:

(i)   dissociation of the soluble complexes at the root surface;
(ii)   diffusion in the soil aggregates; and
(iii) dissolution/desorption from the solid surface.

In this chapter, some recent developments regarding the bioavailability issue are discussed. These studies include speciation measurement, speciation modelling, ion competition for uptake and transport-controlled uptake. The general objective of these studies is to combine the measurement and modelling methods to develop mechanistic approaches to understand bioavailability of ions in terrestrial and aquatic systems.

8.2 Donnan membrane technique

8.2.1 Speciation methods

Measurement can provide direct information on the speciation of chemicals. The partitioning of chemicals between the solid and dissolved phase is often determined by measuring the total amount in both the solid and the solution phase. Total concentration of metal ions in the soil is often measured after digestion or extraction with strong acid. The dissolved ion concentration in the soil solution is measured by extracting soil porewater or extracting the soil with water or electrolyte solutions. To distinguish various chemical forms present in the solid phase is experimentally difficult. The often used step-wise extraction methods can give some information of the binding strength of the ions in the solid phase, but explicit interpretation of the step-wise extraction data in terms of related chemical forms in each step can be misleading.

For metal ions in the solution phase, further speciation measurement is possible with certain techniques. Some of these techniques can distinguish between 'free+labile' and 'inert' fractions of metals, like stripping voltammetry (Xue and Sigg, 2002), diffusive gradients in thin-films (DGT) (Davison and Zhang, 1994 and see Chapter 9) and gel-integrated micro electrodes (GIME) (Pei et al., 2000). Some other techniques can measure the free ion activity in the solution, like ion-selective electrodes (ISE), permeation liquid membrane technique (PLM) (Buffle et al., 2000) and the Donnan membrane technique (DMT) (Temminghoff et al., 2000).

8.2.2 DMT principle

The DMT uses a cation exchange membrane to separate the sample solutions (donor) from a blank solution (acceptor). The negative charge carried by the membrane forms an electrostatic barrier for the neutral and negatively charged species but favours the transport of the positively charged species from the donor solution to the acceptor solution. At equilibrium for the cationic species, the activity of the free cations in the sample solution can be derived from the total ion concentration measured in the acceptor solution according to the Donnan membrane equilibrium principle (Donnan, 1925; Helfferich, 1962), which says that at the Donnan membrane equilibrium, the activity ratios (corrected for charge) of the ions in the solution on the two sides of the membrane are equal:

$$\left(\frac{a_{i,\text{donor}}}{a_{i,\text{acceptor}}}\right)^{1/zi} = \left(\frac{a_{j,\text{donor}}}{a_{j,\text{acceptor}}}\right)^{1/zj} \tag{8.1}$$

where a is the activity of the ion (M), i and j refers to respectively the cation under concern and a reference cation and z the charge of the ion. Subscripts donor and acceptor refer respectively to the donor solution and the acceptor solution. One of the advantages of the DMT approach, in comparison with other methods, is that the disturbance of the sample equilibrium and interference from other components are minimal.

8.2.3 DMT set-up

Fitch and Helmke (1989) have designed a set-up in which the volume of the acceptor solution is very small (200 μl). The small volume of the solution limits the possibilities for further analyses. Temminghoff et al. (2000) adjusted this set-up towards a more efficient solution flow and a higher acceptor volume in order to use it in combination with ICP-MS and/or ICP-AES. Therefore, many elements can be measured in a single sample. The DMT cell designed by

Temminghoff et al. (2000) contains two chambers separated by the cation exchange membrane. The donor (sample) solution and the acceptor solution are circulated through respectively one of these chambers by pumping. It has been shown that with this set-up the Donnan membrane equilibrium for monovalent and bivalent free cations can be obtained in 1–2 days (Temminghoff et al., 2000). This type of DMT cell (see Fig. 8.1), which has been referred to as the lab DMT cell, has been applied to measure the free cation activity in solutions containing inorganic or organic ligands such as EDTA, nitrilo tri acetic acid (NTA) and humic acid (Temminghoff et al., 2000; Oste et al., 2002; Kalis et al., 2006b).

For application of the technique to manure and for *in situ* measurement in rivers and lakes, another type of the DMT cell, the so-called field DMT cell has been developed (Van der Stelt et al., 2005). In the field DMT cell, the acceptor chamber is separated from the outside solution via two cation exchange membranes on both sides. The advantage of the field cell is that it can be applied directly in manure sludge or natural water and that no pumps are required. For the *in situ* measurement in rivers and lakes (see Fig. 8.2), the natural movement of the water can serve as the mixing force. The field DMT cell has been applied to measure nitrogen speciation in manure (Van der Stelt et al., 2005) and metal ion speciation in rivers and lakes (Kalis et al., 2006b). When the DMT method is used *in situ* to measure free metal ion concentrations in natural waters, the field cells are attached to a floating life buoy. By using this life buoy, measurements at different depths below the water surface are possible.

Figure 8.1. *Laboratory measurement with lab using DMT cells.*

Figure 8.2. In situ *measurement field DMT cells.*

To further minimize the disturbance of the equilibrium in the soil samples, a set-up has been developed in which a column containing the soil samples is connected to the lab DMT cell (Weng et al., 2001b). A solution is circulated through the soil column and the donor chamber of the DMT cell (Fig. 8.3). The soil solid phase provides buffering for the solution phase and therefore the disturbance of the equilibrium in the soil samples due to the measurement can be reduced. Using this set-up, Weng et al. (2001b) have measured simultaneously the free ion activity of Cu^{2+}, Cd^{2+}, Ni^{2+}, Pb^{2+} and Zn^{2+} in 2 mM $Ca(NO_3)_2$ solution in equilibrium with various soil samples.

8.2.4 Detection limit

Since the metal ion concentration in the acceptor solution of the DMT is measured with ICP-MS, the detection limit of the DMT for free metal ions is therefore controlled by the detection limit of the ICP-MS, which is in the range of 10^{-7}–10^{-11} M for most metal ions. However, in natural systems, free metal ion concentrations are often lower than this level and reduction of the detection limit of the DMT is therefore required.

One way to lower the detection limit of the DMT is to add ion complexing ligands in the receiving solution (acceptor) to accumulate the ions of concern to a measurable concentration. As far as the amount of the chosen ligand and the complexation constants is known, the free ion concentration in the acceptor can be calculated using the total ion concentration in the acceptor. When Donnan membrane equilibrium is obtained, the free ion concentration in the sample can

Figure 8.3. *A DMT cell connected to a soil column.*

be derived from the free ion concentration in the acceptor based on the Donnan membrane equilibrium principle (Eq. (8.1)). However, to be able to apply such an approach, the condition of the (pseudo) Donnan membrane equilibrium has to be obeyed and the time needed to reach the equilibrium is in practice an important issue.

8.2.5 Ion transport and the DMT

To understand the transport kinetics of ions through the cation exchange membrane in the DMT system, a numerical transport model has been developed (Weng et al., 2005b). In this model, diffusion of the dissolved species in the solution phase and the diffusion of the species that are adsorbed to the membrane are accounted for. Electrochemical reactions are assumed to be instantaneous and chemical equilibrium in each phase (donor, acceptor, membrane) is always maintained. With the same set of model parameters calibrated with experimental data, the transport of K^+, Mg^{2+}, Al^{3+} and NO_3^-, which represent respectively mono-, di- and trivalent cations and monovalent anions, can be described well (Fig. 8.4).

The model simulation shows that, ion transport in the DMT can be controlled by diffusion in the donor and/or in the acceptor solution or by the diffusion in

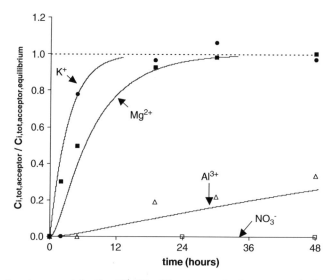

Figure 8.4. *Ion transport in the DMT without complexing agents in both donor and acceptor (2 mM Ca background). Symbols are data and lines are model simulations (from Weng et al., 2005b).*

the membrane itself. Weng et al. (2005b) have proposed a method to determine the relative flux of ion diffusion in the solution phase and in the membrane. The comparison suggests that (1) an increase in degree of ion complexation in the solution, (2) a decrease in the ion valence, and (3) an increase in the background salt concentration, favour the transition from solution diffusion-controlled to a membrane diffusion-controlled transport.

The increase in the equilibrium time due to the addition of ligands to the acceptor depends on:

(i)   When the diffusion in the acceptor solution is the rate-limiting step, adding ligands to the acceptor will not significantly influence the time needed to reach equilibrium, providing that the transport flux in the acceptor solution does not exceed that in the donor; and

(ii) When the diffusion in the membrane is the rate-limiting process, the time needed to reach the Donnan membrane equilibrium will increase with the increase of amount of ligands added to the acceptor.

For the second situation, waiting for the Donnan membrane equilibrium may take a very long time, which is not practical.

8.2.6 Equilibrium and kinetic mode of DMT

In natural samples like surface waters and soil solutions, a significant fraction of metal ions can be complexed with DOM. The DOM-bound species is more important for the metals such as Cu and Pb, which have a high affinity for DOM, than for metals such as Zn, Ni and Cd, which have a relatively weaker affinity for DOM. Generally, in natural samples, the fraction of DOM-bound metals is larger for Cu and Pb than for Zn, Ni and Cd. This means that, for Cu and Pb, the flux in the donor solution is often larger than that in the membrane, whereas for Zn, Ni and Cd this is not the case. When no ligand is present in the acceptor solution, for all these metals, often the diffusion in the acceptor solution is the rate-limiting step and the Donnan membrane equilibrium can be approached in 1–2 days. When ligands are added to the acceptor solution, depending on the situation transport can be controlled by diffusion in the donor solution or in the membrane.

Based on the above analysis, various approaches for measuring the free metal ion concentration in natural samples using the DMT have been proposed (Weng et al., 2005b). (1) When the free ion concentration in the sample is above the detection limit of the ICP-MS (situation 1), there is no need to add ligands to the acceptor solution. The Donnan membrane equilibrium can be reached in 1–2 days and the measurement can be based on the Donnan membrane equilibrium principle. (2) When the free ion concentration in the sample is below the detection limit of the ICP-MS and if the ratio of the total to free ion concentration in the sample is relatively small, for instance for Cd, Ni and Zn close to a neutral pH, a complexing ligand with not too high affinity and at a relatively low concentration can be added to the acceptor. Donnan membrane equilibrium can be expected within a few days time and the free ion concentration in the sample can be calculated based on the equilibrium. (3) For metal ions with a free ion concentration lower than the detection limit of the ICP-MS and a relatively high degree of ion complexation in the sample, for instance Cu and Pb in surface water at near neutral or slightly alkaline pH, a ligand with a high affinity can be added at relatively high concentration to the acceptor solution. This treatment will results in a membrane diffusion-limited transport process, and the free ion concentration in the sample can be derived from the ion flux measured, therefore there is no need to wait for the Donnan membrane equilibrium. The free ion concentration in the sample can be calculated from the total ion concentration in the acceptor measured at a certain time interval using either the numerical model developed or the approximate analytical solutions (Weng et al., 2005b):

$$C_{i,donor} = \frac{V_{acceptor}}{A_e} \frac{\delta_m}{D_{i,m} B^{Z_i}} \frac{C_{i,tot,acceptor}}{t} \qquad (8.2)$$

where C_i is the free ion concentration of ion i in solution (M), $C_{i,total}$ the total concentration of ion i in solution (M), $V_{acceptor}$ the volume of the acceptor

Table 8.1. *Heavy metal concentrations measured in 2003 in the River Rhine near Wageningen.*

	Cd	Ni	Pb	Cu	Zn
Total (−log M)	−9.29	−7.23	−8.95	−7.12	−6.55
Free (−log M)	−10.80	−8.52	−12.11	−10.94	−7.45

solution (m^3), δ_m the thickness of the membrane, $D_{i,m}$ the diffusion coefficient of ion i in the membrane, B the Boltzmann factor in the membrane phase, A_e the effective surface area of a membrane (m^2), t the time (S), and Z_i the charge of ion i.

The kinetic mode of DMT has been tested in the lab and in the field. Free Cu^{2+} concentrations in synthetic solutions containing NTA measured with this approach are in close agreement with those obtained from speciation calculation (Weng et al., 2005b). To measure the free ion activity in surface water, 30 mg/l purified humic acid was added to the acceptor as the complexing ligand for heavy metal ions. Experiments showed that in 2 or 3 days most metal ions have reached the (pseudo) Donnan membrane equilibrium. However, for Cu and sometimes Pb, the equilibrium was not reached in this time interval and the free metal ion concentrations were then derived according to the kinetic approach. Free metal ion concentrations have been measured, among others, in the River Rhine. In Table 8.1, the total and free heavy metal concentration measured in the River Rhine near Wageningen, the Netherlands are given, in which the free ion activity was measured with either the equilibrium mode of DMT (Cd, Ni, Zn, Pb) or the kinetic mode of the DMT (Cu). In a field campaign in which the metal speciation in several rivers and lakes in Europe was measured with various techniques, the free ion activities measured with the DMT are in good agreement with those measured with PLM (Kalis et al., 2006b).

8.3 Multisurface models

8.3.1 Speciation models

Speciation measurements are costly and sometimes very difficult or even not possible (for instance ion speciation in the rhizosphere). For purposes such as the risk assessment and land use planning, prediction of the effects of various changes in land management on the ion speciation is required. Speciation models can be used and various types of models have been developed to calculate chemical speciation or the distribution of chemicals over all relevant forms.

For inorganic chemicals, the most simple model is possibly the solid-solution partitioning or distribution coefficient approach (K_p or K_d), which is the ratio of the concentration in the solid phase to that in the solution phase. Soil characteristics, i.e. pH, cation exchange capacity (CEC), clay content, organic matter content, etc. vary greatly. The variation of soil properties leads to a large range of K_p values measured for one element. For the same metal ions, the K_p value can vary several orders of magnitude (Anderson and Christensen, 1988; Sauve et al., 2000, 2003). Choosing a certain value from this big range leads to big uncertainties in the prediction.

In some other models, empirical pedo-transfer functions are applied that calculate the free or the dissolved ion concentration from its total concentration as a function of some soil characteristics. These functions are often derived by (multiple or step-wise) regression and often in the form of simple linear functions or log-linear functions. The variables used in these functions vary. The most commonly used variables include pH, CEC, organic matter content, DOM content, clay content, etc. When calibrated using certain datasets, these models can give satisfactory description of the same dataset. However, the empirical nature of these models limits their application to other soils that are not included in the calibration. The complexity of the composition and its relation with the solution chemistry of the system hamper the application of the empirical relations to a wider range of samples.

A third type of models can be referred to as mechanistic or process-based geochemical models. In this approach, speciation is calculated based on the chemical processes that are involved using the thermodynamic relations and constants that can best describe these reactions. In the last decades, the development of computer codes in the geochemistry field has greatly increased our ability to carry out sophisticated speciation calculations. Examples of these codes are: MINTEQA2, PHREEQC, ECOSAT, WHAM and ORCHESTRA (Parkhurst et al., 1990; Allison et al., 1991; Keizer and Van Riemsdijk, 1994; Tipping, 1994; Meeussen, 2003). For all these programs, calculations of common inorganic aqueous species are relatively straightforward. However, it has been shown that binding of ions to colloidal particles such as natural organic matter, alumino-silicate clay minerals and metal (hydr)oxides are in many cases important in controlling the speciation of these ions in the environment. In the past, due to the lack of appropriate ion adsorption models, the adsorption of ions has been calculated in some of these geochemistry codes using empirical functions, which is one of the reasons that has led to unsatisfactory results in applying the geochemical models to real samples (Mouvet and Bourg, 1983; Hesterberg et al., 1993). Incorporation of more advanced ion adsorption models for natural colloidal particles can largely improve the predictive capability of these models.

8.3.2 Ion adsorption models

Steady progress has been made in the last decades with respect to the development of surface complexation models (SCMs) for ion binding to geo-colloids. Ion binding to humic substances is a very complicated process due to the combination of chemical heterogeneity, electrostatic interaction, variable stoichiometry of the binding and the competition between various cations and protons for the reactive binding sites. Extensive research over more than a decade has resulted in the well-accepted non-ideal competitive adsorption (NICA)–Donnan model for ion binding that can handle this complexity (De Wit et al., 1990, 1993a, 1993b; Koopal et al., 1994; Benedetti et al., 1995, 1996; Kinniburgh et al., 1996, 1999; Milne et al., 2001, 2003; Koopal et al., 2005). In the same time period models V and VI have been developed by Tipping and Hurley (1992) and Tipping (1994, 1998, 2002) that can also describe competitive ion binding to natural humic materials. Both the NICA model and models V and VI consider that there are two main types of reactive groups on the humics, i.e. the carboxylic type and phenolic type of groups. The heterogeneity of each type of groups is treated differently in the two models. In the NICA model, the distribution of the binding affinity of each type of groups is considered to be continuous, whereas in models V and VI, the distribution is considered discrete. The electrostatic effects on the binding are taken into account in both models with a Donnan-type approach.

In terms of their abundance, the (hydr)oxides of iron, aluminium and manganese are the most important metal (hydr)oxide minerals in soils, sediments and water. They are important colloidal particles in the environment and can bind various cations and anions. A series of SCMs have been developed and applied to describe the charge and ion binding to metal (hydr)oxides. More recently, insight in the structure and nature of the adsorbed species has become available from detailed spectroscopic work. From the modelling point of view, the challenge is to develop sound thermodynamic models that make use of the knowledge gained from spectroscopic studies. The CD-MUSIC model has been developed for this purpose (Hiemstra et al. 1989a, 1989b, 1996; Hiemstra and Van Riemsdijk 1996). The model is based on the structure of the mineral surface, the structure of the adsorbed species and the structure of the electrostatic potential profile in the vicinity of the mineral–water interface.

It has been shown that both the NICA–Donnan model and the CD-MUSIC model can describe ion binding to purified humics or synthetic metal (hydr)oxides particles well. For these models, intensive research has been carried out and model parameters have become available for many ions by model application to experimental dataset of purified humic acids and fulvic acids or synthetic metal (hydr)oxides. The developments of these models together with the availability of model parameters have made it possible to include the (semi)mechanistic ion adsorption models in the geochemical equilibrium calculations.

8.3.3 Multi Surface Model

Attempts have been made to apply these ion adsorption models to understand and to predict ion speciation in environmental systems. Examples are the development and applications of the WHAM (Windermere humic aqueous acid) model (Tipping and Hurley, 1992; Tipping, 1994, 1998, 2002) and the Multi Surface Model. The WHAM model and the Multi Surface Model are two groups of process-based geochemical models in which the models IV–VI or the NICA–Donnan model is used respectively to calculate ion adsorption to natural organic matter.

In the Multi Surface Model proposed by Weng et al. (2001a), ion binding to three types of solid particles, i.e. organic matter, iron (hydr)oxides and clay are considered important for cation speciation in soil. In this research, the organic matter content in the soil samples was measured with loss-on-ignition at 550°C. The total amount of amorphous or crystalline iron (hydr)oxides was derived from respectively the oxalate extractable iron content and the difference between the DCB and oxalate extractable iron. Clay content was determined by sieve and pipet method. The reactivity or site density of these particles is estimated as follows. Because illite is the dominant type of clay mineral in the soils of that particular study, the permanent charge density of the clay particles in the soil is assumed to be equal to that of illite. The charge density of illite was taken from the literature. The soil CEC is mainly determined by ion binding to both clay and organic matter. The site density of organic matter was therefore calculated from the difference of the measured CEC of the soil and the charge present on the clay. It was further assumed that humic acid is the reactive fraction of the organic matter that carries the estimated charge of the organic matter. From the calculated charge of generic humic acid for the pH of the soil sample, the reactive humic acid content is estimated as 31% of the total organic matter. The surface area of amorphous iron (hydr)oxides was assumed to be equal to that of hydrous ferric oxide (HFO) with $600 \, m^2/g$, and that of the crystalline iron (hydr)oxides to be equal to goethite with $100 \, m^2 g$.

Ion binding to organic matter was calculated with the NICA–Donnan model using generic model parameters (Milne et al., 2001, 2003). Ion binding to the crystalline and amorphous iron (hydr)oxides was calculated with respectively the CD-MUSIC model and the generalized two-layer model (Dzombak and Morel, 1990) with model parameters derived for synthetic analogues. Ion binding to clay is assumed to be pH independent and was calculated with a Donnan model. The competition of Al^{3+} ion to the binding was taken into account by assuming that its activity was controlled by gibbsite solubility. Competition with Fe^{3+} ion was not considered. For simplicity, the effects of particle interactions on ion binding were neglected. The predicted metal ion activity in the soil solution of Cu^{2+}, Cd^{2+}, Zn^{2+}, Ni^{2+} and Pb^{2+} was compared to that measured with the DMT. The results show that the agreement is good for Cu, Cd, Zn and Ni. For Pb, the model overestimated the metal activity. Examples of the results are given in Figure 8.5 for Cu and Cd.

Besides the prediction of the free ion activity, the Multi Surface Model can also estimate the relative importance of various types of reactive surface in the ion adsorption. In the study of Weng et al. (2001a), the studied soils are sandy soils in the acidic pH range. For these soils, the model calculations show that soil organic matter is an important sorbent for Cu, Cd, Ni and Zn, especially for Cu. For ions which have a weaker affinity for organic matter than Cu, for instance Cd, Ni and Zn, adsorption to clay minerals can be significant especially for conditions of low pH, low organic matter content and high metal ion loading. According to the model, Cu, Cd, Zn and Ni adsorbed on iron (hydr)oxides is less than 6% of their total amount bound in these samples. The model estimates that iron (hydr)oxides are not important sorbents for Cu, Cd, Zn and Ni in these soils, which have relatively low pH, low content of iron (hydr)oxides compared to soil organic matter and clay silicates.

The usefulness of the Multi Surface Model has also been shown in a study on Ni toxicity to plants grown in three different soils (Weng et al., 2004). In this

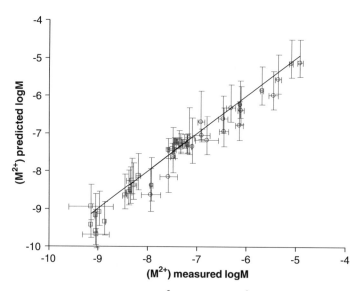

Figure 8.5. *Comparison between free Cu^{2+} ($\bigcirc$) and Cd^{2+} ($\square$) activity in soil solution measured with DMT and calculated with the Multi Surface Model (line is 1 to 1 line, results from Weng et al., 2001a).*

work, a pot experiment was carried out, in which oats were grown in soils spiked with Ni and adjusted to pH 4.7–7. Three types of soils were used, a sandy soil, a clay soil and a sandy soil relatively rich in organic matter and iron (hydr)oxides. The organic matter content in the sandy soil and the clay soil is the same, which is half of the value in the sandy soil that is rich in organic matter and iron (hydr)oxides. The results show that at the same pH, Ni toxicity in the sandy soil and the clay soil is larger than in the sandy soil that is rich in organic matter and iron (hydr)oxides. Using the Multi Surface Model, Ni adsorption to the three types of soils was predicted. The prediction shows that in the pH range of the experiment, Ni adsorption is dominated by the binding to organic matter. Because the sandy soil and the clay soil contain the same amount of organic matter, the binding capacity of these two soils for Ni at the same pH is comparable. Owing to the relatively weak binding to the clay, the large clay content in the clay soil does not contribute significantly to Ni speciation. The large amount of organic matter in the third soil explains the much lower Ni toxicity in this soil compared to the other two. Model calculation (Fig. 8.6) shows that Ni binding to iron (hydr)oxides has a stronger pH dependency than that of organic matter. It can be predicted from these results that binding to oxides will be more important at higher pH.

Similar modelling approach has been adopted by Schroder et al. (2005), Cances et al. (2003), Dijkstra et al. (2004) and Lumsdon (2004). In the study by Schroder et al. (2005), dissolved metal concentration in a river flood plain was

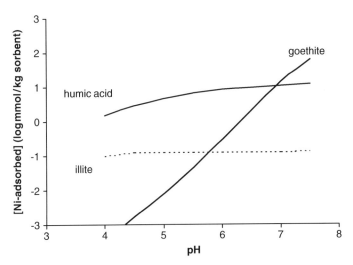

Figure 8.6. *Model prediction of Ni adsorption to individual reactive solids (modified from figure in Weng et al., 2004).*

predicted. In the model, 50% of the solid soil organic matter was assumed as humic acid and 40% of the DOM was assumed as fulvic acid. Metal ion binding to both the solid and DOM was modelled with the NICA–Donnan model. Reasonable agreement with the data was found for Cu and Cd, but the dissolved Zn and Pb concentration was significantly overestimated. In the study by Dijkstra et al. (2004), heavy metal leaching from contaminated soils over a wide range of pH (pH 0.4–12) was simulated. In their application of the Multi Surface Model, humic acid and fulvic acid content in the solid soil and soil solution were measured directly. The results show that the model prediction of the leaching of Cu, Cd, Ni, Zn and Pb was in general adequate, and sometimes excellent. Good agreement between the model predictions and measurements were also found by Cances et al. (2003) for Cu, Cd, Zn and Pb and by Lumsdon (2004) for Cd and Al.

8.4 Competition between ions for uptake

Quantitative approach: Many soil factors influence the bioavailability of elements. In a complex system, the observed effects of soil characteristics on the bioavailability of ions are the net results of the combination of various mechanisms. To be able to analyse the mechanistic factors of the macroscopic effects, a quantitative approach is required that takes into account all processes involved. Nederlof and van Riemsdijk (1995) and Plette et al. (1999) proposed a simplified quantitative model to interpret the bioavailability of heavy metals in terrestrial and aquatic systems. Similar to the BLM approach, accumulation or sorption of metal ions by biota present in a complex system like soil can be regarded as the result of competition for binding of that metal ion by all reactive components, including the biota. The free metal ion in the solution is considered the key chemical species that is involved in the binding to all ligands and the extended Freundlich equation is chosen to describe the adsorption because it leads to simplified analytical expression that gives insight in how certain effects operate. The Freundlich-type equations that are used to describe ion binding to the various components, for instance the plant root and the soil, can be easily combined and the resulting equation can be used to illustrate the net effects of certain factors on ion bioavailability. The authors suggested that the concept of the quantitative model might be a valuable tool in predicting the relative change in availability due to changes in environmental conditions like the pH. The concept explains the apparent paradox that the availability or toxicity of a metal in a nutrient solution generally increases with increasing pH, whereas its availability in the soil generally decreases with increasing pH as will be further illustrated below.

8.4.1 Competition between Ni and protons

Soil pH has been found the most important soil characteristic modulating the availability of heavy metals. The effects of pH on metal toxicity depend on the type of media in which the biota is exposed to the metal. It is common to find that the toxicity of metal ions for plants growing in soils decreases with increasing pH (Wallace et al., 1977; Lexmond, 1980; Smith, 1994), whereas the increase of pH often intensified the biological effects of the metals when the plants are grown in a nutrient solution (Crooke et al., 1954; Hatch et al., 1988). The pH can influence the bioaccumulation of metals in at least two ways: by affecting metal speciation both in the solution and in the solid phase and by influencing the biouptake. A biotic surface, for instance a plant root, may have a pH dependency in terms of binding that is in the same direction as for the soil surfaces (increased binding at increased pH), but generally differs in the magnitude of the effect. This will imply that the metal ion distribution is shifting at increasing pH either towards the biotic system or to the soil surface, depending on the relative magnitude of the effects on the biota and soil, which may result in a higher or lower toxicity of the metal ion.

In a study on Ni toxicity to plants, pH effects on Ni uptake were studied with a solution culture experiment (Weng et al., 2003). In this experiment, oats were grown in nutrient solutions containing various amounts of Ni and the pH was adjusted to 4.0, 5.0, 6.0 and 7.0. After 3 weeks, the dry biomass of oats was measured and the reduction of the biomass was used as an indicator of the toxicity of Ni. The results show that the biomass of oats was reduced at relatively high Ni concentrations. The EC_{50} of Ni in the nutrient solution decreases with the increase of the pH (Fig. 8.7), which means that the toxicity of Ni increases at higher pH. After digestion with strong acid, Ni concentration in the root and shoot part of oats was measured. In line with the pH effects on Ni toxicity, Ni uptake in the biomass also increases with the increase of the pH. The pH-dependent Ni uptake can be described well with the extended Freundlich equation as proposed by Nederlof and van Riemsdijk (1995) and Plette et al. (1999):

$$\log[\text{Ni-shoot}] = 1.15 + 0.65 \log[\text{Ni}^{2+}] + 0.11 \text{pH} \tag{8.3}$$

in which Ni in the shoot is in mmol/kg, $[\text{Ni}^{2+}]$ is the free Ni^{2+} concentration in the nutrient solution in mM, which is calculated from the total Ni concentration taking into account the inorganic complexes that are relevant. Eq. (8.3) says that at the same free Ni^{2+} concentration, Ni uptake by oats will increase with increasing pH.

The pH effects on the uptake of metal ions can be explained as the results of competition of protons (H^+) with the metal ions for the binding sites on the

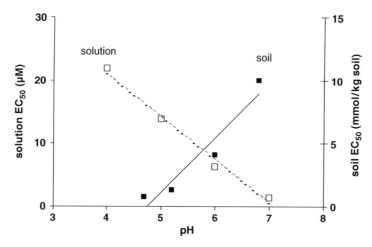

Figure 8.7. *EC$_{50}$ of Ni in the nutrient solution and Ni in soil (results from Weng et al., 2003).*

organisms. At a higher pH, the competition of protons is weaker due to a lower activity of protons than at lower pH, which leads to more binding of the metal ions and therefore the bioavailability/toxicity of metal ions is intensified. Similar pH effects on metal ion bioavailability to organisms grown in aqueous media have been found previously for microorganisms, algae and plants. For instance, Lexmond and Van der Vorm (1981) have found that the toxicity of Cu to maize grown in a nutrient solution was intensified at higher pH compared to that at lower pH. The binding of Cu to the maize root increases substantially with the increase of pH (Nederlof and Van Riemsdijk, 1995). Increased adsorption or uptake of Cu at higher pH was also found for algae, fungi and yeast (Huang and Morehart, 1990, 1991; Xue and Sigg, 1993; Plette et al., 1995).

In the same study on the toxicity of Ni (Weng et al., 2003), a pot experiment was also carried out in which Ni salt was spiked to a sandy soil. The pH of the soil was adjusted to 4 pH levels, in the range of 4.7–6.8. Oats were grown in these pots in a greenhouse for 6 weeks. The results show that at the same total concentration of Ni in the soil, the reduction of dry weight of the biomass is less at a higher pH. The net pH effects on Ni toxicity in the soil are opposite to that in the nutrient solution (Fig. 8.7).

To quantify the pH effects on Ni speciation in the soil, one desorption experiment was carried out. In this experiment, Ni concentration in 0.01 M CaCl$_2$ extraction of the soil samples used in the pot experiment was measured. Free Ni^{2+} concentration in these soils in equilibrium with 0.01 M CaCl$_2$ was also measured using the DMT method in connection with soil columns. Based on these results, a Freundlich equation was derived that describes the pH-dependent Ni adsorption to this soil:

$$\log[\text{Ni-adsorbed}] = -1.35 + 0.71 \log[\text{Ni}^{2+}] + 0.70\text{pH} \qquad (8.4)$$

in which Ni adsorbed is in mmol/kg, free Ni^{2+} in mM. The uptake equation (Eq. (8.3)) and the adsorption equation (Eq. (8.4)) can be combined, which leads to

$$\log[\text{Ni-shoot}] = 3.41 + 0.92 \log[\text{Ni-adsorbed}] - 0.53\text{pH} \qquad (8.5)$$

This derivation shows that due to a stronger pH effect on Ni speciation than on the uptake, the net pH effect on Ni uptake of oats grown in the soil is that Ni uptake should decrease at higher pH. This explains the apparent contradictory results of pH effects on Ni toxicity of plants grown in the nutrient solution and in the soil. The predicted Ni uptake with Eq. (8.5) was compared to the measured results and the agreement is reasonable (Weng et al., 2003). This exercise illustrates that using the quantitative approach more insight can be gained into the mechanisms that control the bioavailability of chemicals in a complex system.

8.4.2 Competition between Al and Ca

Besides protons, competition effects between metal ions for the binding to biotic ligands have also been observed. In a study of Al^{3+} binding to isolated cell wall material of tomato root, it was found that increased Al adsorption to the cell wall material leads to a release of Ca that was initially adsorbed (Postma et al., 2005). When plotting the amount of Al bound to that of Ca bound at a constant pH, a linear relationship is found. The slope of this linear relationship is the same for pH 3.5 and 4.0, which is related to an Al/Ca molar exchange ratio of 1:0.9 at both pH levels. The authors have chosen to use an ion exchange model, the Gaines–Thomas model to describe the competitive adsorption of ions to the cell wall materials. Because protons are always present in the solution, competition between Al/H and Ca/H is also included in the model, in addition to the competition between Al/Ca. The simple exchange model can give a reasonable description of both the adsorption of Al and Ca simultaneously, but some discrepancy remains. Not a single set of model parameters can be obtained that can describe the binding of Ca both in the presence and absence of Al over the whole pH range. The results indicate that there might be more than one type of reactive surface site present in the cell wall material. The relative affinity of H, Ca and Al on these different sites may vary, which can explain the discrepancy found in the model by considering only one type of sites.

8.4.3 Competition between heavy metal ions

In a study on metal ion uptake by plants, Kalis et al. (2006a) have proposed a four-step method for the analysis of metal ion uptake (Fig. 8.8): (i) The metal concentration in the soil solution is related to the total metal content of the soil. (ii) The metal adsorption to the root surface is related to the metal concentration in the soil solution. (iii) The metal content in the roots is related to the adsorption of metal ions to the root surface. (iv) The metal content in the shoots is related to the metal content in the roots.

In a solution culture experiment, perennial ryegrass was grown in nutrient solutions containing Cu, Cd, Ni, Zn and Pb (Kalis et al., 2006a). Humic acid was added to the nutrient solution to simulate DOM in soil solutions. The hypothesis to be tested was that humic acid lowers the free metal ion concentration and, therefore, reduces the metal uptake and, finally, the metal content of the plant. The free metal ion concentrations in the nutrient solutions were measured with the DMT (Temminghoff et al., 2000) and labile metal concentrations with the DGT (Davison and Zhang, 1994). The metal fraction which is adsorbed at the root surface is estimated using a one-minute washing procedure with $0.01\,M$ Na_2H_2EDTA (Slaveykova and Wilkinson, 2002). Even though this method is rather arbitrary in distinguishing between metal adsorbed at the root surface and metal taken up into the root cells, it is linearly related to the metal uptake by the roots (Kalis et al., 2006a) and therefore a good tool to study metal interactions at the root surface. Essential metal ions, such as Cu and Zn, as well

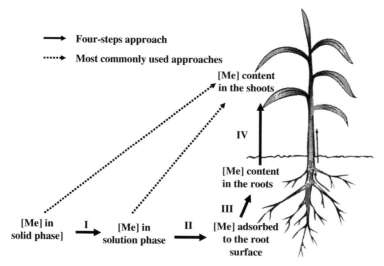

Figure 8.8. *Four-step approach for studying metal uptake by plants, indicated by the black arrows. Dotted arrows indicate the approach of many previous studies.*

as non-essential metal ions, such as Cd, Pb and Ni, were simultaneously present in the studied systems.

The metal content of the shoots depends on the metal content of the roots, either giving a saturation maximum (essential elements) or a linear relationship (non-essential elements). The metal content of the roots is a linear function of the adsorption of metals on the root surface. At a total metal concentration of 1 μM, metals that bind strongly to humic acid (Cu, Pb, and Fe) show a decrease in metal binding by the roots in the presence of humic acid. However, metals that bind weakly to humic acid, like Cd, Ni and Zn show an increase in metal binding by the roots in the presence of humic acid (Fig. 8.9). At 0.1 μM total metal concentration, this effect is less pronounced. These results suggest that complexation of cations such as Cu and Pb with high affinity for (dissolved) organic matter may lead to increased uptake of cations with low affinity for organic matter (Ni, Zn and Cd) because of competition between cations for binding to the organic matter and competition at the root surface.

Assuming that there is equilibrium between the speciation in solution and the competitive binding of the metal ions to the root surface, the presence of humic acid decreases adsorption at the root surface for the metal ions that are bound strongly by humic acid. The relative increase in binding to the root surface for the metal ions that are weakly bound to humic acid, especially at the higher total metal ion concentration, cannot be explained from the effect of the humic acid

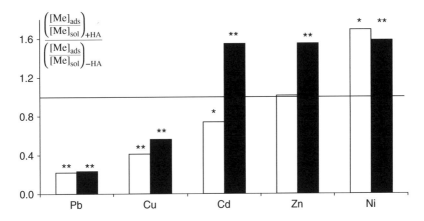

Figure 8.9. *Ratio of the relative metal adsorption (adsorbed/total concentration) to the root surface in the presence of humic acid (HA) to that in the absence of humic acid. Relative metal adsorption is defined as the ratio of absolute metal adsorption to the root surface and the total metal concentration in solution. White bars represent $[Me]_t = 0.1$ μM; black bars represent $[Me]_t = 1$ μM; asterisks indicate an influence of humic acid on metal adsorption (*p < 0.10; **p < 0.05) and Me = metal.*

on the free or labile metal concentrations of the individual metal ions. The results can be explained by taking competitive binding of the metal ions to the roots into account. The competitive binding is affected by the ratio of the affinity constants for a pair of metal ions to a particular biotic site type and by the ratio of the free metal ion concentrations in solution. The ratio of the affinities of a biotic ligand will not be affected by the presence of the humic acid, whereas the free metal ion ratio of a weakly binding metal to a strongly binding metal will increase in the presence of humic acid because of the difference in binding behaviour, which can explain the increased binding of the weakly bound metal ions in the presence of humic acid. The extent of this competitive effect will be more noticeable at higher loading of the reactive sites with metal ions, and this explains why the effect is strongest at the highest total metal ion concentration. The importance of loading of the reactive sites is mostly noticeable for Cd binding by the roots. At $0.1\,\mu M$ Cd, humic acid decreases the Cd binding at the root surface, whereas at $1\,\mu M$ Cd, humic acid increases Cd binding at the root surface (Fig. 8.9).

The Pb and Cu content in the plant roots can be described as a function of the free and labile metal concentration (Fig. 8.10a). However the Cd, Ni and Zn content in the plant shoots depends on the competition between cations for binding at the root surface rather than on the free metal ion concentration alone (Fig. 8.10b). So, especially for weakly binding metals, a multicomponent model may be required when metal uptake by plants is described. In natural soils and water, metals always exist in a multicomponent system. Therefore, competition between cations should be considered when the risks of metal-polluted soils in the environment are assessed. Competition with calcium ions may also play a role for ions like Ni and Zn.

8.5 Ion transport and bioavailability

8.5.1 Intra-aggregate transport

As has been discussed in the introduction, when the supply flux of the chemical from the soil or solution to the organisms is smaller than the potential uptake flux, the kinetics of the supply of the chemical is the factor that limits the bioavailability. This is often the case for phosphorus in soils. Owing to the combination of a relatively high requirement of P by plants and very often a very low concentration in the solution, P is soon depleted in the soil solution in the rhizosphere. To maintain the P flux to the plant root, P in the soil solution needs to be replenished by P adsorbed on the soil particles and inside in soil aggregates. The release of P from the soil particles to the soil solution depends on both the reaction of P with the solid phase and the diffusion process through the pores in the aggregates. The importance of intra-aggregate diffusion as the rate-limiting

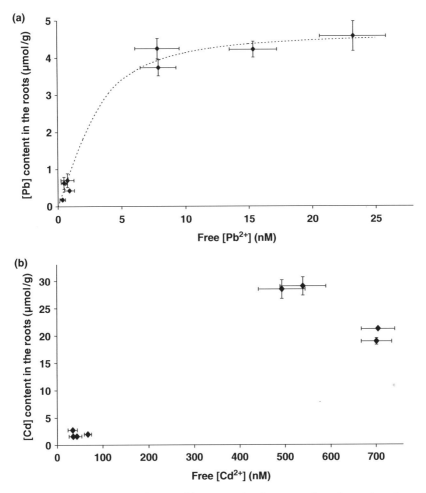

Figure 8.10. *Lead (a) and cadmium (b) content in the roots plotted as a function of the measured free Pb^{2+} (a) and Cd^{2+} (b) concentration in solution. The dotted line represents the fitted Langmuir–Freundlich isotherm. Error bars represent the standard deviation (Fig. 8.10a).*

step in the mass transfer of reactive solutes between sorption sites inside porous sorbents and bulk solution was also demonstrated for various ions (Trivedi and Axe, 2000; Lin and Wu, 2001). To understand this process, a reactive transport model is required that can describe both the adsorption and the transport.

In a study of P bioavailability in soil (Koopmans et al., 2004a, 2004b), slow release of P from the soil aggregates was found in a desorption experiment. Readers are directed to Chapter 14 for an overview of techniques for assessing nutrient bioavailability in soils. The kinetics of P release depends on the speed of shaking. Vigorous shaking leads to a faster equilibrium than weak shaking. The authors explained the slow release of P as a result of P diffusion from the soil aggregates to the bulk solution. A simple Langmuir model was used to describe P adsorption to the soil and the model parameters were optimized by fitting the model to the adsorption data at equilibrium. The reactive transport of P was simulated with a diffusion model in which the soil aggregates were assumed to be porous spheres with a uniform radius. The porosity in the aggregates was fitted, which is 3.5% of the total volume of the aggregates. For the weak shaking in 1:10 0.01 M $CaCl_2$, calculation by assuming the radius of aggregates to be 1 mm gives a good description of the data. For vigorous shaking, the radius has to be reduced to 0.1 mm to obtain good agreement between the calculation and the data. Vigorous shaking causes abrasion of soil particles leading to the exposure of new surface sites and greater P desorption. Vigorous shaking thus leads to a shorter diffusion distance causing a faster P release. The 0.1 mm radius is in agreement with the size of micro soil aggregates (Six et al., 2000), which are the most stable soil aggregates.

The soil samples used in the P desorption experiment were taken from a pot experiment of Koopmans et al. (2004b), in which ryegrass was cropped over 978 days without P addition. Both the P content in the soil and in the ryegrass was analysed during the experiment at 19–47 days intervals. The total P in the soil was measured in an acid ammonium oxalate extraction. With the reactive transport model described above, the authors estimated the P supply rate from the soil at various stages of the experiment. In this calculation, the radius of the soil aggregates was assumed to be 1 mm, the same as in the simulation of the P desorption at mild shaking. The P concentration in the bulk solution was fixed at zero. The P flux calculated in this way can be considered the maximum P supply flux in the soil, because not all the aggregates would be in close contact with the root surface in the pot experiment. Using the total P in the soil measured at various stages of the pot experiment, the maximum P supply rate was calculated with the model. This calculated supply rate was compared to the plant uptake rate, which was calculated from the P content measured in the ryegrass from the experiment. The ratio of the maximum supply rate and uptake rate, which was referred to as the DBI (dynamic bioavailability index), was used as a simple tool to assess the bioavailability of P in the soil. The results show that at the beginning of the pot experiment the DBI was about 10 and it decreased to about 2 after about 300 days, after which the DBI remained at a value of about 2. The plant uptake rate after 300 days decreased continuously, which indicates that at a DBI of about 2, the supply of P is not sufficient to meet the optimal demand of the plant and the soil is most likely the factor limiting P uptake, whereas in the beginning of the experiment the soil was not the limiting factor (very high DBI).

8.5.2 *Effects of root exudates*

As a response to phosphorus deficiency, a number of plant species excrete small organic acids from their roots (Dinkelaker et al., 1989; Hoffland et al., 1989; Gerke and Meyer, 1995; Li et al., 1997). The exudation of small organic acids can enhance the concentration of P in solution due to increase of the dissolution of P-minerals or due to competition with P for adsorption (Fig. 8.11). In a model exercise carried out by Geelhoed et al. (1999), the effects of root exudates on P mobility were studied. In this model, the competitive adsorption of P and citrate on goethite-coated sand was calculated with the CD-MUSIC model, which has been tested with the experimental data of P and citrate adsorption to goethite. The plant root was represented in the model with a cylinder with a radius of 0.15 mm. The root is considered to be a zero sink for $H_2PO_4^-$, the phosphate species taken up by plants. The rate of citrate exudation is kept constant and the value of the exudation rate was taken from the literature. The model simulation was carried out for either constant pH (pH 5) or considering the change of pH due to H^+ release with citrate exudation. Results show that exudation of citrate increased the bioavailability of P. The ratio of available P with and without citrate exudation is larger when the loading of P on goethite is lower. Acidification due to the co-release of protons reduced the effects of citrate on P bioavailability.

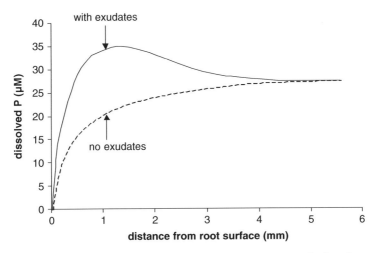

Figure 8.11. *Effects of root exudation on the concentration of phosphorus in soil solution (lines are hand-drawn to illustrate conceptually the effects of root exudates).*

8.6 Outlook for future research

To be able to understand and to predict the factors that influence the bioavailability of chemicals, a process-based approach is required. Chemical speciation of elements is a key process influencing bioavailability of ions. Both measuring and modelling methods have been developed to determine the speciation of chemicals. The DMT method is a promising technique that can measure the free ion concentration/activity. So far, the DMT has mainly been used to measure the speciation of cations. In principle, this technique can also be applied to anions using an anion exchange membrane. By adding ligands to the acceptor, the detection limit of the DMT can be decreased. To measure ion speciation based on the ion transport kinetics in the DMT is a relatively recent development. More testing of this approach is needed before the optimal use of this method can be applied.

Much progress has been made with respect to the development of SCMs for reactive particles such as natural organic matter and metal (hydr)oxides. These models strive to represent as close as possible the physical–chemical nature of the surface, surface sites and the surface complexes. The advantage of these mechanistically oriented models is that they can in principle be applied to multicomponent dynamic systems and are not limited to assume constant solution conditions such as pH and ionic strength. The compilation of a large amount of model parameters and the incorporation of these models into geochemical computer codes have made the application of these advanced SCMs feasible. However, these models and model parameters are derived from results with model particles. Better methods are required that can estimate the reactive surface area of such particles for natural samples. The interactions between different types of particles, like oxides and humics will influence the surface properties of both particles and also ion adsorption to the consortium may deviate from the simple sum. Recently, a new model framework, the LCD (ligand and charge distribution) model has been proposed (Filius et al., 2000; Weng et al., 2005a, 2006). The aim of this model is to describe the adsorption of humic particles to the oxides, and simultaneously, the adsorption of the small ions to both particles. At present, the LCD model is still under development and further improvement and testing is needed before application to natural samples becomes practical.

The BLM model is often considered a steady-state model. However, the model concept can be extended to be able to deal with kinetically controlled bioavailability issues. The slow kinetics that may control the supply rate of chemicals can be due to a slow reaction rate (dissolution, desorption, dissociation, etc) or due to transport processes that can limit the availability. The fundamental speciation approaches discussed above can be combined with the structure of soil aggregates in a reactive transport model. This more sophisticated approach is expected to be a very useful tool in the field of bio-availability research and the application to risk assessment in the near future.

Acknowledgement

We would like to acknowledge the Research Directorate General of the Commission of the European Commission for funding the BIOSPEC project (EVK1-CT-2001-00086).

Reference

Allison, J.D., Brown, D.S., Novo-Gradac, K.J., 1991. MINTEQA2/PRODEFA2-A Geochemical Assessment Model for Environmental Systems: Version 3.0 User's Manual. EPA/600/3-91-021, Environmental Protection Agency, Athens.

Anderson, P.R., Christensen, T.H., 1988. Distribution coefficients of Cd, Co, Ni, and Zn in soils. J. Soil Sci. 39, 15–22.

Antunes, P.M.C., Berkelaar, E.J., Boyle, D., Hale, B.A., Hendershot, W., Voigt, A., 2006. The biotic ligand model for plants and metals: technical challenges for field application. Environ. Toxicol. Chem. 25, 875–882.

Benedetti, M.F., Milne, C.J., Kinniburgh, D.G., Van Riemsdijk, W.H., Koopal, L.K., 1995. Metal-ion binding to humic substances – application of the nonideal competitive adsorption model. Environ. Sci. Technol. 29, 446–457.

Benedetti, M.F., Van Riemsdik, W.H., Koopal, L.K., 1996. Humic substances considered as a heterogeneous Donnan gel phase. Environ. Sci. Technol. 30, 1805–1813.

Buffle, J., Parthasarathy, N., Djane, N.K., Matthiasson, L., 2000. Permeation liquid membranes for field analysis and speciation of trace compounds in waters. In: J. Buffle, G. Horvai (Eds.), *In Situ* Monitoring of Aquatic Systems. Chemical Analysis and Speciation. Wiley, Chichester, pp. 407–493.

Campbell, P.G.C., 1995. Interaction between trace metals and aquatic organisms: a critique of the free-ion activity model. In: A. Tessier, D.R. Turner (Eds.), Metal Speciation and Bioavailability in Aquatic Systems. Wiley, Chichester, pp. 45–102.

Cances, B., Ponthieu, M., Castrec-Rouelle, M., Aubry, E., Benedetti, M.F., 2003. Metal ions speciation in a soil and its solution: experimental data and model results. Geoderma 113, 341–355.

Crooke, W.M., Hunter, J.G., Vergnano, O., 1954. The relationship between nickel toxicity and iron supply. Ann. Appl. Biol. 41, 311–324.

Davison, W., Zhang, H., 1994. *In situ* speciation measurements of trace components in natural-waters using thin-film gels. Nature 367, 546–548.

De Wit, J.C.M., Van Riemsdijk, W.H., Koopal, L.K., 1993a. Proton binding to humic substances. 1. Electrostatic effects. Environ. Sci. Technol. 27, 2005–2014.

De Wit, J.C.M., Van Riemsdijk, W.H., Koopal, L.K., 1993b. Proton binding to humic substances. 2. Chemical heterogeneity and adsorption models. Environ. Sci. Technol. 27, 2015–2022.

De Wit, J.C.M., Van Riemsdijk, W.H., Nederlof, M.M., Kinniburgh, D.G., Koopal, L.K., 1990. Analysis of ion binding on humic substances and the determination of intrinsic affinity distributions. Anal. Chim. Acta 232, 189–207.

Dijkstra, J.J., Meeussen, J.C.L., Comans, R.N.J., 2004. Leaching of heavy metals from contaminated soils: an experimental and modeling study. Environ. Sci. Technol. 38, 4390–4395.

Dinkelaker, B., Romheld, V., Marschner, H., 1989. Citric-acid excretion and precipitation of calcium citrate in the rhizosphere of white lupin (*Lupinus albus* L.). Plant Cell Environ. 12, 285–292.

Donnan, F.G., 1925. The theory of membrane equilibrium. Chem. Rev. 1, 73–79.

Dzombak, D.A., Morel, F.M.M., 1990. Surface Complexation Modeling: Hydrous Ferric Oxide. Wiley, New York.

Filius, J.D., Lumsdon, D.G., Meeussen, J.C.L., Hiemstra, T., Van Riemsdijk, W.H., 2000. Adsorption of fulvic acid on goethite. Geochim. Cosmochim. Acta 64, 51–60.

Fitch, A., Helmke, P.A., 1989. Donnan equilibrium graphite-furnace atomic-absorption estimates of soil extract complexation capacities. Anal. Chem. 61, 1295–1298.

Geelhoed, J.S., Van Riemsdijk, W.H., Findenegg, G.R., 1999. Simulation of the effect of citrate exudation from roots on the plant availability of phosphate adsorbed on goethite. Eur. J. Soil Sci. 50, 379–390.

Gerke, J., Meyer, U., 1995. Phosphate acquisition by red mustard on a humic podzol. J. Plant Nutr. 18, 2409–2429.

Hassler, C.S., Slaveykova, V.I., Wilkinson, K.J., 2004. Some fundamental (and often overlooked) considerations underlying the free ion activity and biotic ligand models. Environ. Toxicol. Chem. 23, 283–291.

Hatch, D.J., Jones, L.H.P., Burau, R.G., 1988. The effect of pH on the uptake of cadmium by 4 plant-species grown in flowing solution culture. Plant Soil 105, 121–126.

Helfferich, F.G., 1962. Ion Exchange. McGraw-Hill, New York.

Hesterberg, D., Bril, J., Delcastilho, P., 1993. Thermodynamic modeling of zinc, cadmium, and copper solubilities in a manured, acidic loamy-sand topsoil. J. Environ. Qual. 22, 681–688.

Hiemstra, T., De Wit, J.C.M., Van Riemsdijk, W.H., 1989a. Multisite proton adsorption modeling at the solid-solution interface of (hydr)oxides – a new approach. 2. Application to various important (hydr)oxides. J. Colloid Interface Sci. 133, 105–117.

Hiemstra, T., Van Riemsdijk, W.H., 1996. A surface structural approach to ion adsorption: the charge distribution (CD) model. J. Colloid Interface Sci. 179, 488–508.

Hiemstra, T., Van Riemsdijk, W.H., Bolt, G.H., 1989b. Multisite proton adsorption modeling at the solid-solution interface of (hydr)oxides – a new approach. 1. Model description and evaluation of intrinsic reaction constants. J. Colloid Interface Sci. 133, 91–104.

Hiemstra, T., Venema, P., Van Riemsdijk, W.H., 1996. Intrinsic proton affinity of reactive surface groups of metal (hydr)oxides: the bond valence principle. J. Colloid Interface Sci. 184, 680–692.

Hoffland, E., Findenegg, G.R., Nelemans, J.A., 1989. Solubilization of rock phosphate by rape. 2. Local root exudation of organic-acids as a response to P-starvation. Plant Soil 113, 161–165.

Huang, C.P., Morehart, A.L., 1990. The removal of Cu(II) from dilute aqueous solutions by Saccharomyces cerevisiae. Water Res. 24, 433–439.

Huang, C.P., Morehart, A.L., 1991. Proton competition in Cu(II) adsorption by fungal mycelia. Water Res. 25, 1365–1375.

Kalis, E.J.J., Temminghoff, E.J.M., Weng, L.P., 2006a. Effects of humic acid and competing cations on metal uptake by Lolium perenne. Environ. Toxicol. Chem. 25, 702–711.

Kalis, E.J.J., Weng, L.P., Dousma, F., Temminghoff, E.J.M., Van Riemsdijk, W.H., 2006b. Measuring free metal ion concentrations in situ in natural waters using the Donnan membrane technique. Environ. Sci. Technol. 40, 955–961.

Keizer, M.G., Van Riemsdijk, W.H., 1994. ECOSAT: Equilibrium Calculation of Speciation and Transport. Manual Program, Agricultural University of Wageningen, Wageningen.

Kinniburgh, D.G., Milne, C.J., Benedetti, M.F., Pinheiro, J.P., Filius, J., Koopal, L.K., Van Riemsdijk, W.H., 1996. Metal ion binding by humic acid: application of the NICA–Donnan model. Environ. Sci. Technol. 30, 1687–1698.

Kinniburgh, D.G., Van Riemsdijk, W.H., Koopal, L.K., Borkovec, M., Benedetti, M.F., Avena, M.J., 1999. Ion binding to natural organic matter: competition, heterogeneity, stoichiometry and thermodynamic consistency. Colloids Surf. A Physicochem. Eng. Asp. 151, 147–166.

Koopal, L.K., Saito, T., Pinheiro, J.P., van Riemsdijk, W.H., 2005. Ion binding to natural organic matter: general considerations and the NICA–Donnan model. Colloids Surf. A –Physicochem. Eng. Asp. 265, 40–54.

Koopal, L.K., van Riemsdijk, W.H., Dewit, J.C.M., Benedetti, M.F., 1994. Analytical isotherm equations for multicomponent adsorption to heterogeneous surfaces. J. Colloid Interface Sci. 166, 51–60.

Koopmans, G.F., Chardon, W.J., de Willigen, P., van Riemsdijk, W.H., 2004a. Phosphorus desorption dynamics in soil and the link to a dynamic concept of bioavailability. J. Environ. Qual. 33, 1393–1402.

Koopmans, G.F., Chardon, W.J., Ehlert, P.A.I., Dolfing, J., Suurs, R.A.A., Oenema, O., van Riemsdijk, W.H., 2004b. Phosphorus availability for plant uptake in a phosphorus-enriched noncalcareous sandy soil. J. Environ. Qual. 33, 965–975.

Lexmond, T.M., 1980. The effect of soil pH on copper toxicity to forage maize grown under field conditions. Neth. J. Agric. Sci. 28, 164–183.

Lexmond, T.M., Van der Vorm, P.D.J., 1981. The effect of pH on copper toxicity to hydroponically grown maize. Neth. J. Agric. Sci. 29, 217–238.

Li, M.G., Shinano, T., Tadano, T., 1997. Distribution of exudates of lupin roots in the rhizosphere under phosphorus deficient conditions. Soil Sci. Plant Nutr. 43, 237–245.

Lin, T.F., Wu, J.K., 2001. Adsorption of arsenite and arsenate within activated alumina grains: equilibrium and kinetics. Water Res. 35, 2049–2057.

Lumsdon, D.G., 2004. Partitioning of organic carbon, aluminium and cadmium between solid and solution in soils: application of a mineral-humic particle additivity model. Eur. J. Soil Sci. 55, 271–285.

Meeussen, J.C.L., 2003. ORCHESTRA: an object-oriented framework for implementing chemical equilibrium models. Environ. Sci. Technol. 37, 1175–1182.

Milne, C.J., Kinniburgh, D.G., Tipping, E., 2001. Generic NICA–Donnan model parameters for proton binding by humic substances. Environ. Sci. Technol. 35, 2049–2059.

Milne, C.J., Kinniburgh, D.G., van Riemsdijk, W.H., 2003. Generic NICA–Donnan model parameters for metal-ion binding by humic substances. Environ. Sci. Technol. 37, 958–971.

Mouvet, C., Bourg, A.C.M., 1983. Speciation (including adsorbed species) of copper, lead, nickel and zinc in the Meuse River – observed results compared to values calculated with a chemical-equilibrium computer-program. Water Res. 17, 641–649.

Naidu, R., Rogers, S., Gupta, V.V.S.R., Kookana, R.S., Bolan, N.S., Adriano, D., 2003. Bioavailability of metals in the soil-plant environment and its potential role in risk assessment: An overview. In: R. Naidu, S. Rogers, V.V.S.R. Gupta, R.S. Kookana, N.S. Bolan and D. Adriano (Eds.), Bioavailability and its Potential Role in Risk Assessment. Science Publishers, New York, pp. 21–59.

Nederlof, M.M., van Riemsdijk, W.H., 1995. Effect of natural organic matter and pH on the bioavailability of metal ions in soils. In: P.M. Huang, J. Berthelin, J.M. Bollag, W.B. McGill and A.L. Page (Eds.), Environmental Impact of Soil Component Interactions: Volume 2: Metals, Other Inorganics, and Microbial Activities. CRC Press Inc., Boca Raton, pp. 75–86.

Oste, L.A., Temminghoff, E.J.M., Lexmond, T.M., van Riemsdijk, W.H., 2002. Measuring and modeling zinc and cadmium binding by humic acid. Anal. Chem. 74, 856–862.

Pagenkopf, G.K., 1983. Gill surface interaction-model for trace-metal toxicity to fishes – role of complexation, pH, and water hardness. Environ. Sci. Technol. 17, 342–347.

Paquin, P.R., Gorsuch, J.W., Apte, S., Batley, G.E., Bowles, K.C., Campbell, P.G.C., Delos, C.G., Di Toro, D.M., Dwyer, R.L., Galvez, F., Gensemer, R.W., Goss, G.G., Hogstrand, C., Janssen, C.R., McGeer, J.C., Naddy, R.B., Playle, R.C., Santore, C., Schneider, U., Stubblefield, W.A., Wood, C.M., Wu, K.B., 2002. The biotic ligand model: a historical overview. Comp. Biochem. Physiol. C Toxicol. Pharmacol. 133, 3–35.

Parkhurst, D.L., Thorstenson, D.C., Plummer, L.N., 1990. PHREEQC – a computer program for geochemical calculations. Water Res. Inv., US Geology Survey, pp. 80–96.

Pei, J.H., Tercier-Waeber, M.L., Buffle, J., 2000. Simultaneous determination and speciation of zinc, cadmium, lead, and copper in natural water with minimum handling and artifacts, by voltammetry on a gel-integrated microelectrode array. Anal. Chem. 72, 161–171.

Plette, A.C.C., Nederlof, M.M., Temminghoff, E.J.M., van Riemsdijk, W.H., 1999. Bioavailability of heavy metals in terrestrial and aquatic systems: a quantitative approach. Environ. Toxicol. Chem. 18, 1882–1890.

Plette, A.C.C., van Riemsdijk, W.H., Benedetti, M.F., Van der Wal, A., 1995. pH dependent charging behavior of isolated cell-walls of a Gram-positive soil bacterium. J. Colloid Interface Sci. 173, 354–363.

Postma, J.W.M., Keltjens, W.G., van Riemsdijk, W.H., 2005. Calcium-(organo)aluminum-proton competition for adsorption to tomato root cell walls: experimental data and exchange model calculations. Environ. Sci. Technol. 39, 5247–5254.

Sauve, S., Manna, S., Turmel, M.C., Roy, A.G., Courchesne, F., 2003. Solid-solution partitioning of Cd, Cu, Ni, Pb, and Zn in the organic horizons of a forest soil. Environ. Sci. Technol. 37, 5191–5196.

Sauve, S., Norvell, W.A., McBride, M., Hendershot, W., 2000. Speciation and complexation of cadmium in extracted soil solutions. Environ. Sci. Technol. 34, 291–296.

Schroder, T.J., Hiemstra, T., Vink, J.P.M., van der Zee, S.E.A.T.M., 2005. Modeling of the solid-solution partitioning of heavy metals and arsenic in embanked flood plain soils of the rivers Rhine and Meuse. Environ. Sci. Technol. 39, 7176–7184.

Six, J., Paustian, K., Elliott, E.T., Combrink, C., 2000. Soil structure and organic matter: I. Distribution of aggregate-size classes and aggregate-associated carbon. Soil Sci. Soc. Am. J. 64, 681–689.

Slaveykova, V.I., Wilkinson, K.J., 2002. Physicochemical aspects of lead bioaccumulation by *Chlorella vulgaris*. Environ. Sci. Technol. 36, 969–975.

Smith, S.R., 1994. Effect of soil-pH on availability to crops of metals in sewage sludge-treated soils. 2. Cadmium uptake by crops and implications for human dietary-intake. Environmental Pollut. 86, 5–13.

Temminghoff, E.J.M., Plette, A.C.C., Van Eck, R., van Riemsdijk, W.H., 2000. Determination of the chemical speciation of trace metals in aqueous systems by the Wageningen Donnan membrane technique. Anal. Chim. Acta 417, 149–157.

Tipping, E., 1994. WHAM – a chemical equilibrium model and computer code for waters, sediments, and soils incorporating a discrete site/electrostatic model of ion-binding by humic substances. Comput. Geosci. 20, 973–1023.

Tipping, E., 1998. Humic ion-binding model VI: an improved description of the interactions of protons and metal ions with humic substances. Aquat. Geochem. 4, 3–48.

Tipping, E., 2002. Cation Binding by Humic Substances. Cambridge University Press, Cambridge.

Tipping, E., Hurley, M.A., 1992. A unifying model of cation binding by humic substances. Geochim. Cosmochim. Acta 56, 3627–3641.

Trivedi, P., Axe, L., 2000. Modeling Cd and Zn sorption to hydrous metal oxides. Environ. Sci. Technol. 34, 2215–2223.

Van der Stelt, B., Temminghoff, E.J.M., van Riemsdijk, W.H., 2005. Measurement of ion speciation in animal slurries using the Donnan membrane technique. Anal. Chim. Acta 552, 135–140.

Van Leeuwen, H.P., 1999. Metal speciation dynamics and bioavailability: inert and labile complexes. Environ. Sci. Technol. 33, 3743–3748.

Wallace, A., Romney, E.M., Cha, J.W., Soufi, S.M., Chaudhry, F.M., 1977. Nickel phytotoxicity in relationship to soil pH manipulation and chelating agents. Commun. Soil Sci. Plant Anal. 8, 757–764.

Weng, L.P., Koopal, L.K., Hiemstra, T., Meeussen, J.C.L., van Riemsdijk, W.H., 2005a. Interactions of calcium and fulvic acid at the goethite–water interface. Geochim. Cosmochim. Acta 69, 325–339.

Weng, L.P., Lexmond, T.M., Wolthoorn, A., Temminghoff, E.J.M., van Riemsdijk, W.H., 2003. Phytotoxicity and bioavailability of nickel: chemical speciation and bioaccumulation. Environ. Toxicol. Chem. 22, 2180–2187.

Weng, L.P., Temminghoff, E.J.M., van Riemsdijk, W.H., 2001a. Contribution of individual sorbents to the control of heavy metal activity in sandy soil. Environ. Sci. Technol. 35, 4436–4443.

Weng, L.P., Temminghoff, E.J.M., van Riemsdijk, W.H., 2001b. Determination of the free ion concentration of trace metals in soil solution using a soil column Donnan membrane technique. Eur. J. Soil Sci. 52, 629–637.

Weng, L.P., van Riemsdijk, W.H., Hiemstra, T., 2006. Adsorption free energy of variable-charge nanoparticles to a charged surface in relation to the change of the average chemical state of the particles. Langmuir 22, 389–397.

Weng, L.P., van Riemsdijk, W.H., Temminghoff, E.J.M., 2005b. Kinetic aspects of Donnan membrane technique for measuring free trace cation concentration. Anal. Chem. 77, 2852–2861.

Weng, L.P., Wolthoorn, A., Lexmond, T.M., Temminghoff, E.J.M., van Riemsdijk, W.H., 2004. Understanding the effects of soil characteristics on phytotoxicity and bioavailability of nickel using speciation models. Environ. Sci. Technol. 38, 156–162.

Xue, H.B., Sigg, L., 1993. Free cupric ion concentration and Cu(II) speciation in an eutrophic lake. Limnol. Oceanogr. 38, 1200–1213.

Xue, H., Sigg, L., 2002. In: T.F. Rozan, M. Taillefert (Eds.), Environmental Electrochemistry: Analysis of Trace Element Biogeochemistry. ACS, Symposium Series, Washington, p. 336.

Developments in Soil Science, volume 32
Ravendra Naidu (Editor)
© 2008 Elsevier B.V. All rights reserved

Chapter 9

DGT MEASUREMENTS TO PREDICT METAL BIOAVAILABILITY IN SOILS

P.S. Hooda and H. Zhang

9.1 Introduction

The mobility of metals largely determines the environmental impact of metal-contaminated soils, as it is the mobile metal species, which largely control metal uptake by plants or other biota as well as metal leaching through the soil. It is, therefore, important that the tests that are used for the purpose of assessing metal mobility should not only be reliable but should also be capable of detecting the changes accurately in metal mobility under field conditions. Historically, soil metal extractions with chemical reagents such as DTPA, ETPA, NH_4NO_3 and $CaCl_2$ have widely been used to assess metal bioavailability. Numerous chemical extractants with varying concentrations have been used and proposed, however, they tend to extract some operationally defined metal fractions, with no particular regard to soil-specific controls on metal bioavailability (Hooda, 2003; Naidu et al., 2003; see Chapter 20). The conventional soil testing procedures, which invariably involve soil sampling, drying, grinding and treating with chemical solutions crucially modify the soil field conditions. Risk assessments of metal-contaminated soils require a comprehensible protocol for testing metal bioavailability (see Chapter 4 and 25). Such a test should ideally be applicable with minimum perturbation of the soil, without disrupting the equilibrium between solid and solution phases that is sensitive to prevailing conditions. It has clearly been shown that measurement of metal concentrations present in soil solution can be a more rational and sensitive approach for evaluating metal bioavailability (McGrath et al., 1999) compared to the chemical-extraction based procedures. However, methods which involve separation of solution and solid phases can potentially disrupt the distribution between the two phases. Furthermore, as plants continually take up metals, local depletion in solution may induce replenishment from the soil, effectively giving the plant access to the labile components associated with the solid phase. Therefore measurement of the labile metal species available to plants, i.e. bioavailable metal, in principle, is best made using *in situ* procedures which do not involve the separation of solution from the solid phase.

The bioavailability of metals is related to their flux into the plant which is dependent on both their concentration in soil solution and their transport rate through the soil. Where there is active removal of metals, the local soil solution equilibrium concentration is further dependent on the re-supply from the solid phase. Quantitative interpretation of this flux of metals from the solid phase is central to considerations of soil testing for bioavailability purposes, yet it is not assessed by conventional soil testing procedures. Several studies have investigated the bioavailability (e.g. Cooperband and Logan, 1994; Yang et al., 1991) of nutrients and metals in soils using ion-exchange resin directly or resin impregnated membranes and resin imbedded in capsules or dialysis bags (see Chapter 14 for information on nutrient bioavailability). All these procedures respond to a flux from the soil to the resin-sink, but the geometry of these sinks with respect to the soil and the diffusion layer between soil and resin-sink is so poorly defined that this flux from soil solution to the sink cannot be quantitatively interpreted.

Davison and Zhang (1994) developed an *in situ* technique capable of quantitatively measuring labile metal species in waters. Further work on this technique, known as DGT (diffusive gradients in thin-films), has shown its application for the measurement of labile metal species in soils (Hooda et al., 1999). The application of DGT for the purpose of assessing metal bioavailability has been developed on the premise that in conventional methods of testing soil solution (1) metal speciation may change during sampling and extraction and (2) the kinetics of metal re-supply from solid phase to solution are not considered. This chapter describes the DGT technique and illustrates its application for the measurements of bioavailable metals in soil.

9.2 Theoretical background

The DGT device consists of two membranes: an ion-permeable hydrogel of thickness Δg, which is backed by a layer of an ion-exchange resin (Chelex-100), embedded in a hydrogel that acts as a sink (Fig. 9.1). The two membranes are held together in an assembly which ensures that only the ion-permeable membrane is exposed to the soil. Transport of simple metal ions from the soil to the resin membrane is determined by the concentration gradient through the known thickness, Δg, of the ion-permeable membrane (Fig. 9.1). Metals ions diffusing through the ion-permeable membrane are rapidly immobilised by the resin membrane. Provided the resin-sink is not saturated and the solution concentration remains constant, a linear concentration gradient is established in the ion-permeable membrane. DGT units are deployed in the soil and the mass of metal, M, accumulated during a given period of deployment is measured. If the metal supply to DGT is solely from the soil solution its initial concentration,

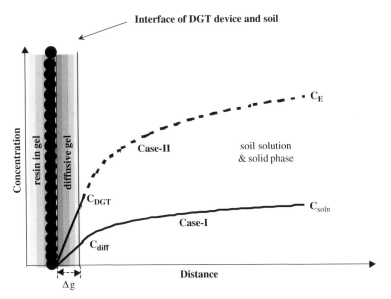

Figure 9.1. *Schematic cross section through the DGT device and adjacent soil solution, showing mean concentrations during the deployment: a solid line represent the case of diffusional re-supply only (Case-I) and a dotted line for the general case when there is some re-supply from the solid phase (Case-II). The dotted line represents the hypothetical concentration in the soil solution assuming re-supply by diffusion only (after Zhang et al., 2001).*

C_{soln}, near to the surface of the device will deplete to C_{diff} at any deployment time (Case-I, Fig. 9.1). After the initial few minutes a pseudo steady state is established at the soil–DGT interface, provided the diffusion coefficient in the diffusive membrane remains constant, as previously found (Zhang and Davison, 1999). The concentration gradient through the diffusive membrane is then linear (Harper et al., 1998), with flux, F (moles cm^{-2} s^{-1}), given by Fick's First Law (Eq. (9.1)).

$$F = DC_{diff}/\Delta g \tag{9.1}$$

Where, C_{diff} (moles cm^{-3}) is the instantaneous concentration at the DGT surface, D (cm^{-2} s^{-1}) the diffusion coefficient in the diffusive membrane of thickness, Δg. The value of D in the diffusive membrane has been measured using diffusion cell (Zhang and Davison, 1999) and they are provided on the DGT Research website (www.dgtresearch.com). The mean flux during the DGT deployment can be represented by the same equation, where C_{diff} then represents the mean concentration at the DGT–soil interface. The mean flux, F, can be calculated from the measured accumulated mass of metal, M (moles), the deployment time, t (s) and the exposed area of the diffusive membrane, A (cm^2):

$$F = M/(At) \qquad (9.2)$$

Consequently the mean metal concentration at the soil–DGT interface for the diffusion-only case (Case-I, Fig. 9.1) can be calculated using the following equation:

$$C_{diff} = M\Delta g/(DAt) \qquad (9.3)$$

The ratio of the mean concentration at the soil–DGT interface due to re-supply by diffusion only to the initial soil solution concentration, C_{soln}, can be expressed as R_{diff}.

$$R_{diff} = C_{diff}/C_{soln} \qquad (9.4)$$

The value of R_{diff} can be determined by a numerical solution of the diffusion equations (Harper et al., 1998), together with the geometry of the DGT device, deployment time and soil tortuosity. For example, for a device with $A = 3.14\,\text{cm}^2$, $\Delta g = 0.93\,\text{mm}$, $t = 24\,\text{h}$ and D (20°C) in the gel for Cu $= 5.42 \times 10^{-6}\,\text{cm}^{-2}\,\text{s}^{-1}$, the value of R_{diff} was 0.08 (Zhang et al., 2001). This is based on the diffusion coefficient in the soil $(4.25 \times 10^{-6}\,\text{cm}^2\,\text{s}^{-1})$ being lower due to tortuosity compared to its value in the diffusive membrane which is the same as in water.

The depletion of metal at the soil–DGT interface may, however, induce its re-supply from the solid phase. The measured accumulated mass of metal will, therefore, be greater due to the contribution from the solid phase compared to the case where metal transport to the DGT device is solely through diffusional supply. Consequently, the mean concentration at the soil–DGT interface, C_{DGT}, as shown in Figure 9.1 (Case-II), is greater than that supplied solely by diffusion from the solution, i.e. C_{diff}. The mean interfacial concentration, C_{DGT}, is then given by Eq. (9.5).

$$C_{DGT} = M\Delta g/(DAt) \qquad (9.5)$$

The C_{DGT} concentration reflects supply from both solution and solid phases, which is maintained approximately constant provided that the kinetically labile reservoir is not significantly depleted close to the DGT device (Harper et al., 1998). In this general case of re-supply contributions from both solution and solid phases, the initial soil solution concentration, C_{soln}, does not denote the effective concentration that is experienced by a sink like DGT or plant roots. This comprehension led Zhang et al. (2001) to propose a concept of effective concentration, C_E, which incorporates both initial soil solution concentration and its enhancement from the solid phase. Just as C_{diff} can be changed to C_{soln} by dividing by R_{diff}, an effective solution concentration, C_E, can be obtained by dividing C_{DGT} by R_{diff} (Eq. (9.6)).

$$C_E = C_{\text{DGT}}/R_{\text{diff}} \qquad (9.6)$$

C_E is the concentration that is required in the soil solution to supply the same amount of metal accumulated solely by diffusion as in Case-II, conceptualised in Figure 9.1. The DGT technique therefore provides direct measurements of the effective concentration. As metal uptake by plants can lower soil solution concentrations in the vicinity of plant roots, C_E is expected to provide a more complete assessment of bioavailable metals. C_E is principally determined by the soil properties, including soil metal concentration, and is independent of the surface area of the DGT device. C_E will, however, increase with deployment time and decrease with increasing thickness of the gel membrane, Δg. This is because the solely diffusional supply progressively decreases with time. The overall supply, however, decreases more slowly when there is solid phase supply. Therefore C_E is conditional on both the deployment time and thickness of the gel membrane, Δg, which must always be quoted (Zhang et al., 2001).

It is an important feature of DGT that it perturbs the soil-solution system in a known way while simultaneously recording the response. The maximum possible flux to the DGT device (Eq. (9.1)) is determined by the DGT geometry and the pore water concentration. The term DGT demand is used where it is necessary to indicate the size of the perturbation imposed by the introduction of a sink for metals.

As the DGT-measured mass or flux is dependent on the concentration gradient between the soil solution and resin membrane, therefore manipulating the thickness of the diffusive membrane can create a variable DGT demand for metal ions. A series of DGT devices can be deployed with increasing thickness, Δg, of the diffusive membrane to create decreasing concentration gradients (Fig. 9.1). According to the theory for a steady-sate situation (Eq. (9.2)), the DGT-measured mass or flux should decrease as the thickness of the diffusive gel membrane increases (Zhang and Davison, 1995). Therefore, a linear plot of DGT-measured mass or flux against $1/\Delta g$ implies that the rate of re-supply of metal ions at the DGT–soil interface (including solid phase contributions) is the same as their removal rate from the soil solution regardless of their removal rate. A deviation from linearity, on the other hand, would imply that in-soil diffusion and any solid phase associated re-supply are unable to satisfy the DGT demand for metal ions when thin diffusive membranes are used (Zhang et al., 1998).

9.3 Experimental procedures

9.3.1 DGT assembly

The DGT device (Fig. 9.2) is a simple tight-fitting piston, which consists of a backing cylinder and a front cap with a 2-cm diameter window. This window, with an area equal to A helps to maintain a close contact with the soil. The shape and dimension of the DGT device, however, can be different depending upon the design. A layer of resin (Chelex) membrane is placed on the top of the backing cylinder with the side containing the resin facing outward and then a layer of ion-permeable membrane is placed directly on top of it. Adherence of soil particles to the diffusive membrane is prevented by placing a 100-μm thick, 0.45 μm pore size Millipore cellulose nitrate membrane on the top of the two layers. This cellulose membrane behaves like an extension of the diffusive membrane (Zhang and Davison, 1995); Δg therefore includes the thickness (mm) of the ion-permeable membrane plus that of the cellulose membrane. Once the membranes are arranged on top of the cylinder (Fig. 9.2), the device is assembled by pressing down the front cap tightly until a good seal is formed on the membrane surface as well as making sure no air is entrapped between the membranes.

9.3.2 Diffusive and resin membranes

The procedure for preparing resin and diffusive gel membranes was first developed by Davison and Zhang (1994). A polyacrylamide hydrogel solution composed of 15% (v/v) acrylamide (Boehringer) and 0.3% (v/v) patented agarose-derived cross linker (DGT Research Ltd., UK) is used in both resin and ion-permeable gels. Freshly prepared 70 μl ammonium persulphate (10% w/w) initiator and 20 μl of TEMED (tetramethylethylenediamine) catalyst are added to every 10 ml of gel solution. This solution is immediately cast between two acid clean glass plates separated by plastic spacers of appropriate thickness and is

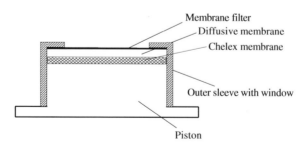

Figure 9.2. *Schematic representation of a section through the cylindrically shaped DGT device, which shows the configuration of the two membranes within the assembly.*

allowed to set in an oven ($42 \pm 2°C$) for at least 45 min until it is fully set. The thickness of the spacers can be varied in order to prepare ion-permeable membranes of desired thickness. To prepare the resin membrane, 2 g of ion-exchange resin Chelex-100 (Na-saturated, 100–200 wet mesh) is added for every 10 ml of gel solution. The resin membrane then can be prepared following the same procedure as for the ion-permeable membrane. All membranes (ion-permeable and resin) are hydrated in ultra pure MilliQ (double de-ionised) water for at least 24 h before use. This hydration step allows them to expand to a stable dimension. A fully hydrated ion-permeable membrane contains about 95% (w/w) water and is stored in 0.01 M $NaNO_3$ solution, and the resin membrane is stored in MilliQ water.

9.3.3 DGT deployment and retrieval

The DGT work so far has generally been limited to its application to soils (< 2 mm soil fraction) under laboratory conditions or alongside plants grown in pots. One critical issue for DGT deployment is that the soil should be moist enough (field capacity or above moisture content) in order to maintain contact with the membrane (Hooda et al., 1999). Field application of DGT although yet to be tested is possible as long as the soil is moist enough.

DGT is deployed by placing the device on the surface of the soil and pushing gently into the surface, making sure that there is no air entrapped and that the membrane is in good contact with the soil. The deployment can vary from few hours to several days, but a deployment time of 16–24 h is generally sufficient. After deploying for a given time, t, the DGT units are retrieved from the soil and are jet-washed with MilliQ water to wash away all the soil particles that may be adhered on the filter membrane. The resin membranes are retrieved and placed into clean plastic vials. For metal elution, the retrieved membranes are completely immersed in 1 ml 1 M HNO_3, by adding to each vial for 24 h (Zhang and Davison, 1995).

Following elution and appropriate dilution, the concentrations of trace metals can be analysed by inductively coupled plasma–mass spectrometry (ICP-MS) or any other suitable analytical equipment (e.g. graphite furnace atomic absorption spectrometry). The DGT accumulated masses of metals in the resin membrane are calculated using Eq. (9.7).

$$M = C_e(V_{HNO_3} + V_{rm})/f_e \qquad (9.7)$$

Where C_e is the concentration of metals (μg l) in the elution solution, V_{HNO_3} the volume of HNO_3 added to the resin membrane, V_{rm} the volume of the resin membrane, typically 0.15 ml, and f_e the elution factor for the metals, typically 0.8 (Zhang and Davison, 1995).

9.4 Example case studies of DGT application

The technique of DGT has been successfully used to measure the *in situ* concentrations of metals in natural waters (Zhang and Davison, 1995) and their fluxes within sediments (Zhang et al., 1995). The bioavailability of metals in soil is dependent on both their concentrations in the soil solution and their rate of transport through the soil (Hooda et al., 1999). This section describes example case studies of DGT application in soils for its general performance as well its ability to predict metal bioavailability.

9.4.1 Metal uptake by DGT

The effect of deployment time on metal uptake by DGT was investigated by deploying DGT devices with 0.4 mm diffusive gel membrane. The soil was acidic (pH 5.0), had a clay loam texture and a maximum water holding capacity (MWHC) of 53%. Soil samples (<2 mm) equivalent to 50 g dry mass were packed into PVC containers before hydrating them to the required water content and allowing them to equilibrate for 24 h at $20 \pm 1 °C$ in an incubator. In this case study, DGT units were deployed in saturated soil samples (53% water) kept at $20 \pm 1 °C$ for varying periods of up to 30 days (Hooda et al., 1999). Example results for Co, Cu and Pb are presented in Figure 9.3.

Initially the response was linear, but after two days the rate of accumulation decreased significantly (Fig. 9.3). The results suggest that after an initial period of deployment the DGT device depleted the available metal adjacent to its surface to the extent that re-supply from the solid phase together with in-soil diffusion were unable to balance the potential DGT demand. Consequently, the concentration of metal ions at the soil–gel interface decreased, causing a drop in the diffusion gradient through the gel membrane (see Fig. 9.1). For shorter deployment time periods, the DGT-accumulated metal mass increases linearly with time, and soil moisture in excess of the MWHC can prolong this liner response by facilitating in-soil metal transport by diffusion (Hooda et al., 1999).

The mass of metal accumulated by DGT as a function of time (Fig. 9.3) clearly shows that this device effectively operates as a sink, and is therefore somewhat analogous to root uptake. The length of time it can be deployed depends on the capacity of the resin-gel.

9.4.2 Soil to solution re-supply kinetics

Depletion of metals in soil solution, as a consequence of plant uptake, results in the re-supply from the solid phase. DGT can provide information about this solid phase to solution re-supply of metals. DGT devices with different diffusive gel thickness can be deployed to create various diffusion gradients and hence

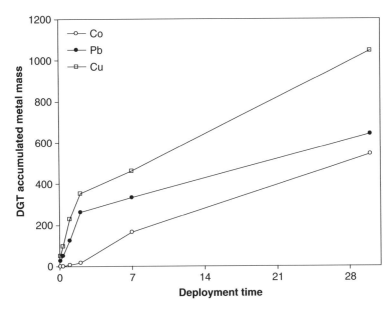

Figure 9.3. *DGT-accumulated mass of Co, Cu and Pb (in ng) measured as a function of deployment time (days). The soil was maintained at MWHC water content (53%) throughout the deployment. The thickness of the diffusive membrane in the DGT probes used was 0.4 mm (after Hooda et al., 1999).*

different rates of demand for metal ions (Fig. 9.1). In this case study, DGT devices with different diffusive membrane thickness from 0.13 to 2.13 mm were deployed in an arable soil. Prior to DGT deployment the soil was equilibrated at 80% moisture content, which was raised to 120% to obtain soil slurry (Zhang et al., 1998). Fluxes from solution to the DGT device were calculated (Eq. (9.2)). Example plots of metal flux against $1/\Delta g$ are presented in Figure 9.4.

All the plots show similar trends. For low values of $1/\Delta g$, corresponding to thicker gel membranes, there is a near linear increase in flux with $1/\Delta g$. This suggests that the flux from soil to solution is able to maintain a steady-state concentration in the soil solution, providing the DGT demand is kept relatively low by having thicker gel membranes. That is, the metal concentration in the soil solution is well buffered by rapid re-supply from the solid phase. When thinner gel membranes are used, the DGT-demanded fluxes are increased. If the demanded flux exceeds the maximum possible soil to solution flux, a steady state is no longer maintained and the concentration at the solution–DGT interface will decrease with time. The DGT-measured flux, which is the average value for the whole deployment period, will then be smaller than it would have been for the well-buffered case.

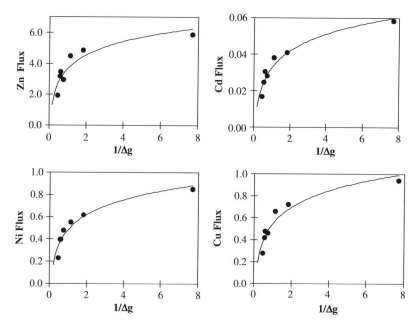

Figure 9.4. *DGT-measured fluxes of Ni, Zn, Cd and Cu (in $pg\,cm^{-2}\,s^{-1}$) plotted against the reciprocal of diffusive gel membrane thickness (in mm) for an arable soil amended with sewage sludge (after Zhang et al., 1998).*

Hooda et al. (1999) found similar relationship between DGT-measured metal fluxes and the thickness of the diffusive membrane, Δg. They further observed that DGT-measured metal fluxes are influenced by soil moisture content. The response was maximal for moisture contents between field capacity (FC) and MWHC (42 and 53%, respectively). For moisture contents greater than the MWHC, the DGT-measured fluxes decreased regardless of the gel membrane thickness, consistent with a dilution-associated decrease in metal concentrations in soil solution (Hooda et al., 1999). The results also showed unexpectedly lower response at soil moisture content lower than FC (37%) regardless of the thickness of the membrane, which suggest that diffusional transport may be restricted in gel or/and soil at low moisture contents. This unique feature of DGT helps to investigate the detailed characteristics of metal transport from soil to solution and how these can be affected by various soil properties (e.g. texture, structure, moisture, pH, organic matter and metal content).

9.4.3 Soil moisture effects on DGT measurements and metal plant uptake

Bioavailability of metals in a given soil is dependent on both their concentrations in the soil solution and their rate of transport through the soil.

Measurements of DGT flux (soil-to-DGT sink) of metals ions should therefore provide a better understanding of their bioavailability. To investigate this, DGT devices were deployed in soils with various soil moisture contents from 27 to 106%. DGT devices with 0.8-mm thick gel membranes were deployed for 24 h, following the experimental procedure and conditions as described in section 9.4.1. The fluxes were strongly influenced by soil moisture contents as shown in the example for Cd (Fig. 9.5). The lowest DGT-measured flux was for the soil with the least moisture content. The Cd flux increased as soil moisture content increased from 27 to 42% and decreased when soil moisture content exceeded the MWHC (53%). The diffusion coefficient of metal ions within the ion-permeable gel used in these studies is effectively the same as in water (Zhang and Davison, 1995). Ion diffusion through soil, however, is dependent on tortuosity of the diffusion pathway. The increase in DGT-measured metal ion fluxes, in this case study, with increasing soil moisture content may be partly due to changes in the tortuosity of the diffusion pathway. The reduced DGT response at small moisture contents may be due to pockets of air in the soil reducing the effective surface area of the DGT membrane (Hooda et al., 1999).

The decrease in DGT-measured fluxes when soil moisture exceeded the MWHC of the soil can be explained by smaller soil solution concentrations due to the dilution effect (Fig. 9.5). The soil solution concentrations were generally at their maximum when the moisture level was at its smallest level of 27%. The concentrations decreased as the moisture level increased, particularly for moisture contents exceeding 80% of MWHC, suggesting the soil solution concentrations reflected a simple dilution effect (Hooda et al., 1999; Davison et al., 2000). This effect of soil moisture on DGT-measured metal fluxes should be reflected by plant metal uptake if the controlling process in soil for the uptake is diffusion. To investigate this, a small plant uptake experiment using the same soil was conducted. Subsamples of the soil equivalent to 250 g dry soil mass were maintained gravimetrically at 27, 32, 37, 42 and 48% moisture contents. Six replicate pots were used for each soil moisture content, three each for seeding cress (*Lepidium sativum* L.) and parallel deployment of DGT probes. During the 18-day experiment, DGT deployments were made consecutively, each lasting for 48 h. Mean DGT metal fluxes obtained during this 18-day period are compared with the metal concentration in the whole cress plants harvested at the end of the experiment (Davison et al., 2000). The integrated DGT flux increased with moisture content, as did the amount of metal in the plants (Fig. 9.6). The general increase in metal uptake with soil moisture is consistent with other work on nutrient uptake by plants (Singh et al., 1998). The results though based on a limited work demonstrate that the response of fluxes measured by DGT

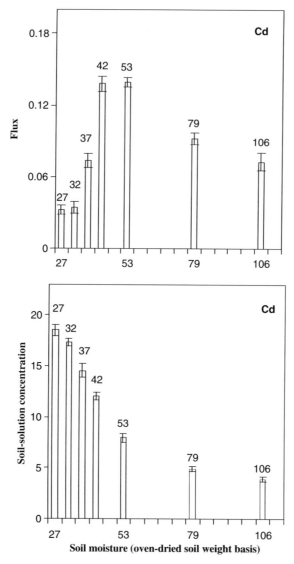

Figure 9.5. *Dependence of DGT-measured Cd fluxes (pg cm^{-2} s^{-1}) and soil solution concentrations (μg dm^{-3}) on soil moisture content expressed as % (w/w), with MWHC water content being 53%.*

to changes in moisture content is similar to the response of plant uptake. The results from this one simple experiment should not be over-interpreted, but it is instructive to consider the mechanistic similarities of uptake by DGT and plants.

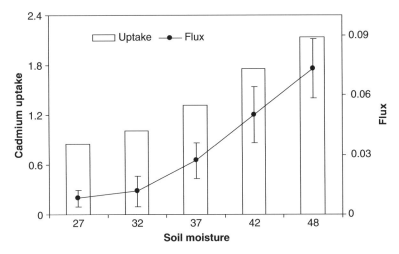

Figure 9.6. *Dependence of the measured mass of Cd in whole cress plants ($\mu g\,g^{-1}$) and the mean DGT fluxes ($pg\,cm^{-2}\,s^{-1}$) during the 18 days on soil moisture content (%, w/w). The maximum soil water content is equivalent to 90% of MWHC.*

9.4.4 DGT-measured effective soil solution concentration and plant uptake

Risk assessment of metal-contaminated soils has largely involved the use of total soil concentration or extractions with chemical solutions, with only limited success (Hooda, 2003). Recently Zhang et al. (2001) developed a new concept of measuring effective soil solution concentration, C_E (Eq. (9.6)) as a means of predicting plant metal uptake. This case study demonstrates the effectiveness of DGT as a tool to predict bioavailable Cu. The work involved 29 soils from diverse sources (Zhang et al., 2001).

Lepidium heterophyllum was grown in replicated pots each containing 500 g soil, with soil moisture being maintained at 80% field capacity. The pots were kept in a glasshouse with the following conditions: 14 h/10 h day/night, 20 °C/12 °C day/ night temperatures. Four weeks after germination, the above ground plant material was harvested, rinsed with deionised water, dried at 80 °C for 24 h and analysed for total Cu content using an inductively coupled plasma atomic emission spectrometry (ICP-AES). Soil solution was extracted at the end of the experiment, after the soil had been raised to 100% field capacity for one week. At the same time DGT units were deployed directly on the surface of each pot. Concentrations of total soluble metals in the soil solutions were determined by ICP-AES and free Cu^{2+} activities in the extracted soil solutions were determined using a Cu^{2+} ion selective electrode (Zhang et al., 2001). Soil from each pot was also sampled and extracted with 0.05 M EDTA (pH 7.0) for 'bioavailable' Cu before being analysed by ICP-AES.

Copper concentration in the plants was plotted against the four soil measurements of Cu (Fig. 9.7). For the whole data the C_E accounted for 98% of the observed variation in plant concentrations. At lower concentrations the relationship was still linear with a high r^2 value of 0.74. On the other hand, the relationships for the whole concentration range against free Cu^{2+} activity, Cu in soil solution and Cu extracted by EDTA were non-linear. For lower concentration ranges while soil solution concentrations were only poorly related ($r^2 = 0.27$)

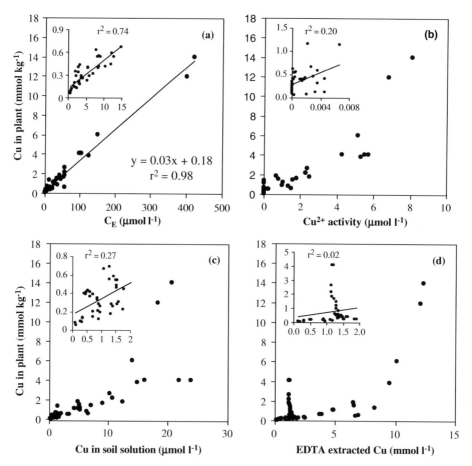

Figure 9.7. *Plots of Cu concentrations in above ground plant tissue versus C_E (a), free Cu^{2+} activity (b), soil solution concentration (c) and EDTA extracted Cu (d) for all individual pots. Insets show the lower concentration range. The linear regression equation and correlation coefficient are shown for the whole data plotted against C_E and the correlation coefficients are shown for all the insets (after Zhang et al., 2001).*

to plant Cu content, Cu^{2+} activity or EDTA extracted Cu had no relationship (Fig. 9.7). Clearly C_E best predicted the Cu concentration in plant shoots across a wide range of concentrations and soil types. The results indicate that C_E, rather than free Cu^{2+} activity in soil solution determines uptake of Cu by *L. heterophyllum*. The good linear relationship of plant uptake and C_E at low concentrations of Cu in soil solution further demonstrates the sensitivity of DGT and its potential use across wide range of soil Cu contents, including deficient conditions (Zhang et al., 2001).

A similar study assessed the bioavailability of Zn, Cd, Pb and Cu in 13 metal-contaminated soils from diverse sources and compared several techniques for predicting metal uptake by wheat (*Triticum aestivum* L.). The metal concentrations in wheat shoots were compared with total soil metal concentrations, total soluble metal (soil solution), free metal activities in soil pore waters, 0.01 M $CaCl_2$-extractable metal concentrations, E values measured by isotopic dilution and effective concentrations, C_E, measured by DGT. While relationships tended to be metal specific, DGT-measured C_E generally best predicted metal uptake by wheat shoots, particularly for Cd and Zn (Nolan et al., 2005). The results of this study support earlier findings where the C_E best predicted Zn uptake by *Lepidium sativum* (Zhang et al., 2004), and Cu uptake by two plant species, *Elsholtzia splendens* and *Silene vulgaris* (Song et al., 2004).

The work clearly shows that kinetically labile solid phase metal contributes to plant uptake, which, unlike soil solution, free metal activity or isotopic dilution is included in the DGT measurement. The DGT technique has proved a consistently better predictor of plant available metal since it provides a measure of the labile metals in soil solution as well as their kinetically labile pool associated with the soil solid phase. The application of DGT technique has largely been limited to soils which were previously homogenised either in laboratory or in pot experiments. Clearly field application of DGT is important in order to further establish and validate the technique. However, recent work by Nowack et al. (2004) suggests that the technique is capable of providing reliable prediction of soil–plant metal relationships under field conditions. They compared DGT application in a soil which was homogenised or undisturbed. While DGT and soil solution measurements differed between the homogenised and undisturbed soil, the uptake of Cu and Zn by ryegrass (*Lolium perenne*) was better predicted by DGT-measured C_E than the soil solution chemistry.

9.5 Conclusion

The case studies of DGT application presented in this chapter show that it is a robust technique, capable of producing highly reliable measurements. DGT, with its plastic assembly offers the possibility of a simple test procedure for soils to assess risks associated with metal-contaminated soils. The case studies clearly show that DGT measurement reflects concentration in the soil solution, the rate of re-supply from soil to solution and the transport rate of metal ions through the soil. These are all factors, which influence uptake of metal ions by plants. The kinetic perturbation of metals in the soil system by DGT is therefore similar to that occurs during plant uptake, as seen in the plant uptake studies. The new concept of effective soil solution concentration that can be obtained easily using DGT measurements may provide a major step forward in assessing metal bioavailability. Further work however is required over a wider range of field conditions with different metals, plant species and water contents.

References

Cooperband, L.R., Logan, T.J., 1994. Measuring in situ changes in labile soil phosphorus with anion-exchange membranes. Soil Sci. Soc. Am. J. 58, 105–114.

Davison, W., Hooda, P.S., Zhang, H., Edwards, A.C., 2000. DGT measured fluxes as surrogates for uptake of metals by plants. Adv. Environ. Res. 3, 550–555.

Davison, W., Zhang, H., 1994. In-situ speciation measurements of trace components in natural waters using thin-film gels. Nature 367, 546–548.

Harper, M.P., Davison, W., Zhang, H., Tych, W., 1998. Kinetics of metal exchange between solids and solutions in sediments and soils interpreted from DGT measured fluxes. Geochim. Cosmochim. Acta 62, 2757–2770.

Hooda, P.S., 2003. A special issue on metals in soils: editorial foreword. Adv. Environ. Res. 8, 1–3.

Hooda, P.S., Zhang, H., Davison, W., Edwards, A.C., 1999. Measuring bioavailable trace metals by diffusive gradients in thin films (DGT): soil moisture effects on its performance in soils. Eur. J. Soil Sci. 50, 285–294.

McGrath, S.P., Knight, B., Kilham, K., Preston, S., Paton, G.I., 1999. Assessment of the bioavailability of heavy metals in soils amended with sewage sludge using a chemical speciation technique and lux-based biosensor. Environ. Toxicol. Chem. 18, 659–663.

Naidu, R., Rogers, S., Gupta, V.V.S.R., Kookana, R.S., Bolan, N.S., Adriano, D., (2003). Bioavailability of metals in the soil-plant environment and its potential role in risk assessment: An overview. In: R. Naidu, S. Rogers, V.V.S.R. Gupta, R.S. Kookana, N.S. Bolan and D. Adriano (Eds.), Bioavailability and its Potential Role in Risk Assessment. Science Publishers, New York, pp. 21–59.

Nolan, A.L., Zhang, H., McLaughlin, M.J., 2005. Prediction of zinc, cadmium, lead, and copper availability to wheat in contaminated soils using chemical speciation, diffusive gradients in thin films, extraction, and isotopic dilution techniques. J. Environ. Qual. 34, 496–507.

Nowack, B., Koehler, S., Schulin, R., 2004. Use of diffusive gradients in thin films (DGT) in undisturbed field soils. Environ. Sci. Technol. 38, 1133–1138.

Singh, R., Chandel, J.D., Bhandari, A.R., 1998. Effect of soil moisture regime on plant, growth, fruiting, quality and nutrient uptake by mango (*Magnifera indica*). J. Indian Agric. Sci. 68, 135–138.

Song, J., Zhao, F.J., Luo, Y.M., McGrath, S.P., Zhang, H., 2004. Copper uptake by *Elsholtzia splendens* and *Silene vulgaris* and assessment of copper phytoavailability in contaminated soils. Environ. Pollut. 128, 307–315.

Yang, J.E., Skogley, E.O., Schaff, B.E., 1991. Nutrient flux to mixed-bed ion-exchange resin: temperature effects. Soil Sci. Soc. Am. J. 55, 762–767.

Zhang, H., Davison, W., 1995. Performance characteristics of diffusion gradients in thin-films for the in-situ measurement of trace metals in aqueous solution. Anal. Chem. 67, 3391–3400.

Zhang, H., Davison, W., 1999. Diffusional characteristics of hydrogels used in DGT and DET techniques. Anal. Chim. Acta 398, 329–340.

Zhang, H., Davison, W., Knight, B., McGrath, S., 1998. In situ measurements of solution concentrations and fluxes of trace metals in soils using DGT. Environ. Sci. Technol. 32, 704–710.

Zhang, H., Davison, W., Miller, S., Tych, W., 1995. In situ resolution measurements of fluxes of Ni, Cu, Fe, and Mn and concentrations of Zn and Cd in porewaters by DGT. Geochim. Cosmoschim. Acta 59, 4181–4192.

Zhang, H., Lombi, E., Smolders, E., McGrath, S., 2004. Zinc availability in field contaminated and spiked soils. Environ. Sci. Technol. 38, 3608–3613.

Zhang, H., Zhao, F.J., Sun, B., Davison, W., McGrath, S.P., 2001. A new method to measure effective soil solution concentration predicts copper availability to plants. Environ. Sci. Technol. 35, 2602–2607.

Developments in Soil Science, volume 32
Ravendra Naidu (Editor)
© 2008 Elsevier B.V. All rights reserved

Chapter 10

ORGANIC CONTAMINANT SPECIATION AND BIOAVAILABILITY IN THE TERRESTRIAL ENVIRONMENT

S.C. Wilson and R. Naidu

10.1 Introduction

Organic contaminants are ubiquitous in the terrestrial environment. Many of the compounds considered environmental pollutants are both persistent and exhibit toxic, carcinogenic, mutagenic and teratogenic potential. Over the past 10 years they have received increased scientific and legislative attention worldwide as the risk they may pose to ecosystems, including plants and humans, has been recognized.

Legislation and guidance determining management and acceptable levels for organic contaminants in soil is often based on total soil concentration. Sometimes this guidance encompasses whole compound groups, such as total polycyclic aromatic hydrocarbons (PAHs), rather than individual species (NEPM, 1999). It has become increasingly clear that the environmental abundance, distribution, bioavailability and ecotoxicity of organic pollutants can be better understood in terms of individual chemical species. The toxicity, environmental behaviour and bioavailability of organic compounds is known to be strongly dependent on compound structure and associated physical and chemical properties. Therefore, total soil concentration (in particular for a group of compounds) may not represent the portion of compound/component compound that may be posing a risk in terms of mobility or availability to plants, microorganisms and higher organisms (Alexander, 2000).

Although the role of chemical species in metal behaviour and bioavailability has been well recognized for many years and extensively researched (Naidu et al., 2003; Sauve, 2003) this has not been the case for organic pollutants. Species specific assessment is critical to the prediction of actual risk posed by organic pollutants to environmental systems. This is limited by our understanding of individual organic chemical species present in soil, factors affecting their abundance in soil and differences between their environmental fate, behaviour and bioavailability (Chapter 2).

10.1.1 Definition of chemical species

A chemical species is defined from IUPAC accepted terms as a specific form of a chemical element, such as a molecular or complex structure or oxidation state

(Templeton et al., 2000; Caruso et al., 2003). Organic chemicals are generally described collectively by the functional groups they contain. Within each chemical group, individual chemical species can encompass a wide range of molecular masses and both chemical and chiral structures.

Analytical techniques used for the identification of chemical species both in solution and associated with solid phases are advanced for metalloids and metals and relationships with bioavailability relatively well documented (Sauve, 2003; Sanz-Medel, 1998). Methods for the identification of certain organic pollutants in soils are also relatively well established, for example, the polychlorinated biphenyls (PCBs) and PAHs (USEPA, 1986), although optimization is required for certain environmental matrices. Methods are usually complex, time consuming and require a combination of analytical techniques involving separation followed by specific species detection (Motelay-Massei et al., 2004). However, reliable and reproducible methods for the analysis and detection of numerous organic chemicals found in environmental matrices, including many pesticides, their metabolites and chiral species, are still being developed (Pepich et al., 2005).

10.2 Speciation, toxicity and abundance in soils

Organic contaminants found in soil belong to a wide range of different chemical groups. The compounds for which there is probably the greatest concern in soil because of long half-lives and persistence are the persistent organic pollutant groups (POPs), hydrophobic, often chlorinated compounds (Harrad, 2001; United Nations Environment Program on Persistent Organic Pollutants). Whilst certain soils receive direct inputs of these chemicals accidently or deliberately such as at contaminated sites or during pesticide usage to agricultural land, on a global scale, direct soil contamination occurs on only a small proportion of land area. Concentrations of many of the persistent pollutants are detected in most soils, however, even in remote areas, as a result of atmospheric deposition (Hassanin et al., 2005; Meijer et al., 2003).

Organic compound groups in soil may exist either as:

(a)   non-ionic molecules;
(b)   charged ions; or
(c)   molecules that exist as ions or unionized species depending on the soil
        environment.

Within each chemical group individual species may:

(a)   have different molecular mass and different chemical formulae; and
(b)   exist as different isomers:

(i) Structural isomers have the same chemical formulae and molecular mass but differ in the ratio or sequence of bonding of their atoms. Structural isomers in general have different physical and chemical properties;

(ii) Stereoisomers have the same molecular mass, chemical formula and sequence of covalent bonds, but differ in the relative disposition of their atoms in space so that they cannot be superimposed on each other but are mirror images of each other. These isomers are said to be chiral and are called enantiomers.

Environmental fate and behaviour of organic compounds is strongly dependent on the compound physical and chemical properties such as polarity, volatility, solubility, hydrophobicity and ionization potential (Sensi, 1992; see Chapter 2). These in turn are governed by compound structure and composition. In the following section we illustrate the range of individual chemical pollutant species that can be found in soil using a number of important groups of environmental organic pollutants as examples. We describe the influence of chemical species on compound toxicity and physico-chemical properties and also review the relative abundance of the different species in soils where data is available as an illustration of species that may be important for inclusion in risk assessment for ecosystems. In later sections we discuss in detail species influence on compound bioavailability.

10.2.1 Non-ionic compounds

Many groups of important soil pollutants exist in soil only as non-ionic species irrespective of the soil environment. The compounds are relatively non-polar due to the atomic constituents, functional groups and configurations of the component molecules. They also tend to be relatively hydrophobic with high octanol–water partition coefficients (K_{ow}), have low solubilities, and are also resistant to biological and abiotic breakdown compared to more polar species. However, physico-chemical and biological properties of the different molecules within each group can vary significantly.

10.2.1.1 Polychlorinated dibenzo-*p*-dioxins and polychlorinated dibenzofurans (PCDD/Fs)

PCDD/Fs are a byproduct of combustion and also the manufacture of compounds such chlorophenoxyalkanoic acids (2,4-D and 2,4,5-T) and pentachlorophenol (Masunaga et al., 2001). They are tricyclic, planar compounds with a basic chemical structure which is illustrated in Figure 10.1. Between one and four of the H atoms on each ring structure can be substituted with Cl atoms enabling a total of 75 PCDD and 135 PCDF individual species. Each individual PCDD/F is referred to as a congener. The congeners with identical chemical formulae (same

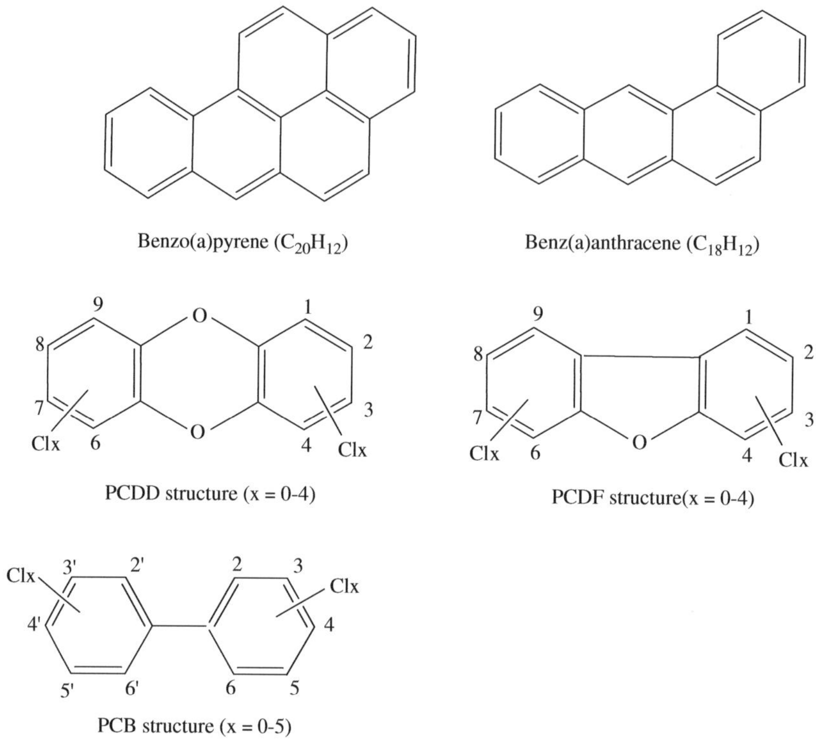

Benzo(a)pyrene ($C_{20}H_{12}$) Benz(a)anthracene ($C_{18}H_{12}$)

PCDD structure (x = 0-4) PCDF structure(x = 0-4)

PCB structure (x = 0-5)

Figure 10.1. *Structures of certain non-ionic organic pollutants.*

number of chlorine atoms) are isomers of each other. Each group of isomers (one for each degree of chlorination) constitutes a homologue group with eight homologue groups possible. The individual compounds become more hydrophobic, less soluble and less volatile with increasing degree of chlorination (Table 10.1).

Considerable variation is evident in the toxicity of individual congers. Toxicity depends on extent and positioning of Cl substitution. The 2,3,7,8-substituted congeners are considered the most toxic to animals. 2,3,7,8-T$_4$CDD is considered to be the most toxic member of this group and the degree of toxicity with additional chlorination decreases to OCDD. Toxicity equivalent factors (TEFs) have been assigned for all the 17 2,3,7,8-substituted PCDD/F congeners as a measure of their toxicity relative to 2,3,7,8-T$_4$CDD which has been universally assigned a TEF of 1. The TEF values devised by the World Health Organisation (Van den Berg et al., 1998) are presented in Table 10.2 which illustrates the variation in toxicity of different PDDD/F congeners.

Significant differences are observed in PCDD/F patterns in different soils (differences are also evidenced when compared to other POPs such as PCBs). This occurs because of the variety of different potential and ongoing sources,

Table 10.1. *Physical-chemical properties for a selection of individual compounds.*

Organic compound	RMW	Solubility	V_p mean	H_c mean	log K_{ow} mean	log K_{oc} mean
PAHs						
Naphthalene	128.2	31	36.8	4.30E+01	3.37	3.14
Benzo(a)anthracene	228.3	0.011	0.000606	5.81E−01	5.91	5.33
Benzo(a)pyrene	252.3	0.0038	2.13E−05	4.60E+01	6.04	6.24
PCBs						
PCB101 (2,2′,4,5,5′)	326.4	0.0165	0.00285	2.16E+01	6.3	6.3
PCB 138 (2,2′,3,4, 4′,5)	360.9	0.002	0.000169	2.49E+01	6.69	6.24
PCB 153 (2,2′,3,5,6,6′)	360.9	0.0026	0.00012	1.76E+01	6.76	6.58
PCB 180 (2,2′,3,4,4′,5,5′)	395.3	0.000509	0.000126	3.42E+01	7.13	6.36
PCDDs						
2,3,7,8-TCDD	322	0.0000193	0.000118	3.34	6.8	6.2
OCDD	460	7.4E−08	9.53E−07	6.84E−01	8.2	6.97
OC pesticides						
p,p′ DDT	354.5	0.003	0.00002	2.36E+00	6	5.61[a]
p,p′ DDE	318	0.04	0.001	7.95E+00	5.7	5.31[a]
p,p′ DDD	320	0.05	0.001	6.40E−01	5.5	5.11[a]
α-HCH	291	1	0.003	8.70E−01	3.81	3.42[a]
β-HCH	291	0.1	0.00004	1.20E−01	3.8	3.41[a]
γ-HCH	291	6.5	0.003	1.30E−01	3.8	3.03[a]
δ-HCH	291	8	0.002	7.30E−02	4.1	3.71[a]
Atrazine	214.7	30	0.00004	2.90E−03	2.4	2.01[a]

Note: Data based on values reported mainly from Mackay et al. (1991, 1992a, b, 1997). RMW ($g\,mol^{-1}$); Solubility $g\,m^{-3}$ or ($mg\,L^{-1}$) at 25°C; V_p, vapour pressure in Pa at 25°C; H_c, Henry's constant in $Pa\,m^3\,mol^{-1}$; K_{ow}, octanol−water partition coefficient; K_{oc}, organic carbon partition coefficient.
[a]No vaules reported in literature. Value calculated from $K_{oc} = 0.41\,K_{ow}$.

many combustion related, apparent production by natural processes (Gaus et al., 2001), and the wide range in physico-chemical properties of the different species (see Table 10.1). A selection of contemporary background surface soils collected from a number of locations around the world showed a common PCDD/F level and homologue pattern in Western Europe (also common to industrial regions) (Green et al., 2004; Hassanin et al., 2005). Near non-detect or low levels were detected at remote areas. In less remote sites increasing concentrations of PCDD homologues with increasing level of chlorination were reported to a maximum abundance of OCDD. However, the PCDFs increased in abundance with decreasing levels of chlorination. The pattern in soils from Australia and Thailand were distinct and dominated by OCDD with lower levels of hexa- and hepta-CDDs. PCDF concentrations were at or below levels of detection. The data suggest an alternative source of PCDD/Fs in Australia and Thailand than those in Europe. In general, these results show that there is a prevalence of hepta-CDDs and octa-CDD in background surface soils.

Table 10.2. *WHO-TEF values (humans/mammals) assigned to PCDD/Fs and PCB congeners.*

Congener	TEF value
Dioxins	
2,3,7,8-TCDD	1
1,2,3,7,8-PeCDD	1
1,2,3,4,7,8-HxCDD	0.1
1,2,3,6,7,8-HxCDD	0.1
1,2,3,7,8,9-HxCDD	0.1
1,2,3,4,6,7,8-HpCDD	0.01
OCDD	0.0001
Furans	
2,3,7,8-TCDF	0.1
1,2,3,7,8-PeCDF	0.05
2,3,4,7,8-PeCDF	0.5
1,2,3,4,7,8-HxCDF	0.1
1,2,3,7,8,9-HxCDF	0.1
1,2,3,6,7,8-HxCDF	0.1
2,3,4,6,7,8-HxCDF	0.1
1,2,3,4,6,7,8-HpCDF	0.01
1,2,3,4,7,8,9-HpCDF	0.01
OCDF	0.0001
PCBs	
Non-ortho	
3,3′4,4′-TeCB (PCB #77)	0.0001
3,4,4′,5-TeCB (PCB #81)	0.0001
3,3′,4,4′,5-PeCB (PCB #126)	0.1
3,3′,4,4′,5,5′-HxCB (PCB #169)	0.01
Mono-ortho	
2,3,3′,4,4′-PeCB (PCB #105)	0.0001
2,3,4,4′,5-PeCB (PCB #114)	0.0005
2,3′,4,4′,5-PeCB (PCB #118)	0.0001
2′,3,4,4′,5-PeCB (PCB #123)	0.0001
2,3,3′,4,4′,5-HxCB (PCB #156)	0.0005
2,3,3′,4,4′,5′-HxCB (PCB #157)	0.0005
2,3′,4,4′,5,5′-HxCB (PCB #167)	0.00001
2,3,3′,4,4′,5,5′-HpCB (PCB #189)	0.0001

Source: Taken from Dyke and Stratford (2002).

10.2.1.2 PCBs

PCBs were first synthesized in 1929 and because of their unusually high chemical stability and electrical resistance were used in a wide range of industrial applications including dielectric fluids in capacitors and transformers, hydraulic fluids, fire retardants and plasticizers (Jones et al., 1991). Most commercial applications used a mixture of PCBs in a chemical formulation known as an Arochlor with different arochlors containing different abundance and composition of component PCB compounds. Because of concern regarding environmental persistence and effects, by the end of the 1970s PCB use and production was ceased in most countries. However, the legacy of PCBs already released into the environment and control of further releases remains a particular concern.

The molecular structure of a PCB consists of a biphenyl with 10 positions where Cl substitution may occur, numbered as illustrated in Figure 10.1. This gives rise to 10 possible homologue groups and 209 possible individual congeners. The definitions between congeners, isomers and homologue groups are identical to those described above for PCDD/Fs. A IUPAC numbering system (Schulz et al., 1989; Ballschmiter and Zell, 1980) is widely used to assign numbers to the individual 209 PCBs.

Considerable specifies specific variations exist in congener toxicity both to humans and animals. The PCBs that have no chlorine substitution at the *ortho* (2,2′,6,6′) positions, the so-called coplanar, "dioxin-like" PCBs are considered most biologically active and toxic, notably 3,3′,4,4′,5-PeCB (No #126), followed by 3,3′,4,4′,5,5′-HxCB (#169) and by 3,3′,4,4′-TeCB (#77) (Safe, 1990). This is because they are able to adopt the coplanar configuration of the dioxins necessary to bind with an intercellular protein, the Ah receptor, considered to be one of the important factors controlling the toxic effect of dioxins. However, mono-*o*-substituted and certain di-*ortho*-substituted PCBs (PCB #170 and PCB #180) also exhibit a toxic response. Those PCBs assigned a TEF by WHO are listed in Table 10.2. Species specific variations in physico-chemical properties also occur for the PCBs. Compounds with increasing degree of chlorination exhibit greater lipophilicity and hydrophobicity, lower solubility and lower volatility (Table 10.1).

The abundance of different PCB species varies widely between different soils and depends on the PCB formulations used. A global survey of 191 background surface soils (0–5 cm) for PCBs and HCB showed differences of up to four orders of magnitude in abundance between sites, with the highest concentrations found in Greenland and mainland Europe (Meijer et al., 2003). Background soil concentrations were strongly influenced by proximity to source region and soil organic matter content. Many of the highest PCB concentrations measured were in the northern hemisphere at latitudes 30–60°N (considered the global source region) or in the organic matter rich soils just to the north of this latitude. The estimated global soil burdens of individual congeners and homologue groups are presented in Table 10.3. Uncertainties in the estimates withstanding, Table 10.3 shows that lighter congeners/homologues are dispersed in soils globally more extensively, with the heavier congeners concentrating in the source zone. Overall, the PCB soil burden generally increases with degree of chlorination to a maximum for the hexa-chlorinated PCBs. In general, remote sites are characterized by a greater proportion of lower molecular weight PCBs whereas industrial and urban sites show higher proportions of the higher molecular weight PCBs (Motelay-Massei et al., 2004).

Table 10.3. *Estimated global surface soil PCB distribution of selected PCBs in background soils (t).*

Congener/homologue group	Global	90°S–30°N	30–60°N	60–80°N
PCB 28	190	50	74	66
PCB 52	210	48	65	94
PCB 101	170	43	88	40
PCB 118	400	83	220	92
PCB 138	530	26	350	150
PCB 153	1220	200	760	260
PCB 180	580	120	360	93
Sum tri (18,31,28)	315	80	130	100
Sum tetra (52,49,44,70)	360	85	180	99
Sum penta (95,90/101,99,87,110,123,118)	1580	230	910	450
Sum hexa (151,149,153/132,141,138,158)	3360	450	2100	790
Sum hepta (187,183,180,170)	1340	220	890	230
Sum octa (199,203,194)	260	32	190	47

Source: Meijer et al. (2003).

10.2.1.3 PAH species

PAHs are released to the environment through a wide range of past and current anthropogenic activities such as production and combustion of petroleum and fossil fuels. They consist of two or more fused benzene rings with the number of possible isomers increasing with the number of rings in the compound. Several thousand possible PAH species are possible although routinely only the 16 assigned by the USEPA as priority pollutants are analysed in soil samples (Keith and Telliard, 1979). Some environmentally significant PAHs are illustrated in Figure 10.1. Again the toxicity of the individual PAH varies with chemical species but some are known to be acutely toxic, in particular benzo(a)pyrene, which shows greater toxicity that other five ringed isomers. The toxicity of the PAHs is related to metabolic transformation by organisms to metabolites that can disrupt DNA and RNA. Other important carcinogens include the benzofluoranthenes, benzo(a)anthracene, dibenzo(a,h)anthracene and indeno(1,2,3-cd)pyrene (Harrad, 2001). Higher molecular weight PAHs show greater hydrophobicity, lower volatility, lower solubility and longer half-lives in soil in general (Table 10.1).

The occurrence of many PAH species in soil has been widely detected (Chen et al., 2005). They are common pollutants at contaminated sites and in soils generally although the abundance and occurrence of each individual chemical species varies widely depending on the source activity (Motelay-Massei et al., 2004). Generally PAHs from petrogenic sources (traffic emissions, petrol stations) are lower molecular weight compounds whilst pyrogenic sources (coal tar on gas works sites) are abundant in high molecular weight PAHs (Wilson and Jones, 1993). Readers are encouraged to Chapter 23 for a detailed consideration of bioavailability and the influence that it has on the efficiency of PAH bioremediation.

10.2.1.4 Polychlorinated naphthalenes (PCNs) and polybrominated diphenyl ethers (PBDEs)

PCNs are a group of compounds with a number of possible different compounds and isomers depending on the chlorination of the naphthalene ring structure. They have similar properties to PCBs and have been used in similar applications but are not prohibited in most countries. In addition to leaking electrical components, sources of PCNs to the environment include thermal emissions, chlor-alkali processes and PCB associations. Because of the naphthalene ring within the compound structure, PCNs are planar compounds exhibiting toxicity similar to that of dioxins and have significant TEQ contributions in environmental samples including soil. Many PCN congeners have been detected in the limited soil samples analysed for these compounds (Meijer et al., 2001b).

PBDEs are used in large quantities as fire retardants. They have attracted attention in recent years because some PBDEs have similar physicochemical, bioaccumulative and toxicological properties to the POPs. Again a number of different individual species are possible depending on Br substitution of the phenyl rings. The major technical products used are penta-BDE and octa-BDE which comprise varying proportions of a number of different congeners. A survey of background soils from the UK and Norway reported that congeners 47, 99, 153 and 145, major constituents of the penta-BDE technical mixture, typically dominated, although a wide range of other congeners were present (Hassanin et al., 2004). Congener 183, a marker for octa-DBE was also present in some samples. This data suggested that PBDEs may be quite stable in the terrestrial environment.

10.2.2 Ionizable compounds

A number of common soil pollutant groups exist as non-ionic species in soil but can also ionize to form cationic or anionic species depending on their acidity/basicity and the pH of the soil environment. We discuss chlorophenols as an example of this category of soil pollutants although numerous other pesticides are also included here (Gevao et al., 2000). For example, chlorophenoxyacetic acid pesticides (2,4-D and 2,4,5-T) form anionic compounds in some environments, whereas the basic triazine herbicide group of pesticides, including atrazine and simazine, can ionize to cationic species depending on the soil environment (Fig. 10.2).

Chlorophenols (CPs) have been used for a wide range of domestic, agricultural and domestic purposes, including pesticide use, and have an evident role in the formation of dioxins and furans (Folke and Lindgaard-Jorgensen, 1985; Paasivirta et al., 1992; Oberg and Rappe, 1992). They are monocyclic aromatic

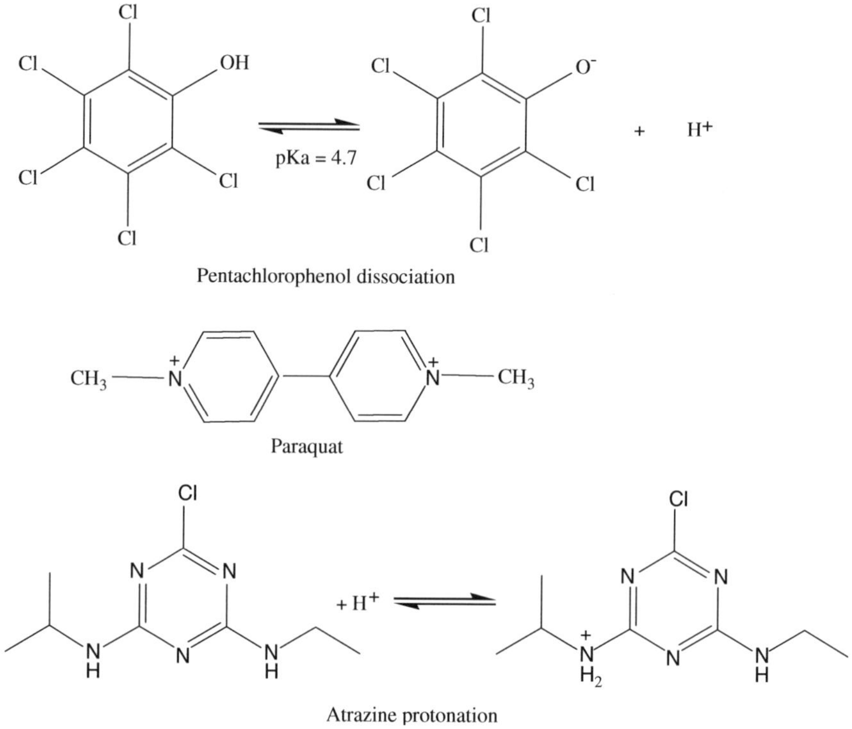

Pentachlorophenol dissociation

Paraquat

Atrazine protonation

Figure 10.2. *Structures of certain ionizable and charged organic pollutants.*

phenol compounds with 1 to 5 chlorine atoms substituted onto the benzene ring. This substitution creates 19 individual species: 3 monochlorophenols (MCP), 6 dichlorophenols (DCP), 6 trichlorophenols (TCP), 3 tetrachlorophenols (TetraCP) and pentachlorophenol (PCP). Five CPs are recognized as a particular concern for soil pollution and exposure due to greater distribution and use. Pentachlorophenol has been used extensively as a fungicide, particularly in wood preservation (Crosby, 1981). 2,4,6-TCP and 2,3,4,6-TetraCP have also been used in mixtures to prevent fungal/microbial infection of wood and to preserve manufactured goods. 2,4-DCP and 2,4,5-TCP are used as raw materials in the manufacture of the chlorophenoxyacetic acid herbicides 2,4-D and 2,4,5-T and as such may be present in the finished product, whilst 2,4,5-TCP has been used in the past specifically as an agricultural pesticide (Ahlborg and Thunberg, 1980).

Chlorophenols are acutely toxic to aquatic and terrestrial organisms with toxicity generally increasing with degree of chlorination (Hobbs et al., 1993). Factors affecting the environmental behaviour and bioavailability of non-ionized CPs are essentially the same as those of the non-ionized organic compound groups described previously. However, chlorophenols are weak hydrophobic acids and as such can dissociate to form phenolate anion (Fig. 10.2). The extent of ionization depends on the individual CP acidity and the system pH. pKa values for chlorophenols range from 5 to 9 with the more highly chlorinated compounds ionizing at lower pH. In fact, some of higher chlorinated CPs including pentachlorophenol exist in many soils as the phenolate ion because soil pH generally exceeds the dissociation constant (Bellin et al., 1990). The implications for compound fate and behaviour in soils, and thence bioavailability, as a result of the presence of the charged species are profound and must be considered in any risk assessment. Factors affecting the bioavailability of the charged species to different organisms are quite different to those of the uncharged species. For example, sorption of PCP has been shown to be stronger than sorption of the phenolate ion (Lagas, 1988; Amiri et al., 2004), and reduced availability for degradation and leaching has been observed in acidic soils because of sorption of the undissociated species (Bellin et al., 1990). However, rapid degradation of the phenolate ion in sludge amended calcareous soils precluded uptake by lettuce, fescue, carrot and chilli pepper (Bellin and O'Connor, 1990).

10.2.3 Charged compounds

A small group of soil contaminants exist soley as charged species and as such their environmental behaviour and bioavailability must be considered separately from that of the non-ionic and ionizable compounds described above. For example, the bipyridium family of pesticides include diquat and paraquat, commonly used commercial herbicides, and exist as planar cationic species in soil (see Fig. 10.2). Most of the environmental fate information published for these pesticides concerns paraquat which is extremely persistent in many soils as a result of strong sorption to soil mainly through associations with interlayer exchange sites of clay minerals (Mordaunt et al., 2005). This renders it essentially biologically unavailable to plants and soil organisms. Less that 0.1% (usually much less) of applied paraquat is present in soil solution and paraquat in soil solution is known to be degraded rapidly (Bromilow, 2003). There is no evidence of reactivation of adsorbed paraquat residues due to desorption under a variety of land management practices with concentrations plateauing rather than continuing to accumulate, even at abnormally high repeated applications. Effects on plants or soil organisms have not been reported at typical application rates (Roberts et al., 2002).

We will not discuss charged species further in this chapter but have included them here as an illustration of the importance of assessing the bioavailability of specific compound species present in the soil environment for accurate risk assessment purposes.

10.2.4 Chiral species

A number of important soil contaminants are chiral. This includes up to 25% of pesticides (Williams, 1996), including several constituents of technical chlordane (*cis*-chlordane, *trans*-chlordane, heptachlor), the major isomer of technical hexachlorocyclohexane (HCH), α-HCH, one of the minor DDT (dichlorodiphenyl-1, 1,1-trichloroethane) isomers, *o,p'*-DDT, and its metabolite, *o,p'*-DDE (dichlorodiphenyl-1,1-dichloroethylene). The 19 PCB congeners with three or four *ortho* chlorine atoms also exist as pairs of stable enantiomers at ambient temperatures (Fig. 10.3) (Wong and Garrison, 2000). Because of the high cost of separation, often both enantiomers are released into the environment as a racemic mixture in the commercially marketed product although only one enantiomer may have the desired application properties.

Physical and chemical processes such as volatilization, sorption and abiotic transformations are not considered enantioselective in environmental systems (Zipper et al., 1998), although this cannot be ruled out for sorption onto organic matter because components of organic matter have chiral centres and may be capable of selectively binding enantiomers and affecting availability. However, biological processes, including those involved in toxicity, biological degradation, organism uptake and metabolism reactions, can be enantiomer selective with one enantiomer reacting preferentially over the other. Different enantiomers can therefore exhibit different toxicities and biological effects. For example, the (−) isomer of *o,p'*-DDT has been identified as the active estrogen mimic with the (+) isomer showing negligible activity to human estrogen receptor (Hoekstra et al., 2001).

Enantioselective degradation of a specific enantiomer in the environment results in the persistence of another (Law et al., 2004). This process will alter the enantiomer ratio in soils with either positive or negative outcomes. Sometimes the more toxic enantiomer persists with greater potential for uptake to plants and higher organisms. Alternatively the more toxic enantiomer is degraded.

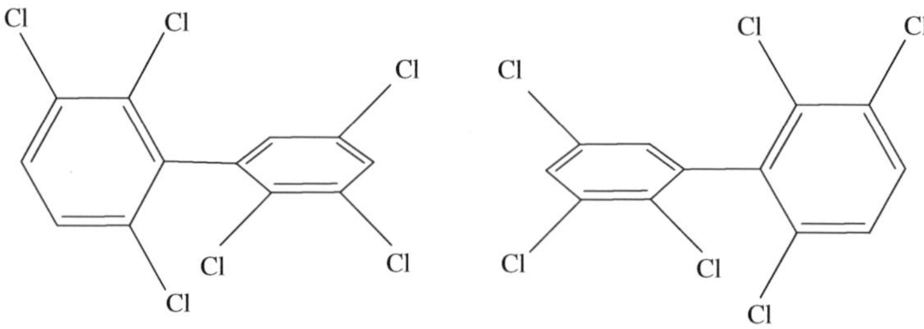

Enantiomers of PCB 135 (2,2',3,3',5,6'-HCB)
(taken from Wong and Garrison, 2000)

Figure 10.3. *Enantiomers of PCB 135 (2, 2', 3, 3', 5, 6'-HxCB). Source: Wong and Garrison (2000).*

Therefore, risk assessment based on the racemic mix is not always accurate. Changing land use has been shown to alter the enantioselectivity in soils (Lewis et al., 1999) although factors affecting this change and determining the specific enantioselectivity by microbial populations are not well understood.

There have been a number of recent attempts to understand the enantiomeric soil compositions for certain soil pollutants primarily for the purposes of atmospheric pollution source apportionment and to understand better air–soil exchanges (Leone et al., 2001; Robson and Harrad, 2004). In Alabama soils, enantiomeric fractions of o,p'-DDT ranged considerably, attributed to differences in microbial populations. However, all soils examined showed preferential loss of the (+) *trans*-chlordane and (–) *cis*-chlordane isomers but enrichment of (+) heptachlor-exo-epoxide and (+) oxychlordane (Wiberg et al., 2001). The enantiomeric composition for α-HCH was close to racemic. Results from UK soils show similar enantiomeric selectivity for these pesticides (Meijer et al., 2001a). Robson and Harrad (2004) investigated enantiomeric fractions of chiral PCBs 95, 136 and 149, commonly found soil contaminants. They found that in topsoil appreciable enrichment of one enantiomer for PCB 95 and the (–) enantiomer for PCBs 136 and 149 occurred. Processes determining this enrichment were not elucidated.

Historically environmental studies have negated to consider differences in the toxicities, biological effects and environmental fate of individual enantiomers of chiral compounds (Magrans et al., 2002). There is a dearth of information on differences in enantiomer uptake in plants and higher organisms. However, in view of the discussion above this is a prerequisite to accurate risk assessment for these compounds.

10.2.5 Pesticides

Certain important pesticides are discussed separately here as an illustration of the potential complexity in terms of considering speciation for accurate risk assessment of many organic compounds. Pesticides encompass a wide range of different compound groups which exhibit all the speciation possibilities described in the preceding sections (Gevao et al., 2000). Examples include the non-ionic organochlorines (DDT, HCH, endosulfan), organophosphates (malathion, diazinon) and carbamates (aldicarb, carbofuran), the ionizable triazines (atrazine, simazine), and the charged bipyridyls (diquat, paraquat). Transformation products of many of the pesticide compounds may also be present in soil at higher levels than the pesticide itself and may themselves exist as different species that are toxic and have potential to adversely affect the environment in terms of persistence, bioaccumulation or increased mobility (Andreu and Pico, 2004). These must be considered in any risk assessment associated with the parent compound.

One example is DDT. Since the 1940s DDT has been used worldwide as an active insecticide in agriculture but also for the control of diseases such as malaria and typhus. However, DDT and also its metabolites DDE and DDD (dichlorodiphenyl-1,1-dichloroethane) (Fig. 10.4) were found to be toxic, stable and highly persistent in the environment (Moore and Walker, 1964). *p,p'*-DDE is known to be a potent androgen receptor antagonist (Kelce et al., 1995). DDD has been reported to have a broader antimicrobial spectrum and to be more toxic to microorganisms than DDT and DDE (Megharaj et al., 1999, 2000). As a result the use of DDT was banned in most countries through the 1970s and 1980s (USEPA, 1972). A number of additional metabolites may also be formed under specific conditions including dichlorodiphenyl-1-chloroethylene (DDMU) and dichlorodiphenylacetic acid (DMA), a polar metabolite formed under anaerobic conditions that has been found to be very persistent and easily leachable (Heberer and Dunnbier, 1999).

Approximately 80% of the DDT technical mixture used in applications was made up of the *para*-substituted *p,p'*-DDT isomer with the *o,p'*-DDT chiral isomer

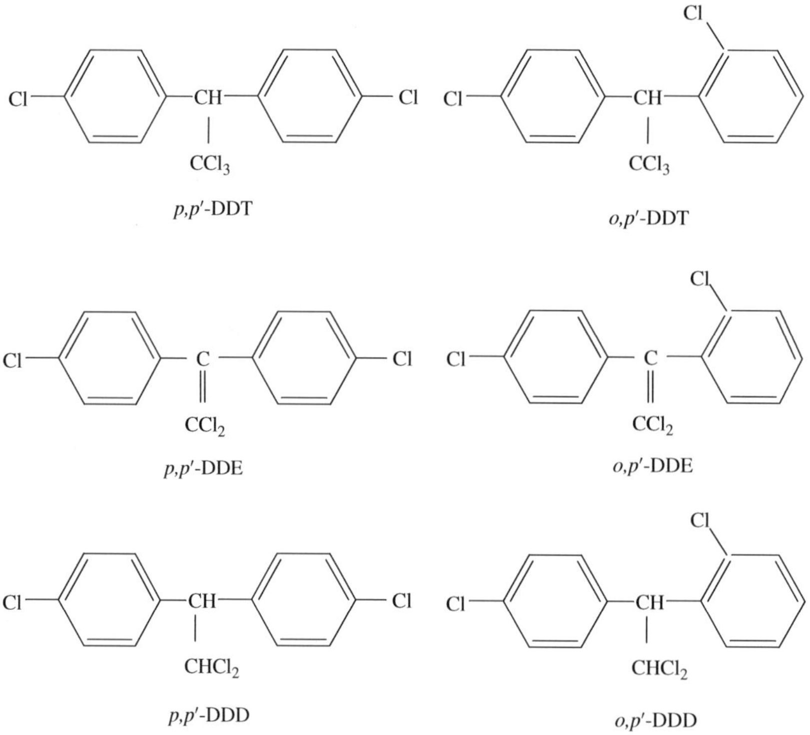

Figure 10.4. *Structure of DDT, DDE and DDD isomers.*

forming a minor component. In Alabama soils, concentrations of the chiral and non chiral species declined at similar rates (Harner et al., 1999) although in soils from Tianjin, China, results indicated that *p,p'*-DDT may have been more diffi-cult to degrade (Gong et al., 2004a). The different metabolite isomers formed (*p,p'*-DDE and *o,p'*-DDE; *p,p'*-DDD and *o,p'*-DDD) strongly depend on environ-ment although DDE is more persistent than DDD or the parent compound DDT and is more likely to be formed under aerobic conditions (Meijer et al., 2001a). In general a high proportion of DDE indicates older residues although parent compound and metabolite concentrations in soil vary depending on proximity to source, time since application and soil properties (Chen et al., 2005; Harner et al., 1999). Only recently have environmental studies considered differences in the fate of different structural and enantiomeric isomers of both the parent compound and its metabolites.

Another complex example is hexachlorocyclohexane (HCH), the term used collectively for the eight (α, β, γ, δ, ε, η, θ) isomers depicted in Figure 10.5. The α-isomer exists in two enantiomeric forms. The γ-isomer, lindane, is the only isomer with strong pesticidal properties. Commercial formulations typically contain 60–70% of the α-isomer, 5–12% β-isomer, 1–12% γ-isomer and 6–10% δ-isomer although refined products with higher proportions of lindane have been used. HCH isomers have different physical and chemical properties as a result of the different axial and equatorial positions of the Cl atoms, and different isomers exhibit different toxicities and have different environmental fates (Willett et al., 1998). All isomers have relatively high water solubilities and moderately high vapour pressures with the α- and γ-isomers being most volatile.

All the HCH isomers are acutely toxic to mammals and also exhibit chronic effects. The α, β and γ-isomers have all been implicated as possible or probable human carcinogens. However, the α-isomer shows the most carcinogenic activity but the β-isomer shows evidence of endocrine disrupting activity in mammals and fish. This is the most stable and environmentally persistent of the isomers because all Cl atoms are in equatorial positions. The most common isomers found in the environment are α, β and γ-isomers. The α-isomer is predominant in air and oceans but the β-isomer is the most abundant isomer accumulating in human fat and has been found in soil at greatest abundance (Gong et al., 2004b; Chen et al., 2005). Loss rates for the different isomers from soil vary depending on the environmental conditions although the order of rate of loss in an agricultural plot treated with the technical mixture was α > γ > δ ≫ β (Chessells et al., 1988). There is some evidence for the isomerization of γ- to α-HCH which may contribute to the α-isomer accumulation (Walker et al., 1999).

These two examples illustrate the complexities but also importance of including all compound speciation in any assessment of bioavailability for risk assessment purposes.

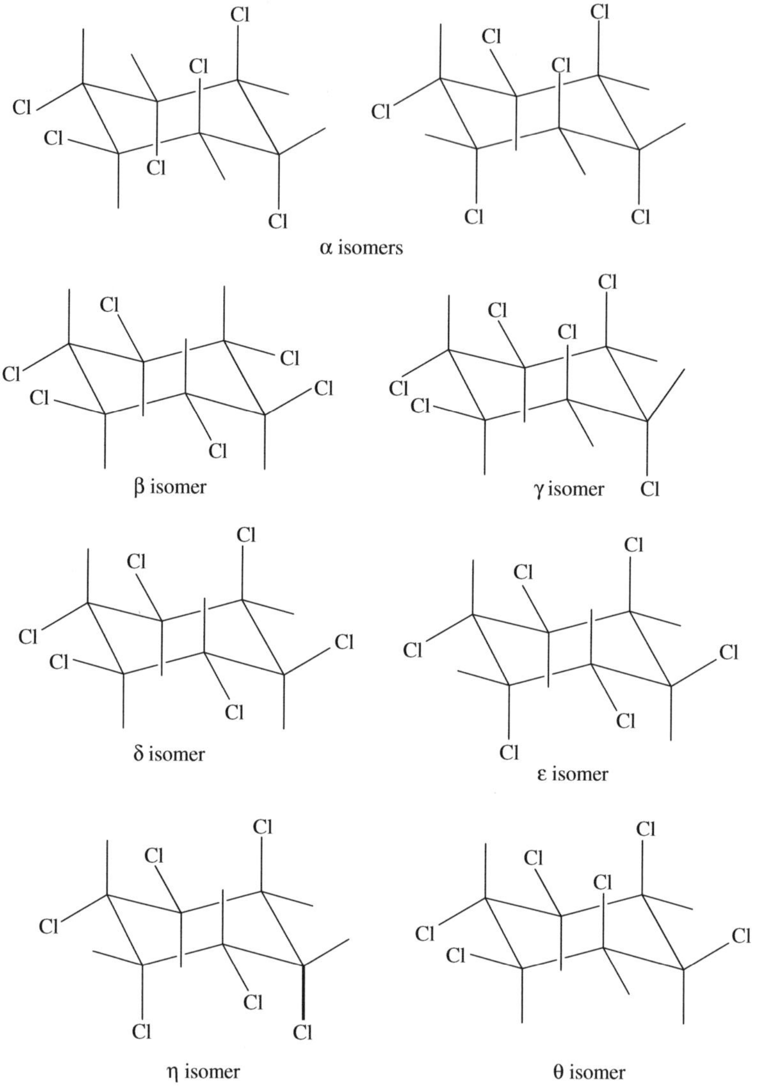

Figure 10.5. *Structures of HCH isomers. Source: Willett et al. (1998).*

10.3 Bioavailability of organic compounds in soil – definitions and measurement

When a compound enters soil it is affected by a number of different processes which determine its ultimate fate. It may be biodegraded, leached, volatilized, undergo abiotic degradation or it may be sorbed onto soil particles (Fig. 10.6). Thus, a portion of the compound added to the soil is potentially bioavailable,

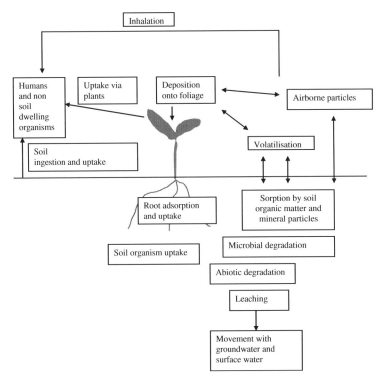

Figure 10.6. *Processes determining organic compound fate in soil.*

that is, it has the potential to be microbially degraded or taken up by plants and other organisms. The actual amount that is taken up/degraded depends on the relative importance of the other soil process occurring and also a variety of different factors affecting bioavailability and extent of the other processes including soil type, time and chemical physico-chemical properties.

In the first part of this section we discuss the concept and definitions of bioavailability as they refer to organic compounds in soil. We then assess the methods available and used to measure bioavailability with specific consideration given to species-specific assessment.

10.3.1 Defining bioavailability

Over recent years there has been an obvious lack of clarity and general agreement on the definition of bioavailability, particularly when referring to contaminants in soil (Adelaide, 2004; Chapter 2, this book). Different definitions used in different disciplines and differences in availability to different organisms have compounded the confusion (Semple et al., 2004). However, incorporation of bioavailability within regulatory frameworks is gaining acceptance (ISO/FDIS 17402, 2007) and consequently standardization of definitions is necessary.

Bioavailability is process dependent as described by Ehlers and Luthy (2003) when they summarized the major U.S. National Research Council (NRC, 2002) report which did not define bioavailability but bioavailability processes "as the individual physical, chemical and biological interactions that determine the exposure of organisms to chemicals associated with soils and sediments" (Fig. 10.7 and Chapter 3 this book). The general concept of bioavailability as defined by ISO/ 11074 (2005): "bioavailability is the degree to which chemicals present in the soil may be absorbed or metabolized by human or ecological receptors or are available for interaction with biological systems" has some limitations in terms of the temporal and species dependence as discussed in detail by Harmsen et al. (2005).

Semple et al. (2004) attempted to address these limitations by defining two components of the bioavailable compound: (i) the bioavailable component, compound that is freely available to cross an organism's cellular membrane from one medium the organism inhabits at a given time. After transfer has occurred, storage, transformation, assimilation or degradation can occur, processes distinct from transfer from the medium and (ii) the bioaccessible component, compound that is available to cross an organism's cellular membrane from the environment, if the organism has the access to the chemical. However, the chemical could either be physically removed from the organism for example occluded in organic matter, or bioavailable only after a period of time. Therefore, bioaccessibility encompasses what is actually bioavailable now plus what is potentially bioavailable as depicted

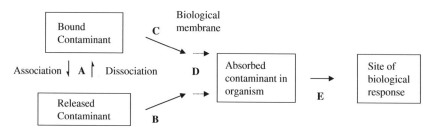

A release of solid-bound contaminant
B transport of released contaminant
C transport of bound contaminant
D uptake across physiological membrane
E incorporation into living organism
A, B, C and D are defined as bioavailability processes but not E , because soil and sediment no longer play a role.

Figure 10.7. *Processes determining compound bioavailability in soil. Source: Ehlers and Luthy (2003).*

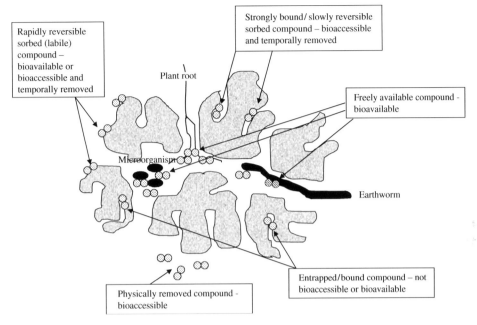

Figure 10.8. *Conceptual diagram of bioavailable and bioaccessible contaminant fractions in soil. Source: Semple et al. (2004).*

in Figure 10.8. These terms are gaining acceptance for terrestrial systems. Readers with interest in the definition of bioavailability are directed to Chapter 3.

Although the essence of definitions will be debated further it is obvious that organic compound bioavailability is intimately related to the compound itself as well as other factors.

10.3.2 Measuring bioavailability

In practical terms an operational definition is necessary. The only direct measure of bioavailability is to use target organisms leading to an operational definition of "bioavailability is the uptake (resulting in accumulation or an effect) of a component by a specified organism during a fixed period of time from a unit mass of soil" (Harmsen et al., 2005). However, this is neither acceptable in the case of humans nor possible nor practicable for many organisms. In the past few years a considerable amount of research has been devoted to develop methods that realistically measure the bioavailability of organic contaminants in soil. Both chemical and biological tests have been used (Lanno et al., 2004; Harmsen et al., 2005; Reid et al., 2000a, 2000b; Semple et al., 2003) although there is an absence of standardized procedures (Naidu et al., 2003).

Biological assays are often imprecise, slow and expensive resulting in greater effort to determine more representative chemical techniques. Also biological methods commonly target the chemical present in the aqueous phase which is generally assumed to be that component directly available for intra-cellular uptake although this is not always the case (Shaw et al., 2000). Chemical tests often appear to estimate the bioaccessible component rather than the bioavailable fraction as defined by Semple et al. (2004). Therefore, any test must be considered as a tool with which to give an indication of the bioavail-ability to a target organism and must be closely correlated with biological assays and field measurements for real effects and not developed independently.

10.3.3 Biological methods
10.3.3.1 Microbial

A number of microbiological methods have been used and standardized to measure biodegradability and as such bioavailability if the biodegradable fraction is also considered to be the bioavailable fraction (Richterich et al., 1998; OECD, 1993). Mineralization assays are also relatively well used where ^{14}C-labelled parent compound converion to $^{14}CO_2$ gives a measure of the microbially avail-able fraction of contaminant (Hatzinger and Alexander, 1995; Semple et al., 2003). Microbial toxicity tests have also been applied but these do not necessarily measure the bioavailable compound faction (Jacobs et al., 1993).

A recent technique in microbial measurement is the use of microbial biosensors. These have found widespread application in environmental monitor-ing and toxicity assessment and have advantages in terms of sensitivity, cost and time compared to the use of higher organisms (Steinberg et al., 1995). A biosensor such as a microorganism and an output signal (electrochemical, optical) are used to measure changing response to pollution or environmental conditions. Genetic modification can enhance detection of the response of the biosensor and the use of reporter genes for bacterial bioluminescence (lux) has found widespread application (Palmer et al., 1998). Lux marked bacteria produce light continuously, with light output directly proportional to metabolic activity. Any chemical induced metabolic stress results in a decline in luminescence directly proportional to the toxicant concentration. These tests appear to be effective for a range of hydrophobic organic pollutants (Boyd et al., 1997; Sousa et al., 1998; Ren and Frymier, 2005) although there has been some conflicting evidence of toxicity assessment (Reid et al., 1998).

A further development which shows promise for the detection of pollutant specific bioavailabiltiy is when lux genes are fused to genes involved in a particular response creating a more specific biosensor. In effect, the biolumines-cence is turned on and proportionally quantifiable in the presence of a target pollutant. This has been used successfully for a number of organic pollutants including naphthalene and certain PCBs (Layton et al., 1998; Heitzer et al., 1992, Nivens et al., 2004). Although these tests have been applied to soil extracts and slurries, there are problems associated with the need for solvents to extract the compounds from soils which impact the biosensor and also effect compound bioavailability (Reid et al., 2000a).

10.3.3.2 Earthworms

Earthworms are commonly used as a means of assessing bioavailability for a wide range of organic xenobiotics including pesticides (Meharg, 1996; Morrison et al., 2000, Kelsey et al., 2005), PAHs (Jager et al., 2003a), PCBs, chlorophenols and chlorobenzenes (Van Gestel and Ma, 1988; Jager et al., 2003b). Earthworms are ideally suited for assessing bioavailability because of their intimate contact with the soil and soil solution and their consumption of large soil volumes. Bioavailability can be measured in terms of bioaccumulation but also in terms of toxicological bioavailability (Lanno et al., 2004).

However, tests with earthworms are limited to chemicals that actually bioaccumulate to measurable levels (log K_{ow} > 4). They can also at times be temporally limited because of slow desorption from soil (Hu et al., 2005) and problems due to species sensitivity have also been reported. Furthermore, uptake by earthworms varies between different species so that results from one study cannot be extrapolated to another (Jager et al., 2005).

10.3.3.3 Whole animal

Whole animal uptake studies have been used for some hydrophobic contaminants for example PCBs in sludge amended soil in cows (McLachlan et al., 1990; McLachlan, 1993; Thomas et al., 1999). Obviously this form of assessment has limitations in terms of time and expense and uptake will depend on animal health and habits. Simulations of the gastrointestinal tract (Van de Wiele et al., 2004) and also in vivo extraction of polluted sediments using digestive fluid removed from target organisms (Weston and Mayer, 1998) have been used as a more rapid and practicable form of measurement. However, these obviously have limitations in terms of the inherent assumptions necessary.

10.3.3.4 Plants

Plants have been extensively used to assess the phytoavailable contaminant fraction for a number of different hydrophobic chemicals (Schroll et al., 1994; Mattina et al., 2000). Plant concentrations can be related to soil concentrations by the use of predictive relationships but uptake ultimately depends on plant species and also on specific environmental conditions so results from one study cannot be assumed for any other environment (Briggs et al., 1982).

10.3.4 Chemical assessment
10.3.4.1 Solvent extraction

Chemical soil extraction techniques have historically aimed to measure the total soil contaminant concentration by exhaustively extracting soil using techniques such as soxhlet extraction. However, a plethora of recent information illustrates that exhaustive extraction gives little indication of actual compound bioavailability and can result in inappropriate estimation of exposure and risk to susceptible populations (Kelsey and Alexander, 1997). Consequently, a number of mild solvent extractants have been used (as detailed in Table 10.4) that give good correlations with contaminant bioavailability for specific species. However, extractants are compound selective and also target species specific so that any chemical extractant method must be developed and correlated for the specific target species.

Table 10.4. *Solvent extractants used to assess chemical bioavailability.*

Compound	Species	Extraction	Author
Atrazine	Pseudomonas R	Methanol:water (1:1)	Kelsey et al. (1997)
	Earthworm *E. foetida*	Methanol:water (9:1)	Kelsey et al. (1997)
Phenanthrene	Pseudomonas R	*n*-butanol	Kelsey et al. (1997) Liste and Alexander (2002)
	Pseudomonas PpG7	*n*-butanol	Liste and Alexander (2002)
	Earthworm *E. foetida*	*n*-butanol	Kelsey et al. (1997)
Pyrene	Pseudomonas R	*n*-butanol	Liste and Alexander (2002)
	Pseudomonas PpG7	*n*-butanol	Liste and Alexander (2002)
Chrysene	Earthworm *E. foetida*	*n*-butanol	Liste and Alexander (2002)
Anthracene, chrysene, pyrene, benzo(a)pyrene	Earthworm *E. foetida*	Tetrahydrofuran	Tang et al. (2002)
DDT+metabolites	Wheat roots	*n*-hexane	Tao et al. (2004)
PAHs and PCBs 110,118,138,153 (not PCBs 8,20,28,52)	*Earthworm Lumbricus terrestris* L.	Methanol:water (1:1)	Krauss et al. (2000a)
USEPA PAHs (not acenaphthylene)	Degradation	Ethanol, acetone–water, toluene–water	Thiele-Bruhn and Brummer (2004)

10.3.4.2 Solid-phase extraction

Pore water analysis can provide a reasonable approximation of bioavailability for certain compounds particularly for microbes (Ronday, 1997). However, pore water analysis is limited by compound solubility in water and mass transfer kinetics of the sorbed compound from the soil. Solid-phase extraction techniques, therefore, use water as the solvent together with a solid-phase extraction matrix that strongly sorbs the compound from the aqueous phase to maintain a mass transfer gradient. The pore water concentration remains low because the compound is adsorbed by the solid phase and therefore all the available compounds will diffuse into the aqueous phase and be adsorbed by the solid phase in a fixed period of time. These techniques offer advantages in terms of less disruption of organic matter, and similar aqueous phase contaminant concentrations to those observed in the field compared to solvent extraction techniques. However, if organism uptake routes are not exclusively from pore water, bioavailability may be underestimated.

Solid-phase extractants used for PAHs, PCBs, chlorobenzenes and DDT and its metabolites include Tenax TA beads (Cornelissen et al., 1998b; Morrison et al., 2000), XAD resin (Lei et al., 2004; Hawthorne et al., 2002), polydimethylsiloxane solid-phase microextraction fibres (Van der Wal et al., 2004a, 2004b) and C18 membranes (Krauss and Wilcke, 2001; Tang et al., 2002; Tang et al., 1999). These have shown good correlations with biodegradation and accumulation in earthworms.

A recent development is the use of hydroxypropyl-β-cyclodextrins, readily soluble molecules with a polar exterior and hydrophobic interior cavity for hydrophobic contaminant adsorption. These have been used successfully as microbe mimics to assess the microbial bioavailability of mainly PAHs to date (Reid et al., 2000b; Cuypers et al., 2002).

10.3.5 Other technologies

Worth mentioning here are a number of other techniques which are finding use for specific applications. Thermal desorption has been used to elucidate the desorption kinetics of sorbed contaminants, with the rapidly desorbing fraction of compound being assumed as the biologically available fraction (Cornelissen et al., 1997). Supercritical fluid extraction has been used to extract PAHs from manufactured gas plant soils (Hawthorne et al., 2002) and individual PCB congeners from sediments (Nilsson and Bjorklund, 2005). Good correlations with pollutant specific bioavailability in field soils have been observed. Selective oxidation of organic matter has also been applied to remove PAHs from expanded organic matter rather than condensed organic matter assuming those associated with the expanded organic matter are bioavailable (Cuypers et al., 2000).

To summarize, a wide range of different techniques have been used to assess organic compound bioavailability in soils with limited standardization between tests. Many have been applied only in the laboratory to spiked systems (albeit aged) using often a limited range of compound species. These do not necessarily mimic bioavailability in long-term field contaminated soil for a multi-component mixture of contaminants where synergistic or antagonistic effects of complex, heterogeneous mixtures may occur (Tang et al., 2002; Hawthorne et al., 2002). Moreover, most of the biological and chemical test validations have used microbes or earthworms as the target species. Although no one single test can encompass all ecological species, the choice of target organism must have particular relevance for the ecosystem for which risk assessment is being undertaken. It is not appropriate to extrapolate data from one study to another (McConnell et al., 1984).

10.4 Factors influencing compound bioavailability

If the contaminant in soil is not lost by processes such as leaching and volatilization, then the fraction that is bioavailable is determined by the rate of mass transfer of the compound to the target organism and the intrinsic ability of the organism to metabolize/uptake the compound (Bosma et al., 1997). This will depend on the organism itself but is also governed by the chemical, its concentration in soil (Wang and Jones, 1994) and a number of other factors that have varying degrees of importance depending on the contaminant and soil environment. These are discussed in the following section with an emphasis on the effects of compound species.

10.4.1 Slow desorption, aging and bound residues

The major soil process determining compound bioavailability is sorption to the soil solid phase (organic and mineral, see Chapter 2). Hydrophobic organic contaminants in soil exhibit a biphasic desorption with an initial phase of rapid desorption followed by a phase of much slower release. An understanding of the processes contributing to the slowly desorbing fraction is critical to assessing the proportion of bioavailable fraction of contaminant in soil. Two models have been proposed to describe the mechanisms of slow sorption/desorption in soil (Pignatello and Xing, 1996; Xing and Pignatello, 1997; Xing et al., 1996).

(i)   Diffusion in organic matter, conceptualized to consist of two regions, the rubbery (expanded) region where diffusion is linear, faster and reversible, and the glassy (condensed) region which contains holes where slow, non-linear, reversible sorption and entrapment may also occur and accounts for desorption retardations.

(ii)   Sorption-retarded pore diffusion where molecular diffusion in pore water is retarded by sorption on pore walls which may or may not be organic matter, influenced by the particle radii, the pore tortuosity and on the constrictivity of pores.

Both mechanisms probably occur simultaneously with relative importance depending on the soil environment. When present above trace levels (0.1–0.5%), SOM is the predominant sorbent for slow desorptive behaviour of hydrophobic organic compounds (Cornelissen et al., 1998a).

It is well established that over time a greater proportion of compound in soil becomes less extractable and less available for uptake or degradation than freshly added compound reducing exposure and thus toxicity and risk (Alexander, 2000; Morrison et al., 2000). This process, known as aging, results in more compounds becoming strongly associated with the soil over time. It has been attributed to a number of different mechanisms. These include formation of strong bonds with soil components (Gevao et al., 2000) and increasing "sequestration" of xenobiotics into the three-dimensional soil solid phase over time as described above, by slow diffusion into micropores of the organic and to a lesser extent mineral phases and partitioning into organic matter (Hatzinger and Alexander, 1995; Tang et al., 1998; Luthy et al., 1997; Cornelissen et al., 1998a; Pignatello and Xing, 1996). The sequestered molecules become less accessible to organisms and microorganisms and therefore less bioavailable. This has been demonstrated for plants, microorganisms and earthworms (Kelsey et al., 1997; White et al., 1997; Macleod and Semple, 2000; Johnson et al., 2002). The bioavailability decreases with time but reaches a value below which a further decline is no longer detectable and at this point degradation is limited by the rate of contaminant transfer (Fig. 10.9). The period to reach this value and the final amount available varies among soils and compounds.

Bound residues in soil represent that part of compound that persists in soil after extraction and may be formed by a number of different soil–compound interactions including sequestration, covalent bond formation and chemisorption (especially for ionizable or ionic compounds), depending on the compound (Gevao et al., 2000; Barraclough et al., 2005). In general, the formation of bound residues reduces the compound bioavailability and bioaccessibility significantly.

The decrease in compound extractability and bioavailability associated with aging, sequestration and bound residues has important implications for risk assessment in long-term contaminated soils. Therefore, an understanding of governing factors is critical but is not yet fully explored. Compound properties such as hydrophobicity (Cornelissen et al., 2000), molecular size, shape, constituent atoms and possibly even chirality, may all have an important effect. Northcott and Jones (2001) showed that the amount of PAH sequestered in

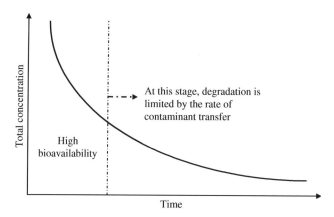

Figure 10.9. *Effect of time on contaminant bioavailability.*

organic matter was related to molecular properties, with relative amount sequestered increasing with compound K_{ow}. Sequestration was particularly important for PAHs composed of four or more aromatic rings. Kottler and Alexander (2001), however, observed that K_{ow}, molecular length and molecular-connectivity index for 21 PAHs were not well correlated with compound sequestration in soil. Formation of non-extractable PAH residues (fluoranthene and benzo(a)pyrene) in surface soil was observed to be rapid compared to PCBs (congeners 28 and 52) in a outdoor lysimeter study (Doick et al., 2005). This was explained by planar PAHs having stronger sorption potentials than the non-planar PCBs. We know also that PAHs sorb more strongly to soils than PCBs of similar hydrophobicity (Krauss et al., 2000a, 2000b) and thus may be more strongly sequestered. A study of the bound fraction of six major use pesticides (atrazine, dicamba, isoproturon, lindane, paraquat and trifluralin) with different properties showed significant differences between bound fraction and potentially available fraction in laboratory microcosms over time (Mordaunt et al., 2005). Bhandari et al. (1997) showed that the extent of irreversible binding of a range of chlorophenol compounds to a sandy surface soil was largely dependent on the initial aqueous concentration although differences in sorptive behaviour were exhibited by phenols of different chlorine substitution. Compound concentration effects on the sequestration of pyrene and phenanthrene have been observed (Chung and Alexander, 1999). However, competitive effects of compound mixtures have not yet been evaluated.

The effect of soil type and environmental conditions on bioavailability, soil–compound interactions and aging has been well documented and reviewed (Chung and Alexander, 1998; Hatzinger and Alexander, 1995; Reid et al., 2000a). Soil texture and composition determine the physical and chemical properties such as pH, aeration, organic matter content, soil inorganic constituents and water content, which are all important in determining soil–compound interactions and the extent of compound sequestration. Chung and Alexander (2002) showed that phenanthrene and atrazine sequestration were highly correlated with organic C, nanoporosity and CEC. pH can change the structure of humic acid and higher concentrations of DDT and its metabolites have been observed in lower pH soils (Gong et al., 2004a). Increased solvent non-extractable residues of pyrene have been observed in biologically active soils compared to sterile soils although the mechanisms for this increased sequestration are not well understood but can possibly be attributed to changes in soil chemistry or structure (Guthrie and Pfaender, 1998; Macleod and Semple, 2003).

Although the soil inorganic constituents are important (particularly in terms of pore size and structure for sorption-retarded pore diffusion), in many soils the amount and nature of organic matter is the most important parameter affecting compound sequestration (Hatzinger and Alexander, 1995; Cornelissen et al., 1998a; Nam et al., 1998; Bogan and Sullivan, 2003). In soils with greater than 0.1% organic carbon partitioning into the organic matter is the predominant sorption process as discussed above (Pignatello and Xing, 1996). Therefore, soils with higher organic carbon contents have a higher capacity to sorb organic contaminants (Macleod and Semple, 2002, 2003). The composition of the organic matter is also important. Macleod and Semple (2003) found that the majority of non-extractable ^{14}C-pyrene was associated with the humin fraction of organic matter rather than the humic or fulvic fractions. Both the water soluble (WSOC) and non-water soluble (Tao and Lin, 2000; Tao et al., 2004) organic matter pools have also been shown to be important to the bioavailability of a range of pollutants (Shaw et al., 2000). Furthermore, organic carbon in soils, particularly contaminated soils, may comprise a range of different materials such as coal, coke and soot. Sorption of PAHs and PCBs onto these particles has been shown to be greater than onto natural organic carbon (Ghosh et al., 2003).

10.4.2 Target organisms

Contaminant uptake mechanisms and rates vary between and within target species and will depend on development stage, health (Wang and Jones, 1994), season and history of contaminant exposure and the mechanism by which the organism acquires and utilizes the compound from the soil. Key factors for development of a soil microbial populations ability to degrade organic contaminants have been discussed by Semple et al. (2003). However, Guerin and Boyd (1992) observed that mineralization of soil-sorbed naphthalene by two bacterial species was different (18% compared to 32%). Numerous other studies have reported differences in uptake between different species of earthworm (Johnson et al., 2002; Yu et al., 2005; Kelsey et al., 2005). Differences in uptake between earthworms and bacteria for soil associated phenanthrene (Kelsey et al., 1997; White et al., 1997; Jager et al., 2005) and biphenyl (Feng et al., 2000) may be explained by the fact that smaller microorganisms can enter micropores. However, microorganisms can only access contaminants in the aqueous phase, whereas earthworms can access them both in the solution and solid phase by ingestion and desorption as the soil passes through the gut. Additionally, some species may have mechanisms to facilitate desorption (Tang et al., 1998a) or to overcome the binding for aged compound.

It is therefore critical for any risk assessment to define the ecosystem of interest so that the relevant target species can be identified. Extrapolation from one species to another is not appropriate and will introduce error in the risk estimation.

10.5 Bioavailability and uptake in organisms

The fraction of compound that is potentially bioavailable in soil may not actually represent that fraction taken up or metabolized because of the importance of other loss processes in soil or intrinsic inability of the organism to take up/metabolize the compound depending on the compound properties. This is an important consideration for risk assessment and as a result compound concentrations in different plants and organisms are often used as an actual measure of the bioavailable component. In this section, therefore, we review differences in the reported uptake of compound species by plants and organisms focusing on compound specific differences.

10.5.1 Plants

The amount of compound taken up by plants is often used as a measure of the phytoavailable fraction, particularly in site specific risk assessment in the contaminated land arena (Nathaniel et al., 2004). However, plant uptake is not always necessarily directly related to that part of the compound available to soil organisms and animals. That withstanding plants are an important exposure route for animals and humans to soil contaminants and uptake of by plants must be considered in any risk assessment.

The four main pathways by which chemicals in soil can enter plants are well established (Topp et al., 1986).

1. Root uptake followed by translocation;
2. Uptake of vapour from the air surrounding the plant by the above ground plant parts;
3. Uptake by external contamination of shoots by soil and dust, followed by retention in the cuticle or cuticle penetration; and
4. Uptake and transport in plant oil cells.

Extent of compound uptake, and the significance of each pathway, is determined by the chemical and its physicochemical properties and also the plant species itself (Topp et al., 1986; Ryan et al., 1988; Scheunert et al., 1994; Lunney et al., 2004). For example, bioconcentration factors for HCH in wheat plants were significantly higher than those in tobacco plants (Sotiriou et al., 1994), and higher in carrots than maize, oat and barley (Schroll and Scheunert, 1992). There is also evidence for preferential uptake of the α-HCH isomer by wheat and maize/pigeon pea from soil where the β- and γ-isomers persisted for longer periods (Singh et al., 1991).

Nash and Beall (1970) observed many years ago differences in the uptake of certain persistent compounds. They reported that the main route of DDT to soybean plants was from foliage absorption of vapour but that dieldrin, endrin and heptachlor were taken up primarily by roots and translocation. Chlordane can also be translocated from soil to vegetation (Mattina et al., 2000). More recently, the uptake of eight xenobiotic chemicals with very different physico-chemical properties in a closed laboratory system showed significant compound variation. There was both foliar and root uptake of hexachlorobenzene and OCDD but no compound translocation, exclusively root uptake and translocation of terbuthylazine, but predominantly foliar uptake of volatile species such as chlorobenzene (Schroll et al., 1994). Translocation is governed by properties such as log K_{ow} and water solubility and is generally seen for compounds with low log $K_{ow} < 4.5$ (Burken and Schnoor, 1997; Briggs et al., 1982; Ryan et al., 1988). However, if the half-life of the compound is short (less than 10 days) then the compound is likely to be lost from the system before being taken up by plants. Compounds with high vapour pressures ($H_c > 10^{-4}$) can be taken up in the vapour phase if not lost to the atmosphere (Ryan et al., 1988). O'Connor et al. (1990) reported that in alkaline soils, the ionization of chlorophenols to the anionic form resulted in increased availability for degradation such that little was taken up by plants.

The predominant pathway for foliage contamination of non-ionic lipophilic organic contaminants for most plants is uptake of volatilized soil contaminant from the air (Travis and Hattemer-Frey, 1988; Fries and Marrow, 1981; Wilson and Meharg, 2003). For soil-sorbed compounds this is limited by strong sorption to soil constituents and compound volatility. Root uptake has been observed (Wild et al., 2005; Wang and Jones, 1994) but again is limited by strong compound adsorption to soil and inefficient transfer to plant roots. Translocation to above ground plant parts is frequently negligible as demonstrated for OCDD, PAHs and chlorobenzenes in carrots (Wild and Jones, 1992; Schroll and Scheunert, 1993; Wang and Jones, 1994). However, even within the same group of compounds there are differences in extent and route of uptake which can be related to compound properties (Scheunert et al., 1994). Wild and Jones (1992) reported that

70% of the total PAH burden in carrots grown in sludge amended soil was in the peel and lower molecular weight compounds were comparatively enriched and dominated the peel total PAH load. Also in sludge amended soils, Wang and Jones (1994) showed that penetration of chlorobenzenes (CB) through carrot peel to the core was restricted, except for dichlorobenzenes. The resistance increased with the CB molecular weight. Among the CBs, tetrachlorobenzenes were taken up most effectively by carrot foliage and root peel, while trichlorobenzenes were taken up most effectively by carrot cores. Similar results have been observed for CB uptake in grass planted in spiked soil. Wilson and Meharg (2003) reported that 1,2,4-trichlorobenzene spiked to soil was absorbed to the greatest extent by roots and tillers compared to HCB and 1,2-dichlorobenzene.

A soil-to-plant uptake route for lipophilic compounds, such as PCDD/Fs and DDT compounds, has been demonstrated for some members of the Cucurbitaceae family as well as certain other food crops (Hulster et al., 1994; Mattina et al., 2000; Lunney et al., 2004). This uptake can vary within the family (White, 2002; White et al., 2003) but shows promise for use in phytoremediation (Lunney et al., 2004). Depending on the binding mechanisms for molecular transport within the plants, certain enantiomers of chiral compounds may be preferentially translocated. Comparison of the soil-to-plant uptake and air-to-plant uptake routes for cis-chlordane (CC), trans-chlordane (TC) (chiral molecules present as racemates in the technical formulation) and trans-nonachlor (major components of the chlordane technical mixture) in zucchini plants showed little enantioselectivity in movement from air to leaf (Lee et al., 2003; White et al., 2002). However, the mechanisms responsible for transport of the chlordane residues from the soil across the root boundary and through the stem to the leaves appeared to be enantioselective with preferential increase in different enantiomers of cis- and trans-chlordane in roots and stems. This, however, was not observed for uptake of DDT enantiomers by pumpkin and zucchini (Lunney et al., 2004).

10.5.2 Microorganisms

Understanding the availability of organic compounds to microorganisms is important because this determines the potential for biodegradation which in turn determines persistence in the environment but also potential for transformations and availability to plants and higher organisms. Much research work has aimed to investigate the availability of organic chemicals for microbial degradation, and methods of enhancing bioavailability, for example by the use of surfactants (Pignatello and Xing, 1996; Zheng and Obbard, 2002). As discussed above the availability of organic chemicals to microorganisms varies significantly between different soils, over time and between different species of microorganism (Guerin and Boyd, 1992; Feng et al., 2000) with some species seeming to overcome the limitations to bioavailability assumed due to compound sorption (Hatzinger and Alexander, 1995). It is generally assumed that microorganisms can only access contaminant in the aqueous phase with bioavailability limited by desorption of compound from the solid particle and diffusion to the organism.

Certain compounds are easier to degrade than others. Wilson and Jones (1993) reviewed the bioremediation of PAHs and found that although a number of different species could mineralize PAHs with four rings or less, co-metabolism was the main mechanism by which those with greater than four rings were degraded and fewer bacterial species had the capability to use four-ring PAHs as their sole carbon and energy source (Walter et al., 1991). However, the white-rot fungus (*Phanerochaete chrysosporium*) has proved successful for degradation of certain recalcitrant compounds including the PAHs (Potin et al., 2004). Zeddel et al. (1993) reported isomer specific bioavailability of PCBs to two species of white-rot fungus with the higher chlorinated species being less available.

Differential rates of degradation of the HCH isomers (α, β, γ, δ) have been reported but the order of isomer degradation rates differs in different studies, possibly illustrating the different bacterial strains used. In some, the β-isomer was the most recalcitrant but in others a different isomer (Siddique et al., 2002; Gupta et al., 2000).

Enantioselective degradation of chiral pesticides in soil by soil microorganisms results in the persistence of one of the enantiomers with a corresponding change in the composition of the component enantiomers from the technical mixture (often the racemic mixture) (Mattina et al., 1999). However, the selectivity may change depending on the soil conditions and land use and has been attributed to the presence of specific degrading organisms (Lewis et al., 1999). The understanding of the populations responsible for the enantioselective degradation and factors affecting their activity is limited.

10.5.3 Soil earthworms

Earthworms are ideally suited for assessing the risk to ecosystems of soil contaminants in view of their intimate contact with soil, assimilating compounds both by ingestion and skin contact (Lanno et al., 2004). They also form an important component for transfer of soil contaminants to higher trophic levels within ecosystems (Harris et al., 2000).

Earthworms are known to take up many hydrophobic compounds (Kelsey et al., 2005; Jager et al., 2003a, 2003b). That component of soil contaminant that is bioavailable to earthworms depends on many factors including the soil, the earthworm species, the compound itself and also compound residence time in soil as discussed above. A decrease in compound bioavailability to earthworms as a result of compound aging has been reported for a number of different compounds including PAHs (Conrad et al., 2002) and pesticides (Morrison et al.,

2000; Gevao et al., 2001). However, non-extracatable aged soil-sorbed compounds can become more available in the presence of earthworms possibly because passage through the gut results in release of soil-sorbed compounds after interaction with digestive enzymes. Also the presence of worms has been shown to retard the bound residue formation and therefore facilitate an increase in the bioavailable fraction of compound in soil (Gevao et al., 2001).

Compounds are taken up by earthworms via (i) passive diffusion from soil solution through the outer membrane or (ii) resorption of the compounds from soil material gassing through the gut. Factors affecting the relative contribution of both pathways are not well understood but include earthworm species as well as compound residence time in soil. Dietary uptake was most important for aged pyrene in soil (Conrad et al., 2002) compared to absorption through the skin possibly because conditions in the gut facilitate the desorption of the strongly bound compounds from soil. The importance of uptake by each route also differs for different compounds. Exposure of *Eisenia andrei* to soil artificially spiked with tetrachlorobenzene, hexachlorobenzene and PCB 153 showed that the contribution of the gut route increased with hydrophobicity and dominated for PCB 153 (Jager et al., 2003b). It seems that uptake through the gut is dominant for very hydrophobic compounds such as PCDD/Fs with log $K_{ow} > 6$. For less hydrophobic compounds absorption through the skin is probably the predominant uptake route.

A study of the uptake of 20 PAHs and 12 PCBs from 25 field contaminated soils by *Lumbricus terrestris L.* showed that the average soil-to-biota accumulation factors for di- and tetra-chlorinated biphenyls were about 100 times higher and for penta- to hepta-chlorinated biphenyls about 10 times higher than those of PAHs with comparable K_{ow} values (Krauss et al., 2000a). Explanations included stronger sorption of PAHs to soil compared to PCBs and/or a slower desorption rate of the PAHs so that soil solution was not replenished for earthworm uptake. These explanations also hold for decreasing uptake with increasing degree of chlorination of PCBs. However, more limited membrane permeability of the more highly chlorinated compounds was also suggested based on experimental observations. Larsen et al. (1992) also found that the bioconcentration of PCBs in *Lumbriscus rubellus* decreased with increasing degree of chlorination, and with decreasing number of chlorine atoms substituted at the ortho position.

Use of compound K_{ow} in determining bioavailability to earthworms is not always successful. When PCBs and chlorobenzenes were freshly added to soil, K_{ow} was important for determining uptake (Belfroid et al., 1995). However, for aged field contaminated soil the bioavailability was independent of compound K_{ow}. This has also been observed for PAHs (Krauss et al., 2000a) and phthalates (Hu et al., 2005).

There is limited data available on the uptake of different compound isomers and enantiomers by earthworms other than those discussed above and only a few studies on a limited number of pesticides (Morrison et al., 2000; Gevao et al., 2001).

10.5.4 Humans and grazing/foraging animals

As well as direct soil ingestion by humans, soil ingestion by grazing animals is a potentially important route of human exposure to soil-bourne organic contaminants through consumption of meat and dairy products, particularly in areas with high soil contaminant concentrations (Beck et al., 1996). Obviously the amount of compound bioavailable to the animal or human will depend on adsorption by the gastrointestinal tract as well as the component in the soil that is available for assimilation. There is evidence that compound specific assimilation of soil-sorbed contaminants does occur.

A number of different animal species have been used to study adsorption from the gastrointestinal tract but notably for only a few groups of xenobiotics (McLachlan et al., 1990; Stephens et al., 1995). Again studies with animals that are not specifically the defined target organisms must be treated with care. Isomer dependent gastrointestinal uptake of PCDD/Fs by cows was shown to be inversely related to PCDD/F chlorination, and lactating cows adsorb the 2,3,7,8-substituted PCDD/Fs preferentially, compared to other congeners in the some homologue group (McLachlan et al., 1990). For PCBs, adsorption was shown to be constant up to a compound log K_{ow} of between 5.5 and 6.5, after which absorption through the digestive tract declines with increasing K_{ow} (McLachlan, 1993; Thomas et al., 1999).

Chickens have also been used as models for grazing animals but have shown some differences compared to absorption by cows (Stephens et al., 1995). PCDD/F bioaccumulation in chickens, as for cows, is tissue dependent and congener-specific with the lower chlorinated congeners showing the greatest degree of absorption (Schuler et al., 1997). This suggests that the passage of hydrophobic contaminants across the intestine is limited by molecular size and water solubility. However, absorption of congeners was not dependent on Cl substitution. There was similar absorption of congeners with the same degree of substitution, and no preferential adsorption of 2,3,7,8-substituted compounds as observed for cows. In chickens no real differences were observed for absorption of coplanar PCBs with different degrees of substitution although only a limited range of PCBs was studied (Pirard and De Pauw, 2005). Uptake by chickens from aged and freshly spiked soils was not significantly different (Stephens et al., 1995). This contrasts with many other studies that show a significant effect of time on contaminant bioavailability (Alexander, 2000; Morrison et al., 2000).

Information on differences in uptake of soil-sorbed enantiomers to animals is limited (Harris et al., 2000). Selective enrichment of the (+) or (−) α-HCH enantiomer has been observed in different animal species and even within tissues of the same species (Willett et al., 1998). The reasons for enantiomer enrichment differences in different environmental media may include different enantiomer degradation rates, microbial populations selective to one isomer or enantiomeric accumulation that creates enantiomer excesses in certain tissues.

10.6 Implications for risk based corrective action

Chemical speciation and bioavailability are increasingly recognized as two contaminant attributes that may underpin risk characterization and remediation of contaminated sites. While not readily acceptable to many regulators, there is now significant shift amongst regulators worldwide for possible inclusion of speciation-bioavailability and toxicity relations in regulatory policies. However, the role of chemical speciation during assessment of bioavailability especially for organic contaminants is yet to be fully appreciated. Indeed, regulatory bodies in Australia recognize the different toxicities of such inorganic substances as arsenic, chromium and vanadium although no emphasis is currently being placed on the different species of organic contaminants. Recent findings of enantio-selective biochemical processes, especially in the case of pesticides, further highlight the need for speciation considerations for bioavailability assessment.

References

Adelaide, 2004. Proceedings from the 3rd International Workshop on Chemical Bioavailability in the Terrestrial Environment. Adelaide, South Australia, 12–15 Sept.

Ahlborg, V.G., Thunberg, T.M., 1980. Chlorinated phenols: occurrence, toxicity, metabolism and environmental impact. Crit. Rev. Toxicol. 7, 1–35.

Alexander, M., 2000. Ageing, bioavailability and overestimation of risks from environmental pollution. Environ. Sci. Technol. 34, 4259–4265.

Amiri, F., Rahman, M.M., Bornick, H., Worch, E., 2004. Sorption behaviour of phenols on natural sandy aquifer material during flow-through column experiments: the effect of pH. Acta Hydroch. Hydrob. 32, 214–224.

Andreu, V., Pico, Y., 2004. Determination of pesticides and their degradation products in soil: critical review and comparison of methods. Trends Anal. Chem. 23, 772–789.

Ballschmiter, K., Zell, M., 1980. Analysis of PCBs by glass capillary gas chromatography. Fresen. Z. Anal. Chem. 302, 20–31.

Barraclough, D., Kearney, T., Croxford, A., 2005. Bound residues: environmental solution or future problem? Environ. Pollut. 133, 85–90.

Beck, A.J., Johnson, D.L., Jones, K.C., 1996. The form and bioavailability of non-ionic organic chemicals in sewage sludge-amended agricultural soils. Sci. Total Environ. 185, 125–149.

Belfroid, A.C., van den Berg, M., Seinen, W., Hermens, J., van Gestel, K., 1995. Uptake, bioavailability and elimination of hydrophobic compounds in earthworms (*Eisenia andrei*) in field-contaminated soil. Environ. Toxicol. Chem. 14, 605–612.

Bellin, C.A., O'Connor, G.A., 1990. Plant uptake of pentachlorophenol from sludge amended soils. J. Environ. Qual. 19, 598–602.

Bellin, C.A., O'Connor, G.A., Jin, Y., 1990. Sorption and degradation of pentachlorophenol in sludge amended soils. J. Environ. Qual. 19, 603–608.

Bhandari, A., Novak, J.T., Burgos, W.D., Berry, D.F., 1997. Irreversible binding of chlorophenols to soil and its impact on bioavailability. J. Environ. Eng. 123, 506–513.

Bogan, B.W., Sullivan, W.R., 2003. Physicochemical soil parameters affecting sequestration and mycobacterial biodegradation of polycyclic aromatic hydrocarbons in soil. Chemosphere 52, 1717–1726.

Bosma, T.N.P., Middeldorp, P.J.M., Schraa, G., Zehnder, A.J.B., 1997. Mass transfer limitation of biotransformation: quantifying bioavailability. Environ. Sci. Technol. 31, 248–252.

Boyd, E.M., Meharg, A.A., Wright, J., Killham, K., 1997. Assessment of toxicological interactions of benzene and its primary degradation products (catechol and phenol) using lux-modified bacterial bioassay. Environ. Toxicol. Chem. 16, 849–856.

Briggs, G.G., Bromilow, R.H., Evans, A.A., Williams, M., 1982. Relationship between lipophilicity and root uptake and translocation of non-ionised chemicals by barley. Pestic. Sci. 13, 495–504.

Bromilow, R.H., 2003. Paraquat and sustainable agriculture. Pestic. Manag. Sci. 60, 340–349.

Burken, J.G., Schooner, J.L., 1997. Uptake and metabolism of atrazine by poplar trees. Environ. Sci. Technol. 31, 1399–1406.

Caruso, J.A., Klaue, B., Michalke, B., Rocke, D.M., 2003. Group assessment: elemental speciation. Ecotoxicol. Environ. Safety 56, 32–44.

Chen, L., Ran, Y., Xing, B., Mai, B., He, J., Wei, X., Fu, J., Sheng, G., 2005. Contents and sources of polycyclic aromatic hydrocarbons and organochlorine pesticides in vegetable soils of Guangzhou, China. Chemosphere 60, 879–890.

Chessells, M.J., Hawker, D.W., Connell, D.W., Papajcsik, I.A., 1988. Factors influencing the distribution of lindane and isomers in soil of an agricultural environment. Chemosphere 17, 1741–1749.

Chung, N., Alexander, M., 1998. Differences in sequestration and bioavailability of organic compounds aged in dissimilar soils. Environ. Sci. Technol. 32, 855–860.

Chung, N., Alexander, M., 1999. Effect of concentration on sequestration and bioavailability of two polycyclic aromatic hydrocarbons. Environ. Sci. Technol. 33, 3605–3608.

Chung, N., Alexander, M., 2002. Effect of soil properties on bioavailability and extractability of phenanthrene and atrazine sequestered in soil. Chemosphere 48, 109–115.

Conrad, A.U., Comber, S.D., Simkiss, K., 2002. Pyrene bioavailability: effect of sediment-chemical contact time on routes of uptake in an oligochaete worm. Chemosphere 49, 447–454.

Cornelissen, G., Van Noort, P.C.M., Parsons, J.R., Govers, H.A.J., 1997. Temperature dependence of slow adsorption and desorption kinetics of organic compounds in sediments. Environ. Sci. Technol. 31, 454–460.

Cornelissen, G., Van Noort, P.C.M., Govers, H.A.J., 1998a. Mechanism of slow desorption of organic compounds from sediments: a study using model sorbents. Environ. Sci. Technol. 32, 3124–3131.

Cornelissen, G., Rigterink, H., Ferdinandy, M.M.A., Van Noort, P.C.M., 1998b. Rapidly desorbing fractions of PAHs in contaminated sediments as a predictor of the extent of bioremediation. Environ. Sci. Technol. 32, 966–970.

Cornelissen, G., Hassell, K.A., van Noort, P.C.M., Kraaij, R., van Ekeren, P.J., Dijkema, C., de Jager, P.A., Govers, H.A.J., 2000. Slow desorption of PCBs and chlorobenzenes from soils and sediments: relations with sorbent and sorbate characteristics. Environ. Pollut. 108, 69–80.

Crosby, D.G., 1981. Environmental chemistry of pentachlorophenol. Pure Appl. Chem. 53, 1051–1080.

Cuypers, C., Grotenhuis, T., Joziasse, J., Rulkens, W., 2000. Rapid persulfate oxidation predicts PAH bioavailability in soils and sediments. Environ. Sci. Technol. 34, 2057–2063.

Cuypers, C., Pancras, T., Grotenhuis, T., Rulkens, W., 2002. The estimation of PAH bioavailability in contaminated sediments using hydroxypropyl-β-cyclodextrin and Triton X-100 extraction techniques. Chemosphere 46, 1235–1245.

Doick, K.J., Klingelmann, E., Burauel, P., Jones, K.C., Semple, K.T., 2005. Long-term fate of polychlorinated biphenyls and polycyclic aromatic hydrocarbons in an agricultural soil. Environ. Sci. Technol. 39, 3663–3670.

Dyke, P.H., Stratford, J., 2002. Changes to the TEF schemes can have significant impacts on regulation and management of PCDD/F and PCB. Chemosphere 47, 103–116.

Ehlers, L., Luthy, R., 2003. Contaminant bioavailability in soil and sediment. Environ. Sci. Technol. 37, 295A–302A.

Feng, Y., Park, J.-H., Voice, T.C., Boyd, S.A., 2000. Bioavailability of soil-sorbed biphenyl to bacteria. Environ. Sci. Technol. 34, 1977–1984.

Folke, J., Lindgaard-Jorgensen, P., 1985. Organics in wheat and rye straw pulp bleaching and combined effluents. (1). Chemical characterization and biodegradation studies. Toxicol. Environ. Chem. 10, 1–24.

Fries, G.F., Marrow, G.S., 1981. Chlorophenyl movement from soil to soybean plants. J. Agric. Food Chem. 29, 757–759.

Gaus, C., Brunskill, G.J., Webber, R., Papke, O., Muller, J.F., 2001. Historical PCDD inputs and their source implications from dated sediment cores in Queensland (Australia). Environ. Sci. Technol. 35, 4597–4603.

Gevao, B., Semple, K.T., Jones, K.J., 2000. Bound residues in soils: a review. Environ. Pollut. 108, 3–14.

Gevao, B., Mordaunt, C., Semple, K.T., Piearce, T.G., Jones, K.C., 2001. Bioavailability of non-extractable (bound) residues to earthworms. Environ. Sci. Technol. 35, 501–507.

Ghosh, U., Zimmerman, J.R., Luthy, R.G., 2003. PCB and PAH speciation amoung particle types in contaminated harbor sediments and effects on PAH bioavailability. Environ. Sci. Technol. 37, 2209–2217.

Gong, Z.M., Tao, S., Xu, F.L., Dawson, R., Liu, W.X., Cui, Y.H., Cao, J., Wang, X.J., Shen, W.R., Zhang, W.J., Quig, B.P., Sun, R., 2004a. Level and distribution of DDT in surface soils from Tianjin, China. Chemosphere 54, 1247–1253.

Gong, Z.M., Xu, F.L., Dawson, R., Cao, J., Liu, W.X., Li, B.G., Shen, W.R., Zhang, W.J., Qu, N.R., Sun, R., Tao, S., 2004b. Residues of hexachlorocyclohexane isomers and their distribution characteristics in soils in the Tianjin area, China. Arch. Environ. Contam. Toxicol. 46, 432–437.

Green, N.J.L., Hassanin, A., Johnston, A.E., Jones, K.C., 2004. Observations on historical, contemporary, and natural PCDD/F. Environ. Sci. Technol. 38, 715–723.

Guerin, W.F., Boyd, S.A., 1992. Differential bioavailability of soil-sorbed naphthalene to two bacterial species. Appl. Environ.l Microbiol. 58, 1142–1152.

Gupta, A., Kaushik, C.P., Kaushik, A., 2000. Degradation of hexachlorocyclohexane (HCH; $\alpha,\beta,\gamma,\delta$) by bacillus circulans and bacillus brevis isolated from soil contaminated with HCH. Soil Biol. Biochem. 32, 1803–1805.

Guthrie, E.A., Pfaender, F.K., 1998. Reduced pyrene bioavailability in microbially active soils. Environ. Sci. Technol. 32, 501–508.

Harmsen, J., Rulkens, W., Eijsackers, H., 2005. Bioavailabilty: concept for understanding or tool for predicting? Land contamination and reclamation 13, 161–171.

Harner, T., Wideman, J.L., Jantunen, L.M.M., Bidleman, T.F., Parkhurst, W.J., 1999. Residues of organochlorine pesticides in Alabama soils. Environ. Pollut. 106, 323–332.

Harrad, S., 2001. The environmental behaviour of persistent organic pollutants. In: R.M. Harrison (Ed.), Pollution: Causes, Effects and Control. 4th Ed.. Royal Society of Chemistry, Cambridge, UK.

Harris, M.L., Wilson, L.K., Elliott, J.E., Bishop, C.A., Tomlin, A.D., Henning, K.V., 2000. Transfer of DDT and metabolites from fruit orchard soils to American robins (*Turdus migratorius*) twenty years after agricultural use of DDT in Canada. Arch. Environ. Contam. Toxicol. 39, 204–220.

Hassanin, A., Breivik, K., Meijer, S.N., Steinnes, E., Thomas, G.O., Jones, K.C., 2004. PBDEs in European background soils: levels and factors controlling their distribution. Environ. Sci. Technol. 38, 738–745.

Hassanin, A., Lee, R.G.M., Steinnes, E., Jones, K.C., 2005. PCDD/Fs in Norwegian and UK soils: implications for sources and environmental cycling. Environ. Sci. Technol. 39, 4784–4792.

Hatzinger, P.B., Alexander, M., 1995. Effect of aging of chemicals in soil on their biodegradability and extractability. Environ. Sci. Technol. 29, 537–545.

Hawthorne, S.B., Poppendieck, D.G., Grananski, C.B., Loehr, R.C., 2002. Comparing PAH availability from manufactured gas plant soils and sediments with chemical and biological tests. 1. PAH release during water desorption and supercritical carbon dioxide extraction. Environ. Sci. Technol. 36, 4795–4803.

Heberer, T., Dunnbier, U., 1999. DDT metabolite bis(chlorophenyl)acetic acid: the neglected environmental contaminant. Environ. Sci. Technol. 33, 2346–2351.

Heitzer, A., Webb, O.F., Thonnard, J.E., Sayler, G.S., 1992. Specific and quantitative assessment of naphthalene and salicylate bioavailability by using a bioluminescent catabolic reporter bacterium. Appl. Environ. Microbiol. 58, 1839–1846.

Hobbs, S.J., Howe, P.D., Dobson, S., 1993. Environmental hazard assessment: pentachlorophenol. TSD/10 Toxic Substances Division, Directorate for Air, Climate and Toxic Substances. Department of the Environment, UK.

Hoekstra, P.F., Burnison, B.K., Neheli, T., Muir, D.C.G., 2001. Enantiomer specific activity of *o,p'*-DDT with the human estrogen receptor. Toxicol. Lett. 125, 75–81.

Hu, X.-Y., Wen, B., Zhang, S., Shan, X.-Q., 2005. Bioavailability of phthalate congeners to earthworms (*Eisenia fetida*) in artificially contaminated soils. Ecotoxicol. Environ. Safety 62, 26–34.

Hulster, A., Muller, J.F., Marschner, H., 1994. Soil-plant transfer of polychlorinated dibenzo-p-dioxins and dibenzofurans to vegetables of the Cucumber family (cucurbitaceae). Environ. Sci. Technol. 28, 1110–1115.

ISO/11074, 2005. Soil Quality – Vocabulary. International Organization for Standardization.

ISO/FDIS 17402, 2007. Soil Quality – Guidance for the selection and application of methods for the assessment of bioavailability of contaminants in soil and soil materials. International Organization for Standardization.

Jacobs, M.W., Coates, J.A., Delfino, J.J., Britton, G., Davis, W.M., Garcia, K.L., 1993. Comparison of sediment extract Microtox toxicity with semivolatile organic priority pollutant concentrations. Arch. Environ. Contam. Toxicol. 24, 461–468.

Jager, T., Baerselman, R., Dijkman, E., De Groot, A.C., Hogendoorn, E.A., De Jong, A., Kruitbosch, J.A.W., Peijnenburg, W.J.G.M., 2003a. Availability of polycyclic aromatic hydrocarbons to earthworms (*eisenia andrei, oligochaeta*) in field polluted soils and soil-sediment mixtures. Environ. Toxicol. Chem. 22, 767–775.

Jager, T., Fleuren, R.H.L.J., Hogendoorn, E.A., De Korte, G., 2003b. Elucidatiing the routes of exposure for organic chemicals in the earthworm *Eisenia Andrei* (*Oligochaeta*). Environ. Sci. Technol. 37, 3399–3404.

Jager, T., Van der Wal, L., Fleuren, R.H.L.J., Barendregt, A., Hermens, J.L.M., 2005. Bioaccumulation of organic chemicals in contaminated soils: evaluation of bioassays with earthworms. Environ. Sci. Technol. 39, 293–298.

Johnson, D.L., Jones, K.C., Langdon, C.J., Piearce, T.G., Semple, K.T., 2002. Temporal changes in earthworm availability and extractability of polycyclic aromatic hydrocarbons in soil. Soil Biol. Biochem. 34, 1363–1370.

Jones, K.C., Burnett, V., Duarte-Davidson, R., Waterhouse, K.S., 1991. PCBs in the environment. Chemistry in Britain, May, p. 435.

Keith, L.H., Telliard, W.A., 1979. Priority pollutants 1. A perspective view. Environ. Sci. Technol. 13, 416–423.

Kelce, W.R., Stone, C.R., Laws, S.C., Gray, L.E., Kemppainen, J.A., Wilson, E.M., 1995. Persistent DDT metabolite DDE is a potent androgen receptor antagonist. Nature 375, 581–585.

Kelsey, J.W., Alexander, M., 1997. Declining bioavailability and inappropriate estimation of risk of persistent compounds. Environ. Toxicol. Chem. 16, 582–585.

Kelsey, J.W., Kottler, B.D., Alexander, M., 1997. Selective chemical extractants to predict bioavailability of soil-aged organic chemicals. Environ. Sci. Technol. 31, 214–217.

Kelsey, J.W., Colino, A., White, J.C., 2005. Effect of species differences, pollutant concentration, and residence time in soil on the bioaccumulation of 2,2,bis-(p-chlorophenyl)-1,1-dicloroethylene by three earthworm species. Environ. Toxicol. Chem. 24, 703–708.

Kottler, B.D., Alexander, M., 2001. Relationship of properties of polycyclic aromatic hydrocarbons to sequestration in soil. Environ. Pollut. 113, 293–298.

Krauss, M., Wilcke, W., 2001. Biomimetric extraction of PAHs and PCBs from soil with octadecyl-modified silica disks to predict their availability to earthworms. Environ. Sci. Technol. 35, 3931–3935.

Krauss, M., Wilcke, W., Zech, W., 2000a. Availability of polycyclic hydrocarbons (PAHs) and polychlori-nated biphenyls (PCBs) to earthworms in urban soils. Environ. Sci. Technol. 34, 4335–4340.

Krauss, M., Wilcke, W., Zech, W., 2000b. Polycyclic aromatic hydrocarbons and polychlorinated biphenyls in forest soils: depth distribution as indicator of different fate. Environ. Pollut. 110, 79–88.

Lagas, P., 1988. Sorption of chlorophenols in the soil. Chemosphere 17, 205–216.

Lanno, R., Wells, J., Conder, J., Bradman, K., Basta, N., 2004. The bioavailability of chemicals in soil for earthworms. Ecotoxicol. Environ. Safety 57, 39–47.

Larsen, B., Pelusio, H.K., Paya-Perez, A., 1992. Bioavailability of polychlorinated biphenyl congeners in soil to earthworm (L. Rubellus) system. Int. J. Environ. Anal. Chem. 46, 149–162.

Law, S., Bidleman, T.F., Martin, M.J., Ruby, M.V., 2004. Evidence of enantioselective degradation of α-hexachlorohexane in groundwater. Environ. Sci. Technol. 38, 1633–1638.

Layton, A.C., Muccini, M., Ghosh, M.M., Sayler, G.S., 1998. Construction of a bioluminescent reported strain to detect polychlorinated biphenyls. Appl. Environ. Microbiol. 64, 5023–5026.

Lee, W.-Y., Iannucci-Berger, W.A., Eitzer, B.D., White, J.C., Mattina, M.I., 2003. Plant uptake and translocation of air-borne chlordane and comparison with the soil-to-plant route. Chemo-sphere 53, 111–121.

Lei, L., Suidan, M.T., Khodadoust, A.P., Tabak, H.H., 2004. Assessing the bioavailability of PAHs in field-contaminated sediment using XAD-2 assisted desorption. Environ. Sci. Technol. 38, 1786–1793.

Leone, A.D., Amato, S., Falconer, R.L., 2001. Emission of chiral organochlorine pesticides from agricultural soils in the Cornbelt Region of the U.S.. Environ. Sci. Technol. 35, 4592–4596.

Lewis, D.L., Garrison, A.W., Wommack, K.E., Whittemore, A., Steudler, P., Melillo, J., 1999. Influence of environmental changes on degradation of chiral pollutants in soils. Nature 401, 898–901.

Liste, H.-H., Alexander, M., 2002. Butanol extraction to predict bioavailability of PAHs in soil. Chemosphere 46, 1011–1017.

Lunney, A.I., Zeeb, B.A., Reimer, K.J., 2004. Uptake of weathered DDT in vascular plants: potential for phytoremediation. Environ. Sci. Technol. 38, 6147–6154.

Luthy, R.G., Aiken, G.R., Brusseau, M.L., Cunningham, S.D., Gchwend, P.M., Pignatello, J.J., Reinhard, M., Traina, S.J., Webber Jr., W.J., Westall, J.C., 1997. Sequestration of hydrophobic organic contaminants by geosorbents. Environ. Sci. Technol. 31, 3341–3347.

Mackay, D., Shiu, W.Y., Ma, K.C., 1991. Illustrated Handbook of Physical-Chemical Properties and Environmental Fate for Organic Chemicals. Volume 1: Monocyclic Hydrocarbons, Chloro-benzenes and PCBs. Lewis Publishers, Michigan, USA.

Mackay, D., Shiu, W.Y., Ma, K.C., 1992a. Illustrated Handbook of Physical-Chemical Properties and Environmental Fate for Organic Chemicals. Volume I: Monocyclic Hydrocarbons, Chlor-obenzenes and PCBs. Lewis Publishers, Michigan, USA.

Mackay, D., Shiu, W.Y., Ma, K.C., 1992b. Illustrated Handbook of Physical-Chemical Properties and Environmental Fate for Organic Chemicals. Volume II: Polynuclear Aromatic Hydrocarbons, Polychlorinated Dioxins and Dibenzofurans. Lewis Publishers, Michigan, USA.

Mackay, D., Shiu, W.Y., Ma, K.C., 1997. Illustrated Handbook of Physical-Chemical Properties and Environmental Fate for Organic Chemicals. Volume V: Pesticides. Lewis Publishers, Michigan, USA.

Macleod, C.J.A., Semple, K.T., 2002. The adaptation of two similar soils to pyrene catabolism. Environ. Pollut. 119, 357–364.

Macleod, C.J.A., Semple, K.T., 2003. Sequential extraction of low concentrations of pyrene and formation of non-extractable residues in sterile and non-sterile soils. Soil Biol. Biochem. 35, 1443–1450.

Macleod, C.J.A., Semple, K.T., 2000. Influence of contact time on extractability and degradation of pyrene in soils. Environ. Sci. Technol. 34, 4952–4957.

Magrans, J.O., Alonso-Prados, J.L., Garcia-Baudin, J.M., 2002. Importance of considering pesticide stereoisomerism-proposal of a scheme to apply Directive 91/414/EEC framework to pesticide active substances manufactured as isomeric mixtures. Chemosphere 49, 461–469.

Masunaga, S., Takasuga, T., Nakanishi, J., 2001. Dioxin and dioxin-like PCB impurities in some Japanese agrochemical formulations. Chemosphere 44, 873–885.

Mattina, M.I., Iannucci-Berger, W., Dtkas, L., Pardus, J., 1999. Impact of long-term weathering, mobility and landuse on chlordane residues in soil. Environ. Sci. Technol. 33, 2425–2431.

Mattina, M.I., Iannucci-Berger, W., Dykas, L., 2000. Chlordane uptake and its translocation in food crops. J. Agric. Food Chem. 48, 1909–1915.

McConnell, E.E., Lucier, G.W., Rumbaugh, R.C., Albro, P.W., Harvan, D.J., Hass, J.R., Harris, M.W., 1984. Dioxin in soil: bioavailability after ingestion by rats and guinea pigs. Science 420, 1077–1079.

McLachlan, M.S., Thoma, H., Reissinger, M., Hutzinger, O., 1990. PCDD/F in an agricultural food chain. Part 1. PCDD/F mass balance for a lactating cow. Chemosphere 20, 1013–1020.

McLachlan, M.S., 1993. Mass balance of polychlorinated biphenyls and other organochlorine compounds in a lactating cow. J. Agric. Food Chem. 41, 474–480.

Megharaj, M., Boul, H.L., Thiele, J.H., 1999. Effect of DDT and its metabolites on soil algae and enzymatic activity. Biol. Fert. Soils 29, 130–134.

Megharaj, M., Kantachote, D., Singleton, I., Naidu, R., 2000. Effects of long-term contamination of DDT on soil microflora with special reference to soil algae and algal transformation of DDT. Environ. Pollut. 109, 35–42.

Meharg, A.A., 1996. Bioavailabiltiy of atrazine to soil microbes in the presence of the earthworm *lumbricus terrestrius* (L.). Soil Biol. Biochem. 28, 555–559.

Meijer, S.N., Halsall, C.J., Harner, T., Peters, A.J., Ockenden, W.A., Jones, K.C., 2001a. Organochlorine pesticide residues in archived UK soil. Environ. Sci. Technol. 35, 1989–1995.

Meijer, S.N., Harner, T., Helm, P.A., Halsall, C.J., Johnston, A.E., Jones, K.C., 2001b. Polychlorinated naphthalenes in U.K. soils: time trends, markers of source and equilibrium status. Environ. Sci. Technol. 35, 4205–4213.

Meijer, S.N., Ockenden, W.A., Sweetman, A., Breivik, K., Grimalt, J.O., Jones, K.C., 2003. Global distribution and budget of PCBs and HCB in background surface soils: implications for sources and environmental processes. Environ. Sci. Technol. 37, 667–672.

Moore, N.W., Walker, C.H., 1964. Organochlorine insecticide residues in wild birds. Nature 207, 1072–1073.

Mordaunt, C.J., Gevao, B., Jones, K.C., Semple, K.T., 2005. Formation of non-extractable pesticide residues: observations on compound differences, measurement and regulatory issues. Environ. Pollut. 133, 25–34.

Morrison, D.E., Robertson, B.K., Alexander, M., 2000. Bioavailability to earthworms of aged DDT, DDE, DDD and dieldrin in soil. Environ. Sci. Technol. 34, 709–713.

Motelay-Massei, A., Ollivon, D., Garban, B., Teil, M.J., Blanchard, M., Chevreuil, M., 2004. Distribution and spatial trends of PAHs and PCBs in soils in the Seine River basin, France. Chemosphere 55, 555–565.

Naidu, R., Rogers, S., Gupta, V.V.S.R., Kookana, R.S., Bolan, N.S., Adriano, D., 2003. Bioavailability of metals in the soil-plant environment and its potential role in risk assessment: An overview. In: R. Naidu, S. Rogers, V.V.S.R. Gupta, R.S. Kookana, N.S. Bolan and D. Adriano (Eds.), Bioavailability and its Potential Role in Risk Assessment. Science Publishers, New York, pp. 21–59.

Nam, K., Chung, N., Alexander, A., 1998. Relationship between organic matter content of soil and the sequestration of phenanthrene. Environ. Sci. Technol. 32, 3785–3788.

Nash, R.G., Beall Jr., M.L., 1970. Chlorinated hydrocarbon insecticides: root uptake versus vapor contamination of soybean foliage. Science 168, 1109–1111.

Nathaniel, P., McCaffrey, C., Ogden, R., Foster, N., Gillett, A., Haynes, D., 2004. Uptake of arsenic by vegetables for human consumption: a study of Wellingborough allotment plots. Land Contam. Reclam. 12, 219–238.

NEPM (1999) National Environment Protection Measure, 1999. Assessment of Site Contamination, National Environment Protection Council, Australia.

Nilsson, T., Bjorklund, E., 2005. Selective supercritical fluid extraction as a tool for determining the PCB fraction accessible for uptake by chironomid larvae in limnic sediment. Chemosphere 60, 141–146.

Nivens, D.E., McKnight, T.E., Moser, S.A., Osbourn, S.J., Siumpson, M.L., Sayler, G.S., 2004. Bioluminescent bioreporter integrated circuits: potentially small rugged and inexpensive whole-cell biosensors for remote environmental monitoring. J. Appl. Microbiol. 96, 33–46.

Northcott, G.L., Jones, K.C., 2001. Partitioning, extraction and formation of nonextractable PAH residues in soil. 1. Compound differences in aging and sequestration. Environ. Sci. Technol. 35, 1103–1110.

NRC (2002) National Research Council. Bioavailability of Contaminants in Soils and Sediments: Processes, Tools and Applications. National Acadamies Press, Washington, DC.

Oberg, L.G., Rappe, C., 1992. Biochemical formation of PCDD/Fs from chlorophenols. Chemosphere 25, 49–52.

O'Connor, G.A., Lujan, J.R., Jin, Y., 1990. Adsorption, degradation and plant availability of 2,4-dinitrophenol in sludge-amended calcareous soils. J. Environ. Qual. 19, 587–593.

OECD, 1993. Organisation for Economic Cooperation and Development. Guidelines for Testing of chemicals. Section 3 – Degradation and Accumulation. Paris (1981), updated OECD Paris 1993.

Paasivirta, J., Tenhola, H., Palm, H., Lammi, R., 1992. Free and bound chlorophenols in kraft pulp bleaching effluents. Chemosphere 24, 1253–1258.

Palmer, G., McFadzean, R., Killam, K., Sinclair, A., Paton, G.I., 1998. Use of lux-based biosensors for rapid diagnosis of pollutants in arable soils. Chemosphere 36, 2683–2697.

Pepich, B.V., Prakash, B., Domino, M.M., Dattilio, T.A., Munch, D.J., Price, E.K., 2005. Development of USEPA Method 527 for the analysis of selected pesticides and flame retardants in the UCMR survey. Environ. Sci. Technol. 39, 4996–5004.

Pignatello, J.J., Xing, B., 1996. Mechanisms of slow sorption of organic chemicals to natural particles. Environ. Sci. Technol. 30, 1–11.

Pirard, C., De Pauw, E., 2005. Uptake of polychlorodibeno-*p*-dioxins and polychlorodibenzofurans and coplanar polychlorobiphenyls in chickens. Environ. Int. 31, 585–591.

Potin, O., Rafin, C., Veignie, E., 2004. Bioremediation of aged polycyclic aromatic hydrocarbons(PAHs)-contaminated soil by filamentous fungi isolated from the soil. Int. Biodeterioration Biodegrad. 54, 45–52.

Reid, B.J., Semple, K.T., Macleod, C.J., Weitz, H., Paton, G.I., 1998. Feasibility of using prokaryote biosensors to assess acute toxicity of polycyclic aromatic hydrocarbons. FEMS Microbiol. Lett. 169, 227–233.

Reid, B.J., Jones, K.C., Semple, K.T., 2000a. Bioavailability of persistent organic pollutants in soils and sediments – a perspective on mechanisms, consequences and assessment. Environ. Pollut. 108, 103–112.

Reid, B.J., Stokes, J.D., Jones, K.C., Semple, K.T., 2000b. Nonexhaustive cyclodextrin-based extraction technique for the exaluation of PAH bioavailability. Environ. Sci. Technol. 34, 3174–3179.

Ren, S., Frymier, P.D., 2005. Toxicity of metals and organic chemicals evaluated with bioluminescence assays. Chemosphere 58, 543–550.

Richterich, K.R., Berger, H., Steber, J., 1998. The two phase closed bottle test – a suitable method for the determination of ready biodegradability of poorly soluble compounds. Chemosphere 37, 319–326.

Roberts, T.R., Dyson, J.S., Lane, M.C.G., 2002. Deactivation of the biological activity of paraquat in the soil environment – a review. J. Agric. Food Chem. 50, 3623–3631.

Robson, S., Harrad, S., 2004. Chiral PCB signatures in air and soil: implications for atmospheric source apportionment. Environ. Sci. Technol. 38, 1662–1666.

Ronday, R., 1997. Centrifugation method for soil pore water assessment of the bioavailability of organic chemicals in soil. Commun. Soil Sci. Plant Anal. 28, 777–785.

Ryan, J.A., Bell, R.M., Davidson, J.M., O'Connor, G.A., 1988. Plant uptake of non-ionic organic chemicals from soils. Chemosphere 17, 2299–2323.

Safe, S., 1990. Polychlorinated biphenyls (PCBs), dibenzofurans (PCDFs) and related compounds: environmental and mechanistic considerations which support the development of toxic equivalency factors (TEFs). Crit. Rev. Toxicol. 21, 51–88.

Sanz-Medel, A., 1998. Trace element analytical speciation in biological systems: importance, challenges and trends. Spectrochim. Acta, Part B, 53197–53211.

Sauve, S., 2003. The role of chemical speciation in bioavailability. In: R. Naidu, V.V.S.R. Gupta, S. Rogers, R.S. Kookana, N.S. Bolan and D.C. Adrino (Eds.), Bioavailability, Toxicity and Risk Relationships in Ecosystems. Science Publishers, USA.

Scheunert, I., Topp, E., Attar, A., Korte, F., 1994. Uptake pathways of chlorobenzenes in plants and their correlation with n-octanol/water partition coefficients. Ecotoxicol. Environ. Safety 27, 90–104.

Schroll, R., Scheunert, I., 1992. A laboratory system to determine separately the uptake of organic chemicals from soil by plant roots and by leaves after vaporization. Chemosphere 24, 97–108.

Schroll, R., Scheunert, I., 1993. Uptake pathways of octachlorodibenzo-p-dioxin from soil by carrots. Chemosphere 26, 1631–1640.

Schroll, R., Bierling, B., Cao, G., Dorfler, U., Lahaniati, M., Langenbach, T., Scheunert, I., Winkler, R., 1994. Uptake pathways of organic chemicals from soil by agricultural plants. Chemosphere 28, 297–303.

Schuler, F., Schmid, P., Schlatter, Ch., 1997. The transfer of polychlorinated dibenzo-p-dioxins and dibenzofurans from soil into eggs of foraging chicken. Chemosphere 34, 711–718.

Schulz, D.E., Petrick, G., Duinker, J.C., 1989. Complete characterization of polychlorinated biphenyl congeners in commercial Aroclor and Clophen mixtures by multidimensional gas chromatography-electron capture detection. Environ. Sci. Technol. 23, 852–859.

Semple, K.T., Morris, A.W.J., Paton, G.I., 2003. Bioavailability of hydrophobic organic contaminants in soils: fundamental concepts and techniques for analysis. Eur. J. Soil Sci. 54, 809–818.

Semple, K.T., Doick, K.J., Jones, K.C., Burauel, P., Craven, A., Harms, H., 2004. Defining bioavailability and bioaccessibility of contaminated soil and sediment is complicated. Environ. Sci. Technol. 229A–331A.

Sensi, N., 1992. Binding mechanisms of pesticides to soil humic substances. Sci. Total Environ. 123/124, 63–76.

Shaw, L.J., Beaton, Y., Glover, L.A., Killham, K., Osborn, D., Meharg, A.A., 2000. Bioavailability of 2,4-dichlorophenol associated with soil water-soluble humic material. Environ. Sci. Technol. 34, 4721–4726.

Siddique, T., Okeke, B.C., Arshad, M., Frankenberger Jr., W.T., 2002. Temperature and pH effects on biodegradation of hexachlorocyclohexane isomers in water and a soil slurry. J. Agric. Food Chem. 50, 5070–5076.

Singh, G., Kathpal, T.S., Spencer, W.F., Dhankar, J.S., 1991. Dissipation of some organochlorine insecticides in cropped and uncropped soil. Environ. Pollut. 70, 219–239.

Sotiriou, N., Moza, P., Scheunert, I., Feicht, E.A., 1994. Uptake and fate of 14C-hexachlorobenzene in tobacco and wheat plants. Fresenius Environ. Bull. 3, 175–180.

Sousa, S., Duffy, C., Weitz, H., Glover, A.L., Bar, E., Henkler, R., Killam, K., 1997. Use of a lux-modified bacterial biosensor to identify constraints to bioremediation of BTEX – contaminated sites. Environ. Toxicol. Chem. 17(6), 1039–1045.

Steinberg, S.M., Poziomek, E.J., Engelmann, W.M., Rogers, K.R., 1995. A review of environmental applications of bioluminescent measurements. Chemosphere 30, 2155–2197.

Stephens, R.D., Petreas, M.X., Hayward, D.G., 1995. Biotransfer and bioaccumulation of dioxins and furans from soil: chickens as a model for foraging animals. Sci. Total Environ. 175, 253–273.

Tang, J., Carroquino, M.J., Robertson, B.K., Alexander, M., 1998. Combined effect of sequestration and bioremediation in reducing the bioavailability of polycyclic aromatic hydrocarbons in soil. Environ. Sci. Technol. 32, 3586–3590.

Tang, W.-C., White, J.C., Alexander, M., 1998a. Utilisation of sorbed compounds by microorganisms specifically isolated for that purpose. Appl. Microbiol. Biotechnol. 49, 117–121.

Tang, J., Robertson, B.K., Alexander, M., 1999. Chemical extraction methods to estimate bioavailability of DDT, DDE and DDD in soil. Environ. Sci. Technol. 33, 4346–4351.

Tang, J., Liste, H.-H., Alexander, M., 2002. Chemical assays of availability to earthworms of polycyclic aromatic hydrocarbons in soil. Chemosphere 48, 35–42.

Tao, S., Lin, B., 2000. Water soluble organic carbon and its measurement in soil and sediment. Water Res. 34, 1751–1755.

Tao, S., Guo, L.Q., Wang, X.J., Liu, W.X., Ju, T.Z., Dawson, R., Cao, J., Xu, F.L., Li, B.G., 2004. Use of sequential ASE extraction to evaluate the bioavailability of DDT and its metabolites to wheat roots in soils with various organic carbon contents. Sci. Total Environ. 320, 1–9.

Templeton, D.M., Ariese, F., Cornelis, R., Danielsson, G., Muntau, H., van Leeuwan, H.P., Lobinski, R., 2000. Guidelines for terms related to chemical speciation and fractionation of elements: definitions, structural aspects and methodical approaches. Pure Appl. Chem. 72, 1453–1470.

Thiele-Bruhn, S., Brummer, G.W., 2004. Fractionated extraction of polycyclic aromatic hydrocarbons (PAHs) from polluted soils: estimation of PAH fraction degradable through bioremediation. Eur. J. Soil Sci. 55, 567–578.

Thomas, G.O., Sweetman, A.J., Jones, K.C., 1999. Input–output balance of polychlorinated biphenyls in a long term study of lactating dairy cows. Environ. Sci. Technol. 33, 104–112.

Topp, E., Scheunert, I., Attar, A., Korte, F., 1986. Factors affecting the uptake of 14C labeled organic chemicals by plants from soil. Ecotoxicol. Environ. Safety 11, 219–228.

Travis, C.C., Hattemer-Frey, H.A., 1988. Uptake of organics by aerial plant parts. Chemosphere 17, 277–283.

USEPA (1972) Public Notices 71-1, 71-2 and 71-5, June 14. Washington, DC, USEPA.

USEPA (1986). Test Methods for Evaluating Solid Wastes, Physical/Chemical Methods. 3rd Ed. USEPA Publication SW-846.

Van Gestel, C.A.M., Ma, W.C., 1988. Bioaccumulation of chlorophenols in earthworms in relation to bioavailability in soil. Ecotoxicol. Environ. Safety 5, 289–297.

Van de Wiele, T.R., Verstraete, W., Siciliano, S.D., 2004. Polycyclic aromatic hydrocarbon release from a soil matrix in the in vitro gastrointestinal tract. J. Environ. Qual. 33, 1343–1353.

Van den Berg, M., Birnbaum, L., Bosveld, A.T.C., Brunström, B., Cook, P., Feeley, M., Giesy, J.P., Hanberg, A., Hasegawa, R., Kennedy, S.W., Kubiak, T., Larsen, J.C., Van Leeuwen, F.X.R., Liem, A.K.D., Nolt, C., Peterson, R.E., Poellinger, L., Safe, S., Schrenk, D., Tillitt, D., Tysklind, M., Younes, M., Waern, F., Zacharewski, T., 1998. Toxic equivalency factors (TEFs) for PCBs, PCDDs and PCDFs for humans and wildlife. Environ. Health Perspect. 106, 775–792.

Van der Wal, L., van Gestel, C.A.M., Hermens, J.L.M., 2004a. Solid phase microextraction as a tool to predict internal concentrations of soil contaminants in terrestrial organisms after exposure to a laboratory standard soil. Chemosphere 54, 561–568.

Van der Wal, L., Jager, T., Fleuren, R.H.L.J., Barendregt, A., Sinnige, T.L., Van Gestel, C.A.M., Hermens, J.L.M., 2004b. Solid phase microextraction to predict bioavailability and accumulation of organic micropollutants in terrestrial organisms after exposure to a field contaminated soil. Environ. Sci. Technol. 38, 4842–4848.

Walker, K., Vallero, D.a., Lewis, R.G., 1999. Factors influencing the distribution of lindane and other hexachlorocyclohexanes in the environment. Environ. Sci. Technol. 33, 4373–4378.

Walter, U., Beyer, M., Klein, J., Rehm, H.-J., 1991. Degradation of pyrene by Rhodococcus sp. UM1. Appl. Microbiol. Biotechnol. 34, 671–676.

Wang, M.-J., Jones, K.C., 1994. Uptake of chlorobenzenes by carrots from spiked and sewage sludge-amended soil. Environ. Sci. Technol. 28, 1260–1267.

Weston, D.P., Mayer, L.M., 1998. Comparison of in vitro digestive fluid extraction and traditional in vivo approaches as measures of polycyclic aromatic hydrocarbon bioavailability from sediments. Environ. Toxicol. Chem. 17, 830–840.

White, J.C., 2002. Differential bioavailability of field-weathered *p,p'*-DDE to plants of the *Cucurbita* and *Cucumis* genera. Chemosphere 49, 143–152.

White, J.C., Kelsey, J.W., Hatzinger, P.B., Alexander, A., 1997. Factors affecting sequestration and bioavailability of phenanthrene in soils. Environ. Toxicol. Chem. 16, 2040–2045.

White, J.C., Mattina, M.I., Eitzer, B.D., Iannucci-Berger, W., 2002. Tracking chlordane compositional and chiral profiles in soil and vegetation. Chemosphere 47, 639–646.

White, J.C., Wang, X., Gent, M.P.N., Isnnucci-Berger, W., Eitzer, B.D., Schultes, N.P., Arienzo, M., Mattine, M.I., 2003. Subspecies-level variation in the phytoextraction of weathered *p,p'*-DDE by *Cucurbita pepo*. Environ. Sci. Technol. 37, 4368–4373.

Wiberg, K., Harner, T., Wideman, J.L., Bidleman, T.F., 2001. Chiral analysis of organochlorine pesticides in Alabama soils. Chemosphere 45, 843–848.

Wild, E., Dent, J., Thomas, G.O., Jones, K.C., 2005. Direct observation of organic contaminant uptake, storage, and metabolism within plant roots. Environ. Sci. Technol. 39, 3695–3702.

Wild, S.R., Jones, K.C., 1992. Polynuclear aromatic hydrocarbon uptake by carrots grown in sludge amended soil. J. Environ. Qual. 21, 217–225.

Willett, K.L., Ulrich, E.M., Hites, R.A., 1998. Differential toxicity and environmental fates of hexachlorocyclohexane isomers. Environ. Sci. Technol. 32, 2197–2207.

Williams, A., 1996. Opportunities for chiral chemicals. Pestic. Sci. 46, 3–9.

Wilson, S.C., Jones, K.C., 1993. Bioremediation of soil contaminated with polycyclic aromatic hydrocarbons (PAHs). A review. Environ. Pollut. 81, 229–249.

Wilson, S.C., Meharg, A.A., 2003. Investigation of organic xenobiotic transfers, partitioning and processing in air-soil-plant systems using a microcosm apparatus. Part II: Comparing the fate of chlorobenzenes in grass planted soil. Chemosphere 53, 583–591.

Wong, C.S., Garrison, A.W., 2000. Enantiomer separation of polychlorinated biphenyl atropisomers and polychlorinated biphenyl retention behaviour on modified cyclodextrin capillary gas chromatography columns. J. Chromatogr. A 866, 213–220.

Xing, B., Pignatello, J.J., 1997. Dual-mode sorption of low-polarity compounds in glassy poly(vinyl chloride) and soil organic matter. Environ. Sci. Technol. 31, 792–799.

Xing, B., Pignatello, J.J., Gigliotti, B., 1996. Competitive sorption between atrazine and other organic compounds in soils and model sorbents. Environ. Sci. Technol. 30, 2432–2440.

Yu, Y.L., Wu, X.M., Li, S.N., Fang, H., Tan, Y.J., Yu, J.Q., 2005. Bioavailability of butachlor and myclobutanil residues in soil to earthworms. Chemosphere 59, 961–967.

Zeddel, A., Majcherczyk, A., Hutterman, A., 1993. Degradation of polychlorinated iphenyls by white rot fungi *Pleurotus ostreatus* and *Trametes versicolor* in a solid state system. Toxicol. Environ. Chem. 40, 255–266.

Zheng, Z., Obbard, J.P., 2002. Polycyclic aromatic hydrocarbon removal from soil by surfactant solubilisation and *Phanerochaete chrysosporium* oxidation. J. Environ. Qual. 31, 1842–1847.

Zipper, C., Suter, M.J.-F., Haderlein, S.B., Gruhl, M., Kohler, H.-P.E., 1998. Changes in the enantiomeric ratio of (R)- to (S)-Mecoprop indicate in situ biodegradation of this chiral herbicide in a polluted aquifer. Environ. Sci. Technol. 32, 2070–2076.

C: Bioavailability and ecotoxicity of contaminants

Ecotoxicology is the study of the fate and effects of toxic agents in ecosystems. This includes the fate of contaminants in ecosystems in relation to interactions with biota while integrating the effects of contaminants from the cellular population level thus extending to entire ecosystem. As such, ecotoxicology is an interdisciplinary subject requiring knowledge on chemistry, biology, and molecular science, including the mechanisms of toxicity and associated ecological processes. Soil microorganisms play an important role in the mineralization of organic matter, nutrient cycling and thereby in maintaining the fertility of soils. Chemical analysis alone is not adequate to estimate the ecotoxic impact of soil contaminants. Moreover, bioassays play a crucial role in the understanding of ecosystem functioning, but only as a tool, and more comprehensive studies of contaminated sites should be performed to understand the influence of chemical speciation and environmental factors on the toxicity of environmental pollutants. The use of bioassays is extremely important for risk assessment of complex mixtures of pollutants, primarily as a tool to demonstrate combined toxicity (integration effects) which cannot be derived from exposure limits of pollutants separately. A multidisciplinary approach is therefore necessary to assess and understand those mechanisms of speciation and bioavailability of pollutants which finally dictate ecosystem effects. The following chapters extend Section C, and here focus on (i) the use of chemical and biological (bioassays) tools in estimating the bioavailability and ecotoxicity of metal and organic contaminants, (ii) interaction of mixed contaminants, (iii) long-term impact of contaminants on soil biota, and (iv) statistical considerations in ecotoxicity estimations.

Developments in Soil Science, volume 32
Ravendra Naidu (Editor)

Chapter 11

BIOAVAILABILITY AND TOXICITY OF CONTAMINANT MIXTURES TO SOIL BIOTA

M. Megharaj and R. Naidu

11.1 Introduction

Global industrialization coupled with modern agriculture has resulted in the production and emissions of various xenobiotic compounds into the environment. Consequently, soil pollution has become a serious problem both nationally and internationally. Soil pollution (urban, industrial and diffuse broad acre) is due to a variety of causes. The most important sources include: (a) former coal gas facilities, (b) petrol stations, (c) pesticide residues and toxic metals in agricultural soils, (d) some defence sites, (e) chemical manufacturing and storage facilities, (f) mine site soils and (g) landfills.

Most of the published literature on ecotoxicology deals with single contaminants. However, the reality is that contaminants exist as mixtures and the toxicity due to mixed contaminants can be vastly different and more severe than single contaminants. Soil and sediment, the ultimate repository/sink for most contaminants, is generally encountered not with a single pollutant, but with mixtures of several contaminants (e.g., organic–organic (pesticides plus their metabolites), organic–inorganic (DDT-As), inorganic–inorganic (sewage sludge containing several heavy metals)). Often the parent chemicals exist simultaneously with their degradation products. About 40% of hazardous waste sites listed on National Priority List of the US Environmental Protection Agency (US EPA) have been stated as co-contaminated (Sandrin et al., 2000). The commonly occurring organic co-contaminants at many contaminated sites around the world include petroleum hydrocarbons, polyaromatic hydrocarbons, chlorinated solvents and pesticides. The common inorganic contaminants found at US EPA superfund sites include As, Ba, Cd, Cr, Pb, Hg, Ni and Zn (Sandrin and Maier, 2003). Assessing the impact of and risk posed by the mixed contaminants is difficult when compared to single contaminants, and as such is one of the most important aspects of practical ecotoxicology. In soil environments under natural conditions organisms are often exposed to complex mixtures of pollutants. However, ecotoxicity assessments are generally restricted to individual contaminants and rarely are mixtures tested. Determining the toxicity of mixed contaminants from a polluted site is not a simple task, as neither the synergistic effects nor the original bioavailability of the contaminant(s) responsible for the toxic effect can be established. The compounds may be metabolically altered to nontoxic or more toxic metabolites, or may exert a direct toxic action on biota and populations. Bacteria and other free living microorganisms may also be subjected to synergistic or antagonistic effects. In certain environments toxicants may cause shifts in microbial populations and this may adversely affect ecosystem function/process. There are two broad approaches used for determining ecotoxicity of mixed contaminants. The first is to determine the toxicity of multiple contaminant systems on organisms through toxicity testing, while the second is to analyse and understand the interactions of various contaminants in the environment (see Chapters 2) in order to determine the effects on chemical activity and bioavailability (see Chapters 3). These two approaches complement each other and are the best means of assessing the impact of complex mixtures of soil contaminants. The field of terrestrial ecotoxicology requires a multidisciplinary approach involving the disciplines such as soil science, environmental chemistry, microbiology, ecology, biology, physiology, biochemistry, engineering and economics. Bioavailability and toxicity are key parameters which need to be adequately understood prior to estimating the risk posed by the contamination or any remedial activity undertaken. The results of quantitative toxicological tests play a key role in identification and evaluation of risk, and thereby increase remediation options for contaminated sites. With risk-based land management gaining more acceptance as a remediation strategy for contaminated sites (see Chapter 4), ecotoxicology plays a major role in identifying risks posed by contamination. However, currently there is a considerable controversy with regard to ecological risk assessment concerning which species and which toxicity tests to apply. No single assay is adequate and a suite of tests are necessary to accurately identify the risk caused by contaminants. This has been clearly demonstrated by Megharaj et al. (2000a, 2000b) while studying the toxicity of total petroleum hydrocarbons on soil biota and their activities.

In this chapter we discuss the toxicity of co-contaminants to soil biota and implications to remediation. Also we provide few case studies from our research into toxicity of co-contaminants.

11.2 Impact on microbiological processes

Elevated concentrations of contaminants at contaminated sites may negatively affect potentially important microbiological processes involved in soil health. A considerable wealth of literature exists on the effect of contaminants on soil biological processes (Giller et al., 1998; Megharaj et al., 1998, 1999a, 1999b, 2000a, 2000b; Naidu et al., 2001; Vig et al., 2003). The adverse impact of metals and organic contaminants such as petroleum hydrocarbons and pesticides on decomposition of organic matter, nutrient transformations, methanogenesis, microbial density and diversity, and soil enzymes have been well studied (Babich and Stotzky, 1985; Bardgett and Saggar, 1994; Burkhardt et al., 1993; Edvantoro et al., 2003; Giller et al., 1998; Kamaludeen et al., 2003; Megharaj et al., 1998, 1999a, 1999b, 2000a, 2000b; Naidu et al., 2001; Vig et al., 2003).

The toxicity of heavy metals such as Cd, in terms of the mineralization of ^{14}C-acetate to ^{14}C–CO_2, to Gram negative *Pseudomonas putida* strain MT 2 increased with increasing pH of soil pore water in an experimental model system (Vanbeelen and Fleurenkemila, 1997). Plette et al. (1996) reported a similar increase in the metal toxicity with increasing pH to soil bacteria in a nutrient medium. A close relationship existed between the amount of metals that can be bound by the organism and the amount of metal that potentially can cause an effect. Metal toxicity was closely related to the binding of metal ions to membrane proteins, prior to membrane transport. Evidence suggested that speciation and bioavailability determined the metal content of the system and the effects of this metal on the biota in the system.

11.3 Pollutant tolerance and adaptation of microorganisms

The toxicity of contaminants to soil organisms depends primarily on soil properties, characteristics of the contaminant and physiology of the organism. The effects of pollutants on non-target microorganisms are difficult to evaluate due to the differential sensitivity of microorganisms, complexity of the population dynamics, multitude of species and conditions used for toxicity assays. Microrganisms may differ in their response to the stress caused by a pollutant (organic or inorganic) as toxicity is related to the biochemistry and physiology of the organism. Microorganisms within species of the same genus or within strains of the same species can also show differences in their sensitivity to pollutants. Giller et al. (1998) demonstrated that *Rhizobium meliloti* was less sensitive to heavy metals than *Rhizobium leguminosarum* and *Rhizobium loti*.

Microorganisms differ in their tolerance mechanisms toward specific contaminants. These mechanisms often include the binding of metals by cell wall or by proteins and extracellular polymers, formation of insoluble metal sulphides, volatilization and enhanced export from cell (Hughes and Poole, 1989). The microbes also influence the metal availability by changing the pH, metal valence, chelation and other mechanisms (Francis, 1990). Microbes are also capable of sorbing metals (Kurek et al., 1982; Beveridge, 1986) which may lead to increased proportions of metal in the solid phase (Brown et al., 1994).

In field situations, heavy metal contamination is often caused by a long-term exposure or repeated exposure to a metal, generally at sub-lethal concentrations.

Such long-term exposure leads to a shift in populations due to the elimination of sensitive populations, adaptation of select microbial communities or increased tolerance of some communities. Such a shift may not be reflected in the total microbial activity measurements and hence microbial activities in conjunction with species diversity should be monitored in order to assess the real impact in risk assessments. Evidence from field studies also suggests a change in the genetic structure of the soil microbial community under long-term metal stress (Giller et al., 1998). Adaptation of the organism to contaminated environment may also play an important role in survival of that organism. For example, a bacterial strain *Sphingomonas* sp. RW1 which survived poorly in a dioxin/dibenzofuran contaminated soil survived well and degraded the compounds upon its adaptation to that soil (Megharaj et al., 1997).

The proportion of Cd-tolerant Gram negative bacteria increased in soils subjected to long-term applications of sewage sludge containing multi-metals such as Cd, Cu, Pb and Zn (Barkay et al., 1985). Increased metal tolerance of soil bacterial communities was observed over time in agricultural soil artificially exposed to Cu, Cd, Zn or Ni, have been reported in literature (Díaz-Raviña et al., 1994; Díaz-Raviña and Bååth, 1996). In heavy metal-polluted soils, the population of microorganisms sensitive to the metal decreases whereas that of tolerant microorganisms increases, resulting in a decrease in the diversity of soil microorganisms (Doleman, 1986). Heavy metals added to the soil can produce a shift from a prokaryote- to eukaryote-dominated population (Doelman, 1986). Available data suggest that Gram negative bacteria are generally more resistant to cadmium and other heavy metals such as copper, mercury, lead and zinc than Gram positive bacteria (Doelman, 1986).

11.4 Degradation products may be more toxic than parent chemicals

Generally in contaminated environments the contaminant in combination with its products, especially where the degradation does occur, can exert more toxic effect than the parent compound (Megharaj et al., 1990). DDT (1,1,1-trichloro-2,2-bis (*p*-chlorophenyl)ethane) and its principle metabolites, DDE (1,1-dichloro-2,2-bis(*p*-chlorophenyl)ethylene) and DDD (1,1-dichloro-2,2-bis(*p*-chlorophenyl)ethane) are widespread environmental contaminants but little information is available concerning their effects on non-target microflora and their activities in long-term contaminated soils. Hence, the impact of long-term DDT pollution of soil was investigated using different criteria such as chemical analysis of DDT residues (DDT, DDE and DDD), microbial biomass, dehydrogenase activity, viable counts of bacteria and fungi and density and diversity of algae (Megharaj et al., 2000a). The chemical analysis of contaminated soils revealed DDT as the predominant compound present followed by its degradation products, DDE

(19–27% of total DDT residues) and DDD (3–13% of total DDT residues) in all the 3 levels of contaminated soils. DDD was found only in minor amounts in low-level contaminated soil compared to medium- and high-level contaminated soils. The presence of DDE and DDD in contaminated soils indicates DDT transformation was occurring in the soil. Viable counts of bacteria and algae declined with increasing DDT contamination while fungal counts, microbial biomass and dehydrogenase activity increased in medium-level contaminated soil (27 mg DDT residues kg^{-1} soil). All the tested parameters were greatly inhibited in high-level contaminated soil (34 mg DDT residues kg^{-1} soil). The predominant algae present in the uncontaminated soil belonged to 2 species of Chlorophyceae (*Chlorococcum* sp. and *Chlorella* sp.) and 2 species of Cyanobacteria (*Anabaena* sp. and *Nostoc* sp.). In high contaminated soil only 1 species of algae, i.e. *Chlorococcum* sp. was present whereas in low and medium contaminated soil 2 species of algae, *Chlorococcum* sp. and *Chlorella* sp. were present. The cyanobacteria were apparently absent in the contaminated soils. Thus the species composition of algae and cyanobacteria was altered in contaminated soils and sensitive species were eliminated in the medium and high contaminated soils suggesting that these organisms could be useful as bioindicators of pollution. A greater toxicity was observed in the medium and high polluted soils where DDD was present as significant portion of the DDT residues suggesting an additive interaction among the DDT residues. Given the fact that DDD was found to have a broader antimicrobial spectrum and more toxic to microorganisms than DDT and DDE (Ko and Lockwood, 1968; Megharaj et al., 1999a, 1999b) the observed greater toxicity in the medium and high contaminated soils where a substantial amount of DDD was present can be explained.

11.5 Toxicity at long-term total petroleum hydrocarbon contaminated site – No single bioassay is adequate to assess the impact of several contaminants

Petroleum hydrocarbons, a mixture of several aliphatic and aromatic hydrocarbons with varying carbon chain length are widespread environmental pollutants. Although biodegradation of petroleum hydrocarbons has been the subject of numerous investigations, information on their toxicity to microorganisms in soil is limited, with virtually no work conducted on soil algae. We carried out a screening experiment for total petroleum hydrocarbons (TPH) and their toxicity to soil algal populations, microbial biomass and soil enzymes (dehydrogenase and urease) in a long-term TPH-polluted site with reference to an adjacent unpolluted site (Megharaj et al., 2000b). Microbial biomass, soil enzyme activity and microalgae declined in medium- to high-level (5,200–21,430 mg kg^{-1} soil) TPH-polluted soils, whereas low-level (2,120 mg kg^{-1} soil) pollution stimulated the algal populations and showed no effect on microbial biomass and enzymes. However, inhibition of all the tested parameters was more severe in soil considered to have medium-level pollution than in soils that were highly polluted. This result could not be explained by chemical analysis alone. Of particular interest was an observed shift in the species composition of algae in polluted soils with elimination of sensitive species in the medium to high polluted soils. Also, an algal growth inhibition test carried out using aqueous eluates prepared from polluted soils supported these results. Furthermore, changes in species composition of algae proved to be more sensitive than microbial biomass and soil enzyme activity measurements. Since different pollutants affect living organisms at different levels and in many ways, it has been recognized now that a battery of bioassay tests utilizing several different biological tests should be used in any monitoring program. Chemical analyses in conjunction with bioassays are necessary in order to detect accurately the environmental pollutants since bioassays report not only the presence of a chemical but also on its bioavailability and biological effects. Thus contaminant toxicity to microorganisms has often been regarded as a direct measure of bioavailability. The results of this experiment suggest that chemical analysis alone is not adequate for toxicological estimations and should be used in conjunction with bioassays. The application of algae in bioassays as pollution indicators can be justified by the fact that these organisms are the primary producers, ubiquitous in occurrence, located at the bottom of the food chain and are therefore the key organisms in sustaining a healthy environment. Furthermore, algae are sensitive to a variety of toxicants (both organics and heavy metals), display a rapid physiological response and have short generation times.

11.6 Toxicity of organic (atrazine) and inorganic (copper) combination to soil biota

Although literature on the effect of metals on organic pollutant biodegradation is scant, available studies indicate that heavy metals have the potential to inhibit organic pollutant biodegradation. Heavy metals such as Cu, Cd, Hg, Zn, and Cr (III) inhibited 2,4-dichloro-phenoxyacetic acid when inoculated with lake sediments (Said and Lewis, 1991). Also, Cd at $60\,mg\,kg^{-1}$ soil has been shown to be inhibitory to 2,4-dichlorophenoxyacetic acid (2,4-D) degradation in soil when inoculated with the 2,4-D degrading bacterium, *Alcaligenes eutrophus* JMP 134 (Roane et al., 2001). Bioaugmentation of As methylating fungi to cattle dip soil contaminated with As ($1390\,mg\,As\,kg^{-1}$) accelerated the rate of As volatilization, however, the As volatilization rate by these fungi was low when the same amount of As is present in combination with DDT at $194\,mg\,kg^{-1}$

(Edvantoro et al., 2003). DDT appears to inhibit the rates of microbial formation of arsine in the contaminated dip soils.

The results of our study on the toxicity and bioavailability of atrazine and copper combinations to soil biota showed a greater toxicity to microbial biomass carbon (5 mg atrazine kg^{-1} and 50 mg Cu kg^{-1}) and lesser algal species diversity than the contaminants present alone. However, no appreciable effect on the activities of phosphatase and sulfatase were observed. Application of Cu at 10 mg kg^{-1} soil had no effect on the persistence of atrazine (at 2 mg kg^{-1} soil) in terms of its half life in soil (30 days). However, Cu at 50 mg kg^{-1} soil increased the half life of atrazine (spiked at 2 mg kg^{-1} soil) to 110 days. This clearly demonstrates that elevated levels of Cu are detrimental to organic-degrading bacteria resulting in prolonged persistence of the organic chemical. In an another experiment where pore waters from 3 soils and algal growth medium were spiked with various combinations of Cu and atrazine and subjected to algal growth bioassay using *Chlorococcum* sp. soil pore waters exhibited antagonistic interaction, but not in algal growth medium. The predominant form of copper in soil pore waters was found to be Cu–DOM complex whereas in growth medium Cu existed as free ion. This clearly suggests that the toxicity observed by conducting the tests using defined laboratory growth media may differ from the real environmental toxicity and caution is required in extrapolating these results to real soil conditions.

11.7 Approaches for bioremediation of co-contaminated soils

Bioremediation of co-contaminated soils with toxic metals and organics is considered to be a difficult task considering the fact that the presence of metals is known to inhibit the organic-degrading microorganisms. In United States alone, over 40% of the Superfund sites are co-contaminated with heavy metals (Sandrin et al., 2000). In Australia, cattle dip sites are contaminated with both arsenic and DDT residues (Edvantoro et al., 2004).

Bioaugmentation with focus on using metal-resistant microorganisms for promoting the degradation of organic contaminants by microorganisms is an emerging area of research in the remediation of co-contaminated sites. Many metal-resistant microorganisms are known to detoxify metals such as Cd, Hg and Se (Roane et al., 1996; Stephen et al., 1999). On the other hand, toxicity of metals to bacteria can inhibit the biodegradation of organic contaminants in co-contaminated soil (Said and Lewis, 1991). However, in a dual bioaugmentation strategy by co-inoculation with metal-detoxifying and organic-degrading bacteria, four metal-resistant (metal-detoxifying) bacterial isolates were used to reduce the soluble concentrations of Cd and thereby allow the degradation of 2,4-D by a Cd-sensitive 2,4-D-degrading bacterium, *Ralstonia eutropha* JMP134 in a co-contaminated soil (Roane et al., 2001). None of the four metal-resistant bacteria could degrade 2,4-D, but each of these isolates supported the degradation of 500 µg 2,4-D ml^{-1} by *R. eutropha* JMP134. Cadmium detoxification by the metal-resistant bacteria involved a plasmid-independent intracellular mechanism. Likewise, a metal-complexing surfactant, rhamnolipid, was used to reduce the toxicity of Cd and then promote the degradation of an organic contaminant, naphthalene by a *Burkholderia* sp. in a model co-contaminated soil (Sandrin et al., 2000).

Waste sites co-contaminated with polycyclic aromatic hydrocarbons (PAHs) and heavy metals are common and widespread in many countries, particularly industrialized ones. White rot fungi are known to oxidize highly condensed PAHs up to 6 aromatic rings with low water solubility (Bumpus et al., 1985; Wolter et al., 1997), while the ability of PAH-degrading bacteria is confined to more soluble PAHs (Cerniglia and Heitkamp, 1989). However, heavy metals such as Cd are toxic to both bacteria and white rot fungi. The highly competitive white rot fungus *Pleurotus ostreatus* could degrade PAH molecules up to 5 to 6 aromatic rings in non-sterile soil both in the presence (up to 100 µg kg^{-1}) and absence of Cd (Baldrian et al., 2000). At 500 µg Cd kg^{-1}, the degradation of PAHs by soil microflora was not affected, while degradation by the white rot fungus was inhibited. Cadmium in the soil was inhibitory to the activity of ligninolytic enzymes, laccase and Mn-dependent peroxidase produced by the fungus, even though degradation rates of PAHs was not altered.

Yet another approach for bioremediation of heavy metals such as Cd is to use hydrogen sulphide generating system for precipitation of Cd and reducing its toxicity in the metal-contaminated and co-contaminated soil. A Cd-resistant *Klebsiella planticola* strain, with an ability to grow at high Cd levels (up to 15 mM CdCl$_2$), effectively precipitated CdS in a medium containing thiosulphate (Sharma et al., 2000). Recently, using DNA recombinant technology, a novel bacterium has been constructed to overproduce sulphide from thiosulphate and then precipitate Cd as a complex of Cd and sulphur (Bang et al., 2000). Transfer of this sulphate-reducing gene to other Cd-sensitive organic pollutant degrading microorganisms could help to remediate the soil co-contaminated with both Cd and organics.

11.8 Conclusion

Approximately 40% of the hazardous sites around the world are co-contaminated with heavy metals(loid)s in addition to organic contaminants. The presence of heavy metals has been shown to be inhibitory to biodegradation of organic chemicals by exerting toxicity to organic-degrading microorganisms. Although considerable literature exists on the toxicity of heavy metals to soil microorganisms studies on toxicity of metals to organic biodegradation are limited. More studies are required on the interaction mechanisms of inorganic–organic co-contaminants. Antagonistic or synergistic toxic effects caused by complex pollutant mixtures such as TPHs, and pesticide combinations (As and DDT as in the cattle dip sites), are difficult to predict by analytical chemistry alone. Results suggested the use of chemical analysis in conjunction with bioassays is necessary for toxicity testing. Therefore, risk-based remediation, particularly in long-term contaminated sites, should be based on toxicological assays in conjunction with chemical assays. The use of metal-resistant microorganisms to promote the degradation of organic pollutants in a soil contaminated with both organic and inorganic pollutants could be an emerging area of research.

Acknowledgement

We thank Dr Michael Beer for critical reading of the manuscript.

References

Babich, H., Stotzky, G., 1985. Heavy metal toxicity to microbe mediated ecological processes – a review and potential application to regulatory policies. Environ. Res. 36, 111–137.

Baldrian, P., Wiesche in der, C., Gabriel, J., Nerud, F., Zadrazil, F., 2000. Influence of cadmium and mercury on activities of ligninolytic enzymes and degradation of polycyclic aromatic hydrocarbons by *Pleurotus ostreatus* in soil. Appl. Environ. Microbiol. 66, 2471–2478.

Bang, S-W., Clark, D.S., Keasling, J.D., 2000. Engineering hydrogen sulfide production and cadmium removal by expression of the thiosulfate reductase gene (*phsABC*) from *Salmonella enterica* serovar typhimurium in *Escherichia coli*. Appl. Environ. Microbiol. 66, 3939–3944.

Bardgett, R.D., Saggar, G., 1994. Effects of heavy metal contamination on the short term decomposition of [^{14}C]glucose in a pasture soil. Soil Biol. Biochem. 26, 727–733.

Barkay, T., Tripp, S.C., Olson, B.H., 1985. Effect of metal-rich sewage sludge application on the bacterial communities of grasslands. Appl. Environ. Microbiol. 49, 333–337.

Beveridge, T.J., 1986. The immobilization of soluble metals by bacteria walls. Biotechnol. Bioeng. Symp. 16, 127–139.

Brown, S., Chaney, R., Angle, J., Baker, A., 1994. Phytoremediation potential of *Thlaspi caerulescens* and bladder campion for zincand cadmium-contaminated soil. J. Environ. Qual. 23, 1151–1157.

Bumpus, J.A., Tien, M., Wright, D., Aust, S.D., 1985. Oxidation of persistent environmental pollutants by a white rot fungus. Science 228, 1434–1436.

Burkhardt, C., Insam, H., Hutchinson, T.C., Reber, H.H., 1993. Impact of heavy metals on the degradative capabilities of soil bacterial communities. Biol. Fertil. Soils 16, 154–156.

Cerniglia, C.E., Heitkamp, M.A., 1989. Microbial degradation of polycyclic aromatic hydrocarbons (PAH) in the aquatic environment. In: U. Varanasi (Ed.), Metabolism of Polycyclic Aromatic Hydrocarbons in the Aquatic Environment. CRC Press, Boca Raton, pp. 41–68.

Díaz-Raviña, M., Bååth, E., 1996. Development of metal tolerance in soil bacterial communities exposed to experimentally increased metal levels. Appl. Environ. Microbiol. 62, 2970–2977.

Díaz-Raviña, M., Bååth, E., Frostegård, Å., 1994. Multiple heavy metal tolerance of soil bacterial communities and its measurement by a thymidine incorporation technique. Appl. Environ. Microbiol. 60, 2238–2247.

Doelman, P., 1986. Resistance of soil microbial communities to heavy metals. In: V. Jensen, A. Kjoller and L.H. Sorensen (Eds.), Microbial Communities in Soil. Elsevier, London and New York, pp. 369–383.

Edvantoro, B.B., Naidu, R., Megharaj, M., Singleton, I., 2003. Changes in microbial properties associated with long-term arsenic and DDT contaminated soils at disused cattle dip sites. Ecotoxicol. Environ. Saf. 55, 344–351.

Edvantoro, B.B., Naidu, R., Megharaj, M., Merrington, G., Singleton, I., 2004. Microbial formation of volatile arsenic in cattle dip site soils contaminated with arsenic and DDT. Appl. Soil Ecol. 25, 207–217.

Francis, A.J., 1990. Microbial dissolution and stabilization of toxic metals and radionuclides in mixed wastes. Experientia 46, 840–851.

Giller, K.E., Witter, E., McGrath, S.P., 1998. Toxicity of heavy metals to micro-organisms and microbial processes in agricultural soils: a review. Soil Biol. Biochem. 30, 1389–1414.

Hughes, M.N., Poole, R.K., 1989. Metals and microorganisms. Chapman and Hall, New York, USA, pp. 1–37.

Kamaludeen, S.P.B., Megharaj, M., Naidu, R., Singleton, I., Juhasz, A.L., Hawke, B.G., Sethunathan, N., 2003. Microbial activity and phospholipid fatty acid pattern in long-term tannery wastes contaminated soil. Ecotoxicol. Environ. Saf. 56, 302–310.

Ko, W.H., Lockwood, J.L., 1968. Conversion of DDT to DDD in soil and the effect of these compounds on soil microorganisms. Can. J. Microbiol. 14, 1069–1073.

Kurek, E., Czaban, J., Bollag, J.-M., 1982. Sorption of cadmium by microorganisms in competition with other soil constituents. Appl. Environ. Mircobiol. 43, 1011–1015.

Megharaj, M., Boul, L.H., Thiele, J.H., 1999b. Persistence and toxicity of DDT and its metabolites toward native algal populations and enzymatic activities in soil. Biol. Fertil. Soils 29, 130–134.

Megharaj, M., Kantachote, D., Singleton, I., 2000a. Effects of long-term contamination of DDT on soil microflora with special reference to soil algae and algal metabolism of DDT. Environ. Pollut. 109, 35–42.

Megharaj, M., Prabhakar Rao, A., Rao, A.S., Venkateswarlu, K., 1990. Interaction effects of carbaryl and its hydrolysis product 1-naphthol towards three isolates of microalgae from rice soil. Agric. Ecosyst. Environ. 31, 293–300.

Megharaj, M., Singleton, I., Kookana, R., Naidu, R., 1999a. Persistence and effects of fenamiphos on native algal populations and enzymatic activities in soil. Soil Biol. Biochem. 31, 1549–1553.

Megharaj, M., Singleton, I., McClure, N.C., 1998. Effect of pentachlorophenol pollution towards microalgae and microbial activities in soil from a former timber processing facility. Bull. Environ. Contam. Toxicol. 61, 108–115.

Megharaj, M., Singleton, I., McClure, N.C., 2000b. Influence of petroleum hydrocarbon contamination on microalgae and microbial activities in a long-term contaminated soil. Arch. Environ. Contam. Toxicol. 38, 439–445.

Megharaj, M., Wittich, R.M., Blasco, R., Pieper, D.H., Timmis, K.N., 1997. Superior survival and degradation of dibenzo-*p*-dioxin and dibenzofuran in soil by soil-adapted *Sphingomonas sp* strain RW1. Appl. Microbiol. Biotechnol. 48, 109–114.

Naidu, R., Krishnamurti, G.S.R., Bolan, N.S., Wenzel, W., Megharaj, M., 2001. Heavy metal interactions in soils and implications for soil microbial biodiversity. In: M.N.V. Prasad (Ed.), Metals in the Environment: Analysis by Biodiversity. Marcel Dekker, New York, pp. 401–432.

Plette, A.C.C., Van Riemsdijk, W.H., Benedetti, M.F., 1996. Competitive binding of protons, calcium, cadmium and zinc to isolated cell walls of a gram-positive soil bacterium. Environ. Sci. Technol. 30, 1902–1910.

Roane, T.M., Pepper, I.L., Miller, R.M., 1996. Microbial remediation of metals. In: R.L. Crawford, D.L. Crawford (Eds.), Bioremediation: Principles and Applications. Cambridge University Press, Cambridge, pp. 312–340.

Roane, T.M., Josephson, K.L., Pepper, I.L., 2001. Microbial cadmium detoxification allows remediation of co-contaminated soil. Appl. Environ. Microbiol. 67, 3208–3215.

Said, W.A., Lewis, D.L., 1991. Quantitative assessment of the effects of metals on microbial degradation of organic chemicals. Appl. Environ. Microbiol. 57, 1498–1503.

Sandrin, T.R., Chech, A.M., Maier, R.M., 2000. A rhamnolipid biosurfactant reduces cadmium toxicity during biodegradation of naphthalene. Appl. Environ. Microbiol. 66, 4585–4588.

Sandrin, T.R., Maier, R.M., 2003. Impact of metals on the biodegradation of organic pollutants. Environ. Health Perspect. 111, 1093–1101.

Sharma, P.K., Balkwill, D.L., Frenkel, A., Vairavamurthy, M.A., 2000. A new *Klebsiella planticola* strain (Cd-1) grows anaerobically at high cadmium concentrations and precipitates cadmium sulfide. Appl. Environ. Microbiol. 66, 3083–3087.

Stephen, J.R., Chang, Y.J., Macnaughton, S.J., Lowalchuk, G.A., Leung, K.T., Flemming, C.A., White, D.C., 1999. Effect of toxic metals on indigenous soil β-subgroup proteobacterium ammonia oxidizer community structur and protection against toxicity by inoculated metal-resistant bacteria. Appl. Environ. Microbiol. 65, 95–101.

Vanbeelen, P., Fleurenkemila, A.K., 1997. Influence of pH on the toxic effects of zinc, cadmium, and pentachlorophenol on pure cultures of soil microorganisms. Environ. Toxicol. Chem. 16, 146–153.

Vig, K., Megharaj, M., Sethunathan, N., 2003. Bioavailability and toxicity of cadmium to microorganisms and their activities in soil: a review. Adv. Environ. Res. 8, 121–135.

Wolter, M., Zadrazil, F., Martens, R., Bahadir, M., 1997. Degradation of eight highly condensed polycyclic aromatic hydrocarbons by *Pleurotus* sp. Florida in solid wheat straw substrate. Appl. Microbiol. Biotechnol. 48, 398–404.

Developments in Soil Science, volume 32
Ravendra Naidu (Editor)

Chapter 12

BIOAVAILABILITY IN SOIL: THE ROLE OF INVERTEBRATE BEHAVIOUR

J. Roembke

12.1 Introduction

Bioavailability is usually defined as a concept that describes the interaction between a chemical and an organism (see Chapter 3 for an overview of the definition of bioavailability). Individual parts of this interaction (e.g. the route of entry, the time of exposure, the type and concentration of the chemical, as well as the properties of the surrounding soil) are well-investigated or are at least identified as being important. The biological receptor with which a specific chemical interacts has also been addressed in recent research. However, from a biological point of view, it seems that the behaviour of individual organisms has been too rarely taken into consideration in this context. In fact, there are various ways in which an organism can actively influence its relationship with a chemical: directly through its own behaviour, or indirectly by changing its environment (e.g. the soil properties).

First, indications on how organisms can change their environment in a way that affects bioavailability of chemicals came from marine studies with polychaete worms living in self-created tubes (Forbes and Forbes, 1994). The exposure of these worms varies greatly, depending whether they stay in their tubes or whether they move on or through the sediment. Depending on the source of the tube material and its properties, exposure of the worms may be higher or lower than the situation in the sediment.

The relationship between a chemical and an organism can be simplified as follows: if there is no uptake (or only a small one), there is no toxicity. This paper focuses on the question: how can organisms avoid considerably or even completely the uptake of chemicals from soil in order to avoid harm? Invertebrates living in the soil or on the soil surface are used as an example; "higher" animals like mammals or plants will not be used owing to wide differences in exposure as well as in physiology. In addition, soil invertebrates, mainly oligochaetes (e.g. earthworms and enchytraeids) plus several groups of arthropods (e.g. springtails or isopods) are ecologically the most important animal groups in the soil. Several species in these groups are considered to be "ecosystem engineers" since "they affect the availability of resources to other organisms through modifications of the physical environment" (Lavelle et al., 1997). Oligochaetes and arthropods act as representatives for the whole soil community and, consequently, have been used as test organisms in soil ecotoxicology (Table 12.1). However, in these tests traditionally acute (mainly mortality) or chronic (growth, biomass, reproduction) endpoints have been used, because an effect on these endpoints clearly affects the population or even the whole ecosystem. In addition, these endpoints can be measured relatively easily.

Despite intensively discussed in the scientific literature (e.g. Spurgeon et al., 2002), neither physiological, biochemical nor genetic endpoints have been used in internationally standardised tests with soil invertebrates. Mainly the difficulty of extrapolating such data to higher levels of biological organisation (i.e. populations or the whole ecosystem) is responsible for this situation (Scott-Fordsmand and Weeks, 1998). In addition, the behaviour of the individual test organisms is usually not reflected in the standard laboratory tests. For example, when a test chemical is mixed into the soil of a test vessel the organisms cannot avoid being exposed. This worst-case scenario (however understandable in the interest of being on the "safe side" when assessing the risk of a chemical) rarely occurs in field situations.

To summarise the situation both from a scientific as well as from a practical point of view (i.e. the design of laboratory tests) there is a need to acknowledge that soil invertebrates can be active players for assessing bioavailability. However, the whole topic is too broad to be treated in one paper. Processes like the movement of chemicals through the soil food web, from invertebrates up to vertebrates like birds and mammals, will not be handled in detail. Any discussion of this topic should also take into consideration that recommendations are ultimately needed on how invertebrate behaviour can be incorporated into existing assessment schemes of the environmental risk of chemicals.

Table 12.1. *Short overview on existing standardised ecotoxicological test systems with soil oligochaetes and arthropods.*

Test species	Endpoint	Guideline
Oligochaetes		
Eisenia fetida, E. andrei	Mortality	OECD, 1984
Eisenia fetida, E. andrei	Reproduction	ISO, 1998; OECD, 2004b
Enchytraeus albidus	Mortality, reproduction	ISO, 2003a; OECD, 2004a
Arthropods		
Folsomia candida	Mortality, reproduction	ISO, 1999a
Hypoaspis aculeifer	Mortality, reproduction	Under development

In this paper, examples of the interactions between soil organisms and chemicals are described and their importance for the overall concept of bioavailability is discussed. In particular, the role of earthworms as ecologically the most important and, at the same time, best-studied soil invertebrates will be highlighted. Following that, the practical consequences of invertebrate behaviour in the presence of contaminants in soil are emphasised by describing a newly developed ecotoxicological earthworm test method, which uses avoidance behaviour as an endpoint. Finally, recommendations for how to use such a test as well as a list of research issues are given.

12.2 Invertebrates as active players in the context of soil contamination

12.2.1 General considerations

In Figure 12.1, an overview of interactions between soil properties, chemicals in the soil and soil-inhabiting invertebrates is given. They can be described

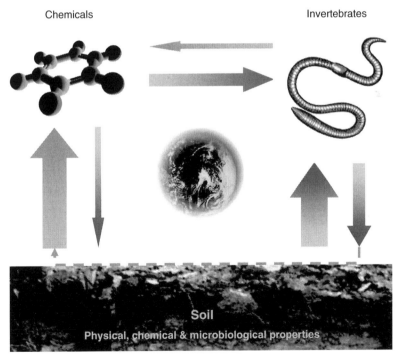

Figure 12.1. *Schematic view of the interactions among soil properties, invertebrates and chemicals in the belowground compartment.*

as follows:

- *Soil – invertebrates*: an intensive two-way interaction takes place. For example, the distribution of species depends significantly on soil properties such as pH, amount of organic matter or texture (NOT on soil type). This relationship differs in its extent in the various invertebrate groups: for example, a soft-bodied oligochaete worm is much more affected by soil properties like pH than an arthropod with a chitin integument (biotic factors like predation or competition also play a role). However, soil invertebrates and in particular large species like earthworms influence soil formation processes through their burrowing activity or the form of the litter layer by influencing organic matter breakdown. When these organisms create micro-habitats (e.g. large burrows) for other species or change the soil properties considerably, they are called "ecosystem engineers" and parts of the soil compartment are named after them (e.g. the drilosphere (=worm habitat) or the termitosphere (=termite habitat); Lavelle et al., 1992).
- *Chemical – soil*: The influence of soil properties on the fate (i.e. metabolism, degradation, adsorption/desorption) of chemicals in soil is fairly well-studied, mainly for heavy metals and for those compounds which require registration before being marketed (pesticides in particular). For example, many organic chemicals are more quickly (and/or differently) metabolised in soils with a high microbial biomass, while the adsorption of others depends greatly on the amount and the composition of the clay fraction of the soil (Domsch, 1992; Hutson and Roberts, 1990). In fact, the partitioning (and, thus, the bioavailability) of some chemicals, in particular heavy metals, can be predicted quite accurately in different soils if the main properties of these soils are taken into consideration (Peijnenburg et al., 1999).
- *Invertebrates – chemicals*: this interaction is usually seen as a one-way process: depending on its concentration a chemical may have some effects on an organism, assuming that both come into contact and/or the compound is taken up into the body. Various ecotoxicological test systems are trying to determine this pathway as accurately as possible. On lower levels of the tiered testing strategy usually acute short-term tests are used for the assessment of single chemicals. In these tests, individual organisms are exposed to the test substance in a "worst-case" scenario: i.e. they cannot avoid coming into contact with the test substance and they do not have the time to build up defence mechanisms. In higher-tier tests (e.g. Terrestrial Model Ecosystems running four months (Knacker et al., 2004)), the exposure situation becomes more realistic but with the exception of the few field tests standardised so far (e.g. the earthworm field test; ISO, 1999b), the organisms (particularly large ones) are still forced to stay exposed to the test substance.

Summarising the current situation, the various ways soil invertebrates can influence chemicals in the soil are seldom taken into consideration. In the following, two different types of active interaction between a chemical and a soil invertebrate are listed.

How soil invertebrates can influence indirectly *the potential effects of a chemical*:

The degradation of a chemical can be enhanced and/or modified either physically (by bioturbation) or biologically (by enhancing microbial activity) by soil invertebrates. In the first case, a chemical applied to a soil and adsorbed at the soil surface can be transported to deeper layers by the burrowing activity of earthworms (e.g. Farenhorst et al., 2000). In deeper soil layers, the chemical is either metabolised more slowly (e.g. it cannot be degraded by photolysis anymore) or just the opposite happens, since it is moved to an environment with conditions more favourable for degrading micro-organisms (e.g. more constant moisture). The fate of Atrazine in earthworm middens is an example for the second possibility (Akhouri et al., 1997). Soil invertebrates also can directly affect microbial populations which are responsible for the degradation of many organics, For example, enchytraeids or collembolans keep micro-organisms in their optimum growth stage by that their grazing activity. Finally, burrowing activity also influences the leaching behaviour of many chemicals, since in particular the deep, stable vertical burrows of large earthworms are known to be responsible for groundwater contamination by soil-applied chemicals (Edwards et al., 1990; Steenhuis et al., 1990).

How soil invertebrates can influence directly *the potential effects of a chemical*:

- Chemicals can be metabolised and/or degraded internally (e.g. in the gut) or externally by soil invertebrates (e.g. Stenersen, 1992). This process is a by-product of the feeding activity of the animals, in particular those which feed their way indiscriminately through the soil. While it is known that digestive enzymes are responsible for this effect, little research has been done regarding how and to what extent this process is ecotoxicologically relevant under field conditions. In this context, it is scientifically interesting to distinguish whether such metabolisation is conducted in the invertebrate tissue itself or by microbes living in the gut of the animal.
- Soil invertebrates can store a chemical (in particular heavy metals) in their body in such a way that it does not affect their physiology. The best-known examples are the chloragog cells of oligochaetes like earthworms or enchytraeids. At least some species of the mainly aquatic oligochaete family Tubificidae can store potentially harmful chemicals in posterior segments which can then be shed (Lucan-Bouché et al., 1999). Disregarding biomagnification within the food chain, the chemicals are not available during the storage period for other organisms or for abiotic metabolisation processes.

- Soil invertebrates can adapt themselves to toxic chemicals physiologically and/
  or by developing genetic defence mechanisms when being exposed for longer
  periods of time to certain chemicals. Well-known examples are the adaptation
  of earthworms and collembolans to heavy metals when living at sites with a
  long history of mining (e.g. Posthuma, 1990). This process can be economically
  devastating in the case of the development of resistance by pest species to
  insecticides. Just a few decades appear sufficient to change soil invertebrate
  populations so they are significantly less sensitive to these contaminants.

The most efficient way to influence the potential effects of a chemical is to
avoid any contact which could lead to harm of the organism. In the case of soil
invertebrates, two different ways are possible:

- Some soil invertebrates have the ability to use defence mechanisms against
  toxic chemicals. For example, earthworms can excrete mucus (a protein-rich
  slime with anti-bacterial properties), which covers the whole outer surface of
  the worm. This mucus can act as a barrier – but due to its highly adsorptive
  structure it can also enhance exposure to a chemical. In some cases the
  contact with a chemical can also increase the mucus production so much
  that the worm becomes highly stressed. Theoretically, a protective structure
  (i.e. with foreign material) could be built up as a defence mechanism. However, good examples of this possibility are not known for soil invertebrates.
- Upon realising that there is a potentially toxic chemical in its immediate
  surrounding, the animal can move away from the source. The pre-condition
  is that the organism has the ability to detect the potentially toxic chemical
  early enough to react. This repellent effect is often used in the case of insect
  pests (e.g. for protecting people against mosquitoes). Despite various
  observations of avoidance behaviour in laboratory and field situations, it
  has not been well-studied for soil invertebrates.

In the following, only the last type of interaction between chemicals and soil
invertebrates will be further discussed.

12.2.2 *Avoidance behaviour in soil arthropods*

As mentioned earlier, there are several examples of avoidance behaviour among
soil invertebrates. Relatively well-studied is the reaction of species living on the
soil surface, in the litter layer or even on leaves of herbs or other plants, since these
organisms can easily be observed, even under field conditions. For example, the
walking behaviour of spiders is quite different, depending on whether they are
moving on a clean leaf or on a leaf which has been treated with a pesticide (Jagers
Op Akkerhuis, 1994). In the latter case, the insects try to minimise the contact
between their legs and the treated leaf surface by moving more quickly. When
carabid beetles, wolf spiders or isopods are kept in a test vessel with contaminated

soil, they try to escape from the vessel, and if this is not possible, end up either in hyperactivity or (except spiders) huddle one above another in a corner (pers. observation). In the case of carabid beetles or isopods, these movements have been assessed by using automatic video-tracking systems (Bayley et al., 1997).

The best-studied case of avoidance behaviour of soil arthropods is the effect of heavy metals, pesticides and contaminated soils on collembolans (Sjögren, 1997; Heupel, 2002; Natal-da-Luz et al., 2004). Using experiences with five different springtail species and the fungicide Betanal (a.i. phenmedipham), the latter author proposed to standardise a laboratory test with avoidance behaviour as the measurement endpoint. At least in the case of the species *Onychiurus armatus* the sensitivity of the avoidance test exceeded that of mortality and reproduction tests with Betanal. Similar results have been found in tests with the species *Folsomia candida* and *Sinella communis* (the latter from Australia) when exposed to copper and Phenol. However, no avoidance up to 1000 mg cadmium/kg soil dry weight (dw) was found in tests with *Sinella communis*, while the EC50 for reproduction is much lower 50.1 mg/kg soil (dw) (Greenslade and Vaughan, 2003).

12.2.3 Avoidance behaviour in oligochaetes

For the first time, Tomlin (1992) discussed oligochaetes, in particular lumbricid earthworms but also potworms (Enchytraeidae), as well-suited for the assessment of chemical stress using behavioural endpoints. This conclusion arises partly because these soft-bodied organisms live in close contact with soil particles and soil pore water. In addition, they possess a huge number of chemical sense organs all over their body surface, but especially in the head (Fig. 12.2) and – in a somewhat lower number – the tail region (Edwards and Bohlen, 1996; Römbke and Schmidt, 1999).

Oligochaetes are known to react very sensitively to a wide range of chemicals (Didden and Römbke, 2001; Edwards and Bohlen, 1992). One of the first examples involving a behavioural endpoint was observed by Cook et al. (1980), who found no acute effects of DDT on earthworms, but surface casting was significantly reduced, thus indicating a repellence or avoidance reaction.

There are further reasons why earthworms are so well-suited for the investigation and use of avoidance behaviour:

a. Earthworms are relatively large organisms which usually (but not always) live in the soil. Their activity can be directly (e.g. when moving on the surface) or indirectly (e.g. by observing their feeding or defecating behaviour) assessed in the field (Edwards and Bohlen, 1996);

b. Most of the lumbricid species can be classified in one of the three ecological groups, which can be distinguished morphologically or by their ecological preferences including the way they behave and move (Bouché, 1977; Lee, 1985). Due to this classification, it is relatively easy to predict which

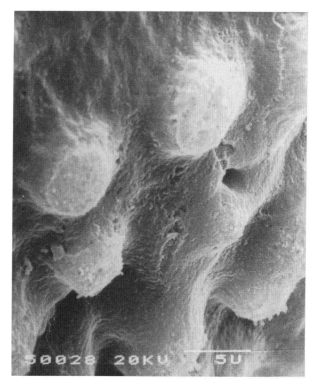

Figure 12.2. *Chemical sense organs in the head region of the enchytraeid species* Cognettia sphagnetorum *(diameter: 5 μm).*

ecological consequences an effect of the behaviour of a certain earthworm species might have. These three groups are defined as follows:

- *epigeics (litter inhibiters)* species that live above the soil surface, typically in the litter layers of forest soils, making no burrows. Often small, reddish-coloured, quick-moving animals with short life cycles, which are under high predation pressure. Survive drought usually in the cocoon stage.

- *endogeics (mineral dwellers)* species that inhabit mineral soil, making horizontal non-permanent burrows, mainly in the uppermost 10–15 cm of soil. Often whitish, slow-moving animals of variable size and intermediate longevity and life-cycle length, they experience relatively low predation pressure from surface-dwelling animals; they enter diapause in response to drought.

- *anecics (vertical burrowers)* species that live in permanent vertical burrows in mineral soil layers (up to 3 m deep). Often large worms, darkly coloured on the dorsal side (at least the anterior body), able to withdraw rapidly, but are usually slow-moving animals. These are relatively long-lived species with long life cycles; they experience under high predation pressure when at the surface but are protected in the burrows. They survive drought usually in a quiescent stage.

The behavioural differences among these three groups lead to quite different exposure situations, in particular in combination of different chemical formulations (e.g. sprayed versus granular pesticides). However, due to its ecological importance only the last group (anecics) will be discussed in detail in the next section.

12.2.4 Example: anecic earthworms

Earthworm species belonging to the anecics (vertical burrowers) are ecologically very important and are thus called "ecosystem engineers". The best-known example is the night crawler *Lumbricus terrestris* (belonging to the family Lumbricidae), originally a native of Central and Northern Europe, but due to anthropogenic activities now a peregrine species, i.e. distributed in temperate regions world-wide (Blakemore, 2002). Surface mating and feeding make this species especially susceptible to pesticides applied to crop sites (Tomlin, 1992). By building mucus-lined permanent burrows, *L. terrestris* creates areas in which chemicals may either accumulate or be more quickly degraded in comparison to the surrounding soil. In the latter case, the burrows are a "safe haven" for the worms, e.g. after application of Atrazine (Akhouri et al., 1997).

After application of pesticides like Benomyl, strange behaviour of earthworms has been observed in the field (Mather and Christensen, 1998). *L. terrestris* was particularly involved, since this species often moves on the soil surface in search of food or for reproduction (Fig. 12.3). An enhanced crawling

Figure 12.3. *Two adult* Lumbricus terrestris *reproducing on the soil surface. Note that both individuals stick with their tails in their burrows.*

activity, especially during daytime, is negative for the worms due to higher risk (light, predators). In addition, a decrease in feeding activity by the worms has a negative impact on litter decomposition (i.e. organic matter breakdown), one of the most important ecological services of soil organisms. For example, the fungicide Benomyl causes an accumulation of litter in orchards, since the feeding of *L. terrestris* on the apple leaves was strongly affected (Kennel, 1990).

12.3 Practical use of soil invertebrate behaviour: the earthworm avoidance test

Based on observations outlined in the previous section, the question arises how to use these experiences in practical ecotoxicology. The most promising approach is the performance of behavioural tests.

From the 1980s behavioural responses of earthworms to pesticides and heavy metals were recorded in the literature (Cook et al., 1980; Wentsel and Guelta, 1988). However, it took some time before this observation was transformed into a proposal for an ecotoxicological laboratory test (e.g. Yeardley et al., 1996; Slimak, 1997; Stephenson et al., 1998). Later still, the negative impact of chemicals on earthworm behaviour was accepted as ecologically relevant by the International Organisation for Standardisation (ISO), which is currently discussing the draft of a test guideline (ISO, 2007). Despite the fact that anecic species like *L. terrestris* are ecologically most relevant, the common compost worm (*Eisenia fetida, E. andrei*) is most often used in these tests due to its higher practicability as in the standard acute and chronic effect tests (cf. Table 12.1). Details of the new test are summarised in Table 12.2.

In short, the test principle can be described as follows (ISO, 2007) adult earthworms are exposed simultaneously to a control soil and a contaminated soil or a soil containing test substances. Test soil and control soil are placed into each test vessel and the earthworms are thus presented with a choice between the test soil and the control soil. Two test-vessel designs are available: (I) a two-section test vessel and (II) a six-section test vessel (Fig. 12.4). After an incubation period of 2 days, the number of worms is determined in all sections of the vessels.

Avoidance tests with various soils and different contaminations have been performed often in recent years. These indicated the worms are relatively

Table 12.2. *Short characterisation of the earthworm avoidance test (ISO, 2007).*

Name	Earthworm avoidance test
Species	*Eisenia fetida, E. andrei*
Substrate	Artificial soil or field soils like LUFA 2.2
Duration	2 days
Parameter	Behaviour (avoidance response)
Design	NOEC, ECx, limit test (number of vessels depends on the design; 5 replicates)
Test vessels	Either with two or six sections (Fig. 12.4)

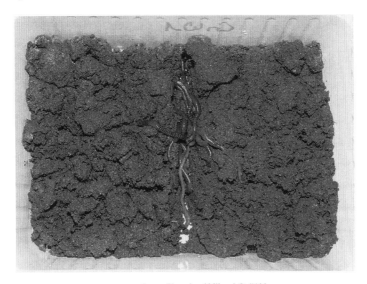

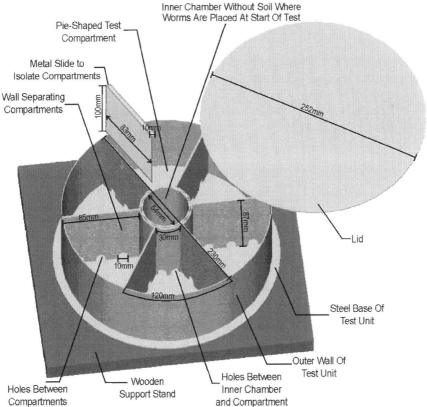

Figure 12.4. *Design of the ISO Earthworm Avoidance Test: two and six section design (ISO, 2007).*

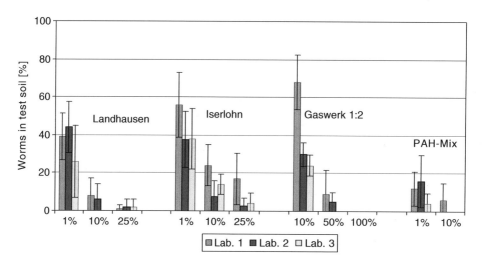

Figure 12.5. *Percentage of* Eisenia fetida *in four contaminated soils mixed with a control soil (LUFA 2.2) in a dilution series and tested in three laboratories (Hund-Rinke et al., 2003).*

insensitive towards different properties between soils, i.e. different field soils can be used for this kind of test (Hund-Rinke and Wiechering, 2001). In fact, out of a range of Central European soils only acid forest soils are clearly avoided. Up to now it has been shown that the test is suitable for mineral oil, polyaromatic hydrocarbons, TNT, lead, manganese, zinc, copper sulphate, petroleum hydrocarbons (crude oil) and mixtures consisting of several heavy metals, KCl, NH_4Cl, benomyl, carbendazim, lambda-cyhalothrin, mancozeb and complex hydrocarbon mixtures such as amines and glycol products and condensate (ISO, 2007; Garcia, 2004; Schaefer, 2004).

As an example, also for the reproducibility of the test, the percentage of *Eisenia fetida* in four contaminated soils (control soil: LUFA 2.2) tested in three laboratories is shown in Figure 12.5 (Hund-Rinke et al., 2003). Generally speaking, the sensitivity of this screening method is comparable to the much more time-consuming (but ecologically more relevant) earthworm reproduction test. Amendments may be included in the ISO guideline, allowing for example the use of enchytraeids in the same test design (Amorim et al., 2005).

12.4 Avoidance tests in environmental risk assessment

In the European Union, the risk of chemicals for the soil is assessed by comparing exposure and effect concentrations in a tiered system of tests, starting with simple acute tests and ending with very complex field studies (e.g. EU, 1991). Any new test system must be designed in such a way that an effect concentration, either as a NOEC or an ECx, can be determined. Technically, the Earthworm Avoidance Test as described by ISO (2004) fulfils this requirement (in fact, EC50 values for several pesticides have already been published; e.g. for Benomyl and Carbendazim (Garcia, 2004)). However, the specific pros and cons of this test must be appropriately evaluated before recommending its use for the environmental risk assessment of chemicals.

Pros of avoidance tests:

– Compared with other tests using soil invertebrates this test is quick, cheap and easy to perform.
– Its sensitivity towards a wide range of chemicals is usually in the same order of magnitude as the earthworm reproduction test, or is at least equal to that of the acute test.

Cons of avoidance tests:

– The sensitivity is species-specific and chemical-specific (e.g. the endogeic earthworm *Aporrectodea caliginosa* is unable to detect organophosphate pesticides at field-relevant concentrations (Hodge et al., 2000)).
– Avoidance behaviour can be affected by soil properties (e.g. low pH).

Putting the issue of chemical-specificity aside for a moment, the avoidance test seems to fulfil important criteria for an ecotoxicological standard test. However, due to the much higher ecological relevance of reproduction tests this test can only be discussed as an alternative to the existing acute test. In light of this, the sublethal test could be used as a screening test in order to decide whether further tests are necessary (Schaefer, 2004).

This conclusion is especially true for the use of the avoidance tests as a tool to assess the environmental risk of potentially contaminated field soils. As mentioned earlier, the existing experiences support its use for testing soil samples with an often complex mixture of contaminants. Effects of such soils are either compared with a standard control soil (e.g. Organisation for Economic Co-Operation and Development (OECD) artificial soil or LUFA 2.2) or with a site-specific soil with the same properties as the test soil but without contamination (ISO, 2003b). Such a use of avoidance tests was already proposed by Ferguson et al. (1998) while discussing the scientific basis of risk assessment of contaminated sites in Europe.

12.5 Outlook

Several issues remain to be clarified before an avoidance test can be recommended without restrictions for regulatory purposes. The most important of these issues are:

a. The usefulness of species other than the compost worm currently mentioned in the ISO draft guideline has to be checked (e.g. enchytraeids, but mainly arthropods). For example, the walking behaviour of arthropods like collembolans and carabid beetles has to be investigated using PC-controlled video systems.

b. Why certain chemicals are "detectable" by earthworms and others not, remains to be investigated. The test can only be recommended if non-detectable chemicals are an exception – otherwise the test results would be too unreliable.

c. Up to now, differences between earthworm species and between life-stages are not known in detail. For example, adult worms can escape from contaminants but juveniles are probably much less able to avoid contaminants.

Assuming that these issues can be resolved in the near future, the use of an (earthworm) avoidance test as a screening tool for the risk assessment of single chemicals spiked into standard soils or of potentially contaminated soil samples is recommended.

References

Akhouri, N.M., Kladivko, E.J., Turco, R.F., 1997. Sorption and degradation of atrazine in middens formed by *Lumbricus terrestris*. Soil Biol. Biochem. 29, 663–666.

Amorim, M.J., Römbke, J., Soares, A.M.V.M., 2005. Avoidance behaviour of *Enchytraeus albidus*: effects of benomyl, carbendazim, phenmedipham and different soil types. Chemosphere 59, 501–510.

Bayley, M., Baatrup, E., Bjerregaard, P., 1997. Woodlouse locomotion behaviour in the assessment of clean and contaminated field sites. Environ. Toxicol. Chem. 16, 2309–2314.

Blakemore, R., 2002. Cosmopolitan earthworms – an eco-taxonomic guide to the peregrine species of the world. VermEcology, Australia (426 pp + 80 figures).

Bouché, M.B., 1977. Strategies lombriciennes. Ecol. Bull. (Stockholm) 25, 122–132.

Cook, A.G., Critchley, B.R., Critchley, U., Perfect, T.J., Yeadon, R., 1980. Effects of cultivation and DDT on earthworm activity in a forest soil in the subhumid tropics. J. Appl. Ecol. 17, 344–357.

Didden, W.A.M., Römbke, J., 2001. Enchytraeids as indicator organisms for chemical stress in terrestrial ecosystems. Ecotoxicol. Environ. Saf. 50, 25–43.

Domsch, K.-H., 1992. Pestizide im Boden. Mikrobieller Abbau und Nebenwirkungen auf Mikroorganismen. VCH, Weinheim, 575p.

Edwards, C.A., Bohlen, P.J., 1992. The effects of toxic chemicals to earthworms. Rev. Environ. Contam. Toxicol. 125, 23–99.

Edwards, C.A., Bohlen, P.J., 1996. Biology of Earthworms., 3rd Ed, Chapman & Hall, London, 276 pp.

Edwards, W.M., Shipitalo, M.J., Owens, L.B., Norton, L.D., 1990. Effect of *Lumbricus terrestris* burrows on hydrology of continuous no-till corn fields. Geoderma 46, 73–84.

EU (European Union) 1991 Council Directive Concerning the Placing of Plant Protection Products on the Market No. 91/414/EEC. Brusseles, Belgium.

Farenhorst, A., Topp, E., Bowman, B.T., Tomlin, A.D., 2000. Earthworms and the dissipation and distribution of atrazine in the soil profile. Soil Biol. Biochem. 32, 23–33.

Ferguson, C., Darmendrail, D., Freier, K., Jensen, B.K., Jensen, J., Kasamas, H., Urzelai, A., Vegter, J., 1998. Risk Assessment for Contaminated Sites in Europe. Vol. 1. Scientific Basis, LQM Press, Nottingham.

Forbes, V.E., Forbes, T.L., 1994. Ecotoxicology in Theory and Practice. Ecotoxicology Series No. 2. Chapman & Hall, London, 247 pp.

Garcia, M.V.B., 2004. Effects of Pesticides on Soil Fauna: Development of Ecotoxicological Test Methods for Tropical Regions. ZEF Series No. 19. University of Bonn, Germany, 281 pp.

Greenslade, P., Vaughan, G.T., 2003. A comparison of Collembola species for toxicity testing of Australian soils. Pedobiologia 47, 171–179.

Heupel, K., 2002. Avoidance response of different Collembolan species to Betanal. Eur. J. Soil Biol. 38, 273–276.

Hodge, S., Webster, K.M., Booth, L., Hepplethwaite, V., O'Halloran, K., 2000. Non-avoidance of organophosphate insecticides by the earthworm *Aporrectodea caliginosa* (Lumbricidae). Soil Biol. Biochem. 32, 425–428.

Hund-Rinke, K., Römbke, J., Achazi, R., Warnecke, D., 2003. Avoidance test with *E. fetida* as indicator for the habitat function of soils – results of a laboratory comparison test. J. Soils Sediments 3, 7–12.

Hund-Rinke, K., Wiechering, H., 2001. Earthworm avoidance test for soil assessments. An alternative for acute and reproduction tests. J. Soils Sediments 1, 15–20.

Hutson, D.H., Roberts, T.R., 1990. Environmental Fate of Pesticides. Progress in Pesticide Biochemistry Toxicology Vol. 7. Wiley, Chichester, 286 p.

ISO (International Organization for Standardization), 1998. Soil Quality-Effects of Pollutants on Earthworms (*Eisenia fetida*)-Part 2: Determination of Effects on Reproduction. ISO 11268-2, Geneva, Switzerland.

ISO (International Organization for Standardization), 1999a. Soil Quality-Inhibition of Reproduction of Collembola (*Folsomia candida*) by Soil Pollutants. ISO 11267, Geneva, Switzerland.

ISO (International Organization for Standardization), 1999b. Soil Quality – Effects of Pollutants on Earthworms-Part 3: Guidance on the Determination of Effects in Field Situations. ISO 11268-3, Geneva, Switzerland.

ISO (International Organization for Standardization), 2003a. Soil Quality-Effects of Pollutants on Enchytraeidae (*Enchytraeus* sp.). Determination of Effects on Reproduction and Survival. ISO 16387, Geneva, Switzerland.

ISO (International Organization for Standardization), 2003b. Soil Quality-Guidance on the Ecotoxicological Characterization of Soils and Soil Materials. ISO 15799, Geneva, Switzerland.

ISO (International Organization for Standardization), 2005. Draft: Soil Quality-Avoidance Test for Evaluating the Quality of Soils and the Toxicity of Chemicals. Test with Earthworms (*Eisenia fetida/andrei*). ISO 17512, Geneva, Switzerland.

Jagers Op Akkerhuis, G., 1994. Effects of walking activity and physical factors on the short-term toxity of deltamethrin spraying in Adult Epigeal Money Spiders (Linyphiidae). In: M.H. Donker, H. Eijsackers and F. Heimbach (Eds.), Ecotoxicology of Soil Organisms. CRC Press, Boca Raton, pp. 323–338.

Kennel, W., 1990. The role of the earthworm *Lumbricus terrestris* in integrated fruit production. Acta Hortic. 285, 149–156.

Knacker, T., Van Gestel, C.A.M., Jones, S.E., Soares, A.M.V.M., Schallnass, H.-J., Förster, B., Edwards, C.A., 2004. Ring-testing and field-validation of a Terrestrial Model Ecosystem (TME) – an instrument for testing potentially harmful substances: conceptual approach and study design. Ecotoxicology 13, 9–28.

Lavelle, P., Bignell, D., Lepage, M., 1997. Soil function in a changing world: the role of invertebrate ecosystem engineers. Eur. J. Soil Biol. 33, 159–193.

Lavelle, P., Spain, A.V., Blanchart, E., Martin, A., Martin, S., 1992. Impact of soil fauna on the properties of soils in the humid tropics. In: P.A. Sanchez, R. Lal (Eds.), Myths and Science of Soils in the Tropics. ASA Madison, Wisconsin, pp. 157–185.

Lee, K.E., 1985. Earthworms – their ecology and relationships with soils and land use. Academic Press, Sydney, Australia, 411 pp.

Lucan-Bouché, M.-L., Biagianti-Risbourg, S., Arsac, F., Fernet, G., 1999. An original decontamination process developed by the aquatic oligochaete *Tubifex tubifex* exposed to copper and lead. Aquat. Toxicol. 45, 9–17.

Mather, J.G., Christensen, O.M., 1998. Earthworm surface migration in the field: influence of pesticides using Benomyl as test chemical. In: S.C. Sheppard, J.D. Bembridge, M. Holmstrup and L. Posthuma (Eds.), Advances in Earthworm Ecotoxicology. SETAC Press, Pensacola, USA, pp. 327–340.

Natal-da-Luz, T., Ribeiro, R., Sousa, J.P., 2004. Avoidance tests with Collembola and earthworms as early screening tools for site specific assessment of polluted soils. Environ. Toxicol. Chem. 23, 2188–2193.

OECD (Organisation for Economic Co-Operation and Development), 1984. Guideline for Testing of Chemicals No. 207. Earthworm Acute Toxicity Test, Paris.

OECD (Organisation for Economic Co-Operation and Development), 2004a. Guideline for Testing of Chemicals No. 220. Enchytraeidae Reproduction Test, Paris.

OECD (Organisation for Economic Co-Operation and Development), 2004b. Guideline for Testing of Chemicals No. 222. Earthworm Reproduction Test, Paris.

Peijnenburg, W.J.G.M., Posthuma, L., Zweers, P.G.P.C., Baerselman, R., de Groot, A.C., van Veen, R.P.M., Jager, T., 1999. Prediction of metal bioavailability in Dutch field soils for the oligochaete *Enchytraeus crypticus*. Ecotoxicol. Environ. Saf. 43, 170–186.

Posthuma, L., 1990. Genetic differentiation between populations of *Orchesella cincta* (Collembola) from heavy-metal contaminated sites. J. Appl. Ecol. 27, 609–622.

Römbke, J., Schmidt, M., 1999. REM documentation of putative cuticular sense organs of enchytraeids. In: R.M. Schmelz and K. Sühlo (Eds.), Newsletter on Enchytraeidae. University of Osnabrueck, Osnabrueck, Vol. 6, pp. 15–20.

Schaefer, M., 2004. Assessing 2,4,6-trinitrotoluene (TNT)-contaminated soil using three different earthworm test methods. Ecotoxicol. Environ. Saf. 57, 74–80.

Scott-Fordsmand, J., Weeks, J.M., 1998. Review of selected biomarkers in earthworms. In: S.C. Sheppard, J.D. Bembridge, M. Holmstrup and L. Posthuma (Eds.), Advances in Earthworm Ecotoxicology. SETAC Press, Pensacola, USA, pp. 173–189.

Sjögren, M., 1997. Dispersal rates of Collembola in metal polluted soil. Pedobiologia 41, 506–513.

Slimak, K.M., 1997. Avoidance response as a sublethal effect of pesticides on *Lumbricus terrestris* (Oligochaeta). Soil Biol. Biochem. 29, 713–715.

Spurgeon, D., Svendsen, C., Hankard, P., Weeks, J., Kille, P., Fishwick, S., 2002. Review of Sublethal Ecotoxicological Tests for Measuring Harm in Terrestrial Ecosystems. Environment Agency, Bristol.

Steenhuis, T.S., Staubitz, W., Andreini, M.S., Surface, J., Richard, T.L., Paulsen, R., Pickering, N.B., Hagermann, J.R., Geohring, L.D., 1990. Preferential movement of pesticides and tracers in agricultural soils. J. Irrigation Drainage Engin. 116, 50–66.

Stenersen, J., 1992. Uptake and metabolism of xenobiotics by earthworms. In: P. Greig-Smith, H. Becker, P.J. Edwards and F. Heimbach (Eds.), Ecotoxicology of Earthworms, Intercept, Andover, pp. 129–138.

Stephenson, G.L., Kaushik, A., Kaushik, A., Kaushik, N.K., Solomon, K.R., Steele, T., Scroggins, R.P., 1998. Use of an avoidance-response test to assess the toxicity of contaminated soils to earthworms. In: S.C. Sheppard, J.D. Bembridge, M. Holmstrup and L. Posthuma (Eds.), Advances in Earthworm Ecotoxicology. SETAC Press, Pensacola, USA, pp. 67–81.

Tomlin, A.D., 1992. Behaviour as a source of earthworm susceptibility to ecotoxicants. In: P. Greig-Smith, H. Becker, P.J. Edwards and F. Heimbach (Eds.), Ecotoxicology of Earthworms, Intercept, Andover, pp. 116–125.

Wentsel, R.S., Guelta, M.A., 1988. Avoidance of brass powder-contaminated soil by the earthworm *Lumbricus terrestris*. Environ. Toxicol. Chem. 7, 241–243.

Yeardley, R.B., Lazorchak, J.M., Lazorchak, J.M., Gast, L.C., 1996. The potential of an earthworm avoidance test for evaluation of hazardous waste sites. Environ. Toxicol. Chem. 15, 1532–1537.

Developments in Soil Science, volume 32
Ravendra Naidu (Editor)

Chapter 13

RELATIONSHIP BETWEEN BIOCHEMICAL ACTIVITY AND METAL CONCENTRATION IN SOIL AMENDED WITH SEWAGE SLUDGE

Tom W. Speir

13.1 Introduction

Heavy metals are toxic to living organisms primarily because of their protein binding capacity and hence their ability to inhibit enzyme catalysis, to precipitate cytoplasm proteins and to disrupt cellular membrane structures. There have been many studies to demonstrate toxicity to soil microorganisms (see Chapters 11–13 and 18) and inhibition of soil enzyme-catalysed processes caused by the amendment of soils with heavy metal salts (e.g., Juma and Tabatabai, 1977; Liang and Tabatabai, 1978; Lighthart et al., 1983; Doelman and Haanstra, 1984, 1986; Speir and Ross, 2002). Studies of this type have helped lead to the recognition that soil microorganisms are far more sensitive to heavy metal stress than other soil organisms or plants (Giller et al., 1998). Consequently, tests of microbial function and soil biochemical processes are currently being used to assess soil quality and 'health' in circumstances where soils are exposed to heavy metals. Such ecotoxicological tests have been suggested as appropriate determinants of soil metal limits in regulations pertaining to the application of sewage sludge on land (e.g., Stadelmann and Santschi-Fuhrimann, 1987; Scott-Fordsmand and Pedersen, 1995). However, as is found in investigations across the soil science spectrum, the vast array of experimental conditions used and the inherent complexity and variety of soils make it almost impossible to draw conclusions about critical heavy metal concentrations from the literature.

This chapter presents an overview based on previous reviews (Smith, 1991, 1996; McGrath et al., 1995; Giller et al., 1998) and recent literature to assess the relationship between biochemical activity and metal concentrations in soil amended with sewage sludge and attempt to address the following questions:

(a) What are the short-term (acute) effects of adding metal-contaminated sewage sludge to land and how/why do they differ from soil amendment with heavy metal salts?
(b) What are the long-term (chronic) effects of adding metal-contaminated sewage sludge to land and how/why do they differ from the acute effects?

(c)   What do adverse effects on soil biochemical properties mean and should
      we be concerned?

13.2 Acute effects of sewage sludge on soil biochemical properties

13.2.1 Natural sludge without heavy metal amendments

Aware of previous investigations that demonstrated inhibitory effects of heavy
metal salts on soil enzyme activities, and of the prevalence of trace metals in
sewage sludge, Frankenberger et al. (1983) measured the effect on soil urease
activity of soil amendment with sludge. They found that urease activity was
lower than that in unamended soil at the lowest sludge loading rates (2.2 and
$8.9 \, g \, kg^{-1}$), but activity was invariably greater at higher rates of amendment
$(22–100 \, g \, kg^{-1})$. This was attributed to extremely high metal concentrations in
the sludges inhibiting urease at low amendment rates, but stimulating microbial
activity and urease synthesis by the additional organic matter and nutrients
supplied by the sludge at high amendment rates. To support this contention,
they found that the order of increasing urease activity generally followed that of
sludge organic matter content. Subsequently, short-term incubation studies of
soils amended with low rates of sewage sludge $(5.6–12.3 \, g \, kg^{-1})$ have
demonstrated, more often than not, that soil biological activity was enhanced
(e.g., Bonmati et al., 1985; Chander et al., 1995; Vieira, 2001; Debosz et al., 2002;
García-Gil et al., 2004). In these studies, however, the results are not necessarily
at variance with those of Frankenberger et al. (1983), because the heavy metal
concentrations of the sludges used were generally much lower. At slightly
higher sludge application rates (up to $44.8 \, g \, kg^{-1}$), Chander et al. (1995) found
that, initially, soil microbial biomass was enhanced in proportion to the amount
of sludge added, i.e., similarly to Frankenberger et al. (1983), biological activity
was proportional to the amount of organic matter added. Even at very much
higher rates (up to 35% w/w), Wong et al. (1998, 2000) found that soil
respiration increased with increasing amount of sludge applied, but on
incubation for 70 days, the respiratory activity had declined in all treatments
to levels similar to those of the control soils.

In contrast to the above studies, Reddy et al. (1987) showed that moderate
applications (equivalent to $40–120 \, t \, ha^{-1}$) of relatively uncontaminated sludge
had adverse effects on soil enzyme activity after 40 days' incubation in pots
growing soya bean. This was generally attributed to increased amounts of
DTPA-soluble heavy metals in the amended soils. The authors provided
insufficient information to convert their application rates to $g \, kg^{-1}$ and allow
comparison with other investigations. However, the low metal contents of the
sludge used and the rather low DTPA-soluble metal concentrations suggest that
other factors may have contributed to the observed enzyme inhibition. In a

subsequent study using the same soil and sewage sludge, Reddy and Faza (1989) found that dehydrogenase activity increased in all treatments over four days' incubation. However, activity was always lower in sludge-amended soil than in the control and decreased with increasing loading rate (again 40–$120\,t\,ha^{-1}$). This result was again attributed to the heavy metals in the sludge. Interpretation of dehydrogenase activity, especially after treatment of soil with a nutrient- and C-rich amendment, is, however, fraught with difficulties. Bremner and Tabatabai (1973) and Rossel et al. (1997) have reported that alternative electron acceptors, such as nitrate and humic materials, which are both potential constituents of sewage sludge, interfere with the dehydrogenase assay. In addition, Chander and Brookes (1991b) ascribed a similar interference to Cu and, in light of these methodological problems, Speir and Ross (2002) suggested that dehydrogenase cannot be relied upon as a measure of biological activity in sludge-amended soils.

Abd-Alla et al. (1999), using a pot-trial, studied the effect of very high application rates of sewage sludge on nodulation, nitrogen fixation and plant growth of three legumes. The sludge used was relatively highly contaminated by Cu and Zn (1000 and $2500\,mg\,kg^{-1}$, respectively) and at application rates up to 30% w/w, nodulation, nitrogenase activity and growth were significantly increased. However, at higher rates (40 and 50%), significant inhibition of these properties was observed. The authors concluded that these adverse effects were due to the impact of Cu and Zn on *Rhizobium*. Considering the sensitivity shown by some strains of this legume symbiont, it is not surprising that it is adversely affected at metal concentrations of 400–500 and 1000–$1250\,mg\,kg^{-1}$ Cu and Zn, respectively, imposed by this unrealistically high sludge application rate.

The principal aim of most of the above investigations was to examine the benefits (stemming from the organic matter and/or nutrients) of applying sewage sludge to soil. The experiments were short-term because the beneficial effects are to a large extent transient – nutrients are taken up by plants, or they are leached or immobilised, organic matter decomposes. Generally, the experimental design did not allow metal effects on soil biochemical processes to be assessed, either because there were no effects, or because their impacts were hidden by the general enhancement of biological activity. There are two ways that these limitations can be overcome: spike the sewage sludge with heavy metals to ensure much higher concentrations; and/or, continue the incubations with sludge for a much longer time to ensure transient effects have passed.

13.2.2 Heavy metal-enriched, or -spiked, sludge

Coppola et al. (1988) conducted a pot-trial using two contrasting soils amended with sewage sludge that had been spiked with $CdSO_4$ to give soil Cd concentrations of 0, 2, 4, 8 and 16 mg kg^{-1}. Although the CdSO$_4$ had been added to the sludge two weeks beforehand to allow 'equilibration', plant uptake of Cd was very high (50–70 µg g^{-1}), suggesting that a significant proportion was present in a highly available form. Microbial respiration, measured at the end of the crop growing period, was unaffected by Cd, as also was microbial biomass (by ATP extraction) at all Cd concentrations except at the highest rate in one of the soils. However, increasing Cd concentration in both soils markedly reduced non-symbiotic N$_2$-fixation. In a study comparing the effect of increasing concentrations of PbCl$_2$ (0–500 µg Pb g^{-1} soil), with and without sewage sludge (0.5%), Dar (1997) found that the presence of sludge had no effect on the extent of inhibition of C mineralisation (respiration) or reduction of microbial biomass. In effect, this was a metal salt amendment study and the sewage sludge was essentially irrelevant. Similarly, Moreno et al. (2002, 2003) amended soil and soil–sludge mixtures (5% sludge) with Cd and Ni salts to achieve final metal concentrations up to 8000 mg kg^{-1} soil and found very strong inhibitory effects on dehydrogenase activity, ATP content, microbial respiration and microbial biomass. These studies were, again, essentially metal salt amendment experiments, although the authors concluded that the presence of sewage sludge did provide some protection to soil microbial functions.

Khan and Scullion (1999) amended soil with sewage sludges (4% w/w) that had been spiked with metal salts one week before. The salts were added in combination, i.e., Cu and Ni, Cu and Zn, and Ni and Zn, to achieve soil metal concentrations of approximately 75 and 125% of current EC limits (Cu 135, Ni 75, Zn 300 mg kg^{-1} soil). Soil respiration declined in all treatments over seven weeks' incubation, but after the first week, respiration was higher in the presence of metals than in their absence at each assay time. In contrast, increasing metal input generally reduced microbial biomass C. There was a marked reduction in soil pH in the latter stages of the incubation and this was probably responsible for an increase of available Ni and Zn (CaCl$_2$-extractable) observed at week seven. Indeed, at this time the proportions of total metal removed by this mild extractant were up to 20 and 28% of Ni and Zn, respectively. This is an extraordinarily high proportion and emphasises that amending sludge with metal salts does not truly simulate a sludge that has achieved a metal-enriched state via the normal sewage treatment process. In a similarly designed experiment, but using individual metal-spiked sludge to give metal concentrations of Cd 75, Cu 800, Ni 240, Pb 900 and Zn 900 mg kg^{-1} soil, Khan and Scullion (2002) again found that soil respiration often responded positively to increasing metal concentrations. However, there was no response to Cd at either three or seven weeks' incubation, nor to Ni at seven weeks', and the response to Zn was small. Microbial biomass C and N were generally reduced by increasing metal input, but the responses were not significant for

Cd and Pb (biomass C), and Cd and Pb (seven weeks') and Zn (three weeks') (biomass N). The authors concluded that microbial populations and processes in soil can be influenced by metals in sludges at input rates close to, or for Zn within, current EC limits. In marked contrast to the above two studies, Rost et al. (2001) found that amending soil with spiked sludge to give soil Zn concentrations up to $800\,\mathrm{mg\,kg^{-1}}$ markedly reduced respiration, but had minimal effect on microbial biomass C. Unlike in the previous studies, this resulted in a decline in the ratio of these two properties, the metabolic quotient $q\mathrm{CO_2}$, which is contrary to what would be expected in a microbial population under stress (Killham, 1985). Protease enzyme activity and arginine ammonification were also largely unaffected by increasing Zn concentrations (Rost et al., 2001).

It is apparent from the above studies that short-term incubation of soils treated with metal-spiked sewage sludges often result in adverse effects on soil biochemical processes, in contrast to similar incubations with 'natural' sludges. However, in most, if not all instances, this type of experimental approach is little different from applying metal salts to the soils – the soil is 'shock-dosed' with high concentrations of metals that are considerably more available than those in sewage sludge. Metal salt amendment experiments may have a place in mechanistic studies, but their relevance to assessment of the consequences of land application of sewage sludge is highly debatable. This was recognised many years ago (Logan and Chaney, 1983; Chaney et al., 1987) and has been further emphasised by Smith (1991) and by Giller et al. (1998). Heavy metals in sewage sludge are invariably strongly complexed with mineral and organic components of the sludge and are, therefore, considered to be far less biologically available than metals in salts. In addition, sludge application rates are generally set at levels that do not allow rapid accumulation of metals. Both the type of the microbial response and its sensitivity may be very different and the response obtained in the laboratory may bear little relation to the response seen in the field (Giller et al., 1998). Renella et al. (2002), in a metal salt amendment study, were unable to simulate long-term effects of heavy metals on microbial biomass and processes. They found that the microbial biomass that develops under conditions of chronic metal toxicity responds differently to heavy metals than that subjected to acute exposure. They concluded that this confirmed the assertion of Giller et al. (1998).

In an attempt to determine the minimum soil concentrations of individual metals added in sewage sludge affecting the soil microbial biomass and its activity, Chander et al. (1995) used metal-enriched sludges prepared by Adams and Sanders (1984). The sludges were the product from a pilot sewage treatment plant in which the raw sewage was spiked with metal salts (Cu, Ni and Zn, individually) before treatment. Using sludges prepared in this way should have overcome the high metal availability issue common to spiked-sludge experiments, and neutral salt (0.5 M K_2SO_4) extraction results confirmed that only a low proportion of each metal was extractable at the commencement of the incubation period. However, the amount removed by this extractant increased markedly during a 64-week incubation, suggesting that the metals were initially chelated by the sludge organic matter but became more available as this decomposed during incubation. Sludge was applied at rates from 40–160 t ha^{-1} equivalent, which gave soil metal concentrations of, Cu 150–540, Ni 80–280 and Zn 340–1160 mg kg^{-1}, respectively. Sludge application increased microbial biomass to the same extent, irrespective of whether the sludges were metal-enriched or not and biomass was greatest (4.5-fold higher) at the highest loading rate. During the 64-week incubation, microbial biomass declined exponentially in all treatments and large applications of high metal sludges resulted in final biomasses that were smaller than those given low metal sludge or no sludge. Zinc and Cu at twice the EU limits decreased microbial biomass by about 20% and Ni at four times the limit decreased it by 15%. Dehydrogenase activity and N-mineralisation were also adversely affected in the high metal treatments after 64-weeks' incubation. The approach used in this experiment may have overcome reservations about the form and availability of metals in the sludge-treated soils, and has also involved a much longer incubation period than other studies. However, the shock effect on soil organisms of a sudden high metal exposure, as opposed to a gradual build-up (Giller et al., 1998), remains an unresolved issue in such an investigation. The results suggested that microbial biomass becomes severely impaired as the sludge organic matter is mineralised, but it is debatable, with these high single-dose applications of metals, whether the microbial populations could have had time to adapt (Giller et al., 1998).

13.3 Chronic effects of sewage sludge on soil biochemical properties

Early research indicated that there was little risk to the microbial processes involved in C and N mineralisation in soils from the application of sewage sludge to land (Smith, 1996). These particular processes are now known to be relatively robust and it is only since the development of new techniques to measure the size and activity of the soil microbial biomass that adverse effects attributable to long-term application of sewage sludge have been observed (Smith, 1996).

Brookes and McGrath (1984) provided the first real evidence that the heavy metals supplied to soil during long-term application of sewage sludge have an adverse effect on soil biochemical processes. They showed that the microbial biomass, measured by the fumigation–incubation technique of Jenkinson and Powlson (1976) and by ATP extraction (Tate and Jenkinson, 1982), was substantially reduced in a loamy sand soil from plots at the Woburn (UK) Experimental Farm that had received sewage sludge or sludge-compost between 1942 and 1961, and no amendments except inorganic fertilisers subsequently. They also found that soil respiration was not affected, meaning that the biomass-specific respiration or respiratory quotient (qCO_2) increased in the sludge plots – the stress response postulated by Killham (1985). Brookes and McGrath (1984) suggested that, either less microbial biomass is produced per unit of substrate input in the presence of toxic metals, or that the microbial populations do not live as long in sludge-treated soils. They also examined soils (sandy loam) from plots at Luddington that had received sludge individually enriched in Zn, Cu, Ni or Cr. They found reduced microbial biomass in plots treated with Cu- and Ni-sludges, but in soils from the Zn- and Cr-sludge plots, soil microbial biomasses were in the same order as those in soils given uncontaminated sludge or fertiliser. They concluded that these results suggested that Zn and Cr were not responsible for the toxic effects observed at Woburn. Subsequently, Brookes et al. (1986) showed that development of colonies of blue-green algae (cyanobacteria) and rates of nitrogen fixation were retarded and reduced, respectively, in a high metal soil (sludge-treated plot) from Woburn, compared with those from a low metal soil (farm-yard manure-treated plot). These effects occurred at total metal concentrations about (Zn and Cu), less than (Ni) and considerably greater than (Cd), the 1986 EU soil limits (CEC, 1986). Using soils from the same Woburn plots, McGrath et al. (1988) and Giller et al. (1989) found poor clover growth and ineffective white nodules that only contained *Rhizobium* strains unable to fix nitrogen. Moreover, unless a very large inoculum of effective *R. leguminosarum* bv *trifolii* was applied to soils held moist for two months before sowing with white clover, no N_2-fixation was observed in re-inoculated soils. These adverse effects again occurred at metal concentrations close to the EU guidelines, with a 50% reduction in nitrogen fixation occurring at total metal concentrations of (mg kg^{-1}) 334 Zn, 99 Cu, 27 Ni and 10 Cd (McGrath et al., 1988).

The above series of experiments provided the first strong evidence that, after the confounding effects of sewage sludge organic matter have diminished with time, changes in the structure and function of soil microbial populations can be observed. These changes occurred in soils that had received 'normal' sludge applications, whereby metal concentrations have built up gradually over time. Seemingly, if the microbial populations have adapted (Giller et al., 1998), this adaptation has involved reduction of total microbial biomass and diminished functionality of some important populations. Moreover, this has occurred at total metal concentrations disturbingly close to recommended soil limit values. However, apart from some indication from the Luddington result (Brookes and McGrath, 1984) that Zn and Cr may be of less concern than Cu and Ni, there was little progress towards determining which metal, or combination of metals, was responsible for the apparent toxicity. Since then, there have been a considerable number of investigations of microbial and biochemical properties of soils from long-term experimental sites containing plots previously treated with sewage sludge in the UK, Germany and Sweden. The purpose of these studies was to confirm or refute the Woburn results and/or to try to determine which metal (or combination) was causing the adverse effects. Because of the importance of nitrogen fixation, and the apparent sensitivity of the process and the organisms responsible for it, a number of these investigations have focused on *Rhizobium* and on free-living organisms, cyanobacteria and heterotrophs (e.g., Mårtensson and Witter, 1990; Lorenz et al., 1992; Smith and Giller, 1992; Chaudri et al., 1993; Dahlin et al., 1997; Munn et al., 1997; Smith, 1997; Obbard, 2001). Most have been extensively reviewed by others (McGrath et al., 1995; Giller et al., 1998) and, for this reason and because specific soil microbial populations fall outside the scope of this review, I will return my focus to soil biochemical properties.

Chander and Brookes (1991a) re-visited the Luddington site (sandy loam soil, 15% clay) sampled by Brookes and McGrath (1984) and another site (Lee Valley, stony silt loam soil, 21% clay) that had received identical one-off applications ($125\,t\,ha^{-1}$ dry solids) of Zn, Cu, Ni or Cr-enriched sewage sludge 22 years before. In spite of receiving identical sludge treatments, there were marked differences in soil metal concentrations (especially Zn) between the two sites and at Luddington, between replicate plots. This was attributed to physical movement of soil across plot boundaries during many years of cultivation. At both sites, Cu and Zn at about 2–3 times EC soil limit concentrations caused an accumulation of soil organic matter and a reduction of microbial biomass. In contrast, Ni also at 2–3 times EC limits, had no effect on organic matter or microbial biomass. The authors concluded that the most probable order of metal toxicity to the microbial biomass (in terms of the EC limits) is Cu > Zn ≫ Ni or Cd. Subsequently, Chander and Brookes (1993) investigated the relationships between total metals, $CaCl_2$-extractable metals and soil microbial biomass in soils from a UK experimental farm (Gleadthorpe) that had received metal-spiked sewage sludge eight and four years beforehand. The soil, a sandy loam (9% clay) had, in 1982, received sludges ($200\,t\,ha^{-1}$ dry solids) spiked individually with Zn, Cu and Ni and with combinations of Zn+Cu and Zn+Ni, each over a range of metal concentrations. In 1986, some plots received a further addition of sludge naturally contaminated with high levels of Zn and/or Cu to boost metal concentrations. Soil Ni concentrations never exceeded EC limits, but those of Zn and Cu exceeded limit values by up to 2.3 and 4.9 times, respectively. Addition of uncontaminated sludge and Ni-contaminated sludge increased microbial biomass C by about 8%, compared to amounts in soil that had never received sludge. Similarly, slight elevations of Zn and Cu also increased biomass C, but at all concentrations exceeding the soil limit values (Zn 300 and Cu 140 mg kg^{-1}) biomass C was reduced. At the highest concentrations of Zn alone (705 mg kg^{-1}) and Cu alone (690 mg kg^{-1}) biomass C was reduced by 36 and 51%, respectively. Of major significance was that Zn and Cu in combination had a much greater adverse effect on biomass C than did the individual metals at similar concentrations. In contrast, a combination of Zn and Ni did not decrease biomass C. This may be because in these latter combinations, neither metal exceeded the soil limit concentrations. The authors also calculated microbial biomass C as a percentage of total soil C and these results further emphasised the effects of metals added singly or in combination (Fig. 13.1). Moreover, whereas biomass C generally comprises about 1–4% of total soil organic C, and in the control soil and that treated with uncontaminated or slightly contaminated sludge the value was about 1.5–1.6% (i.e., in the normal range), the percentage was often less than half of this in the soils containing high metal concentrations (Fig. 13.1). Chander and Brookes (1993) concluded that the ratio of soil biomass C to total organic C provides a sensitive indicator of the effects of heavy metals on the soil microbial biomass. Perhaps the one note of caution about this study emerges from the CaCl$_2$-extractable metal

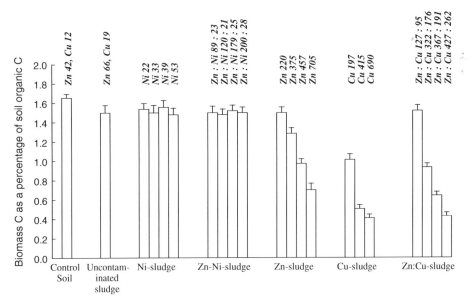

Figure 13.1. *Microbial biomass C expressed as a percentage of total soil organic C in Gleadthorpe soils (standard errors shown). Values given in italics text are the total soil metal concentrations ($\mu g\,g^{-1}$ soil). (Reprinted from Chander and Brookes (1993) with permission from Elsevier.)*

concentrations. For the soil treated with uncontaminated sludge, $CaCl_2$ extracted 23% of the total Zn, whereas in soils where total Zn exceeded EU limit concentrations, 38–42% was extracted. Similarly for Cu, only 1% was extracted from the uncontaminated sludge-treated soil, but up to 9% was extracted from soils with total Cu above limit concentrations. This indicated that the metals in soil that received the spiked sludge were somewhat more available than those in soil that received the natural sludge, even though the system had had eight years to equilibrate. In addition, this study again raises the issue of the capacity for the microbial populations to adapt after 'shock-dosing' with one application of highly contaminated sewage sludge, as discussed by Giller et al. (1998).

The field trials at Braunschweig in Germany provided strong evidence, through the study of *Rhizobium*, that Zn was the metal likely to be causing adverse effects at both this site and at the Woburn site (Chaudri et al., 1993; McGrath et al., 1995). There are two trials at Braunschweig, one under long-term arable management and one converted from woodland to arable in the 1950s. The soil type and texture (silty loam) are the same at both sites, but they differ in pH (ex-woodland >1 pH unit lower) and in organic C content (comparable sludge treatments have more C at the ex-woodland site) (Chaudri et al., 1993). Plots were treated with liquid sewage sludge (anaerobically digested and undigested) at rates of 100 or $300 \, m^3 \, ha^{-1}$ (approx. 5 and $15 \, t \, ha^{-1}$) annually from 1980. After the first year, the sludge was either 'low metal' or 'high metal'; the latter obtained by spiking low metal sludge with salts of Cd, Cr, Cu, Hg, Ni, Pb and Zn. Obbard et al. (1994) measured dehydrogenase activity in the presence of added glucose – potential dehydrogenase, indicative of the physiologically active microbial biomass (Rossel and Tarradellas, 1991). At the time of sampling, only Zn in the high-rate high metal sludge plots exceeded the EC soil limit concentration ($300 \, mg \, kg^{-1}$). Significantly higher substrate-induced dehydrogenase activities were found in soils from both trials that had received $100 \, m^3 \, ha^{-1}$ sludge compared with the respective control soils, irrespective of whether the sludge had been spiked or not. However, application of digested sludge at the higher rate resulted in dehydrogenase activities that were significantly lower than those of the control soils, with little difference found between the effects of high and low metal sludges. In contrast, undigested sludge (applied to the long-term arable site only) stimulated activity at both low and high application rates, presumably because of the greater abundance of readily digestible organic matter it contains. The authors concluded that, although selected heavy metals may suppress substrate-induced dehydrogenase activity at high concentrations in sludge-amended soils, the level and type of sludge addition had a more significant effect in influencing activity than the concentration of heavy metals present.

At about the same time as the above study, Fließbach et al. (1994) investigated the microbial biomass and activity of the soils of the Braunschweig trials. Again, in their samples, the only occasions that soil limit concentrations were exceeded were for Zn in the high-rate, high metal treatments. Compared to the unsludged control soils, sludge application caused an increase in microbial biomass C, but this beneficial effect was less pronounced in soils receiving high metal sludge. In the $300\,m^3\,ha^{-1}$ treatments the reduction in the high metal plots averaged 26%. The microbial C/organic C ratio, previously postulated to be a sensitive indicator of adverse effects of heavy metals (Chander and Brookes, 1993), was always lower in the high metal plots than in those receiving the same application rate of low metal sludge – up to 32% lower at the highest contamination of the former woodland site. Microbial respiration increased substantially in the plots receiving $300\,m^3\,ha^{-1}\,y^{-1}$, this being especially marked in the former woodland soil. In both trials, the increase was greater in the high metal than in the low metal soils. This, coupled with the decrease of microbial biomass C in the high metal plots, resulted in a markedly higher qCO_2, especially in the high metal plots receiving the highest sludge application rates. Fließbach et al. (1994) also showed that the fungal contribution to substrate-induced respiration increased with sludge application in both trials; from 78% in the unsludged arable soil to 96% in the $300\,m^3$ high metal soil, and from 71 to 97% in the equivalent old woodland soils. The authors concluded that long-term effects of exposure to heavy metals may result in a less efficient and heavy metal tolerant soil microbial community, that is mainly composed of fungi. They were also concerned that ecosystem structures did not appear to be protected by the EC soil metal limit concentrations, since heavy metal-induced changes were detectable even below these concentration limits. Subsequently, Witter et al. (2000) revisited the Braunschweig sludge-treated arable plots and confirmed that the microbial biomass C/organic C ratios were significantly reduced in the $300\,m^3\,ha^{-1}$ treatments. They were also able to demonstrate that there was a progressive development of metal tolerance with increased metal loading.

Witter and colleagues also conducted a series of studies on soils from a field experiment at Ultuna, Uppsala, in Sweden, that had received a number of amendments, including sewage sludge application every second year from 1956 (Witter et al., 1993; Witter and Dahlin, 1995; Dahlin and Witter, 1998). When sampled in 1988, only Cu, Ni and Zn concentrations (125, 35 and $230\,mg\,kg^{-1}$, respectively) approached EU limit values. They found that sewage sludge was one of three amendments that considerably reduced soil pH and this, to an extent, confounded their interpretation of effects on soil microbial biomass, which was also reduced in these treatments (Witter et al., 1993). They considered that the lower microbial biomass content in the sewage sludge-treated soil could have been due to low degradability of the sludge C, or to the somewhat lower pH, or to the elevated heavy metal concentrations imposing a stress on the microbial population. Liming the soil did not increase the size of the microbial biomass and it was concluded that there was a negative effect of higher heavy metal concentrations in the sludge-amended soil. As with previous studies (Chander and Brookes, 1991c), this was attributed to a reduced substrate-utilisation efficiency of the soil microbial biomass. Witter et al. (1993) also found that soil respiration in the sewage sludge-amended soil was somewhat lower than expected on the basis of the soil's C content. This was attributed again to the heavy metals present in the soil, but as mentioned previously and probably not given due regard, this could as easily be due to sewage sludge C being less degradable than the C supplied in the other amendments used in the experiment. Subsequently, Witter and Dahlin (1995), using labelled glucose and straw additions to soils from the same plots, showed that more glucose C had respired from the sewage sludge-amended soil than from fertilised treatments or a farmyard manure treatment, but that the sum of respired and microbial biomass-incorporated glucose C did not differ between soils. This suggested that in the sludge-treated soil a greater proportion of glucose was used for energy. The straw amendment data showed, in contrast, that less straw-C was respired in the sludge-treated soil and that the sum of incorporated and respired straw-C was also lower, suggesting a reduced capacity to metabolise more complex substrates. In a further experiment on the same site, Dahlin and Witter (1998) concluded that higher maintenance requirements most likely contributed to the smaller microbial biomass in the sewage sludge-amended soil. They were, however, unable to demonstrate that reduced substrate-utilisation efficiency also contributed.

Dahlin et al. (1997) measured, along with other properties, the size of the microbial biomass, its specific respiration rate, basal respiration and glucose-induced respiration at another Swedish trial at Brunnby, where sewage sludge had been added every fifth year from 1966–1989 at rates of 0, 5, 10 and 20 dry $t\,ha^{-1}$. Even at the highest application rate, heavy metal concentrations were relatively low, with mean concentrations: Cd 0.78, Cr 31, Cu 60, Ni 23, Pb 25 and Zn 117 mm kg^{-1} soil. Nevertheless, the only property that was not affected by past sludge application was basal respiration. The changes were, however, moderate, compared to those found at the Woburn site, which is encouraging considering that the metal concentrations were much lower than at Woburn. The most contaminated soil at Brunnby showed a 15% reduction in the ratio of microbial biomass C to soil organic C and a 20% increase in specific respiration rate (c.f., a 50% decrease and a doubling, respectively, at Woburn, Brookes and McGrath, 1984). In addition, microbial growth upon addition of glucose was characterised by a near doubling of the lag time before onset of exponential growth and a higher specific growth rate compared with that in the control soil.

The most disturbing aspect of this investigation was that, even though statistically significant effects on soil microbial properties were only seen at the highest rate of sludge application, several properties were apparently affected even at the low rates of sludge addition. They correlated significantly with soil Cu concentration, with the particularly close fit of microbial growth characteristics after glucose addition suggesting a gradual change as Cu concentrations increased, rather than a threshold level for metal toxicity.

Kelly et al. (1999) investigated soil metal concentrations and microbial communities in a plot that had received a single large dose (244 t ha^{-1}) of dewatered highly contaminated sewage sludge 18 years before sampling. The soil was a silty clay loam, with metal contents Cd 45, Cr 512, Cu 341, Ni 159, Pb 337 and Zn 1506 mg kg^{-1}. The soil metal concentrations in this US study were generally very much higher than EU limit values, but were mainly within acceptable ranges for that country (USEPA, 1993). Microbial biomass C, although lower in the sludge-treated plot than in an adjacent control, was not reduced significantly. Decreased rates of BIOLOG colour development were found in samples from the sludge-treated plot, indicating reduced substrate-utilisation efficiency. In addition, dehydrogenase activity was markedly and significantly reduced (20-fold) in the sludge-treated plot. However, as mentioned previously, but not discussed by Kelly et al. (1999), there are problems with the interpretation of dehydrogenase activity in sludge-treated soils and/or soils contaminated with Cu (Chander and Brookes, 1991b; Speir and Ross, 2002). Consequently, there is evidence, but not particularly strong evidence, of reduced soil biological activity and efficiency, from this study of a soil contaminated by relatively high levels of sewage sludge-borne heavy metals.

In a Japanese study of a volcanically derived sandy loam soil, Kunito et al. (2001) found no suppression of microbial activities (microbial biomass and nine enzyme activities) by the addition of sewage sludge that raised total Cu and Zn to 144 and 216 mg kg^{-1}, respectively. Sludge had been applied twice annually since 1978. Microbial biomass C and four enzyme activities increased, mostly not significantly, and the remaining enzyme activities decreased, all not significantly, with sewage sludge application. However, the ratios of enzyme activity to microbial biomass tended to decrease with sludge application, suggesting that trace metals caused some adverse effects on soil microorganisms. These ratios, as with that between microbial biomass C and organic C, have been suggested to be sensitive indicators of environmental stress resulting from soil contamination (Tscherko and Kandeler, 1997; Landi et al., 2000). However, there was a marked difference in pH between the control plot and the sludge-treated plot and regression analysis was unable to determine whether the changes that occurred were a result of this or of the metals in the sludge.

Parat et al. (2005) also found that microbial biomass C had been increased by long-term sewage sludge application (every two years from 1974 to 1993) in a sandy soil sampled six years after sludge application ceased. The greatest increase occurred in plots receiving the highest rates of sludge, and where total Zn concentrations had increased to $> 800\,\mathrm{mg\,kg^{-1}}$. When microbial biomass was expressed as a proportion of total C, there were no significant differences between control and sludge-amended plots. The authors suggest that the absence of effects at such high Zn concentrations can be attributed to most of the metal existing in low availability oxide and residual forms. However, this is only a partial explanation, because exchangeable soil Zn was relatively high $(67\,\mathrm{mg\,kg^{-1}})$ and perhaps the results indicate that the soil microorganisms have adapted to the higher trace metal concentrations (Parat et al., 2005).

13.4 What is the relationship between metal concentration and soil biochemical activity?

Over the past 30 years there have been many demonstrations of the adverse effects of heavy metal salts on a large number of soil biochemical properties, including indices of microbial activity and biomass, organic matter and nitrogen dynamics and soil enzyme activities. These effects have generally been measured over short-term incubations – unrealistic scenarios for environmental impact assessment. However, such investigations cannot be dismissed as irrelevant in the context of the long-term fate and effects of heavy metals in sewage sludge, because they provide valuable insights into the factors that influence the behaviour of metals in soil. Speir and Ross (2002) argue that inhibition of enzyme activity in heavy metal-contaminated soil should reflect the bioavailability of the metals, since the mechanisms that are protecting soil enzymes are likely to be the same mechanisms limiting metal uptake by plants and soil organisms. In the longer term, contaminated site assessment has shown that the effects of heavy metal contamination on soil biochemical properties are long lasting (e.g., Tyler, 1974; Bardgett et al., 1994; Kuperman and Carreiro, 1997). However, there is insufficient evidence to determine whether soil-related protective mechanisms are equally effective when the biochemical properties are reduced by repression and diminished microbial growth, as they are by direct inhibition.

Addition of sewage sludge to land introduces a major confounding factor not present in metal amendment studies, or at most sites contaminated by heavy metals – a large amount of nutrient-rich, readily degradable organic matter. There is no question that the presence of this organic matter totally dominates the subsequent behaviour of the soil microbial populations, at least in the short-term. Consequently, short-term studies have, as reviewed above, almost always shown positive responses to the organic matter and nutrients that mask any possible adverse effects due to heavy metals. However, there is almost certainly another factor involved as well – greatly reduced bioavailability of metals that have entered the waste stream in low concentrations and been concentrated in sludge during the sewage treatment process. This reduced availability results from complexation, precipitation and sorption reactions, as well as other interactions, with the sludge organic and mineral matrix. Studies with metal-spiked sludge have tended to emphasise the low availability of metals that have passed as a 'normal' burden through wastewater treatment plants, because the added metals are demonstrably more bioavailable, at least in the short-term. Consequently, adverse effects on biochemical properties have been found, in spite of the added organic matter.

Clearly, of most importance to ecosystem functioning are the long-term consequences of sewage sludge application to land. Given the issues raised above, what then might be expected to happen under long-term sludge application and, especially, what might be expected after sludge application has ceased? The answer depends on what happens to the heavy metals after the mineralisable sludge organic matter has disappeared – does the metal bioavailability change, or are there always sufficient soil organic matter and reactive mineral components to keep the metals effectively inert? Results from some, but not all, long-term field trials suggest that there have been changes in microbial community structure, reductions in microbial biomass, increased microbial stress and reduced capacity to carry out certain key functions (e.g., N_2-fixation, C-mineralisation). However, in several of these trials sludge is still being applied and others involve the use of metal-spiked sludge, making comparisons very difficult. The Woburn trials probably represent the benchmark, strongly indicating that adverse effects on soil biochemical properties, microbial populations and microbial functions occur at metal concentrations around or below the 1986 EU limit values. Although Zn has been strongly implicated as the key contaminant, there is no proof that it is. Indeed, early work at Luddington suggested that Zn was unimportant (Brookes and McGrath, 1984) and others have provided evidence that Cd is the critical element at least for some soil organisms (Smith and Giller, 1992). The importance of soil type, and hence soil properties that might influence metal availability, has not emerged from the long-term studies, except that Woburn is a loamy sand soil, which might be expected to be less protective than heavier-textured soils. It is noteworthy that, in studies on other sandy soils, e.g., Kunito et al. (2001) and Parat et al. (2005) no effects attributable to heavy metals were observed.

In conclusion, it is obvious that there is no clear relationship between metal(s) concentration and biochemical activities in sewage sludge-amended soils. Short-term studies cannot give us useful data because they are not appropriate and, moreover, they are confounded by sludge organic matter. Long-term field trials are appropriate, but were never set up to answer questions about adverse effects of heavy metals – they were components of soil fertility trials. Issues that have not yet been comprehensively addressed in field trials are:

(1)   What are the effects of soil type, soil properties and soil management?
(2)   What is the influence of sewage treatment process on the character of the sludge produced and how does this influence the long-term fate of the metals?
(3)   What are the critical soil concentrations of each of the major metal contaminants in sewage sludge and how are they influenced by factors raised in questions (1) and (2)?

In recent times, purpose-designed long-term field trials have been set up in different parts of the world in attempts to provide answers to these questions.

References

Abd-Alla, M., Yan, F., Schubert, S., 1999. Effect of sewage sludge application on nodulation, nitrogen fixation, and plant growth of faba bean, soybean and lupin. J. Appl. Bot. 73, 69–75.

Adams, TMcM., Sanders, J.R., 1984. The effect of pH on the release to solution of zinc, copper and nickel from metal-loaded sewage sludge. Environ. Pollut. 8B, 85–99.

Bardgett, R.D., Speir, T.W., Ross, D.J., Yeates, G.W., Kettles, H.A., 1994. Impact of pasture contamination by copper, chromium and arsenic timber preservative on soil microbial properties and nematodes. Biol. Fertil. Soils 18, 71–79.

Bonmati, M., Pujola, M., Sana, J., Soliva, M., Felipo, M.T., Garau, M., Ceccanti, B., Nannipieri, P., 1985. Chemical properties, populations of nitrite oxidizers, urease and phosphatase activities in sewage sludge-amended soil. Plant Soil 84, 79–91.

Bremner, J.M., Tabatabai, M.A., 1973. Effect of some inorganic substances on TTC assay of dehydrogenase activity in soils. Soil Biol. Biochem. 5, 385–386.

Brookes, P.C., McGrath, S.P., 1984. Effects of metal toxicity on the size of the soil microbial biomass. J. Soil Sci. 35, 341–346.

Brookes, P.C., McGrath, S.P., Heijnen, C., 1986. Metal residues in soils previously treated with sewage-sludge and their effects on growth and nitrogen fixation by blue-green algae. Soil Biol. Biochem. 18, 345–353.

CEC–Commission of the European Communities, 1986. Council directive (86/278/EEC) on the protection of the environment, and in particular of the soil, when sewage sludge is used in agriculture. Off. J. Eur. Comm. 181, 6–12.

Chander, K., Brookes, P.C., 1991a. Effects of heavy metals from past applications of sewage sludge on microbial biomass and organic matter accumulation in a sandy loam and silty loam U.K. soil. Soil Biol. Biochem. 23, 927–932.

Chander, K., Brookes, P.C., 1991b. Is the dehydrogenase assay invalid as a method to estimate microbial activity in Cu-contaminated and non-contaminated soils? Soil Biol. Biochem. 23, 909–915.

Chander, K., Brookes, P.C., 1991c. Microbial biomass dynamics during the decomposition of glucose and maize in metal-contaminated and non-contaminated soils. Soil Biol. Biochem. 23, 917–925.

Chander, K., Brookes, P.C., 1993. Residual effects of zinc, copper and nickel in sewage sludge on microbial biomass in a sandy loam. Soil Biol. Biochem. 25, 1231–1239.

Chander, K., Brookes, P.C., Harding, S.A., 1995. Microbial biomass dynamics following addition of metal-enriched sewage sludges to a sandy loam. Soil Biol. Biochem. 27, 1409–1421.

Chaney, R.L., Bruins, R.F.J., Baker, D.E., Korcak, R.F., Smith, J.E., Cole, D., 1987. Transfer of sludge-applied trace elements to the food chain. In: A.L. Page, T.J. Logan and J.A. Ryan (Eds.), Land Application of Sludge Food Chain Implications (Ed.). Lewis Publishers Inc, Chelsea, Michigan, pp. 67–99.

Chaudri, A.M., McGrath, S.P., Giller, K.E., Rietz, E., Sauerbeck, D.R., 1993. Enumeration of indiginous *Rhizobium leguminosarum* biovar *trifolii* in soils previously treated with metal-contaminated sewage sludge. Soil Biol. Biochem. 25, 301–309.

Coppola, S., Dumontet, S., Pontonio, M., Basile, G., Marino, P., 1988. Effect of cadmium-bearing sewage sludge on crop plants and microorganisms in two different soils. Agric. Ecosyst. Environ. 20, 188–194.

Dahlin, S., Witter, E., 1998. Can the low microbial biomass C- to-organic C ratio in an acid and a metal contaminated soil be explained by differences in the substrate utilization efficiency and maintenance requirements? Soil Biol. Biochem. 30, 633–641.

Dahlin, S., Witter, E., Mårtensson, A., Turner, A., Bååth, E., 1997. Where's the limit? Changes in the microbiological properties of agricultural soils at low levels of metal contamination. Soil Biol. Biochem. 29, 1405–1415.

Dar, G.H., 1997. Impact of lead and sewage sludge on soil microbial biomass and carbon and nitrogen mineralization. Bull. Environ. Contam. Toxicol. 58, 224–240.

Debosz, K., Petersen, S.O., Kure, L.K., Ambus, P., 2002. Evaluating effects of sewage sludge and household compost on soil physical, chemical and microbiological properties. Appl. Soil Ecol. 19, 237–248.

Doelman, P., Haanstra, L., 1984. Short-term and long-term effects of cadmium, chromium, copper, nickel, lead and zinc on soil microbial respiration in relation to abiotic soil factors. Plant Soil 79, 317–327.

Doelman, P., Haanstra, L., 1986. Short- and long-term effects of heavy metals on urease activity in soils. Biol. Fertil. Soils 2, 213–218.

Fließbach, A., Martens, R., Reber, H.H., 1994. Soil microbial biomass and microbial activity in soils treated with heavy metal contaminated sewage sludge. Soil Biol. Biochem. 26, 1201–1205.

Frankenberger Jr., W.T., Johanson, J.B., Nelson, C.O., 1983. Urease activity in sewage sludge-amended soils. Soil Biol. Biochem. 15, 543–549.

García-Gil, J.C., Plaza, C., Senesi, N., Brunetti, G., Polo, A., 2004. Effects of sewage sludge amendment on humic acids and microbiological properties of a semiarid Mediterranean soil. Biol. Fertil. Soils 39, 320–328.

Giller, K.E., McGrath, S.P., Hirsch, P.R., 1989. Absence of nitrogen fixation in clover grown on soil subject to long-term contamination with heavy metals is due to survival of only ineffective *Rhizobium*. Soil Biol. Biochem. 21, 841–848.

Giller, K.E., Witter, E., McGrath, S.P., 1998. Toxicity of heavy metals to microorganisms and microbial processes in agricultural soils: a review. Soil. Biol. Biochem. 30, 1389–1414.

Jenkinson, D.S., Powlson, D.S., 1976. The effect of biocidal treatments on metabolism in soil, V. Method for measuring soil biomass. Soil Biol. Biochem. 8, 209–213.

Juma, N.G., Tabatabai, M.A., 1977. Effects of trace elements on phosphatase activity in soils. Soil Sci. Soc. Am. J. 41, 343–346.

Kelly, J.J., Häggblom, M., Tate III, R.L., 1999. Effects of the land application of sewage sludge on soil heavy metal concentrations and soil microbial communities. Soil Biol. Biochem. 31, 1467–1470.

Khan, M., Scullion, J., 1999. Microbial activity in grassland soil amended with sewage sludge containing varying rates and combinations of Cu, Ni and Zn. Biol. Fertil. Soils 30, 202–209.

Khan, M., Scullion, J., 2002. Effect of metal (Cd, Cu, Ni, Pb or Zn) enrichment of sewage-sludge on soil micro-organisms and their activities. Appl. Soil Ecol. 20, 145–155.

Killham, K., 1985. A physiological determination of the impact of environmental stress on the activity of microbial biomass. Environ. Pollut. (Ser. A) 38, 283–294.

Kunito, T., Saeki, K., Goto, S., Hayashi, H., Oyaizu, H., Matsumoto, S., 2001. Copper and zinc fractions affecting microorganisms in long-term sludge-amended soils. Biores. Technol. 79, 135–146.

Kuperman, R.G., Carreiro, M.M., 1997. Soil heavy metal concentrations, microbial biomass and enzyme activities in a contaminated grassland ecosystem. Soil Biol. Biochem. 29, 179–190.

Landi, L., Renella, G., Moreno, J.L., Falchini, L., Nannipieri, P., 2000. Influence of cadmium on the metabolic quotient, L-:D-glutamic acid respiration ratio and enzyme activity:microbial biomass ratio under laboratory conditions. Biol. Fertil. Soils 32, 8–16.

Liang, C.N., Tabatabai, M.A., 1978. Effects of trace elements on nitrification in soils. J. Environ. Qual. 7, 291–293.

Lighthart, B., Baham, J., Volk, V.V., 1983. Microbial respiration and chemical speciation in metal-amended soils. J. Environ. Qual. 12, 543–548.

Logan, T.J., Chaney, R.L., 1983. Utilization of municipal wastewater and sludge on land-metals. In: A.L. Page, T.L. Gleason III, J.E. Smith Jr., I.K. Iskander and J.E. Sommers (Eds.), Proceedings of the 1983 Workshop, on Utilization of Municipal Wastewater and Sludge on Land. University of California, Riverside, pp. 235–326.

Lorenz, S.E., McGrath, S.P., Giller, K.E., 1992. Assessment of free-living nitrogen fixation activity as a biological indicator of heavy metal toxicity in soil. Soil Biol. Biochem. 24, 601–606.

Mårtensson, A.M., Witter, E., 1990. Influence of various soil amendments on nitrogen-fixing soil microorganisms in a long-term field experiment, with special reference to sewage sludge. Soil Biol. Biochem. 22, 977–982.

McGrath, S.P., Brookes, P.C., Giller, K.E., 1988. Effects of potentially toxic metals in soil derived from past applications of sewage sludge on nitrogen fixation by *Trifolium repens* L. Soil Biol. Biochem. 20, 415–424.

McGrath, S.P., Chaudri, A.M., Giller, K.E., 1995. Long-term effects of metals in sewage sludge on soils, microorganisms and plants. J. Indust. Microbiol. 14, 94–104.

Moreno, J.L., Hernández, T., Pérez, A., García, C., 2002. Toxicity of cadmium to soil microbial activity: effect of sewage sludge addition to soil on the ecological dose. Appl. Soil Ecol. 21, 149–158.

Moreno, J.L., Pérez, A., Aliaga, A., Hernández, T., 2003. The ecological dose of nickel in a semiarid soil amended with sewage sludge related to the unamended soil. Water Air Soil Pollut. 143, 289–300.

Munn, K.J., Evans, J., Chalk, P.M., Morris, S.G., Whatmuff, M., 1997. Symbiotic effectiveness of *Rhizobium trifolii* and mineralisation of legume nitrogen in response to past amendment of a soil with sewage sludge. J. Sustain. Agric. 11, 23–37.

Obbard, J.P., 2001. Ecotoxicological assessment of heavy metals in sewage sludge amended soils. Appl. Geochem. 16, 1405–1411.

Obbard, J.P., Sauerbeck, D., Jones, K.C., 1994. Dehydrogenase activity of the microbial biomass in soils from a field experiment amended with heavy metal contaminated sewage sludges. Sci. Total Environ. 142, 157–162.

Parat, C., Chaussod, R., Lévêque, J., Andreux, F., 2005. Long-term effects of metal-containing farmyard manure and sewage sludge on soil organic matter in a fluvisol. Soil Biol. Biochem. 37, 673–679.

Reddy, G.B., Faza, A., 1989. Dehydrogenase activity in sludge amended soil. Soil Biol. Biochem. 21, 327.

Reddy, G.B., Faza, A., Bennett Jr., R., 1987. Activity of enzymes in rhizosphere and non-rhizosphere soils amended with sludge. Soil Biol. Biochem. 19, 203–205.

Renella, G., Chaudri, A.M., Brookes, P.C., 2002. Fresh additions of heavy metals do not model long-term effects on microbial biomass and activity. Soil Biol. Biochem. 34, 121–124.

Rossel, D., Tarradellas, J., 1991. Dehydrogenase activity of soil microflora: significance in ecotoxicological tests. Environ. Toxic. Water Qual. 6, 17–33.

Rossel, D., Tarradellas, J., Bitton, G., Morel, J-L., 1997. Use of enzymes in soil ecotoxicology: a case for dehydrogenase and hydrolytic enzymes. In: J. Tarradellas, J. Bitton and D. Rossel (Eds.), Soil Ecotoxicology. Boca Raton, CRC Press, pp. 179–206.

Rost, U., Joergensen, R.G., Chander, K., 2001. Effects of Zn enriched sewage sludge on microbial activities and biomass in soil. Soil Biol. Biochem. 33, 633–638.

Scott-Fordsmand, J. J., Pedersen, M. B., 1995. Soil quality criteria for selected inorganic compounds. Arbejdsrapport fra Miljøstyrelsen Working Report No. 48, Danish Environmental Protection Agency, Ministry of Environment and Energy, Copenhagen, pp. 200.

Smith, S.R., 1991. Effects of sewage sludge application on soil microbial processes and soil fertility. In: B.A. Stewart (Ed.), Advances in Soil Science, Vol. 16. Springer-Verlag, New York, pp. 191–212.

Smith, S.R. (Ed.), 1996. Effects of PTEs on soil fertility and natural ecosystems. In: Agricultural Recycling of sewage Sludge and the Environment. CAB International, Wallingford, pp. 119–153.

Smith, S.R., 1997. *Rhizobium* in soils contaminated with copper and zinc following the long-term application of sewage sludge and other organic wastes. Soil Biol. Biochem. 29, 1475–1489.

Smith, S.R., Giller, K.E., 1992. Effective *Rhizobium leguminosarum* biovar *trifolii* present in five soils contaminated with heavy metals from long-term applications of sewage sludge of metal mine spoil. Soil Biol. Biochem. 24, 781–788.

Speir, T.W., Ross, D.J., 2002. Hydrolytic enzyme activities to assess soil degradation and recovery. In: R.G. Burns, R.P. Dick (Eds.), Enzymes in the Environment: Activity, Ecology and Applications. Marcel Dekker Inc, New York, pp. 407–431.

Stadelmann, F.X., Santschi-Fuhrimann, E., 1987. Beitrag zur Abstützung von Schwermetall-Richtwerten im Boden mit Hilfe von Bodenatmungsmessungen. Eidgenössische Forschungsanstalt für Agrikulturchemie und Umwelthygenie (FAC), Liebefeld-Bern.

Tate, K.R., Jenkinson, D.S., 1982. Adenosine triphosphate measurements in soil: an improved method. Soil Biol. Biochem. 14, 331–335.

Tscherko, D., Kandeler, E., 1997. Ecotoxicological effects of fluorine deposits on microbial biomass and enzyme activity in grassland. Eur. J. Soil Sci. 48, 329–335.

Tyler, G., 1974. Heavy metal pollution and soil enzymatic activity. Plant Soil 40, 303–311.

US EPA—United States Environment Protection Agency, 1993. Part 503 — Standards for the use and disposal of sewage sludge. Federal Register 58, 9387–9404.

Vieira, R.F., 2001. Sewage sludge effects on soybean growth and nitrogen fixation. Biol. Fertil. Soils 34, 196–200.

Witter, E., Dahlin, S., 1995. Microbial utilization of [U-^{14}C]-labelled straw and [U-^{13}C]-labelled glucose in soils of contrasting pH and metal status. Soil Biol. Biochem. 27, 1507–1516.

Witter, E., Gong, P., Bååth, E., Marstorp, H., 2000. A study of the structure and metal tolerance of the soil microbial community six years after cessation of sewage sludge applications. Environ. Toxicol. Chem. 19, 1983–1991.

Witter, E., Mårtensson, A.M., Garcia, F.V., 1993. Size of the soil microbial biomass in a long-term field experiment as affected by different N-fertilizers and organic manures. Soil Biol. Biochem. 25, 659–669.

Wong, J.W.C., Lai, K.M., Fang, M., Ma, K.K., 1998. Effect of sewage sludge amendment on soil microbial activity and nutrient mineralization. Environ. Internat. 24, 935–943.

Wong, J.W.C., Lai, K.M., Fang, M., Ma, K.K., 2000. Soil biology of low grade landfill soil with sewage sludge amendment. Environ. Technol. 21, 1233–1238.

D: Bioavailability of nutrients and agrichemicals

A review of literature shows that unlike highly contaminated soils that have formed the focus of much bioavailability research, only limited attention has been directed to 'bioavailability of nutrients and agrichemicals'. This section focuses on techniques currently being used for assessing nutrient and other agrichemical bioavailability in soils including a critical review of gaps and new areas of research. A special aspect of the impact of nutrients and agrichemicals is that their environmental impact is through non-point source pollution. Thus a key issue is, and increasingly will be, the scaling up of local results to the catchment scale.

Developments in Soil Science, volume 32
Ravendra Naidu (Editor)

Chapter 14

TECHNIQUES FOR ASSESSING NUTRIENT BIOAVAILABILITY IN SOILS: CURRENT AND FUTURE ISSUES

M.J. Hedley

14.1 Introduction

One hundred and fifty years of research by soil chemists and agronomists has attempted to determine those portions of chemical compounds in soils that readily release essential elements (*nutrients*) for uptake by the roots of food and fibre producing crops (Russell, 1973; Soon, 1985). The techniques used have included studies of X-ray diffraction to identify minerals in soils (Dixon and Weed, 1989), dissolution and extraction to measure the amounts and solubility of minerals and organic matter in soils, exchange reactions of charged nutrient ions with reactive surfaces of minerals and soils to estimate (sometimes using radioisotope tracers) the immediate reserve of available nutrients in soils, aerobic and anaerobic incubations to assess nutrient release from soil organic matter through the hydrolytic activities of soil micro-organisms and studies of the processes facilitating and constraining the transport of nutrients through soils to roots (Barber, 1984; Tinker and Nye, 2000).

There are 17 elements essential for plant growth that can be derived by uptake from the soil solution. Nutrient availability in soil is therefore a broad subject dealing with complex chemical reactions involved in the functioning of soil–water–plant–atmosphere systems. There are some excellent in-depth reviews dealing with the fundamental plant root zone processes involved in nutrient acquisition by plants (e.g. Marschner, 1998; Tinker and Nye, 2000; Huang and Germida, 2002; Trolove et al., 2003).

This review discusses some selected techniques used to quantify nutrient acquisition by plant roots, addresses soil tests and isotopic exchange studies (see Chapter 7) as indicators of potential bioavailability and briefly covers computer simulation of dynamic nutrient cycling processes. In terms of developing techniques to assess nutrient availability most attention in past years has been directed at developing soil tests through extensive studies correlating the amounts of nutrients released to weak chemical extractants with crop yield, or with the amounts of nutrients taken up by crop plants (Westerman, 1990; Peverill et al., 1999).

A number of general conclusions can be drawn from this research that are pertinent to the dogma driving development of techniques for assessing nutrient bioavailability. These conclusions are best summarised by a brief general discussion of the current conceptual models of nutrient availability in soil.

14.2 Conceptual models of nutrient availability in soil

Nutrients are primarily taken up (some actively) as low-charge-density ions and sometimes as uncharged small low-molecular-weight complexes (including chelated species; Marschner, 1998) from the thin, soil–water film (solution phase) that provides connectivity between the plant root and the complex organo-mineral fabric (solid phase) of the soil (Fig. 14.1). The simple organic and inorganic ions are immediately replaced from diffuse layers of ions attracted to neutralise the variably charged surfaces of the soil solid phase. There is a great variation (even within elements) in the chemical form, reactivity and solubility of nutrients associated with the solid-phase surfaces, which comprise inorganic, crystalline and amorphous precipitates, complex organo-metallic polymers and living organisms. Over the longer-term there is release of nutrients to the

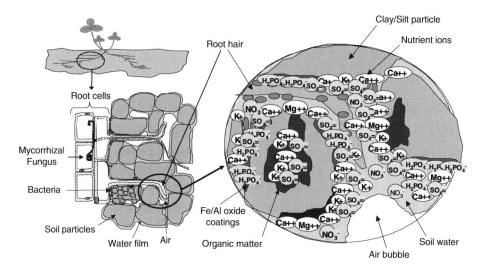

Figure 14.1. *A cartoon representing the concept of the growth of plant roots in larger soil macropores with root hairs exploring water-filled soil micropores. Nutrient ions taken up from the micropore solution by the root hair may be replaced by cations on negatively charged clay and organic matter surfaces and anions adsorbed to Fe and Al oxide surfaces.*

solution phase involving different degrees of desorption, exchange and dissolution from the solid mineral phase and biochemical (enzymatic hydrolysis, oxidation-reduction) mineralisation of the organic phase.

14.2.1 Chemical forms
14.2.1.1 Soil solution

For a nutrient to be considered 'immediately-bioavailable' to plants and microorganisms it must be in a water-soluble form (ion, small molecule or dissolved gas) that can rapidly move through protein transporter channels in the plant, or microbial, cell membranes (e.g. Maser et al., 2002).

Immediately-available water-soluble nutrients are in four major forms: discrete ions, outer sphere ion associations or complexes, called ion pairs, and inner-sphere metal–ligand complexes and metal chelates (Sparks, 2003). In soil solution, discrete cations and anions are hydrated, surrounded by spheres of co-ordinated water molecules (e.g. cations K^+, Na^+, Ca^{2+}, Mg^{2+} and anions $H_2PO_4^-$, SO_4^{2-}, Cl^-, NO_3^-). These major cations and anions do not induce any of the co-ordinated water molecules to donate protons and the hydrated ion carries the same charge as the central co-ordinating anion, or cation. Net positively charged hydrated-cations may be attracted to and associate with net negatively charged hydrated-anions to form ion pairs, which retain the co-ordinated sphere of water molecules around each ion (outer-sphere complex). Alternatively, the cation and anion may be so strongly attracted that water molecules are displaced and an ionic, or a single covalent bond, is formed to produce an inner-sphere, metal–ligand complex (e.g. $CaSO_4^0$ and $CaHCO_3^+$). If the complex is created with two or more covalent bonds then a stronger complex is formed called a metal–chelate (e.g. tris(oxalato)cobaltate $[Co(C_2O_4)_3]^{3-}$). Polycarboxylic acids including amino acids form some of the strongest soluble complexes with metals. The major nutrient cations have small ion size and high positive charge. They behave as hard Lewis acids (Sparks, 2003), forming ionic solids and are present predominantly as discrete ions in soil solution (Table 14.1).

The major nutrient anions (Table 14.1) have high negative charge. They behave as hard Lewis bases (Sparks, 2003), forming ionic solids and also are present predominantly as discrete ions in soil solution (Table 14.1). Ion pairs and metal–ligand complexes of associated major nutrient cations and anions are often minor components of the ion species in soil solution.

Most of the minor-nutrient elements are transition metals (e.g. cations Mn^{2+}, Fe^{2+}, Co^{2+}, Cu^{2+}). They behave as soft Lewis acids, predominantly forming metal–ligand complexes, or chelates, through covalent bonds with soft Lewis bases such as amino-N and sulphydryl-S groups of soluble organics. The stabilities of many of these complexes are pH dependent but at near neutral soil

Table 14.1. *Major solid-phase forms of nutrients in soil, their release process to soil solution and the major species in a mildly acid to neutral soil solution*[a].

Solid reserve form	Release process	Major solution species in mildly acid-neutral soils
Nitrogen		
Predominantly in organic matter	Microbial, hydrolytic extracellular enzymes	Free hydrated ions
Interlayer NH_4^+ in 2:1 layer clays	Slow cation exchange	NH_4^+ NO_3^-
Surface exchangeable NH_4^+	Rapid cation exchange	
Sulphur		
Predominantly in organic matter	Microbial hydrolytic extracellular enzymes	75% free hydrated ion, SO_4^{2-}
Adsorbed to oxide surfaces	Ligand exchange	
Sulphides (e.g. Fe S) and sulphur in anoxic soils	Microbial oxidation	25% metal ligand with Ca^{2+} Mg^{2+} (e.g. $CaSO_4^0$)
Gypsum (calcium sulphate)	dissolution	
Phosphorus		
Organic matter	Microbial, hydrolytic extracellular enzymes	90% free hydrated ions, $H_2PO_4^-$ HPO_4^{2-}
Adsorbed and bound by ligand exchange to oxide surfaces and $CaCO_3$	Ligand exchange	10% metal ligand complex (e.g. $FeH_2PO_4^+$)
Primary apatites	dissolution	
Potassium, Calcium and Magnesium		
Silicate minerals (e.g. feldspars micas and 2:1 clays)	Dissolution, slow cation exchange	100% free hydrated ion, K^+
Carbonates and sulphates	dissolution	90% free hydrated ion, Ca^{2+} Mg^{2+}
Surface exchangeable cations	Cation exchange	10% metal ligand complex (e.g. $CaHCO_3^+$)
Iron, Manganese, Zinc, Cobalt and Copper		
Precipitated as hydroxyoxides	Chelation, reduction	70% free metal, Zn^{2+} 30% metal ligand complex (e.g. $ZnSO_4^0$ $ZnHCO_3^+$)
Organo-metal-oxide complexes	Dissociation, chelation	99% metal-organic ligand complex, Cu-SOM
Exchangeable	Cation exchange, desorption	70% free metal, Mn^{2+} Fe^{2+}; 30% ligand complex (e.g. Fe SO_4^0, $FeHCO_3^+$ Fe-organic ligand Fe-SOM)
Co^{2+} adsorption on manganese-oxide complexes	Desorption, chelation, reduction	Hydrated ion, Co^{2+}; Co-organic ligand complex
Boron and Molybdenum		
Adsorbed on Fe, Al, and Mn oxides and other clay minerals	Desorption	100% H_3BO_3 free acid, hydrated anion, MoO_4^{2-}; 10% metal ligand complex (e.g. Fe MoO_4^+)

[a]Compiled from Sparks (2003), Wild (1988) and Sposito (1983).

solution pH, these complexes can be the dominant ion species of minor-nutrients in soil solution (Table 14.1).

14.2.1.2 Soil solution–soil solid phase equilibria (rapid)

In aerobic soils, the solution phase can be considered to be the thin, soil–water film that provides connectivity (Fig. 14.1) between the plant root and the complex, organo-mineral fabric of the soil (solid phase). In the solution phase, the concentrations (activities) of charged ion species and polar molecules are buffered, in the short-term (seconds/minutes), by a rapid dynamic equilibrium with a larger pool of the same, or related, species adsorbed (or precipitated) onto oppositely charged solid-phase surfaces. A decrease in the concentration of solution species, caused by uptake by plant roots and micro-organisms, disturbs the equilibria, resulting in desorption of ion species from the solid phase, or dissolution of precipitates. An increase in the concentration of solution species caused by addition of a soluble fertiliser to the soil also will disturb these equilibria resulting in adsorption of ion species to the solid phase, or surface precipitation. Thus, adsorption–desorption and precipitation–dissolution equilibria are capable of buffering the concentration of solution-phase species within narrow ranges.

The concentrations of the dominant major cations (Ca^{2+}, K^+, Na^+ and Mg^{2+}) in soil solution are predominantly buffered by non-specific adsorption, or exchange, of the hydrated ion on negatively charged, soil mineral or organic matter surfaces. Net negative charges on soil particle surfaces are generated by isomorphous substitution of lower valence metals for silicon (Si^{4+}) and aluminium (Al^{3+}) in silicate and alumina sheets of secondary clay minerals, dissociation of carboxylic and phenolic functional groups of soil organic matter, or cell walls of soil bacteria and fungi in surface biofilms. Soils containing secondary clays with 2:1 layer structures (e.g. weathering products of mica, such as illite or vermiculite, with two silicate sheets sandwiching an alumina sheet) generate the highest levels of negative charge, from both isomorphic substitution in silicate and alumina sheets and inclusion of lower valence metals in the interlayer spaces. These clays also develop high specific surface areas through weathering and expansion of the interlayers. Highly charged clays attract decomposing organic residues as well as metal cations. The surface complexes responsible for cation exchange in topsoils are therefore heterogenous mixtures of clay minerals, metal oxides and the end products of organic matter decomposition, such as humic and fulvic acids, and the decomposer micro-organisms themselves. While conceptual molecular models of metal–humic–clay structures have been simulated (Schulten et al., 1998), the actual molecular structure of these surfaces are unknown.

The species and concentrations of the transition metals (e.g. cations Mn^{2+}, Fe^{2+}, Co^{2+}, Cu^{2+}) in soil solution are influenced by more complex solid–solution phase equilibria partly because either their oxidation state may change depending on the redox potential of the soil or partly because their hydrated ions in soil solution behave as weak acids, donating protons from the co-ordinated water molecules. Under aerobic conditions (high redox potentials) at near neutral pH free metal ion concentrations in soil solution are low because the higher oxidation states of these metals readily form low solubility hydrous oxide polymers, or precipitates, on soil surfaces. In addition, they readily form metal–ligand complexes with soluble organic matter in soil solution that may then be adsorbed onto soil surfaces. The dissolution of the hydrous oxides and the stability of the metal–organic ligands are pH sensitive. Therefore, concentrations of soil solution species of these transition metals are much more sensitive to small changes in soil pH, oxygen diffusion rates and factors that influence the concentration of the soluble organic matter (Sparrow and Uren, 1987).

The anions (NO_3^-, HCO_3^-, SO_4^{2-}, $H_2PO_4^-$, HPO_4^{2-} and Cl^-) are mainly adsorbed on to positively charged soil surfaces as hydrated ions. The positively charged surface may be a layer of non-specially sorbed cations, whose positive charge is not fully neutralised by the soil surface negative charge. In addition, hydrous oxide surfaces of Al, Fe and Mn will carry net positive charge in soils with near neutral pH or lower. To these surfaces, $H_2PO_4^-$, HPO_4^{2-} and SO_4^{2-} may be specifically adsorbed as they form inner sphere complexes with the metal ions of the hydrous oxide surface. The quantity of species in a position to contribute to rapid exchange, desorption and dissolution reactions have been measured relatively easily by desorption–dilution (dissolution–dilution) and ion exchange experiments (Allen et al., 2001; Barrow, 1993). Radioactive or stable isotopes of essential elements have been used extensively in ion exchange studies to measure the pool size of exchangeable nutrients in soils (Di et al., 1997).

14.2.1.3 Soil solution–soil solid phase equilibria (slow)

Although the bulk of research has been on the fast abiotic reactions buffering nutrient concentrations in soil solution, the slower reactions involved in mineral dissolution and the decomposition of organo-mineral surface coatings have importance in the long-term (months/years) supply of plant-available nutrients. This is because the size of the cycling (exchangeable) pool of a nutrient in soil is usually small relative to the amounts of non-exchangeable nutrient in inorganic soil minerals and recalcitrant soil organic matter. With soil components that have slow rates of dissolution, or decomposition, the surface area in contact with the soil solution becomes a rate-limiting factor. Thus, minerals and organo-mineral complexes in the silt and clay size fractions of soils (large surface areas per unit weight or volume) are the important constituents to focus on for slow nutrient release.

Most of soil N and S and half the soil P commonly occurs in organic compounds (Wild, 1988). Mineralisation rates are highly variable depending not only on soil moisture and temperature regimes but also on their physical accessibility to decomposer organisms (see review by Guggenberger and Haider, 2001). Any solubilisation of organic N, P and S will require the action of a suite of extracellular enzymes to depolymerise complex organic materials and finally release smaller organic or inorganic ions (e.g. Kandeler et al., 2002).

The major nutrient cation Ca^{2+} is predominantly found in aluminium-silicate minerals such as feldspars and amphiboles and contributions from carbonates and apatite, depending upon the soil's pH and weathering stage. Mg^{2+} is released mainly from ferromagnesian minerals such as biotite, serpentine, hornblende and olivine in weakly weathered parent materials but in soils it is present in secondary clay minerals such as chlorite, vermiculite, illite and montmorillonite. Slow-release reserves of K^+ are mainly the clay weathering products of micas (e.g. illite) and feldspars (Dixon and Weed, 1989). Trace element nutrients are associated with these clay-sized, organo-mineral complexes. In particular, the transition metals will form complexes with other dominant mixed Al–Fe and Mn metal oxide surfaces and soil organic matter (Andersen et al., 2002; Li et al., 2003). Mn oxides can be responsible for significant Co^{2+} adsorption and occlusion, particularly if the pH rises (Quantin et al., 2002).

The range of combinations of soil mineral–metal oxide–organic matter complexes is diverse but the processes inducing their dissolution can be classi-fied into six categories (Table 14.1). (1) Dissolution when the product of the ion activities in soil solution falls below the solubility product of the solid (e.g. Hinsinger and Jaillard, 1993). (2) Ion exchange between ions fixed within clay lattices and higher concentrations of counter ions in solution (e.g. Moritsuka et al., 2002). (3) Acid hydrolysis (e.g. Hinsinger et al., 1993) with roots and soil micro-organisms being the main source of carbonic acid (respiration) and excreted organic acids (Huang and Germida, 2002). (4) Reduction (Blaylock and James, 1994; Quantin et al., 2002; Marschner et al., 2003). (5) Chelation by organic acids and siderophores (Dakora and Phillips, 2002; Huang and Germida, 2002). (6) Decomposition catalysed by hydrolytic enzymes (Kandeler et al., 2002).

14.2.2 Biophysical constraints

Soil water is the transport medium which enables nutrient ions, chelates and molecules to pass from the soil to the plant root. Soil moisture tension controls the number of water-filled pores and the cross-sectional area of soil pores that are available for solute transport (Tinker and Nye, 2000). In a moist soil ($> -10\,kPa$ moisture tension), $0.05\,mm$ diameter pores will be water-filled. As the soil dries out these medium sized pores empty, leaving a thin water film on their walls and only the micro-pores remain water-filled. This has the consequence of decreasing the number of pathways for solute transport and effectively increases the path length for solutes to travel to the root. The decreased volumetric water content in the soil also decreases the habitat size and activity of microbial decomposers. Soil moisture content therefore has a marked effect on nutrient bioavailability by influencing nutrient transport rates and the mineralisation of nutrients form soil organic matter. Microbial and plant root processes making nutrients bioavailable are temperature dependent as well as being spatially dependent on the architecture of the soil's water-filled pore space. Although roots are mostly thought of as sinks for nutrients and water, they are also sources of carbon substrates, providing the energy for micro-organism growth and the microbially mediated processes of nutrient transformation in soil. Plant root growth and death with its supporting cast of micro-organisms leads to the formation of stable soil aggregates. While these aggregates stay intact, pools of exchangeable and solid-phase nutrients within the aggregates remain isolated from the active available nutrient pool.

At a greater scale, biophysical factors constraining nutrient availability are influenced strongly by the physical management of the soil fabric and by the seasonal patterns of plant growth and nutrient demand (dictated by plant species and climate). Winter drainage events can cause marked reductions in nitrate concentrations in the plant root zone, particularly when soil temperatures are low, limiting the replenishment of nitrate from organic matter mineralisation.

14.2.3 Description of actual or potential nutrient bioavailablity

Currently, we do not have good mechanistic descriptions of the actual bioavailability of nutrients in soils beyond one season. The introductory discussion makes it clear that actual nutrient bioavailability is spatially and temporally dynamic and influenced by unpredictable climatic events. A dynamic explanation is required that only mathematical modelling can offer but some of the soil biological and physical processes are complex and currently defy precise mathematical description (Tinker and Nye, 2000). The static (one point in time) techniques for measuring nutrient bioavailability in soils are simply indicators of potential nutrient bioavailability that may, or may, not be realised by the biological system.

Past research has, however, identified for most nutrients the key processes at the soil–water–plant interface that have the greatest influence on plant nutrient uptake (Marschner, 1998; Tinker and Nye, 2000). This knowledge has been useful in the development of techniques for assessing the potential bioavailability of nutrients in soil. Current research and development of these techniques can be characterised into three areas. (1) Applied tools for assessing and removing soil mineral constraints to crop production. (2) Applied tools for assessing soil management threats to water quality. (3) Research tools for building a fundamental understanding of the quantity, form and dynamics of plant-available nutrients in soils.

14.3 Applied tools for assessing and removing soil mineral constraints to crop production

14.3.1 Soil testing and crop response trials

A dominant research goal has been to develop soil tests that can extract from soils amounts of nutrients that are strongly correlated with crop growth response to increasing amounts of nutrient applied in single-factor field trials. Successful research develops mathematical algorithms (calibrations) to explain the soil test/plant growth (harvested yield) response function and the relationship between the amount of nutrient applied to the soil and the change in soil test values (Fig. 14.2). These two pieces of information plus the value of the crop at harvest can be used to calculate estimates of fertiliser application rates which are appropriate for sustainable, or profitable, farming.

The general term for this area is 'soil testing' and there are several excellent manuals including those edited by Westerman (1990) and Peverill et al. (1999), which cover many procedures currently conducted. In this paper, the aim is to present a general overview of the progress that has been made in soil testing procedures and highlight some areas of future research interest.

14.3.1.1 The interpretive value of soil tests depends on adherence to standardised procedures

Expanding the statements of McLaughlin et al. (1999), there are three areas of soil testing that must be rigorously standardised for soil testing procedures to be effective in their purpose of identifying the degree of a soil nutrient constraint and determining the appropriate application rate of fertiliser to alleviate that constraint. These areas are:

(i) A standardised soil sampling protocol that takes into consideration general soil condition (temperature, moisture content, stage of ground preparation or stage of crop growth), the position of different soil types in the landscape, the appropriate soil sampling depth for the crop, features of

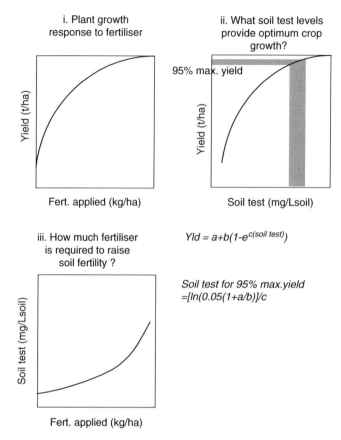

Figure 14.2. *Steps in calibrating soil test values to crop yield response in single factor nutrient field trials. (i) Different fertiliser rates are applied to a crop grown at a nutrient responsive site and yield measurements are recorded. (ii) Plots are soil tested and yields are expressed against soil test values. A Mitscherlich curve is fitted to explain variation in yield with soil test. Soil test to achieve 95% of maximum yield calculated. (iii) Soil test values are plotted against fertiliser applied to obtain values for expected soil test change per unit fertiliser applied.*

the farming system (headland turning areas, shelter belts, stock management, fertiliser history and application method) and landscape (slope, erosion, water availability) that create fertility gradients (Fig. 14.3), or, patches with distinct differences in soil nutrient concentration (James and Wells, 1990; Brown, 1993, 1999). The objective of the soil testing programme influences the soil sampling strategy. To monitor the effectiveness of a corrective fertiliser programme over time a series of marked sampling transects on each soil type is required so that sampling sites are easily found in subsequent years. This minimises the influence of

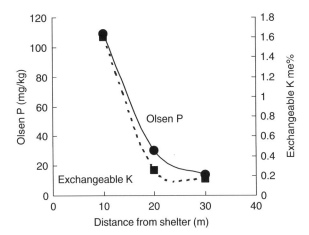

Figure 14.3. *Soil fertility gradient caused by stock sheltering adjacent to trees (data from Matthew et al., 1988).*

the soil spatial variability on test values. To identify the yield potential of different paddocks, or soil types, targeted random sampling may be the best approach. After sampling, moist soil should be chilled and/or air dried before prompt dispatch to the analytical laboratory. Air drying is standard, and results in marked increases in extractable P and S, partly due to the death and lysis of soil micro-organisms (Sparling et al., 1985). Leaving samples under a warm moist condition may accelerate mineralisation of N and S (see reviews by Brown, 1993, 1999).

(ii) Standardised analysis protocols that deliver highly reproducible results and can report at least a 5–10-fold change in the soil test value over the range of soil nutrient concentrations that produce between 10 and 100% relative crop yield (a concept demonstrated in Fig. 14.2, achieved in Fig. 14.5 but not in Fig.14. 4).

(iii) Calibration with an extensive range of crop responses to nutrient availability. These may have to be repeated with contrasting climates and soil types (Dahnke and Olsen, 1990). Even with one crop calibration curves can vary with season and soil type (e.g. pasture response curves to available phosphate, Sinclair et al. (1997)).

14.3.1.2 Laboratory protocols

Routinely used analytical protocols for assessing the bioavailability of nutrients in a soil sample almost exclusively involve extractions using chemical reagents that are weak acids, alkalis, chelating agents, or oxidants (Table 14.2).

In most cases, the reagents were selected originally because their solubilising action on soil nutrients matched concepts held by the method developers of the

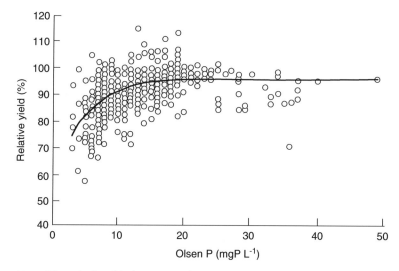

Figure 14.4. *The relationship between relative yield of pasture and Olsen P test in each treatment across 17 trial sites in New Zealand (Sinclair et al., 1997).*

nutrient acquisition processes of plants. The laboratory protocols for P and exchangeable cations (Table 14.2) recognise that most soils have relatively large pools of exchangeable forms of these nutrients and the dominant method of extraction is competitive desorption, or exchange, with a counter ion (Table 14.1).

For some N and S tests, decomposition and release of N and S from organic matter is simulated by anaerobic and aerobic incubations (Table 14.2). The organic-S test developed by Watkinson and Kear (1996) is one of the few sulphur tests in routine laboratory use that assesses the potential of soils to mineralise S. For the Cu, strong chelating agents (DTPA and EDTA; Table 14.2) are used to release Cu from soil organic matter complexes (Table 14.1).

It is fortuitous that the amount of nutrient extracted by some reagents (expressed as kg P/hectare plough layer) is approximately equivalent to the amount taken up by a crop in a growing season (e.g. P availability measured by the Olsen (Olsen et al., 1954) or Bray (Bray and Kurtz, 1945)). This fuels the common misconception among growers and farmers that soil tests measure the size of the plant-available pool of nutrients in a soil. This is of course incorrect. Soil tests simply provide an index of nutrient availability that under some conditions correlates well with plant growth response to that nutrient.

The majority of extractive tests were developed and calibrated before rhizosphere studies had demonstrated that a range of soil nutrients forms can be depleted by plant roots (see later this chapter) and before isotopic exchange studies had indicated the size of exchangeable pools of nutrients in soils

Table 14.2. *A list of selected soil tests for routine analysis of the bioavailability of nutrients in soils.*

Soil test (reference)	Soil: solution ratio	Extractants, pH (at 20–25°C unless stated)	Extraction time
Phosphate			
Olsen (Olsen et al., 1954)	1:20	0.5 M Sodium bicarbonate, pH 8.5	30 min
Colwell (Colwell., 1963)	1:20	0.5 M Sodium bicarbonate, pH 8.5	16 h
Bray 1 (Bray and Kurtz, 1945)	1:7	0.03 M Ammonium fluoride plus 0.025 M hydrochloric acid, pH 3	1 min
Mehlich (Mehlich, 1953)	1:4	0.025 M Sulphuric acid plus 0.05 M hydrochloric acid, pH 1.0	5 min
Egnér (Egnér et al., 1960)	1:20	0.1 M Ammonium lactate plus 0.4 M Acetic acid	4 h
Resin (Saggar et al., 1999)	1:100	Anion plus cation exchange resin membrane, pH of soil	16 h
Exchangeable cations			
NH₄OAc (Rayment and Higginson, 1992)	1:50	Leaching with 1 M ammonium acetate, pH 7	2–3 h
Mehlich (Mehlich, 1953)	1:4	0.025 M Sulphuric acid plus 0.05 M hydrochloric acid, pH 1.0	5 min
Resin (van Raij, 1998)		Anion plus cation exchange resin membrane, pH of soil	16 h
Qian and Schoenau (2002)			in field
Sulphate			
MCP extractable sulphate (Rayment and Higginson, 1992)	1:5	0.01 M Ca(H₂PO₄)₂, pH 4	17 h
Sulphate plus extractable organic sulphur Watkinson and Kear (1996)	1:5	0.02 M K(H₂PO₄), pH 4	30 min
Nitrogen			
Anaerobic incubation (Keeney, 1982)		5 g Soil incubated with 12.5 ml water in sealed tube at 30°C. Extraction 2 M KCl	2 weeks
Aerobic incubation (Keeney, 1982)		5 g Soil+15 g quartz sand +3 ml water incubated at 30°C. Extraction 2 M KCl	2 weeks
Hot KCl-hydrolysable (Gianello and Bremner, 1986)	1.5:10	2 M KCl, sealed digestion at 100°C	4 h
NaOH-hydrolysable (Wang and Li, 1991)	4:10	Sealed jar at 40°C to capture NH₃ released	24 h
Extractable NO₃ to depth Qian and Schoenau (1995)		Anion-exchange membrane sticks placed in soil	2 weeks
Trace elements			
Copper			
Brennan and Best (1999)		DTPA (diethylenetriaminepentaacetic acid) or EDTA (ethylenediaminetetraacetic acid)	
Manganese		0.01 M CaCl₂	
Sheppard and Bates (1982), Uren (1999)			
Boron		Water, 100°C	20 min
Hot water extractable Haddad and Kaldor (1982)			

(see later this chapter). New soil P tests have been developed and compared with isotopic dilution studies (e.g. Delgado and Torrent, 2001) and may offer better descriptions of plant available nutrient pools across a wider range of soil conditions than some established tests. A small number of calibrations have been completed for these tests (e.g. resin test for phosphorus; van Raij and Quaggio, 1990; van Raij, 1998; Saggar et al., 1999). However, calibration of soil tests to crop yield responses in field trials is extremely expensive and provides the major hindrance to new tests being adopted in favour of existing calibrated tests.

14.3.1.3 Interpretive calibration

Among the major nutrients, research has produced reproducible and robustly calibrated soil tests only for P (e.g. Moody and Bolland, 1999) and exchangeable K on some soils (e.g. Gourley, 1999). The reasons for this limited range, which does not include nitrogen and sulphur, are probably two fold. Firstly, the plant-available ionic forms of phosphate and exchangeable cations are in equilibrium with significantly large reserves (Table 14.1) of the same ionic forms in diffuse layers attracted to neutralise the variably charged surfaces of the organo-mineral surfaces of the soil fabric. These reserves can be extracted by solutions of exchangeable counter ions, or ligands (Table 14.2). On the other hand, reserves of nitrogen and sulphur in soils are mostly present as insoluble organic compounds. For ammonium and nitrate ions in soil solution to be replenished there must be enzymatic hydrolysis of soil organic matter by soil organisms. These processes, mostly performed by the extracellular enzymes of bacteria and fungi, are particularly sensitive to microbial growth patterns, therefore the rates of nutrient release are very sensitive to soil temperature and moisture conditions. To complicate matters the nitrate and sulphate, the products of organic N and S mineralisation, are relatively easily leached from many soils. The following illustrates the problem. Wang et al. (1999) showed that the amounts of soil N mineralised in anaerobic and aerobic incubation techniques that exchange with solution $^{15}NH_4^+$ are strongly correlated to amounts of total soil N, hot KCl extractable N and NaOH extractable N (Table 14.3).

Thus, all these laboratory protocols for extractable N are good indices of potentially mineralisable N. Yet estimates of N supply derived using some of these tests (plus NO_3^- accumulated in the profile) on soil sampled from either 0–15 cm (the plough layer) or 0–60 cm (the whole wheat root zone) explain less than 23% of the variation in N uptake by wheat over different landscape units in one paddock (Table 14.4; Walley et al., 2002). The main reason for the poor prediction of N uptake by the wheat crop appeared to be small differences in elevation in a rolling landscape that influenced the soil moisture regime during

Table 14.3. *Linear relationships between three indices of soil nitrogen availability and the amounts of N mineralised in anaerobic and aerobic incubations of 19 topsoils from Eastern Australia (data derived from Wang et al., 2001).*

Gross mineralisable N (mg N kg^{-1} soil)		Total soil nitrogen	NaOH extractable N	Hot KCl extractable N
Anaerobic incubation	Intercept	7	28	47
Range 25–250	Slope	0.044	0.51	2.9
	r^2	0.69	0.71	0.41
Aerobic incubation	Intercept	7	4	12
Range 25–250	Slope	0.041	0.61	3.8
	r^2	0.91	0.87	0.57

Table 14.4. *Pearson correlations between measure soil characteristics and N accumulation in a spring wheat crop within 100 sampling points of a 300 m transect across rolling terrain including Typic Haploborolls and Typic Aquolls (Walley et al., 2002).*

Parameter measured	Pearson correlation with N uptake by spring wheat	
A horizon depth	0.41	
Relative elevation	−0.40	
	Soil depth analysed	
	0–15 cm	0–60 cm
Spring moisture	0.38	0.28
Total carbon	0.44	0.33
Total nitrogen	ns	0.30
Mineral N	ns	ns
Aerobic mineralisable N	0.37	nd
Hot KCl extractable N	0.48	0.47
Anion exchange resin	0.32	0.32

ns–not significant; nd–not determined.

crop growth but also during soil formation. Topographically induced differences in moisture regime affected both the amount of potentially mineralisable N measured in the laboratory test and the extent of mineralisation that took place leading up to, and during, crop growth.

Also unpredictable seasonal variations in climate seriously compromise the ability of soil N tests to predict both the extent of soil N mineralisation and plant uptake of mineralised N. Although multiple regression models can be used to identify important factors, they cannot combat seasonal variation. Introducing spatially and temporally (daily time step) precise models of soil temperature and water regime with precision mapping of the potentially mineralisable soil N pool could solve these problems.

Single factor production response 'calibration scales' for soil P tests are also not without problems in their predictive value (Hedley et al., 1995). Sinclair et al. (1997) reported the relative pasture production response curves for Olsen extractable P in New Zealand, derived from the NZMAF, National Series of phosphate trials. Pasture dry matter maximum yield potentials (P deficiency removed) were highly variable among the trial sites mainly due to differences caused by factors other than P availability influencing plant growth. Sinclair et al. (1997) normalised pasture yield responses to P availability by calculating relative yield response at each site. This was achieved by expressing nutrient (P in this case) limited plant yield (or plant growth response) as a percentage relative to the maximum or non-nutrient limited yield at the site (or field trial). To use this approach non-nutrient limited yield (or maximum yield) must be known, or can be accurately estimated, from the trial treatments. Normalising yield data to % relative yield did not cure the site-to-site and within-site variability in the pasture yield response to soil P status. The best fit asymptotic relationship, employing Olsen P as the predictor, explained less than 27% of the variation in % relative yield across 17 field trial sites and plot treatments (Sinclair et al., 1997; Fig. 14.4). This still remained a single factor approach to describing yield variation and has its limitations. In clover/grass pastures in New Zealand, the two most common constraining factors are soil available water and nitrogen availability.

Moir et al. (2000) also saw large variations in annual accumulated pasture yield response to Olsen P soil test values across 13 trial sites, which differed in initial soil fertility and rainfall regime. Simple soil water balances were used to calculate soil-limited evapotranspiration from the pasture by using daily rainfall and temperature measured at each site, and estimates of regional solar radiation. Pasture growth rates at each site were expressed as the mean amount of dry matter grown per hectare per millimetre of soil-limited evapotranspiration. Moir et al. (2000) demonstrated that for 13 field trial sites, varying in fertility status, a simple natural logarithm function explained 89% of the variation in the legume/grass growth rate per millimetre actual evapotranspiration caused by changes in soil Olsen P status (Fig. 14.5). This empirical use of the growth-limiting plant-available water to take the noise out of site-to-site variation in the calibration of Olsen P both improves the calibration of the soil test and allows improved accuracy in predicting the seasonal pattern of P (and other nutrient) uptake where P availability limits pasture growth. A practical benefit is that the climate-sensitive pasture growth model developed by Moir et al. (2000), which predicts actual plant yield, can be used to test different future climate patterns so that farm investment risk can be assessed.

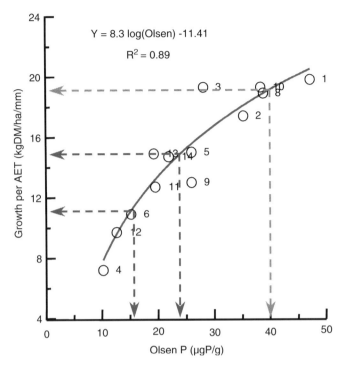

$$Y = 8.3 \log(\text{Olsen}) - 11.41$$
$$R^2 = 0.89$$

Figure 14.5. *Relationship between pasture growth per unit of estimated soil limited evapotranspiration (AET) and Olsen P for 13 grazed pasture sites in New Zealand (Moir et al., 2000).*

14.3.1.4 Developments with multi-nutrient soil tests

A current area of research interest is the development of single tests that deliver multi-element read out. Two key drivers for this research in the US have been the need to reduce the cost of testing large numbers of soil samples for the application of precision farming techniques and the advent of multi-element analysers such as Inductively coupled Plasma–Atomic Absorption Spectrophotometer (ICP-AAS) and Inductively coupled Plasma–Mass Spectrophotometer (ICP-MS) (Jones et al., 1998). The Mehlich 3 (0.2 M CH$_3$COOH, 0.25 M NH$_4$NO$_3$, 0.015 M NH$_4$F, 0.013 M HNO$_3$ and 0.001 M EDTA; Mehlich, 1984) and AD-DTPA (1 M NH$_4$HCO$_3$ and 0.005 M DTPA at pH 7.6; Soltanpour and Schwab, 1977) appear to be the two tests most likely to be adopted (Table 14.5). There is also considerable interest in the use of combined cation–anion exchange resin membranes for multi-element testing (Qian et al., 1992; van Raij, 1998; Qian and Schoenau, 2002).

The lack of calibration data against crop response is one of the main factors constraining widespread adoption of these multi-element tests. Progress is being

Table 14.5. *Multi-nutrient extractants, soil adaptability and elements determined (modified from Jones, 1990).*

Extractant	Soil type	Elements determined
Morgan	All acid soils and soil-less mixes	P, K, Ca, Mg, Cu, Fe, Mn, Zn, NO_3, NH_4, SO_4, Al, As, Hg, Pb
Wolf-Morgan	All acid soils and organic soils	P, K, Ca, Mg, Cu, Fe, Mn, Zn, Al, NO_3, NH_4
Mehlich No. 1	Acid sandy soils	P, K, Ca, Mg, Na, Mn, Zn
Mehlich No. 3	All acid soils	P, K, Ca, Mg, Na, B, Cu, Fe, Mn, Zn
AB DTPA	Alkaline soils	P, K, Na, Fe, Mn, Zn, As, Cd, NO_3
Ion exchange resin Beads Van Raij (1998)	All soils	P, K, Na, Ca, SO_4^-, NO_3^-, NH_4^+
Membrane McLaughlin et al. (1993), Qian et al. (1992), Qian and Schoenau, 2002	All soils	P, K, Na, Ca, Mg, SO_4^-, NO_3^-, NH_4^+

made by regression analysis of the amounts of nutrients extracted by multi-element tests against standard tests, which have in the past been calibrated against crop yield response. Studies by Qian et al. (1992), van Raij (1998) and McLaughlin et al. (1993) show linear correlations between ion exchange resin extractable nutrients and standard tests for exchangeable K, Ca and Mg. Previous calibrations also exist for extractable P (Saggar et al., 1999) and extractable SO_4 (Searle, 1992). Extensive research into in-field calibration of ion exchange resin membranes has taken place in Canada (Qian and Schoenau, 2002) and New Zealand (Saggar et al., 1999), but little progress has been made elsewhere.

14.4 Applied tools for assessing soil management threats to water quality

14.4.1 Soil testing and nutrient transfer to waters

The main objective for this area of environmental research is to develop soil tests capable of identifying soil patches in the landscape that can potentially contribute to undesirable nutrient enrichment of drainage water, leading to nuisance algal growth in surface water. The main emphasis has been on soil phosphate loss to surface waters.

It is impossible to remove the phosphorus constraint to the biological productivity of agricultural soils without increasing the biological productivity of waters draining those soils. Dissolved phosphorus (DP) concentrations rising above 0.3×10^{-9} M will stimulate accelerated growth of algae and other aquatic plants (Sharpley and Rekolainen, 1997). The DP may enter the surface water directly, or can be released from particulate P (P rich sediment or organic matter) carried in surface run-off or subsurface drainage (Sharpley et al., 1995). Patches in the landscape that receive fertiliser, animal manure or sewage sludge can be expected to increase DP in run-off.

Excessive application of fertiliser alone, or in combination with manure from housed animals, has created large areas of P-enriched surface soils in Europe (Sibbesen and Runge-Metzger, 1995) and in the US (Sharpley et al., 1994; Sharpley et al., 1995). In these regions diffuse run-off and drainage from agricultural land is now seen as the major contributor to elevated DP in surface waters.

New Zealand (Sharpley et al., 1977), UK (Heckrath et al., 1995) and US (Pote et al., 1996; McDowell and Sharpley, 2001) research shows that as soil P status increases, so does the concentration of P in run-off and drainage P water, albeit in some soils we may not be able to measure that increase (Fig. 14.6). A number of researchers have established critical threshold soil test values by using the break point method (Heckrath et al., 1995; Hesketh and Brookes, 2000; McDowell and Sharpley, 2001) to establish a soil test value above which accelerated P loss in drainage waters may occur (Fig. 14.6).

14.4.1.1 Simulating the pattern of change in drainage water P concentrations

Johnston and Polton (1993), Heckrath et al. (1995) and Hesketh and Brookes (2000) demonstrated that the relationship between Olsen soil test P values and P released to soil suspensions (1.5 w/v) in 0.01 M $CaCl_2$, simulated the relationship between Olsen soil test P and drainage water P concentrations from those soils (Fig. 14.7). Break points occurred at similar soil test values. McDowell

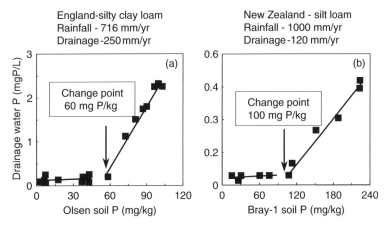

Figure 14.6. *Relationships between soil test values and dissolved inorganic P (DIP) in drainage water (a, Heckrath et al., 1995; b, Sharpley et al., 1977). Redrawn from Hedley and Sharpley (1998).*

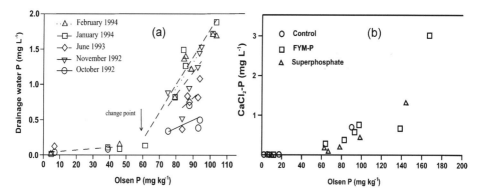

Figure 14.7. *The similarity of the relationship between Olsen P values and (a) drainage water P concentrations (Heckrath et al., 1995) and (b) 0.01 M CaCl₂ extractable P (Johnston and Polton, 1993). Redrawn from Hesketh and Brookes (2000).*

and Sharpley (2001) studied a number of soils and catchment areas to demonstrate a strong linear relationship between 0.01 M CaCl2 extractable P and the average drainage water P concentrations (Fig. 14.8).

14.4.1.2 Estimating P desorption potential from topsoils

In recent years research has focussed on developing techniques for identifying soil patches enriched in P that would contribute elevated amounts of biologically available P to surface waters (Sharpley et al., 1994; Sims, 1998). Sharpley (1996) and co-workers (Pote et al., 1996) have successfully used ion exchange resins and Fe oxide coated filter paper (Menon et al., 1997) to estimate potentially desorbable inorganic P (Pi) from sediment samples. The amounts of soil Pi desorbed to Fe-oxide filter paper, or P extracted by 0.1 M NaOH, were very strongly correlated with the growth of three P-limited algae on sediment as the sole source of P (Fig. 14.9). The Fe-oxide impregnated filter paper technique is seen to simulate desorption of P from sediment for uptake by aquatic plants and correlates well with P extracted from soil by common soil P tests (Sharpley, 1994).

There has been widespread 'lobbying' from scientists for soil testing laboratories to offer these new 'environmental soil tests' for improved phosphorus management of soils (Sharpley et al., 1994; Sims, 1998). It is also recognised, however, that if 'normal' soil test/plant growth response curves (discussed 'previous' section) had been used to maximise crop yield but avoid excess application of fertiliser and manure, the P enrichment of arable soils would not be as extensive. This concept, illustrated in Fig. 14.10, has been proposed for European soils by Higgs et al. (2000). Thus, it would appear that current and future research will be directed at determining acceptable agronomic soil P test

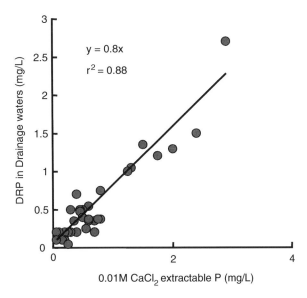

Figure 14.8. *Relationship between 0.01 M CaCl₂ extractable P from surface soils and the average drainage water P concentrations in a number of different catchments. Redrawn from McDowell and Sharpley (2001).*

critical values above which soils must not be raised. Sharpley and Tunney (2000) present recommended values for the soil tests used in some American states. These authors also report that the EPA in Ireland have adopted a threshold of 15 mg P kg^{-1} Morgan's P (approximately 60 mg P kg^{-1} Olsen P) above which there can be no land application of manure.

14.5 Research tools for building a fundamental understanding of the quantity, form and dynamics of plant-available nutrients in soils

14.5.1 Isotope exchange studies

Research using isotope exchange techniques to determine the amounts of readily and slowly available nutrients in soils (Table 14.6) and the rates of nutrient transfer between soil forms and uptake by plants (see recent papers by Frossard et al., 1994; Di et al., 1997; Hamon et al., 2002) has continued for more than 50 years.

14.5.1.1 *E* and *L* value determinations

Isotopic dilution studies offer the potential to determine the size of the pool of nutrients on the soil solid phase that is in 'quasi' equilibrium with ions in soil solution without chemical extraction (IAEA, 1976). Readers are directed to

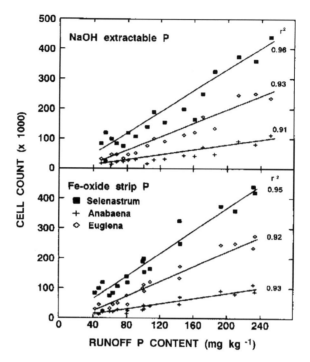

Figure 14.9. *Relationships between the growth of three algal species on suspended sediment and the amounts of P extracted from the sediment by Fe-oxide impregnated filter papers and 0.1 M NaOH extraction (redrawn from Sharpley et al., 1994).*

Chapter 7 for the use of isotopic dilution techniques for the assessment of rhizosphere processes, redox sensitive species and to natural attenuation.

The exchange theory to interpret the data was based on simple models of surface adsorption (McAuliffe et al., 1948). It was assumed that nutrient X has an exchangeable pool in the soil (X_E) that is the sum of the amount of nutrient in soil solution (X_L, mg $X \cdot kg^{-1}$soil) plus an amount of readily exchangeable nutrient on the soil surface (X_S, mg $X \cdot kg^{-1}$soil),

$$^A X_E = {}^A X_L + {}^A X_S \qquad (14.1)$$

and that (X_L and X_S) readily exchange.

$$^A X_L \rightleftharpoons {}^A X_S \qquad (14.2)$$

The native isotope of X in the soil is $^A X$. To determine unknown $^A X_E$, a known activity of isotope $^B X_T$ (Becquerels of radio activity or atom % excess of stable

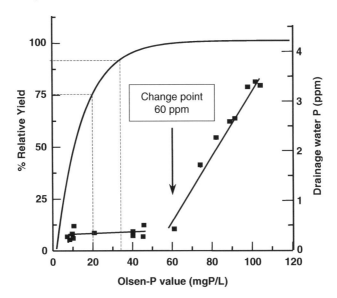

Figure 14.10. *A conceptual view of the relationship between optimum Olsen P values for obtaining high relative pasture yield and likely drainage water P concentrations on a soil with medium to low P retention (concept derived from Higgs et al., 2000).*

isotope) is added in a trace amount of $^A X_T$ and is well mixed with a sample of soil that has an aqueous phase, which can be sampled (moist incubation or slurry suspension of soil). It is assumed that the chemically similar isotope $^B X$ rapidly equilibrates with the total new exchangeable pool of $(^A X_E + ^A X_T)$.

As a 'quasi' equilibrium exists, the specific activity of the soil solution $^B X_L / ^A X_L$ will also equal the specific activity of the new exchangeable pool $^B X_T /$ $(^A X_E + ^A X_T)$. Providing the soil solution can be sampled, $^B X_L$ and $^A X_L$ can be determined by scintillation counting (or mass spectrometry) and chemical analysis, respectively.

$$\frac{^B X_T}{^A X_E + ^A X_T} = \frac{^B X_L}{^A X_L} \tag{14.3}$$

The original exchangeable pool is calculated as

$$^A X_E = \frac{^B X_T \, ^A X_L}{^B X_L} - ^A X_T \tag{14.4}$$

When the exchangeable pool of nutrient is determined in laboratory incubations it has been termed the E value (Russell et al., 1954).

Table 14.6. *A summary of selected studies using isotopic dilution to measure the size and transformations of the exchangeable pool of nutrients in soil.*

Isotopic exchange method	Bioavailability application	Definition and example references
Single point in time measurements		
E value	Determination of labile pool or exchangeable pool of nutrient in soil in exchange with isotopic species added to the soil solution	Russell et al. (1954), Frossard et al. (1994), Di et al. (1997), Buhler et al. (2003)
L value	Determination of the pool of nutrients made exchangeable with isotopic species added to soil solution by rhizosphere effects	Larsen (1952), Frossard et al. (1994), Di et al. (1997), Buhler et al. (2003)
(L–E) values	Crude estimate of enhanced dissolution of non-exchangeable nutrients by rhizosphere effects	Frossard et al. (1994), Hedley et al. (1982)
Inverse dilution with E or L values	Estimate of fertiliser dissolution or mineralisation of organic residue	Hood (2000), Di et al. (1997)
Kinetic studies		
ΔE_t values	Exchangeable nutrient pool and soil nutrient buffer capacity	Di et al. (1994b, 2000)
Inverse dilution ΔE_t	Rates of fertiliser dissolution or rates of nutrient immobilisation or loss from soils	Di et al. (1994b, 2000)
ΔL_T	Rates of enhanced dissolution in the rhizosphere and nutrient immobilisation in soils	Di et al. (2000)

If plant growth (P) is used to take up nutrient X from soil solution then the specific activity of the nutrient taken up ($^{B}X_P/^{A}X_P$) is also assumed to be the same as the specific activity of the exchangeable pool and can be used in Eq. (14.5) to calculate the size of the exchangeable nutrient pool in soils. This has been termed L value (Larsen, 1952). If a plant is grown from seed, then the isotopic marker taken up by the plant $^{B}X_P$ can also exchange with nutrient contained in the seed ($^{A}X_W$). Alternatively leakage of an unknown fraction of nutrient from the seed can be exchanged with ^{B}X remaining in the soil. The degree of isotopic exchange with nutrients in the seed is therefore uncertain, but becomes important when the amount of nutrient taken up by plants $^{A}X_P$ is small compared to $^{A}X_W$. The largest error will be caused by not accounting for seed dilution of $^{B}X_P$. Therefore, it is useful to calculate two L values, the first considering full exchange with seed nutrients,

$$L \text{ value} \quad ^{A}X_E = \frac{^{B}X_T{}^{A}X_P}{^{B}X_P} \quad \text{(including full exchange with seed nutrients)} \quad (14.5)$$

or no exchange with seed nutrients,

$$L \text{ value} \quad {}^{A}X_{E} = \frac{{}^{B}X_{T}({}^{A}X_{P} - {}^{A}X_{W})}{{}^{B}X_{P}} \quad \text{(no exchange with seed nutrients)} \quad (14.6)$$

The results of such calculations using P uptake by *Brassica napus* seedlings from a ^{32}P labelled soil that had weak P buffering characteristics are shown in Fig. 14.11. The maximum uncertainty in L value is larger during early stages of plant growth when seed P comprised a large proportion of total plant P.

The complex chemical and physical nature of most solid-phase nutrient species in soils (see introduction, weakly and strongly adsorbed inorganic forms, discrete sand, silt and clay mineral phases, organo-mineral coatings, interlayer charge balancing ions in layered silicate clays) creates a wide range of forms of nutrients in soils that have a differing exchangeability and solubility. This is reflected in E values, which tend to increase over the short-term (minutes/hours) as the period of isotope equilibration with the soil increases (Frossard et al., 1994). L values are also likely to increase over a period of days (or weeks) as rhizosphere processes such as acidification, or enhanced hydrolytic enzyme activity, solubilise additional pools of non-exchangeable nutrient (e.g. Hedley et al., 1982). The choice of plant species and the conditions of plant growth can cause significant differences between E and L values determined on the same soil samples. This is particularly so when soils contain significant non-exchangeable pools of acid-soluble inorganic phosphate and

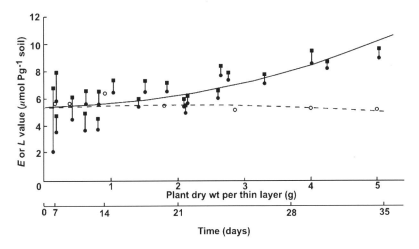

Figure 14.11. *A comparison of changes in the isotopically exchangeable pool of P in the rhizosphere soil of* B. Napus *(L value, solid line) with E values (dashed line) determined in incubated, unplanted soil (redrawn from Hedley et al., 1982).*

organic phosphate. For example Hedley et al. (1982) showed that solubilisation of acid soluble soil P in the P deficient rhizosphere of *B. napus* was
associated with L values becoming increasingly larger that E values (Fig. 14.11).

14.5.1.2 Kinetic studies

Fardeau et al. (1985) used a standard soil slurry incubation to examine the
relationship between increasing E value and time of incubation (1–100 min). The
proportion of radioactive $^{32}PO_4$ ions (r_t/R) remaining in solution with time (t)
could be explained with a decay series,

$$\frac{r_t}{R} = \frac{r_1}{R}\left\{ t + \left(\frac{r_1}{R}\right)^{(1/n)} \right\}^{-n} + \frac{r_\infty}{R} \tag{14.7}$$

where R, r_1 and r_∞ are the solution activity of ^{32}P at zero time, after 1 min and at
infinity, respectively. n is the slope of the regression $ln\,(r_t/R)$ against $ln\,(t)$ for
($t \leq 100$ min) and is calculated from the $r_{1\,min}$ measurement and an assumed
value for r_∞/R. It is assumed that the P exchange equilibria is at steady state so
that after infinite equilibration r_∞/R is equal to the ratio of the amount of P in
soil solution (Cp)/total amount of inorganic phosphorus (Pi) in the soil. Once n
is determined from experimental work with short equilibration times, r_t/R can
be predicted for longer periods of exchange that may simulate a crop growing
season. An estimate of the exchangeable pool (E_t) can be calculated from

$$E_t = \frac{RC_P}{r_t} \tag{14.8}$$

Frossard et al. (1994) used this method to calculate a theoretical E value for a
time equivalent ($E_{13\ weeks}$) to the period of growth of common bent grass
(*Agrostis capillaris* L.) on a range of P fertilised and unfertilised soils. L values
determined from P uptake by bent grass between 8 and 13 weeks generally
agreed with the $E_{13\ week}$ value. Frossard et al. (1994) concluded that the
similarity between E and L values confirmed that plant available P, at least, is P
that will exchange in the long-term in laboratory slurry incubations. The authors
note, however, that there are several circumstances where error of accuracy can
occur in determining E values. The most common problem is low, or non-
detectable, solution phosphate concentrations in strongly weathered soils with
high soluble concentrations of interfering Si ions (Frossard et al., 1994). In
tropical soils with almost non-detectable solution phosphate concentration E
values become unreliable (Buhler et al., 2003). Salcedo et al. (1991) prescribe the
use of an anion exchange resin to concentrate P from the extracted soil solution
and allow colorimetric determination in the absence of interfering soluble (Si).
More recently, Maertens et al. (2004) have proposed that the resin membranes

are included in the slurry incubation. The authors suggest that the presence of resin membrane will not induce dissolution of non-exchangeable P forms. This result is not consistent with the results of Trolove et al. (1996), who demonstrated accelerated dissolution of rock phosphate when resin membranes were equilibrated with soils of decreasing pH. Such resin-enhanced dissolution may be considered an appropriate dilution of the isotopic tracer in studies that try to simulate dissolution of minerals in the rhizosphere.

14.5.1.3 Inverse dilution to measure dissolution and mineralisation rates

The technique of inverse isotopic dilution (IAEA, 1976) is becoming popular again to determine the solubilisation of normally non-exchangeable fertiliser materials (Di et al., 1995), the mineralisation of organic residues (Hood, 2001) and the bioavilability of nutrients in biosolids (Frossard et al., 1996a, 1996b). In one option of this technique, the exchangeable nutrient pool $^A X_E$ in the soil is first equilibrated with the tracer $^B X_T$ until a 'quasi' exchange equilibrium is established. As in Eq. (14.4), the specific activity of the sampled soil solution in equilibrium with the exchangeable nutrient pool is used to calculate $^A X_E$. Several 'labelled soil' replicates are prepared. The 'fertilising' material ($^A X_F$), a fertiliser, or organic residue is then mixed with half the replicates of labelled soil. As a fraction $\propto$ of the nutrient is released from the 'fertilising' material, it enters the exchangeable pool, diluting the specific activity of the pool to $^B X_T / (^A X_E + \propto \ ^A X_F)$ to give a new 'fertilised' soil solution-specific activity of $^B X_{LF} / ^A X_{LF}$ which can be measured

$$\therefore \ ^A X_E + \propto \ ^A X_F = \frac{^B X_T \ ^A X_{LF}}{^B X_{LF}} \tag{14.9}$$

But in the unfertilised replicate

$$^A X_E = \frac{^B X_T \ ^A X_L}{^B X_L} \tag{14.10}$$

substituting Eq. (14.10) in Eq. (14.9) the amount of fertiliser material released

$$\propto \ ^A X_E = \ ^B X_T \left(\frac{^A X_{LF}}{^B X_{LF}} - \frac{^A X_L}{^B X_L} \right) \tag{14.11}$$

Montenegro and Zapata (2002) used this technique to assist the selection of plant species that are effective users of low solubility rock phosphates. Using ^{32}P labelled soil fertilised with unlabelled phosphate rock, they identified lower shoot-specific activities in some *B. napus* cultivars that were able to achieve greater solubilisation of phosphate rock than others.

For nutrients that are involved in inorganic fixation in soils, or immobilisation into non-exchangeable organic forms, or are leached from field soils, the addition of the fertiliser material may change the rate of some of these processes and error will arise because the rate at which isotope is removed from the exchangeable pool differs in fertilised and unfertilised soils. Di et al. (1994a, 1994b) used the principles for isotopic dilution for a pulse-labelled, single, exchangeable pool with an input rate (fertiliser dissolution and/or organic matter mineralisation) and an output rate (plant uptake and/or fixation or soil organic matter synthesis or leaching) to demonstrate how more information on nutrient bioavailability (and retention in soils) can be gained if a series of measurements of the specific activity of the exchangeable pool are made during plant growth, or during incubation of fertiliser with soil. Kinetic studies of isotope exchange equilibria will be most useful in determining the dissolution rates and decomposition rates of fertilisers, manures and biosolids added to soil. Di et al. (2000) discuss the application of kinetic studies of isotope dilution to measurement of nitrogen, phosphorus and sulphur transformations in soils and explain their uses and limitations.

14.5.1.4 Evaluating whether soil tests extract exchangeable forms of nutrients

Isotopic dilution studies using soil incubations (E value) plus plant growth (L values) carried out in conjunction with chemical extraction of the labelled soils, allow researchers to determine whether the chemical soil testing procedure draws nutrients from the same sized exchangeable pool as the plant. Aigner et al. (2002) used the kinetic technique developed by Fardeau et al. (1985) to show a strong correlation between E values determined by a 1 min slurry incubation and soil P extracted by the Fe strip method (Menon et al., 1997). In addition, the specific activities of P extracted by the Fe strip and in the $E_{1\ min}$ soil solution were similar, indicating the soil test P was drawn from the same sized exchangeable pool. Ayoub et al. (2003) grew the hyperaccumulator *Thlaspi caerulescens, Taraxum officinale* and *Hordeum vugare L.* on two soils of contrasting Zn content (one had been used for sewage sludge disposal) that had been uniformly labelled with ^{67}Zn. All plants had similar L values on both soils and E and L values were of similar magnitude. This indicated that *T. caerulescens* took up Zn from the same exchangeable pool as the other plant species and its hyperaccumulation activity was not associated with increased solubilisation of the soil Zn pool. The respective isotopically exchangeable Zn pools in each soil made up 67% (high Zn soil) and 13% (low Zn soil) of their total Zn content. A range of reagents (0.1 M $NaNO_3$, 0.01 M $CaCl_2$, 0.43 M CH_3COOH and 0.05 M EDTA) extracted lower amounts of Zn with higher specific activities of Zn than sampled by the plants. This demonstrated that whereas amounts of extractable nutrient from a soil may correlate well with plant uptake, the nutrient pool sourced by the plant may be different to that extracted by chemical reagents.

14.5.2 Rhizosphere studies

The product of the concentrations of nutrients in the soil solution and the volume of water transpired by plants provides an empirical evaluation of whether the transpirational flow (mass flow) of water from soil through the plant is capable of supplying the nutrients accumulated in the shoot mass. Using the average values for nutrient concentrations in maize at harvest, a crop water use value of approximately 300 mm and average soil solution nutrient concentration for mid-western USA topsoils, Barber et al. (1962) estimated that the major nutrient ions Ca, Mg and SO_4 would be oversupplied at the root surface, while K, P and most of the trace elements would be undersupplied. Experiments with seedling root growth in thin layers of soil labelled with the radioisotopes allowed autoradiographs to be developed, which confirmed concentration depletion zones around roots for ^{33}P, K (surrogate ^{86}Rb), ^{99}Mo and ^{65}Zn (see reviews by Barber, 1984 and Tinker and Nye, 2000). In the depletion zone, the majority of ion movement was by diffusion down a concentration gradient to the root surface. Mathematical models of nutrient supply to plant by mass flow and diffusion from an exchangeable pool of nutrients in the soil gives a reasonable explanation of plant nutrient uptake in well fertilised soils (see reviews by Barber, 1984 and Tinker and Nye, 2000). However, in nutrient deficient soils, nutrient uptake often exceeds that predicted by mass flow, diffusion and exchange/desorption (Tinker and Nye, 2000). Accelerated solubilisation of non-exchangeable nutrient pools in the rhizosphere can account for these differences. Once techniques were developed to produce rhizosphere soil in quantities permitting chemical analysis (e.g. Grinsted et al., 1982; Kuchenbuch and Jungk, 1982) then it was relatively easy to chemically fractionate the rhizosphere soil to characterise the forms of nutrients being solublised (e.g. Hedley et al., 1982). The added advantage of the technique developed by Kuchenbuch and Jungk (1982), which created a planar rhizosphere (Fig. 14.12), was that soil at increasing distances from the rhizoplane could be sampled and analysed for evidence of root-induced change. Several researchers have used techniques like this to study the mobilisation of soil and fertiliser phosphorus (e.g. Hedley et al., 1982, 1994; Gahoonia and Nielsen, 1992), potassium (Hinsinger and Jaillard, 1993; Hylander et al., 1999), copper (Alam et al., 2001) and zinc (Kirk and Bajita, 1995) by root-induced processes.

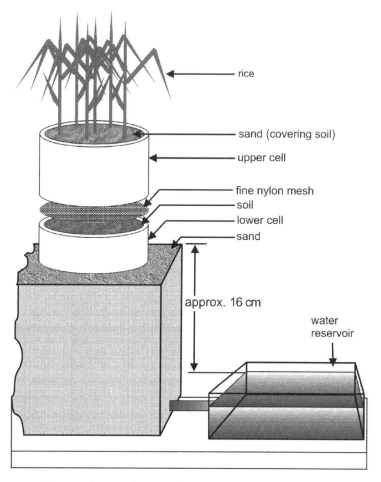

Figure 14.12. *Root study container used to create planar-rhizosphere soil in which P depletion profiles can be measured in the lower cell at increasing distances from the root plane at the 24 μm mesh (after Kuchenbuch and Jungk, 1982).*

14.5.2.1 Rhizosphere-induced P mobilisation

The extra information on nutrient fate and bioavailability in soil that can be gained by combining rhizosphere studies with chemical fractionation is illustrated in studies with upland rice (Hedley et al., 1994). After fertilising a Philippine Ultisol with monocalcium phosphate (MCP) only 5–6% of the added soluble Pi remained in the easily extracted pool (cation and anion exchange resin extractable Pi). The majority, 75–82%, of the added P was found strongly adsorbed to Fe and Al oxides (NaOH extractable Pi fraction). Despite the strong adsorption of P to Fe and Al oxides, upland rice was able to take up most of the

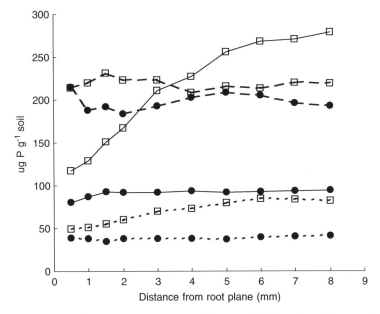

Figure 14.13. *The profiles of NaOH-extractable organic P (— —) and inorganic P (———) and 0.5 M H$_2$SO$_4$-extractable P (- - -) in Cavinti soil either unfertilised (•) or fertilised (□) with monocalcium phosphate at 240 μg P g^{-1} soil (data from Hedley et al., 1994).*

recently added Pi in the slice of soil adjacent to the rice rhizoplane (Fig. 14.13). This solubilisation of strongly adsorbed P in the upland rice rhizosphere can be explained if the Fe-oxide surface is dissolved by the chelating action of root-released citric acid (Kirk, 1999).

The forms of P depleted in the rhizosphere of plants are dependent upon the stage of soil weathering, the forms of P that have accumulated from fertiliser application (Hedley et al., 1995) and the plant species. Recently, Hinsinger (2001) and Trolove et al. (2003) reviewed these factors. In low P status soils some plant species show significant rhizosphere-induced mineralisation of soil organic P (e.g. Tarafdar and Jungk, 1987; Gahoonia and Nielsen, 1992; Zoysa et al., 1998; George et al., 2002). It is not clear why some species are able to mobilise more organic P than others when grown in the same soil. The answer obviously lies with the suite of hydrolytic enzymes that are induced in the rhizosphere (Trolove et al., 2003), but the mechanisms of enzyme induction require further investigation.

14.5.2.2 Rhizosphere-induced trace element mobilisation

Since the pioneering studies of Marschner and others (see reviews by Marschner and Romheld, 1994; Marschner, 1998) on the role of siderophores and phytosiderophores in the efficient plant uptake of iron from soils, similar processes have been found to be important in Zn (vonWiren et al., 1996) and probably Cu (Tao et al., 2003) uptake from soil by plant roots. As yet there is still no clear model of Cu uptake that explains the accumulation of Cu in plants from soil. In agricultural soils it is common for more than 99% of the total dissolved Cu concentration to be strongly bound to dissolved organic carbon (DOC) (Römkens, 1998). Inside the xylem sap of the plant root, copper is strongly complexed with amino acids (e.g. histidine (His) and nicotianamine (NA); Liao et al., 2000a). The question arises: how is Cu transported from a natural soil solution across the root cell wall? Are the protein transporters in the cell membrane responsible for the transfer of free Cu^{2+} ions, or organically complexed Cu? Studying soil solution behaviour in a pH stat combining a free copper ion electrode provides a useful technique for studying the nature of the dominant soluble Cu species (Liao et al., 2000a, 2000b). Liao et al. (2000b) found that similar to amino acid–Cu complexes (Liao et al., 2000a), soil solution organic matter complexes probably have unique pH dependent stability relationships (Jeffery and Uren, 1983; Liao et al., 2000b). Different pH dependent stability for soil solution Cu complexes may help explain why two soils (Ashhurst stony silt loam and Wairoa pumice soil) can have similar amounts total Cu complexed in soil solution yet only one soil (Wairoa pumice soil) grows Cu deficient plants (Fig. 14.14). Liao et al. (2000b) proposed the following concepts to explain why these differences in Cu uptake by plants may occur in soils with similarly low soil solution free Cu^{2+} concentrations.

In the soil solution of Ashhurst stony silt loam at pH 5.6– 6.0 (Fig. 14.15), 99% of all Cu in soil solution in the bulk soil is bound to soil organic matter (Cu-SOM) and there is less than $10\,\mu g\ L^{-1}$ of free Cu^{2+} in bulk soil solution (Figs. 14.15 and 14.16). The Cu–SOM complexes may move to the rhizosphere soils by mass flow or diffusion. When soluble Cu–SOM complexes enter the rhizosphere and approach the rhizoplane (Fig. 14.16), particularly near the root elongation zone or zone of root hair proliferation, the pH is likely to be lower than the bulk soil, causing the Cu–SOM to dissociate and to release more free Cu^{2+} ions. To achieve passive uptake of free Cu^{2+} ions via transporter channels in the root cell membrane, the pH at the rhizoplane in the Ashhurst stony silt loam has to decrease to $<$pH 5.1 and the plant root xylem be at $>$ pH 5.6 (Fig. 14.16). In this way root surface Cu^{2+} concentrations of 2.5–$3\,\mu g\ Cu^{2+}\ L^{-1}$ may allow diffusion of free Cu^{2+} to xylem sap free Cu^{2+} concentrations of 1.6–$3.2\,\mu g\ Cu^{2+}\ L^{-1}$ (i.e. a concentration gradient of free Cu^{2+} is created across the cell wall). The free Cu^{2+} is taken up and then loaded into the xylem, where pH is in the range 5.5–6.5 and Cu forms strong complexes with NA and His, and is transported to shoots via xylem sap (Fig. 14.16).

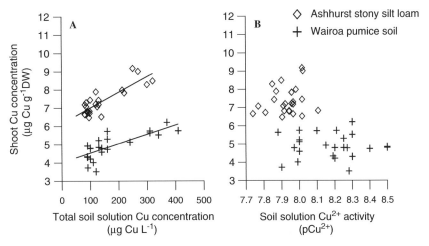

Figure 14.14. *The relationship between total soil solution Cu concentration and shoot Cu concentrations (A), and the lack of a relationship between the soil solution Cu^{2+} activity and shoot Cu concentrations (B) for ryegrass grown in pots of Ashhurst stony silt loam and Wairoa pumice soil (Liao et al., 2000b).*

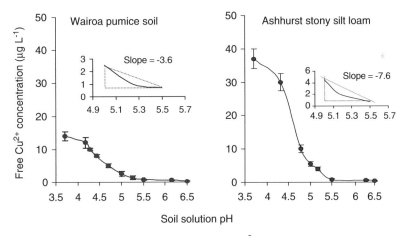

Figure 14.15. *The relationship between free Cu^{2+} concentration and pH of soil solutions from pots treated with $50\,kg\,Cu\,ha^{-1}$ as $CuSO_4 \cdot 5H_2O$ (Liao et al., 2000b).*

Passive uptake of free Cu^{2+} ions in the Wairoa pumice soil is less likely to occur because a rhizosphere pH of <5.0 would be required.

If pH does not change across the rhizosphere/root boundary as described above, then the conditions represented in (Fig. 14.16) would suggest that root surface, active carrier binding sites for Cu must have association constants for

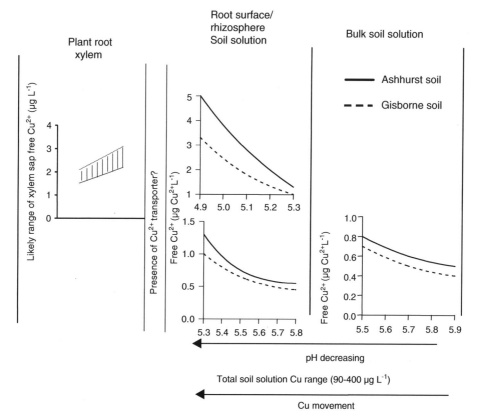

Figure 14.16. *A conceptual model showing the free Cu²⁺ concentrations in solution as Cu moves from the bulk soil solution of pH 5.5 to 5.9 to the lower pH rhizoplane and is transferred across the root cell wall to the xylem sap Liao et al. (2000b).*

Cu greater than the soluble soil organic matter at the same pH. Root-exuded mucilage is unlikely to present a barrier to this process since reported association constants for mucilage-Cu are only 3.4–5.4 (log K_a) (Mench et al., 1987, 1988; Morel et al., 1986), compared to soil organic matter of 7.8–11.3 (Stevenson, 1994) and xylem amino acids (His and NA) of 17.5–18.6 (May et al., 1977; Stephan et al., 1996).

Further rhizosphere and soil solution studies will be required to test whether the concepts described above do represent the mechanism of Cu uptake from low copper status soils.

14.5.3 Computer simulation

Conceptual models of nutrient uptake by plants, or whole nutrient cycles and experimentally determined rate-limiting processes have been translated into mathematical algorithms and numerical solutions for time-dependent processes (see reviews of nitrogen and phosphorus models by Benbi and Richter (2003) and Tinker and Nye (2000), respectively). Computer simulation models are required to test theories and interpret experimental results concerning the mechanisms of nutrient availability in soils, however, their greatest value is in simulating the temporal dynamics of nutrient availability caused by seasonal crop growth (e.g. Greenwood et al., 2001), discrete drainage or run-off events (e.g. El-Sadek et al., 2003) and long-term land management (Tabachow et al., 2001; El-Sadek et al., 2003). Successful simulation of nutrient bioavailability in soil over time leads to the development of models that provide practical management advice to farmers and growers (e.g. Greenwood et al., 2001).

Simulation models have been usefully applied to solve problems at every scale of nutrient bioavailability. Soil solution speciation models such as MINTEQA2 (Allison et al., 1991) are invaluable for characterising the division of a total soluble nutrient concentration into species of cations, anions and ion pairs and complexes. This is important if the interactions between soil solution and adsorbing soil surface are to be described to model such processes as nutrient uptake by plants and leaching of nutrients from soils.

Also at the detailed scale are simulation models of rhizosphere processes. Kirk (1999) has developed a model for the upland rice rhizosphere, where root-released citrate is generated, diffuses into the soil to chelate iron from the Fe-hydrous oxide surface. P adsorbed on the hydrous oxide surface is solublised and diffuses back to the root surface. The model was used by Kirk (1999) to explain the soil P depletion profiles measured by Hedley et al. (1994) in the upland rice rhizosphere (Fig. 14.13). This modelling development arose because the upland rice P uptake was larger than could be explained by a simpler model that simulated only the desorption and diffusive flux of phosphate from bulk soil to the outer shell of a P depleted rhizosphere. The latter is the process of nutrient ion uptake simulated by the Barber–Cushman model (Barber and Cushman, 1981; Barber, 1984), which has been employed to simulate P and K uptake by plants (Barber, 1984). According to Greenwood et al. (2001), this model has limited application to field grown crops (e.g. Lu and Miller, 1994) because it does not simulate the effect of climatic conditions and soil nutrient supply on the growth of shoot and root mass. Recently, Greenwood et al. (2001) have introduced a mechanistic model for P uptake by vegetable crops that addresses some limitations of the Barber–Cushman type model and aims to provide practical advice to growers on rates of starter P fertiliser to apply given different initial soil P statuses. The model is very detailed in the mechanistic way it describes the dissolution of granular starter fertiliser and the diffusive flux of P to newly generated segments of root as crop growth proceeds. Root growth is a function of crop growth that is responsive to P supply from a soil in which soil volumetric moisture content (controlled by a daily soil water balance model), soil test status and P buffer capacity influence the rate of P diffusion to the root segments. Model outputs of yield, P off-take in the crop and final soil P test status provide a useful readout for the environmental scientist wishing to predict P run-off risks (Karpinets et al., 2004), or grower wishing to maximise profits by improving fertiliser use efficiency.

To give useful practical advice on the likelihood that farm practices cause nutrient accumulation or depletion, models need not be complex in their simulation of detailed nutrient cycling processes, but can be used to give an overview of the nutrient transformations within an agroecosystem and its nutrient gains and losses. The model Overseer Nutrient Budgets (Ledgard et al., 1999) achieves this for a number of farming systems by using an annual mass balance of nutrient inputs in fertiliser, feed, atmospheric N fixation etc. and nutrient losses in product, drainage and gaseous loss. In New Zealand, the model is gaining wide use with fertiliser company advisory officers who wish to identify farm practices that are causing excessive accumulation of soil nutrients. An Overseer nutrient budget for a dairy farm paddock used for land treatment of dairy shed and feed pad effluent is shown in Table 14.7. The large amount of maize silage fed to cows on the feed pad (supplement) allows a high stocking rate and adds to the farm nutrient loading. The high stocking rate creates more effluent (in proportion to time spent on the feeding pad and dairy shed) requiring land treatment and the rate of annual N, P and K accumulation in the soil is excessive (Table 14.7). Instructions to the farmer to cease fertiliser application on this block and increase the block area will partly remedy the problem. The model can be used to explore different management scenarios that have the potential to solve the nutrient accumulation problem.

14.6 Future directions

Research into nutrient bioavailability is normally justified using the following categories: (1) the academic challenge of understanding and describing the fundamental biochemical and biophysical processes involved in nutrient cycling in terrestrial systems, (2) the practical challenge of improving soil fertility advice to by farmers and growers to overcome shortages in regional food production and, in other regions, improving the efficiency of nutrient use in soil–plant systems to maximise crop production profitability and minimise food and fibre costs, and (3) the practical challenge of identifying nutrient-enriched terrestrial environments that pose a threat to drainage water quality.

In category (1) further studies are required on rhizosphere processes, particularly for immobile trace elements and the role of root released compounds in major element nutrition. Plant and animal deficiencies of

Table 14.7. *A nutrient input–output balance predicted by the Overseer Nutrient Budgets model for a block of grazed pasture (19 ha) receiving farm dairy effluent and feed pad effluent on a 153 ha New Zealand dairy farm (carrying 3.6 cows per hectare and using maize silage to supplement on pasture grazing).*

	N	P	K	S	Ca	Mg	Na
Inputs				kg/ha/y			
Fertiliser	66	60	63	56	97	19	0
Effluent added	458	57	430	55	84	45	22
Atmospheric (legume/rain)	25	0	4	8	6	13	80
Irrigation	0	0	0	0	0	0	0
Slow-release	0	3	3	0	0	0	2
Supplements	39	7	44	4	32	19	12
Outputs				kg/ha/y			
Milk/Meat	112	20	25	7	28	2	7
Excreta transfer	70	8	68	10	11	7	3
Hay silage	0	0	0	0	0	0	0
Gaseous loss	104	0	0	0	0	0	0
Drainage/run-off	65	4	119	141	51	10	68
Accumulation in soil organic matter or fixation	239	42	0	−35	0	0	0
Change in available pool	0	54	331	0	129	77	38

Cu, Co, Zn and Se still cannot be diagnosed adequately from our current understanding of these elements indexes of plant availability in the soil. If we can improve methods for maintaining soil solution characteristics as 'near native' as possible then solution speciation studies in bulk and rhizosphere soil could provide improved knowledge of trace element release from soil and uptake by roots. Where appropriate, it may be profitable to combine isotopic labelling of the soil prior to plant growth and then utilise electrophoretic (Strobel, 2001), or ion exchange chromatographic methods (Collins et al., 2001), to identify the charged species present. It would be useful to combine the rhizosphere studies with the screening of plant species (or cultivars) for efficient nutrient uptake processes.

Currently in category (2), there is still considerable research being conducted that focuses only on the laboratory development of improved extraction methods for use as indices of nutrient availability in soils. Refining laboratory protocols to give improved definition of bioavailable nutrients in a laboratory soil suspension may not best serve the agricultural communities' investment in science. It needs to be recognised that the large errors created by field spatial and temporal variability in exchangeable nutrient pool sizes and plant growth responses will limit the advantages gained by improvements in laboratory protocol. Rapid remotely sensed field methods (Robert, 2002) are required for estimating the spatial component of nutrient variability in the field. These field methods will offer a faster and more economic way to direct annual precision application of fertiliser to a variable yield landscape. Standard soil testing procedures should be constrained to permanent transects for monitoring whether the general fertiliser policy is causing soil fertility to increase or decline. In reality, a visit to the farming community in most countries will reveal that much is to be gained by simply causing more farmers to soil test as part of a fertility management plan. In addition, training on the approved field protocols for field sampling soil cores will be a major improvement in increasing the knowledge base about nutrient availability status.

In category (3), current research has made good progress in developing tests that indicate whether soil or sediment samples are likely to enrich drainage and run-off waters with P. It will have to be left to the soil carbon and nitrogen modellers to improve risk-assessment procedures for nitrate leaching. There is an urgent need for research techniques that are able to track agricultural run-off into rivers and lakes, where its nutrient contribution to algal and periphyton growth can be assessed. In the meantime, the environmental risk of nutrient enriched soil will be assessed either on the agronomic soil test status or specifically designed soil P saturation indices. Critical values for these tests are probably enforced in regional environmental protection plans in several developed countries. Protocols for sampling and undertaking these tests and their calibrations will have to be robust and stand the test of several cases in environmental law courts. Such developments are likely to take soil tests into an 'era of certainty' for which they are unlikely to be fit.

Although the fundamental knowledge requirement for all categories is similar, our current mechanistic understanding of the processes involved is incomplete and the majority of progress in categories 2 and 3 is made by observation, correlation and empirical modelling. Developments in increasing computing power mean that the goal of making a completely mechanistic nutrient cycling model capable of interpreting and predicting the nutrient sustainability of an ecosystem is looking feasible, particularly when recent progress with crop nutrient uptake models (e.g. Greenwood et al., 2001) and nitrogen cycling models (Tabachow et al., 2001; El-Sadek et al., 2003) are considered. The two major challenges for experimenters and modellers are to fully simulate nutrient cycling in field soil growing conditions and to produce models that involve multi-nutrient cycling rather than single element cycling.

References

Aigner, M., Fardeau, J.C., Zapata, F., 2002. Does the Pi strip method allow assessment of the available soil P? Comparison against the reference isotope method. Nutr. Cycl. Agroecosys. 63, 49–58.
Alam, S., Kamei, S., Kawai, S., 2001. Metal micronutrients in xylem sap of iron-deficient barley as affected by plant-borne, microbial, and synthetic metal chelators. Soil Sci. Plant Nutr. 47, 149–156.

Allen, D.G., Barrow, N.J., Bolland, M.D.A., 2001. Comparing simple methods for measuring phosphate sorption by soils. Aust. J. Soil Res. 39, 1433–1442.

Allison, J.D., Brown, D.S., Novo-Gradac, K.J., 1991. MINTEQA2/PRODEFA2, A Geochemical Assessment Model for Environmental Systems. Version 3.11. ESEPA, Athens, GA.

Andersen, M.K., Raulund-Rasmussen, K., Hansen, H.C.B., Strobel, B.W., 2002. Distribution and fractionation of heavy metals in pairs of arable and afforested soils in Denmark. Eur. J. Soil Sci. 53, 491–502.

Ayoub, A.S., McGaw, B.A., Shand, C.A., Midwood, A.J., 2003. Phytoavailability of Cd and Zn in soil estimated by stable isotope exchange and chemical extraction. Plant Soil 252, 291–300.

Barber, S.A., 1984. Soil nutrient bioavailability, a mechanistic approach. Wiley, New York, 398pp.

Barber, S.A., Cushman, J.H., 1981. Nitrogen uptake model for agronomic crops. In: J.K. Iskander (Ed.), Modeling Waste Water Renovation- Land Treatment. Wiley Interscience, New York, pp. 382–409.

Barber, S.A. Walker, J.M., Vasey, E.H., 1962. Principles of ion movement through the soil to the plant root. Transactions of International Society of Soil Science, Commisions IV and V. International Soil Science Conference, Soil Bureau, Lower Hutt, New Zealand, pp. 121–124.

Barrow, N.J., 1993. Mechanisms of reaction of zinc with soil and soil components. In: A.D. Robson (Ed.), Zinc in soils and plants. Proceedings of the International Symposium held at The University of Western Australia, 27–28 September, 1993, Kluwer Academic Publishers, Dordrecht, Netherlands, pp. 15–31.

Benbi, D., Richter, J., 2003. A critical review of some approaches to modelling nitrogen mineralization. Biol. Fert. Soils 35, 168–183.

Blaylock, M.J., James, B.R., 1994. Redox transformations and plant uptake of selenium resulting from root–soil interactions. Plant Soil 158, 1–12.

Bray, R.H., Kurtz, L.T., 1945. Determination of total, organic, and available forms of phosphorus in soils. Soil Soc. 59, 39–45.

Brennan, R.F., Best, E., 1999. Copper. In: K.I. Peverill, L.A. Sparrow and D.J.A. Reuter (Eds.), Soil analysis interpretation manual. Australian Soil and Plant Analysis Council, CSIRO Publishing, Collingwood, Australia, pp. 263–280.

Brown, A.J., 1993. A review of soil sampling for chemical-analysis. Aust J. Exp. Agric. 33, 983–1006.

Brown, A.J., 1999. Soil sampling and sample handling for chemical analysis. In: K.I. Peverill, L.A. Sparrow and D.J.A. Reuter (Eds.), A soil analysis interpretation manual. Australian Soil and Plant Analysis Council, CSIRO Publishing, Collingwood, Australia, pp. 35–53.

Buhler, S., Oberson, A., Sinaj, S., Friesen, D.K., Frossard, E., 2003. Isotope methods for assessing plant available phosphorus in acid tropical soils. Eur. J. Soil Sci. 54, 605–616.

Collins, R.N., Onisko, B.C., McLaughlin, M.J., Merrington, G., 2001. Determination of metal–EDTA complexes in soil solution and plant xylem by ion chromatography-electrospray mass spectrometry. Environ Sci. Technol. 35, 2589–2593.

Colwell, J.D., 1963. The estimation of phosphorus fertilizer requirements of wheat in southern New South Wales by soil analysis. Aust. J. Exp. Agric. Anim. Husb. 3, 190–197.

Dahnke, W.C., Olsen, R.A., 1990. Soil test correlation, calibration and recommendation. In: R.L. Westerman (Ed.), Soil Testing and Plant Analysis, 3rd Ed. Book series No. 3. Soil Science Society of America Inc., Madison, WI. pp. 45–72.

Dakora, F.D., Phillips, D.A., 2002. Root exudates as mediators of mineral acquisition in low nutrient environments. Plant Soil 245, 35–47.

Delgado, A., Torrent, J., 2001. Comparison of soil extraction procedures for estimating phosphorus release potential of agricultural soils. Commun. Soil Sci. Plant Anal. 32, 87–105.

Di, H.J., Condron, L.M., Frossard, E., 1997. Isotope techniques to study phosphorus cycling in agricultural and forest soils, a review. Biol. Fertil. Soils 24, 1–12.

Di, H.J., Cameron, K.C., McLaren, R.G., 2000. Isotopic dilution methods to determine the gross transformation rates of nitrogen, phosphorus, and sulfur in soils, a review of the theory, methodologies, and limitations. Aust. J. Soil Res. 38, 213–230.

Di, H.J., Harrison, R., Campbell, A.S., 1994a. Assessment of methods for studying the dissolution of phosphate fertilizers of differing solubility in soil. I. An isotopic dilution method. Fert. Res. 38, 1–9.

Di, H.J., Harrison, R., Campbell, A.S., 1994b. Assessment of methods for studying the dissolution of phosphate fertilizer of differing solubility in soil. III. The relationship between rate of dissolution and plant response. Fert. Res. 38, 19–27.

Di, H.J., Harrison, R., Campbell, A.S., 1995. An isotope injection technique to measure agronomic effectiveness of different phosphate sources in undisturbed soils. Aus. J. Exp. Agric. 35, 489–493.

Dixon, J.B., Weed, S.B., 1989. Minerals in Soil Environments. , 2nd Ed., Madison,, WI, 1244pp.

Egnér, H., Riehm, H., Domingo, W.R., 1960. Untersuchungen über die chemische Bodenanalyse als Gundlage für die Beurteilung des Nährtof-fzustandes der Böden. Chemishe Extraktions-methoden zur phosphor – und kaliumbestimmung. Kungl. Lantbr. Hôgsk. Ann. 26, 199–215.

El-Sadek, A., Oorts, K., Sammels, L., Timmerman, A., Radwan, M., Feyen, J., 2003. Comparative study of two nitrogen models. J. Irrig. Drain. Eng.-ASCE 129, 44–52.

Fardeau, J.C., Morel, C., Jappé, J., 1985. Cinétique d'échange des ions phosphate dans les systémes sol-solution. Vérification expérmentale de l'équation théorique. Comptes reduc de l'Académie des Sciences Paris 300, Série III 8, pp. 371–376.

Frossard, E., Fardeau, J.C., Brossard, M., Morel, J.L., 1994. Soil isotopically exchangeable phosphorus, a comparison between E and L values. Soil Sci. Soc. Am. J. 58, 85–86.

Frossard, E., Sinaj, S., Dufour, P., 1996a. Phosphorus in urban sewage sludges as assessed by isotopic exchange. Soil Sci. Am. J. 60, 179–182.

Frossard, E., Sinaj, S., Zhang, L.M., Morel, J.L., 1996b. The fate of sludge phosphorus in soil–plant systems. Soil Sci. Soc. Am. J. 60, 1248–1253.

Gahoonia, T.S., Nielsen, N.E., 1992. The effects of root-induced pH changes on the depletion of inorganic and organic phosphorus in the rhizosphere. Plant Soil 143, 185–191.

George, T.S., Gregory, P.J., Robinson, J.S., Buresh, R.J., 2002. Changes in phosphorus concentrations and pH in the rhizosphere of some agroforestry and crop species. Plant Soil 246, 65–73.

Gianello, C., Bremner, J.M., 1986. Comparison of chemical methods of assessing potentially available organic nitrogen in soil. Comm. Soil Sci. Plant Anal. 17, 215–236.

Gourley, C.J.P., 1999. Potassium. In: K.I. Peverill, L.A. Sparrow and D.J.A. Reuter (Eds.), Soil Analysis Interpretation Manual. Australian Soil and Plant Analysis Council, CSIRO Publishing, Collingwood, Australia, pp. 229–243.

Greenwood, D.J., Karpinets, T.V., Stone, D.A., 2001. Dynamic model for the effects of soil P and fertilizer P on crop growth, P uptake and soil P in arable cropping, Model description. Ann. Bot. 88, 279–291.

Grinsted, M.J., Hedley, M.J., White, R.E., Nye, P.H., 1982. Plant-induced changes in the rhizosphere of rape (Brassica napus var. Emerald) seedlings. I. pH change and the increase of P concentration in the soil solution. New Phytol. 91, 19–29.

Guggenberger, G., Haider, K.M., 2001. Effect of mineral colloids on biogeochemical cycling of C, N, P, and S in soil. In: P.M. Huang, J.M. Bollag and N. Senesi (Eds.), Interactions between soil particles and microorganisms, impact on the terrestrial ecosystem. Wiley, Chichester, UK, pp. 267–322.

Haddad, K.S., Kaldor, C.J., 1982. Effect of parent material, natural available soil boron, and applied boron and lime on the growth and chemical composition of lucerne on some acidic soils of the Central Tablelands of New South Wales. Aust. J. Exp. Agric. Animal Hus. 22, 317–323.

Hamon, R.E., Bertrand, I., McLaughlin, M.J., 2002. Use and abuse of isotopic exchange data in soil chemistry. Aust. J. Soil. Res. 40, 1371–1381.

Heckrath, G., Brookes, P.C., Poulton, D.R., Goulding, K.W.T., 1995. Phosphorus leaching from soils containing different. J. Environ. Qual. 24, 904–910.

Hedley, M.J., Sharpley, A.N., 1998. Strategies for global nutrient cycling. In: L. Currie (Ed.), Long-term Nutrient needs for New Zealand's Primary Industries: Global Supply, Production

Requirements, and Environmental Constraints, Palmerston North, New Zealand: The Fertilizer and Lime Research Centre, Massey Univ., pp. 70–95.

Hedley, M.J., White, R.E., Nye, P.H., 1982. Plant-induced changes in the rhizosphere of rape (*Brassica napus* var. Emerald) seedlings. III. Changes in L. value, soil phosphate fractions and phosphatase. New Phytol. 215, 45–56.

Hedley, M.J., Kirk, G.J.D., Santos, M.B., 1994. Phosphorus efficiency and the forms of soil phosphorus utilised by upland rice cultivars. Plant Soil 158, 53–62.

Hedley, M.J., Mortvedt, J.J., Bolan, N.S., Syers, J.K., 1995. Phosphorus fertility management in agroecosystems. In: H. Tiessen (Ed.), Phosphorus in the Global environment. Wiley, New York, pp. 59–93. http://www.icsu-scope.org/downloadpubs/scope54/TOC.htm.

Hedley, M.J, Sharpley, A.N., 1998. Strategies for global nutrient recycling. In: L.D. Currie and Loganathan, P. (Eds.), The Long-Term Nutrient Needs for New Zealand's Primary Industries, Global Supply, Production Requirements and Environmental Constraints. Occasional Report No. 11. Fertiliser and Lime Research Centre, Massey University, Palmerston North, pp. 189–214.

Hesketh, N., Brookes, P.C., 2000. Development of an indicator for risk of phosphate leaching. J. Environ. Qual. 29, 105–110.

Higgs, B., Johnston, A.E., Salter, J.L., Dawson, C.J., 2000. Some aspects of achieving sustainable phosphorus use in agriculture. J. Environ. Qual. 29, 80–87.

Hinsinger, P., 2001. Bioavailability of soil inorganic P in the rhizosphere as affected by root-induced chemical changes, a review. Plant Soil 237, 173–195.

Hinsinger, P., Elsass, F., Jaillard, B., Robert, M., 1993. Root-induced irreversible transformation of a trioctahedral mica in the rhizosphere of rape. J. Soil Sci. 44, 535–545.

Hinsinger, P., Jaillard, B., 1993. Root-induced release of interlayer potassium and vermiculitization of phlogopite as related to potassium depletion in the rhizosphere of ryegrass. J. Soil Sci. 44, 525–534.

Hood, R., 2001. Evaluation of a new approach to the nitrogen-15 isotope dilution technique, to estimate crop N uptake from organic residues in the field. Biol. Fertil. Soils 43, 156–161.

Hood, R., Merckx, R., Jensen, E.S., Powlson, D., Matijevic, M., Hardarson, G., 2000. Stimating crop N uptake from organic residues using a new approach to the 15N isotope dilution technique. Plant Soil 223, 33–44.

Huang, P.M., Germida, J.J., 2002. Chemical and biological processes in the rhizosphere, metal pollutants. In: P.M. Huang, J.-M. Bollag and N. Sensi (Eds.), Interactions between Soil Particles and Microorganisms. Wiley, Chichester, UK, pp. 382–425.

Hylander, L.D., Ae, N., Hatta, T., Sugiyama, M., 1999. Exploitation of K near roots of cotton, maize, upland rice, and soybean grown in an Ultisol. Plant Soil 208, 33–41.

IAEA., 1976. Tracer manual on crops and soils. Technical report Series No. 171, International Atomic Energy Agency, Vienna, Austria.

James, S.W., Wells, K.L., 1990. Soil sample collection and handling, technique based on source and degree of field variability. In: R. L. Westerman (Ed.), Soil Testing and Plant Analysis, 3rd Ed. Book series No. 3. Soil Science Society of America Inc., Madison, WI, pp. 25–44.

Jeffery, J.J., Uren, N.C., 1983. Copper and zinc species in the soil solution and effects of soil pH. Aust. J. Soil. Res. 21, 479–488.

Johnston, A.E., Poulton, P.R., 1993. The role of phosphorus in crop production and soil fertility: 150 years of field experiments at Rothamsted, UK. In: J.J. Schultz (Ed.), Phosphate Fertilisers and the Environment. International Fertilizer Development Center, Muscle Shoals, AL, pp. 45–63.

Jones, J.W., Tsuji, G.Y., Hoogenboom, G., Hunt, L.A., Thornton, P.K., Wilkens, P.W., Imamura, D.T., Bowen, W.T., Singh, U., 1998. Decision support system for agrotechnology transfer. In: G.Y. Tsuji, G. Hoogenboom and P.K. Thornton (Eds.), Understanding Options for Agricultural Production. Kluwer Academic Publishers, Dordrecht, The Netherlands, pp. 157–177.

Kandeler, E., Marschner, P., Tscherko, D., Gahoonia, T.S., Nielsen, N.E., 2002. Microbial community composition and functional diversity in the rhizosphere of maize. Plant Soil 238, 301–312.

Karpinets, T.V., Greenwood, D.J., Ammons, J.T., 2004. Predictive mechanistic model of soil phosphorus dynamics with readily available inputs. Soil Sci. Soc. Am. J. 68, 644–653.

Keeney, D.R., 1982. Nitrogen-availability indices. In: A.L. Page, R.H. Miller and D.R. Keeney (Eds.), Methods of Soil Analysis, Part 2 Agronomy 9 2nd Ed., ASA, SSSA, Madison, WI, pp. 711–733.

Kirk, G.J.D., 1999. A model of phosphate solubilisation by organic anion excretion from plant roots. Eur. J. Soil Sci. 50, 369–378.

Kirk, G.J.D., Bajita, J.B., 1995. Root-induced iron oxidation, pH changes and zinc solubilization in the rhizosphere of lowland rice. New Phytol. 131, 129–137.

Kuchenbuch, R., Jungk, A., 1982. A method for determining concentration profiles at the soil–root interface by thin slicing rhizospheric soil. Plant Soil 68, 391–394.

Larsen, S., 1952. The use of ^{32}P in studies on the uptake of phosphorus by plants. Plant Soil 4, 1–10.

Ledgard, S.F, Williams, P.H., Broom, F.D., Thorrold, B.S., Wheeler, D.M., Willis, V.J., 1999. OVERSEER^{TM} – a nutrient budgeting model for pastoral farming, wheat, potatoes, apples and kiwifruit. In: L.D. Currie, M.J. Hedley, D.J. Horne and P. Loganathan (Eds.), Best Soil Management Practices for Production. Occasional Report No. 12. FLRC, Massey University, Palmerston North, pp. 143–152.

Li, J., Rate, A.W., Gilkes, R.J., 2003. Fractionation of trace elements in some non-agricultural Australian soils. Aust. J. Soil. Res. 41, 1389–1402.

Liao, M.T., Hedley, M.J., Woolley, D.J., Brooks, R.R., Nichols, M.A., 2000a. Copper uptake and translocation in chicory (Cichorium intybus L. cv Grasslands Puna) and tomato (Lycopersicon esculentum Mill. cv Rondy) plants grown in NFT system. II. The role of nicotianamine and histidine in xylem sap copper transport. Plant Soil 223, 243–252.

Liao, M.T., Hedley, M.J., Woolley, D.J., Loganathan, P., Brooks, R.R., Nichols, M.A., 2000b. Knowledge to explain the "unexplainable" Cu uptake by plants from copper-fertilised soils. In: L.D. Currie and P. Loganathan (Eds.), Proceedings of the Workshop, Soil Research, A Knowledge Industry for Land-based Exporters, Occasional Report No. 13, February 9–10, Fertilizer and Lime Research Centre, Massey University, Palmerston North, pp. 279–289.

Lu, S., Miller, M.H., 1994. Prediction of phosphorus uptake by field-grown maize with the Barber–Cushman model. Soil Sci. Soc. Am. J. 58, 852–857.

Maertens, E., Thijs, E., Smolders, E., Degryse, F., Cong, P.T., Merckx, R., 2004. An anion resin membrane technique to overcome detection limits of isotopically exchanged P in P-sorbing soils. Eur. J. Soil Sci. 55, 63–69.

Marschner, H., 1998. Soil–root interface, biological and biochemical processes. In: P.M. Huang, D.C. Adriano, T.J. Logan and R.T. Chekai (Eds.), Soil Chemistry and Ecosystem Health, SSA Special Publication 52, Soil Science Society of America, Madison, WI, pp. 191–233.

Marschner, P., Fu, Q.-L., Rengel, Z., 2003. Manganese availability and microbial populations in the rhizosphere of wheat genotypes differing in tolerance to Mn deficiency. J. Plant Nutr. Soil Sci. 166, 712–718.

Marschner, H., Romheld, V., 1994. Strategies of plants for acquisition of iron. Plant Soil 165, 261–274.

Maser, P., Gierth, M., Schroeder, J.I., 2002. Molecular mechanisms of potassium and sodium uptake in plants. Plant Soil 247, 43–54.

Matthew, C., Tillman, R.W., Hedley, M.J., Thompson, M., 1988. Observations on the relationship between soil fertility, pasture botanical composition and pasture growth rate; for a North Island lowland pasture. Proceedings of the New Zealand Grasslands Association 49, 141–145.

May, P.M., Linder, P.W., Williams, D.R., 1977. Computer simulation of metal-ion equilibria in biofluids, models for the low-molecular-weight complex distribution of calcium (II), magnesium (II), manganese (II), iron (III), copper (II), Zinc (II) and lead (II) ions in human blood plasma. J. Chem. Soc., 588–595.

McAuliffe, C.D., Hall, N.S., Dean, L.A., Hendricks, S.B., 1948. Exchange reactions between phosphates and soil, hydroxylic surfaces of soil minerals. Soil Sci. Soc. Am. Proc. 12, 119–123.

McDowell, R.W., Sharpley, A.N., 2001. Approximating phosphorus release from soils to surface run-off and subsurface drainage. J. Environ. Qual. 30, 508–520.

McLaughlin, M.J., Lancaster, P.A., Sale, P.W.G., Uren, N.C., Peverill, K.I., 1993. Use of cation/anion exchange membrane for multi-element testing of acidic soils. Plant Soil 155/156, 223–226.

McLaughlin, M.J., Reuter, D.J., Rayment, G.E., 1999. Soil testing – principles and concepts. In: K.I. Peverill, L.A. Sparrow and D.J. Reuter (Eds.), A soil analysis interpretation manual. Australian Soil and Plant Analysis Council, CSIRO Publishing, Collingwood, Australia, pp. 1–21.

Mehlich, A., 1953. Determination of P, Ca, Mg, K and NH$_4$. Division Mimeo, North Carolina Soil Test, Raleigh, NC.

Mehlich, A., 1984. Mehlich 3 soil test extractant, a modification of Mehlich 2 extractant. Commun. Soil Sci. Plant Anal. 15, 1409–1416.

Mench, M., Morel, J.L., Guckert, A., 1987. Metal binding properties of high molecular weight soluble exudates from maize (Zea mays L.) roots. Biol. Fert. Soils 3, 165–169.

Mench, M., Morel, J.L., Guckert, A., Guillet, B., 1988. Metal binding with root exudates of low molecular weight. J. Soil Sci. 39, 51–527.

Menon, R.G., Chien, S.H., Chardon, W.J., 1997. Iron oxide-impregnated filter paper (Pi-test). II. A review of its application. Nutr. Cycl. Agroecosys. 47, 7–18.

Moir, J.L., Scotter, D.R., Hedley, M.J., Mackay, A.D., 2000. A climate driven soil fertility dependent, pasture production model. N.Z. J. Agric. Res. 43, 491–500.

Montenegro, A., Zapata, F., 2002. Rape genotypic differences in P uptake and utilization from phosphate rocks in an Andisol of Chile. Nutr. Cycl. Agroecosys. 63, 27–33.

Moody, P.W., Bolland, M.D.A., 1999. Phosphorus. In: K.I. Peverill, L.A. Sparrow and D.J. Reuter (Eds.), A Soil Analysis Interpretation Manual. Australian Soil and Plant Analysis Council, CSIRO Publishing, Collingwood, Australia, pp. 187–220.

Morel, J.L., Mench, M., Guckert, A., 1986. Measurement of Pb^{2+}, Cu^{2+} and Cd^{2+} binding with mucilage exudates from maize (Zea mays L.) roots. Biol. Fertil. Soils 2, 29–34.

Moritsuka, N., Yanai, J., Kosaki, T., 2002. Depletion of nonexchangeable potassium in the maize rhizosphere and its possible releasing processes. Paper no. 2233, 17th World Congress of Soil Science, Bangkok.

Olsen, S.R., Cole, C.V., Watanabe, F.S., Dean, L.A., 1954. Estimation of available phosphorus in soils by extraction with sodium bicarbonate. US Department of Agriculture Cicular 939. US Government Printing Office, Washington DC, USA, p. 19.

Peverill, K.I., Sparrow, L.A., Reuter, D.J., 1999. A soil analysis interpretation manual. Australian Soil and Plant Analysis Council, CSIRO Publishing, Collingwood, Australia, 369pp.

Pote, D.H., Daniel, T.C., Sharpley, A.M., Moore, P.A., Edwards, D.R., Nichols, D.J., 1996. Relating extractable soil phosphorus to phosphorus losses in run-off. Soil Sci. Soc. Am. J. 60, 855–859.

Qian, P., Schoenau, J.J., 1995. Assessing nitrogen mineralization from soil organic matter using anion exchange membranes. Fert. Res. 40, 143–148.

Qian, P., Schoenau, J.J., 2002. Practical applications of ion exchange resins in agricultural and environmental soil research. Can. J. Soil Sci. 82, 9–21.

Qian, P., Schoenau, J.J., Huang, W.Z., 1992. Use of ion exchange membranes in routine soil testing. Commun. Soil Sci. Plant Anal. 23, 1791–1804.

Quantin, C., Becquer, T., Rouiller, J.H., Berthelin, J., 2002. Redistribution of metals in a New Caledonia Ferralsol after microbial weathering. Soil Sci. Soc. Am. J. 66, 1797–1804.

Rayment, G.E., Higginson, F.R., 1992. Australian laboratory handbook of soil and water chemical methods. Inkata Press, Melbourne.

Robert, P.C., 2002. Precision agriculture: a challenge for crop nutrition management. Plant Soil 247, 143–149.

Römkens, P.F.A.M., Dolfing, J., 1998. Effect of Ca on the solubility and molecular size distribution of DOC and Cu binding in soil solution samples. Environ. Sci. Technol. 32, 363–369.

Russell, E.W., 1973. Soil Conditions and Plant Growth, 10th Ed., Longman, London, UK.

Russell, R.S., Rickson, J.B., Adams, S.N., 1954. Isotope equilibria between phosphates in soil and their significance in the assessment of fertility by tracer methods. J. Soil Sci. 5, 85–105.

Saggar, S., Hedley, M.J., White, R.E., Perrott, K.W., Gregg, P.E.H., Cornforth, I.S., Sinclair, A.G., 1999. Development and evaluation of an improved soil test for phosphorus, 3, field comparison of Olsen, Colwell and Resin soil P tests for New Zealand pasture soils. Nutr. Cycl. Agroecosys. 55, 35–50.

Salcedo, I.H., Bertino, F., Sampaio, E.V.S.B., 1991. Reactivity of phosphorus in northeastern Brazilian soils assess by isotopic dilution. Soil Sci. Soc. Am. J. 55, 140–145.

Schulten, H.R., Leiweber, P., Schnitzer, M., 1998. Analytical pyrolysis and computer modelling of humic and soil particles. In: P.M. Huang, N. Senesi and J. Buffle (Eds.), Structure and Surface Reactions of Soil Particles. Wiley, New York, pp. 282–320.

Searle, P.L., 1992. The extraction of sulphate and mineralisable sulphur from soil with an anion exchange membrane. Commun. Soil Sci. Plant Anal. 23, 2087–2095.

Sharpley, A.N., 1996. Availability of phosphorus in manured soils. Soil Sci. Soc. Am. J. 60, 1459–1466.

Sharpley, A.N., Hedley, M.J., Sibbesen, E., Hillbricht-Ilkowska, A., House, W.A., Ryszkowski, L., 1995. Phosphorus transfers fro terrestrial to aquatic ecosystems. In: H. Tiessen (Ed.), Phosphorus in the Global Environment. Wiley, New York, pp. 171–201.

Sharpley, A.N., Rekolainen, S., 1997. Phosphorus in agriculture and its environmental. In: H. Tunney, O.T. Carton, P.C. Brookes and A.E. Johnston (Eds.), Phosphorus loss from soil to water. CAB International, Wallinford, NY, pp. 1–53.

Sharpley, A.N., Sims, J.T., Pierzynski, G.M., 1994. Innovative soil phosphorus availability indices, assessing inorganic phosphorus. Prospects for improving nutrient recommendations, SSSA Special Publication 40, pp. 115–141.

Sharpley, A.N., Tillman, R.W., Syers, J.K., 1977. Use of laboratory extraction data to predict losses of dissolved inorganic phosphate in surface runoff and tile drainage. J. Environ. Qual. 6, 33–36.

Sharpley, A.N., Tunney, H., 2000. Phosphorus research strategies to meet agricultural and environmental challenges for the 21st Century. J. Environ. Qual. 29, 176–181.

Sheppard, S.C., Bates, T.E., 1982. Selection of a soil extraction and multiple regession model to predict plant available manganese. Commun. Soil Plant Anal. 13, 1095–1113.

Sibbesen, E., Runge-Metzger, A., 1995. Phosphorus balance in European agriculture: status and policy options. In: H. Tiessen (Ed.), Phosphorus in the Global Environment: Transfers, Cycles, and Management. Wiley, New York, pp. 43–57.

Sims, J.T., 1998. Phosphorus soil testing, innovations for water quality protection. Commun. Soil Sci. Plant Anal. 29, 1471–1489.

Sinclair, A.G., Johnstone, P.D., Smith, L.C., Roberts, A.H.C., O'Connor, M.B., Morton, J.D., 1997. Relationship between pasture dry matter yield and soil Olsen P from a series of long-term field trials. N.Z. J. Agric. Res. 40, 559–567.

Soltanpour, P.N., Schwab, A.P., 1977. A new soil test for simultaneous extraction of macro- and micro-nutrients in alkaline soils. Commun. Soil Sci. Plant Anal. 8, 195–207.

Soon, Y.K., 1985. Soil Nutrient Availability, Chemistry and Concepts. Van Nostrand Reinhold Soil Science Series. Van Nostrand Reinhold, New York, 353pp.

Sparks, D.L., 2003. Environmental Soil Chemistry, 2nd Ed., Academic Press, New York, 352pp.

Sparling, G.P., Whale, K.N., Ramsay, A.J., 1985. Quantifying the contribution from the soil microbial biomass to the extractable P-levels of fresh and air-dried soils. Aust. J. Soil Res. 23, 613–621.

Sparrow, L.A., Uren, N.C., 1987. The effect of temperature and soil pH on the oxidation and reduction of manganese in acidic soils. Soil Biol. Biochem. 19, 143–148.

Sposito, G., 1983. The chemical forms of trace metals in soils. In: I. Thornton (Ed.), Applied Environmental Geochemistry, Academic Press, London, pp. 123–170.

Stephan, U.W., Schmidke, I., Stephan, V.W., Scholz, G., 1996. The nicotianamine molecule is made-to-measure for complexation of metal micronutrients in plants. Biometals 9, 84–90.

Stevenson, F.J., 1994. Humus Chemistry: Genesis, Composition, Reactions. 2nd Ed., John Wiley and Sons, New York, 433pp.

Strobel, B.W., 2001. Influence of vegetation on low-molecular-weight carboxylic acids in soil solution – a review. Geoderma 99, 169–198.

Tabachow, R.M., Peirce, J.J., Richter, D.D., 2001. Biogeochemical models relating soil nitrogen losses to plant-available N. Environ. Eng. Sci. 18, 81–89.

Tao, S., Chen, Y.J., Xu, F.L., Cao, B.G.Li., 2003. Changes of copper speciation in maize rhizosphere soil. Environ. Pollut. 122, 447–454.

Tarafdar, J.C., Jungk, A., 1987. Phosphatase activity in the rhizosphere and its relation to the depletion of soil organic phosphorus. Biol. Fertil. Soil 3, 199–204.

Tinker, P.B., Nye, P.H., 2000. Solute Movement in the Rhizosphere. Topics in Sustainable Agronomy. Oxford University Press, Oxford, UK, 444pp.

Trolove, S.N., Hedley, M.J., Caradus, J.R., Mackay, A.D., 1996. Uptake of phosphorus from different sources by Lotus pendunculatus and three genotypes of Trifolium repens. 2. Forms of phosphate utilised and acidification of the rhizosphere. Aust. J. Soil Res. 34, 1027–1040.

Trolove, S.N., Hedley, M.J., Kirk, G.J.D., Bolan, N.S., Loganathan, P., 2003. Progress in selected areas of rhizosphere research on P acquisition. Aust. J. Soil Res. 41, 471–499.

Uren, N.C., 1999. Manganese. In: K.I. Peverill, L.A. Sparrow and D.J. Reuter (Eds.), A Soil Analysis Interpretation Manual. Australian Soil and Plant Analysis Council, CSIRO Publishing, Collingwood, Australia, pp. 287–294.

van Raij, B., 1998. Bioavailable tests, alternatives to standard soil extractions. Commun. Soil Sci. Plant Anal. 29, 1553–1570.

van Raij, B., Quaggio, J.A., 1990. Extractable phosphorus availability indexes as affected by liming. Comm. Soil Sci. Plant Anal. 21, 1267–1276.

VonWiren, N., Marschner, H., Romheld, V., 1996. Roots of iron-efficient maize also absorb phytosiderophore-chelated zinc. Plant Physiol. 111, 1119–1125.

Walley, F., Yates, T., van Groenigen, J.-W., van Kessel, C., 2002. Relationships between soil nitrogen availability indices, yield, and nitrogen accumulation of wheat. Soil Sci. Soc. Am. J. 66, 1549–1561.

Wang, S.J., Fox, D.G., Cherney, D.J.R., Klausner, S.D., Bourquin, L.D., 1999. Impact of dairy farming on well water nitrate level and soil concentrations of phosphorus and potassium. J. Dairy Sci. 82, 2164–2169.

Wang, W., Smith, C.J., Chalk, P.M., Chen, D., 2001. Evaluating chemical and physical indices of nitrogen mineralization capacity with an unequivocal reference. Soil Sci. Soc. Am. J. 65, 368–376.

Wang, W.J., Li, R.G., 1991. Evaluation of the methods for determining available nitrogen in soil on Hebei plain. Chinese J. Soil Sci. 22, 263–266.

Watkinson, J.H., Kear, M.J., 1996. Sulfate and mineralisable organic sulfur in pastoral soils of New Zealand. 2. A soil test for mineralisable organic sulfur. Aust. J. Soil Res. 34, 405–412.

Westerman, R.L., 1990. Soil Testing and Plant Analysis, 3rd Ed. Soil Science Society of America, Book Series No. 3, SSSA, Madison, WI.

Wild, A. (Ed.) 1988. Russells Soil Conditions and Plant Growth. 11 Ed.. Longman Scientific and Technical, London, U.K, p. 991.

Zoysa, A.K.N., Loganathan, P., Hedley, M.J., 1998. Phosphate rock dissolution and transformation in the rhizosphere of tea (Camellia sinensis L.) compared with other plant species. Eur. J. Soil Sci. 48, 477–486.

Developments in Soil Science, volume 32
Ravendra Naidu (Editor)

Chapter 15

THE ROLE OF INHIBITORS IN THE BIOAVAILABILITY AND MITIGATION OF NITROGEN LOSSES IN GRASSLAND ECOSYSTEMS

J. Singh, S. Saggar, N.S. Bolan and M. Zaman

15.1 Introduction

Managed grasslands have a high demand for nitrogen (N) for plant growth. These grassland soils need a continuous supply of N inputs from various sources to meet animal feed demand and sustain productivity. Addition of N to soils not only increases plant productivity but also results in increased nitrate (NO_3^-) leaching and release of gaseous N such as ammonia (NH_3) and nitrous oxide (N_2O). Recent sharp increase in fertiliser N inputs to intensively managed and grazed grasslands has rekindled the debate on its impact on atmospheric, terrestrial and aquatic environments. There has been increasing interest in the use of nitrogen inhibitors in mitigating environmental impacts of N losses through leaching and gaseous emissions from excretal, fertiliser and effluent N inputs. This chapter gives an overview of the sources of N input to grazed grasslands, the dynamics of N in grassland soils, environmental impacts of N losses; discusses the role of inhibitors in improving N bioavailability and mitigating N losses; identifies the gaps and the limitations from the existing information and concludes by presenting the main research needs to devise mitigation strategies with inhibitors.

15.2 Issues

Grasslands worldwide occupy 117 million km^2 of vegetated lands and provide forage for over 1800 million livestock units and wildlife (World Resources Institute, 2000). These managed grasslands are highly productive, with increased pasture production being the major goal for the pastoral farmers for higher per hectare animal productivity. The fertility of grassland (pasture) soils can be substantially altered by grazing animals mainly through the deposition of dung and urine and their subsequent transformation and transport in soils (Haynes and Williams, 1993; Saggar et al., 1990a, 1990b). In legume-based pastures, N is derived from biological fixation of atmospheric N (BNF), through the addition of manures and fertilisers, and the uneven deposition of animal excreta. In non-legume-based pastures, such as grass pastures in Europe, most N is derived from fertiliser and manure application. Although in legume-based pastures, most of the N is derived from BNF, a small amount of fertiliser N is traditionally added during the early spring season, mainly to overcome the deficiency caused by the slow rates of BNF and mineralisation of soil organic matter. Nitrogen is extremely dynamic in grazed pastoral soils, always changing or moving. Nitrogen is the major nutrient element that most strongly regulates pasture production but N is also a major contributor to environmental degradation.

The global N fertiliser demand is expected to grow at an average annual rate of 1.7% per annum, reaching 94.6 million tonnes N in 2008. It is estimated that New Zealand agricultural systems receive an annual input of about 3 million tonnes N, with 1.58 million tonnes from animal excreta, 0.9–1.1 million tonnes from BNF, 0.33 million tonnes from fertilisers, and about 0.01–0.015 million tonnes from atmospheric deposition (Saggar, 2004). Recently fertiliser N inputs to New Zealand grazed pastures have increased sharply, and this increase is expected to continue in the foreseeable future. N fertiliser use in 1990 was 59,265 tonnes; the estimates for 2003–04 were 348,000 tonnes (Ministry for the Environment, 2005). This increased use of reactive-N benefits society, but it also represents a significant cost to society through increased nitrate (NO_3^-) leaching and enhanced ammonia (NH_3), nitrous oxide (N_2O) emissions. The increasing fertiliser N input to grazed pastures has rekindled the debate on its impact on atmospheric, terrestrial and aquatic environments (Parliamentary Commissioner for the Environment Report, 2004).

Nitrogen is an important plant nutrient and its loss affects both the quality and quantity of feed and animal production. An increase in NO_3^- concentration in groundwater resulting from leaching has been linked to increasing incidences of NO_3^- toxicity in human and livestock (i.e., methaemoglobinaemia) (Gupta et al., 1999; Bolan et al., 2004a). Nitrate leaching is one of the biggest issues facing the New Zealand agriculture sector at present. Similarly, grazed pastures are identified as an important source of NH_3 and nitrous oxide (N_2O). While NH_3 is implicated in acid rain, N_2O is involved in ozone depletion and global warming (i.e., greenhouse gas).

Both NO_3^- leaching and gaseous emissions are considered a greater issue for dairying compared to sheep and beef farms. Results of studies conducted by Saggar and associates (Saggar, 2004; Saggar et al., 2004a, 2005a, 2005b) show a 5- to 10-fold increase in nitrous oxide (N_2O) emissions in grazed pasture compared with ungrazed pasture, and also a much higher N_2O emission factor for dairy-grazed than sheep-grazed soils. Therefore, the increasing amounts of N going on to hill country could increase both leaching and gaseous emissions of N and create further problems in the future.

Various approaches have been used to improve the bioavailability of N, and mitigate the economic and environmental impacts of N losses. One such approach is the use of nitrification and urease inhibitors (NIs and UIs). Recently in New Zealand there has been increasing interest in the use of commercially formulated NIs and UIs (e.g., EcoN, N-Care and SustaiN[1]) to reduce the loss of N through leaching and gaseous emissions, and enhance plant productivity.

The aim of this chapter is to discuss the role of inhibitors in the bioavailability of N. As N exists in many different inorganic and organic forms in soils and these N forms undergo several transformations, an understanding of N dynamics can help to illustrate the importance of N bioavailability and its fate in the environment. This chapter therefore summarises the various areas of this topic including: (i) a brief summary of N inputs and dynamics in grazed pastures; (ii) an outline of the environmental impacts of N losses; (iii) an illustration of the role of inhibitors in improving the N bioavailability and mitigating N losses; (iv) a brief description of research on the use of inhibitors in New Zealand and (v) conclusions presenting the main research needs.

15.3 Sources of nitrogen input in grazed pastures

In grazed pastures, N is derived from BNF, through the addition of manures and fertilisers, and the deposition of animal excreta. In many countries, including Australia, New Zealand and parts of North America and Europe, the use of legume-based pasture is the most common grazing management practice. In BNF, the N_2 atom is biochemically reduced from its most oxidised state (N_2) to its most reduced form (NH_4^+):

$$N_2 + 8H + 8e^- + 16ATP \text{ (energy)} \xrightarrow{\text{Nitrogenase}} 2NH_3 + H_2 + 16ADP + 16P_i \quad (15.1)$$

This biochemical reaction is performed exclusively by prokaryotes (a large range of nitrogen-fixing bacteria such as *Rhizobium* and cyanobacteria), using an enzyme complex termed nitrogenase.

The amount of BNF in legume-based pastures depends on a number of factors including legume species, soil and climatic conditions, nutrient supply and grazing management. High levels of available phosphorus in soils are essential for maintaining both the presence and N_2-fixing activity of legumes in pastures and for supplying N to these pastures. Similarly, adequate levels of other nutrients, such as sulphur and molybdenum in particular, are required. For example, the largest annual estimates of $680 \, \text{kg N ha}^{-1}$ for white clover/ryegrass

---

[1]EcoN: Ravensdown Fertiliser Co-operative Ltd.; N-Care: Ballance AgriNutrients Ltd.; SustaiN: Summit-Quinphos Ltd.

pasture were obtained using soils of low N status and under optimum conditions for N fixation (Ledgard, 1995). Work by Sears and co-workers in the 1950s and 1960s suggested white clover-based pastures were capable of fixing $500-700 \, \text{kg N ha}^{-1} \text{y}^{-1}$ (Sears et al., 1965), whereas measurements from the 1970s onwards suggest annual BNF rates in the range $65-392 \, \text{kg N ha}^{-1}$ are more common for grass/clover pastures in New Zealand (Crush, 1987; Goh and Williams, 1999; Ledgard et al., 1990; Ledgard et al., 1996; Ledgard and Steele, 1992). These measurements also indicate annual BNF averaging from about $185 \, \text{kg N ha}^{-1}$ for sheep farms to $200-250 \, \text{kg N ha}^{-1}$ for more intensively managed dairy farms (White, 1989). BNF rates in the range $100-300 \, \text{kg N ha}^{-1} \text{y}^{-1}$ are therefore common for grass/clover pastures in New Zealand (Ledgard et al., 1990). It is estimated that New Zealand agricultural systems receive an annual N input of 0.9–1.1 million tonnes through BNF (Saggar, 2004).

Legumes tend to use soil N when the concentration of inorganic N is high, which results in less BNF. Consequently, application of fertiliser N to legume crops or pastures generally reduces the amount of BNF in soils (Table 15.1). Addition of fertiliser N causes an initial decrease in BNF, as the legume is suppressed and uptake of fertiliser N occurs. While reviewing the data on the effect of fertiliser N on BNF from various clover-based field experiments, Ledgard et al. (2001) reported an annual decrease of up to 75%. For each kg of fertiliser N applied, BNF decreased by between 0.3 and 0.7 kg N. The data in Table 15.1 suggest, with increasing N addition, BNF by clover continues to decrease, and the percent decrease varies from 20 to 75% depending on the time of application and grazing management. The impact of increasing fertiliser application on BNF is depicted in Figure 15.1. Furthermore, in intensively grazed systems, most of the ingested N (75–95%) is excreted, mostly in urine. This has a major direct effect on BNF by altering soil N status.

Nitrogen fertilisers are used widely in the grass-based intensive pasture production systems of Europe and North America. Pure grass pasture often responds linearly up to $200-400 \, \text{kg N ha}^{-1} \text{y}^{-1}$, and application rates in this range are common (Whitehead, 1995). Where pastures are cut for conservation,

Table 15.1. *Effect of N fertiliser in reducing biological N fixation (BNF) in New Zealand pastures.*

Fertiliser N $(\text{kg N ha}^{-1} \text{y}^{-1})$	Biological N fixation $(\text{kg N ha}^{-1} \text{y}^{-1})$	Decrease (%)	Reference
0, 390	111, 47	58	Ledgard et al. (1996)
0, 100	100, 70	30	Crush (1987)
0, 200, 400	210, 170, 70	19, 67	Ledgard (1995)
0, 200, 400	154, 99, 39	36, 75	Ledgard et al. (2001)

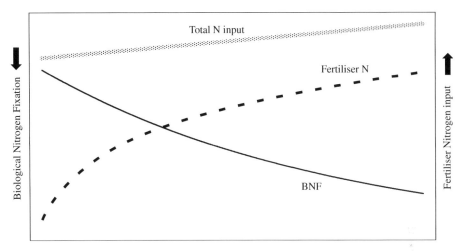

Figure 15.1. *The influence of increased nitrogen fertiliser application on biological N fixation (BNF) in legume-based pastures.*

large quantities of nutrients are removed and the optimum N rate can be greater than that under grazed swards, where N is returned to pasture in the form of animal excreta. In legume-based pastures, a small amount of N fertiliser has traditionally been added during the winter/spring period, mainly to offset the low level of biological N fixation during this period. However, there has recently been a sharp increase in the use of N fertilisers in grazed pastures. This is attributed to a number of reasons: (i) extra feed can be produced throughout the year to increase the stocking rate, achieve early calving, extend lactation later into autumn and make more high-quality silage to feed later in the lactation; (ii) feed obtained from N fertiliser application can be used to replace more expensive feed supplements and (iii) the productivity and the profitability of the farm can be increased by fertiliser N application. The recent arrival of the clover weevil in New Zealand will also induce more fertiliser N use unless effective control measures can be found. More recently, there has been increasing interest in the use of N fertilisers in the hill country.

In grazed pastures, a substantial amount of N is recycled through the direct deposition of animal excreta. Usually between 5 and 35% of the N in pasture protein is converted into animal protein (i.e., milk and meat), and the remaining N is excreted in dung and urine. The proportion of total N intake excreted and its partition between urine and faeces are dependent on the type of animal, the intake of dry matter and the N concentration of the diet. For sheep and cattle, faecal excretion of N is usually about $0.8\,g\ N\ 100\,g^{-1}$ of dry matter consumed, regardless of the N content of the feed (Whitehead, 1995). The majority of the

N is excreted in urine and the proportion of N in the urine increases with increasing N content of the diet.

The concentration of N in urine may vary from 1 to $20\,g\,N\,L^{-1}$ because of factors such as N content in the diet and the volume of water consumption, but it is normally in the range $8–15\,g\,N\,L^{-1}$. The proportion of urine N present as urea increases with an increase in N intake. Typically, over 70% of the N in urine is present as urea; the rest consists of amino acids and peptides (Haynes and Williams, 1993). The bulk of the N in faeces is in organic forms. About 20–25% of faecal N is water-soluble, 15–25% is undigested dietary N, and the remaining 50–65% is present in bacterial cells (Oenema et al., 1997). New Zealand's 5.32 million dairy cows and 4.5 million beef cattle daily excrete around $300,000\,m^3$ of dung and 180 million m^3 urine. It is estimated that annually in New Zealand about 70 million m^3 of effluent are being generated from dairy sheds, 4 million from piggery farms, and 50 million m^3 from meat-processing plants (Saggar et al., 2004b).

The estimated global amount of N voided by animals ranges between 80 and 130 million tonnes per year, and is as large as, or larger than, the global annual N fertiliser consumption of about 90 million tonnes. In New Zealand, however, the animals void almost 5 times more N (1.5 million tonnes of N) than the N fertiliser input (0.34 million tonnes).

15.4 Nitrogen dynamics in pasture soils

To understand the biochemical mechanisms involved in mitigating N losses through the use of inhibitors, it is important to understand the dynamics of N in soils. A detailed description of the biotic and abiotic N transformations is given in Bolan et al. (2004b). A simplified version of the transformation of N in a legume-based pasture is presented in Figure 15.2. The N transformations in soils include: mineralisation, immobilisation, nitrification, denitrification, NH_3 volatilisation, NH_4^+ fixation and NO_3^- leaching. While the first four reactions involve soil micro-organisms (biotic), the last three involve only chemical/physical processes (abiotic).

15.4.1 Mineralisation

The mineralisation process involves the conversion of plant-unavailable organic forms into plant-available inorganic forms by soil micro-organisms. The process includes aminization and ammonification reactions. Aminization is a microbial process in which heterotrophic micro-organisms first hydrolyse the macro-molecules of organic N compounds, (e.g., proteins, into simple N compounds, such as amines and amino acids). For example, when blood and bone fertiliser, which contains protein as the major N compound, is added to pasture soils,

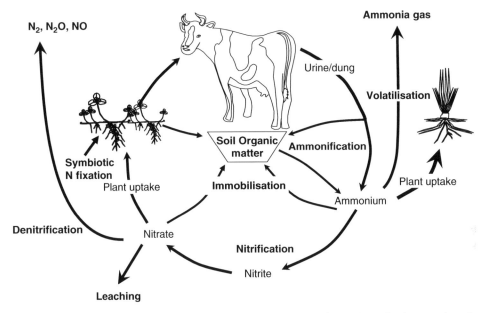

Figure 15.2. *Schematic representation of nitrogen transformations in legume-based pastures (Dr. M.J. Hedley, pers. comm.).*

it first undergoes aminization reactions (Eq. (15.2)).

$$\text{Proteins (in Blood and Bone)} \rightarrow \text{amines} (R - NH_2) + CO_2 \qquad (15.2)$$

Ammonification is a biological process in which a group of micro-organisms converts amines and amino acids into NH_4^+ ions. For example, urea $(CO(NH_2)_2)$ in animal urine and fertilisers undergoes the ammonification reaction releasing NH_4^+ ions (Eq. (15.3)). This process is also known as "urea hydrolysis" and is carried out in the presence of the urease enzyme in the soil. Urease is a powerful enzyme produced by practically all microbial and plant species. The ammonification process also releases hydroxyl (OH^-) ions and hence the pH around the urea granules or urine spots in soil increases resulting in alkaline conditions. The build up to high NH_4^+ ion concentration and the elevation of pH during the ammonification reaction provide ideal conditions for ammonia volatilisation to occur. Thus, the addition of urease inhibitors, which retard urea hydrolysis, is likely to reduce ammonia volatilisation (see below).

$$CO(NH_2)_2 \xrightarrow{\text{urease}} 2NH_4^+ + 2OH^- + CO_2 \qquad (15.3)$$

15.4.2 Nitrification

The biological conversion of NH_4^+ to NO_3^- is known as nitrification (Eq. (15.4)). Broadly, the nitrification reaction involves a two-step process in which the NH_4^+ ions are first converted (oxidised) into nitrite (NO_2^-) and then into NO_3^-. A sequence of reactions involved in the first step of ammonium oxidation to nitrite. Since the rate of conversion of NO_2^- to NO_2^- is faster than the conversion of NH_4^+ to NO_2^-, it is unlikely that NO_2^-, which is toxic to plants, accumulates under most soil and climatic conditions. The nitrification process produces H^+ ions, thereby decreasing the pH. The NO_3^- ions produced through the oxidation of NH_4^+ ions are subject to leaching and gaseous emission losses. Thus, the addition of NIs that retards the nitrate production is likely to reduce the loss of N through these two processes (see below).

$$2NH_4^+ + 4O_2 \rightarrow 2NO_3^- + 4H^+ + 2H_2O \qquad (15.4)$$

15.4.3 Immobilisation

Immobilisation is a microbial process in which the plant-available NH_4^+ and NO_3^- ions are converted to plant-unavailable organic N. For example, the addition of carbon (C)-rich substances such as maize stubble and cereal straw in arable soils promotes immobilisation and reduces N availability to plants. Of major concern from a practical point of view is the amount of C relative to N (i.e., C:N ratio) in the decomposing organic matter. Problems arise when the N content of the decomposing organic matter is small, because microbes may become deprived of N and compete with plants for the available N in soil. Thus, the addition of plant residues with a high C:N ratio induces immobilisation of soil N by micro-organisms, thereby decreasing the amount of plant-available soil N.

15.4.4 Denitrification

In waterlogged soils, some micro-organisms obtain their oxygen from NO_3^-, resulting in the reduction of NO_3^-. The reduction of NO_3^- proceeds in a series of steps, producing NO_2^-, nitric oxide (NO), N_2O and N_2 gas (Eq. (15.5)). Denitrification results not only in the loss of a valuable plant nutrient but also in the release of N_2O (a potent greenhouse gas), which is also implicated in the destruction of atmospheric ozone.

$$NO_3^- \rightarrow NO_2^- \rightarrow NO \rightarrow N_2O \rightarrow N_2 \qquad (15.5)$$

15.4.5 Ammonium fixation

Ammonium, being a cation, is strongly retained on cation exchange sites, whereas NO_3^-, being an anion, is very weakly adsorbed onto the soil particles.

Nitrate moves with water, and subsequent NO_3^- leaching not only results in the loss of a valuable nutrient but also causes ground water pollution. A high NO_3^- concentration in drinking water is toxic, especially to infants, and has been linked with "blue baby syndrome" (methaemoglobinaemea). The World Heath Organisation has therefore stipulated a safe upper limit (11.3 mg NO_3–N L^{-1} or 50 mg NO_3 L^{-1}) in drinking water.

Although leaching losses occur from both fertiliser N and urine N, a number of studies have shown that in grazed pastures, the latter provides the major pathways for NO_3^- leaching (Di and Cameron, 2002a, 2002b). The release of mineral N from faeces results in elevated concentrations of mineral N in the soil below the dung patch. The high concentrations of NO_3^- in dung patches (e.g., 90–130 mg N kg^{-1}) (Ryden, 1986) can also be a significant source of both NO_3^- leaching and gaseous losses of N_2O and N_2 from grazed pastures.

15.4.6 Ammonia volatilisation

Ammonium ions in an alkaline medium dissociate into gaseous NH_3, which is subjected to volatilisation losses (Eq. (15.6)). Ammonia volatilisation occurs when the soil pH is high (>7.5). In the case of urea application and urine deposition, the initial increase in soil pH through the ammonification process (Eq. (15.2)) is likely to result in NH_3 volatilisation.

$$NH_4^+ + OH^- \rightarrow NH_3 \uparrow H_2O(pK_a 7.6) \qquad (15.6)$$

15.5 Environmental impact of N losses

Loss of N, occurring mainly through NH_3 volatilisation, biological denitrification and NO_3^- leaching, has both economic and environmental implications (Bolan et al., 2004b). In the context of environmental pollution and global climate change, while NO_3^- leaching attracts attention because of its potential human and animal health effects, and water pollution, gaseous N, such as NH_3, N_2O and NO, cause concern because of their radiative or chemical effects on the atmosphere. Since 1900, the global anthropogenic use of reactive forms of N has increased from less than 5 to approximately 20 million tonnes N in 1950 to almost 150 million tonnes N in 1996, and is expected to approach 190 million tonnes N by 2020. This newly reactive N is derived from production of synthetic fertilisers, from the increased production of crops that fix N biologically, and from fossil fuel consumption (Mosier and Kroeze, 2000). This increased use of reactive-N benefits the society, but it also represents a significant cost to society through increased NO_3^- leaching and enhanced emissions of NO_x (pronounced "knox", sum of NO and NO_2), NH_3, N_2O and deposition of NO_y (sum of knox plus all other oxidised forms of N such as HNO_3 and peroxyacetyl nitrate in the atmosphere) and NH_x (Mosier et al., 2001).

In New Zealand, both NO_3^- leaching and gaseous emissions are considered more important issues for dairying than for sheep and beef farms. The environmental effects of NO_3^- leached to groundwater and other waterways and the potential damage to soils are a major concern to the farming industry, the scientific community, and the society. The accumulation of NO_3^- in the environment results mainly from non-point source leaching and runoff from the over-application of N fertilisers, voided urine and dung, and from poorly or untreated effluents and sewage. In addition, NO_3^--containing wastes are produced by many industrial processes including paper and munitions manufacturing. As agriculture is implicated in the NO_3^- pollution problem, farmers and rural communities are the most affected. Environment Waikato data suggest the quality of about 10% of the groundwater in the livestock farming area of the region is below World Health Organisation drinking water standards (Anon, 2005). The declining water quality of Lake Taupo and the Rotorua Lakes in New Zealand has been linked to the land use within the catchment and the export of N from farming and other sources. Farming has been shown to be a major contributor to the algal blooms occurring in Lake Rotoiti in New Zealand (Ministry for the Environment, 2003). High concentrations of NO_3^- in lakes, rivers and estuaries can result in eutrophication and algal blooms, and links have also been made between high NO_3^- and toxicity in fish eggs, amphibian eggs and tadpoles (Agriculture and Agri-Food Canada, 2003). Health impacts on humans and animal life occur from drinking water, and/or eating foods high in NO_3^- such as vegetables. Nitrate consumption at high concentrations has been linked to adult ailments such as gastric cancer, ulceration of the mouth and/or stomach lining; it is especially linked to infants, causing the condition known as methaemoglobinaemia, also called "blue baby syndrome". Nitrates have a direct, caustic effect on the lining of the gut if consumed in large quantities. Signs of poisoning include diarrhoea, salivation and abdominal pain. Nitrate is converted in the gut to NO_2^-, which then combines with haemoglobin to form methemoglobin, thus decreasing the ability of the blood to carry oxygen. Infants are more susceptible to NO_3^- toxicity than older children or adults. Although NO_3^- is not always toxic to animals, NO_3^- toxicity in grazing animals is likely to occur when they ingest water and forage that are high in NO_3^- concentration (Bolan et al., 2004a). Ruminants are more susceptible to NO_3^- toxicity than simple-stomached animals because rumen microbes enhance the reduction of NO_3^- to NO_2^- in the digestive tract. The toxicity symptoms include trembling, staggering gait, rapid respiration and prostration. Affected animals cease to eat and soon collapse and die without convulsions. Losses of weight and milk production and non-infectious abortion have been noted as sublethal effects in dairy cattle. Uncertainty exists about the level of NO_3^- ingestion that is considered the minimal lethal dose. Studies have indicated that 7.6–9.0 g NO_3^-–N 100 kg^{-1} body weight is lethal to animals. Assuming that the average daily pasture intake by dairy cattle is 4% of the body weight, pasture with a NO_3^-–N content of more than 0.21% is likely to be toxic to animals with a live body weight of 300 kg (Table 15.2) (Bolan et al., 2004a). It is important to remember that the pasture NO_3^- levels above which toxicity occurs depend on the rate of pasture ingestion by the animal. Careful use of N fertiliser, an awareness of plant factors, the effect of weather and cattle management can help reduce losses from NO_3^- poisoning.

Agriculture is also one of the major sources of gaseous N emissions that result from increased N fertiliser use, animal excreta and organic manures, or N fixed by legumes, thereby polluting the environment. For example, NH_3 affects visibility, aerosol chemistry, health and climate, as it causes acidification and eutrophication when deposited in soil and water. It also acts as a neutralising agent for acidic aerosols, besides affecting vegetation and forming NO_3^-. Ammonia has a short lifetime in the atmosphere but it can act as a secondary source of NO and N_2O, which are directly or indirectly involved in global warming. Nitrous oxide accounts for 2–4% of total Global Warming Potential (GWP) (Watson et al., 1992). In the last few decades, the concentration of N_2O in the atmosphere has progressively increased at an annual rate of 0.2–0.3% as a result of human activities (Prinn et al., 1990; Rasmussen and Khalil, 1986), and about 70% of the anthropogenic N_2O increase is attributed to agriculture (Watson et al., 1992).

It is generally recognised that the use of chemical N fertiliser is the most important contributor to N_2O emissions from agricultural soils worldwide. It is estimated that about 1.5 million tonnes of N is injected annually into the atmosphere as N_2O as a result of fertiliser application, which represents about

Table 15.2. *Level of nitrate in feed and animal response (from Bolan et al., 2004b).*

% NO_3–N	% $NO_3^†$	Comment or animal response
0.0–0.06	0.0–0.27	Safe to feed. Use caution with pregnant or young animals at the upper level
0.07–0.12	0.28–0.54	Generally safe when fed with balanced ration; for pregnant animals limit nitrate feed to 1/2 of daily dry matter intake
0.13–0.21	0.55–0.92	Limit to 1/4 of the total daily ration; ration should be well fortified with energy, minerals and vitamins
>0.21	>0.92	Toxic. Extreme caution should be used. A well-mixed feed or pelleting of the feed will reduce sorting by animals. Amount of dilution with other feeds depends on nitrate level

†% NO_3 = % NO_3–N × 4.43

44% of the anthropogenic input and about 13% of the total annual input of N_2O to the atmosphere (Watson et al., 1992). BNF and animal manures are the other major contributors to N_2O atmospheric input. BNF can be nitrified and denitrified in the same way as fertiliser N, thus resulting in N_2O emission. Legumes can thus increase N_2O emissions by a factor of 2 to 3 compared with non-legume pastures (Duxbury et al., 1982). Addition of animal waste/manure to soil supplies additional quantities of C and N, promotes microbial activity, and may release substantial amounts of N_2O (Beauchamp, 1997). A 5- to 10-fold increase in N_2O was observed in grazed pasture compared with ungrazed pasture (Saggar et al., 2004a), suggesting in grazed pastures it is animal excreta deposited in the form of dung and urine that provide high concentrations of available N and C, and are the principal source of N_2O production.

15.6 Inhibitors in nitrogen cycle

Nitrogen inhibitors are the compounds used in controlling N dynamics in soils to reduce N losses. These can be grouped into two categories: (i) urease inhibitors (UIs); and (ii) nitrification inhibitors (NIs). The general theory for using NIs and UIs is they will slow N turnover by slowing the oxidation of N to NO_3^-, causing N to stay in the more immobile form of NH_4^+. The UIs are used to control urea hydrolysis and the subsequent ammonification process through their effect on urease enzyme. The NIs are used to control the oxidation of ammonium (NH_4^+) ions to nitrite (NO_2^-) ions (Figure 15.3).

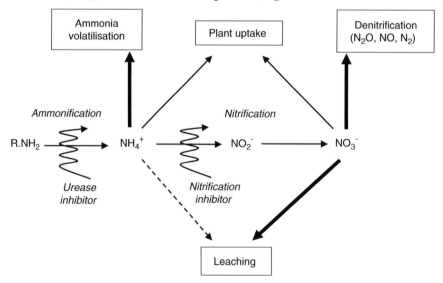

Figure 15.3. *Mechanisms of ammonification and nitrification, and the role of inhibitors in controlling nitrogen transformations and losses.*

Inhibitors do not inhibit nitrification indefinitely, but usually between 4 and 10 weeks depending upon soil temperature and pH. These include both specific and non-specific compounds. The specific inhibitors tend to control micro-organisms/enzymes involved in specific biochemical reactions, e.g., enzymes involved in the ammonification (UIs) and nitrification (NIs) processes, whereas the non-specific inhibitors tend to have a blanket effect on microbial community in soils. Non-specific inhibitors include many agricultural pesticides, such as herbicides (e.g., Monuran), fungicides (e.g., Terrazole), insecticides (e.g., BHC) and fumigants (e.g., Telone) that affect the activities of soil micro-organisms, including ammonification and nitrification processes. Martens and Bremner (1984, 1993) studied the effects of 46 herbicides and 15 insecticides applied at the rate of $5 \mu g\, g^{-1}$ soil and 17 fungicides applied at the rate of $1 \mu g\, g^{-1}$ soil, on urea hydrolysis and nitrification of urea N in two fine-textured and two coarse-textured soils. They found none of the herbicides and fungicides retarded urea hydrolysis in any of the four soils. However, ten herbicides (amitrole, 2,4-D amine, chlorpropham, dinoseb, propanil, propham, acifluorfen, diclopfop methyl, fenoxaprop ethyl and tridiphane) and five insecticides (carbaryl, lindane, trimethacarb, diazinon and fenitrothion) retarded nitrification of N in the two coarse-textured soils. Among the 17 fungicides studied, one (maneb) retarded urea hydrolysis in all four soils, and seven (anilazine, benomyl, chloranil, captan, maneb, mancozeb and thiram) in two coarse-textured soils. The fungicides, with the exception of benomyl, fenaminosulf, folpet, metalaxyl, metham-sodium, PCNB and teraazole, retarded nitrification in one of the coarse-textured soils.

Most of the chemical compounds used as inhibitors affect the growth and proliferation of micro-organisms, thereby inhibiting the nutrient cycling processes in soils. A major concern, therefore, is the environmental fate of such inhibitors when repeatedly applied to soil. Some inhibitors are inhibitory to plant growth at concentrations that effectively inhibit nitrification.

Although many specific inhibitors are found to specifically block a particular enzyme system in the N transformation reaction sequence, when used in high concentration these compounds may act as general biocides, indicating specificity is often linked to concentration. Furthermore, depending on the concentration, these chemicals can either kill micro-organisms (i.e., biocidal action) or temporarily inhibit microbial function (i.e., biostatic action). However, complete inhibition of any N transformation process is seldom achieved with the use of these chemicals.

15.6.1 Urease Inhibitors

UIs slow the conversion of urea to NH_4^+ by inhibiting the urease enzyme, which reduces NH_4^+ concentration in the soil solution and hence lowers the potential for NH$_3$ volatilisation and seedling damage. Slowing the hydrolysis of urea allows more time for it to diffuse away from the application site or for rain or irrigation to dilute urea and NH$_4^+$ concentration at the soil surface and increase its dispersion in the soil subsequently retaining NH$_3$ in the soil.

Urease activity

The urea in cattle urine and in fertiliser is usually hydrolysed within a few days by an enzyme termed "urease", which is present in many plants and plant litter (Freney and Black, 1988) and in most species of bacteria, yeast and fungi. The enzyme catalyses the hydrolysis of urea to NH$_4^+$ (Eq. (15.3)) and carbamate ions which decompose to CO$_2$ and NH$_3$.

The active site of urease contains two nickel (II) atoms linked by a carbamate bridge. Two imidazole N atoms are bound to each Ni atom; a carboxylate group and a water molecule fill the remaining coordination site of the metal ion. The ability to hydrolyse urea is found to vary from 17 to 70 % for soil bacteria and from 78 to 98% for soil fungi (Lloyd and Sheaffe, 1973). Although soil urease is considered to be of microbial origin there is evidence that some soil urease activity may be derived from plants (Frankenberger and Tabatabai, 1982). However, there is no direct evidence for the production of urease by plant roots.

The urease activity of soils is associated with organic matter (Kissel and Cabrera, 1988; O'Toole et al., 1982; Reynolds et al., 1985): as the organic matter content of soil decreases with depth, so too does urease activity (Bremner and Mulvaney, 1978; Mulvaney and Bremner, 1981). Urease activity is greater in grassland than in cultivated soils (O' Toole et al., 1985; Reynolds et al., 1985; Whitehead and Raistrick, 1993), which probably relates to differences in organic matter and microbial activity.

Hydrolysis of urea is temperature dependent and increases with soil temperature over the range 0–40°C (Vlek and Carter, 1983), though slight hydrolysis has been detected at sub-zero temperatures (Bremner and Mulvaney, 1978). It is also affected by urea concentration, soil water and soil pH. The optimum pH for urea hydrolysis is between 6.0 and 7.0 (Kissel and Cabrera, 1988).

Mechanism of inhibition of urease

Thousands of chemicals have been tested as potential inhibitors of soil urease activity, for use with urea fertilisers. These can be classified according to their structures or according to their binding modes with urease, and mostly fall into three groups: (i) reactive organic or inorganic compounds (e.g., alk(en)yl thiosulfinate, aydroquinone, p-benzoquinone) that react with sulfhydryl (mercapto) groups in the urease enzyme; (ii) metal-chelating compounds (e.g., caprylohydroxamic acid, acetohydroxamic acid) that cause inhibition due to complex formation with one of the Ni atoms at the active site of urease and (iii) competitive inhibitors (e.g., hydroxyurea, phosphoroamides, phenyl phosphorodiamidate (PPDA), N-(*n*-butyl)phosphorothioic triamide (NBPT) that resemble urea molecule (structural analogue), and bind to the active site of urease enzyme. Amtul et al. (2002) divided UIs into (i) substrate-analogue inhibitors, and (ii) non-substrate-like or mechanism-based inhibitors, depending on their binding modes.

Substrate-analogue inhibitors have structural similarities to urea and inhibit urease by competing for the same active site on the enzyme. Thiourea, methylurea, hydroxyl urea and numerous hydroxamic acids are the main examples of the substrate-analogue UIs.

Non-substrate-analogue inhibitors do not have any close structural similarity with urea, but they interfere with the enzyme's catalysis mechanism leading to enzyme inactivation. These compounds are also called "mechanism-based" inhibitors, e.g., imidazoles and sulphydryl reagents like *p*-chloromercuribenzoate, polyhydric phenols, aminocresols (Rodgers, 1984) and quinones (e.g., *p*-benzoquinone, 2,5-demethylbenzoquinone).

A number of UIs have been studied and tested over the last 30 years, but the following two groups have gained importance during the last few years as potent UIs:

(i) *Hydroxamic acids*: Hydroxamic acid [R–CONH–OH, R–C(OH)=NOH] (HXA) derivatives characterised by a terminal O=C–NHOH functionality were discovered by Kobashi et al. (1962). Since then a range of hydroxamic acids have been designed and examined (Gale and Atkins, 1969; Kobashi et al., 1980; Nervig and Kadis, 1976). The best studied hydroxamate and the prototype of this class of inhibitors is acetohydroxamic acid (AHA), which inhibits ureases from *Clostridium sordelli, Escherichia coli, Morganella morganii, Proteus mirabilis, Proteus vulgaris, Providencia rettgeri, Staphylococcus aureus* (Rosenstein et al., 1981) and many other micro-organisms, as well as ureases from soil (Pugh and Waid, 1969). AHA is a stable synthetic lead molecule, weakly acidic and highly soluble in water, which structurally resembles urea. Hydroxamic acids are effective metal chelates and their mechanism of inhibition involves binding to the metal ions of the active site of enzyme.

(ii) *Phosphorodiamidates*: The synthetic phosphorodiamidates are more potent than HXA and can be successfully used to inhibit the urease activity of ureolytic bacteria in soil (Bremner et al., 1986; Byrnes et al., 1983; Kobashi et al., 1985; Liao and Raines, 1985; Martens and Bremner, 1984; Rao and Ghai, 1986). The strong interaction between urease and phosphoroamide compounds may result from the electrostatic stabilisation and structural similarity to the tetrahedral geometry that may mimic an intermediate state in enzymatic catalysis. Many compounds have been studied and evaluated (Broadbent et al., 1985; Martens and Bremner, 1984; Mulvaney and Bremner, 1981; O'Connor and Hendrickson, 1987), though most have shown limited potential as fertiliser amendment due to problems of low effectiveness, lack of sustained action or lack of stability in fertiliser. NBPT is currently the most promising and effective at low concentrations when mixed with urea (Bremner and Chai, 1986; Joo et al., 1987). NBPT is not an active UI and must be converted in the soil to its oxygen analogue *N*-(*n*-butyl) phosphoric triamide (BNPO), which is the actual UI (Christianson et al., 1990). The conversion of NBPT to its oxygen analogue *N*-(*n*-butyl) phosphoric triamide (NBPTO) is rapid, occurring within minutes/hours in aerobic soils (Byrnes and Freney, 1995), but it can take several days in the floodwater of tropical soils. NBPTO forms a tridentate ligand with the urease enzyme, blocking the active site (Manunza et al., 1999).

15.6.2 Nitrification inhibitors

NIs have been used in agriculture to improve fertiliser efficiency and crop yields and to minimise denitrification and/or leaching losses of NO_3^-, by maintaining applied fertiliser N in the soil as NH_4^+–N (Bronson et al., 1991; Smith et al., 1989; Yadvinder Singh and Beauchamp, 1989). They can reduce emissions of N_2O directly by reducing the fraction of NH_4^+–N oxidised to NO_3^- and therefore the N_2O loss associated with nitrification before crop uptake, or indirectly by reducing the amount of NO_3^- substrate available for denitrification (Aulakh et al., 1984; Bronson et al., 1992).

Nitrification process

Nitrification usually refers to chemolithotrophic nitrification, but heterotrophic nitrification also exists. Heterotrophic nitrification is the oxidation of any reduced form of N, including organic N. Fungi are considered to be most efficient heterotrophic nitrifiers (Killham, 1986). Chemolithotrophic nitrification is mediated by micro-organisms belonging to the family *Nitrobacteriaceae* (Watson et al., 1989). These organisms drive energy from the oxidation of NH_4^+ and NO_2^-, and can use CO_2 as a sole C source (Hooper et al., 1997). The oxidation of NH_4^+ and NO_2^- is mediated primarily by two separate groups of autotrophic bacteria: (i) ammonia-oxidising bacteria, belonging to the genera *Nitrosomonas*, *Nitroso-coccus*, *Nitrosospira* and *Nitrosolobus* (Bock et al., 1991); and (ii) nitrite-oxidising bacteria, belonging to the genera *Nirobacter*, *Nitrococcus*, *Nitrospira* and *Nitrospina*. Of the nitrite-oxidising bacteria, only *Nitrobacter* has been detected in soils (Bock et al., 1991). Ammonium oxidation is often thought to be the rate-limiting step in autotrophic nitrification (De Boer and Kowalchuck, 2001). The main factors that affect nitrification are soil temperature, moisture, pH and the substrates NH_4^+, O_2 and CO_2 (Stevenson, 1982). *Nitrosomans europea* has been used in most physiological research regarding chemolithotrophic ammonia oxidation. Ammonia oxidation is mediated by two enzymes, ammonia monooxygenase and hydroxylamine oxidoreductase (Hooper et al., 1997). Ammonia monooxygenase is located in the cytoplasimic membrane and converts NH_4^+ to hydroxylamine, and hydroxylamine oxidoreductase is located in the periplasm and converts hydroxylamine to NO_2^- (Wood, 1986). It is evident from molecular techniques that representatives of genus *Nitrosospira*, and especially of cluster 3, are dominant ammonia-oxidising bacteria in fertilised soils (Kowalchuck and Stephen, 2001). The heterotrophic nitrifying bacteria (*Paracoccus dentrificans, Thiosphaera panto-tropha, Psedomonas putida* and *Alcaligenes faecalis*) possess ammonia- and hydroxalamine-oxidising enzymes that have strong similarities with those of autotrophic nitrifiers (Kuenen and Robertson, 1994; Moir et al., 1996).

NIs are chemicals designed to slow this process, reducing the risk that N will be lost through leaching and denitrification. Most of the NIs inactivate the ammonia mono-oxygenase. Many substances can potentially inhibit the nitrification reactions. Metals are particularly strong inhibitors of the reactions: when exposed to more than one inhibitor, the extent of inhibition increases greatly.

Among the large number of chemicals reported as NIs only eight (NP, nitrapyrin or N- Serve [2-chloro-6-(tri-chloromethyl)pyridine]; AM [2-amino-4-chloro-6-methylpyrimidine]; DCD [dicyanamide]; ST [2-sulfanil-amodo thia-zole]; TU [thiourea]; Dwell [5-ethoxy-3-trichloromethyl-1,2,4-thiadiazole]; MBT [2-mercaptobenzothiazole] and acetylene [C_2H_2]) have been widely tested. So far only two (NP, DCD) have gained acceptance for practical use. NP is seldom effective because of sorption on soil colloids, hydrolysis to 6-chloropicolinic acid and loss by volatilisation; it is also corrosive, explosive and toxic to plants. DCD is expensive for large-scale use in agriculture, and high application rates (25 kg DCD ha^{-1}) are required for significant inhibition (Merino et al., 2002). A new NI, DMPP or ENTEC (3,4-dimethylpyrazol phosphate), effective at low concentrations of 0.5–1.0 kg active compound ha^{-1}, has recently been developed in Germany (Zerulla et al., 2001).

Mechanism of inhibition of nitrification

In general, specific NIs are the compounds that retard oxidation of NH_4^+ to NO_2^-, without affecting subsequent oxidation of NO_2^- to NO_3^-. For example, a specific inhibitor such as DCD acts through its effect on cytochrome oxidase involved in the oxidation of hydroxylamine to NO_2^- during the nitrification process. The length of this effect is a function of the concentration of DCD in the product and the frequency of application. The non-specific inhibitors affect all enzymes in the same way. Non-specific methods of inhibition include any physical or chemical changes that ultimately denature the protein portion of the enzyme and are therefore irreversible, e.g., benzotriazoles (Bz), used as a corrosion inhibitor for decades, is an effective NI in soils under warm climate.

DCD, the dimeric form of cyanamide with relatively high water solubility ($23 \, g \, L^{-1}$ at $13°C$), is receiving renewed interest, as it can move with fertilisers in the soil and can be dissolved in liquid manures (Amberger, 1989). It also contains about 65% N, is non-volatile, degrades to CO_2, NH_3 and H_2O, and thus acts as a slow release N fertiliser. It is a bacteriostatic, non-toxic, chemical with LD_{50} of $10 \, g \, kg^{-1}$ body weight, which is about 3 times higher than NaCl (Amberger, 1989). DCD inhibits the first stage of nitrification, the oxidation of NH_4^+ to NO_2^- (Eq. (15.7)) specifically affecting *Nitrosomonas europea* (Zacherl and Amberger, 1990). Presumably this effect is due to reaction of the CN group of DCD with sulfhydrl or heavy metal groups of the bacteria's respiratory enzyme.

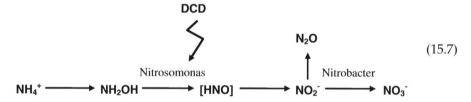

$$(15.7)$$

The bioactivity and effectiveness of NIs may depend on many factors such as soil organic matter, soil pH and soil temperature (Di and Cameron, 2004a; Irigoyen et al., 2003; Prasad and Power, 1995; Rodgers et al., 1985). Generally organic matter reduces the effectiveness of NIs (McCarty and Bremner, 1989, 1990) either by stimulating microbial activity that results in faster degradation of inhibitors (Slangen and Keerkhoff, 1984) or by reducing the bioactivity of inhibitors through absorption on the organic matter. DCD applied at $100 \, \mu g$ DCD–N g^{-1} soil was only moderately effective in an organic soil (40% organic C) (Sahrawat et al., 1987). DCD is influenced by humic and fulvic acid in the organic matter, e.g., when shaken with purified humic and fulvic acid for 24 h, both the amino and nitrile ends of DCD were sorbed to the humic materials in organic matter (Jacinthe and Pichtel, 1992). The addition of undecomposed organic matter drastically reduced the effectiveness of NIs (Puttanna et al., 1999).

DCD decomposes more slowly in strongly acidic soils than in slightly acidic soils. Only 4.1% of DCD was mineralised in acid soils (pH 4–4.3) in 60 days, compared with 48% in a near-neutral (pH 6.8) soil (Rodgers et al., 1985). The addition of lime (increasing the soil pH from 5.4 to 8.3) generally decreases NIs effectiveness (Puttanna et al., 1999), due to increased nitrifier activity and increased general microbial activity (Slangen and Keerkhoff, 1984) that rapidly biodegrades NIs. However, Bz and DCD showed resistance to degradation compared with the other two inhibitors (*o*-nitrophenol and *n*-nitroaniline).

One of the most important factors controlling the persistence of NIs is temperature (Keeney, 1980; Zourarakis and Killorn, 1990). NIs are more effective in laboratory incubations at temperatures well below optimal for nitrification (Bundy and Bremner, 1973). This effect is likely to be the result of a combination of greater inhibitor persistence due to slow degradation and/or slow volatilisation and low nitrification activity. Vilsmeier (1980) found that 0.67 mg DCD–N $100 \, g^{-1}$ soil was degraded in 60 days to 0.6 mg at 8°C and to 0.1 mg at 20°C. Rapid reduction in nitrification inhibition by several inhibitors including DCD with increased temperature from 20 to 30°C is common (McCarty and Bremner, 1989; Puttanna et al., 1999; Di and Cameron, 2004a), due to their faster degradation.

15.7 Bioavailability of N with inhibitors

Plants take N both as NH_4^+ and NO_3^-. Synchronizing plant N uptake with the released NH_4^+ or NO_3^- by controlling the rate at which urea (in applied urine or fertiliser) is hydrolysed to NH_4^+ and its subsequent oxidation to NO_3^- and the temporary rise in soil pH is critical to minimise gaseous and leaching losses of N. Most plants prefer NO_3^- over NH_4^+; however, the rate of uptake of NH_4^+ is often found to be greater than that of NO_3^-, especially at low temperatures.

Since plant roots can absorb both NH_4^+ and NO_3^- ions, ammonification and nitrification processes markedly influence the N absorption efficiency by plants, mainly by controlling the concentrations of these ions in soil solution. It has often been shown that while UIs decrease the concentration of NH_4^+ ions, NIs increase the concentration of NH_4^+ ions and decrease NO_3^- ions (Fig. 15.3). Thus by controlling nitrification, it is possible not only to increase the N absorption efficiency by plants but also to minimise the N loss by leaching and volatilisation. To decrease the N loss, chemical fertilisers with NIs have been developed, but their application is very limited in developing countries because of their high cost. Certain tropical grass species, such as *Brachiaria humidicola*, have been shown to inhibit the nitrification process by suppressing the growth of ammonium-oxidising bacteria, accumulating NO_3^- in the soil, and enhancing N absorption (see http://ss.jircas.affrc.go.jp/kanko/newsletter/nl1999/No.18/03ishikawa.htm).

The application of DCD with urine has been found to reduce soil NO_3^- production but increase the NH_4^+ concentration at all the soil depths (0–500 mm) compared with urine alone (Cookson and Cornforth, 2002). The capacity of NIs to preserve N in NH_4^+ form depends on several factors, such as soil temperature (Guiraud and Marol, 1992; Zerulla et al., 2001), soil humidity (Grundmann et al., 1995) and treatment doses (Rajbanshi et al., 1992). The metabolic degradation of NI–DCD follows a linear kinetic equation (Rajbanshi et al., 1992) so does the increase in soil NH_4^+ concentration (Irigoyen et al., 2003), and its bioavailability. The decreased NO_3^- concentration found in lettuce leaves with the addition of DCD (Montemurro et al., 1998) has been attributed to high levels of NH_4^+ in soil. Application of NIs like NP and DCD caused a significant increase in NH_4^+ through mineralisation when applied alone to grasslands (Rodgers and Ashworth, 1982). The alkaline N fertilisers, such as NH_3 and urea, have been found to generally nitrify faster than acid-forming fertilisers such as ammonium sulphate (Abbasi et al., 2003; Keeney, 1980). Hence the relative inhibition by NIs could be lower with the alkaline-forming fertilisers, at least in acid and neutral soils.

UI (NBPT) delays urea hydrolysis and thus keeps the N in urea form (Bremner et al., 1991; Wang et al., 1991; Watson et al., 1994b). Although UIs have little impact on nitrification (Bundy and Bremner, 1973; Bremner et al., 1986), NBPT has been shown to decrease NO_2^- and NO_3^- accumulation in soil compared with unamended urea (Bremner and Chai, 1989), suggesting an association with reduced rates of nitrification (Watson et al., 1994b).

The assimilation of N by plants is a complex biochemical process involving a series of N assimilatory enzymes, and is beyond the scope of this chapter. For a detailed review on these processes please refer to Stewart et al. (1980) and Lea (1993).

Briefly, in the presence of sufficient sun light as a source of energy, N assimilatory enzymes (NO_3^- reductase) in plants rapidly reduce NO_3^- to NH_3, which is then assimilated into glutamine and glutamate (Fig. 15.4). Glutamine, glutamate and organic acids arise from carbohydrate metabolism then serve as N donors in the biosynthesis of amines, amides and essentially all amino acids and nucleic acids. The amino acids thus serve as building blocks for synthesis of proteins. Thus NO_3^- reduction occurs both in aerial portions (shoots and leaves) and in roots of plants; however, most reduction occurs in shoot. The relative importance of these two sites of NO_3^- conversion is considered most important.

It is evident from Figure 15.4 that the rate of NH_4^+ assimilation is faster than that of NO_3^- as the former is directly incorporated into organic compounds. In

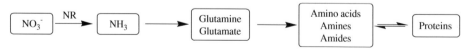

Figure 15.4. *Schematic representation of nitrogen metabolism in plants (NR denotes nitrate reductase).*

order to maintain a charge balance, plant uptake of NH_4^+ and NO_3^- affects the pH of rhizosphere by releasing either hydrogen (H^+) or hydroxyl (OH^-) ions. The release of such ions by plants also affects the uptake of other anions and cations. For example, NH_4^+ absorption reduces the uptake of cations such as calcium (Ca^{++}), magnesium (Mg^{++}) and potassium (K^+) and increases uptake of anions such as phosphate and sulphate. NO_3^- uptake reduces the absorption of anions.

A number of studies carried out in different agricultural systems reported that the application of N fertiliser with UIs or NIs improves the bioavailability of N, resulting in increased dry matter yield and N uptake (Watson et al., 1998; Xu et al., 2002; Zaman et al., 2005). Such increases are always attributed to the delayed urea hydrolysis by UIs and to NH_4^+ retention by NIs. However, there is little information in the literature on the type of N taken up by plants after the application of N inhibitors and their effects on the biochemical processes of N assimilation. Recently Zaman et al. (2005) reported 20%, 17% and 15% increase in pasture production from urea applied with UI (NBPT), NI (DCD) and UI+NI, respectively. Similar increases in dry matter production have been obtained from urine applied with UIs and NIs. Cameron et al. (2005) also reported treating urine patches with DCD may increase pasture production by 15%. Application of DCD ($15\,kg\,ha^{-1}$) with urine increased pasture yield by an average of 49% in the autumn and by 18% in the spring urine treatment (Di and Cameron, 2002a, 2002b). DCD applied twice with urine plus urea resulted in a 15% to 33% increase in pasture yields and a 24% increase in N uptake (Di and Cameron, 2004c). Cameron et al. (2005) also reported that treating urine patches with DCD may increase pasture production by 15%. This higher dry matter yield and N uptake with inhibitors can be attributed to the retention of applied N as mineral N or organic N in the soil profile, which subsequently becomes available for pasture plants.

15.8 Effect of inhibitors on N losses

Many research trials have confirmed inhibitors are effective in delaying the conversion of either urea to NH_4^+ (UIs) or NH_4^+ to NO_3^- (NIs) (Fig. 15.3). The majority of research indicates the application of UIs to soils with fertiliser urea or urine reduces NH_3 volatilisation while the application of NIs reduces NO_3^- leaching and N_2O emissions. Some studies also show NIs increase NH_3 volatilisation (Davies and Williams, 1995; Nastri et al., 2000).

Treating urea with UI (NBPT) reduces NH_3 loss from surface applications (Bremner et al., 1991; Clay et al., 1990; Carmona et al., 1990). Laboratory (Carmona et al., 1990; Vittori-Antisari et al., 1996) and field studies (Rawluk et al., 2001; Watson et al., 1994a) have shown increased inhibition of urease activity with an increasing rate of UI that followed the law of diminishing returns

(Watson et al., 1994b). NBPT can reduce NH_3 volatilisation significantly in urea, with concentrations as low as 0.005% (w/w) (Carmona et al., 1990). Christianson et al. (1990) observed 68% inhibition of urea hydrolysis at 0.01% NBPT (w/w) and NH_3 losses 1.5–3 times lower when the rate was increased to 0.1%. The optimum concentration of NBPT for temperate grassland soils to inhibit urea hydrolysis is 0.1% of urea (w/w) (Watson et al., 1994b). However, it has been observed that NBPT is less effective at higher temperatures (Bremner et al., 1991) and in soils with high levels of organic carbon (Carmona et al., 1990; Wang et al., 1991). A soil incubation study, using a wide range of soil types, indicated the effectiveness of NBPT in lowering NH_3 volatilisation was the greatest in soils with a high pH and low buffering capacity (Watson et al., 1994a). As these were the soil conditions leading to high NH_3 loss from unamended urea, NBPT has the potential to improve the efficiency of urea for temperate grassland. There is little evidence of any long-term adverse effect on grass production or reduced efficacy with repeated applications of NBPT amended urea over a period of 3 years (Watson et al., 1998). Results of more recent studies on the effect of UIs on NH_3 volatilisation are summarised in Table 15.3.

UIs have little effect on nitrification. Although NBPT has been shown to decrease NO_2^- and NO_3^- accumulation in soil (Bremner and Chai, 1989; Watson et al., 1994a), this is probably due to the slow formation of exchangeable NH_4^+ caused by the inhibition of urea hydrolysis (Vittori-Antisari et al., 1996).

In New Zealand, there has recently been increasing interest in the use of NIs to mitigate environmental impacts of N losses through leaching and gaseous emissions from animal excreta and effluent application (Table 15.4). Di and Cameron (2002b) found the application of DCD following two urine applications ($1000\,kg\,N\,ha^{-1}$) reduced N_2O emissions by 82%. Williamson and Jarvis (1997) obtained 74% reduction in N_2O emissions in a short-term study (37 days) where DCD was applied to urine ($60\,kg\,N\,ha^{-1}$). The new inhibitor DMPP ($1\,kg\,ha^{-1}$) reduced N_2O emissions by 60% in autumn and by 48% in spring when applied to a grassland after slurry application (Merino et al., 2005).

About 60% reduction in NO_3^- leaching from grazed pasture soils, including animal urine patches with DCD, has been reported with soil lysimeters (Di and Cameron, 2004c) using a free-draining shallow stony soil. As NO_3^- leaching is accompanied by counter cations, e.g., calcium, potassium and magnesium, the leaching of these cations was also reduced by the NI (Di and Cameron, 2004b).

There is some evidence that both UIs and NIs may have detrimental effects on plant leaves, e.g., transient leaf tip scorch with UIs, and DCD phytoxicity under certain weather condition (Belastegui Macadam et al., 2003; Bremner, 1995; Prasad and Power, 1995; Watson, 2000). However, the benefits of inhibitors in reducing N losses and increasing pasture production would appear to outweigh these short-term detrimental effects. These same trials showed a wide range of

Table 15.3. *Selected references on the effect of urease inhibitors (UIs) in nitrogen economy.*

Inhibitor (dose)	N source (kg N ha^{-1})	Reduction in N losses	Type of crop	Effect on dry matter yield	Country	Reference
UI–NBPT (0.05% w/w)	Urea (100)	83% decrease in NH$_3$ loss	Perennial ryegrass pasture	9% increase	Ireland	Watson et al. (1994b)
UI–NBPT (0.05%, 0.10%, 0.15% w/w)	Urea (100)	75–81% (sandy loam) 75–85% (clay loam) decrease in NH$_3$ loss	–	–	Canada	Rawluk et al. (2001)
UI–NBPT (0.25% w/w)	Urea (120)	89% (sandy loam) 47% (clay loam) decrease in NH$_3$ loss	Wheat crop	No significant effect	Italy	Gioacchini et al. (2002)
UI–Agrotain (0.1% w/w)	Urea / Urine (600)	27% decrease in NH$_3$ loss / 23% decrease in NH$_3$ loss	Ryegrass-clover pasture	Not monitored	New Zealand	Singh et al. (2003, 2004)
UI–HQ (0.3% w/w)+NI–DCD (5% w/w)	Urea	62% decrease in N$_2$O emissions	Rice crop	35–37% increase	Belgium	Xu et al. (2002)

Table 15.4. *Selected references on the effect of nitrification inhibitors (NIs) in nitrogen economy.*

Inhibitor (dose)	N source (kg N h^{-1})	Reduction in N losses	Type of crop	Effect on dry matter yield	Country	Reference
NI–DCD (25 kg ha^{-1})	CAN (80)	42% decrease in N$_2$O	Perennial ryegrass pasture	–	Spain	Merino et al. (2002)
	cattle slurry (85)	60% decrease in N$_2$O				
NI–DCD (12 kg ha^{-1})	Dairy farm effluent (1100)	18% reduction in NO$_3^-$–N leaching	Ryegrass pasture	19.2% increase	New Zealand	Williamson et al. (1998)
NI–DCD (25 kg ha^{-1})	Cattle urine (450)	36% reduction in NO$_3$–N content (0–100 mm)	Ryegrass-clover pasture	No significant effect	New Zealand	Cookson and Cornforth (2002)
NI–DCD	Cattle urine (1000)	76% (autumn) 42% (spring) reduction in NO$_3^-$–N leaching	Ryegrass-clover pasture	30% increase	New Zealand	Di and Cameron (2002a)
NI–DCD (6 kg ha^{-1})	Urea (60)	–	Wheat crop	22–25% increase	USA	Rao and Popham (1999)
NI–DCD (12.5 kg ha^{-1}) Nitrapyrin (7.5 kg ha^{-1})	Urea (120)	56–58%	Perennial rye grass- spring barley	–	Scotland	McTaggart et al. (1997)
	Urea (120)	40%				

economic returns, depending upon soil type, drainage, time of application and environmental conditions. The greatest likelihood of N losses is from coarse-textured or poorly drained soils; it is in these situations that the use of inhibitors would be most economical. However, inhibitors do not work as well in coarse-textured soils as in these soils urea and NH_4^+ ions have a tendency to move away from the inhibitor with rainfall or irrigation (University of Illinois).

Studies on the effect of inhibitors on N economy (Table 15.4) have shown the inhibitory action of these chemicals depends on their persistence and bioactivity in soils, which in turn are affected by the intrinsic properties of the compound, soil properties and climatic conditions. The half-lives of inhibitors may vary from a few days to several weeks, depending on the nature of the compound, rate of application, soil type, pH and season (soil temperature). The ideal inhibitor for use in agriculture should:

- specifically block an enzymatic reaction (e.g., NIs should block ammonium oxidation to nitrite, but not nitrite oxidation to nitrate, during the nitrification process);
- remain in close contact with N compounds (e.g., UIs must move with urea molecules that are not readily absorbed by soil; whereas NIs must be close to NH_4^+ ions that are readily retained by soil);
- not adversely affect other beneficial soil organisms and higher plants;
- remain effective in the soil for several weeks after N input through fertiliser addition and excretal deposition;
- not to be toxic to animals and humans at the levels used to inhibit nitrification effectively and
- cost effective to use.

The ultimate goal of any inhibitor is to increase the efficiency of N use. For an economic benefit to occur, the N saved from leaching and gaseous losses by using the inhibitors would have to result either in an increase in pasture production, with a value greater than the cost of the inhibitors, or in a reduction in fertiliser input. The economic benefits of reduced environmental pollution and future damage to our environment from leaching and gaseous emissions are of higher significance over the long-term than the productivity gains. The value of inhibitors in reducing N losses from N fertilisers and increasing crop yields is well established in arable soils. The inhibitors are also reported to increase pasture production. The increase in stocking rates needed to utilise this extra pasture may, however, enhance emissions of other greenhouse gases. Results of a recent desktop study (de Klein and Monaghan, 2005), demonstrated that the use of NIs had a limited effect on total greenhouse gas emissions reduction, compared with the reduction in N_2O emissions, due to an increase in both CH_4 and CO_2 emissions from the farm system.

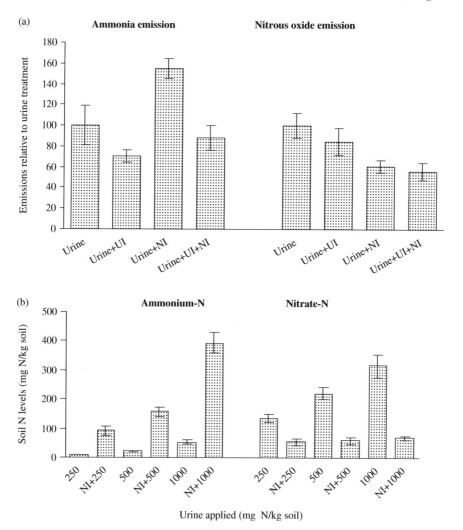

Figure 15.5. *Effect of inhibitors applied with urine on ammonia and nitrous emissions. (Nitrous oxide quantitatively measured by static chamber method and ammonia qualitatively measured by passive absorption method. Error bars are 2 standard deviation, n = 3).*

Lysimeter studies by Di and Cameron (2002a, 2002b, 2003, 2004b, 2004c) showed DCD reduced NO_3^--leaching and N_2O emissions from urine and urea applications. Results of a current PhD study on the impact of the inhibitors on N dynamics (Fig. 15.5a, b) confirmed DCD reduced soil N_2O emissions and NO_3^-

leaching from cattle urine but also suggested DCD increased NH_3 volatilisation and potential NH_4^+–N leaching.

Understanding the soil and plant processes controlling DCD decomposition, the variable response in different soils, and the impact DCD has in causing changes in the N transformations and N cycle is the focus of our current research. This will allow us to develop simple assays and models to monitor and simulate the degradation of these inhibitors in various soil types, and enhance our understanding of the impact of these inhibitors on the bioavailability of N in managed grassland ecosystems.

15.9 Conclusions

Application of UIs or NIs with N fertiliser, animal urine and animal slurries improves the bioavailability of N, resulting in increased dry matter yield and N uptake. Such increases result from the delayed urea hydrolysis by UIs and NH_4^+ retention by NIs. Certain tropical grass species such as *Brachiaria humidicola* appear to inhibit the nitrification process by suppressing the growth of ammonia-oxidising bacteria, accumulating NO_3^- in the soil and enhancing N absorption. However, there is little information in the literature on the type of N taken up by plants after the application of inhibitors or on their effects on the biochemical processes of N assimilation. These inhibitors may have a detrimental effect on plant leaves under certain weather condition but the benefits of inhibitors in increasing pasture production and reducing N losses would appear to prevail over these short-term detrimental effects.

Loss of N, occurring mainly through NH_3 volatilisation, biological denitrification and NO_3^- leaching, has both economic and environmental implications. Therefore, the economic benefits of reduced environmental pollution and future damage to our environment as a result of the use of N inhibitors are of higher significance to the productivity gains over the long term.

The value of inhibitors in mitigating N losses would depend on their rate of biodegradation and persistence in soils. Currently, there is a strong debate in New Zealand on the effectiveness of the NI, DCD, in mitigating N loss. However, it is difficult to devise mitigation strategies from the existing information because the key soil and environmental factors influencing DCD efficiency are poorly understood. Furthermore, there is little information on the long-term impact these inhibitors will have in altering the N cycle of grazed pasture systems, and on the issues of toxicity.

Furthermore, a quantitative understanding of the interrelations between N_2O and NH_3 emissions, and NO_3^- and NH_4^+ leaching is central to understanding how pasture systems behave and respond to inhibitors and to determine the effectiveness of land-management strategies to reduce overall N losses. Mitigation strategies neglecting these interrelations may be suboptimal. For example, there are already claims that NIs lead to increased NH_3 volatilisation and have the potential to enhance NH_4^+ leaching. However, there are limited quantitative field data available to assess accurately the NH_3 volatilisation and NH_4^+ leaching contribution of NIs.

References

Parliamentary Commissioner for the Environment, 2004. Growing for good – intensive farming, sustainability and New Zealand's environment. Parliamentary Commissioner for the Environment, Wellington, New Zealand, 236 p. http://www.pce.govt.nz.

University of Illinois, undated. This Land: 50 ways farmers can protect their groundwater. College of Agricultural, Consumer and Environmental Sciences, University of Illinois Urbana-Champaign. http://www.thisland.uiuc.edu/50ways/50ways_5.html.

Abbasi, M.K., Shah, Z., Adams, W.A., 2003. Effect of the nitrification inhibitor nitrapyrin on the fate of nitrogen applied to a soil incubated under laboratory conditions. J. Plant Nutr. Soil Sci. 166, 513–518.

Agriculture and Agri-Food Canada, 2003. The health of our water: Towards a sustainable agriculture in Canada, Ch.7d. http://www.res2.agr.gc.ca/publications/hw/07d_e.htm.

Amberger, A., 1989. Research on dicyandiamide as a nitrification inhibitor and future outlook. Commun. Soil Sci. Plant Anal. 20, 1933–1955.

Amtul, Z., Atta-ur-Rahman, Siddiqui, R.A., Choudhary, M.I., 2002. Chemistry and mechanism of urease inhibition. Curr. Med. Chem. 9, ?323–1348.

Anon, 2005. Advice on hand for farmers on nitrate leaching from soil. Ctry.-Wide 27, 34.

Aulakh, M.S., Rennie, D.A., Paul, E.A., 1984. Acetylene and N-serve effects upon N_2O emissions from NH_4^+ and NO_3^- treated soils under aerobic and anaerobic conditions. Soil Biol. Biochem. 16, 351–356.

Beauchamp, E.G., 1997. Nitrous oxide emissions from agricultural soils. Can. J. Soil Sci. 77, 113–123.

Belastegui Macadam, X.M., del Prado, A., Merino, P., Estavillo, J.M., Pinto, M., Gonzalez-Murua, C., 2003. Dicyandiamide and 3,4-dimethyl pyrazole phosphate decrease N_2O emissions from grassland but dicyandiamide produces deleterious effects in clover. J. Plant Physiol. 160, 1517–1523.

Bock, E., Koops, H.-P., Harms, H., Ahlers, B., 1991. The biochemistry of nitrifying organisms. In: J.M. Shively, L.L. Barton (Eds.), Variations in Autotrophic Life. Academic Press, London, pp. 171–200.

Bolan, N.S., Saggar, S., Kemp, P.D., 2004a. Nitrate toxicity in grazing animals. In: Proceedings of the New Zealand Institute of Primary Industry Management and Rural Valuers Conference, Hamilton, New Zealand, pp. 154–166.

Bolan, N.S., Saggar, S., Luo, J.F., Bhandral, R., Singh, J., 2004b. Gaseous emissions of nitrogen from grazed pastures: processes, measurements and modelling, environmental implications, and mitigation. Adv. Agron. 84, 37–120.

Bremner, J.M., McCarty, G.W., Yeomans, J.C., Chai, H.S., 1986. Effects of phosphoroamides on nitrification, denitrification, and mineralization of organic nitrogen in soil. Commun. Soil Sci. Plant Anal. 17, 369–384.

Bremner, J.M., McCarty, G.W., Higuchi, T., 1991. Persistence of the inhibitory effects of phosphoroamides on urea hydrolysis in soils. Commun. Soil Sci. Plant Anal. 22, 1519–1526.

Bremner, J.M., 1995. Recent research on problems in the use of urea as a nitrogen fertiliser. Fertil. Res. 42, 321–329.

Bremner, J.M., Chai, H.S., 1986. Evaluation of N-butyl phosphorothioic triamide for retardation of urea hydrolysis in soil. Commun. Soil Sci. Plant Anal. 17, 337–351.

Bremner, J.M., Chai, H.S., 1989. Effects of phosphoroamides on ammonia volatilization and nitrite accumulation in soils treated with urea. Biol. Fertil. Soils 8, 227–230.

Bremner, J.M., Mulvaney, R.L., 1978. Urease activity in soils. In: R.G. Burns (Ed.), Soil Enzymes. Academic Press, London, pp. 149–196.

Broadbent, F.E., Nakashima, T., Chang, G.Y., 1985. Performance of some urease inhibitors in field trials with corn. Soil Sci. Soc. Am. J. 49, 348–351.

Bronson, K.F., Touchton, J.T., Hauck, R.D., Kelley, K.R., 1991. N-15 recovery in winter-wheat as affected by application timing and dicyandiamide. Soil Sci. Soc. Am. J. 55, 130–135.

Bronson, K.F., Mosier, A.R., Bishnoi, S.R., 1992. Nitrous-oxide emissions in irrigated corn as affected by nitrification inhibitors. Soil Sci. Soc. Am. J. 56, 161–165.

Bundy, L.G., Bremner, J.M., 1973. Inhibition of nitrification in soils. Soil Sci. Soc. Am. Proc. 37.

Byrnes, B.H., Freney, J.R., 1995. Recent developments in the use of urease inhibitors in the tropics. Fertil. Res. 42, 251–259.

Byrnes, B.H., Savant, N.K., Craswell, E.T., 1983. Effect of a urease inhibitor phenyl phosphor-odiamidate on the efficiency of urea applied to rice. Soil Sci. Soc. Am. J. 47, 270–274.

Cameron, K.C., Di, H.J., Moir, J., Roberts, A., Pellow, R., Christie, R., 2005. Treating grazed pasture soil with a nitrification inhibitor "ECO-N" to decrease nitrate leaching. In: L.D. Currie and J.A. Hanly (Eds.), Developments in Fertiliser Application Technologies and Nutrient Management. Fertliser and Lime Research Centre, Massey University, Palmerston North, New Zealand, pp. 93–103.

Carmona, G., Christianson, C.B., Byrnes, B.H., 1990. Temperature and low concentration effects of the urease inhibitor N-(n-butyl) thiophosphoric triamide (Nbtpt) on ammonia volatilization from urea. Soil Biol. Biochem. 22, 933–937.

Christianson, C.B., Byrnes, B.H., Carmona, G., 1990. A comparison of the sulfur and oxygen analogs of phosphoric triamide urease inhibitors in reducing urea hydrolysis and ammonia volatilization. Fertil. Res. 26, 21–27.

Clay, D.E., Malzer, G.L., Anderson, J.L., 1990. Ammonia volatilization from urea as influenced by soil-temperature, soil-water content, and nitrification and hydrolysis inhibitors. Soil Sci. Soc. Am. J. 54, 263–266.

Cookson, W.R., Cornforth, I.S., 2002. Dicyandiamide slows nitrification in dairy cattle urine patches: effects on soil solution composition, soil pH and pasture yield. Soil Biol. Biochem. 34, 1461–1465.

Crush, J.R., 1987. Nitrogen fixation. In: M.J. Baker, W.M. Williams (Eds.), White Clover. CAB International, Wallingford, UK, pp. 185–202.

Davies, D.M., Williams, P.J., 1995. The effect of the nitrification inhibitor dicyandiamide on nitrate leaching and ammonia volatilization – a UK nitrate sensitive areas perspective. J. Environ. Manage. 45, 263–272.

De Boer, W., Kowalchuck, G.A., 2001. Nitrification in acid soils: microorganisms and mechanisms. Soil Biol. Biochem. 33, 853–866.

de Klein, C.A.M., Monaghan, R.M., 2005. The impact of potential nitrous oxide mitigation strategies on the environmental and economic performance of dairy systems in 4 New Zealand catchments. In: A. van Amstel (co-ordinator), Non-CO$_2$ Greenhouse Gases (NCGG-4). Millipress Rotterdam, The Netherlands, pp. 593–600.

Di, H.J., Cameron, K.C., 2002a. Nitrate leaching in temperate agroecosystems: sources, factors and mitigating strategies. Nutr. Cycl. Agroecosys. 64, 237–256.

Di, H.J., Cameron, K.C., 2002b. The use of a nitrification inhibitor, dicyandiamide (DCD), to decrease nitrate leaching and nitrous oxide emissions in a simulated grazed and irrigated grassland. Soil Use Manage. 18, 395–403.

Di, H.J., Cameron, K.C., 2003. Mitigation of nitrous oxide emissions in spray-irrigated grazed grassland by treating the soil with dicyandiamide, a nitrification inhibitor. Soil Use Manage. 19, 284–290.

Di, H.J., Cameron, K.C., 2004a. Effects of temperature and application rate of a nitrification inhibitor, dicyandiamide (DCD), on nitrification rate and microbial biomass in a grazed pasture soil. Aust. J. Soil Res. 42, 927–932.

Di, H.J., Cameron, K.C., 2004b. Effects of the nitrification inhibitor dicyandiamide on potassium, magnesium and calcium leaching in grazed grassland. Soil Use Manage. 20, 2–7.

Di, H.J., Cameron, K.C., 2004c. Treating grazed pasture soil with a nitrification inhibitor, eco-n (TM), to decrease nitrate leaching in a deep sandy soil under spray irrigation – a lysimeter study. N Z J. Agric. Res. 47, 351–361.

Duxbury, J.M., Bouldin, D.R., Terry, R.E., Tate, R.L., 1982. Emissions of nitrous-oxide from soils. Nature 298, 462–464.

Frankenberger, W.T., Tabatabai, M.A., 1982. Amidase and urease activities in plants. Plant Soil 64, 153–166.

Freney, J.R., Black, A.S., 1988. Importance of ammonia volatilization as a loss process. In: J.R. Wilson (Ed.), Advances in Nitrogen Cycling in Agricultural Ecosystems. CAB International, Wallingford, pp. 156–173.

Gale, G.C., Atkins, I.M., 1969. Inhibition of urease by hydroxamic acids. Archives Internationales de Pharmacodynamie et de Therapie 180, 289–298.

Gioacchini, P., Nastri, A., Marzadori, C., Giovannini, C., Antisari, L.V., Gessa, C., 2002. Influence of urease and nitrification inhibitors on N losses from soils fertilized with urea. Biol. Fertil. Soils 36, 129–135.

Goh, K.M., Williams, P.H., 1999. Comparative nutrient budgets of temperate grazed pastures. In: L.O. Fresco, E.M. Smailing and O. Oenema (Eds.), Nutrient Diseuilibria in Global Agroecosystems: Concepts and Case Studies. CAB International, Wallingford, UK, pp. 173–191.

Grundmann, G.L., Renault, P., Rosso, L., Bardin, R., 1995. Differential effects of soil water content and temperature on nitrification and aeration. Soil Sci. Soc. Am. J. 59, 1342–1349.

Guiraud, G., Marol, C., 1992. Influence of temperature on mineralization kinetics with a nitrification inhibitor (mixture of dicyandiamide and ammonium thiosulphate). Biol. Fertil. Soils 13, 1–5.

Gupta, S.K., Gupta, R.C., Seth, A.K., Gupta, A.B., Bassin, J.K., Gupta, A., 1999. Adaptation of cytochrome-b5 reductase activity and methalobinemia in areas witha high nitrate concentration in drinking water. Bull. World Health Organ. 77, 749–755.

Haynes, R.J., Williams, P.H., 1993. Nutrient cycling and soil fertility in the grazed pasture ecosystem. Adv. Agron. 49, 119–199.

Hooper, A.B., Vannelli, T., Bergmann, D.J., Acciero, D., 1997. Enzymology of oxidation of ammonia to nitrite by bacteria. Antonie van Leeuwenhoek 71, 59–67.

Irigoyen, I., Muro, J., Azpilikueta, M., Aparicio-Tejo, P., Lamsfus, C., 2003. Ammonium oxidation kinetics in the presence of nitrification inhibitors DCD and DMPP at various temperatures. Aust. J. Soil Res. 41, 1177–1183.

Jacinthe, P.A., Pichtel, J.R., 1992. Interaction of nitrapyrin and dicyandiamide with soil humic compounds. Soil Sci. Soc. Am. J. 56, 465–470.

Joo, Y.K., Christians, N.E., Bremner, J.M., 1987. Effect of N-(normal-butyl) thiophosphoric triamide (Nbpt) on growth-response and ammonia volatilization following fertilization of kentucky bluegrass (Poa-Pratensis L) with urea. J. Fertil. Issues 4, 98–102.

Keeney, D.R., 1980. Factors affecting the persistence and bioactivity of nitrification inhibitors. In: M. Stelly (Ed.), Nitrification Inhibitors: Potentials and Limitations. Vol. 38, American Society of Agronomy, Madison, WI, pp. 33–46.

Killham, K., 1986. Heterotrophic nitrification. In: J.I. Prosseer (Ed.), Nitrification. Vol. 20, IRL Press, Oxford, UK, pp. 117–126.

Kissel, D.E., Cabrera, M.L., 1988. Factors affecting urea hydrolysis. In: B.R. Bock and D.E. Kissel (Eds.), Ammonia Volatilization from Urea Fertilizers. Bulletin Y-206, National Fertilizer Development Center, Tennessee Valley Authority, Muscle Shoals, Alabama, pp. 53–66.

Kobashi, K., Hase, J.I., Uehara, K., 1962. Specific inhibition of urease by hydroxamic acids. Biochim. Biophys. Acta 62, 380–383.

Kobashi, K., Munakata, K.I., Takebe, S., Hase, J.I., 1980. Therapy for uro lithiasis by hydroxamic acids. II. Urease inhibitory potency and urinary excretion rate of hippurohydroxamic acid derivatives. J. Pharmacobio-dyn. 3, 444–450.

Kobashi, K., Takebe, S., Numata, A., 1985. Specific-inhibition of urease by n-acylphosphoric triamides. J. Biochem. 98, 1681–1688.

Kowalchuck, G.A., Stephen, J.R., 2001. Ammonia-oxidising bacteria: a model for molecular ecology. Annu. Rev. Microbiol. 55, 485–529.

Kuenen, J.G., Robertson, L.A., 1994. Combined nitrification-denitrification processes. FEMS Microbiol. Rev. 15, 109–117.

Lea, P., 1993. Nitrogen metabolism. In: P.J. Lea, R.C. Leegood (Eds.), Plant Biochemistry and Moleculer Biology. Wiley, New York, pp. 155–180.

Ledgard, S.F., Brier, G.J., Upsdell, M.P., 1990. Effect of clover cultivar on production and nitrogen-fixation in clover-ryegrass swards under dairy-cow grazing. N Z J. Agric. Res. 33, 243–249.

Ledgard, S.F., Steele, K.W., 1992. Biological nitrogen fixation in mixed legume/grass pastures. Plant Soil 141, 137–153.

Ledgard, S.F., 1995. Leaching rife with high nitrogen application. In: H. Dunn (Ed.), New Zealand Dairy Exporter. Vol. 70, New Zealand Dairy Exporter Ltd., Wellington, pp. 20–21.

Ledgard, S.F., Sprosen, M.S., Steele, K.W., 1996. Nitrogen fixation by nine white clover cultivars in grazed pasture, as affected by nitrogen fertilization. Plant Soil 178, 193–203.

Ledgard, S.F., Sprosen, M.S., Penno, J.W., Rajendram, G.S., 2001. Nitrogen fixation by white clover in pastures grazed by dairy cows: temporal variation and effects of nitrogen fertilization. Plant Soil 229, 177–187.

Liao, C.F.H., Raines, S.G., 1985. Inhibition of soil urease activity by amido derivatives of phosphoric and thiophosphoric acids. Plant Soil 85, 149–152.

Lloyd, A.B., Sheaffe, M.J., 1973. Urease activity in soils. Plant Soil 39, 71–80.

Manunza, B., Deiana, S., Pintore, M., Gessa, C., 1999. The binding mechanism of urea, hydroxamic acid and N-(n-butyl)-phosphoric triamide to the urease active site. A comparative molecular dynamics study. Soil Biol. Biochem. 31, 789–796.

Martens, D.A., Bremner, J.M., 1984. Effectiveness of phosphoroamides for retardation of urea hydrolysis in soils. Soil Sci. Soc. Am. J. 48, 302–305.

Martens, D.A., Bremner, J.M., 1993. Influence of herbicides on transformations of urea nitrogen in soil. J. Environ. Sci. Health B 28, 377–395.

McBride, M.B., 1994. Environmental Soil Chemistry. Oxford University Press, New York.

McCarty, G.W., Bremner, J.M., 1989. Laboratory evaluation of dicyandiamide as a soil nitrification inhibitor. Commun. Soil Sci. Plant Anal. 20, 2049–2065.

McCarty, G.W., Bremner, J.M., 1990. Persistence of effects of nitrification inhibitors added to soils. Commun. Soil Sci. Plant Anal. 21, 639–648.

McTaggart, I.P., Clayton, H., Parker, J., Swan, L., Smith, K.A., 1997. Nitrous oxide emissions from grassland and spring barley, following N fertiliser application with and without nitrification inhibitors. Biol. Fertil. Soils 25, 261–268.

Merino, P., Estavillo, J.M., Graciolli, L.A., Pinto, M., Lacuesta, M., Munoz-Rueda, A., Gonzalez-Murua, C., 2002. Mitigation of N_2O emissions from grassland by nitrification inhibitor and Actilith F2 applied with fertilizer and cattle slurry. Soil Use Manage. 18, 135–141.

Merino, P., Menendez, S., Pinto, M., Gonzalez-Murua, C., Estavillo, J.M., 2005. 3,4-dimethylpyrazole phosphate reduces nitrous oxide emissions from grassland after slurry application. Soil Use Manage. 21, 53–57.

Ministry for the Environment, 2003. Report to the Minister for the Environment on Lake Rotioti and other Rotorua Lakes. Ministry for the Environment, Welington. http://www.mfe.govt.nz/issues/water/rotorua-lakes/minister-report.html.

Ministry for the Environment, 2005. New Zealand's Greenhouse Gas Inventory Report, 1990–2003 and Common Reporting Format Tables, April 2005. Ministry for the Environment, Wellington. http://www.climatechange.givt.nz.

Moir, J.W.B., Crossman, L.C., Spiro, S., Richardson, D.J., 1996. The purification of ammonia-oxygenase from Paracoccus denitrificans. FEBS Lett. 387, 71–74.

Montemurro, F., Capotorti, G., Lacertosa, G., Palazzo, D., 1998. Effects of urease and nitrification inhibitors application on urea fate in soil and nitrate accumulation in lettuce. J. Plant Nutr. 21, 245–252.

Mosier, A.R., Kroeze, C., 2000. Potential impact of global atmospheric N_2O budget of the increased nitrogen input required to meet future global food demands. Chemosphere – Glob. Change Sci. 2, 465–473.

Mosier, A., Bleken, M.A., Chaiwanakupt, P., Ellis, E.C., Freney, J.R., Howarth, R.B., Matson, P.A., Minami, K., Naylor, R., Weeks, K.N., Zhu, Z.L., 2001. Policy implications of human-accelerated nitrogen cycling. Biogeochemistry 52, 281–320.

Mulvaney, R.L., Bremner, J.M., 1981. Control of urea transformations in soils. In: E.A. Paul, J.N. Ladd (Eds.), Soil Biochemistry. Vol. 5, Marcel Dekker, New York, pp. 153–196.

Nastri, A., Toderi, G., Bernati, E., Govi, G., 2000. Ammonia volatilization and yield response from urea applied to wheat with urease (NBPT) and nitrification (DCD) inhibitors. Agrochimica 44, 231–239.

Nervig, R.M., Kadis, S., 1976. Effect of hydroxamic acids on growth and urease activity in Cornebacterium renale. Can. J. Microbiol. 22, 544–551.

Nommik, H., Vathras, G.L., 1982. Retention and fixation of ammonium and ammonia in soils. In: F.J. Stevenson (Ed.), Nitrogen in Agricultural Soils. Vol. 22, American Society of Agronomy, Madison, WI, pp. 123–171.

O'Connor, M.J., Hendrickson, L.L., 1987. Effect of phenylphosphorodiamidate on ammonia volatilization as affected by soil temperature and rate and distribution of urea. Soil Sci. Soc. Am. J. 51, 1062–1066.

O' Toole, P., Morgan, M.A., McGarry, S.J., 1985. A comparative study of urease activities in pasture and tillage soils. Commun. Soil Sci. Plant Anal. 16, 759–773.

Oenema, O., Velthof, G.L., Yamulki, S., Jarvis, S.C., 1997. Nitrous oxide emissions from grazed grassland. Soil Use Manage. 13, 288–295.

O'Toole, P., Morgan, M.A., McAleese, D.M., 1982. Effects of soil properties, temperature and urea concentration on patterns and rates of urea hydrolysis in some Irish soils. Ir. J. Agric. Res. 21, 185–197.

Prasad, R., Power, J.F., 1995. Nitrification inhibitors for agriculture, health, and the environment. Adv. Agron. 54, 233–281.

Prinn, R., Cunnold, D., Rasmussen, R., Simmonds, P., Alyea, F., Crawford, A., Fraser, P., Rosen, R., 1990. Atmospheric emissions and trends of nitrous-oxide deduced from 10 years of Ale-Gauge data. J. Geophys. Res. 95, 18369–18385.

Pugh, K.B., Waid, J.S., 1969. The influence of hydroxamates on ammonia loss from an acid loamy sand treated with urea. Soil Biol. Biochem. 1, 195–206.

Puttanna, K., Gowda, N.M.N., Rao, E., 1999. Effect of concentration, temperature, moisture, liming and organic matter on the efficacy of the nitrification inhibitors benzotriazole, o-nitrophenol, m-nitroaniline and dicyandiamide. Nutr. Cycl. Agroecosys. 54, 251–257.

Rajbanshi, S.S., Benckier, G., Ottow, J.C.G., 1992. Effects of concentration, incubation temperature, and repeated applications on degradation kinetics of dicyandiamide (DCD) in model experiments with a silt loam soil. Biol. Fertil. Soils 13, 61–64.

Rao, D.L.N., Ghai, S.K., 1986. Effect of phenylphosphorodiamidate on urea hydrolysis, ammonia volatilization and rice growth in an alkali soil. Plant Soil 94, 313–320.

Rao, S.C., Popham, T.W., 1999. Urea placement and nitrification inhibitor effects on growth and nitrogen accumulation by no-till winter wheat. Crop Sci. 39, 1115–1119.

Rasmussen, R.A., Khalil, M.A.K., 1986. Atmospheric trace gases - trends and distributions over the last decade. Science 232, 1623–1624.

Rawluk, C.D.L., Grant, C.A., Racz, G.J., 2001. Ammonia volatilization from soils fertilized with urea and varying rates of urease inhibitor NBPT. Can. J. Soil Sci. 81, 239–246.

Reynolds, C.M., Wolf, D.C., Armbruster, J.A., 1985. Factors related to urea hydrolysis in soils. Soil Sci. Soc. Am. J. 49, 104–108.

Rodgers, G.A., 1984. Inhibition of soil urease activity by aminocresols. Plant Soil 79, 155–157.

Rodgers, G.A., Wickramasinghe, K.N., Jenkinson, D.S., 1985. Mineralization of dicyandiamide, labelled with N^{15}, in acid soils. Soil Biol. Biochem. 17, 253–254.

Rodgers, G.A., Ashworth, J., 1982. Use of nitrification inhibitors to improve recovery of mineralized nitrogen by winter wheat. J. Sci. Food Agric. 33, 1229–1236.

Rosenstein, I.J., Hamiltonmiller, J.M., Brumfitt, W., 1981. Role of urease in the formation of infection stones – comparison of ureases from different sources. Infect. Immun. 32, 32–37.

Ryden, J.C., 1986. Gaseous losses of nitrogen from grassland. In: H.G. van Meer, J.C. Ryden and G.C. Ennik (Eds.), Nitrogen Fluxes in Intensive Grassland Systems. Martinus Nijhoff Publishers, Dordrecht.

Saggar, S., Hedley, M.J., Gillingham, A.G., Rowarth, J.S., Richardson, S., Bolan, N.S., Gregg, P.E.H., 1990a. Predicting the fate of fertilizer sulfur in grazed hill country pastures by modeling the transfer and accumulation of soil-phosphorus. N Z J. Agric. Res. 33, 129–138.

Saggar, S., Mackay, A.D., Hedley, M.J., Lambert, M.G., Clark, D.A., 1990b. A nutrient-transfer model to explain the fate of phosphorus and sulfur in a grazed hill-country pasture. Agric. Ecosyst. Environ. 30, 295–315.

Saggar, S., 2004. Changes in nitrogen dynamics of legume-based pastures with incresed nitrogen fertiliser use: Impacts on New Zealnd's nitrous oxide emissions inventory. N Z Soil News 52, 110–117.

Saggar, S., andrew, R.M., Tate, K.R., Hedley, C.B., Rodda, N.J., Townsend, J.A., 2004a. Modelling nitrous oxide emissions from New Zealand dairy grazed pastures. Nutr. Cycl. Agroecosys. 68, 243–255.

Saggar, S., Bolan, N.S., Bhandral, R., Hedley, C., Luo, J., 2004b. Emissions of methane, ammonia and nitrous oxide from animal excreta deposition and farm effluent application in grazed pastures. N Z J. Agric. Res. 47, 513–544.

Saggar, S., Giltrap, D.J., Hedley, C.B., Lambie, S.M., Tate, K.R., Li, C., 2005a. Nitrous oxide emissions from grazed perennial pastures in New Zealand. In: A. van Amstel (co-ordinator), Non-CO2 Greenhouse Gases (NCGG-4). Millipress Rotterdam, The Netherlands, pp. 113–121.

Saggar, S., Giltrap, D.J., Li, C., Hedley, C.B., Tate, K.R., Lambie, S.M., 2005b. Modelling nitrous oxide emissions from grazed grasslands in New Zealand. In: F.P.O' O' Mara, R.J. Wilkins, L.T. Mannetje, D.K. Lovett, P.A.M. Rogers and T.M. Boland (Eds.), XX international Grasslands Congress. Wageningen Academic Publishers, The Netherlands, p. 566.

Sahrawat, K.L., Keeney, D.R., Adams, S.S., 1987. Ability of nitrapyrin, dicyandiamide and acetylene to retard nitrification in a mineral and an organic soil. Plant Soil 101, 179–182.

Sears, P.D., Goodall, V.C., Jackman, R.H., Robinson, G.S., 1965. Pasture growth and soil fertility. VIII. The influence of grasses, white clover, fertilizers and the return of herbage clippings on pasture production of an impoverished soil. N Z J. Agric. Res. 8, 270–283.

Singh, Y., Beauchamp, E.G., 1989. Nitrogen transformations near urea in soil – effects of nitrification inhibition, nitrifier activity and liming. Fertil. Res. 18, 201–212.

Singh, J., Bolan, N., Saggar, S., 2003. A method for simultaneous measurement of ammonia volatilization and nitrous oxide emissions from intact soil cores. In: L.D. Currie and J.A. Hanly (Eds.), Tools for Nutrient and Pollutant Management: Applications to Agriculture and Environmental Quality. Occasional Report No. 17, Fertilizer and Lime Research Centre, Massey University, Palmerston North, New Zealand, pp. 224–231.

Singh, J., Saggar, S., Bolan, N., 2004. Mitigating gaseous losses of N from pasture soil with urease and nitrification inhibitors. In: Proceedings of the Australia New Zealand SuperSoil 2004 Conference, Sydney, Australia, 6–9 December. http://www.regional.org.au/au/asssi.

Slangen, J.H.G., Keerkhoff, P., 1984. Nitrification inhibitors in agriculture and horticulture: a literature review. Fertil. Res. 5, 1–76.

Smith, K.A., Crichton, I.J., McTaggart, I.P., Lang, R.W., 1989. Inhibition of nitrification by dicyandiamide in cool temperate conditions. In: J.A. Hansen, K. Henriksen (Eds.), Nitrogen in Organic Wastes Applied in Soils. Academic Press, London, pp. 289–303.

Stevenson, F.J., 1982. Nitrogen in Agricultural Soils. American Society of Agronomy, Crop Science Society of America, Soil Science Society of America, Madison, WI.

Stewart, G.P., Mann, A.F., Fentem, P.A., 1980. Enzymes of glutamate formation: gluatamate dehydrogenase, glutamine synthetase, glutamate synthase. In: B.F. Miflin (Ed.), The Biochemsitry of Plants. Vol. 5, Academic Press, New York, pp. 271–327.

Vilsmeier, K., 1980. Effect of temperature on the breakdown of dicyandiamide in the soil. Zeitschrift Fur Pflanzenernahrung Und Bodenkunde 143, 113–118.

Vittori-Antisari, L., Marzadori, C., Gioacchini, P., Ricci, S., Gessa, C., 1996. Effects of the urease inhibitor N-(n-butyl) thiophosphoric triamide in low concentrations on ammonia volatilization and evolution of mineral nitrogen. Biol. Fertil. Soils 22, 196–201.

Vlek, P.L.G., Carter, M.F., 1983. The effect of soil environment and fertilizer modifications on the rate of urea hydrolysis. Soil Sci. 136, 56–63.

Wang, Z., Van Cleemput, O., Liantie, L., Baert, L., 1991. Effect of organic matter and urease inhibitors on urea hydrolysis and immobilization of urea nitrogen in an alkaline soil. Biol. Fertil. Soils 11, 101–104.

Watson, S.W., Bock, E., Harms, H., Koops, H.-P., Hooper, A.B., 1989. Nitrifying bacteria. In: J.T. Staley, M.T. Bryant, N. Pfennig and J.D. Holt (Eds.), Bergeys Manual of Systematic Bacteriology. Vol. 3, Williams and Wilkins, Baltimore, pp. 1808–1834.

Watson, R.T., Meiro Filho, L.C., Sanhueza, E., Janetos, A., 1992. Sources and sinks. In: J.T. Houghton, B.A. Callander and S.K. Varney (Eds.), Climate Change 1992, the Supplementary Report to the IPCC Scientific Assessment. Cambridge University Press, Cambridge, pp. 25–46.

Watson, C.J., Miller, H., Poland, P., Kilpatrick, D.J., Allen, M.D.B., Garrett, M.K., Christianson, C.B., 1994a. Soil properties and the ability of the urease inhibitor N-(n-Butyl) thiophosphoric triamide (Nbtpt) to reduce ammonia volatilization from surface-applied urea. Soil Biol. Biochem. 26, 1165–1171.

Watson, C.J., Poland, P., Miller, H., Allen, M.B.D., Garrett, M.K., Christianson, C.B., 1994b. Agronomic assessment and N-15 recovery of urea amended with the urease inhibitor Nbtpt (N-(n-Butyl) thiophosphoric triamide) for temperate grassland. Plant Soil 161, 167–177.

Watson, C.J., Poland, P., Allen, M.B.D., 1998. The efficacy of repeated applications of the urease inhibitor N-(n-butyl) thiophosphoric triamide for improving the efficiency of urea fertilizer utilization on temperate grassland. Grass Forage Sci. 53, 137–145.

Watson, C.J., 2000. Urease activity and inhibition – principles and practice. In: The International Fertiliser Society Proceedings No. 454.

White, R.E., 1989. Nitrogen transformations, fixation and losses – a brief introduction and overview. In: R.E. White, L.D. Currie (Eds.), Proceedings Nitrogen in New Zealand Agriculture and Horticulture. Fertiliser and Lime Research Centre, Massey University, Palmerston North, pp. 83–87.

Whitehead, D.C., 1995. Grassland Nitrogen. CAB International, Wallingford, UK.

Whitehead, D.C., Raistrick, N., 1993. The volatilization of ammonia from cattle urine applied to soils as influenced by soil properties. Plant Soil 148, 43–51.

Williamson, J.C., Jarvis, S.C., 1997. Effect of dicyandiamide on nitrous oxide flux following return of animal excreta to grassland. Soil Biol. Biochem. 29, 1575–1578.

Williamson, J.C., Taylor, M.D., Torrens, R.S., Vojvodic-Vukovic, M., 1998. Reducing nitrogen leaching from dairy farm effluent-irrigated pasture using dicyandiamide: a lysimeter study. Agric. Ecosyst. Environ. 69, 81–88.

Wood, P.M., 1986. Nitrification as a bacterial energy source. In: J.I. Prosser (Ed.), Nitrication, Special Publication of the Society for General Microbiology. Vol. 20, IRL Press, Oxford, UK, pp. 39–64.

World Resources Institute (2000) World Resources 2000–2001. World Resources Institute, Washington, DC, pp. 389.

Xu, X.K., Boeckx, P., Van Cleemput, O., Zhou, L.K., 2002. Urease and nitrification inhibitors to reduce emissions of CH_4 and N_2O in rice production. Nutr. Cycl. Agroecosys. 64, 203–211.

Zacherl, B., Amberger, A., 1990. Effect of the nitrification inhibitors dicyandiamide, nitrapyrin and thiourea on *nitrosomonas europaea*. Fertil. Res. 22, 37–44.

Zaman, M., Nguyen, L., Blennerhassett, J.D., Quin, B.F., 2005. Increasing the utilisation of urea fertilisers by pasture. In: L.D. Currie, J.A. Hanly (Eds.), Developments in Fertiliser Application Technologies and Nutrient Management. Vol. 18, Fertiliser and Lime Research Centre, Massey University, Palmerston North, pp. 276–284.

Zerulla, W., Barth, T., Dressel, J., Erhardt, K., von Locquenghien, K.H., Pasda, G., Radle, M., Wissemeier, A.H., 2001. 3,4-Dimethylpyrazole phosphate (DMPP) – a new nitrification inhibitor for agriculture and horticulture – an introduction. Biol. Fertil. Soils 34, 79–84.

Zourarakis, D., Killorn, R., 1990. The efficacy of 2 nitrification inhibitors at high-temperature in 2 Iowa soils. Soil Sci. 149, 185–190.

Developments in Soil Science, volume 32
Ravendra Naidu (Editor)
© 2008 Elsevier B.V. All rights reserved

Chapter 16

ASSESSMENT OF PHOSPHORUS BIOAVAILABILITY FROM ORGANIC WASTES IN SOIL

B. Fuentes, M. de la Luz Mora, N.S. Bolan and R. Naidu

16.1 Introduction

In rural areas, where economic activities are driven by the export of raw materials, and food and fibre products, cultivation and livestock production have increased and patterns of land use have changed significantly. Intensive livestock production and agricultural farming practices tend to "mine" many nutrients from soil, causing modifications in the nutrient flux and thereby resulting in decreased soil fertility, mainly due to the deficiencies in P, N, K, Ca and Mg, and the presence of phytotoxic elements such as soluble Al and Mn, particularly in acid soils (Borie and Rubio, 1999; Mora et al., 1999; Haynes and Mokolobate, 2001; Mora et al., 2004; Mora et al., 2006).

To maintain soil fertility, it is necessary to add nutrients to agricultural systems. Farmers use mineral fertilizers, lime, organic waste, compost and various management techniques to maintain soil fertility. Nowadays recovering nutrients by reincorporating organic residues in farming systems is a sustainable alternative that can contribute to restoring the natural environmental equilibrium (Unsal and Sozudogru, 2001; Griffin et al., 2003). For this, it is essential to organize and optimize the use of natural resources because, although the recycling of nutrients through agricultural application of organic waste such as biosolids, sludges and manure is an ecological method of nutrient recovery, other factors must be considered such as social acceptability, soil nutrient saturation and contamination issues (Levlin et al., 2002).

The high consumption of raw materials and the generation of organic waste from biological treatment as a consequence of world population increase have led to political decisions in Europe and elsewhere for more sustainable resource use, residue disposal and ecosystem protection. To achieve these aims, strict nutrient discharge limits and waste management strategies have been introduced (Woods et al., 1999).

Phosphorus (P) is an essential nutrient (Mkhabela and Warman, 2005; see Chapters 14) because it is a constituent of macromolecular cell structures and is essential for microorganism, plant and animal growth. In spite of its wide distribution in nature, P is a limited resource (Adnan et al., 2003; Shimamura et al., 2003) and it is deficient in most soils (Vassilev et al., 2001; Soil P exists in various chemical forms, including inorganic P (Pi) and organic P (Po). These compounds have multiple sources of natural and anthropogenic origin, which differ widely in their behaviour and fate in both natural and managed soils (Shand and Smith, 1997; Turner et al., 2003a; Turner et al., 2003b; Hansen, et al., 2004), specifically in relation to bioavailability, as various forms can undergo cycling at different rates, being retained in soils or made available to plants (Chen et al., 2003; De Brouwere et al., 2003; Nwoke et al., 2003).

Orthophosphate ions from Pi and Po differ in their release rates in different soils (Nwoke et al., 2003). This process is influenced by the soil environment: physicochemical characteristics (controlled mainly by soil properties like pH, soluble Al, Fe, Ca and organic matter content, causing dissolution, precipitation, adsorption and desorption reactions); biological reactions (produced by microbial activity involved in P cycling that involves reactions of mineralization and immobilization, that are governed by soluble–insoluble P ratio in soil solution) and soil management practices (Rubaek and Sibbesen, 1993; Celi et al., 2001; Borie and Rubio, 2003; Griffin et al., 2003; Mkhabela and Warman, 2005). The acidity and exchangeable Al, commonly encountered in Chilean soils enhances strong adsorption of P (Mora et al., 2002). Under these conditions microbial activity is reduced and therefore the mineralization of organic P became slow.

The problem of P deficiency can potentially be solved by the progressive return of organic materials to soils. However, the quantity, quality and management are fundamental factors that affect P availability from residues (Kwabiah et al., 2003a). Identification of P in waste (both its organic and inorganic forms) is a fundamental prerequisite to understanding nutrient dynamics in soil-waste systems and the mechanisms responsible for the release of potentially bioavailable forms of P over time. For effective P recovery from organic residues incorporated in soil, it is fundamental to identify the P distribution patterns in residues and subsequently in soil, which will help determine the optimal conditions for P release into plants.

Phosphate removal for recycling from wastewater has been practiced widely and a number of effective and reliable techniques have been employed (Yoshino et al., 2003). In particular, wastewater treatment currently involves the removal of phosphorus by either chemical or biological means (Battistoni et al., 2002; Levlin et al., 2002). Adoption of technologies for P recovery would therefore resolve problems of distribution, inefficient use and transport (Szögi et al., 2006). The cost of P removal treatment can be offset by gaining a recyclable phosphate product and reducing sludge disposal costs and volume (Battistoni et al., 2002). Greaves et al. (1999) indicated that P recovery in a form suitable to be used as fertilizer offers better control of P application and its release rate into the soil, so reducing the need to import P fertilizers and the cost of exporting nutrients to P-deficient areas.

This chapter presents a critical review of the factors that influence the bioavailability of P compounds in the dynamic soil environment, looking particularly at P compounds recovered from organic waste. This is one area that is often overlooked by soil fertility experts and environmental managers.

16.2 Phosphorus compounds in soil environment

Phosphorus is an essential element for the metabolism of living organisms because it is a component of ATP, nucleic acid and phospholipids that compose the cellular membranes. Moreover, it is used in plant energy metabolism in ATP and ADP molecules and intermediate compounds of respiration and photo-synthesis (Taiz and Zeiger, 1998).

Phosphorus in soil comes from both pedogenic and anthropogenic sources, the majority of P being introduced as mineral fertilizers (Bolan et al., 2005) or organic residues; a significant quantity of insoluble and more stable P can accumulate in soils with regular P application (Scherer and Sharma, 2002; Verma et al., 2005). Phosphorus mineral fertilizers include water-soluble (fast release as superphosphate, triple superphosphate, monoammonium phosphate and diammonium phosphate), water-insoluble (slow release as phosphate rock and basic slag) and mixtures of both (partially acidulate phosphate rocks and superphosphate and mixed rock) (Bolan and Duraisamy, 2003; Bolan et al., 2005). The main sources of P from organic sources include manure, crop residues and sludge from different treatments in sustainable waste handling systems.

The most significant P compound in soil in terms of bioavailability is the orthophosphate anion, which is associated with readily accessible short-term availability for plants. For longer-term availability Po compounds can be degraded to release orthophosphate anions. The concentration and distribution of labile and stable Po may vary with season, soil type and vegetation/cropping system (Rubaek and Sibbesen, 1993). However, P bioavailability is affected by external factors such as adsorption reactions, pH, organic matter content and mineral immobilization. Furthermore, the nature of P compounds and their interactions in waste affect the degree of availability of orthophosphate ions. Therefore, the effective availability of P compounds applied to soil from residues depends firstly on the form of P added and, later on, the external factors that contribute to increased or decreased availability over time.

For the speciation of P, bioavailability and transformation over time in soil and organic residue, techniques such as nuclear magnetic resonance (NMR) and isotopic exchange have been developed. These provide an understanding of the chemical form of P compounds and their interaction in both natural and managed soils (Frossard et al., 2002; Turner et al., 2003a; Turner et al., 2003b; Briceño et al., 2004).

16.2.1 Inorganic phosphorus

Plants take up P from soil solutions mainly as orthophosphate anions ($H_2PO_4^-$ or $H_2PO_4^{-2}$ or PO_4^{3-}). Although the relative proportion of these three phosphate anions in soil solution depends on the pH, there has been some debate about the major ionic form in which P is taken up by roots.

The orthophosphate condensation produces polyphosphates, the first being pyrophosphate ion ($P_2O_7^{4-}$), and the second tripolyphosphate ion ($P_3O_{10}^{5-}$), as is shown in the following reactions:

$$2PO_4^{3-} + H_2O \leftrightarrow P_2O_7^{4-} + 2OH^-$$

$$P_2O_7^{4-} + PO_4^{3-} + H_2O \leftrightarrow P_3O_{10}^{5-} + 2OH^-$$

The structural formulas of the acidic form, $H_4P_2O_7$ and $H_5P_3O_{10}$ are:

Pyrophosphoric acid   Tripolyphosphoric acid

Polymeric phosphates hydrolyse in water and their hydrolysis rate is pH dependent. The number of phosphate groups in anionic form determines the interaction force of this molecule with the metal cation present in soil, residue and water. The Pi compounds vary in their pK_a values (Table 16.1). Knowing the pH values allows an estimation of possible interactions with other ions in the solution.

Turner et al. (2003b) identified different P species such as orthophosphate, pyrophosphate and polyphosphate in permanent pasture soils by means of NaOH-EDTA extract. Briceño et al. (2004) determined smaller quantities of pyrophosphate in soil with high P fixation, indicating that these molecules can be present as organic-pyrophosphates, which are products of microbial degradation or else come from persistent and recalcitrant organic matter in soil.

16.2.2 Organic phosphorus

Organic P compounds, considered recalcitrant in soil (Tarafdar and Claassen, 2003) and derived from microorganisms, plant and animal waste, form part of humified materials which are stabilized in the soil matrix. For P from organic

Table 16.1. *Dissociation acid constant of P compound.*

Ionizable hydrogen	pK_a
$H_3PO_4 \leftrightarrow H^+ + H_2PO_4^-$	2.12[a]
$H_2PO_4^- \leftrightarrow H^+ + HPO_4^{2-}$	7.21[a]
$HPO_4^{2-} \leftrightarrow H^+ + PO_4^{3-}$	12.32[a]
$H_4P_2O_7 \leftrightarrow H^+ + H_3P_2O_7^-$	Quite small[b]
$H_3P_2O_7^- \leftrightarrow H^+ + H_2P_2O_7^{2-}$	2.64[b]
$H_2P_2O_7^{2-} \leftrightarrow H^+ + HP_2O_7^{3-}$	6.76[b]
$HP_2O_7^{3-} \leftrightarrow H^+ + P_2O_7^{4-}$	9.42[b]
$H_5P_3O_{10} \leftrightarrow H^+ + H_4P_3O_{10}^-$	Quite small[b]
$H_4P_3O_{10}^- \leftrightarrow H^+ + H_3P_3O_{10}^{2-}$	Quite small[b]
$H_3P_3O_{10}^{2-} \leftrightarrow H^+ + H_2P_3O_{10}^{3-}$	2.30[b]
$H_2P_3O_{10}^{3-} \leftrightarrow H^+ + HP_3O_{10}^{4-}$	6.50[b]
$HP_3O_{10}^{4-} \leftrightarrow H^+ + P_3O_{10}^{5-}$	9.24[b]

[a]From Tan (1993).
[b]From Manahan (1994).

compounds to be available, it needs to be hydrolysed and mineralized by the microbial biomass, which is a fundamental process for the release of orthophosphate ions to plants (Haygarth and Jarvis, 1997; Hayes et al., 2000a; Borie and Rubio, 2003; Oehl et al., 2004) and the maintenance of the P cycle in ecosystems. Through a mineralization process, Po compounds represent an important P source for plants and microorganisms (Thien and Myers, 1992; Makarov et al., 2005) in soil with low levels of bioavailable P (Parfitt et al., 2005) and in unfertilized cropping systems (Linquist et al., 1997).

Knowledge of the chemical nature of Po in soil is fundamental for determining the contribution of these compounds to the bioavailable P pool, for predicting association with soil constituents, susceptibility to microbial attack and physical position inside or outside the soil macro and micro aggregates (Rubaek and Sibbesen, 1993; Borie et al., 1989; Borie and Rubio, 2003; Hansen et al., 2004; Makarov et al., 2005). Po concentration in soils with high fixation capacity is estimated at 654–1942 ppm, representing 49–64% of total phosphorus (Borie et al., 1989).

Depending on the type of bond, Po compounds are classified as: orthophosphate monoester, orthophosphate diester and phosphonates; their bond type is shown in Figure 16.1.

Orthophosphate monoester presents only one covalent bonding where P is linked to the O–C molecule, forming species P–O–C. Some compounds that present this bond type include: *myo*-inositol hexa*kis*phosphate, glucose-6-phosphate, para-nitrophenyl phosphate and nucleotides. Orthophosphate monoesters are also a product of the degradation of the macromolecules phospholipids and nucleic acid. These compounds have been found in several

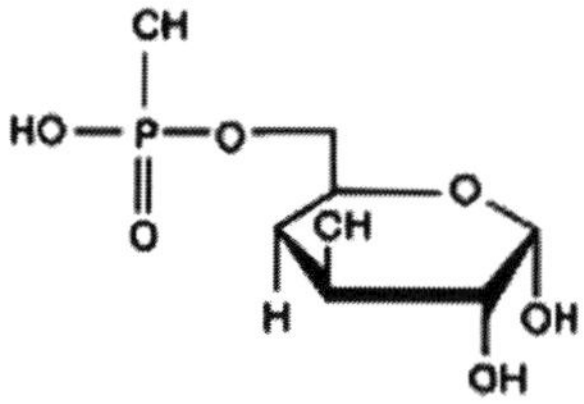

Orthophosphate monoester
D-Glucose-6-phosphate

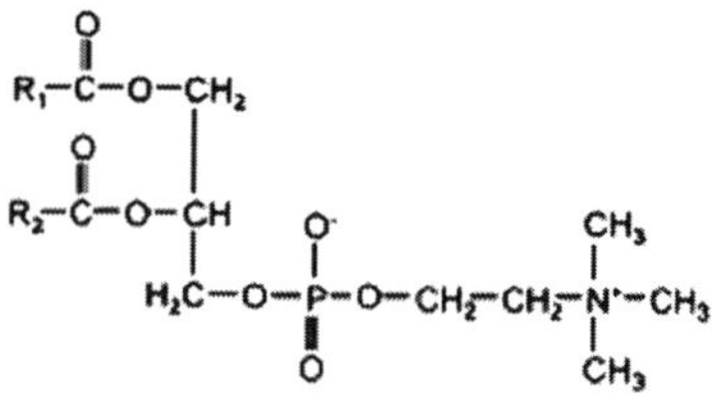

Orthophosphate diester
L-α-Phosphatidyl choline
(lecithin)

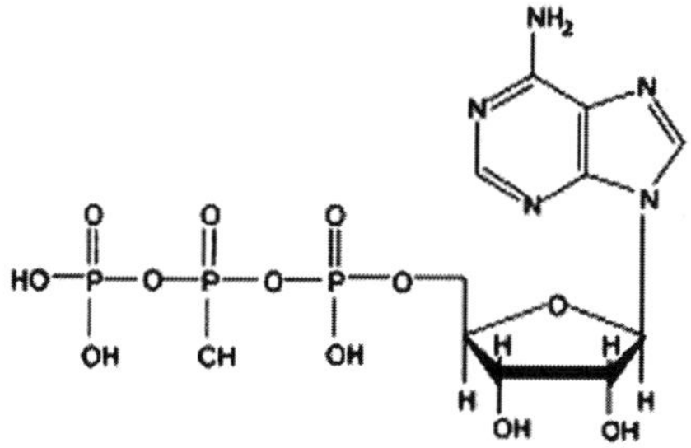

Organic Polyphosphate
(monoester)
Adenosine-5-triphosphate

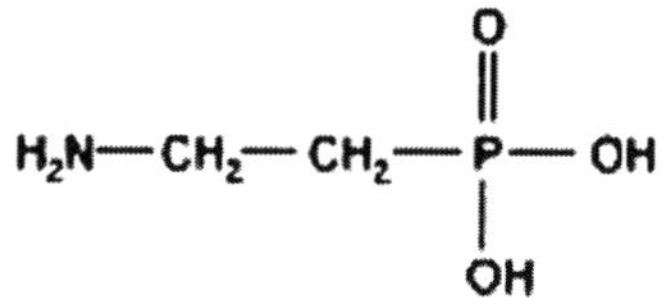

Phosphonate
2-Aminoethyl phosphonic acid

Figure 16.1. *Organic phosphorus compound (from Turner et al., 2005).*

organic residues, mainly manure, compost and sewage sludge, which are potential inositol phosphate sources in soil, when applied to agricultural land (Turner et al., 2002b) and can make their way into waterways from effluent discharges.

Briceño et al. (2004) determined that Pi along with the orthophosphate monoester were the major P chemical species of the variable charge detected in Chilean soil, diester orthophosphate was 5–10% and pyrophosphate was 0–3% of the total P. They also indicate the presence of inositol phosphates in the range 17–64% of total P.

Temporal variation in the amount of monoester orthophosphate depends on soil type, level of fertilization, movement in P fraction (Zhang et al., 1999) and microbial biomass, because it is expected that some microorganisms will degrade and utilize the hydrolysed P (Andlid et al., 2004).

Within monoester phosphate compounds, inositol phosphates are found widely in the natural environment (Turner et al., 2002b; Turner and McKelvie, 2002; Andlid et al., 2004), representing 42–67% of organic P from Chilean volcanic soil (Borie et al., 1989). However, a large knowledge gap exists about origins, chemical forms, bioavailability and the mobility of these compounds in the environment (Turner and McKelvie, 2002; Turner et al., 2002b). Inositol

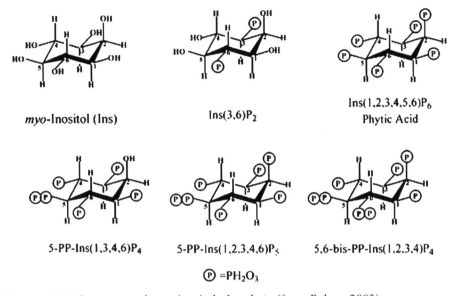

Figure 16.2. *Structures of myo-inositol phosphate (from Raboy, 2003).*

Table 16.2. *Dissociation acid constants of myo-inositol hexakisphosphate.*

	pK											
	pK_1	pK_2	pK_3	pK_4	pK_5	pK_6	pK_7	pK_8	pK_9	pK_{10}	pK_{11}	pK_{12}
From Cosgrove (1980) in Celi et al. (2001)	1.84	1.84	1.84	1.84	1.84	1.84	6.3	6.3	9.7	9.7	9.7	9.7
From Costello et al. (1976) in Turner et al. (2002b)	1.1	1.5	1.5	1.7	2.1	2.1	5.7	6.9	7.6	10.0	10.0	12.0

phosphate belongs to a family of phosphoric esters derived from hexahydroxy-cyclohexane (inositol) (Turner et al., 2002b). These molecules can exist as inositol in various states of phosphorylation. In Figure 16.2, an inositol phosphate molecule with varying phosphorylation is shown, including the more highly phosphorylated inositol compound that contains 7 and 8 mol of P per inositol mol.

The number of substituted phosphate groups on the inositol ring may vary between one and six (Turner et al., 2002b), having 12 ionizable protons (whose dissociation acid constants are indicated in Table 16.2), with 12 coordinate ligands for complexing cations being potential.

Inositol phosphates form complexes with monovalent cations generating monomeric salts such as the sodium inositol phosphate compound commonly known as phytate (Andlid et al., 2004), with polyvalent cation inositol phosphate potentially forming inter and intra molecular bonds, resulting in the formation of polymeric compounds (Dao, 2003).

In general, inositol phosphate is strongly adsorbed in soil (Yadav and Tarafdar, 2003). The intensity of this interaction is a function of negative charge density. Borie et al. (1989) determined that inositol phosphate formed strong associations with humic and fulvic acid. *Myo*-inositol hexakisphosphate has a high anionic charge density, allowing it to form stable complexes in soil (Turner et al., 2002b; Briceño et al., 2004; McDowell and Stewart, 2006) that are protected from microbial degradation and dephosphorylation reactions (Briceño et al., 2004; Hansen et al., 2004) which decrease their availability. Moreover, due to low solubility, they are most abundant in the soil (Yadav and Tarafdar, 2003). The adsorption process occurs through their phosphate groups, similar to the free orthophosphate ion, forming a binuclear complex (Celi et al., 2001) with soil constituents.

Although compound-type inositol phosphates are a dominant component in most soils, other monoester phosphate compounds are probably more important in terms of short-term cycling and P availability for plants due to the high stabilizing capacity of these compounds in the soil structure, which inhibit their degradation. However, the hydrolysis of inositol phosphate to inositol and phosphoric acid is considered an important metabolic process in several bio-systems (Pandey et al., 2001; Turner et al., 2002a).

Within the inositol phosphate, *myo*-inositol hexakisphosphate (phytic acid) is the most abundant form in seeds, in particular in cereal grains (Andlid et al., 2004), roots and tubers where the clear function is storage and phosphate and inositol recovery. Typically, this compound accumulates during seed development and is broken down during germination, about $75 \pm 10\%$ of total P in seed dry weight corresponding to phytic acid (Raboy, 2003).

Orthophosphate diester is a compound where orthophosphate is bound to O–C by two ester linkages forming the C–O–P–O–C molecule. Major diester-P sources in soil and organic waste include nucleic acid and are derived from phosphoprotein and phospholipids (Briceño et al., 2004). Orthophosphate diester can be an important P source for plants, as it has a lower charge density which allows for an ionically smaller interaction with soil constituents compared to orthophosphate monoester, and therefore is more exposed to enzymatic degradation (Taranto et al., 2000; McDowell and Stewart, 2006) and subsequent orthophosphate release. These compounds play an important role in P transformation in ecosystems (Makarov et al., 2002). Borie and Rubio (2003) indicated that in agricultural soil, nucleic acids and phospholipids are of microbial origin and are present in small quantities, the low level possibly indicating that these P compounds are a substrate of rapid cycling; therefore, P lipid and nucleic acid in soil could represent an important intermediate compound for plant and microbial nutrition. Makarov et al. (2005) investigated the organic P composition in plants, bacteria and fungus in soil by means of P-31 NMR, demonstrating that the proportion of P species in bacteria and plants are very similar in base to ratios of monoester and diester orthophosphate in alkaline extract, while fungi differ considerably and present high proportions of monoesters (phospholipid prevalence) and polyphosphate.

Phosphonates correspond to anions of phosphonic acid and are similar to phosphates except that they have a carbon–phosphorus bond (C–P) instead of the carbon–oxygen–phosphorus (C–O–P) linkage. The molecule type is $R-PO(OH)_2$.

Phosphonates are highly water-soluble, poorly soluble in organic solvents and are not volatile (Nowack, 2003). They have a resistance to chemical hydrolysis and thermal descomposition (Lesueur et al., 2005), are chemically stable making possible their use as detergents, plasticizers, coolant additives, pesticides and chelating agents in many industrial applications through their ability to form strong complexes with metals (Barja et al., 2001; Nowack, 2003; Lesueur et al., 2005). Their photodegradation and biodegradation are discussed (Nowack, 2003). Phosphonates can enter the environment in large amounts as pollution from industrial activities (Matys et al., 2001).

The first phosphonate to be identified as naturally occurring was 2-amino-ethilphosphonic acid (Hayes et al., 2000b) which is found in plants and many animals, mostly in cell membranes (Nowack, 2003). Turner et al. (2003b) have identified this compound in soil extract. Several bacteria, such as *E. coli*, *Pseudomonas spp.*, have developed the ability to metabolize phosphonates as a nutrient source (C, P and N) by cleaving the C–P bond to release P (Hayes et al., 2000b; Matys et al., 2001). This process involves the lyase enzymes which degrade the direct C–P bond (Gimsing et al., 2004), but the cell-free activity of C–P lyase has not been properly demonstrated in bacteria (Matys et al., 2001).

Phosphonates adsorb very strongly onto almost all mineral surfaces in the pH range of water (Nowack, 2003). These compound are also adsorbed on soil sediment and sewage sludge. Phosphonate adsorption, photodegradation and biodegradation are modified by the presence of metals due to the formation of soluble and non-soluble complexes (Barja et al., 2001). In general, in all types of compounds, it is possible that the stability of the metal complexes increases with the number of orthophosphate or phosphonic acid groups.

Chemical studies of phosphonates focus on polyphosphonates, in particular aminopolyphosphonates such as glyphosate (*N*-phosphonomethyl-glycine), a broad-spectrum postemergent herbicide (Barja et al., 2001; Gimsing et al., 2004). Their biodegradation in soil and water is dependent on microbial activity and the presence of metal ions (Barja et al., 2001). Phosphonates in soil strongly sorb to Al and Fe oxides. However, in the presence of Ca and Mg they form ternary surface-phosphonate-Ca complexes. For this reason, glyphosate in soil decreases herbicidal activity. In addition, it forms non-soluble compounds by adsorption reaction, which protects it from microbial degradation. Furthermore, as adsorption of glyphosate is linked by means of phosphate groups, it is possible that this herbicide can compete for adsorption sites in soil with free orthophosphate (Gimsing et al., 2004). Nowack (2002) investigated phosphonate adsorption onto activated sludge and humic acid and found that adsorption onto humic acids decreases sharply with increasing pH with negligible adsorption at pH above 6.5 and the adsorption capacity of the sludge for phosphonates at neutral pH can be explained by its amorphous iron oxide content.

In natural systems, phosphonates have not been found with the same frequency as other P organic compounds. Pant et al. (1999) determined that in soil NaOH extract, glucose-6-phosphate, glycerophosphate, nucleoside monophosphates (up to 91%) and polynucleotides (up to 58%) were the major forms of total soil organic P and in lesser proportion determine phosphonate presence (up to 7%). Briceño et al. (2004) identified several P compounds in soil samples (high capacity P fixation): orthophosphate monoester, orthophosphate diester, orthophosphate and pyrophosphate were present more often and in greater quantities than either polyphosphate or phosphonate. Turner et al. (2003b) ascertained that the main P compounds present in NaOH–EDTA extract from permanent pasture soil were orthophosphate monoesters (154–751 mg kg^{-1} soil, equivalent to 29–60% of the total P extracted and equivalent to 71–91% of the Po extracted). Within these compounds they identified *myo*-inositol hexa*kis*phosphate, choline phosphate, glucose-1-phosphate, β-glycerophosphate and phosphatidic acid. To a lesser degree they also found orthophosphate diesters (between 2 and 10% of the total P extracted and 3 and 15% of the Po extracted) identified as DNA and phospholipids. The least-quantified group was phosphonates at 16 mg kg^{-1} soil (equivalent to 3% of total P extracted and 5% Po extracted). This agrees with Zhang et al. (1999), who reported phosphonate quantities <2 mg kg^{-1} soil, indicating a negligible presence.

16.3 Soil factors influencing phosphorus bioavailability

16.3.1 Phosphorus adsorption reactions

Compared with the other nutrients, P is the least mobile and bioavailable under varied soil conditions and is a limiting factor for plant growth. The low bioavailability is due to the strong sorption of orthophosphate ions to soil constituents (Hinsinger, 2001). This is reflected by fixation of phosphate in allophanic soils due to allophane (Borie and Zunino, 1983; De Brouwere et al., 2003).

The evolution of soil may influence the adsorption process. Younger soils are composted by primary minerals and have relatively high pH. Through the course of time, the weathering of primary minerals increases soil acidity leading to the formation of secondary clay minerals dominated by reactive iron and aluminium oxides, which adsorb P compounds more readily. However, in soils where the amorphous non-crystalline minerals are dominant (with great surface area), there is a greater capacity to adsorb phosphate than secondary crystalline clays (Olander and Vitousek, 2005).

The physiochemical processes of soil are determined by surface reaction. It is therefore necessary to study solid-solution interlayer soil to understand and explain their macroscopic behaviour (Briceño et al., 1999). Specific anion adsorption may induce an increase in negative surface charge and/or decrease in positive surface charge (Xu et al., 2003). Mora and Canales (1995a) indicate that the first reaction between phosphate molecules and active surface sites is associated with the mononuclear complex, followed by binuclear complex formations. In acid soils, mineralogy and geochemistry favour strong retention of orthophosphate ions. However, although total P is high, low levels of P are maintained in the soil solution (Iyamuremye et al., 1996a; Hinsinger, 2001). The availability of P to plants grown on acid soils depends largely on the extent to which phosphate ions form slowly soluble complexes or are adsorbed onto mineral surfaces (Mora et al., 2002). The fixation reaction can occur between orthophosphate and Fe or Al hydrous oxides (Nwoke et al., 2003), or between orthophosphate and silicate minerals forming monodentate (labile) and bidentate (irreversible) complexes (Mora and Canales, 1995a; Mora and Canales, 1995b). The acid soil pH contributes to increasing concentrations of exchangeable Al^{3+}, Fe^{3+} and Mn^{2+}, which react readily with phosphate, increasing precipitation as they diminish its bioavailability (Tan, 1993; Iyamuremye et al., 1996a; Iyamuremye et al., 1996c; Haynes and Mokolobate, 2001).

Research in acid soils of southern Chile, indicated that the use of acidifying fertilizers as input, accelerated the acidification process in the soil, increasing dramatically its Al concentration and favouring the adsorption of P (Mora et al., 2004). To solve acidity problems and retention of P in these soils, calcareous amendments have been applied. Mora et al. (1999), affirm that the application of calcitic or dolomitic lime reduces P adsorption capacity in Chilean Andisols due to the increase of negative charge and decrease of surface acidity.

In soils with variable charge minerals, orthophosphate anions are specifically adsorbed and form inner-sphere complexes by replacing coordinated surface –OH and –OH$_2$ groups with the orthophosphate ions (Borie and Rubio, 2003;

Briceño et al., 2004; Jara et al., 2005). These soils can adsorb both anions and cations. Therefore the adsorption can be specific as well as electrostatic (Xu et al., 2003).

In variable charge soils, both positive and negative charges are pH dependent and it is caused by adsorption and desorption of the potential determining of H^+ and OH^- ions (Mora et al., 1999). The point of zero net charge (PZNC) is the pH value where the soil surface will have the same amount of positive and negative charge, resulting in a zero net charge. When pH values are lower than PZNC, the soil surface tends to have a mainly positive charge (Hyun and Lee, 2004). Therefore, at pH values less than PZNC, orthophosphate adsorption is favoured due to the enhancing anion exchange capacity by protonation of pH-dependent surface species and electrostatic attraction between anions and the positively charged compound surface.

In calcareous soils, P fixation and retention also occur due to high amounts of soluble and exchangeable Ca^{2+} resulting from reaction of $CaCO_3$ with orthophosphate anions to form precipitate. Carreira et al. (2006) indicated that pedogenic $CaCO_3$ is the primary geochemical agent in arid ecosystems capable of reducing leaching losses of P, through secondary precipitation of Ca–P minerals and/or strong sorption reaction of P with $CaCO_3$.

16.3.2 Soil solution pH

The distribution of P species in soil is determined primarily by the pH of soil solution (Hinsinger, 2001) due to the variation of proton dissociation from orthophosphoric acid or polyphosphates, which are characterized by pK_a values (Table 16.1) that represent an important property of chemical compounds and indicate their ionization capability. In the normal pH soil range, (Fig. 16.3) $H_2PO_4^-$ and HPO_4^{2-} come from phosphoric acid (Tan, 1993).

The pH of soil solution also influences organic P compounds. Table 16.2 shows pK_a values for *myo*-inositol hexa*kis*phosphate, which indicate that in normal pH-range soil, these compounds have at least six dissociated protons, which confer a high negative charge density to the molecule, causing a strong interaction with soil minerals.

Depending on the charge density, P ions have a tendency to form complex species with several metal cations such as Ca, Mg, Fe and Al with different bond forces forming mono-, di- and multi-dentate bonds.

Increased pH in acid soil is fundamental for increasing P bioavailability because it causes the precipitation of exchangeable Al and Fe, thus reducing the potential for orthophosphate ion precipitation (Iyamuremye et al., 1996c).

When pK_a values for different P species incorporated to soil are known as well as the soil pH, it is possible to interpret the P compound behaviour in terms

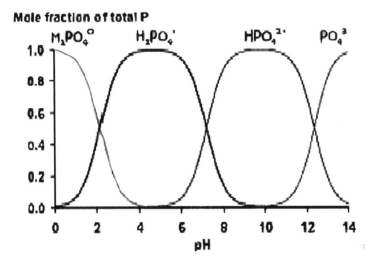

Figure 16.3. *Speciation of orthophosphate ions (expressed as mole fraction of total P) in solution as a function of pH (from Hinsinger, 2001).*

of adsorption reaction. This makes it possible to predict the interaction with the soil mineral and organic matter in order to obtain an overall estimate of P compound bioavailability. However, in soil systems in particular, pK_a values are only a chemical property for understanding the adsorption mechanism. Finally the many interactions in the complex matrix – where oxidation–reduction, hydrolysis, protonation, desprotonation, catalysis, substitution and anabolic reactions occur simultaneously with mineralization and immobilization – must be known in order to understand P bioavailability. The changes in the rizhosphere: root-induced acidification or alkalinization affects P bioavailability (Hinsinger, 2001).

16.3.3 Organic matter content

Organic matter in soil has a support function for nutrient exchange and improves soil fertility by modifying physical, chemical and biological conditions. The total acidity or exchange capacity of humified soil is attributed to the presence of dissociable protons in aromatic and aliphatic carboxylic, phenolic and hydroxyl groups (Tan, 1993). Fe-humate and Al-humate complexes can influence soil behaviour and reactivity (Tan, 1993; Mora and Canales, 1995a; Mora and Canales, 1995b) and generate large buffer capacity in a wide pH range, acquiring a fundamental role in regulating nutrient transport and availability in soil solution (García-Gil et al., 2004). Humic acids are characterized by lower total acidity (5–6 mEq g^{-1}) and lower carboxyl contents than fulvic acid which presents a total acidity of 10–12 mEq g^{-1} (Tan, 1993).

Organic matter adsorption on clays occurs by ligand exchange with the surface hydroxyl groups. These are therefore specifically adsorbed onto the surface, generating a competitive effect on P adsorption (Mora and Canales, 1995a; Mora and Canales, 1995b; Haynes and Mokolobate, 2001; Jara et al., 2005). Mora and Canales (1995a) investigated an organic matter role in P adsorption in the Chilean Andisols, and found that the adsorption of orthophosphate begins to increase when organic matter is extracted from soil, which generates a greater surface area of amorphous minerals and new active sites for adsorption within the soil inorganic matrix. However, Mora and Canales (1995b) studied humic acid capacity for reactivity modification of allophanic soil, and observed that an increase of surface acidity enhances orthophosphate adsorption capacity due to a proton-donation mechanism which favours adsorption orthophosphate reactions.

Other studies have focused on the stimulation of P adsorption, a stabilizing product of Al, Fe surfaces by organic matter. This provides new adsorption sites (De Brouwere et al., 2003) for P compounds. Several investigators have demonstrated a strong correlation between organic carbon and P (Borie et al., 1989; Escudey et al., 2001; Briceño et al., 2004); in this context, Borie et al. (1989) and Borie and Rubio (2003) determined that P in Chilean soil is strongly associated with organic macromolecules, and that this P organic compound (Po) represents a high proportion of total P (49–64%). They also determined that P associated with humic acid ranged from 51 to 68% of Po and P associated with fulvic acid varied from 32–49% of Po. Escudey et al. (2001) determined that Chilean acid soils with high organic carbon content have a higher P in organic form, the P related to humic substances corresponding to 71–93% of organic P in soil. This illustrates the role of organic matter in P retention through the formation of stable complex P-humus and confirms that the ultimate sink for P (Briceño et al., 2004) is not readily available to plants (Borie et al., 1989).

16.3.4 Soil organic acids and its influence on phosphorus bioavailability

Soil contains low molecular weight organic acids which have one or more carboxylic groups (Jones, 1998). Some of these acids are in anionic form such as citrate, oxalate, acetate, malate, isocitrate and tartrate. In soil their quantity varies between 10^{-3} and 10^{-5} M (Kpomblekou-A and Tabatabai, 2003). Organic acids are released by plant root exudates and microorganisms; they can also be products of the degradation of complex organic molecules (Jones, 1998; Geelhoed et al., 1999; Jones et al., 2003; Yadav and Tarafdar, 2003). Along with other low molecular weight compounds such as sugar and amino acid, organic acids represent one of the most labile sources of C in soil. These acids are transitory owing to degradation by microorganisms or complexation with cation but organic acids are generated continuously (Geelhoed et al., 1999; Jones et al., 2003; Xu et al., 2003).

The adsorption of organic acids produces an impact on soil surface charge, thereby allowing metal cation complexation in solution and anion displacement from the soil matrix (Jones, 1998). Xu et al. (2003) showed the negative charge in soil gradually increased with the rise in citrate concentration, whereas the positive charge decreased sharply at first and then changed gradually. This means that the first organic anions added were consumed to neutralize the positive surface charge, and subsequently adsorbed anions created negative surface charge. The extent of metal complexation depends on the nature of the organic acid, the metal characteristic and solution pH (Geelhoed et al., 1999). Jones et al. (2003) determined that monovalent organic acids are weakly adsorbed in the surface soil, whereas divalent and trivalent organic acids are adsorbed to a greater degree and hence form a highly stable complex. Acetic acid (pK_a value 4.75) has one carboxylic group and it alone can form a monodentate complex with Al or Fe from surface soil; whereas oxalate, malate and citrate acids (pK_a values between 1.23 and 3.16) have two or three carboxylic groups that can form multichelate complexes which affect their mobility in soil (Xu et al., 2003).

Organic acid has been related to nutrient mobilization, mainly in P insoluble utilization, and therefore enhances its P bioavailability (Bolan et al., 1994; Ström et al., 2002) with P adsorption decreasing due to competition for active sites for the adsorption and dissolution of light insoluble P compounds, for example, Ca, Fe and Al phosphates (Bolan et al., 1994; Geelhoed et al., 1999; Jones et al., 2003). However, released P depends on factors such as pH solution, organic acid characteristics and soil mineralogy (Bolan et al., 1994; Ström et al., 2002). Bolan et al. (1994), indicated organic acid increases P availability in soils mainly through decreased P adsorption; the carboxylic group number influences P availability in the following order: tricarboxilic > dicarboxilic > mono-carboxilic acid.

However, studies by Jara et al. (2005) have revealed that the adsorption of phosphate in binary and ternary systems of sulphate, citrate (or oxalate) on synthetic and natural allophanes soils at different pH demonstrate that phosphate adsorption was gently inhibited by organic acids (lower than 10 µM). Jones (1998) studied two mechanisms for released P that operate simultaneously: ligand exchange, whereby organic acid directly replaces P on ligand exchange surfaces (Al(OH)$_3$ and Fe(OH)$_3$), and metal ion complexation on the surface, which constitutes the exchange matrix holding the P (Ca^{2+} or Fe^{3+}). Ström et al. (2002) suggested that the P release mechanism with organic acid may be different depending on soil type: in calcareous soil, oxalate mainly releases P contained in Ca–P minerals forming precipitate of Ca-oxalate. By contrast, citrate has greater affinity for Fe^{3+} and Al^{3+} predominantly located in acid soils and from there it releases the P contained in Fe–P and Al–P minerals.

Investigators have proven organic acid release is enhanced by exudate of root plants subjected to P deficiencies such as white lupin, rape, bean and maize. However, an enhanced organic acid exudation in roots exposed to toxic levels of Al^{+3} has been proven, the above-mentioned suggesting a role in the system of plant detoxification or it could be an active mechanism in response to environmental stress (Jones, 1998; Jones et al., 2003). Shen et al. (2002) demonstrated that the higher P uptake in different bean genotypes may be largely explained by the higher exudation of organic acid (citrate was a major component followed by tartrate and acetate); additionally, acid root exudates of P-deficient plants of the four bean genotypes had a higher ability to release P from Al–P and Fe–P in comparison with P-adequate root exudates.

The contribution of organic acid to the enhanced bioavailability of P has been used to increase the P released from fertilizer, soils and P compounds with different degrees of reactivity (Ström et al., 2002; Shen et al., 2002; Kpomblekou-A and Tabatabai, 2003). Kpomblekou-A and Tabatabai (2003) proved the abilities of different organic acid (glycolic, pyruvic, oxalic, malonic, fumaric, salicylic, tartaric, cis-aconitic and citric) to release P from phosphoric rocks with a wide range of reactivity. They indicated that tri-carboxilic acids (cis-Aconitic and citric) and di-carboxilic acids (oxalic, malonic, fumaric and tartaric) are more effective than mono-carboxylic acids (glycolic, pyruvic and salicylic acid) in releasing P from the different phosphoric rocks.

16.3.5 Phosphatase enzymes in soil

Enzymes are biological catalysts of innumerable reactions and are of great agronomic and ecological value. They are released into soil by microorganisms and plant root exudates, participating in the forming and degrading of molecules and contributing to the nutrient cycling of carbon, nitrogenous, phosphorus and sulphur (Pascual et al., 2002). In the soil, the variability, quantity and specificity of enzymes are in relation to soil management and organic waste incorporation (Deng and Tabatabai, 1997).

The cell during normal growth and division produce intra- and extra-cellular enzymes, the former being necessary to exert catalytic action within the cell, a restricted and compartmentalized system (Rao et al., 2000) and the latter being responsible for the hydrolysis of high molecular weight or insoluble substrates, potential toxins destruction and nutrient solubilization (Pascual et al., 2002).

In the nutrient cycle P has different chemical forms, developing complex molecules and being united by different energy bonds. This characteristic in soil solution determines their P transport potential and availability for enzymes (Shand and Smith, 1997). Phosphatase is a general name used to describe a broad group of extra-cellular enzymes, which catalyze the hydrolysis of both phosphate ester bonds and anhydride bonds. These enzymes play a fundamental role in the phosphorus cycle, allowing orthophosphate to be released from organic and inorganic compounds thereby increasing the bioavailable P (Thien and Myers, 1992; Deng and Tabatabai, 1997; Rao et al., 2000).

For the development of enzymatic processes which contribute to increasing the pool of bioavailable orthophosphate ions, several kinds of phosphatases are needed that are selected for different phosphorus compounds. Understanding these processes requires information on the P chemical nature of organic forms (Turner et al., 2002a) and inorganic forms. Phosphatase enzymes studied in soil include phytases, acid and alkaline phosphatase, phospholipase and pyrophosphatase.

Phytases (*myo*-inositol-hexakisphosphate phosphohydrolase, Ec 3.1.3.8) have an established role in the phosphate release from *myo*-inositol-hexakisphosphate. Soil and water microorganisms are capable of hydrolyzing phytic acid through this enzyme (Pandey et al., 2001).

Acid phosphatase (Ec 3.1.3.2) and *alkaline phosphatase* (Ec 3.1.3.1) are orthophosphoric monoester phosphohydrolases (Deng and Tabatabai, 1997; Pant and Warman, 2000) that hydrolyse compound-type orthophosphate monoester.

Phospholipase (phosphatidylcholine cholinephosphohydrolase Ec 3.1.4.3) and **nuclease** (Ec 3.1.31.1) are orthophosphoric diester phosphohydrolases (Ec 3.1.4) (Deng and Tabatabai, 1997; Pant and Warman, 2000), which hydrolyse compound-type orthophosphate diester.

Inorganic pyrophosphatases (pyrophosphate phosphohydrolase Ec 3.6.1) act on acid anhydride bonds in phosphoryl-containing anhydrides (Deng and Tabatabai, 1997).

Pant and Warman (2000) found an increase of up to 49% in the hydrolysis of soluble unreactive phosphorus, when simultaneous incubation of immobilized phospholipase and acid phosphatase was performed, attributing this increase to two consecutive enzymatic reactions, the first of which was the hydrolysis of phospholipids into simple constituents and the second was the hydrolysis of the products of degradation by acid phosphatase.

Approaching phosphatase specificity for P compound hydrolysis, the substrate specificity technique has been implemented to identify different P compounds in different fractions from soil, water and manure (Turner et al., 2002a; He et al., 2004). In this context, Turner et al. (2002a) used this technique to hydrolyse molybdate unreactive phosphorus (MUP) to classify these P compounds into groups with potentially mobile organic P in soil–water extract. These researchers test substrate specificity of phosphatases. Using alkaline phosphomonoesterase they achieved P recovery from condensed compounds and labile orthophosphate monoester (adenosine 5′-triphosphate, glucose-6-phophate, p-nitrophenyl phosphate and tetra sodium pyrophosphate). However, the authors found that phytase releases orthophosphate from inositol hexakisphosphate and also hydrolysed condensed phosphates, labile orthophosphate monoester and orthophosphate diesters. In this context, Shand and Smith (1997) reported that acid phosphatase from wheat germ and phytase from wheat hydrolysed P from a range of substrates including monoesters, diester phosphate, pyrophosphate and polyphosphates, indicating the lack of specificity of these enzymes.

As with other extra-cellular enzymes, phosphatases in soil are affected by environmental conditions such as pH, temperature, presence or absence of inhibitors, strong fixation onto clay or humic colloids through adsorption processes entrapment or co-polymerization. Chemical or conformational changes of enzyme structure occurring upon immobilization (Rao et al., 2000), which could have a positive or negative influence on enzymatic activity. Moreover, soil can be an inhospitable environment for extra-cellular enzymes because they suffer biological and non-biological denaturalization and inactivation can occur (Pascual et al., 2002). Gianfreda and Bollag (1994) determined acid phosphatase catalytic activities (from wheat germ) immobilized on different supports: glass beads, montmorillonite, kaolinite and soil. The recovery of acid phosphatase activity (calculated as percentages of the specific activity of the free enzyme) was 72% on glass beads and 30, 35 and 29% respectively for other supports. Rao et al. (2000) studied immobilized acid phosphatase activity (from potato) on different supports: inorganic support (clays), organic matrix (tannic acid) and organo-mineral complexes showed catalytic features quite different from those of the free enzyme. Huang and Shindo (2000) investigated the effect of the copper on the activity and kinetic properties of acid phosphatase both free and immobilized onto different support. Their result indicated that enzymatic activity of free acid phosphatase decreased significantly from 51.6 to 27.2% with the Cu concentration increase (0.1–0.4 mM Cu). The enzymes immobilized onto soil clays and kaolin mineral showed similar degrees of reductions for their residual activities in presence of Cu. The enzymes immobilized on these surfaces retained 45–50% and 20–25% of the original activity in the presence of 0.1 and 0.4 mM Cu. Pant and Warman (2000), who indicated the optimum pH for achieving maximal enzymatic activity of nuclease, phospholipase, alkaline and acid phosphatase, pH values of 8.9, 7.3, 9.8 and 4.8, respectively. The optimum pH found for the same enzymes but immobilized was 6.0 for acid phosphatase and 7.0 for the other enzymes.

To explain the differences in enzymatic activity of immobilized enzymes, Quiquampoix (1987) proposed a model with two stages to describe the effect of surfaces on enzymes: firstly, adsorption due to attraction forces that act on enzymes and promote surface fixation, and secondly, modification in the conformation of the adsorbed enzyme when the forces of attraction are strong. Thus, strong attraction forces generate modification in enzyme conformation and therefore cause activity modifications. Weak forces allow enzymes to conserve their active structure.

However, investigators have demonstrated that root exudates from plants exposed to orthophosphate-deficient conditions present more phosphatase activity than roots from plants subject to high orthophosphate levels. Yadav and Tarafdar (2001) studied the conditions promoting the maximum release of acid phosphatase from plant root in the presence of different concentrations of Pi (KH_2PO_4) and different Po compound (phytin, lecithin and glycerphosphate). They found the maximum acid phosphatase secretion was observed under P-deficient conditions and phytine treatment, as well as determining that the lowest phosphatase secretion was by plants under Pi-sufficient. The acid phosphatase activity was increased depending on the degree of hydrolysable complex substrates in the following order: phytin > lecithin > glycerophosphate > phosphate (Fig. 16.4). Tarafdar and Claassen (2003) assessed the role of Po in soil solution for wheat plant nutrition in an oxisol that contained P bound to Fe and Al. They observed a reduction of Po concentration in soil solution up to 22, 75 and 90% after 9, 17 and 24 days, respectively. They hypothesized that after the hydrolysis of Po by the phosphatase enzyme; Pi is either taken up by wheat plants or adsorbed by soil components. They also found that if Po concentration decreases in the soil solution, over time the acid phosphatase activity also decreases, indicating that plants secrete acid phosphatase according to the amount of Po present in the system; otherwise, when the Pi concentration of the soil solution increases, the acid phosphatase activity of plant roots decreases.

Efficient microorganism has been identified for phytase, acid phosphatase and alkaline phosphatase production: *Aspergillus niger*, *Aspergillus awamori*, *Emericella nidulans*, *Emericella rugulosa*, *Penicillium simplicissimum* and *Penicillium rubrum* (Yadav and Tarafdar 2003). Several phosphatase enzymes are utilized in industrial processes. Recent research has shown that microbial sources are more promising for the production of phytases on a commercial level. Two strains of *Aspergillus sp.*, *A. niger* and *A. ficcum* have most commonly been employed for this end (Pandey et al., 2001). Vassilev et al. (2001) affirm that the manipulation of microbial populations releases P for plant nutrition and is a promising

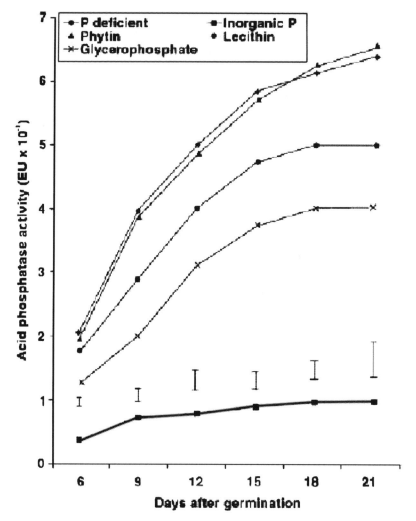

Figure 16.4. *Influence of source P supply, on acid phosphatase secretion by plants with time (from Yadav and Tarafdar, 2001).*

practice for agriculture. This enzyme has been utilized as a diet supplement for livestock to enable optimized orthophosphate release from seeds. Andlid et al. (2004) studied the yeast phytase ability of different *Saccharomyces cerevisiae* genotypes for degrading extra-cellular *myo*-inositol hexa*kis*phosphates and studied the Pi effect on the gene family repression that encoded secretory phosphatase. They showed an efficient repression of phytase activity for all *S. cerevisiae* tested in the presence of extra-cellular Pi.

16.3.6 Organic wastes incorporation and its influence on P bioavailability

The incorporation of organic residues can improve soil conditions (Mokolobate and Haynes, 2002; Pypers et al., 2005) making an increase in P availability possible. Borie et al. (2002) reported that crop residues are commonly applied to the soil as a source of nutrients for plant. This practice fits well with the current world-wide trends towards resource conservation in agriculture and particularly the growing interest of local farmers in alternative agricultural practices, such as organic farming and zero or reduced tillage.

The improvement in P availability following residue amendment depends on residue characteristics which reduce P sorption strength in soils (Haynes and Mokolobate, 2001; Pypers et al., 2005) and P bioavailability compounds incorporated into the soil. Iyamuremye et al. (1996b) indicated that P-rich residues resulted in a significant increase in the P pool available (resin and $NaHCO_3$), readily mineralizable Po ($NaHCO_3$), chemisorbed (NaOH) fractions in the soil.

The main mechanisms (which are simultaneous) involved in enhancing P availability are: orthophosphate ion incorporation, arise in pH accompanied by P solubilization, production and release of organic anions, increased enzymatic activity, incorporation of organic matter and complexation of the exchangeable ions as Al, Fe, Ca and Mn.

Organic matter in residues contains significant quantities of P compounds and during mineralization, orthophosphate is released into the soil solution. In addition, inorganic and organic products are generated during the partial decomposition of organic waste, and humic substances and organic acid can be adsorbed into soil surfaces decreasing the potential P adsorption by blocking sites for the formation of complexes with Al, Fe and Ca, (Iyamuremye et al., 1996b; Haynes and Mokolobate, 2001; Akhtar et al., 2002, Mkhabela and Warman, 2005). Kwabiah et al. (2003b) proved the capacity of plant residues to increase bioavailable P in soil, and demonstrated the direct influence of plant quality material input on labile P. They attributed variations to three factors: changes in the release of P from soluble P fractions depending on plant materials, changes in immobilization rates of P fractions and changes in the adsorption–desorption capacity of the soil amendment. Borie et al. (2002) added plant residues (wheat and lupine) and carried out the appropriate management of arbuscular mycorrhizal symbiosis in an acidic soil from southern Chile. They found that organic amendment increased soil pH, plant performance and mineral acquisition (wheat more than lupine), P availability and arbuscular mycorrhizal development (lupine more than wheat). Mokolobate and Haynes (2002) investigated the effects caused by the incorporation of organic waste in acid soil and they found higher proton consumption capacity, high $CaCO_3$

content, ash alkalinity and increased pH followed in the order, poultry manure > filter cake > household compost > grass residues. The residues decreased exchangeable Al in the same order.

Organic waste incorporation, specifically in compost and manure, increases pH values in mixed waste soil. This pH increase can be associated with specific organic molecules adsorption by ligand exchange with the release of OH^- ions during organic matter mineralization, proton consumption by functional groups associated with organic materials, proton consumption during decarboxylation of organic acid anions and the release of OH^- during reduction reactions (Mokolobate and Haynes, 2002). Iyamuremye et al. (1996a) indicated that manure and alfalfa residues caused the highest increase in pH values, averaging more than 1 unit in soil at 28 days, and also causing exchangeable Al reduction in soil. Pypers et al. (2005) found that organic amendment significantly increased soil pH; soil samples with initial pH of 4.97 and 4.39 increased 0.48–0.67 pH units on the first day of application, then on the third and seventh days after residue application up to a pH readings were 5.95 and 5.72, respectively. However, pH values then decreased at 36 days after residue application. Sharpley et al. (2004) investigated soil with varying manure application histories and found that pH was significantly greater in manured soils than in untreated soils, presenting differences in a range of 1.5–0.4 units, using poultry litter and swine slurry manure, respectively; moreover, organic carbon, exchangeable Ca and total P increased in soil amendment as well. Pypers et al. (2005) determined that Al extractable concentrations in soil were considerably reduced by organic amended (green residues) from 1.6 and 2.3 $cmol_c\,kg^{-1}$ to 0.63 and 0.72 $cmol_c\,kg^{-1}$ on the first day. Linear regression analysis further revealed a significant positive correlation between pH increase and reduction of exchangeable Al, which favour orthophosphate anions in soil solution.

It has been demonstrated that organic waste incorporated into soil has an influence on the composition and enhancement of microbial biomass and therefore produced changes in enzymatic activity (Speir et al., 2004). Oehl et al. (2001a) found that organically cultivated soil revealed higher microbial biomass than conventionally cultivated soil and non-fertilized soil, indicating that microbial P in organically cultivated soil (17.6–16.5 $mg\,kg^{-1}$) was higher than conventionally cultivated and non-fertilizer soil (9.0–8.0 $mg\,kg^{-1}$). García-Gil et al. (2000) determined that the microbial biomass enhanced long-term in soil with organic amendments was mainly due to the microbial biomass in the organic residues and the addition of carbon substrate, which stimulates indigenous soil microbiology. However, these investigators found phosphatase activity was inhibited in organically amended soil when incorporated into municipal solid residue compost, suggesting that phosphatase activity can be inhibited not only by heavy metals, but also by inorganic phosphate feedback, as this enzyme is activated by low P availability in soil. Dee et al. (2003) determined that phosphatase activity was greatly increased due to waste incorporation, suggesting that the rate of synthesis and release by microorganisms or their stability in soil is related to pH.

It is important to highlight that organic acids formed during organic waste decomposition by microbial degradation have a significant effect on the P adsorption–desorption dynamic (Iyamuremye et al., 1996c). This is likely the main mechanism involved in the reduction of P adsorption. These organic anions compete with orthophosphate for sites on soil surfaces and are also able to replace P bound to soil surfaces, producing an increase in P availability (Pypers et al., 2005). Iyamuremye et al. (1996c) reported malic, malonic, maleic, succinic, formic, citric and acetic acid in soil solution samples amended with organic residues and determined that citric acid affects the speciation of Al, Fe and P by forming complexes with these metals, preventing the formation of gibbsite, goethite and P precipitation.

16.3.7 Soil use and phosphorus bioavailability

Land use practices also determine differences in P dynamic, quantities and therefore bioavailability, depending mainly on the cultivation system and long-term fertilizer input. The variation of P fractions in different types of soil can be attributed to a combination of factors including parent material, vegetation cover, fertilizer use, climatic conditions and microbial activity (Rubaek and Sibbesen, 1993; Gale et al., 2000; Pinochet et al., 2001; Chen et al., 2003; De Brouwere et al., 2003). McDowell and Stewart (2006) indicated that in a closed system (native ecosystem with no agricultural development), the P cycling between inorganic and organic forms is determined by factors such as climate, topography, soil parent material, biomass and time. Besides, Taranto et al. (2000) pointed out fertilizer have not been applied the rate at which P is cycled through plant residues and soil organic matter. In agricultural soil, P input and output changes result in an altered P balance and P form. Pinochet et al. (2001) investigated the relation between P fractions and soil evolution in a native forest soil, demonstrating that total P increased (354–$1414\,\mathrm{mg\,kg^{-1}}$) according to the degree of evolution, probably due to factors such as time, temperature, moisture, soil parent material and, in addition to these, the re-incorporation of organic P compounds of different quality and quantity. In all soils studied, contents of Po (53–82%) were found to be higher than inorganic P, especially in more recent soil than in evolved soil. Olander and Vitousek (2005) indicated that mature tropical forests are generally productive without substantial P inputs, suggesting that the forest biota have a lower P demand than crops or else they compete more effectively for P. It is agreed, too, that differences exist between horizons in both P adsorption reactions and P mineralization by microbial activity. Since most available P tends to cycle through the organic horizon where the microbial immobilization is a better competitor for available P than the adsorption process, horizon characteristics may be important in controlling short-term P availability in forest soil.

The excessive and continuous application of P as mineral fertilizer or organic residues in amounts exceeding crop uptake in soil with a large P-fixing capacity (as inner-sphere complexation via a ligand exchange mechanism) will inevitably lead to P accumulation (Akhtar et al., 2002; Indiati and Rossi, 2002; Borie and Rubio, 2003; Staats et al., 2004). Borie and Rubio (2003) determined that cultivated soil presented higher amounts of both total and bioavailable P than uncultivated soil due to P accumulation produced by agronomic soil management, especially phosphate and manure fertilization. Haustein et al. (2000) considered that when soils reach their capacity for PO_4^{3-} adsorption, added orthophosphate ions are retained with low intensity (possibly via electrostatic interaction on variably charged mineral surfaces) and P addition is more easily desorbed and transported, which facilitates P availability or loss. McDowell and Stewart (2006) studied P behaviour in soil for different uses (native, pasture and forest) and determined that P distribution and forms were influenced both by soil type and land use. They found that the total P concentrations were less in forest soil than in pasture soils, which is presumably due to P input via fertilizer and liming in pasture soil. They showed that the inorganic to organic P ratio decreases in the order pasture > forest > native, while the organic C to P ratio decreases from native > forest > pasture. Among all the Po compounds, monoester orthophosphate was identified in the greatest quantities in all soil representing 31–60% of total P, followed by diester orthophosphate at 1–12% of total P. The authors also indicated a decline of the diester to monoester ratio. However, the ratio was generally greater in forest than pasture soils, which were attributed to the labile nature of diester, the mineralization of monoester in forest soil and the increase of monoester in pasture soil due to organic debris.

Changes in agricultural practices produce effects on soil biomass content affecting Po mineralization and P bioavailability. Oehl et al. (2002) pointed out that organic farming can lower or prevent available P accumulation in agricultural topsoils and thereby reduce diffuse P losses to water. Brookes et al. (1984) suggested that biomass P contributions to plant nutrition are much more important in grasslands than in arable soil. This indicates that in areas with less intensive agriculture, less inorganic P fertilizer can be used since P released from the biomass may make a significant contribution to the overall P.

Some varieties of plants are more efficient than other in the uptake of phosphorus and this is another factor that establishes P bioavailability. Some plant species such as white lupin (develops proteoid roots under P-deficient condition) utilize non-labile P effectively in either inorganic or organic forms, which releases citrate to mobilize calcium-bound phosphate; pigeonpea utilizes iron and aluminium-bound phosphate by releasing piscidic acid (Shibata and Yano, 2003). Taranto et al. (2000) compare the effect of roots in two contrasting ecosystems (*Banksia* woodland versus pasture) on transformations and mobility of various P compounds added to soil: RNA, Fe-phytate and phosphoric rock. They determined that most P in RNA added to *Banksia* soil was mineralized and leached as orthophosphate ions within the first two months, attributing this behaviour to root release of organic acids and phosphatase activity and possibly to greater microbial activity in the soil. In pasture soils, RNA-P was mineralized at a slower rate, root growth was rapid and less P was leached, whereas additions of Fe-phytate and rock phosphate released little orthophosphate.

16.4 Fractionation of soil phosphorus

Chemical extraction is based on the use of extract solutions such as water, diluted salt solutions or diluted weak and strong acid solutions, and has been traditionally used to identify P pools in soil and later in wastes like manure, sediments, compost and sludge. This fractionation method is useful for obtaining information about potential availability, mobility by soil erosion or runoff, leaching of P through the soil column and also for quantifying changes in Po, but it cannot identify specific classes of P compounds (Leinweber et al., 1997; Zhang et al., 1999; Gale et al., 2000; Scherer and Sharma, 2002; Akhtar et al., 2002; Frossard et al., 2002; Indiati and Rossi, 2002; Griffin et al., 2003; Chen et al., 2003; McDowell and Sharpley, 2003; McDowell and Stewart, 2006).

In soil, sequential chemical extractions use increasingly strong solutions to recover progressively more recalcitrant P (Linquist et al., 1997) and their interpretation is based on the individual action of extracting, sequencing and relating chemical and biological soil properties to plant availability (Pinochet et al., 2001), and differentiating separate P fractions or pools. The information about these fractions provides knowledge of P sinks and sources in soils to design efficient management operations (Verma et al., 2005). Among P fractionation methodologies, P sink tests including anion exchange resin or iron oxide impregnate are used to determine labile P. These methods do not destroy, or only damage slightly, chemical conditions in the soil. Basically, resins capture orthophosphate present in soil solutions, generating a concentration gradient to release this anion from the soil, and simulating the way plant roots take up Pi from the soil solution (Rubaek and Sibbesen, 1993).

The fractionation traditional method was proposed by Hedley et al. (1982) and current investigations are based on this method.

Soluble P or readily available: the solution utilized for extraction has been water (He et al., 2004; Ajiboye et al., 2004), NH$_4$Cl (Akhtar et al., 2002) anion exchange resin or Fe oxide impregnated strip (Iyamuremye et al., 1996b; Linquist et al., 1997; Gale et al., 2000; Pinochet et al., 2001; Borie and Rubio, 2003; Nwoke et al., 2003) and NaHCO$_3$ (Iyamuremye et al., 1996b; Leinweber et al.,1997; Linquist et al., 1997; Gale et al., 2000; Taranto et al., 2000; Baeza, 2001; Pinochet et al., 2001; Frossard et al., 2002; Borie and Rubio, 2003; Nwoke et al., 2003; Ajiboye et al., 2004; He et al., 2004; McDowell and Stewart, 2006).

Labile P in the solid soil phase: Pi and Po are easily extracted from the mineral surface or, in other words, slightly adsorbed on Fe and Al minerals, which are supposedly labile. The extract solution employed is NaOH (Iyamuremye et al., 1996b; Leinweber et al., 1997; Linquist et al., 1997; Gale et al., 2000; Taranto et al., 2000; Baeza 2001; Escudey et al., 2001; Frossard et al., 2002; Borie and Rubio, 2003; Nwoke et al., 2003; He et al., 2004; Ajiboye et al., 2004).

Non-labile P: resistant Pi, bond Ca forming apatite or octacalcium phosphate. HCl is used as an extracting solution (Iyamuremye et al., 1996b; Linquist et al., 1997; Gale et al., 2000; Pinochet et al., 2001; Akhtar et al., 2002; Frossard et al., 2002; McDowell and Stewart, 2006).

Residual fractions or recalcitrant pool: this is slowly exchangeable or non-exchangeable from the mineral surface. Extraction is subjected to strong digestion with H$_2$SO$_4$/H$_2$O$_2$ (Iyamuremye et al., 1996b; Leinweber et al., 1997; Linquist et al., 1997; Pinochet et al., 2001; Borie and Rubio, 2003).

The conventional methodologies for P fractionation in soil have been used to study P availability in different residues. Table 16.3 summarizes some studies of fractionation in residues. However, He et al. (2003) indicate that conventional methodology for P fractionation in soil is not appropriate for characterizing P in waste like manure, as the physiochemical characteristics of animal manure may differ from soil, and generate different P distribution patterns in soil and manure due to the origin of the mineral and organic matrices, respectively. The authors discovered that a major portion of P in manure is soluble in H$_2$O and NaHCO$_3$, but in soil the major P portion was extracted with NaOH and HCl, adding in the differences in the chemical compositions of soil and manure fractions.

Detailed information, identification and quantification of different forms of P compounds in soil and manures can be obtained by alkaline extraction and determination by ^{31}P NMR spectroscopy (Pant et al., 1999; Zhang et al., 1999; Taranto et al., 2000; Baeza 2001; Cade-Menun et al., 2002; Turner et al., 2003a; Turner et al., 2003b; Makarov et al., 2002; Briceño et al., 2004; Hansen et al., 2004; Makarov et al., 2005; McDowell and Stewart, 2005; McDowell and Stewart, 2006). With this methodology, several compound classes can be identified: e.g., orthophosphate monoesters and diesters, pyrophosphates and phosphonates, in particular nucleic acid, phospholipids and polyphosphates in

Table 16.3. *Fractionation of organic residues mainly manures.*

Residue	Fractionation	References
Dairy manure	Lipid phosphorus (ether–alcohol); total acid soluble P (trichloroacetic acid); residual (H_2SO_4–H_2SeO_3–H_2O_2)	Barnett (1994a, 1994b)
Sewage sludge	NH_4Cl, NaOH, HCl and residual-P	Rydin (1996); Scherer and Sharma (2002)
Liquid pig manure, dry poultry manure	Resin; $NaHCO_3$; NaOH; H_2SO_4; residual	Leinweber et al. (1997)
Swine manure	Deionizer water; $NaHCO_3$; NaOH and HCl	He et al. (2003)
Biosolid, hog, dairy and beef manure	Deionizer water; $NaHCO_3$; NaOH and HCl	Ajiboye et al. (2004)
Compost	Deionizer water; $NaHCO_3$; NaOH; HCl and residual	Frossard et al. (2002); Scherer and Sharma (2002)
Liquid lagoon and solid manure	Exchangeable P ($Mg(NO_3)_2$); P associated with calcium ($NaCH_3COO$); P associated with organic matter (NaClO); Fe oxides (citrate, dithionate and bicarbonate) and residual (HNO_3/H_2SO_4 and HF)	Hansen et al. (2004)
Dairy manure	H_2O or resin; $NaHCO_3$; NaOH and HCl or H_2SO_4	He et al. (2004)
Fresh and dry dung dairy, deer, sheep	H_2O; $NaHCO_3$; NaOH, HCl and residual ($K_2S_2O_8$ and H_2SO_4)	McDowell and Stewart (2005)
Sludge from anaerobic digestion	H_2O, $NaHCO_3$+$Na_2S_2O_4$, NaOH and HCl	Ding et al. (2005)

complex matrices. Toor et al. (2003) pointed out that unreactive P may contain organic and some condensed forms of P, these species were a predominant P fraction and comprised 85–88% of the TP concentration in leachate from grassland soil and through NMR spectroscopy revealed that the dominant type was monoester orthophosphate (67.4%) followed by diester orthophosphate (20.2%).

Nevertheless, NMR presents some inconveniences: P extraction with NaOH-EDTA presents a degree of inefficiency probably due to the large acid P pool which is insoluble in NaOH solutions, and which varies with type of soil and its management (McDowell and Stewart, 2006). Additionally, orthophosphate diester can be susceptible to degradation in NaOH and NaOH-EDTA extract, which would underestimate diester orthophosphate concentration and over-estimate monoester phosphate (Pant et al., 1999; Leinweber et al., 1997; Turner et al., 2003a). Greaves et al. (1999) indicated that chemical identification of organic form is difficult due to the highly reactive nature of the PO_4^{3}, with its propensity for hydrolysis from the organic moiety during extraction.

Currently, several investigators (Traoré et al., 1999; Oehl et al., 2001b; Frossard et al., 2002; Sinaj et al., 2002; Bühler et al., 2003; Oehl et al., 2004) have introduced the technique of isotopic exchange to determine P availability in soil, compost and manure. This technique enables measurement of the amount of orthophosphate that can be transferred from the solid soil to the solution in a given time and therefore provides information on availability.

To quantify P in solution, several investigators (Chapman et al., 1997; Shand and Smith, 1997; Turner et al., 2002a) classify P compounds according to their reactions with molybdate, such as molybdate reactive phosphorus (MRP) and MUP. In addition, these classifications are so complex that dissolved organic phosphorus (DOP) and dissolved condensed phosphorus (DCP) have been incorporated as subclassifications of total P in soil solutions, manures and sludge. P quantified (MRP) is only approximated to inorganic orthophosphate but this could include both free inorganic orthophosphate (PO_4^{3-}) and inorganic orthophosphate bound to organic molecules through cation bridges which may be released under the acidic conditions required for the formation of the molybdenum blue complex (Chapman et al., 1997; Greaves et al., 1999; Turner et al., 2002a). Dao (2003) observed that molybdenum-reactive phosphate determination by colorimetry appeared to overestimate dissolved P in the presence of *myo*-inositol hexakisphosphate, explaining for the higher molybdenum-reactive P estimates that molybdate apparently interacted with the phosphate ester groups of inositol to simulate the phosphomolybdate-antimony complex and the reduction of the complex by ascorbic acid yielded the blue colour that can be quantified as PO_4–P. MUP includes both orthophosphate ions not hydrolysable from organic and condensed P compounds (Shand and Smith, 1997; Turner et al., 2002a).

16.5 Phosphorus in organic residues

With the aim of reutilizing nutrients and diminishing organic residue disposal in landfills, it has been necessary to recycle residues, and agricultural land application is an alternative (Unsal and Sozudogru, 2001; Seyhan and Erdincler, 2003; Balmér, 2004; Mkhabela and Warman, 2005). The advantage of using organic waste in agriculture is that it contains organic matter, macronutrients and microelements which may be recovered and reincorporated into natural cycles (Seyhan and Erdincler, 2003; Balmér, 2004; Wang et al., 2004).

Organic matter in manure generates changes in chemical, physical and biological soil amendment properties (Wienhold, 2005) because it facilitates and increases cationic exchange, increasing pH, improving the soil structure and increasing water-holding capacity and modifier microbial activity. However, it

Table 16.4. *Values of total P in organic residues.*

Organic residue	Total P (g Kg^{-1})	References
Beef manure	2.94	Iyamuremye et al. (1996a)
	4.02	Griffin et al. (2003)
Dairy manure	1.6	García-Gil et al. (2000)
	4.9–8.9	Ebeling et al. (2002)
	4.35	Griffin et al. (2003)
	4.1–18.3	He et al. (2004)
	3.5–9.8	Sharpley et al. (2004)
	11.00 (solid)	Hansen et al. (2004)
	0.05–0.12 (lagoon)	Hansen et al. (2004)
Poultry manure	26.21	Leinweber et al. (1997)
	21	Mokolobate and Haynes (2002)
	23.60	Griffin et al. (2003)
	25.2–27.8	Sharpley et al. (2004)
Swine manure	16.22 (liquid)	Leinweber et al. (1997)
	24.69	Griffin et al. (2003)
	29.7	Sharpley et al. (2004)
Municipal solid waste	5.0 (compost)	García-Gil et al. (2000)
	2.9–5.6	Mkhabela and Warman (2005)
CSUW	2.25–7.19	Frossard et al. (2002)
CSUW and WW	2.09–3.43	Frossard et al. (2002)
Anaerobic sewage sludge	38.3	Akhtar et al. (2002)

Notes: CSUW; compost solid urban waste, WW; woody waste.

may also influence the fixation of nutrients and consequently decrease bioavailability. Furthermore, organic matter behaves as an accumulation phase for heavy metals (Speir et al., 2004). The over-application of nutrients – principally N and P – could pose a risk to the environment. In this area, P species and quantities in organic waste vary and therefore behave differently when applied to soil. Table 16.4 indicates total P values in organic residues.

It has been documented that organic waste incorporation causes changes in Pi and Po soil fraction distribution patterns. Baeza (2001) has investigated the effects of applying fresh cow dung, over P pools in acid Chilean Andisol. P pools in the soil change following the breakdown of cow dung. Labile-P fraction (P–NaHCO$_3$) has a short-term increment after dung application in 0–10 cm depth soil. P–NaHCO$_3$ fraction increased (from 69 to 72 mg kg^{-1}) in 60 days. Then this fraction decreased (10 mg kg^{-1}) due to microbial activity and pH change. From 120 to 180 days the P–NaHCO$_3$ recovered its original level. Scherer and Sharma (2002) indicated that bioavailable P (H$_2$O–P) increased in all treatments when compost, manure and sewage sludge were applied to soil (Table 16.5). However, P bioavailability over time has not been studied sufficiently, as available P varied depending on mineralization immobilization rates and fixation reactions in soil.

Table 16.5. *Influence of long-term application on organic waste with different P fractions in soil (mg kg^{-1} soil).*

P fractions	Soil	Farmyard manure	Compost	Sewage sludge
H$_2$O–P	20.1	54.3	32.2	28.1
NaHCO$_3$–Pi	44.9	105.2	95.6	98.0
NaHCO$_3$–Po	4.4	8.0	6.1	9.1
NaOH–Pi	70.1	152.9	159.5	182.7
NaOH–Po	25.9	32.1	20.2	71.8
HCl–P	1.4	7.4	10.2	7.4
Residual–P	55.3	48.6	48.6	26.8
Total P	222.1	408.5	372.4	423.9

Source: Adapted from Scherer and Sharma (2002).
Notes: The data correspond to different application rates: farmyard manure (10 t ha^{-1}), compost (24.5 t ha^{-1}) and sewage sludge (3.72 t ha^{-1}).

16.5.1 Phosphorus in animal residues

A common and traditional practice has utilized manure as organic fertilizer or soil improver, due to its higher contribution of organic matter and nutrients, mainly nitrogen and phosphorus (Rebollar and Mateos, 1999; Malley et al., 2002; Ajiboye et al., 2004; He et al., 2004; Speir et al., 2004; Wienhold, 2005).

Recently, intensified livestock production has generated an increased production of organic residues, and environmental regulations have been adopted for the application of manures in agriculture (Pandey et al., 2001), as they are a diffuse source of pollution, with the P in particular being directly associated with eutrophication in bodies of water (Haygarth and Jarvis, 1997; Gale et al., 2000; Withers et al., 2001; Ebeling et al., 2002; Hodgkinson et al., 2002; Ajiboye et al., 2004; He et al., 2004). Manure is applied to soils according to the nitrogen needs of the plants. The N/P ratio in manure is lower, which indicates that it contains more P than is needed by most crops. As a result, there is a surplus of P that goes unused (Atia and Mallarino, 2002; Nelson et al., 2003; Wienhold, 2005). Gale et al. (2000) determined that in soils with the highest rates of manure application total P surplus to crop requirements: 210 and 280 kg P ha^{-1}, total P increased in a range of 92–65%, respectively. This increase was registered in all soil fractions. The authors suggested that this increase in labile P creates potential for adverse impacts on water quality. In 1997, Haygarth and Jarvis demonstrated a significant relationship between runoff discharge and total phosphorus loss, indicating that mechanisms such as hydrolysis and Po mineralization and the erosion of particulate P may be important for releasing P from soils.

For adequate P handling of animal residues, it is necessary to know the P compound, dynamics and transformation in waste and soil amendments. DeRouchey et al. (2002) proposed ensuring appropriate management of livestock waste by providing nutrient profiles of various manure types to help livestock operators accurately apply manure to their land. This practice allows crops or forages to utilize nutrients from the manure, so decreasing the need for chemical fertilizer. Currently in development are management practices designed to optimize manure recycling in relation to soil type, presence of drainage systems and timing of manure application in order to avoid potential P loss (Hodgkinson et al., 2002; He et al., 2004). Hodgkinson et al. (2002) demonstrated that pig slurry application before winter drainage in arable alkaline soil resulted in high concentrations of dissolved P (up to $10 \, mg \, L^{-1}$) and total P (up to $75 \, mg \, L^{-1}$) in drain flow.

P in manures can vary widely depending on animal physiology, species and age, composition of diets, duration of manure storage, moisture content and type of bedding material (Atia and Mallarino, 2002; McDowell and Stewart, 2005). Hansen et al. (2004) affirmed that the forms of organic P in manure applied to soil will depend on the nature of the animal manure and type of storage. The authors also found important differences in total P contend among solid and liquid dairy manure of $11 \, g \, kg^{-1}$ and $0.05–0.12 \, g \, kg^{-1}$, respectively. Griffin et al. (2003) determined the chemical composition of beef, dairy, poultry and swine manure, finding total P concentrations equal to 4.05; 4.35; 23.60 and $24.69 \, g \, kg^{-1}$ dry matter (DM). Atia and Mallarino (2002) pointed out that the aforementioned variations, in addition to factors such as reapplication treatment, climate and soil characteristics, can influence P availability. He et al. (2004) determined from 13 dairy manures that total P varied between 4100 and $18,300 \, mg \, kg^{-1}$, phytate was present as the major hydrolysable Po compound, and DNA was low in all P fractions analysed.

However, the main differences in P quantity and species in manures are the product of P content variation in a diet (Barrow, 1975), where P is added as an orthophosphate supplement or a vegetable ingredient that contains P in ATP, nucleic acid, phospholipids, phosphoprotein and phosphoglucides. Among the phosphoglucides, phytic acid was the most important with 60–80% of total P in grain (Rebollar and Mateos, 1999). Monogastric animals, such as pigs and poultry, possess small quantities of phytase to efficiently digest phytate in their feed. Hence it is considered that P from phytate should not be used for these animals, as they excrete large amounts of P in the form of phytic acid (Pandey et al., 2001; Malley et al., 2002). Phytase research has recently intensified with the aim of incorporating it in the livestock diet in order to take advantage of the phytic acid. This would mean a Pi supplementation in feed and consequently a decrease in fecal excretion. Phytase supplementation in animal feed can reduce the amount of P in manure up to approximately 30% (Pandey et al., 2001). Turner et al. (2002b) discussed the agronomic advantages of phytase supplements in animal diets and the environmental benefits. They agreed that phytase supplements can reduce the concentration of total P in manures, but the corresponding changes in manure-P composition may have implications for P transfer to bodies of water; by transforming inositol phosphate (which is adsorbed in the soil) into Pi, it runs off easily into bodies of water. Ebeling et al. (2002) indicated that if supplementation could be reduced to the minimum concentration needed for optimum production, the amount of P in manure and the applications to farmland would also decrease. The authors conclude in their investigation that phosphorus concentration in dairy diets influences the forms and amount of P in manures. Besides, high P diet manure contributes more to runoff than low P diet manure.

P recovery from liquid manure is an attractive technology when the application to land is risky. Greaves et al. (1999) revised the potential for P recovery from animal manures as a method of increasing the sustainability of the global P cycle, indicating that P recovery is likely to require a process for extraction and later conversion to a useful product. Burns et al. (2003) highlight phosphorus precipitation in manure prior to land application as offering the potential to recover excess phosphorus from animal manures and move it to cropping areas that require phosphorus fertilizer input. Szögi et al. (2006) have developed a technology for P recovery from swine manure. These technologies add calcium hydroxide to precipitate P, obtaining sludge with a high moisture content which is rich in calcium phosphate as a final byproduct. It has the potential to be reused as fertilizer, but it would first be necessary to dehydrate it in order to transport it economically.

Nevertheless, for managing technologies that allow optimum recycling of P it is necessary to know the P fractions present in different manure types and their behaviour in soil. Investigators have recently analysed P speciation in these residues and factor that condition P availability. Leinweber et al. (1997) performed P fractionation in manure (liquid pig and chicken) and determined that of total P, 11 and 29% were resin extract, 13 and 10% were $NaHCO_3$ extract, 10 and 3% were NaOH extract, 27 and 17% were H_2SO_4 and 39 and 41% were residual for liquid pig and chicken manure, respectively. McDowell and Stewart (2005) examined P fractions in the dung of pastoral grazing animals (dairy cattle, deer and sheep). Date for fractionation indicated that P in fresh samples was mainly available: P in water (15–36% Pt) and P in bicarbonate (36–45% Pt) and less P was distributed in the more recalcitrant fractions: NaOH (17–19% Pt), HCl (4–11% Pt) and residual (8–10% Pt). Griffin et al. (2003) analysed changes in soil produced by different manure applications (beef, dairy, poultry and swine) and pointed out that all extractable soil P pools changed over time and exhibited significant differences according to the P source, time and interaction were common. Hansen et al. (2004) investigated P forms in manure stored in solid or lagoon, by means of NaOH-EDTA extract and NMR. Their results showed P compounds in solid and liquid manure were similar, indicating that about 30% of the total P is in organic form. The primary forms extracted from total P were orthophosphate (63.3 and 58.4%), pyrophosphate (3.5 and 7.1%), monoester specifically phytic acid (15.6 and 10.8%), another monoester (14.4 and 20.1%) and diester as phospholipids (1.8%) and DNA (0.9 and 1.8%) and phosphonates (0.5%) for solid and lagoon manures. Laboski and Lamb (2003) determined the relative effectiveness of manure and fertilizer for increasing P availability in soil. However, P in manure was always significantly more available than P in fertilizer for the majority of soils, after 1 and 9 months of incubation. This behaviour could be explained because the Po fractions in manure are composed mainly of high molecular weight compounds, probably DNA, polyphosphates and inositol phosphate, which can be adsorbed onto the soil surface and contribute to the release of Pi bound to the surface. Moreover, organic acids produced during microbial degradation compete with P manure for sorption sites on soil, causing P manure to be more available. Scherer and Sharma (2002) indicated a positive effect of farmyard manure on P availability is probably due to the organic anions produced during organic matter decomposition, which can be chelating agents for Fe and Al ions in manure, blocking potential sites for P adsorption. Sharpley et al. (2004) reported that P concentrations in inorganic fractions, organic fractions and residual P were significantly greater in soils receiving manures compared with untreated soil. However, their results revealed a relative increase in inorganic P forms compared with organic P. The overall change in inorganic P comprised 26–57% of total P in untreated soil and increasing from 49 to 80% of total P in treated soils.

16.5.2 Phosphorus in compost

Compost is a simple and economic biotechnology for organic waste treatment, with its high content of organic matter and high nutrient concentration which can be taken up by a crop (Traoré et al., 1999; Garrido et al., 2002; Speir et al., 2004). The most important factor affecting the successful use of compost for agricultural practices is the degree of maturity and stability (both terms are defined as degree of decomposition of organic matter). However, stability is related to the activity level of microbial biomass and compost maturity refers to the rate of decomposition of phytotoxic organic substances (Benito et al., 2003).

Composting is an aerobic process, where biological exothermic oxidation of organic matter is converted into a final stable humified product. Formed by a dynamic and rapid succession of microbial populations, this process can be developed in static piles, rows or reactors. Composting is facilitated under the following conditions: pH between 5 and 11, C/N between 30 and 40, humidity between 40 and 65% (Garrido et al., 2002). Frossard et al. (2002) indicated that understanding compost properties is essential for optimizing its recycling. García-Gil et al. (2004) indicated that the effect of compost on the elemental and functional composition of humic acid in soil is dependent on amendment rate and time application management, which determine beneficial changes in the acid base properties of amended soil and the enhancement of its buffering capacity. This experience highlights the importance of organic matter for maintaining pH, which is directly related to P availability.

The P content in compost is quantified as total P but, within this total, neither the quantities nor the P species that are available is entirely clear. Compost application in intensive cropping system soils often exceeds P uptake owing to an accumulation of P species, which are unavailable or not readily available. It has been determined that high levels of total P in soil amended with compost increased fixed-form P with Fe, Al and Ca (Park et al., 2004).

Although compost has high levels of Pi as orthophosphate, this information is not time-specific, and there is little information with respect to the P dynamic during the composting process. Frossard et al. (2002) determined P fractionation in compost in order to estimate P forms and bioavailability from various soil/compost mixtures. Studies conducted by this investigative group have indicated that only 30–50% of the total Pi in the compost is rapidly plant available, suggesting that a large fraction of Pi compost is present as condensed calcium phosphate such as apatites or octacalcium phosphate, which are not soluble in water, bicarbonate or NaOH. Among Po species present in compost inositol, phosphate and DNA have been identified. Eghball (2003), by fractionating P in composts and uncomposted manure, found that more than 75% of P in both types of manure was Pi. Po was 16% of total P in composted manure and 25% in non-composted manure, which indicates that Po is mineralized during the composting process.

According to the evolution of the composting process, immature compost contains substantial amounts of easily degradable organic compounds like organic acids (Smith and Hughes, 2004) and has a higher microbial biomass and therefore enzymatic activity, which decreases upon the compost reaching maturity and stabilizing. Thus, the maturity or age of the compost can affect P availability. Adler and Sikora (2003) suggested that water-extractable P may increase when soils are amended with biologically active immature compost, due to the generation of organic acids or humic substances which can compete with adsorption sites, and so generate higher available P in amended soil. Parameters such as pH, cation and anion exchange capacity, alkalinity, humification index, enzymatic phosphatase activity and organic acid content can be studied to characterize their P availability. Mkhabela and Warman (2005) studied the influence of compost from municipal solid waste on the P soil dynamic, suggesting that compost may be effective as a mineral fertilizer in increasing and supplying available P for soil, and they found that mixed soil-compost increases pH in the range of 0.2–0.6 units in soil with an initial pH of 5.8. The authors also indicated that P fixation was slightly decreased (78% fixations for control to 70% for compost) with the addition of compost. They explained that organic materials may enhance soil P availability because of the organic acid present in wastes as well as the increasing pH due to C mineralization and hydroxyl ion production; however, the incorporation of basic cations such as Ca and Mg may decrease P availability. Furthermore, higher microbial activity after compost application increases phosphate activity and thus releases P during the decomposition of organic matter. Pascual et al. (2002) proposed that changes in soil phosphatase activity are dependent upon the degree of stabilization of residues. When fresh residues are applied to soil, the phosphatase activity initially increases compared with mature compost; over time (360 days), however, this activity decreases. Enzymatic activity also depends on the quantity of microbial biomass contained in residues. The author further discovered that soil amended with mature compost did not show any such decrease in immobilized phosphatase activity, relating this to the importance of composting to protect enzymes against degradation. Park et al. (2004), however, demonstrated that compost and a mixture of compost and fertilizer increase total organic and available P. The authors also indicated that when P was applied from compost or fertilizer, it can inhibit phosphatase activity and subsequently suppress the mineralization processes, favouring Po accumulation.

Nevertheless, the addition of organic matter to soil can provide new sites for capture of P, depending on soil solution pH. Frossard et al. (2002) indicated that the amount of organic matter in compost controls the concentration of Pi in solution. This could be because organic waste is the original source of P in the compost, has an inhibitory effect on the precipitation of Pi in insoluble forms and favours the mineralization of P. Speir et al. (2004) found an increase in total C and N, cation exchange capacity, total and labile P and exchangeable cations in soil amended with compost from sludge at one and two years after application, indicating potentially enhanced nutrient storage. This retention of cations and anions can be attributed to sites for bonds generated by organic matter.

16.6 Phosphorus compounds recovered from sludges

The application of sludge and biosolids to land leads to problems associated with contamination of waterbodies from excess nutrients and toxicity caused by heavy metals. Bioconversion process technology appears promising in resolving sludge disposal problems (Wang et al., 2004). Maguire et al. (2000) mentioned that in practice, biosolids are usually applied according to crop N requirements and do not exceed load rates for trace elements, generating a P accumulation in soils as total P and bioavailable P.

Sludge application to land has been largely researched and evaluated, and it is necessary to consider methodologies for treatment, management, disposal and reuse which do not create new environmental problems (Woods et al., 1999; Kroiss, 2004).

To prevent nutrient overload – especially P – when domestic biosolids and agroindustrial wastes are applied to land or treated wastewater are disposal into water bodies, alternatives for nutrient reduction have been tested (Nelson et al., 2003; De-Bashan and Bashan, 2004). Greaves et al. (1999) indicated that, in order to enter into a biogeochemical cycle, P compounds obtained from residues must be in an available and non-toxic compound. The most common approach for dealing with phosphate precipitates in wastewater is their use in the industrial process for making fertilizer (De-Bashan and Bashan, 2004) but in most cases these precipitates are applied to a mixture of sludge and soil. However, depending on the type of precipitate deposited, P will be more or less bioavailable.

Thus, it is necessary to quantify the fertilizing value of biosolids as part of P management. This is crucial to identifying the availability of P in biosolids when applied to land (Seyhan and Erdincler, 2003). Scherer and Sharma (2002) determined that total P was higher in sewage sludge than compost and manure, but its percentage contribution of bioavailable P was low compared to other residues, which can be attributed to the low Fe/P ratio in sludge. It is assumed that P may be adsorbed on colloid surfaces with Fe^{3+}, serving as a bridge resulting in a lower P delivery.

In general, P removal from wastewater treatment plants is obtained by using biological removal or physiochemical precipitation to achieve crystallization, using metal cations for the formation of a range of P minerals (Greaves et al., 1999; Battistoni et al., 2002; Levlin et al., 2002; Yoshino et al., 2003). In the two last cases, the removal of P is mediated by converting the ionic form present in wastewater into a solid fraction which can be an insoluble salt precipitate or microbial biomass in sludge (Adnan et al., 2003; Nelson et al., 2003; De-Bashan and Bashan, 2004). Akhtar et al. (2002) determined that total P content in anaerobically digested sewage sludge was approximately $38.3\,g\,kg^{-1}$ and within this total, the bioavailable (that part extractable with iron oxide impregnated filter paper) was $1260\,mg\,kg^{-1}$. However, sludge application to soil increased soluble P more than 10-fold for all studied soils at 20, 60 and 120 days.

However, the P forms present in materials have differing degrees of availability depending on the separation method used because in the residue matrix different and simultaneous processes are generated that involve precipitated forms, adsorption–desorption, microbiological degradation and complexation reaction with humic substances and organic acid.

16.6.1 Chemical precipitation

P precipitation is important for a number of natural, biological and engineering processes, and phosphorus recovery from wastewater through precipitation has received a great deal of attention. In wastewater treatment, precipitating agents for orthophosphate anions are typically $FeCl_3$, $FeSO_4$ $Al(OH)_3$, $MgCl_2$, $CaCO_3$ and $CaCl_2$, which are added at various points in the conventional treatment process to convert soluble phosphate to a particulate form. Phosphorus removal appears the more sustainable process, if it is achieved by combining extraction with the recovery of recyclable quality precipitates (De-Bashan and Bashan, 2004).

The aim of chemical precipitation is to achieve P crystallization and trap soluble P in solid fraction sludge and prevent its discharge into the aqueous effluent, so decreasing enrichment of this nutrient in bodies of water and hence avoiding the risk of eutrophication in lakes and rivers.

Phosphorus removed by chemical precipitation presents technical and economic disadvantages, but the production of recyclable material in the form of phosphate pellets is desirable as raw material for industry or as a slow-release fertilizer (Battistoni et al., 2002; De-Bashan and Bashan, 2004).

During P chemical precipitation from organic effluents, diverse metal cations (mainly Ca^{2+}, Mg^{2+}, Fe^{3+} and Al^{3+}), inorganic ligands or organic acids cause interference with nucleation sites and render the proper crystallization of P minerals difficult (van der Houwen and Valsami-Jones, 2001; Udert et al., 2003). During compound removal, there can also be P adsorption and P precipitation in the solid phase which yields P precipitates with different chemical properties and therefore different behaviour when applied to soil. This indicates that the transfer coefficient between water and sludge differs widely and depends on physical and chemical balances (Kroiss, 2004), which will allow the several chemical P compounds formed during precipitation to be deposited in sludge.

Thermodynamically, the phenomenon of precipitation is produced in supersaturated conditions, where the mineral ion activity product (IAP) exceeds the solubility product (Kps) (Udert et al., 2003). However, in complex systems such as wastewater, complex ions are formed in solution, which have a metal cation centre bound to one or more molecules or ions. Measurement of the tendency of metallic ion to form a complex ion is given by the formation constant or stability constant (K). Thus, the composition of total precipitate mineral mass is a function of the type and concentration of ions, and the pH of the phase equilibrium (Udert et al., 2003).

In soils, the recovery of P from fertilizer and manure applications as measured by bioavailability depends on the redistribution of P fertilizer and manure-derived P. Ajiboye et al. (2004) suggest that P in biosolids from anaerobic digestion is not readily available for plant uptake because the P is usually associated with the more recalcitrant fraction (HCl) and less with the labile fraction (H_2O and $NaHCO_3$), probably because of the Fe, Al and Ca intrinsically present in biosolids and from inorganic chemicals added to wastewater and sludge such as polymers and flocculants. However, these investigators indicated that the decrease in P availability could be beneficial, as it is less likely to be transported by surface runoff. Maguire et al. (2000) indicated that P availability from sewage sludge is lower than that of fertilizer or manure, probably due to the precipitation of phosphates caused by the addition of $Al_2(SO_4)_3$ or $FeCl_3$ to the wastewater.

Precipitation of aluminium–phosphorus compounds: Aluminium hydroxide $Al(OH)_3$ is a strong adsorption agent for orthophosphate and condensed phosphate. A theoretical analysis of $Al(OH)_3$ precipitating agent demonstrates that orthophosphate removal is not achieved via the precipitated $AlPO_4$ when applying conventional alum to the wastewater, but via aluminium hydroxide phosphate precipitation (De-Bashan and Bashan, 2004). Aluminium compounds formed during precipitation depend on the type and quantity of organic matter present in the wastewater. Organic matter inhibits phosphorus removal. Solid aluminium species with different organic compounds are formed depending on solution components. The extent of inhibition grows with increasing concentrations of organic matter (De-Bashan and Bashan, 2004). When these P compounds of aluminium are applied to soil, multiple reactions can occur and it is possible that in acid soil with a high level of Al^{3+}, the solubility of P–Al compounds falls under these conditions because the equilibrium is displaced by precipitate formation.

Calcium phosphate precipitation: The reaction of PO_4^{3-} with Ca^{2+} can form a number of related products depending on reaction conditions. Into the complex chemistry of calcium phosphates, hydroxylapatite (HAP, $Ca_5OH(PO_3)_3$) compounds are the most stable calcium phosphates. There is a range of other calcium phosphates that may precipitate as a precursor to HAP, namely dicalcium phosphate dihydrate (bruscite DCPD, $CaHPO_4 \times 2H_2O$), octacalcium phosphate (OCP, $Ca_4H(PO_4)_3 \times 2.5H_2O$), tricalcium phosphate (TCP, $Ca_3(PO_4)_2$) and amorphous calcium phosphates (ACP) (van der Houwen and Valsami-Jones, 2001).

Van der Houwen and Valsami-Jones (2001) proved the influence of organic acid on phosphate calcium precipitation; they found that when citrate is present, high calcium and phosphate concentrations were necessary for precipitation to occur. The high concentration is partly due to the complex formation with citrate in solution, which results in a lower free ion concentration of calcium. This implies a specific organic mechanism of inhibition, since citrate reacts with calcium and can be adsorbed into colloidal calcium phosphate precipitates and slow down the transformation of these particles into more stable calcium phosphate.

The formation of calcium phosphate could be recycled to the non-agricultural P industry as a starting material for processing (Greaves et al., 1999). However, direct application of pure fertilizers has low reactivity and thus an insufficient release of phosphorus when applied to soil (Vassilev et al., 2001). The addition of Ca in sludge, in order to precipitate phosphate or to alkalinize the sludge, reduces availability of P. However at high pH values, phosphorus makes a very stable compound with heavy metals and its availability to plants is reduced (Seyhan and Erdincler, 2003). Sharpley et al. (2004) indicated that in manured soil, crystalline P-forms were dominated by species such as tricalcium and octacalcium phosphate minerals, as a result of P and Ca being introduced into the system through manure. By contrast, in untreated soil, hydroxapatite was the main mineral form.

Iron phosphate precipitation: Among P removal mechanisms is the precipitation of vivianite ($Fe_2(PO_4)_3 \times 8H_2O$), a non-soluble ferrous phosphate which is achieved by adding $FeSO_4$ to wastewater (De-Bashan and Bashan, 2004). The precipitate sludge generated comprises iron phosphate and iron hydroxide (Woods et al., 1999). In Sweden, chemical precipitation by use of iron salt is the most commonly used method for phosphorus removal because iron salt has a low solubility constant. However, phosphate bound to iron is not used as a raw material in the fertilizer industry, which prefers calcium phosphate because of its easy release of phosphorus (Levlin et al., 2002).

Magnesium ammonium phosphate hexahydrate: The most promising compound for P and N recovery from wastewater plants is magnesium ammonium phosphate (struvite, $MgNH_4PO_4 \times 6H_2O$) (Nelson et al., 2003). Different investigators have studied phosphorus recovery from various supernatants of anaerobic digestion in sewage treatment plants (Chimenos et al., 2003; Yoshino et al., 2003) and from piggery wastes (Wrigley et al., 1992; Nelson et al., 2003). Struvite precipitate occurs in natural form in anaerobic treatment systems causing problems in pipework and equipment (Stratful et al., 2001; Adnan et al., 2003; van Rensburg et al., 2003). Struvite requires that its components be simultaneously available in the wastewater in the molecular ratio $1(Mg^{2+})$: $1(NH^{4+})$:$1(PO_4^{3-})$ molar ratios (Nelson et al., 2003; De-Bashan and Bashan, 2004). Struvite formation requires high pH values between 8.6 and 10.6 (Battistoni et al., 1997; van Rensburg et al. (2003) indicated the critical pH values for struvite precipitation is 7.6. The struvite formation is shown by reaction.

$$HPO_4^{2-} + NH_4^+ + Mg^{2+} + OH^- + 5H_2O \rightarrow MgNH_4PO_4 \times 6H_2O$$

P removal uses struvite from wastewaters leading to the production of a recyclable material in the form of phosphate pellets. This is desirable – particularly if pellets with a low water content and a relatively high P content are obtained (Battistoni et al., 2002) – since they can be used as a slow fertilizer which also contains nitrogen and magnesium (Adnan et al., 2003; Battistoni et al., 2002; Burns et al., 2003; Chimenos et al., 2003; Shimamura et al., 2003; Nelson et al., 2003). Battistoni et al. (2002) have devised a phosphate removal technique by exclusively using the chemical–physical properties of anaerobic supernatants without adding any chemical, but the removal process requires optimized nucleation and precipitation mechanisms in order to force the P to precipitate on sand seed. Wrigley et al. (1992) were able to precipitate struvite crystals from the supernatant of anaerobic digestion from piggery wastes with $MgSO_4$, and they achieved a 75% decrease in initial phosphate concentration in batch assay and 90% in continuous process. Nelson et al. (2003) examined P removal from anaerobic swine lagoon liquid, finding that changes in both pH and/or Mg:P ratios promoted struvite precipitation in this effluent. Besides, struvite precipitation decreased the PO_4–P from an initial concentration of $51\,mg\,L^{-1}$ to $7.6\,mg\,L^{-1}$ in approximately 30 min at pH 9.0 and a Mg:P ratio of 1.2:1.

16.6.2 Biological phosphorus removal

Special attention is given to biological phosphorus removal, rather than the chemical processes from wastewater treatment (Chen et al., 2005). Removing phosphate by biological means can be accomplished through two independent mechanisms: firstly, by directing phosphorus incorporation into biomass in biological sludge, where cell mass proportion growth associated with BOD removal and its P is removed by means of the waste-activated sludge, and secondly, by enhancing the storage capacity of phosphorus as polyphosphate in the microbial biomass-known as enhanced biological phosphorus removal (EBPR) (Greaves et al., 1999; De-Bashan and Bashan, 2004; Mulkerrins et al., 2004; Chen et al., 2005). EBPR is a technique developed for the removal of dissolved phosphorus from wastewater; it is based on biological P uptake in excess of the normal metabolic requirement (Seyhan and Erdincler, 2003). Hence P is accumulated in the form of polyphosphate in microorganisms (Greaves et al., 1999). For this technique, it is not necessary to use chemicals to achieve precipitation (Liu et al., 2005).

The phosphorus accumulating organisms (PAO) include a wide range of aerobic bacteria such as: *Acinetobacter, Aeromonas, Alcaligenes, Bacillus, Citrobacter, Corynebacterium, Flavobacterium, Moraxella, Pseudomonas* (Greaves et al., 1999).

EBPR systems consist of two sequential stages: anaerobic and aerobic (Greaves et al., 1999; Liu et al., 2005). The anaerobic provision zone in the activated sludge bioreactor allows enhanced removal of phosphorus, beyond the need for biomass synthesis (Woods et al., 1999). During anaerobic phase, carbon substrates are taken up from solution and subsequently stored in biomass as polyhydroxyalkanoates (PHA); the necessary energy for this PHA biosynthesis is provided by the degradation of polyphosphate and glycogen stored in the cell, generating a P release to an external medium. In the aerobic phase, where no external carbon source is present, the stored PHA is oxidized and used for cell growth along with the phosphate released in the anaerobic process. In the EBPR process, the aerobic phosphate uptake is higher than the phosphate released in anaerobic face, resulting in net phosphate removal (Chen et al., 2005). Liu et al. (2005) have indicated that phosphorus can be biologically stored or chemically precipitated in microorganisms. Precipitation in the cell can occur due to certain multivalent metal ions, such as Ca^{2+} and Mg^{2+}, which can bond to phosphate to form chemical precipitates of polyphosphate. Chemical precipitation, however, is not attributed to biological phosphorus removal. The authors found that less than 10% of the total P content in granules is in the form of precipitate phosphate. Seyhan and Erdincler (2003) investigated the effect o lime-stabilization on the P availability in EBPR sludges through the yield increase of *Lolium perenne* and found that the maximum yield was attained at 100 ppm P for sludge and limed sludges, whereas in the case of triple superphosphate (TSP), the optimum dosage was 200 ppm P. This can be attributed to the presence of different forms of phosphates in the sludge besides water-soluble phosphorus, which are transformed into available forms through mineralization. The authors also indicated that water content affects P availability and they suggested that P was less available in dried sludges. They added that during the preparation of lime sludge pellets, prepared by lime-stabilization of dehydrated sludge, this could generate the formation of less soluble and more crystalline phosphates when they are dry.

16.7 Conclusions

These studies show that the current trend is to maximize the use of P from residues. This makes necessary a knowledge of the P species, evolution and their interaction in the residues and their subsequent fate in the soil environment. In terms of crop-land application, organic wastes and various organic and inorganic P compounds have different levels of bioavailability, depending on the type and treatment of wastes and the characteristics of the soil. It is important to highlight that the factors that determine P availability in soil should be studied to learn more about this influence on P availability in organic residues. In the course of our investigation, we note that the evolution and dynamic P in organic waste has not been thoroughly studied, especially the stages of maximum P availability in the degradation process related to phosphatase activity and organic acid. This information could help to optimize the use of P from organic residues.

Acknowledgements

Authors thank Grant Project MECESUP FRO 0309 and the International Cooperation Project FONDECYT 7020934.

References

Adler, P.R., Sikora, L.J., 2003. Changes in soil phosphorus availability with poultry compost age. Commun. Soil Sci. Plant Anal. 34(1&2), 81–95.

Adnan, A., Mavinic, D.S., Koch, F.A., 2003. Pilot-scale study of phosphorus recovery through struvite crystallization-examining the process feasibility. J. Environ. Eng. Sci. 2, 315–324.

Ajiboye, B., Akinremi, O.O., Racz, G.J., 2004. Laboratory characterization of phosphorus in fresh and over-dried organic amendment. J. Environ. Qual. 33, 1062–1069.

Akhtar, M., McCallister, D.L., Eskridge, K.M., 2002. Availability and fractionation of phosphorus in sewage sludge-amended soils. Commun. Soil Sci. Plant Anal. 33(13&14), 2057–2068.

Andlid, T.A., Veide, J., Sandberg, A., 2004. Metabolism of extracellular inositol hexaphosphate (phytate) by *Saccharomyces cerevisiae*. Int. J. Food Microbiol. 97, 157–169.

Atia, A.M., Mallarino, A.P., 2002. Agronomic and environmental soil phosphorus testing in soils receiving liquid swine manure. Soil Sci. Soc. Am. J. 66, 1696–1705.

Baeza, G., 2001. Un modelo Para el Reciclaje de Nitrógeno, Fósforo y Azufre en el Sistema Suelo-Planta-Animal en Suelos ácidos. Tesis de Doctorado. Universidad de Santiago, Chile.

Balmér, P., 2004. Phosphorus recovery-an overview of potentials and possibilities. Water Sci. Technol. 49(10), 185–190.

Barja, B.C., Herszage, J., dos Santos Afonso, M., 2001. Iron (III)-phosphonate complexes. Polyhedron 20, 1821–1830.

Barnett, G.M., 1994a. Phosphorus forms in animal manure. Bioresour Technol. 49, 139–147.

Barnett, G.M., 1994b. Manure P fractionation. Bioresour. Technol. 49, 149–155.

Barrow, N.J., 1975. Chemical form of organic phosphate in sheep faeces. Aust. J. Soil. Res. 13, 63–67.

Battistoni, P., De Angelis, A., Prisciandaro, M., Boccadoro, R., Bolzonella, D., 2002. P removal from anaerobic supernatants by struvite crystallization: long term validation and process modelling. Water Res. 36, 1927–1938.

Battistoni, P., Fava, G., Pavan, P., Musacco, A., Cecchi, F., 1997. Phosphate removal in anaerobic liquors by struvite crystallization without addition of chemicals: preliminary results. Water Res. 31(11), 2925–2929.

Benito, M., Masaguer, A., Moliner, A., Arrigo, N., Palma, R.M., 2003. Chemical and microbiological parameters for the characterisation of the stability and maturity of pruning waste compost. Biol. Fertil. Soils 37, 184–189.

Bolan, N.S., Adriano, D.C., Naidu, R., Mora, M.L., Mahimairaja, S., 2005. Phosphorus-trace element interactions in soil-plant systems. In: J. Thomas Sims, A.N. Sharpley, G.M. Pierzynski, D.T. Westermann, M.L. Cabrera, J.M. Powell, T.C. Daniel and P.J.A. Withers (Eds.), Phosphorus: Agriculture and the Environment. Hardcover, 1121 pages, ASA-CSSA-SSSA, Agron. Monogr. 46. USA, USA Publishers. pp. 317–352 (ISBN: 0-89118-157-1; Chapter 10).

Bolan, N.S., Duraisamy, P., 2003. Role of inorganic and organic soil amendments on immobilisation and phytoavailability of heavy metals: a review involving specific case studies. Aust. J. Soil Res. 41(3), 533–555.

Bolan, N.S., Naidu, R., Mahimairaja, S., Baskaran, S., 1994. Influence of low-molecular-weight organic acid on the solubilization of phosphates. Biol. Fertil. Soils 18(4), 311–319.

Borie, F., Redel, Y., Rubio, R., Rouanet, J.L., Barea, J.M., 2002. Interactions between crop residues application and mycorrhizal developments and some soil-root interface properties and mineral acquisition by plants in an acidic soil. Biol. Fertil. Soils 36, 151–160.

Borie, F., Rubio, R., 1999. Effects of arbuscular mycorrhizae and liming on growth and mineral acquisition of aluminum-tolerant and aluminum-sensitive barley cultivars. J. Plant Nutr. 22, 121–137.

Borie, F., Rubio, R., 2003. Total and organic phosphorus in Chilean volcanic soils. Gayana Bot. 60(1), 69–78.

Borie, F., Zunino, H., 1983. Organic matter-phosphorus associations as a sink in P-fixation processes in allophanic soils of Chile. Soil Biol. Biochem. 15, 599–603.

Borie, F., Zunino, H., Martínez, L., 1989. Macromolecule-P associations and inositol phosphates in some Chilean volcanic soils of temperate regions. Commun. Soil Sci. Plant Anal. 20(17&18), 1881–1894.

Briceño, M., Escudey, M., Galindo, G., 1999. Adsorción de cloruro, fosfato y ftalato en fracciones extraídas de un Andisol chileno. Bol. Soc. Chil. Quím. 44, 11–23.

Briceño, M., Escudey, M., Galindo, G., Borchardt, D., Chang, A., 2004. Characterization of chemical phosphorus forms in volcanic soils using ^{31}P-NMR spectroscopy. Commum. Soil Sci. Plant Anal. 35(9&10), 1323–1337.

Brookes, P.C., Powlson, D.S., Jenkinson, D.S., 1984. Phosphorus in the soil microbial biomass. Soil Biol. Biochem. 16(2), 169–175.

Bühler, S., Oberson, A., Sinaj, S., Friesen, D.K., Frossard, E., 2003. Isotope methods for assessing plant available phosphorus in acid tropical soils. Eur. J. Soil Sci. 54, 605–616.

Burns, R.T., Moody, L.B., Celen, I., Buchanan, J.R., 2003. Optimization of phosphorus precipitation from swine manure slurries to enhance recovery. Water Sci. Technol. 48(1), 139–146.

Cade – Menun, B.J., Liu, C.W., Nunlist, R., McColl, J.G., 2002. Soil and litter phosphorus-31 nuclear magnetic resonance spectroscopy: extractants, metals, and phosphorus relaxation times. J. Environ. Qual. 31, 457–465.

Carreira, J.A., Viñegla, B., Lajtha, K., 2006. Secondary $CaCO_3$ and precipitation of P–Ca compounds control the retention of soil P in arid ecosystems. J. Arid Environ. 64, 460–473.

Celi, L., Presta, M., Ajmore-Marsan, F., Barberis, E., 2001. Effects of pH and electrolytes on inositol hexaphosphate interaction with goethite. Soil. Sci. Soc. Am. J. 65, 753–760.

Chapman, P.J., Edwards, A.C., Shand, C.A., 1997. The phosphorus composition of soil solutions and soil leachates: influence of soil – solution ratio. Eur. J. Soil Sci. 48, 703–710.

Chen, C.R., Sinaj, S., Condron, L.M., Frossard, E., Sherlock, R.R., Davis, M.R., 2003. Characterization of phosphorus availability in selected New Zealand grassland soils. Nutr. Cycl. Agroecosys. 65, 89–100.

Chen, Y., Chen, Y.S., Xu, Q., Zhou, Q., Gu, G., 2005. Comparison between acclimated and unacclimated biomass affecting anaerobic-aerobic transformations in the biological removal of phosphorus. Process Biochem. 40, 723–732.

Chimenos, J.M., Fernández, A.I., Villalba, G., Segarra, M., Urruticoechea, A., Artaza, B., Espiell, F., 2003. Removal of ammonium and phosphates from wastewater resulting from the process of cochineal extraction using MgO-containing by-product. Water Res. 27, 1601–1607.

Cosgrove, D.J., 1980. Inositol Phosphates. Studies in Organic Chemistry. Elsevier, Amsterdam, The Netherlands.

Costello, A.J., Glonek, T., Myers, T.C., 1976. ^{31}P nuclear magnetic resonance-pH titrations of *myo*-inositol hexaphosphate. Carbohydr. Res. 46, 159–171.

Dao, T.H., 2003. Polyvalent cation effects on *myo*-inositol hexakis dihydrogenphosphate enzymatic dephosphorylation in dairy wastewater. J. Environ. Qual. 32, 694–701.

De-Bashan, L.E., Bashan, Y., 2004. Recent advanced in removing phosphorus from wastewater and its future used as fertilizer (1997–2003). Water Res. 38, 4222–4246.

De Brouwere, K., Thijs, A., Hens, M., Merckx, R., 2003. Forms and bioavailability of soil phosphorus in temperate forests in southern Chile and Flanders. Gayana Bot. 60(1), 17–23.

Dee, B.M., Haynes, R.J., Graham, M.H., 2003. Changes in soil acidity and the size and activity of the microbial biomass in response to the addition of sugar mill wastes. Biol. Fertil. Soils 37, 47–54.

Deng, S.P., Tabatabai, M.A., 1997. Effect of tillage and residue management on enzyme activities in soils: III. Phosphatases and arylsulfatase. Biol. Fertil. Soils 24, 141–146.

DeRouchey, J.M., Goodband, R.D., Nelssen, J.L., Tokach, M.D., Dritz, S.S., Murphy, J.P., 2002. Nutrient composition of Kansas swine lagoons and hoop barn manure. J. Anim. Sci. 80, 2051–2061.

Ding, L., Wang, X., Zhu, Y., Edwards, M., Glindemann, D., Ren, H., 2005. Effect of pH on phosphine production and the fate of phosphorus during anaerobic process with granular sludge. Chemosphere 59, 49–54.

Ebeling, A.M., Bundy, L.G., Powell, J.M., Andraski, T.W., 2002. Dairy diet phosphorus effects on phosphorus losses in runoff from land-applied manure. Soil. Sci. Soc. Am. J. 66, 284–291.

Eghball, B., 2003. Leaching of phosphorus fractions following manure or compost application. Commun. Soil. Sci. Plant Anal. 34(19&20), 2803–2815.

Escudey, M., Galindo, G., Förster, J.E., Briceño, M., Diaz, P., Chang, A., 2001. Chemical forms of phosphorus of volcanic ash-derived soils in Chile. Commun. Soil. Sci. Plant Anal. 32(5&6), 601–616.

Frossard, E., Skrabal, P., Sinaj, S., Bangerter, F., Traoré, O., 2002. Forms and exchangeability of inorganic phosphate in composted solid organic wastes. Nutr. Cycl. Agroecosys. 62, 103–113.

Gale, P.M., Mullen, M.D., Cieslik, C., Tyler, D.D., Duck, B.N., Kirchner, M., McClure, J., 2000. Phosphorus distribution and availability is response to dairy manure applications. Commun. Soil. Sci. Plant Anal. 31(5&6), 553–565.

García-Gil, J.C., Ceppi, S.B., Velasco, M.I., Polo, A., Senesi, N., 2004. Long-term effects of amendment with municipal solid waste compost on the elemental and acidic functional group composition and pH-buffer capacity of soil humic acids. Geoderma 121, 135–142.

García-Gil, J.C., Plaza, C., Soler-Rovira, P., Polo, A., 2000. Long-term effects of municipal solid waste compost application on soil enzyme activities and microbial biomass. Soil Biol. Biochem. 32, 1907–1913.

Garrido, S.E., Vilchis, J., André, C., García, J., Alvarez, A., Gorostieta, E., 2002. Aerobic thermophilic composting of waste sludge from gelatin-grenetine industry. Resour. Conservat. Recycl. 34, 161–173.

Geelhoed, J.S., van Riemsdijk, W.H., Findenegg, G.R., 1999. Simulation of the effect of citrate exudation from roots on the plant availability of phosphate adsorbed on goethite. Eur. J. Soil Sci. 50, 379–390.

Gianfreda, L., Bollag, J.M., 1994. Effect of soils on the behavior of immobilized enzymes. Soil Sci. Soc. Am. J. 58, 1672–1681.

Gimsing, A.L., Borggaard, O.K., Jacobsen, O.S., Aamand, J., Sorensen, J., 2004. Chemical and microbiological soil characteristics controlling glyphosate mineralization in Danish surface soils. Appl. Soil Ecol. 27, 233–242.

Greaves, J., Hobbs, P., Chadwick, D., Haygarth, P., 1999. Prospect for the recovery of phosphorus from animal manures. Rev. Environ. Technol. 20, 697–708.

Griffin, T.S., Honeycutt, C.W., He, Z., 2003. Changes in soil phosphorus from manure application. Soil Sci. Soc. Am. J. 67, 645–653.

Hansen, J.C., Cade-Menun, B.J., Strawn, D.G., 2004. Phosphorus speciation in manure-amended alkaline soils. J. Environ. Qual. 33, 1521–1527.

Haustein, G.K., Daniel, T.C., Miller, D.M., Moore, P.A., McNew, R.W., 2000. Aluminum-containing residuals influence high-phosphorus soils and runoff water quality. J. Environ. Qual. 29, 1954–1959.

Hayes, J.E., Richardson, A.E., Simpson, R.J., 2000a. Components of organic phosphorus in soil extracts that are hydrolysed by phytase and acid phosphatase. Biol. Fertil. Soils 32, 279–286.

Hayes, V.E.A., Ternan, N.G., McMullan, G., 2000b. Organophosphonate metabolism by a moderately halophilic bacterial isolate. FEMS Microbiol. Lett. 186, 171–175.

Haygarth, P.M., Jarvis, S.C., 1997. Soil derived phosphorus in surface runoff from grazed grassland lysimeters. Water Res. 31(1), 140–148.

Haynes, R.J., Mokolobate, M.S., 2001. Amelioration of Al toxicity and P deficiency in acid soils by additions of organic residues: a critical review of the phenomenon and the mechanisms involved. Nutr. Cycl. Agroecosys. 59, 47–63.

He, Z., Griffin, T.S., Honeycutt, C.W., 2004. Phosphorus distribution in dairy manures. J. Environ. Qual. 33(4), 1528–1534.

He, Z., Honeycutt, C.W., Griffin, T.S., 2003. Comparative investigation of sequentially extracted phosphorus fractions in a sandy loam soil and a swine manure. Commun. Soil. Sci. Plant Anal. 34(11&12), 1729–1742.

Hedley, M.J., Stewart, J.W.B., Chauhan, B.S., 1982. Changes in inorganic and organic soil phosphorus fractions induced by cultivation practices and by laboratory incubations. Soil Sci. Soc. Am. J. 46, 970–976.

Hinsinger, P., 2001. Bioavailability of soil inorganic P in the rhizosphere as affected by root-induced chemical changes: a review. Plant Soil 237, 173–195.

Hodgkinson, R.A., Chambers, B.J., Withers, P.J.A., Cross, R., 2002. Phosphorus losses to surface waters following organic manure applications to drained clay soil. Agric. Water Manage. 57, 155–173.

Huang, Q., Shindo, H., 2000. Effects of copper on the activity and kinetics of free and immobilized acid phosphatase. Soil Biol. Biochem. 32, 1885–1892.

Hyun, S., Lee, L., 2004. Hydrophilic and hydrophobic sorption of organic acids by variable-charge soils: effect of chemical acidity and acidic functional group. Environ. Sci. Technol. 38, 5413–5419.

Indiati, R., Rossi, N., 2002. Phosphorus reserves and availability in soils receiving long-term manure applications. Commun. Soil Sci. Plant Anal. 33(9&10), 1589–1605.

Iyamuremye, F., Dick, R.P., Baham, J., 1996a. Organic amendments and phosphorus dynamics: I. Phosphorus chemistry and sorption. Soil Sci. 161(7), 426–435.

Iyamuremye, F., Dick, R.P., Baham, J., 1996b. Organic amendments and phosphorus dynamics: II. Distribution of soil phosphorus fractions. Soil Sci. 161(7), 436–443.

Iyamuremye, F., Dick, R.P., Baham, J., 1996c. Organic amendments and phosphorus dynamics: III. Phosphorus speciation. Soil Sci. 161(7), 444–451.

Jara, A., Violante, A., Pigna, M., Mora, M.L., 2005. Mutual interactions of sulfate, oxalate, citrate and phosphate on synthetic and natural allophanes. Soil Sci. Soc. Am. J. 70, 337–346.

Jones, D.L., 1998. Organic acid in the rhizosphere – a critical review. Plant Soil 205, 25–44.

Jones, D.L., Dennis, P.G., Owen, A.G., van Hees, P.A.W., 2003. Organic acid behavior in soils – misconceptions and knowledge gaps. Plant Soil 248, 31–41.

Kpomblekou-A, K., Tabatabai, M.A., 2003. Effect of low-molecular weight organic acids on phosphorus release and phytoavailability of phosphorus in phosphate rocks added to soil. Agric. Ecosys. Environ. 100, 275–284.

Kroiss, H., 2004. What is the potential for using the resources in sludge?. Water Sci. Technol. 49(10), 1–10.

Kwabiah, A.B., Palm, C.A., Stoskopf, N.C., Voroney, R.P., 2003b. Response of soil microbial biomass dynamic to quality of plant materials with emphasis on P availability. Soil Biol. Biochem. 35, 207–216.

Kwabiah, A.B., Stoskopf, N.C., Palm, C.A., Voroney, R.P., Rao, M.R., Gacheru, E., 2003a. Phosphorus availability and maize response to organic and inorganic fertilizer inputs in a short term study in western Kenya. Agric. Ecosys. Environ. 95, 49–59.

Laboski, C.A.M., Lamb, J.A., 2003. Changes in soil test phosphorus concentration after application of manure or fertilizer. Soil. Sci. Soc. Am. J. 67, 544–554.

Leinweber, P., Haumaier, L., Zech, W., 1997. Sequential extractions and ^{31}P-NMR spectroscopy of phosphorus forms in animal manures, whole soils and particle-size separates from a densely populated livestock area in northwest Germany. Biol. Fertil. Soils 25, 89–94.

Lesueur, C., Pfeffer, M., Fuerhacker, M., 2005. Photodegradation of phosphonates in water. Chemosphere 59, 685–691.

Levlin, E., Löwén, M., Stark, K., Hultman, , 2002. Effects of phosphorus recovery requirements on Swedish sludge management. Water Sci. Technol. 46(4), 435–440.

Linquist, B.A., Singleton, P.W., Cassman, K.G., 1997. Inorganic and organic phosphorus dynamic during a build-up and decline of available phosphorus in an ultisol. Soil Sci. 162(4), 254–264.

Liu, Y., Lin, Y., Tay, J., 2005. The elemental compositions of P-accumulating microbial granules developed in sequencing batch reactors. Process Biochem. 40(10), 3258–3262.

Maguire, R.O., Sims, J.T., Coale, F.J., 2000. Phosphorus fractionation in biosolids-amended soils: relationship to soluble and desorbable phosphorus. Soil Sci. Soc. Am. J. 64, 2018–2024.

Makarov, M.I., Haumaier, L., Zech, W., 2002. Nature of soil organic phosphorus: as assessment of peak assignments in the diester region of ^{31}P NMR spectra. Soil Biol. Biochem. 34, 1467–1477.

Makarov, M.I., Haumaier, L., Zech, W., Marfenina, O.E., Lysak, L.V., 2005. Can ^{31}P NMR spectroscopy be used to indicate the origins of soil organic phosphates?. Soil Biol. Biochem. 37, 15–25.

Malley, D.F., Yesmin, L., Eilers, R.G., 2002. Rapid analysis of hog manure and manure-amended soils using near-infrared spectroscopy. Soil Sci. Soc. Am. J. 66, 1677–1686.

Manahan, S.E., 1994. Environmental Chemistry. , 6th Ed., Lewis Publishers, USA, 811 pp.

Matys, S.V., Laurinavichius, K.S., Krupyanko, V.I., Nesmeyanova, M.A., 2001. Optimization of degradation of methylphosphonate – analogue of toxic pollutant with direct C-P bond by *Escherichia coli.* Process Biochem. 36, 821–827.

McDowell, R.W., Sharpley, A.N., 2003. Phosphorus solubility and release kinetic as a function of soil test P concentration. Geoderma 112, 143–154.

McDowell, R.W., Stewart, I., 2005. Phosphorus in fresh and dry dung of grazing dairy cattle, deer, and sheep: sequential fraction and phosphorus-31 nuclear magnetic resonance analyses. J. Environ. Qual. 34, 598–607.

McDowell, R.W., Stewart, I., 2006. The phosphorus composition of contrasting soils in pastoral, native and forest management in Otago, New Zealand: sequential extraction and ^{31}P NMR. Geoderma 130(1–2), 176–189.

Mkhabela, M.S., Warman, P.R., 2005. The influence of municipal solid waste compost on yield, soil phosphorus availability and uptake by two vegetable crops grown in a Pugwash sandy loam soil in Nova Scotia. Agric. Ecosys. Environ. 106, 57–67.

Mokolobate, M.S., Haynes, R.J., 2002. Comparative liming effect of four organic residues applied to an acid soil. Biol. Fertil. Soils 35, 79–85.

Mora, M.L., Alfaro, M., Williams, P., Stehr, W., Demanet, R., 2004. Effect of fertiliser input on soil acidification in relation to growth and chemical composition of a pasture, and animal production. Rev. Cienc. Suelo Nutr. Veg. 4(1), 29–40.

Mora, M.L., Alfaro, M.A., Jarvis, S.C., Demanet, R., Cartes, P., 2006. Soil aluminium availability in Andisols of southern Chile and its effect on forage production and animal metabolism. Soil Use Manage. 22(1), 95–101.

Mora, M.L., Baeza, G., Pizarro, C., Demanet, R., 1999. Effect of calcitic and dolomitic lime on physicochemical properties of a Chilean Andisol. Commun. Soil Sci. Plant Anal. 30(3&4), 427–439.

Mora, M.L., Canales, J., 1995a. Interactions of humic substances with allophanic compounds. Commun. Soil. Sci. Plant Anal. 26(17&18), 2805–2817.

Mora, M.L., Canales, J., 1995b. Humin-clay interactions on surface reactivity in Chilean Andisols. Commun. Soil. Sci. Plant Anal. 26(17&18), 2819–2828.

Mora, M.L., Cartes, P., Demanet, R., Cornforth, I.S., 2002. Effects of lime and gypsum on pasture growth and composition on an acid Andisol in Chile, South America. Commun. Soil. Sci. Plant Anal. 33(13&14), 2069–2081.

Mulkerrins, D., Dobson, A.D.W., Colleran, E., 2004. Parameters affecting biological phosphate removal from wastewaters. Environ. Int. 30, 249–259.

Nelson, N.O., Mikkelsen, R.L., Hesterberg, D.L., 2003. Struvite precipitation in anaerobic swine lagoon liquid: effect of pH and Mg:P ratio and determination of rate constant. Biores. Technol. 89, 229–236.

Nowack, B., 2002. Aminopolyphosphonate removal during wastewater treatment. Watet Res. 36, 4636–4642.

Nowack, B., 2003. Environmental chemistry of phosphonates. Water Res. 37, 2533–2546.

Nwoke, O.C., Vanlauwe, B., Diels, J., Sanginga, N., Osonubi, O., Merckx, R., 2003. Assessment of labile phosphorus fractions and adsorption characteristics in relation to soil properties of West African savanna soils. Agric. Ecosys. Environ. 100, 285–294.

Oehl, F., Frossard, E., Fliessbach, A., Dubois, D., Oberson, A., 2004. Basal organic phosphorus mineralization in soils under different farming systems. Soil Biol. Biochem. 36, 667–675.

Oehl, F., Oberson, A., Probst, M., Fliessbach, A., Roth, H.R., Frossard, E., 2001a. Kinetics of microbial phosphorus uptake in cultivated soils. Biol. Fertil. Soils 34, 31–41.

Oehl, F., Oberson, A., Sinaj, S., Frossard, E., 2001b. Organic phosphorus mineralization studies using isotopic dilution techniques. Soil Sci. Soc. Am. J. 65, 780–787.

Oehl, F., Oberson, A., Tagmann, H.U., Besson, J.M., Dubois, D., Mäder, P., Roth, H.-R., Frossard, E., 2002. Phosphorus budget and phosphorus availability in soil under organic and conventional farming. Nutr. Cycl. Agroecosys. 62, 25–35.

Olander, L.P., Vitousek, P.M., 2005. Short-term controls over inorganic phosphorus during soil and ecosystem development. Soil Biol. Biochem. 37, 651–659.

Pandey, A., Szakacs, G., Soccol, C.R., Rodriguez-Leon, J.A., Soccol, V.T., 2001. Production, purification and properties of microbial phytases. Biores. Technol. 77, 203–214.

Pant, H.K., Warman, P.R., 2000. Enzymatic hydrolysis of soil organic phosphorus by immobilized phosphatases. Boil. Fertil. Soils 30, 306–311.

Pant, H.K., Warmam, P.R., Nowak, J., 1999. Identification of soil organic phosphorus by ^{31}P nuclear magnetic resonance spectroscopy. Commun. Soil Sci. Plant Anal. 30(5&6), 757–772.

Parfitt, R.L., Yeates, G.W., Ross, D.J., Mackay, A.D., Budding, P.J., 2005. Relationships between soil biota, nitrogen and phosphorus availability, and pasture growth under organic and conventional management. Appl. Soil Ecol. 28, 1–13.

Park, M., Singvilay, O., Shin, W., Kim, E., Chung, J., Sa, T., 2004. Effect of long-term compost and fertilizer application on soil phosphorus status under paddy cropping system. Commun. Soil Sci. Plant Anal. 35(11&12), 1635–1644.

Pascual, J.A., Moreno, J.L., Hernández, T., García, C., 2002. Persistence of immobilized and total urease and phosphatase activities in a soil amended with organic wastes. Biores. Technol. 82, 73–78.

Pinochet, D., Epple, G., MacDonald, R., 2001. Fracciones de fósforo orgánico e inorgánico en un transecto de suelos de origen volcánico y metamórfico. Rev. Cienc. Suelo Nutr. Veg. 1, 58–69.

Pypers, P., Verstraete, S., Thi, C.P., Merckx, R., 2005. Changes in mineral nitrogen, phosphorus availability and salt-extractable aluminium following the application of green manure residues in two weathered soils of South Vietnam. Soil Biol. Biochem. 37, 163–172.

Quiquampoix, H., 1987. A stepwise approach to the understanding of extracellular enzyme activity in soil II. Competitive effects on the adsorption of a β-D-glucosidase in mixed mineral or organo-mineral systems. Biochimie 69, 765–771.

Raboy, V., 2003. *myo*-Inositol-1, 2, 3, 4, 5, 6-hexakisphosphate. Phytochemistry 64, 1033–1043.

Rao, M.A., Violante, A., Gianfreda, L., 2000. Interaction of acid phosphatase with clays, organic molecules and organo-mineral complexes: kinetic and stability. Soil Biol. Biochem. 32, 1007–1014.

Rebollar, P.G.Y., Mateos, G.G., 1999. El fósforo en nutrición animal. Necesidades, valoración de materias primas y mejora de la disponibilidad. En Curso de Especialización FEDNA: Avances en nutrición y alimentación animal. In: P.G. Rebollar, C. de Blas and Y.G.G. Mateos (Eds.), Fundación Española para el Desarrollo de la Nutrición Animal. Madrid, España.

Rubaek, G.H., Sibbesen, E., 1993. Resin extraction of labile, soil organic phosphorus. J. Soil Sci. 44, 467–478.

Rydin, E., 1996. Experimental studies simulating potential phosphorus release from municipal sewage sludge deposits. Water Res. 30(7), 1695–1701.

Scherer, H.W., Sharma, S.P., 2002. Phosphorus fractions and phosphorus delivery potential of a luvisol derived from loess amended with organic materials. Biol. Fertil. Soils 35, 414–419.

Seyhan, D., Erdincler, A., 2003. Effect of lime stabilization of enhanced biological phosphorus removal sludges on the phosphorus availability to plants. Water Sci. Technol. 48(1), 155–162.

Shand, C.A., Smith, S., 1997. Enzymatic release of phosphate from model substrates and P compounds in soil solution from a peaty podzol. Biol. Fertil. Soils 24, 183–187.

Sharpley, A.N., McDowell, R.W., Kleinman, P.J.A., 2004. Amount, forms, and solubility of phosphorus in soils receiving manure. Soil Sci. Soc. Am. J. 68, 2048–2057.

Shen, H., Yan, X., Zhao, M., Zheng, S., Wang, X., 2002. Exudation of organic acids in common bean as related to mobilization of aluminum- and iron-bound phosphates. Environ. Exp. Bot. 48, 1–9.

Shibata, R., Yano, K., 2003. Phosphorus acquisition from non-labile sources in peanut and pigeonpea with mycorrhizal interaction. Appl. Soil Ecol. 24, 133–141.

Shimamura, K., Tanaka, T., Miura, Y., Ishikawa, H., 2003. Development of a high-efficiency phosphorus recovery method using a fluidized-bed crystallized phosphorus removal system. Water Sci. Techol. 48(1), 163–170.

Sinaj, S., Traoré, O., Frossard, E., 2002. Effect of compost and soil properties on the availability of compost phosphate for withe clover (*Trifolium repens* L.). Nutr. Cycl. Agroecosys. 62, 89–102.

Smith, D.C., Hughes, J.C., 2004. Changes in maturity indicators during the degradation of organic wastes subjected to simple composting procedures. Biol. Fertil. Soils 39, 280–286.

Speir, T.W., Horswell, J., van Schaik, A.P., McLaren, R.G., Fietje, G., 2004. Composted biosolids enhance fertility of a sandy loam soil under dairy pasture. Biol. Fertil. Soil 40, 349–358.

Staats, K.E., Arai, Y., Sparks, D.L., 2004. Alum amendment effects on phosphorus release and distribution in poultry litter-amended sandy soils. J. Environ. Qual. 33, 1904–1911.

Stratful, I., Scrimshaw, M.D., Lester, J.N., 2001. Conditions influencing the precipitation of magnesium ammonium phosphate. Water Res. 35(17), 4191–4199.

Ström, L., Owen, A.G., Godbold, D.L., Jones, D.L., 2002. Organic acid mediated P mobilization in the rhizosphere and uptake by maize roots. Soil Biol. Biochem. 34, 703–710.

Szögi, A.A., Vanotti, M.B., Hunt, P.G., 2006. Dewatering of phosphorus extracted from liquid swine waste. Biores. Technol. 97(1), 183–190.

Taiz, L., Zeiger, E., 1998. Plant Physiology., 2nd Ed., Sinauer Associates, Inc. Publishers, Massachusetts, USA, 792 pp.

Tan, K.H., 1993. Principles of Soil Chemistry., 2nd Ed., Marcel Dekker, USA.

Tarafdar, J.C., Claassen, N., 2003. Organic phosphorus utilization by wheat plants under sterile conditions. Biol. Fertil. Soils 39, 25–29.

Taranto, M.T., Adams, M.A., Polglase, P.J., 2000. Sequential fractionation and characterization (^{31}P-NMR) of phosphorus-amended soils in *Banksia integrifolia* (L.f.) woodland and adjacent pasture. Soil Biol. Biochem. 32, 169–177.

Thien, S.J., Myers, R., 1992. Determination of bioavailable phosphorus in soil. Soil Sci. Soc. Am. J. 56, 814–818.

Toor, G.S., Condron, L.M., Di, H.J., Cameron, K.C., Cade-Menun, B.J., 2003. Characterization of organic phosphorus in leachate from a grassland soil. Soil Biol. Biochem. 35, 1317–1323.

Traoré, O., Sinaj, S., Frossard, E., Van De Kerkhove, J.M., 1999. Effect of composting time on phosphate exchangeability. Nutr. Cycl. Agroecosys. 55, 123–131.

Turner, B., McKelvie, I.D., 2002. A Novel technique for the pre-concentration and extraction of inositol hexakisphosphate from soil extracts with determination by phosphorus-31 nuclear magnetic resonance. J. Environ. Qual. 31, 466–470.

Turner, B.L., Mahieu, N., Condron, L.M., 2003a. Phosphorus-31 nuclear magnetic resonance spectral assignments of phosphorus compounds in soil NaOH-EDTA extracts. Soil Sci. Soc. Am. J. 67, 497–510.

Turner, B.L., Mahieu, N., Condron, L.M., 2003b. The phosphorus composition of temperate pasture soils determined by NaOH-EDTA extraction and solution ^{31}P NMR spectroscopy. Organic Geochem. 34, 1199–1210.

Turner, B.L., Cade-Menun, B.J., Condron, L.M., Newman, S., 2005. Extraction of soil organic phosphorus. Talanta 66, 294–306.

Turner, B.L., McKelvie, I.D., Haygarth, P.M., 2002a. Characterisation of water-extractable soil organic phosphorus by phosphatase hydrolysis. Soil Biol. Biochem. 34, 27–35.

Turner, B.L., Papházy, M.J., Haygarth, P.M., McKelvie, I.D., 2002b. Inositol phosphates in the environment. Phil. Trans. R. Soc. Lond. B. 357, 449–469.

Udert, K.M., Larsen, T.A., Gujer, W., 2003. Estimating the precipitation potential in urine-collecting systems. Water Res. 37, 2667–2677.

Unsal, T., Sozudogru, S., 2001. Description of characteristics of humic substances from different waste materials. Biores. Technol. 78, 239–242.

van der Houwen, J.A.M., Valsami-Jones, E., 2001. The application of calcium phosphate precipitation chemistry to phosphorus recovery: the influence of organic ligands. Environ. Technol. 22, 1325–1335.

van Rensburg, P., Musvoto, E.V., Wentzell, M.C., Ekama, G.A., 2003. Modelling multiple mineral precipitation in anaerobic digester liquor. Water Res. 37, 3087–3097.

Vassilev, N., Vassileva, M., Fenice, M., Federici, F., 2001. Immobilized cell technology applied in solubilization of insoluble inorganic (rock) phosphates and P plant acquisition. Biores. Technol. 79, 263–271.

Verma, S., Subehia, S.K., Sharma, S.P., 2005. Phosphorus fractions in an acid soil continuously fertilized with mineral and organic fertilizers. Biol. Fertil. Soils 41, 295–300.

Wang, J.Y., Stabnikova, O., Tay, S.T.L., Ivanov, V., Tay, J.H., 2004. Biotechnology of intensive aerobic conversion of sewage sludge and food waste into fertilizer. Water Sci. Technol. 49(10), 147–154.

Wienhold, B.J., 2005. Changes in soil attributes following low phosphorus swine slurry applications to no-tillage sorghum. Soil Sci. Soc. Am. J. 69, 206–214.

Withers, P.J.A., Edwards, A.C., Foy, R.H., 2001. Phosphorus cycling in UK agriculture and implications for phosphorus loss from soil. Soil Use Manage. 17, 139–149.

Woods, N.C., Sock, S.M., Daigger, G.T., 1999. Phosphorus recovery technology modeling and feasibility evaluation for municipal wastewater treatment plants. Environ. Technol. 20, 663–679.

Wrigley, T.J., Webb, K.M., Venkitachalm, H., 1992. A laboratory study of struvite precipitation after anaerobic digestion of piggery wastes. Biores. Technol. 41, 117–121.

Xu, R., Zhao, A., Ji, G., 2003. Effect of low-molecular-weight organic anions on surface charge of variable charge soils. J. Colloid Interface Sci. 264, 322–326.

Yadav, R.S., Tarafdar, J.C., 2001. Influence of organic and inorganic phosphorus supply on the maximum secretion of acid phosphatase by plants. Biol. Fertil. Soils 34, 140–143.

Yadav, R.S., Tarafdar, J.C., 2003. Phytase and phosphatase producing fungi in arid and semi-arid soils and their efficiency in hydrolyzing different organic P compounds. Soil Biol. Biochem. 35, 1–7.

Yoshino, M., Yao, M., Tsuno, H., Somiya, I., 2003. Removal and recovery of phosphate and ammonium as struvite from supernatant in anaerobic digestion. Water Sci. Technol. 48(1), 171–178.

Zhang, T.Q., Mackenzie, A.F., Sauriol, F., 1999. Nature of soil organic phosphorus as affected by long-term fertilization under continuous corn (*Zea Mays L.*): a ^{31}P NMR study. Soil Sci. 164(9), 662–670.

Developments in Soil Science, volume 32
Ravendra Naidu (Editor)

Chapter 17

BIOLOGICAL TRANSFORMATION AND BIOAVAILABILITY OF NUTRIENT ELEMENTS IN ACID SOILS AS AFFECTED BY LIMING

N.S. Bolan, J. Rowarth, M. de la Luz Mora, D. Adriano and D. Curtin

17.1 Introduction

Soil acidification is a natural process, which can be accelerated by plant, animal and human activities. In managed ecosystems, such as pastoral and arable farms, soil acidification is caused mainly by the net release of H^+ ions during the transformation and cycling of three major elements, carbon (C), nitrogen (N) and sulphur (S) (Bolan and Hedley, 2003). Acidification particularly impacts on the potential productivity of soils with low pH buffering capacity. For example, in many parts of the world, permanent legume-based pastoral farming and continuous use of acid-forming fertilizers (e.g., ammonium sulphate) in arable cropping have resulted in aluminium (Al) and manganese (Mn) toxicity in soils (Sumner and Yamada, 2002; Mora et al., 2005), thereby limiting their potential for sustainable production.

Soil acidity has detrimental affects on both plants and soil organisms (Robson and Abbott, 1989; Runge and Rode, 1991; Bloom, 2000), and environmental degradation (Bolan et al., 2003a, 2003b) (Fig. 17.1). The activities and functioning of soil organisms generally are reduced, resulting in the inhibition of biological functions, such as N fixation by legumes, and decomposition of organic matter and nutrient re-cycling is slowed (Gregorich and Janzen, 2000). In acid soils, the concentrations of Al and Mn and other heavy metals become toxic to plant growth (Ritchie, 1989; Mora et al., 1999a). Phosphorus (P) and molybdenum (Mo) may become insoluble and unavailable (Mortvedt, 2000), and low pH may result in a reduction in availability of basic essential cations, such as calcium (Ca) and magnesium (Mg) (Sumner et al., 1991).

Acid produced in soils is commonly neutralized by applying agricultural lime (primarily calcium carbonate), thereby maintaining the possibility of sustainable production and environmental protection. Liming enhances the physical, chemical and biological properties of soil through its direct effect on the amelioration of soil acidity and its indirect effect on the mobilization of some of the major and trace element nutrients, immobilization of toxic metals, and improvements in soil structure and hydraulic conductivity. In variable charge

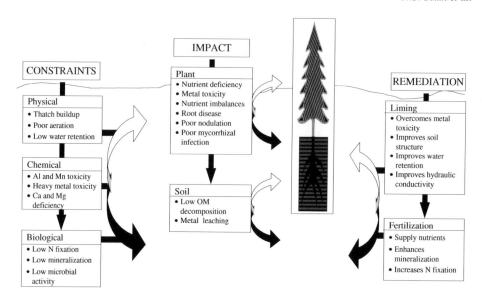

Figure 17.1. *Acidity-related constraints, impact and remediation in soil-plant system.*

soils, liming can also be used as a management tool to manipulate the surface charge, thereby controlling the reactions of nutrient ions and contaminants. Liming has been shown to provide optimum conditions for a suite of biological functions that include N fixation by legumes, mineralization of organic matter and microbial redox reactions (Mora and Barrow, 1996; Haynes and Naidu, 1998; Bolan et al., 1999).

Plant nutrient elements undergo both chemical and biological transformations in soils, which control their im(mobilization) and bioavailability. Biological transformation processes embody biogeochemical reactions that are mediated through microbial activity and include mineralization/immobilization, microbial oxidation/reduction and biomethylation/demethylation (Coleman and Whitman, 2005). Nitrogen (N), phosphorus (P), sulphur (S), boron (B), cobalt (Co), molybdenum (Mo) and selenium (Se) are some of the important nutrient elements that occur in significant quantities in organic forms, and are subject to biological transformation in soils by microorganisms. Biological transformation processes play a significant role in both enhancing their bioavailability and mitigating their impacts on terrestrial, aquatic and atmospheric environments. Some of these biotransformation processes can be manipulated to take advantage of their beneficial effects in remediating contaminated sites (e.g., removal of Se through biomethylation process).

Although the impact of soil acidification on chemical transformation of nutrient and toxic elements has been reviewed by a number of workers (e.g., Reuss, 1986; Sumner et al., 1991; Bolan et al., 2003a, 2003b), no comprehensive review on the effect of liming on biological transformation of nutrient elements in soils in relation to sustainable production and environmental protection has been published. The present review discusses briefly the various processes that generate acid during the cycling of C, N and S in soils, the distribution of nutrient elements in soil organic matter and the biological processes involved in the transformation of these nutrient elements. The detrimental effects of soil acidity on microbial functions, and the beneficial effects of liming in overcoming the problems associated with soil acidity, are examined in relation to the biological transformation of major and trace element nutrients.

17.2 Processes of acid generation in soils

Rowell and Wild (1985) summarized the causes of soil acidification: dissolved CO_2 has the greatest acidifying effect in soils with pH less than approximately 6.5; fertilizers containing ammonium ions or urea acidify soil whether the ions are taken up directly by plants or nitrified first; oxidation of N and S in soil organic matter causes acidification (particularly after deforestation); the acidifying effect of rainfall and dry deposition is due to sulphuric and nitric acids, sulphur dioxide and ammonium ions. The most significant hydrogen ion generating processes in the soil occur during the biogeochemical cycling of C, N and S (Table 17.1). In the case of the C cycle, dissolution of CO_2 to form H_2CO_3 in soil solution and synthesis and dissociation of carboxylic acids produced by plants and microorganisms are the two main sources of H^+ ions (Helyar and Porter, 1989). The assimilation of CO_2 into carboxylic acids in higher plants indirectly acidifies their rhizosphere. In the case of N and S cycles, mineralization of organic matter and the subsequent oxidation of N and S

Table 17.1. *Proton generation processes in C, N and S cycles in managed ecosystem.*

Process	Reaction equation	H^+ (molc/mol)	Equation No.
Carbon cycle			
Dissolution of carbon dioxide	$CO_2 + H_2O \rightarrow H_2CO_3 \rightarrow H^+ + HCO_3^-$	+1	(3.1)
Synthesis of organic acid	Organic C $\rightarrow$ R.COOH $\rightarrow$ R.COO$^-$ + H$^+$	+1	(3.2)
Nitrogen cycle			
N fixation	$2N_2 + 2H_2O + 4R.OH \rightarrow 4R.NH_2 + 3O_2$	0	(3.3)
Ammonium assimilation	$NH_4^+ + R.OH \rightarrow R.NH_2 + H_2O + H^+$	+1	(3.6)
Ammonia volatilization	$NH_4^+ + OH^- \rightarrow NH_3\uparrow + H_2O$	+1	(3.7)
Nitrification	$NH_4^+ + 2O_2 \rightarrow NO_3^- + H_2O + 2H^+$	+2	(3.8)
Sulfur cycle			
Mineralization of organic S	2Organic S $+ 3O_2 + 2H_2O \rightarrow 2SO_4^{2-} + 4H^+$	+2	(3.11)
Oxidation of So	$2So + 2H_2O + 3O_2 \rightarrow 2SO_4^{2-} + 4H^+$	+2	(3.13)

nutrients result in the release of H^+ ions. Leaching of nitrate (NO_3^-) and sulphate (SO_4^{2-}) with charge-balancing basic cations not only results in the depletion of plant-available cations but is also likely to accelerate the acidification process (Oenema, 1990; de Klein et al., 1997; Alewell et al., 2001).

The processes involved in the generation of H^+ during C, N and S cycling in soils have been discussed in detail by Bolan and Hedley (2003) and Tang and Rengel (2003) and can be grouped into two main categories: plant-induced and soil-induced. In managed ecosystems used for agricultural production, regular fertilizer use can also contribute to soil acidification (Mora et al., 1999b; Bloom, 2000; Sumner and Yamada, 2002).

17.2.1 Plant-induced processes

The plant-induced processes involved in the generation of H^+ include the uptake and assimilation of C, N and S. In higher plants, the metabolism of carbohydrates assimilated during the photosynthetic process results in the synthesis of organic acids, such as malic and oxalic acids. These organic acids produce H^+ ions during the subsequent dissociation of some of their carboxyl groups at the cytoplasmic pH of the plants (Raven, 1985). Most of the terrestrial plant species tend to maintain cytoplasmic pH by excreting H^+ ions into the soil solution, thereby acidifying their rhizosphere (Felle, 1988).

Plants use N in three main forms – as an anion (nitrate, NO_3^-), as a cation (ammonium, NH_4^+) or as a neutral N_2 molecule (biological N_2 fixation) (Marschner, 1995; Stockdale et al., 1997). Depending upon the form of N taken up and the mechanism involved in its subsequent assimilation in the plant, excesses of cation or anion uptake may occur (Tang and Rengel, 2003). Plants generally release H^+, OH^- or bicarbonate (HCO_3^-) ions in order to maintain charge balance during the nutrient uptake process. It has been shown that while the uptake of NH_4^+ and N_2 fixation results in a net release of H^+ ions, uptake of NO_3^- can result in a net release of OH^- (Haynes, 1990). Assimilation of NH_4^+ in plants results in the release of H^+ firstly through the deprotonation of NH_4^+. The amino acids and proteins synthesized through the assimilation of NH_4^+ release additional H^+ during their dissociation (Raven, 1985).

Many legumes commonly export H^+ into their rhizosphere when actively fixing N_2 (Haynes, 1983; Liu et al., 1989). Part of the H^+ generated within the legume root comes from the dissociation of the carboxyl groups of amino acids. The quantity of H^+ released during N_2 fixation is a function of C assimilation and therefore depends mainly on the form and amount of amino acids and organic acids synthesized within the plant (Raven, 1985). For example, some tropical legumes do not accumulate cations to the same extent as temperate legumes, thereby resulting in less acidification of the rhizosphere

(Tang et al., 1997). Tropical legumes assimilate part of the fixed N as ureides (allantoin and allantoic acid) that have high pK_a values (e.g., allantoin pK_a 8.96) and are therefore unlikely to dissociate and donate H$^+$ ions under the pH regimes common in the cytoplasm and xylem (Ledgard and Peoples, 1988).

17.2.2 Soil-induced processes

Transformation of C, N and S in soils results in the release of H$^+$ ions. The continuous production of CO_2 through microbial and root respiration increases the concentration of CO_2 in the soil, which dissolves in soil solution resulting in H_2CO_3 (Andrews and Schlesinger, 2001). However, since the solubility of CO_2 in water is negligible below pH 5, the release of CO_2 through respiration is unlikely to cause soil pH to drop below 5 (Jobbágy and Jackson, 2003). Soil microorganisms produce organic acids when the plant litter they are decomposing is rich in organic compounds but low in basic cations. The quantity and diversity of organic acids produced from litter varies depending mainly on the nature of the plant species (Uren, 2001; Bertin et al., 2003).

Nitrification and ammonia (NH_3) volatilization reactions in soils result in the release of H$^+$ ions (Stockdale et al., 1997). Soil microorganisms firstly convert organic N to NH_4^+ (ammonification), and then oxidize NH_4^+ to NO_3^- (nitrification). Whereas the ammonification process results in the release of OH$^-$, the nitrification process results in the release of H$^+$. Combined ammonification and nitrification processes, in theory generate one net mole of H$^+$ for every mole of N transformed. Protons are also produced during the mineralization and subsequent oxidation of S in soil organic matter. However, the net release of H$^+$ during the transformation of N and S depends on the extent of their removal through plant uptake, leaching and immobilization (Helyar, 1976; Logan et al., 1999, 2000; Rothe et al., 2002).

Fertilizer application in managed ecosystems used for agricultural production is one of the major contributors to soil acidification (Tisdale et al., 1985; Sumner and Yamada, 2002). The processes involved in the acidification of soil through fertilizer application have been discussed in detail elsewhere (e.g., Rengel, 2003, and chapters therein). Application of N fertilizers, such as urea and ammonium sulphate, to soils produces H$^+$ mainly during the nitrification of NH_4^+ to NO_3^- and the subsequent leaching of the NO_3^- (Bouman et al., 1995; Bergholm et al., 2003). Dissolution of monocalcium phosphate in superphosphate fertilizers, such as single superphosphate and triple superphosphate, results in the formation of dicalcium phosphate with a release of phosphoric acid close to the fertilizer granules. Similarly, when elemental sulphur (S^o)-based fertilizers are added to soils, oxidation of S^o releases sulphuric acid, which induces considerable acidification of the soil matrix (Barrow, 1971). Reclamation of soils acidified by elemental sulphur requires addition of large quantities of lime and N fertilizer (Nyborg, 1982).

17.3 Biological transformation of nutrients in soils

Nutrients added to soils through fertilizer and manure application undergo biological transformation processes that include mineralization/immobilization and microbial oxidation/reduction reactions. In order to examine the effect of liming on biological transformation of nutrients in acid soils it is important to understand the distribution of nutrient elements in organic forms as well as the biochemical reactions involved in the mineralization/immobilization and microbial oxidation/reduction reactions.

17.3.1 Distribution of nutrients in soils

Nutrient elements occur in soils in various fractions and forms (Shuman, 1991). For example, they can be

(a) in structural components of primary and secondary minerals (e.g., K in feldspar);
(b) precipitated in inorganic forms, including those occluded by Fe, Al and Mn oxides (e.g., occluded P);
(c) complexed by organic matter (e.g., complexed Cu);
(d) incorporated into organic matter including microbial biomass (e.g., biomass N, P and S);
(e) specifically adsorbed onto silicate clay minerals and Fe, Al and Mn oxides (e.g., adsorbed P and Zn);
(f) in exchangeable forms on the exchange sites of silicate clays and organic matter (e.g., exchangeable anions and cations); and
(g) water soluble, as free and complexed ions (e.g., NO_3^- and NH_4^+ in soil solution).

Significant quantities of N, P, S, B, Co, Mo and Se occur in organic forms (Fraction d), especially in soils that are rich in organic matter such as permanent legume based pasture soils. These organic forms are subject to biological transformation in soils. The relative contribution of organic matter in maintaining the supply of bioavailable N pool had been well recognized, leading to the development of soil tests based on mineralizable N pools (Wardle, 1992; Moore et al., 2000; Dersch et al., 2003; Olfs et al., 2004). The importance of organic pools of other nutrient elements, such as P and S, especially in soils rich in organic matter (e.g., pasture soils and forestry) in the bioavailability of these nutrients has also been recognized (Condron et al., 1993; Motavalli et al., 1993; Chepkwony et al., 2001; Prietzel et al., 2001; McDowell, 2003), but has not yet been incorporated into soil tests.

Soil organic N constitutes 90–98% of total N in most soils. The organic N pool includes crude protein, simple protein, amino acids, amino sugars and nucleic acids (Jarvis et al., 1996; Stockdale et al., 1997). The amino acids are present in the soil solution and also bound to clay minerals and humic colloids. Amino acids, being readily decomposed by microorganisms, have only a transient existence in soil. Thus, the amount of amino acids present in the soil solution at any one time represents a balance between synthesis and assimilation by microorganisms (Badalucco and Kuikman, 2001). Amino sugars, or N-containing carbohydrates, occur as structural components of a broad group of substances, for example, the mucopolysaccharides in combination with mucopeptides and mucoproteins. Some of the amino sugar material in soil may exist in the form of an alkali-insoluble polysaccharide referred to as chitin (Paul and Clark, 1996). Nucleic acids, which occur in the cells of all living organisms, consist of individual mononucleotide units joined by a phosphoric acid ester linkage. Small amounts of N have also been extracted from soil in the form of glycerophosphatides, amines and vitamins (Stevenson, 1985a; Blackmer, 2000a, 2000b).

Soil organic P constitutes 30–50% of total P in most soils (Dalal, 1977) although some Chilean Andisols have been reported to have up to 75% of their total P content occurring mainly as monoester phosphate (Briceño et al., 2004; Escudey et al., 2004). Organic P occurs mostly as esters of orthophosphoric acid, and five classes of esters have been identified by ^{31}P-NMR (Newman and Tate, 1980; Condron et al., 1996). These include inositol phosphates, phospholipids, nucleic acids, nucleotides and sugar phosphates. The first three are the dominant groups in most soils and phytic acid is the most common ester of the first group found in soils. Phosphatidylcholine (lecithin) and phosphotidyletha-nolamine are the predominant phospholipids in soils. The two distinct nucleic acids include ribonucleic acid and deoxyribonucleic acid (Stevenson, 1985a; Sharpley, 2000).

The main forms of organic S in soils include S-containing amino acids (cysteine, cystine and methionine), peptides (glutathione and γ-glutamylcys-teine) and sulphonates, in which S is directly bonded to carbon (C–S), and the true organic esters of sulphuric acid (C–O–S), in which S is bonded to oxygen in the form of $C–O–SO_3^-$ linkages. Sulphamates may also be found, in which S occurs in the form of $N–O–SO_3^-$ and $N–SO_3^-$ groups (Saggar et al., 1998).

Some of the boron in soils is associated with organic matter, partly by adsorption (Keren et al., 1985) and partly through the reactions of boric acid with hydroxyl aliphatic acids and aromatic compounds containing *o*-dihydroxy groups (Stevenson, 1985b). Organic matter acts as a major source of plant-available B in many soils, as is evident from a decrease in the availability of B under dry conditions resulting from low microbial activity. Molybdenum is present partly as organically bound Mo (Harter, 1991). Selenium occurs in organic forms partly as organic complexes and partly as Se amino acid synthesized by microorganisms and plants (Gissel-Nielsen et al., 1984; Mayland, 1994). The availability of Se is increased by the presence of Mn oxides, which tend to oxidize selenite to selenate, and possibly by the secretion of exudates in the rhizosphere, which may promote this oxidation (Blaylock and James, 1994). In contrast, the addition of organic matter was found to reduce the uptake of selenate by tall fescue, probably due to an enhanced reduction of selenate to selenite (Ajwa et al., 1998).

17.3.2 Mineralization/immobilization

The mineralization process involves the conversion of plant-unavailable organic forms of nutrients such as N, S, P and some of the trace elements into plant-available inorganic forms by soil microorganisms. Immobilization is the reverse process in which the plant-available nutrients are converted to plant-unavailable forms through incorporation into organic compounds (Fig. 17.2). In most cases mineralization/immobilization processes involve microbial oxidation/reduction reactions, which are discussed in a later section with specific examples. In this section, the general aspects of mineralization/immobilization processes controlling the transformation of N, P and S are discussed.

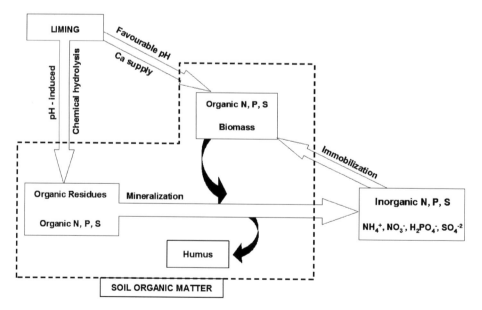

Figure 17.2. *Mineralization/immobilization processes as affected by liming.*

Mineralization/immobilization reactions involved in the cycling of N in soils have been studied in more detail than other nutrients such as P and S. The main reason is that organic N is a major source of plant-available N, which is the most limiting nutrient for both plant growth and microbial activity in most soils. Nitrogen thus controls crop production, the breakdown of organic matter and microbial transformation of other nutrients. In a report (Sparling et al., 2002) from a soil-quality monitoring project (1995–2001) involving 512 different sites and 9 land-use categories, the top 10 cm of soil of dairy and drystock pastures ($n = 269$) was found to contain 5.11 ± 0.09 total N (mg/cm^3). For cropping and horticulture soils ($n = 67$) the total N was 2.73 ± 0.12 (mg/cm^3), and from indigenous forest ($n = 58$) was 3.48 ± 0.16 (mg/cm^3). An early model for estimating nitrogen leaching losses (Di and Cameron, 2000) suggested that 200 kg/ha N would be produced from mineralization annually, and that pasture N uptake would be of the order of 300 kg/ha N. In further work, however, gross N mineralization rates measured under glasshouse conditions were found to be equivalent on an annual basis to between 492 and 1351 kg/ha N (Mishra et al., 2005). Soluble nitrogen in the soil, whether from fertilizer, urine or mineralization of organic matter, in whatever production system, in excess of the ability of plants to assimilate it can result in groundwater contamination (e.g., Crush et al., 1997; Di and Cameron, 2002).

Mineralization of organic N includes aminization and ammonification reactions. Aminization is a microbial process in which the heterotrophic microorganisms hydrolyse the macromolecules of organic N compounds, such as proteins, into simple N compounds, such as amines and amino acids (Eq. (17.1)). Ammonification is a biological process in which a group of microorganisms converts amines and amino acids that are formed by the aminization process and also added through urine and urea fertilizer into NH_4^+ ions (Eq. (17.2)). This process also releases hydroxyl (OH^-) ions and hence the pH around the urea granules or urine spots in soil increases to a maximum of 8. The majority of the NH_4^+ ions produced during the mineralization process undergo microbial oxidation process known as nitrification reaction (see below).

$$\text{Proteins} \rightarrow \text{amines} (R - NH_2) + CO_2 \qquad (17.1)$$

$$CO(NH_2)_2 \rightarrow 2NH_4^+ + 2OH^- + CO_2 \qquad (17.2)$$

Immobilization is a biological process (which could be considered to include plant as well as microbial immobilization) in which the plant-available NH_4^+ and NO_3^- are converted to plant-unavailable organic N. In arable soils, the addition of carbon-rich substances such as maize stubble and cereal straw promotes immobilization, thereby decreasing N availability to plants. The C:N ratio of microbial tissue is approximately 8 (Paul and Clark, 1996). This implies that during the decomposition of organic matter by microorganisms, one unit of N is needed for every eight units of C assimilated. Therefore, when plant residues with a high C:N ratio are added to soils, the lack of N will inhibit the decomposition or the microbes will compete for available N from the soil, resulting in a deficiency of N for the subsequent crop (Jensen, 1997). Immobilization exceeds mineralization when the C:N ratio of the plant residue is greater than approximately 25–30 (Parr and Papendick, 1978), but transient immobilization of soil N even in crop residues having a relatively low C:N ratio (e.g., 15) can contribute to reduced leaching (Jensen, 1997).

Both abiotic and biotic soil processes control P and S dynamics. Abiotic processes include physiochemical reactions such as phosphate and sulphate sorption–desorption and precipitation–dissolution. Biotic processes include the microbial assimilation of phosphate and sulphate into organic forms and the mineralization of plant and animal residues by soil microorganisms through extracellular hydrolytic enzymes, such as phosphatase and sulphatase (Germida and Siciliano, 2001 (this work says that this is what they do – it did not research the function)). Microorganisms and plant roots produce acid phosphatases, bacteria also produce alkaline phosphatases. Phosphatases hydrolyse organic P to inorganic P, thereby making it bioavailable. Similarly, a number of micro-organisms produce sulphatase (e.g., arylsulphatase by *Penicillium* sp.) which hydrolyses organic S such as sulphate esters to inorganic S (Ganeshamurthy and Nielsen, 1990).

The mineralization of P and S in soils during the decomposition of organic matter and the subsequent release of phosphate and sulphate ions occurs in two main phases. The first phase, which takes place within a short period, involves the release of any phosphate or sulphate that is already water-soluble, together with the mineralization of labile organic P and S, while the second phase, which takes place over an extended period, involves the mineralization of more complex organic P and S compounds (Freney et al., 1975).

17.3.3 Microbial oxidation/reduction

Nitrogen, S, Fe, Mn and Se are some of the nutrient elements that are most commonly subjected to microbial oxidation/reduction (redox reactions) reactions. In general, metals are less soluble in their higher oxidation state, whereas in the case of non-metal and metalloids, the solubility and mobility depend on both the oxidation state and the ionic form (cation vs. anion) (Adriano, 2001).

The redox reactions are grouped into two categories, assimilatory and dissimilatory. In assimilatory reactions, microorganisms assimilate only those elements which they need to make protein and body tissue (Brock and Madigan, 1991). Carbon, N, P, S and Se undergo assimilatory redox reactions. The substrate will serve a role in the physiology of the organism by acting as the terminal electron acceptor and permitting growth. Examples include O_2 for aerobes, simple organic molecules for fermentative microbes, sulphate for sulphate-reducing bacteria (e.g., *Desulfovibrio*), nitrate for denitrifying bacteria (e.g., *Flavobacterium*) and CO_2 for methanogenic bacteria. In the dissimilatory reactions, the elemental substrate has no known role in the physiology of the species responsible for the reaction, and represents merely casual reductions coupled to enzymatic or microbial oxidations of some other substrates.

The redox reactions involved in N cycling in soils have been studied in detail in relation to both bioavailability and environmental degradation resulting from nitrate leaching and gaseous emission. Aerobic conditions favour the nitrification of ammonium to nitrate, which may then be susceptible to loss from the soil through leaching. Anaerobic conditions favour the denitrification of nitrate with the loss of gaseous N_2 and N_2O (Tiedje, 1982).

The biological oxidation of NH_4^+ to NO_3^- is known as nitrification (Eq. (17.3)). Ammonium ions are released either indirectly from the ammonification reaction of organic matter, urea and organic forms of N fertilizers or directly from the solubilization of ammonium fertilizers. The nitrification reaction is a two-step process in which the NH_4^+ ions are first converted (oxidized) into nitrite (NO_2^-) and then to NO_3^-. Since the rate of conversion of NO_2^- to NO_3^- is faster than the conversion of NH_4^+ to NO_2^-, it is unlikely that NO_2^-, which is toxic to plants, accumulates under most soil and climatic conditions. The nitrification process produces H^+ ions, thereby decreasing the pH.

$$2NH_4^+ + 4O_2 \rightarrow 2NO_3^- + 4H^+ + 2H_2O \qquad (17.3)$$

Denitrification is the last step in the N cycle, where the fixed N is returned to the atmospheric pool of N_2. Biological denitrification is defined as the dissimilatory reduction of NO_3^- or NO_2^- by essentially anaerobic bacteria producing molecular N_2 or oxides of N when oxygen is limiting (Payne, 1981; Zumft, 1997). Denitrification is carried out by respiratory denitrifiers that gain energy by coupling N oxide reduction to electron transport phosphorylation during anaerobic respiration (Tiedje, 1988). Denitrifying bacteria can be present in nearly all soils and are generally facultative aerobes. The main genera capable of denitrification in soil include *Pseudomonas*, *Bacillus*, *Alculigenes* and *Flavobacterium* (Payne, 1981; Firestone, 1982; Tiedje, 1988).

Many denitrifying bacteria are chemoheterotrophs, i.e., they can use NO_3^- as their primary electron acceptor to obtain energy from organic compounds (heterotrophic denitrification). In addition, some autotrophic organisms can obtain energy by using NO_3^- for oxidation of inorganic compounds, such as elemental sulphur (autotrophic denitrification). As facultative aerobes, denitrifiers can be considered bacteria that prefer to use O_2 as their electron acceptor, and can use NO_3^- as a terminal acceptor of electrons only when O_2 is not available. Under conditions of limited O_2 availability, aerobic respiration can apparently provide the energy needed for synthesis of new enzymes required for NO_3^- reduction.

The general pathway for the reduction of NO_3^- during the denitrification process may be represented by Eq. (17.4) (Payne, 1981; Firestone, 1982). The general requirements for biological denitrification are: (a) the presence of bacteria possessing the metabolic capacity, (b) suitable electron donors such as organic C compounds, (c) anaerobic conditions or restricted O_2 availability and (d) N oxides, NO_3^-, NO_2^-, NO or N_2O as terminal electron acceptors. The process of denitrification is therefore generally promoted under anaerobic conditions, high concentrations of soil NO_3^- and a readily available source of carbon.

$$5(CH_2O) + 4NO_3^- + 4H^+ \rightarrow 2N_2 + 5CO_2 + 7H_2O \qquad (17.4)$$

Sulphur compounds are also subject to microbially mediated oxidation and reduction reactions. In aerobic conditions, metal sulphides (e.g., pyrite) are oxidized microbially to sulphuric acid (Evangelou and Zhang, 1995). Similarly, elemental S, which is used as an important source of slow-release S fertilizer, is oxidized to sulphate ions available for plant uptake (Watkinson and Bolan, 1998). However, severely anaerobic conditions may result in sulphate being converted to insoluble metal sulphides or, in some circumstances, to volatile organic sulphides (Lomans et al., 2002). Since the metal sulphides are insoluble, severely anaerobic conditions may result in elements such as Zn, Cu and Co becoming less available for plant uptake (Barrett and McBride, 2007). Also anaerobic conditions may increase the availability of Fe through changing its oxidation from Fe^{3+} to Fe^{2+} (Ratering and Schnell, 2000), and moderately anaerobic conditions may increase the solubility of other elements, such as Mn, Cu, Co and Mo, associated with the hydrous Fe oxides (Sahrawat, 2005).

Selenium occurs in organic forms, and is therefore subject to microbial redox reactions. Although Fe and Mn do not occur in organic forms in soils, microorganisms mediate their transformation through redox reactions. Redox reactions involving Fe and Mn in particular have generally been attributed to indirect action of microorganisms (e.g., electron donors are thought to be reduced fermentation products), although there has been some evidence suggesting the use of these metals as electron acceptors (Tugel et al., 1986). For example, oxidation of Fe and Mn occurs spontaneously in the absence of microbial activity when reduced environments are exposed to oxygen (Alexander, 1977).

In living systems Se tends to be reduced rather than oxidized. Selenium is reduced under both aerobic and anaerobic conditions. Dissimilatory selenate reduction to Se° is the major biological transformation for remediation of Se oxyanions in anoxic sediments. Selenate [Se(VI)] is more mobile than selenite [Se(IV)] because the later is strongly adsorbed onto soil minerals and organic matter under near neutral pH conditions. When Se(IV) and Se(VI) are introduced into moderately reducing conditions, they are quickly transformed through microbial processes to Se° and/or organic Se compounds. Selenite is readily reduced to the elemental state by chemical reductants, such as sulphide and hydroxylamine, or biochemically by systems, such as gluthione reductase. Since both selenate and nitrate can be used as terminal electron acceptors by many microorganisms, presence of nitrate in the system inhibits the reduction of Se (Frankenberger and Losi, 1995).

17.3.4 Methylation/demethylation

The redox reactions of trace elements such as As, Se and Hg also include methylation/demethylation reactions. Methylation is considered to be the major process of volatilization of As, Hg and Se in soils and sediments, resulting in the release of poisonous methyl gas (Huysmans and Frankenberger, 1991; Zhang et al., 1999; Zhang and Frankenberger, 2000; Ullrich et al., 2001). Although methylation of metals occurs through both chemical (abiotic) and biological processes, biological methylation (biomethylation) is considered to be the dominant process in soils and aquatic environments (Allen, 1991; Gilmour et al., 1992; Kerry et al., 1991; Sadiq, 1997; Azaizeh et al., 1997, 2003). Thayer (2002) suggested that methylation could be considererd in two categories: trans-methylation and fission-methylation. Trans-methylation refers to the transfer of an intact methyl group from one compound (methyl donor) to another compound (methyl acceptor). Fission-methylation refers to the fission of a compound (methyl source), not necessarily containing a methyl group. This is then captured by another compound and the resulting molecule is reduced to a methyl group.

At present there is substantial evidence for the biomethylation of Se in soils and aquatic systems (Frankenberger and Arshad, 2001). Microorganisms in soils and sediments act as biologically active methylators. Organic matter provides the source of methyl donor for both biomethylation and abiotic methylation in soils and sediments.

Selenium biomethylation (Fig. 17.3) is of interest because it represents a potential mechanism for the removal of Se from contaminated environments,

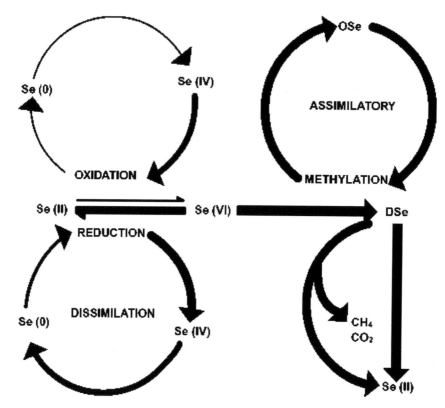

Figure 17.3. *Biotransformation of selenium in soils (key required).*

and it is believed that methylated compounds, such as dimethyl selenide (DMSe) are less toxic than dissolved Se oxyanions. Fungi predominate among the Se methylating microbes in soils, although some bacterial isolates have also been identified. Hydrogen oxidizing methanogens such as *Methanobacterium omelianskii* are involved in the reductive methylation; demethylation is effected by methylotrophic bacteria. Dimethyl selenide can be demethylated in anoxic sediments as well as anaerobically by an obligate methylotroph similar to *Methanococcoides methylutens*.

Five different volatile forms of reduced Se have been detected: hydrogen selenide (H_2Se), methaneselenol (CH_3SeH), dimethyl selenide (CH_3SeCH_3), dimethyl selenenyl sulphide (CH_3SeSCH_3) and dimethyl diselenide (CH_3Se-$SeCH_3$). The relatively high vapour pressure of these compounds enhances the transformation of Se from soils and sediments to aqueous and vapour phases. However, the rapid oxidation of the first two in this list and lower vapour pressure of the last two leave CH_3SeCH_3 as the most significant contributor to atmospheric Se input (Frankenberger and Losi, 1995). Anaerobic demethylation reaction may result in the formation of toxic and reactive H_2Se from less toxic DMSe. Although H_2Se undergoes rapid chemical (possibly bacterial) oxidation under oxic conditions, it can exist for long periods in an aerobic environment. Aerobic demethylation of DMSe will yield selenate, thereby retaining Se in the system.

Methylation has been proposed as a mechanism for detoxification – a biological mechanism for the removal of toxic metals by converting them to methyl derivatives that is subsequently removed by volatilization or extraction with solvents (Frankenberger and Losi, 1995).

17.4 Soil acidity and bioavailability of nutrients

Acidity affects the diversity and activity of microorganisms (Holub et al., 1993), and so can influence the rate of organic matter decomposition. This in turn influences the number of simple organic molecules available for microbial growth and functioning that are necessary for eventual mobilization of N and other constituent elements (P and S) of soil organic matter (Alexander, 1977). In highly acidic conditions, organic matter accumulates giving rise to vast storehouse of nutrients that can be exploited by subsequent liming (Marschner and Wilczynski, 1991; Prescott et al., 2000; Godbold, 2003).

17.4.1 Major nutrients

In many soils, organic matter is the main source of N, P and S and, since mineralization of organic matter is affected by acidity, the release of these nutrients for plant uptake can decrease as acidity increases. Further, in highly weathered acid soils, phosphate and sulphate ions are adsorbed by sesquioxide surfaces, precipitated as Fe and Al phosphate and as Al sulphate (Marsh et al., 1987; Haynes, 1984; Mora et al., 1999c), thereby limiting their accessibility for microbial transformation and uptake by higher plants.

Among microbial nitrogen transformation processes, nitrification has been predicted to be the most sensitive to soil acidity, because it is catalysed by a relatively limited diversity of autotrophic nitrifiers (Firestone et al., 1984). Nitrification is markedly reduced below pH 6 and is undetectable below pH 4.5 (Alexander, 1977). In contrast, ammonification reactions are insensitive to acidity over the range of pH typified in agricultural soils, resulting in the accumulation of NH_4^+–N. Thus, the effect of acidity on the soluble N status of most agricultural soils is limited to nitrification, implying that for plants which use both NH_4^+ and NO_3^-, acidity is unlikely to control the availability of N. However, for certain plant species that are unable to use NH_4^+–N, acidification can result in restricted uptake of N or even NH_4^+ toxicity (Marschner, 1995).

Denitrification is expected to be less sensitive to soil acidity than nitrification simply because it is catalysed by a relatively diverse group of heterotrophic microorganism. While this process has been shown to be inhibited in extremely acid soils (Nägele and Conrad, 1990), the overall rates of the reaction have not been found to be particularly sensitive to slight shifts in soil pH (Firestone, 1982; Bolan et al., 2003a, 2003b). However, soil acidity has often been shown to influence the $N_2O:N_2$ ratio in the gases produced. The proportion of N_2O increases as pH decreases, with N_2O frequently being the dominant product in acid soil (Christensen, 1985; Parkin et al., 1985). However, results of research investigating whether liming can mitigate N_2O fluxes from urine-amended soil, prompted by the contribution of N_2O from agricultural soils, suggested that the soil pH giving the greatest N_2O fluxes derived from nitrification was approximately 5.8–6.1 (Clough et al., 2003); a pH range of approximately 4.7–7.7 was investigated, and the authors did suggest that further research was required under conditions suitable for denitrification to examine the fate of the nitrate pool.

17.4.2 Trace element nutrients

The effect of acidity on biotransformation of trace elements is mediated through its effect on geochemical reactions such as chemical speciation, adsorption/desorption and precipitation/dissolution. In general, while the solubility of trace element metals such as Zn and Cu increases with a decrease in pH, the reverse is true for non-metals such as Mo and Se. However, as Cu is complexed with organic matter, the slow rate of decomposition of organic matter in acid soils decreases the release of Cu (Cavallaro and McBride, 1980; Jeffery and Uren, 1983).

A decrease in pH results in increasing concentrations of soluble Mn and below pH 6, Mn can adversely affect the growth of sensitive crops (Jones and Fox, 1978). However, redox conditions usually influence concentrations of soluble Mn more than pH. Similarly because of low solubility of iron (Fe) even under very acid conditions (Lindsay, 1979), redox reactions are likely to be of greater importance in the rhizosphere in dissolving sufficient Fe to meet the requirements of plants (Marschner, 1995). The solubility of Mo and Se is affected by their speciation in soils, which in turn is controlled largely by pH and Eh.

17.5 Liming and bioavailability of nutrients

In the case of managed ecosystems used for agricultural production, the rate of acid generation can be altered by decreasing the losses of C, N and S from the system, selecting plant species that do not accumulate cation excesses, and selecting the fertilizer type that produce less acid (Gregan et al., 1989; Edmeades, 2003).

Traditionally, liming is the most common practice used to overcome the impact of soil acidification and to create conditions conducive to achieve crop and animal production at the most favourable economic rates (Sumner and Yamada, 2002). To achieve such production on acid soils, it is necessary to apply sufficient lime to (a) eliminate toxicities of Al^{3+} and Mn^{2+}, (b) supply adequate levels of Ca^{2+} and Mg^{2+}, (c) facilitate the use of water, (d) create conditions which maximize the availability and uptake of the essential nutrient and the performance of rhizobial-legume association and (e) create condition which control soil pathogens. Lime recommendations are based on achieving a target pH, base saturation or acid $(Al^{3+}+H^{+})$ saturation to optimize plant growth (Farina and Channon, 1991).

Liming enhances the physical, chemical and biological characteristics of soil through its direct effect on the amelioration of soil acidity and through its indirect effects on the mobilization of plant nutrients, immobilization of toxic heavy metals and the improvements in soil structure and hydraulic conductivity (Haynes and Naidu, 1998) (Fig. 17.4).

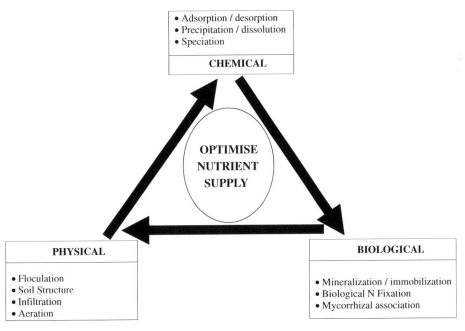

Figure 17.4. *Effect of liming on the physical, chemical and biological properties of soil in relation to nutrient supply.*

Liming has been shown to provide optimum conditions for a number of biological activities that include N fixation (Bååth and Arnebrant, 1994; Wheeler et al., 1997), and mineralization of N, P and S in soils (Anderson, 1998). The enhanced mineralization of these nutrient ions is likely to cause an increase in their concentration in soil solution for plant uptake and for leaching (Lyngstad, 1992; Arnold et al., 1994; Neale et al., 1997). Nitrogen-fixing bacteria in legume plants require Ca, hence liming is likely to enhance N fixation (Muchovej et al., 1986).

Liming is often recommended for the successful colonization of earthworms in pasture soils. The lime-induced increase in earthworm activity may influence the soil structure and macroporosity through the release of polysaccharide and the burrowing activity of earthworm (Springett and Syers, 1984; Robinson et al., 1992). Liming has been shown to cause short-term increases in microbial biomass and soil enzyme activity (Haynes and Swift, 1988). Increased microbial activity and the subsequent production of extracellular polysaccharides which act as a binding agent can increase soil aggregate stability (Kay, 1998; Kay and Angers, 2000).

In summary, liming induces biological transformation of nutrient elements through a number of processes. These include (i) provision of favourable pH conditions for the proliferation of microbial diversity and functioning, (ii) mitigating metal toxicity on biological activities, (iii) increasing the bioavailable carbon content through pH-induced chemical solubilization of soil organic matter and (iv) increasing the bioavailability of nutrient ions through pH-induced desorption.

17.5.1 Major nutrients

Although acidity is recognized as an important regulator of microbial activity and diversity (Paul and Clark, 1996), the agronomic significance of its effect has been difficult to assess. From a review of the literature, Adams and Martin (1984) concluded that mineralization of organic matter occurs over a wide pH range, but the rate decreases below about pH 6. In a liming study in Wisconsin, Dancer et al. (1973) showed that N mineralization was not affected by pH in the range 4.7–6.6, but nitrification was decreased 3–5-fold as pH decreased. Waring and Gilliam (1983) found that liming had little influence on mineralization in acid (pH < 5) Histosols in North Carolina. In contrast, Curtin and Smillie (1986) partly attributed lime-induced yield increase of two grass species to increased availability of N. In a study with 40 soils from Alberta (Canada), Nyborg and Hoyt (1978) found no relationship between N mineralized in 120 days and soil pH. Even so, liming to pH 6.7 almost doubled the amount of N mineralized during a laboratory incubation. In an associated field study, liming increased crop N uptake by 15–42 kg/ha in the first year, but by only 7–10 kg/ha in the second year.

In a laboratory study, Curtin et al. (1998a) found that liming two slightly acid (pH 5.7–5.8) soils increased mineralization of N and C. The effect of liming on mineralization was temporary (approximately 28 days) and was attributed to the release of labile organic matter when soil pH was increased.

Liming has often been shown to enhance the mineralization of organic matter, thereby releasing inorganic plant nutrients such as N, S and P to soil solution. Unless these nutrients are actively taken up by plants they can be leached from the soil profile (Simmons et al., 1996), even in the case of P (e.g., Sims et al., 1998; Toor et al., 2004).

Liming affects both the chemical and microbial transformation of N in soils. In general, NH_4^+–N is nitrified more rapidly on addition of lime due to an increase in the activity of microorganisms involved in nitrification (Lyngstad, 1992; Puttanna et al., 1999). The efficiency of nitrification inhibitors decreases with the addition of lime. This is probably due to an increase in nitrifier activity and also due to an increase in general microbial activity (Slangen and Kerkhoff, 1984), resulting in rapid biodegradation of nitrification inhibitors. All these processes result in an increase in NO_3^- concentration, potentially leading to NO_3^- leaching, especially in the absence of active plant uptake (Kreutzer, 1995).

A number of laboratory and field studies have examined the effects of lime on nutrient transformations in forest soils affected by acid rain. For example, Persson et al. (1990) observed that liming resulted in lower net N mineralization rate in needle litter and mor humus of coniferous soils. The reduction was more pronounced in soils dominated by NH_4^+ than those dominated by NO_3^-. This was attributed to the increase in fixation of NH_4^+ with increasing pH resulting from liming. Similarly, Backman and Klemedtsson (2003) examined the potential nitrification of acid soils 6–7 years after liming at four coniferous forest sites in the central and western parts of southern Sweden. The effect of liming on the total N pool and potential nitrification was limited to the uppermost 10 cm that is rich in N. This suggests that the effect of liming on the potential nitrification is dependent on N availability as influenced by N mineralization, trees' demands for N and atmospheric N inputs.

Borken and Brumme (1997) examined the effect of surface liming of acidic forest soils on CO_2 and N_2O emissions and CH_4 sequestration using closed chambers in two deciduous and two spruce forests. Liming had no clear effect on CO_2 emissions, but resulted in a reduction of N_2O emissions by 9–62% and in an increase of CH_4 uptake by 26–580%. Results of the research reported by Clough et al. (2003), already discussed, indicate that the relationship between pH and N_2O is complex. Conflicting results in the literature have been attributed to antecedent moisture regime, carbon supply, bacterial populations present, temperature, nitrate concentration and soil redox potential (Beauchamp et al., 1989; Firestone and Davidson, 1989; Nägele and Conrad, 1990; Stevens et al., 1998).

Peat soils are generally acidic and require regular lime application in order to achieve sustainable production (e.g., Schmehl et al., 1950; Li et al., 2006 a, 2006b). In an examination of the effect of liming on N and P mineralization in peat soils, Chapman et al. (2003) found that N mineralization was higher in fen peat than in peat bogs, but net P mineralization was negligible in both sites. Adding lime had no effect on the labile N pool, and cumulative N and P mineralization in peat bogs. However, lime additions decreased potentially mineralizable N and cumulative N mineralization in the fen aerobic incubations, but increased cumulative P mineralization in the fen anaerobic incubations. Although both are peatlands, bog and fen nutrient cycles are controlled by different environmental factors including redox conditions that may explain the differences in the effect of lime on mineralization of N and P.

The forest floor is expected to play a major role in determining the total ecosystem response to watershed liming because of its high concentration of nutrients and its high level of microbial activity. Simmons et al. (1996) estimated net N mineralization and net nitrification in a field survey using the buried-bag approach. Both the laboratory and field experiments demonstrated that N mineralization increased at high lime doses (pH 5–6), thereby inducing nitrate leaching. However, nitrate leaching from microcosms with live roots was not affected by liming, suggesting that roots in the forest floor may prevent excess nitrate leaching. Similarly, Kreutzer (1995) showed that liming forest soils caused a loss of about 170 kg N/ha or 14% of the store of the surface humus layer mainly through leaching. The results indicated that liming could aggravate the increasing problem of nitrate contamination of forest groundwater resources which is associated with deposition of atmospheric nitrogen compounds.

Bailey (1995) indicated that liming affects N recovery by grassland swards in at least two different ways, each associated with a different phase in the soil mineralization immobilization turnover cycle. During phases of net N mineralization, liming by raising soil pH stimulates biomass activity and increases the amount of organic N mineralized. In contrast, during phases of net N immobilization, liming by increasing Ca availability in the rhizosphere improves the ability of plants to absorb N, and thus helps them to compete more effectively with the biomass for mineral N.

Liming may either increase or decrease the supply of plant-available P in soils (Haynes, 1984). Increase in P availability resulting from liming has often been attributed to the mitigation of Al-induced inhibitory effect on P uptake and the physico-chemical mobilization of P resulting from increased pH. Liming may also accelerate the rate of organic P mineralization due to increased rates of microbial activity, but the practical significance of this effect remains unclear because of the difficulty in measuring P mineralization rates. At a long-term experimental site in New Zealand, Condron and Goh (1989) attributed declines in organic P in the 0–7.5 cm soil layer between 1971 and 1974 to increased mineralization as a result of liming in 1972. Similarly, Chepkwony et al. (2001) noticed that liming a pasture soil rich in organic P (total P = 816 mg/kg; organic P = 440 mg/kg) caused an increase in the mineralization of organic P, as measured by a decrease in the specific activity of ^{32}P in Bray No 1 and resin extracts. Armstrong and Helyar (1992) noticed that liming of an acid soil had only a minor effect on the distribution of added P between various soil inorganic fractions but resulted in a depletion of the organic fraction of 0.5 M NaHCO$_3$ and 1.0 M HCl, which they attributed to the increase in the bioavailability of organic P due to liming.

There have been relatively few studies in which the so-called *P-sparing effect* of lime (Tillman and Syers, 1982) has been quantified under field conditions. The *P-sparing effect* of lime refers to a situation in which lime increases the actual availability of P to the plant. It excludes situations where plant P uptake is increased as a result of elimination of Al (or Mn) toxicity by liming. An increase in the availability of P can be expected only on soils which have accumulated some reserve (in this case organic P or inorganic "fixed" P) of this element.

Liming also affects P availability through its effect on mycorrhizal fungi (Sano et al., 2002; Johnson et al., 2005). Antibus and Linkins (1992) noticed that liming did not affect the diversity of ectomycorrhizal morphological types, but increased the relative frequency of certain morphotypes. Lime-induced shifts in ectomorphotypes have the potential to alter the mineralization of organic P. Liming reduced soil phosphatase activity. Bååth and Arnebrant (1993), while examining microfungal species composition in coniferous forest soils which had been treated with lime or wood ash, noticed clear differences in species composition due to liming in one site: *Penicillium spinulosum, Oidiodendron* cf. *truncatum, Mortierella* spp., and two sterile taxa decreased in abundance in limed areas, whereas *Geomyces pannorum, Penicillium* cf. *brevicompactum, Trichoderma polysporum* and *Trichosporiella sporotrichioides* increased in isolation frequency.

In pasture soils, liming has been shown to have only a minor effect on the distribution of added phosphate between the various soil inorganic P fractions but results in a depletion in the organic fractions indicating lime-induced mineralization of organic P (Armstrong and Helyar, 1992). Condron et al. (1993) showed that liming altered the potential availability of soil organic matter and organic P, as shown by a marked reduction in the extractability of soil organic P with sodium bicarbonate and sodium hydroxide. Similarly, McDowell (2003) characterized the relative quantity of P forms and potential P loss from a series of grassland soils with a range of Olsen P concentrations (14–120 mg/kg), liming (e.g., pH 4.1–6.6) and fertilization histories. Forms and concentrations of P species such as the distribution of mono- and diester P were a function of soil management, i.e., greater concentrations of these species were noted at lower soil pH.

Applying lime has been shown to increase the concentration of SO_4^{2-} in soil solution (Bolan et al., 1988; Mora et al., 2005) and several reasons have been proposed to explain this (Korentajer et al., 1983). These include (i) SO_4^{2-} mineralized from soil organic matter by microorganisms growing in a more favourable pH environment; (ii) SO_4^{2-} released from organic matter by chemical hydrolysis; (iii) adsorbed SO_4^{2-} released from the soil surface and (iv) SO_4^{2-} released from sparingly soluble Fe and Al hydroxy sulphates, which become more soluble at higher pH.

Leaching of SO_4^{2-} from a New Zealand soil with high S retention capacity increased several-fold when pH was increased from 4.7 to 7.0 by application of $CaCO_3$ (Bolan et al., 1988). In situations where S supply is marginal, enhanced SO_4^{2-} leaching after liming could lead to S deficiency. Decreases in the uptake of S by corn (Zea mays L.) from two limed soils which were subjected to leaching under greenhouse conditions provided some evidence that leaching-induced S losses may lead to S deficiencies in plants grown in limed soils (Korentajer et al., 1983).

Valeur and Nilsson (1993) examined the long-term effect of liming on S mineralization in open (with regular leaching) and closed incubation systems using mor humus soils collected 6 years after dolomitic lime (8800 kg/ha) application to Norway spruce stand in southern Sweden. The open incubation resulted in a higher net S mineralization in both the limed and control humus, and limed humus had a lower net S mineralization regardless of the incubation technique used. More organic S was leached from the open incubation columns when lime had been applied than in the control. Limed humus had the greatest microbial activity irrespective of the incubation technique. Immobilization of sulphate was greater in the limed incubation than in the control. In another study, Valeur et al. (2000) observed that the net mineralization of S in a coniferous soil, 7 years after liming, decreased with increasing liming rates, while the soil respiration as measured by CO_2 release increased at high liming rates. The inverse relationship between net S mineralization and soil respiration indicated an increase in S immobilization at increasing lime application rates.

The effect of three types of liming material differing in Ca^{2+} content [calcium carbonate ($CaCO_3$), dolomite ($CaMg(CO_3)_2$) and magnesium carbonate ($MgCO_3$)] on net S mineralization was examined by Valeur et al. (2000). In the limed treatments, the leaching of dissolved organic carbon (DOC) and the net S mineralization increased with decreasing Ca^{2+} content of the lime and increasing degree of lime dissolution. The accumulated SO_4^{2-} leaching was greater in the limed soil than the control treatment. There were consistently significant positive correlations between the amount of SO_4^{2-} leached and DOC. These results indicate that net S mineralization is strongly related to the solubilization of the organic matter (DOC formation) and that pH and/or Ca^{2+} affect the net S mineralization through their effects on organic matter solubility.

Marschner (1993) measured S release from unsterilized and γ-irradiation sterilized soil samples. In unsterilized soil samples, applying lime to the acid forest topsoil caused an immediate increase in extractable sulphate, but enhanced sulphate release rates were detectable only during the first 4 weeks of incubation. In the sterilized samples, γ-irradiation initially caused an increase in extractable sulphate which was attributed mainly to the destruction of microbial biomass. After the second week, no further differences in extractable sulphate were detectable between the treatments. The data indicated that even five years after applying lime the observed increased sulphate concentrations in the soil solution could be attributed to the mineralization of organic S due to enhanced microbial activity.

17.5.2 Trace element nutrients

Liming acid soils affects the biotransformation of trace element nutrients and their subsequent bioavailability through a number of processes that include (i) pH-induced changes in trace element speciation; (ii) pH-induced changes in trace element sorption/desorption; (iii) competitive effect of lime-derived Ca^{2+} on sorption; (iv) pH-induced mobilization of DOC and its subsequent effect on chelation of trace elements and (v) changes in redox and methylation/demethylation reactions.

With the exception of Mo and Se, bioavailability of most other trace element nutrients decreases with liming mainly due to decrease in the concentration of these elements in soil solution resulting from increased retention. For example, most problems with Fe nutrition are encountered when pH is raised by liming, resulting in a sufficient depression of Fe solubility to limit uptake in crops (Marschner, 1995). This phenomenon is often referred to as "lime-induced iron chlorosis". The effect of pH >6 in lowering free metal ion activities in soils has been attributed to the increase in pH-dependent surface charge on oxides of Fe, Al and Mn (Stahl and James, 1991), chelation by organic matter or precipitation of metal hydroxides (Lindsay, 1979).

Strong complexation of Cu by soil organic matter is believed to be an important factor in explaining why Cu deficiencies are not as prevalent as Zn deficiencies in limed soils, even though the two cations show similar diminution in solubility with increasing pH (Lindsay, 1979). For example, Kreutzer (1995) noticed that the concentration of such metals as Cu and Pb, which form stable organic complexes, increased in the soil solution after liming a Norway Spruce forest soil. Thus, liming may trigger the "chemical time bomb effect" (i.e., solubilization and release of metals) of certain metals and turnover of organic matter plays a crucial part in this lime-induced effect (Mayer, 1993).

Unlike most other trace elements, bioavailability of Mo and Se in soils is greatest under alkaline pH than under acidic condition. Liming has been shown to increase the mobilization of Mo through solubilization of Mo held in organic matter and desorption of adsorbed Mo from silicate clay minerals iron and aluminium oxides (Lang and Kaupenjohann, 1998). Thus, liming acid soils often helps to correct Mo deficiency. Conversely, plant response to application of Mo under field conditions is more effective on acid soils (Adriano, 2001). However, liming can increase the amount of plant-available Mo only on soils which have a reserve of Mo. There are some soils in New Zealand (e.g., some highly weathered "gumland" soils of Northland and deep acid peats in the Waikato) that have no reserve Mo and hence liming is ineffective in raising available Mo. In such cases Mo must be applied as a fertilizer.

Liming of soil has been reported to have a small but significant negative effect on plant uptake of Se (Carlson et al., 1991; Hansson et al., 1994). However, Hamdy and Gissel-Nielsen (1977) noticed that liming of a sandy soil increased Se volatilization by 1.2 fold.

Liming affects biotransformation processes through its effects on the adsorption and speciation of metals. For example, Rogers (1977) observed that the extent of methylation of Hg decreased with an increase in soil pH. This has been attributed to the unavailability of Hg(II) at high pH due to its stronger adsorption and to the reduced supply of methylating organic matter at high pH. Neutral and alkaline environments favour dimethylmercury; under acidic conditions less volatile monomethyl mercury is formed, reducing the release of volatile Hg to atmosphere.

Liming can also influence bioaccumulation of metals by microorganisms. An increase in pH facilitates surface binding of metals (Crout et al., 2006) through pH-induced increases in surface charge. Furthermore, energy dependent metal uptake is frequently pH-dependent and maximum rates are observed between 6 and 7 (Wilkinson et al., 2002).

17.6 Conclusions and future research needs

Soil acidification is a natural process that can be accelerated by farming activities. Under managed farming systems, most of the acidity is generated from C, N and S biogeochemical cycles in soils and from fertilizer application. Soil acidification decreases the availability of major plant nutrients mainly through inhibition of microbial activity, and induces the phytotoxicity of metals, mainly through chemically induced solubilization.

The primary objective of liming soils used for agricultural production is to reduce the concentrations of Al and Mn, which are phytotoxic and are soluble in acid soils, thereby limiting the potential of these soils for production. Liming acid soils also provides optimum conditions for microbial activity, thereby enhancing the rate of decomposition of organic matter and the consequential mobilization of nutrients such as N, P and S and some of the trace elements associated with organic matter for plant uptake.

Liming has been identified in the enhanced mineralization and nitrification of organic N, and has been shown to decrease the amount of P fertilizer required to boost yield in some soils. This reduction in P requirements results directly from an increased solubilization of organic and inorganic soil P and its subsequent uptake and indirectly from an increase in P uptake due to reduced Al and Mn toxicity. Liming is practiced as a management tool to enhance the solubilization of trace elements such as Mo and Se, thereby increasing their bioavailability. Liming can be used to control the volatilization loss of trace elements which are subject to methylation reactions.

Future research should be focussed on ecological liming effects, such as microbial diversity and functioning, dynamics of organic matter and mobilization of nutrient elements from organic matter. The impact of liming on microbial-plant interactions (e.g., mycorrhizal association) should be examined in relation to nutrient cycling and bioavailability. The role of liming in microbially mediated processes that are effective in the bioremediation of metal contaminants in terrestrial and aquatic system needs to be examined under field conditions.

Clearly, soil biology, and the influence upon it of changes in pH, is extremely important in soil function. If we want to understand and know more about how to mitigate pollutants and pesticides, and the pros and cons of different production systems (including no-till, organic and conventional), we must understand more about soil biology. Soil biology drives nutrient availability (through decomposition, mineralization and immobilization), soil aggregation and pesticide degradation. It is the mediator for biogeochemical cycling and understanding more will allow improved efficiency of production systems in terms not only of nutrient uptake but also of plant health through the employment of microorganisms rather than pesticides. It will also allow agriculture to reduce impacts on soil erosion and hence on marine environments.

Results from research in different countries cannot simply be transported to other countries because of the unique combinations of intensive and extensive production, soil types, climate, pests and diseases that exist. Each country needs its own programme in research in soil biology, and also needs funding to maintain long-term trials, and, indeed, to attract the next generation of researchers.

Of further note is that results from soil biology research are not germane to only one industry, and so the research is unlikely to attract industry funding. Soil biology is at the core of national productivity, and so much research must be funded as a national good.

References

Adams, M.A., Martin, J.B., 1984. Liming effects on nitrogen use and efficiency. In: R.D. Hauck (Ed.), Nitrogen in Crop Production. ASA, CSSA, and SSSA, Madison, WI, pp. 417–426.

Adriano, D.C., 2001. Trace Elements in Terrestrial Environments: Biogeochemistry, Bioavailability and Risks of Metals. 2nd Ed., Springer, New York.

Ajwa, H.A., Bañuelos, G.S., Mayland, H.F., 1998. Selenium uptake by plants from soils amended with inorganic and organic materials. J. Environ. Qual. 27, 1218–1227.

Alewell, C., Armbruster, M., Bittersohl, J., Evans, C.D., Meesenburg, H., Moritz, K., Prechtel, A., 2001. Are there signs of acidification reversal in freshwaters of the low mountain ranges in Germany? Hydrol. Earth Syst. Sci. 5(3), 367–378.

Alexander, M., 1977. Introduction to Soil Microbiology. 2nd Ed., Wiley, New York.

Allen, K.N., 1991. Seasonal variation of selenium in outdoor experimental stream–wetland systems. J. Environ. Qual. 20, 865–868.

Anderson, T.H., 1998. The influence of acid irrigation and liming on the soil microbial biomass in a Norway spruce (*Picea abies* [L.] k.) stand. Plant Soil 199(1), 117–122.

Andrews, J.A., Schlesinger, W.H., 2001. Soil CO_2 dynamics, acidification and chemical weathering in a temperate forest with experimental CO_2 enrichment. Global Biogeochem. Cycles 15, 149–162.

Antibus, R.K., Linkins, A.E., 1992. Effects of liming a red pine forest floor on mycorrhizal numbers and mycorrhizal and soil acid-phosphatase activities. Soil Biol. Biochem. 24(5), 479–487.

Armstrong, R.D., Helyar, K.R., 1992. Changes in soil phosphate fractions in the rhizosphere of semi-arid pasture grasses. Aust. J. Soil Res. 30(2), 131–143.

Arnold, G., Vanbeusichem, M.L., van Diest, A., 1994. Nitrogen mineralization and H^+ transfers in a scots pine (*Pinus sylvestris* L.) forest soil as affected by liming. Plant Soil 161, 209–218.

Azaizeh, H.A., Gowthaman, S., Terry, N., 1997. Microbial selenium volatilization in rhizosphere and bulk soils from a constructed wetland. J. Environ. Qual. 26, 666–672.

Azaizeh, H.A., Salhani, N., Sebesvari, Z., Emons, H., 2003. The potential of rhizosphere microbes isolated from a constructed wetland to biomethylate selenium. J. Environ. Qual. 32, 55–62.

Bååth, E., Arnebrant, K., 1993. Microfungi in coniferous forest soils treated with lime or wood ash. Biol. Fertil. Soils 15, 91–95.

Bååth, E., Arnebrant, K., 1994. Growth rate and response of bacterial communities to pH in limed and ash treated soils. Soil Biol. Biochem. 26, 995–1001.

Backman, J.S.K., Klemedtsson, A.K., 2003. Increased nitrification in acid coniferous forest soil due to high nitrogen deposition and liming. Scand. J. Forest Res. 18(6), 514–524.

Badalucco, L., Kuikman, P., 2001. Mineralization and immobilization in the rhizosphere. In: R. Pinton, P. Nannipieri (Eds.), The Rhizosphere. Biochemistry and Organic Substances at the Soil–Plant Interface. Marcel Dekker, New York, pp. 159–196.

Bailey, J.S., 1995. Liming and nitrogen efficiency – some effects of increased calcium supply and increased soil-pH on nitrogen recovery by perennial ryegrass. Commun. Soil Sci. Plant Anal. 26(7–8), 1233–1246.

Barrett, K.A., McBride, M.B., 2007. Dissolution of zinc-cadmium sulfide solid solutions in aerated aqueous suspension. Soil Sci. Soc. Am. J. 71, 322–328.

Barrow, N.J., 1971. Slowly available sulphur fertilizers in south-western Australia. 1. Elemental sulphur. Aust. J. Exp. Agric. Animal Husbandry 11, 211–216.

Beauchamp, E.G., Trevors, J.T., Paul, J.W., 1989. Carbon sources for bacterial denitrification. Adv. Soil Sci. 10, 114–142.

Bergholm, J., Berggren, D., Alavi, G., 2003. Soil acidification induced by ammonium sulphate addition in a Norway spruce forest in Southwest Sweden. Water Air Soil Pollut. 148(1–4), 87–109.

Bertin, C., Yang, X.H., Weston, L.A., 2003. The role of root exudates and allelochemicals in the rhizosphere. Plant Soil 256(1), 67–83.

Blackmer, A.M., 2000a. Bioavailability of nitrogen. In: M.E. Sumner (Ed.), Handbook of Soil Science. CRC Press, Boca Raton, FL, pp. D3–D17.

Blackmer, A.M., 2000b. Hold nitrogen on manured cornfields in 2000. Integr. Crop Manag. Newsl. IC-484(2), 16–17.

Blaylock, M.J., James, B.R., 1994. Redox transformations and plant uptake of selenium resulting from root–soil interactions. Plant Soil 158(1), 1–12.

Bloom, P.R., 2000. Soil pH and pH buffering. Chapter 10. In: M.E. Sumner (Ed.), Handbook of Soil Science. CRC Press, Boca Raton, FL.

Bolan, N.S., Adriano, D., Curtin, D., 2003a. Soil acidification and liming interactions with nutrient and heavy metal transformation and bioavailability. Adv. Agron. 78, 215–272.

Bolan, N.S., Adriano, D.C., Natesan, R., Koo, B.-J., 2003b. Effects of organic amendments on the reduction and phytoavailability of chromate in mineral soil. J. Environ. Qual. 32, 120–128.

Bolan, N.S., Hedley, M.J., 2003. The role of carbon, nitrogen and sulphur in soil acidification. In: Z. Rengel (Ed.), Handbook of Soil Acidification. Marcel Dekker, New York, pp. 29–56.

Bolan, N.S., Naidu, R., Syers, J.K., Tillman, R.W., 1999. Surface charge and solute interactions in soils. Adv. Agron. 67, 88–141.

Bolan, N.S., Syers, J.K., Tillman, R.W., Scotter, D.R., 1988. Effect of liming and phosphate additions on sulfate leaching in soils. J. Soil Sci. 39, 493–504.

Borken, W., Brumme, R., 1997. Liming practice in temperate forest ecosystems and the effects on CO_2, N_2O and CH_4 fluxes. Soil Use Manage. 13, 251–257.

Bouman, O.T., Curtin, D., Campbell, C.A., Biederbeck, V.O., Ukrainetz, H., 1995. Soil acidification from long-term use of anhydrous ammonia and urea. Soil Sci. Soc. Am. J. 59, 1488–1494.

Briceño, M., Escudey, M., Galindo, G., Borchardt, D., Chang, A., 2004. Characterization of chemical phosphorus forms in volcanic soils using ^{31}P-NMR spectroscopy. Commun. Soil Sci. Plant Anal. 35(9&10), 1323–1337.

Brock, T.D., Madigan, M.T., 1991. Biology of Microorganisms. 6th Ed., Prentice-Hall, Englewood Cliffs, NJ.

Carlson, L., Babione, M., Godfrey, P.J., Fowler, A., 1991. Ecological survey of heathlands in Cape Cod National Seashore, MA. Final report to the Cape Cod National Seashore. Contract No. CX-1600-3-0005, Department of Botany, University of Massachusetts, Amherst, MA.

Cavallaro, N., McBride, M.B., 1980. Activities of Cu^{2+} and Cd^{2+} in soil solutions as affected by pH. Soil Sci. Soc. Am. J. 44, 729–732.

Chapman, S.J., Campbell, C.D., Puri, G., 2003. Native woodland expansion: soil chemical and microbiological indicators of change. Soil Biol. Biochem. 35(6), 753–764.

Chepkwony, C.K., Haynes, R.J., Swift, R.S., Harrison, R., 2001. Mineralization of soil organic P induced by drying and rewetting as a source of plant-available P in limed and unlimed samples of an acid soil. Plant Soil 234(1), 83–90.

Christensen, S., 1985. Denitrification in an acid soil: effects of slurry and potassium nitrate on the evolution of nitrous oxide and on nitrate-reducing bacteria. Soil Biol. Biochem. 17, 757–764.

Clough, T.J., Sherlock, R.R., Kelliher, F.M., 2003. Can liming mitigate N_2O fluxes from a urine-amended soil? Aust. J. Soil Res. 41, 439–457.

Coleman, D.C., Whitman, W.B., 2005. Linking species richness, biodiversity and ecosystem function in soil systems. Pedobiologia 49(6), 479–497.

Condron, L.M., Davis, M.R., Newman, R.H., Cornforth, I.S., 1996. Influence of conifers on the forms of phosphorus in selected New Zealand grassland soils. Biol. Fertil. Soils 21, 37–42.

Condron, L.M., Goh, K.M., 1989. Effects of long-term phosphaic fertilizer applications on amounts and forms of phosphorus in soils under irrigated pasture in New Zealand. J. Soil Sci. 40, 383–395.

Condron, L.M., Tiessen, H., Trasarcepeda, C., Moir, J.O., Stewart, J.W.B., 1993. Effects of liming on organic-matter decomposition and phosphorus extractability in an acid humio ranker soil from northwest Spain. Biol. Fertil. Soils 15, 279–284.

Crout, N.M.J., Tye, A.M., Zhang, H., McGrath, S.P., Young, S.D., 2006. Kinetics of metal fixation in soils: measurement and modeling by isotopic dilution. Environ. Toxicol. Chem. 25(3), 659–663.

Crush, J.R., Cathcart, S.N., Singleton, P., Longhurst, R.D., 1997. Potential for nitrate leaching from different land uses in the Pukekohe area. Proc. N. Z. Grasslands Assoc. 59, 55–58.

Curtin, D., Campbell, C.A., Jalil, A., 1998a. Effects of acidity on mineralization: pH-dependence of organic matter mineralization in weakly acidic soils. Soil Biol. Biochem. 30(1), 57–64.

Curtin, D., Smillie, G.W., 1986. Effects of liming on soil chemical characteristics and grass growth in laboratory and long-term field-amended soils. 1. Soil chemistry. Plant Soil 95, 15–22.

Dalal, R., 1977. Soil organic phosphorus. Adv. Agron. 29, 83–117.

Dancer, W.S., Peterson, L.A., Chesters, G., 1973. Ammonification and nitrification of N as influenced by soil pH and previous N treatments. Soil Sci. Soc. Am. Proc. 37(1), 67–69.

de Klein, C.A.M., Monaghan, R.M., Sinclair, A.G., 1997. Soil acidification: a provisional model for New Zealand pastoral systems. N. Z. J. Agric. Res. 40, 541–557.

Dersch, G., Pfeffer, M., Danneberg, O.H., 2003. Determination of the N mineralization potential of different soils by anaerobic incubation as calibrated in a pot experiment. Bodenkultur 54(2), 69–81.

Di, H.J., Cameron, K.C., 2000. Calculating nitrogen leaching losses and critical nitrogen application rates in dairy pasture systems using a semi-empirical model. N. Z. J. Agric. Res. 43, 139–147.

Di, H.J., Cameron, K.C., 2002. Nitrate leaching and pasture production from different nitrogen sources on a shallow stony soil under flood-irrigated dairy pasture. Aust. J. Soil Res. 40(2), 317–334.

Edmeades, D.C., 2003. The long-term effects of manures and fertilisers on soil productivity and quality: a review. Nutr. Cycl. Agroecosyst. 66(2), 165–180.

Escudey, M., Galindo, G., Briceño, M., Chang, A., 2004. Influence of particle size on ^{31}P-NMR analysis of extracts from volcanic ash-derived soils in Chile. J. Chilean Chem. Soc. 49(1), 5–9.

Evangelou, V.P., Zhang, Y.L., 1995. A review: pyrite oxidation mechanisms and acid mine drainage prevention. Crit. Rev. Environ. Sci. Technol. 252, 141–199.

Farina, M.P.W., Channon, P., 1991. A field comparison of lime requirement indices for maize. Plant Soil 134, 127–135.

Felle, H., 1988. Short-term pH regulations in plants. Physiol. Plantarum 74, 583–591.

Firestone, M.K., 1982. Biological denitrification. In: F.J. Stevenson (Ed.), Nitrogen in Agricultural Soils, Agronomy Monograph 22. American Society of Agronomy, Madison, WI, pp. 289–326.

Firestone, M.K., Davidson, E.A., 1989. Microbiological basis of NO and N_2O production and consumption in soil. In: M.O. Andreae, D.S. Schimel (Eds.), Exchange of Trace Gases Between Terrestrial Ecosystems and the Atmosphere. Wiley, Chichester, pp. 7–21.

Firestone, M.K., McColl, J.G., Killham, K.S., Brooks, P.D., 1984. Microbial response to acid deposition and effects of plant productivity. In: R.A. Linthurst (Ed.), Direct and Indirect Effects of Acid Deposition. Butterworths, Stoneham, MA, pp. 51–62.

Frankenberger, W.T. Jr., Arshad, M., 2001. Bioremediation of selenium-contaminated sediments and water. BioFactors 14(1/4), 241–254.

Frankenberger, W.T. Jr., Losi, M.E., 1995. Applications of bioremediation in the cleanup of heavy metals and metalloids. In: H.D. Skipper, R.F. Turco (Eds.), Bioremediation Science and Applications. SSSA, Special Publication No. 43, Madison, WI, pp. 173–210.

Freney, J.R., Melville, G.E., Williams, C.H., 1975. Soil organic matter fractions as sources of plant available sulphur. Soil Biol. Biochem. 7, 217–221.

Ganeshamurthy, A.N., Nielsen, N.E., 1990. Arylsulphatase and the biochemical mineralization of soil organic sulphur. Soil Biol. Biochem. 22, 1163–1165.

Germida, J., Siciliano, S., 2001. Taxonomic diversity of bacteria associated with the roots of modern, recent and ancient wheat cultivars. Biol. Fertil. Soils 33(5), 410–415.

Gilmour, C.C., Henry, E.A., Mitchell, R., 1992. Sulfate stimulation of mercury methylation in freshwater sediments. Environ. Sci. Technol. 26(11), 2281–2287.

Gissel-Nielsen, G., Gupta, U.C., Lamand, M., Westernmarck, T., 1984. Selenium in soils and plants and its importance in livestock and human nutrition. Adv. Agron. 37, 297–459.

Godbold, D.L., 2003. Managing acidification and acidity in forest soils. In: Z. Rengel (Ed.), Handbook of Soil Acidity. Marcel Dekker, New York, pp. 431–448.

Gregan, P.D., Hirth, J.R., Conyers, M.K., 1989. Amelioration of soil acidity by liming and other amendments. In: A.D. Robson (Ed.), Soil Acidity and Plant Growth. Academic Press, New York, pp. 205–264.

Gregorich, E.G., Janzen, H.H., 2000. Decomposition. In: M.E. Sumner (Ed.), Handbook of Soil Science, Section C, Soil Biology and Biochemistry. CRC Press, Boca Raton, FL.

Hamdy, A.A., Gissel-Nielsen, G., 1977. Fractionation of soil selenium. Z. Pflanz. Bodenkunde 6, 697–703.

Hansson, L., Johanson, K., Olin, A., Siman, G., 1994. Method for the determination of selenium in low-selenium grain and the effect of liming on the uptake of selenium by barley and oats. Acta Agric. Scand.B – Soil Plant Sci. 44(4), 193–200.

Harter, R.D., 1991. Micronutrient adsorption–desorption reactions in soils. In: J.J. Mortvedt, F.R. Cox, L.M. Shuman and R.M. Welch (Eds.), Micronutrients in Agriculture, 2nd Ed. SSSA, Madison, WI, pp. 59–88.

Haynes, R.J., 1983. Soil acidification induced by leguminous crops. Grass Forage Sci. 38, 1–11.

Haynes, R.J., 1984. Lime and phosphate in the soil-plant systems. Adv. Agron. 37, 249–467.

Haynes, R.J., 1990. Active ion uptake and maintenance of cation–anion balance: a critical examination of their role in regulating rhizosphere pH. Plant Soil 126, 247–264.

Haynes, R.J., Naidu, R., 1998. Influence of lime, fertilizer and manure applications on soil organic matter content and soil physical conditions: a review. Nutr. Cycl. Agroecosyst. 51, 123–137.

Haynes, R.J., Swift, R.S., 1988. Effects of lime and phosphate additions on changes in enzyme-activities, microbial biomass and levels of extractable nitrogen, sulfur and phosphorus in an acid soil. Biol. Fertil. Soils 6, 153–158.

Helyar, K.R., 1976. Nitrogen cycling and soil acidification. J. Aust. Inst. Agric. Sci. 42, 217–221.

Helyar, K.R., Porter, W.M., 1989. Soil acidification, its measurement and the processes involved. In: A.D. Robson (Ed.), Soil Acidity and Plant Growth. Academic Press, Sydney, pp. 61–101.

Holub, Z., Simonovicova, A., Banasova, V., 1993. The influence of acidification on some chemical and microbiological properties of soil, those determining plant viability. Biologia 48(6), 671–675.

Huysmans, K.D., Frankenberger, W.T., 1991. Evolution of trimethyllarsine by a *Penicillium* sp. isolated from agricultural evaporation pond water. Sci. Total Environ. 105, 13–28.

Jarvis, S.C., Stockdale, E.A., Shepherd, M.A., Powlson, D.S., 1996. Nitrogen mineralization in temperate agricultural soils: processes and measurement. Adv. Agron. 57, 187–235.

Jeffery, J.J., Uren, N.C., 1983. Copper and zinc species in the soil solution and the effects of soil pH. Aust. J. Res. 21, 479–488.

Jensen, E.S., 1997. Nitrogen immobilization and mineralization during initial decomposition of 15N-labelled pea and barley residues. Biol. Fertil. Soils 24, 39–44.

Jobbágy, E.G., Jackson, R.B., 2003. Patterns and mechanisms of soil acidification in the conversion of grasslands to forests. Biogeochemistry 64, 205–229.

Johnson, D., Leake, J.R., Read, D.J., 2005. Liming and nitrogen fertilization affects phosphatase activities, microbial biomass and mycorrhizal colonisation in upland grassland. Plant Soil 271(1/2), 157–164.

Jones, J.P., Fox, R.L., 1978. Phosphorus nutrition of plants influenced by manganese and aluminium uptake from an oxisol. Soil Sci. 126, 230–236.

Kay, B.D., 1998. Soil structure and organic carbon: a review. In: R. Lal, J.M. Kimble, R.F. Follett and B.A. Stewart (Eds.), Soil Processes and the Carbon Cycle. CRC Press, Boca Raton, FL, pp. 169–197.

Kay, B.D., Angers, D.A., 2000. Soil structure. In: M. Sumner (Ed.), Handbook of Soil Science. CRC Press, Boca Raton, FL, pp. 229–276.

Keren, R., Bingham, F.T., Rhoades, J.D., 1985. Plant uptake of boron as affected by boron distribution between the liquid and the solid phases in soil. Soil Sci. Soc. Am. J. 49, 297–302.

Kerry, A., Welbourn, P.M., Prucha, B., Mierle, G., 1991. Mercury methylation by sulfate-reducing bacteria from sediments of an acid stressed lake. Water Air Soil Pollut. 56, 565–575.

Korentajer, L., Byrnes, H.B., Hellums, T.D., 1983. The effect of liming and leaching on sulfur supplying-capacity of soils. Soil Sci. Soc. Am. J. 47, 525–530.

Kreutzer, K., 1995. Effects of forest liming on soil processes. Plant Soil 168/169, 447–470.

Lang, F., Kaupenjohann, M., 1998. Influence of liming and kieserite fertilization on the molybdenum dynamics of forest sites. Forstwissenchaftliches Centralblatt 117(6), 316–326.

Ledgard, S.F., Peoples, M.B., 1988. Measurement of nitrogen fixation in the field. In: J.R. Wilson (Ed.), Advances in N cycling in agricultural ecosystems. Proceedings of the symposium held in Brisbane, Australia, 11–15 May, 1987, CAB International, pp. 351–367.

Li, G.D., Helyar, K.R., Conyers, M.K., Castleman, L.J.C., Fisher, R.P., Poile, G.J., Lisle, C.J., Cullis, B.R., Cregan, P.D., 2006b. Pasture and sheep responses to lime applications in a grazing experiment in a high-rainfall area, south-eastern Australia. II. Liveweight gain and wool production. Aust. J. Agric. Res. 57(10), 1057–1066.

Li, G.D., Helyar, K.R., Welham, S.J., Conyers, M.K., Castleman, L.J.C., Fisher, R.P., Evans, C.M., Cullis, B.R., Cregan, P.D., 2006a. Pasture and sheep responses to lime applications in a grazing experiment in a high-rainfall area, south-eastern Australia. I. Pasture production. Aust. J. Agric. Res. 57(10), 1045–1055.

Lindsay, W.L., 1979. Chemical Equilibria in Soils. Wiley, New York.

Liu, W.C., Lund, L.J., Page, A.L., 1989. Acidity produced by leguminous plants through symbiotic dinitrogen fixation. J. Environ. Qual. 18(4), 529–534.

Logan, K.A.B., Thomas, R.J., Raven, J.A., 1999. Hydrogen ion production and ammonium uptake by two tropical forage grasses. J. Plant Nutr. 22(1), 53–66.

Logan, K.A.B., Thomas, R.J., Raven, J.A., 2000. Effect of ammonium and phosphorus supply on H^+ production in gel by two tropical forage grasses. J. Plant Nutr. 23(1), 41–54.

Lomans, B.P., van der Drift, C., Pol, A., Op den Camp, H.J.M., 2002. Microbial cycling of volatile organic sulfur compounds. Cell. Mol. Life Sci. 59, 575–588.

Lyngstad, I., 1992. Effect of liming on mineralization of soil-nitrogen as measured by plant uptake and nitrogen released during incubation. Plant Soil 144, 247–253.

Marschner, B., 1993. Microbial contribution to sulfate mobilization after mining an acid forest soil. J. Soil Sci. 44(3), 459–466.

Marschner, B., Wilczynski, A.W., 1991. The effect of liming on quantity and chemical composition of soil organic matter in a pine forest in Berlin, Germany. Plant Soil 137(2), 229–236.

Marschner, H., 1995. Mineral Nutrition of Higher Plants. 2nd Ed., Academic Press, London.

Marsh, K.B., Tillman, R.W., Syers, J.K., 1987. Charge relationships of sulphate sorption by soils. Soil Sci. Soc. Am. J. 51, 318–323.

Mayer, R., 1993. Chemical time bombs related to forestry practice – distribution and behavior of pollutants in forest soils. Land Degrad. Rehabil. 4(4), 275–279.

Mayland, H.F., 1994. Selenium in plant and animal nutrition. In: W.T. Frankenberger, S. Benson (Eds.), Selenium in the Environment. Marcel Dekker, New York, pp. 29–45.

McDowell, R.W., 2003. Identification of phosphorus species in extracts of soils with contrasting management histories. Commun. Soil Sci. Plant Anal. 34(7–8), 1083–1095.

Mishra, S., Di, H.J., Cameron, K.C., Monaghan, R., Carran, A., 2005. Gross nitrogen mineralization rates in pastural soils and their relationships with organic nitrogen fractions, microbial biomass and protease activity under glasshouse conditions. Biol. Fertil. Soils 42, 45–53.

Moore, J.M., Klose, S., Tabatabai, M.A., 2000. Soil microbial biomass carbon and nitrogen as affected by cropping systems. Biol. Fertil. Soils 31(3–4), 200–210.

Mora, M.L., Baeza, G., Pizarro, C., Demanet, R., 1999b. Effect of calcitic and dolomitic lime on physicochemical properties of a Chilean Andisol. Commun. Soil Sci. Plant Anal. 30(3/4), 427–439.

Mora, M.L., Barrow, N.J., 1996. The effects of time of incubation on the relation between charge and pH of soil. Eur. J. Soil Sci. 47, 131–136.

Mora, M.L., García, J., Santander, J., Demanet, R., 1999c. Rol de los fertilizantes nitrogenados y fosfatados en los procesos de acidificación de los suelos. Frontera Agrícola 5(1&2), 59–81.

Mora, M.L., Schnettler, B., Demanet, R., 1999a. Effect of liming and gypsum on soil chemistry, yield and mineral composition of ryegrass grown in an acidic Andisol. Commun. Soil Sci. Plant Anal. 30, 1251–1266.

Mora, M.L., Shene, C., Violante, A., Demanet, R., Bolan, N.S., 2005. The effect of organic matter and soil chemical properties on sulfate sorption in Chilean volcanic soil. In: P.M. Huang, A. Violante, J.M. Bollag and P. Vityakon (Eds.), Soil abiotic and biotic interaction and the impact on the ecosystem and human welfare. Science Pulishers, Enfield, pp. 223–244.

Mortvedt, J.J., 2000. Bioavailability of micronutrients. In: M.E. Sumner (Ed.), Handbook of Soil Science. CRC Press LLC, Boca Raton, FL, pp. D71–D86.

Motavalli, P.P., Duxbury, J.M., DeSouza, D.M.G., 1993. The influence of organic soil amendments on sulphate adsorption and sulphur availability in a Brazilian Oxisol. Plant Soil 154(2), 301–308.

Muchovej, R.M.C., Borges, A.C., Novais, R.F., Thiebaut, J.T.L., 1986. Effect of liming level and Ca:Mg ratios on yield, nitrogen content and nodulation of soyabeans grown in acid Cerrado soil. J. Soil Sci. 37(2), 235–240.

Nägele, W., Conrad, R., 1990. Influence of pH on the release of NO and N_2O from fertilized and unfertilized soil. Biol. Fertil. Soils 10, 139–144.

Neale, S.P., Shah, Z., Adams, W.A., 1997. Changes in microbial biomass and nitrogen turnover in acidic organic soils following liming. Soil Biol. Biochem. 29(9–10), 1463–1474.

Newman, R.H., Tate, K.R., 1980. Soil phosphorus characterization by ^{31}P-nuclear magnetic resonance. Commun. Soil Sci. Plant Anal. 11(9), 835–842.

Nyborg, M., 1982. Environmental implications of wind-blown sulphur from sour natural gas processing plant in Alberta. In: A.I. More (Ed.), International Sulphur Conference '82. Vol. 1, The British Sulphur Corporation, Purley Press, London, pp. 455–462.

Nyborg, M., Hoyt, P.B., 1978. Effects of soil acidity and liming on mineralization of soil nitrogen. Can. J. Soil Sci. 58, 331–338.

Oenema, O., 1990. Calculated rates of soil acidification of intensively used grassland in the Netherlands. Fert. Res. 26(1–3), 217–228.

Olfs, H.W., Neu, A., Werner, W., 2004. Soil N transformations after application of N-15 labeled biomass in incubation experiments with repeated soil drying and rewetting. J. Plant Nutr. Soil Sci. – Z. pflanz. bodenkunde 167(2), 147–152.

Parkin, T.B., Sexstone, A.J., Tiedje, J.M., 1985. Adaptation of denitrifying populations to low soil pH. Appl. Environ. Microbiol. 49, 1053–1056.

Parr, J.F., Papendick, R.I., 1978. Factors affecting the decomposition of crop residues by microorganisms. In: W.R. Oschwald (Ed.), Crop Residue Management Systems. American Society of Agronomy, Publication No. 31, Madison, WI, pp. 101–129.

Paul, E.A., Clark, F.E., 1996. Soil Microbiology and Biochemistry. 2nd Ed., Academic press, San Diego.

Payne, W.J., 1981. Denitrification. Wiley, New York.

Persson, T., Wirén, A., Anderson, S., 1990. Effects of liming on carbon and nitrogen mineralization in coniferous forests. Water Air Soil Pollut. 54, 351–364.

Prescott, C.E., Maynard, D.G., Laiho, R., 2000. Humus in northern forests: friend or foe? Forest Ecol. Manage. 133(1/2), 23–36.

Prietzel, J., Weick, C., Korintenberg, J., Seybold, G., Thumerer, T., Treml, B., 2001. Effects of repeated $(NH_4)_2SO_4$ application on sulfur pools in soil, soil microbial biomass, and ground vegetation of two watersheds in the Black Forest/Germany. Plant Soil 230(2), 287–305.

Puttanna, K., Gowda, N.M.N., Rao, E.V.S.P., 1999. Effect of concentration, temperature, moisture, liming, and organic matter on the efficacy of the nitrification inhibitors benzotriazole, o-nitrophenol, m-nitroaniline and dicyandiarnide. Nutr. Cycl. Agroecosyst. 54, 251–257.

Ratering, S., Schnell, S., 2000. Localization of iron-reducing activity in paddy soil by profile studies. Biogeochemistry 48, 341–365.

Raven, J.A., 1985. Regulation of pH and generation of osmolarity in vascular plants: a cost-benefit analysis in relation to efficiency of use of energy, nitrogen and water. New Phytol. 101, 25–77.

Rengel, Z., 2003. Handbook of Soil Acidity. Marcel Dekker, New York.

Reuss, J.O., 1986. Acid Deposition and the Acidification of Soil and Water. Springer, New York.

Ritchie, G.S.P., 1989. The chemical behavior of aluminium, hydrogen and manganese in acid soils. In: A.D. Robson (Ed.), Soil Acidity and Plant Growth. Academic Press, New York, pp. 1–60.

Robinson, Ch., Ineson, P., Piearce, T.G., Rowland, A.P., 1992. Nitrogen mobilization by earthworms in limed peat soils under *Picea sitchensis*. J. Appl. Ecol. 29(1), 226–237.

Robson, A.D., Abbott, L.K., 1989. The effect of soil acidity on microbial activity in soils. In: A.D. Robson (Ed.), Soil Acidity and Plant Growth. Academic Press, New York, pp. 139–166.

Rogers, R.D., 1977. Abiological methylation of mercury in soil. J. Environ. Qual. 6(4), 463–467.

Rothe, A., Huber, C., Kreutzer, K., Weis, W., 2002. Deposition and soil leaching in stands of Norway spruce and European Beech: results from the Höglwald research in comparison with other European case studies. Plant Soil 240(1), 33–45.

Rowell, D.L., Wild, A., 1985. Causes of soil acidification: a summary. Soil Use Manage. 1(1), 32–33.

Runge, M., Rode, M.W., 1991. Effects of soil acidity on plant associations. In: B. Ulrich, M.E. Sumner (Eds.), Soil Acidity. Springer, New York, pp. 183–202.

Sadiq, M., 1997. Arsenic chemistry in soils: an overview of thermodynamic predictions and field observations. Water Soil Pollut. 93(3–4), 117–136.

Saggar, S., Hedley, M.J., Phimsarn, S., 1998. Dynamics of sulfur transformations in grazed pastures. In: D.G. Maynard (Ed.), Sulfur in the Environment. Marcel Dekker, New York, pp 45–94.

Sahrawat, K.L., 2005. Fertility and organic matter in submerged rice soils. Curr. Sci. 88, 735–739.

Sano, S.M., Abbott, L.K., Solaiman, M.Z., Robson, A.D., 2002. Influence of liming, inoculum level and inoculum placement on root colonization of subterranean clover. Mycorrhiza 12(6), 285–290.

Schmehl, W.R., Peech, M., Bradfiled, R., 1950. Causes of poor growth of plants in acid soils and beneficial effects of liming: I. Evaluation of factors responsible for acid-soil injury. Soil Sci. 70, 393–410.

Sharpley, A.N., 2000. Bioavailable phosphorus in soil. In: G.M. Pierzynski (Ed.), Methods for Phosphorus Analysis for Soils, Sediments, Residuals, and Waters. Southern Cooperative Series Bulletin 396, pp. 38–43.

Shuman, L.M., 1991. Chemical forms of micronutrients in soils. In: R.J. Luxmoore (Ed.), Micronutrients in Agriculture. SSSA, Madison, WI, pp. 113–144.

Simmons, J.A., Yavitt, J.B., Fahey, T.J., 1996. Watershed liming effects on the forest floor N cycle. Biogeochemistry 32(3), 221–244.

Sims, J.T., Simard, R.R., Joern, B.C., 1998. Phosphorus loss in agricultural drainage: historical perspective and current research. J. Environ. Qual. 27, 277–293.

Slangen, L.H.G., Kerkhoff, P., 1984. Nitrification inhibitors in agriculture and horticulture: a literature review. Fertilizer Research 5, 1–76.

Sparling, G.P., Stephens, P.R., Schipper, L.A., Bettjeman, W., 2002. Soil quality at regional and national scales in New Zealand. OECD Expert Meeting on Soil Organic Carbon Indicators for Agricultural Land, Ottawa, Canada, 15–18 October. Accessed from http://www.webdomino1.oecd.org.comnet/agr/soil.nsf 17th April 2007, 7pp.

Springett, J.A., Syers, J.K., 1984. Effect of pH and calcium content of soil on earthworm cast production in the laboratory. Soil Biol. Biochem. 16, 185–189.

Stahl, R.S., James, B.R., 1991. Zinc sorption by manganese-oxide-coated sand as a function of pH. Soil Sci. Soc. Am. J. 55, 1291–1294.

Stevens, R.J., Laughlin, R.J., Malone, J.P., 1998. Soil pH affects the process reducing nitrate to nitrous oxide and di-nitrogen. Soil Biol. Biochem. 30, 1119–1126.

Stevenson, F.J., 1985a. Geochemistry of soil humic substances. In: G.R. Aiken, D.M. McKnight, R.L. Wershaw and P. MacCarthy (Eds.), Humic Substances in Soil, Sediment and Water: Geochemistry, Isolation and Characterization. Wiley, New York, pp. 13–52.

Stevenson, F.J., 1985b. Cycles of Carbon, Nitrogen, Phosphorus, Sulphur and Micronutrients. Wiley, New York.

Stockdale, E.A., Gaunt, J.L., Vos, J., 1997. Soil-plant nitrogen dynamics: what concepts are required? Eur. J. Agron. 7(1–3), 145–159.

Sumner, M.E., Fey, M.V., Noble, A.D., 1991. Nutrient status and toxicity problems in acid soils. In: B. Ulrich, M.E. Sumner (Eds.), Soil Acidity. Springer, New York, pp. 149–182.

Sumner, M.E., Yamada, T., 2002. Farming with acidity. Commun. Soil Sci. Plant Anal. 33(15–18), 2467–2496.

Tang, C., McLay, C.D.A., Barton, L., 1997. A comparison of proton excretion of twelve pasture legumes grown in nutrient solution. Aust. J. Exp. Agric. 37(5), 563–570.

Tang, C., Rengel, Z., 2003. Role of plant cation/anion uptake ratio in soil acidification. In: Z. Rengel (Ed.), Handbook of Soil Acidity. Marcel Dekker, New York, pp. 57–81.

Thayer, J.S., 2002. Biological methylation of less-studied elements. Appl. Organomet. Chem. 16(12), 677–691.

Tiedje, J.M., 1982. Denitrification. In: A.L. Page, R.H. Miller and D.R. Keeney (Eds.), Methods of Soil Analysis, Part 2. American Society of Agronomy, Agronomy Monograph No. 9, Madison, WI, pp. 1011–1026.

Tiedje, J.M., 1988. Ecology of denitrification and dissimilatory nitrate reduction to ammonium. In: A.J.B. Zehnder (Ed.), Environmental Microbiology of Anaerobes. Wiley, New York, pp. 179–244.

Tillman, R.W., Syers, J.K., 1982. Potential of lime for improving the efficiency of phosphate fertilizer use. N. Z. Agric. Sci. 16, 90–92.

Tisdale, S., Nelson, W., Beaton, J., 1985. Soil fertility and fertilizers. 4th Ed., Collier McMillan Publishing, London, UK.

Toor, G.S., Condron, L.M., Di, H.J., Cameron, L.C., Sims, J.T., 2004. Assessment of phosphorus leaching losses from a free draining grassland soil. Nutr. Cycl. Agroecosyst. 69(2), 167–184.

Tugel, J.B., Hines, M.E., Jones, G.E., 1986. Microbial iron reduction by enrichment cultures isolated from estuarine sediments. Appl. Environ. Microbiol. 52, 1167–1172.

Ullrich, S.M., Tanton, T.W., Abdrashitova, S.A., 2001. Mercury in the aquatic environment: a review of factors affecting methylation. Crit. Rev. Environ. Sci. Technol. 31(3), 241–293.

Uren, N.C., 2001. Types, amounts, and possible functions of compounds released into rhizosphere by soil-grown plants. In: R. Pinton, Z. Varanini and P. Nannipieri (Eds.), The Rhizosphere: Biochemistry, and Organic Substance at the Soil-Plant Interface. Marcel Dekker, New York, pp. 22–32.

Valeur, I., Andersson, S., Nilsson, S.L., 2000. Calcium content of liming material and its effects on sulphur release in a coniferous forest soil. Biogeochemistry 50, 1–20.

Valeur, I., Nilsson, I., 1993. Effects of lime and two incubation techniques on sulfur mineralization in a forest soil. Soil Biol. Biochem. 25(10), 1343–1350.

Wardle, D.A., 1992. A comparative assessment of factors which influence microbial biomass carbon and nitrogen levels in soil. Biol. Rev. Camb. Philos. Soc. 67(3), 321–358.

Waring, S.A., Gilliam, J.W., 1983. The effect of acidity on nitrate reduction and denitrification in lower coastal plain soils. Soil Sci. Soc. Am. J. 47(2), 246–251.

Watkinson, J.H., Bolan, N.S., 1998. Modeling the rate of elemental sulfur oxidation in soils. In: D.G. Maynard (Ed.), Sulphur in the Environment. Marcel Dekker, New York, pp. 135–172.

Wheeler, D.M., Edmeades, D.C., Morton, J.D., 1997. Effect of lime on yield, N fixation, and plant N uptake from the soil by pasture on three contrasting trials in New Zealand. N. Z. J. Agric. Res. 40(3), 397–408.

Wilkinson, K.J., Slaveykova, V.I., Hassler, C.S., Rossier, C., 2002. Physicochemical mechanisms of trace metal bioaccumulation by microorganisms. Chimia 56, 681–684.

Zhang, Y.Q., Frankenberger, W.T., 2000. Formation of dimethylselenonium compounds in soil. Environ. Sci. Technol. 34(5), 776–783.

Zhang, Y.Q., Frankenberger, W.T., More, J.N., 1999. Effect of soil moisture on dimethylselenide transport and transformation to nonvolatile selenium. Environ. Sci. Technol. 33(19), 3415–3420.

Zumft, W.G., 1997. Cell biology and molecular basis of denitrification. Microbiol. Mol. Biol. Rev. 61(4), 533–616.

E: Tools to assess bioavailability

Traditional methods of assessing soil and water contamination involve exhaustive and often expensive techniques. For this reason, there is now a concerted move towards developing highly sensitive indicators of contaminant bioavailability. Such indicators may include bio or chemi-sensors. This section presents an overview of the tools currently being used to assess contaminant bioavailability, plus the new cutting-edge technique that are currently being developed. The section links up with all the sections presented above.

Developments in Soil Science, volume 32
Ravendra Naidu (Editor)

Chapter 18

CONTAMINANT CONCENTRATIONS IN ORGANISMS AS INDICATORS OF BIOAVAILABILITY: A REVIEW OF KINETIC THEORY AND THE USE OF TARGET SPECIES IN BIOMONITORING

Nico M. van Straalen

18.1 Introduction

Bioavailability is a concept that is hard to define (see Chapter 3). It combines properties of the substance, the organism and the environment, which interact with each other (see Chapters 11–13 and 18). Numerous attempts have been made to define bioavailability in terms of the fractions obtained by chemical extractions of a contaminant from the abiotic matrix (Chapters 7 and 20). While these studies have considerably improved the insight into the mechanisms that control speciation and the fate of substances in the abiotic environment (Salomons and Stigliani, 1995), extrapolations to organisms are often troubled by the great diversity of uptake routes and organism-specific physiologies. From the perspective of the organism, only the fraction that may pass cell membranes is relevant. This fraction may, however, be in dynamic equilibrium with a fraction that is reversibly bound to the environmental matrix, which in its turn, may interact with a strongly bound fraction (Fig. 18.1). It is usually assumed that the partitioning of a contaminant over the different compartments in the environment is at equilibrium. Under this the so-called equilibrium partitioning theory, distributions of chemicals may be predicted from their physico-chemical properties (Di Toro et al., 1991).

One of the most important arguments for consideration of contaminants inside organisms is that internal concentrations are more closely related to the bioavailable fraction than to the total concentration in the environment (Phillips and Segar, 1986). However, measurement of residues in organisms should not be seen as a way to estimating the total concentration in the environment (this can be more easily done by taking samples from the abiotic environment), but as a way to estimate the bioavailable concentration. Correlations with chemically defined fractions can then be sought to explain bioavailability in terms of the chemical species involved.

While simplistic, Figure 18.1 presents the dynamics of contaminants from the time it comes into contact with soil to when it is absorbed by organisms. For

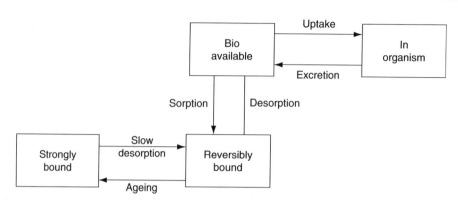

Figure 18.1. *General scheme of the four different compartments over which a contaminant in the environment may be distributed.*

heavy metals the bioavailable fraction involves a variety of different chemical species that are taken up at different rates by different organisms, while in some cases the organism itself may stimulate the uptake by excreting complexing agents or by changing speciation through bioturbating activities (Hare, 1992, Morel, 1997). In the case of organic hydrophobic micropollutants, uptake is not only governed by the soluble concentration of the chemical in the environment, but also by the flux of digested food components across the cell membranes of the gut epithelium which, for highly lipophilic chemicals, contributes significantly to the total uptake (Gobas et al., 1993; Belfroid et al., 1996).

Internal concentrations prove their usefulness most clearly under conditions where the total concentration in the environment does not change, but the bioavailable concentration does. An example is the use of earthworms to measure heavy metal bioavailability in soil. If the ratio between internal and external concentrations were constant, the use of earthworms would provide no more information than abiotic measurements. Soil ecotoxicological studies show, however, that the ratio is sensitive to soil pH due to the fact that acidity of the pore water influences partitioning of metals over reversibly bound and available fractions (Van Gestel et al., 1995). Residue changes in earthworms thus indicate changes in bioavailability. Although some authors have argued otherwise (Martin and Coughtrey, 1982), a linear relationship between internal and total external concentrations is not always a desirable property for bioindicator organisms.

Another reason supporting measurements in biological matrices is that residue data can be used to estimate the ecological risk of the contamination. The risk posed by chemicals does not depend on its total concentrations in the abiotic environment, but on the flux into organisms. Although there may be exceptions to this rule (e.g. some chemicals causing an irritating effect to the skin without being taken up), in the majority of cases the internal concentration determines the effect. If bioindication data are used for risk assessment, residue data should be supplemented with thresholds that indicate the maximal internal concentrations tolerated by the organism. This paper reviews the principles that underlie the use of contaminant residues in organisms as indicators of bioavailability.

18.2 Toxicokinetic interpretation of residues

Rather than a static quantity, the residue of a contaminant in an organism is the result of a dynamic balance between gain and loss processes. The theory of compartment modelling (Jacquez, 1972; Doucet and Sloep, 1992) has been widely used to describe changes in internal concentrations as a function of time. The most common approach is to assume that the organism can be considered as a single compartment and that the loss of the chemical from the organism is proportional to the internal amount. A balance equation describing the process can then be written as:

$$\frac{dQ}{dt} = A(t) - k\,Q(t) \tag{18.1}$$

where Q is the body burden of the chemical (in absolute units, e.g. µg), $A(t)$ the function specifying the rate of uptake (in mass per time, e.g. $\mu g\,d^{-1}$), and k the excretion rate constant (in d^{-1}). As suggested by Moriarty (1984), balance equations in compartment models are preferably expressed in terms of the body burden, rather than the concentration; it is only after solving the mass-specific balance equation that concentration changes are derived by the dividing body burdens by the weight or volume of the organism.

The uptake rate $A(t)$ is often written as being proportional to the external concentration:

$$A(t) = k_1\,C(t)\,V \tag{18.2}$$

where k_1 is the uptake rate constant expressed in d^{-1}, $C(t)$ the external concentration (e.g. in $\mu g\,L^{-1}$), and V the volume of the external compartment from which the organism can take up the substance (in L). In most of the publications that apply toxicokinetic modelling to environmental problems, it is assumed that Eq. (18.2) holds, but the external concentration is usually not considered as a true second compartment of the system, i.e., $C(t)$ is assumed not to decrease due to uptake by the organism or to increase due to excretion (except in the method of Banerjee et al., 1984). In addition, the volume V is usually not considered separately but is included in the rate constant, k_1, which is then expressed in $L\,d^{-1}$. However, if the external compartment is assumed to be constant or to behave autonomously as a function of time, it is more logical to write the uptake term directly as $A(t)$ like in Eq. (18.1), and ignore the formulation in Eq. (18.2).

The differential Eq. (18.1) can be solved analytically for various input functions $A(t)$. Two of these are considered in Figure 18.2. In the case of a block

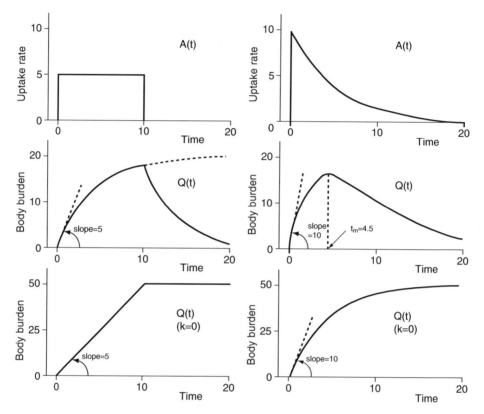

Figure 18.2. *Predicted changes in body burden, $Q(t)$, when an organism can be described by a single linear compartment and is exposed to a contaminant with uptake regime $A(t)$ as shown in the upper left panel (block pulse) and the upper right panel (diluted pulse). For purposes of comparison, the total amount of substance taken up over time (the area under the uptake curve) has been chosen the same in the two cases; it equals at $t_c = 50$ in the block function and $a_m/k_0 = 50$ in the diluted pulse. Changes in body burden according to formulas (18.3) and (18.5) are drawn in the middle panels for $a = 5$, $t_c = 10$, $a_m = 10$, $k = 0.25$, and $k_0 = 0.2$ (arbitrary units). The initial slope of Q at $t = 0$ is $a = 5$, resp. $a_m = 10$. The equilibrium concentration under continuous exposure (a block pulse of long duration) is $a/k = 20$. There is no equilibrium in the diluted pulse exposure; the maximum concentration in the animal is reached at $t_m = (\ln k_0 - \ln k)/(k_0 - k) = 4.46$. The lower panels give the change in body burden when the organism cannot excrete the chemical ($k = 0$); a concentration of 50 is then reached in both cases.*

input, the rate of uptake jumps to a value of a on time zero and jumps back to zero on time t_c. This type of uptake regime is often applied in ecotoxicity experiments where test animals are exposed to a constant concentration for some time and then transferred to a clean environment to depurate. The solution of Eq. (18.1) under these conditions, for $t \geqslant 0$ and assuming $Q(0) = 0$, is:

$$Q(t) = \frac{a}{k}\left[1 - e^{-kt} - S(t_c)\left\{1 - e^{-k(t-t_c)}\right\}\right] \tag{18.3}$$

where $S(t_c)$ is a unit step function: $S(t_c) = 0$ for $t < t_c$, and $S(t_c) = 1$ for $t \geqslant t_c$. The time course of Q predicted by Eq. (18.3) is pictured in Figure 18.2. Body burden initially increases at a rate a, then tends to approach an asymptotic value equal to a/k, but it decreases exponentially after the environment becomes clean again on time t_c. The constant k is also known as the turnover rate of the chemical in the organism and $1/k$ is the average residence time. Eq. (18.3) is used in curve fitting to estimate the parameters a and k from a time series of observed body burdens when test animals have been exposed to a block exposure pulse (e.g. Janssen et al., 1991).

Another type of uptake regime, also shown in Figure 18.2, is a so-called "diluted pulse" (Widianarko and Van Straalen, 1996). This occurs for example in the case of non-persistent pesticides sprayed onto an agricultural field, or an instantaneous discharge in a river, diluted by diffusion. For the diluted pulse, the uptake function can be written as a decreasing exponential for $t \geqslant 0$:

$$A(t) = a_m e^{-k_0 t} \tag{18.4}$$

where a_m is the uptake rate on time zero (in $\mu g\, d^{-1}$), and k_0 the rate constant for the disappearance of the chemical from the environment (in d^{-1}). With a diluted pulse for $A(t)$ in Eq. (18.1), the solution for $Q(t)$ is (Southworth et al., 1978; Landrum et al., 1992; Widianarko and Van Straalen, 1996):

$$Q(t) = \frac{a_m}{k - k_0}\left[e^{-k_0 t} - e^{-kt}\right] \tag{18.5}$$

The curve for $Q(t)$ initially increases at rate a_m, then shows a hump, with a maximum at time t_m (the value of this time depends on k_0 and k), and decreases according to the difference between the two exponentials (Fig. 18.2).

An application of the diluted pulse model to real data is shown in Figure 18.3. Edwards (1976) measured the concentration of the insecticide diazinon in soil and in slugs sampled from a wheat field, which had been treated with a single application early in the season. Diazinon was detected in the slugs for a considerable period, even after the residues in the soil had decayed below the detection limit. Eq. (18.5) was fitted to these data and parameters were estimated by a least squares routine. The half-life of diazinon in slugs is longer than in soil

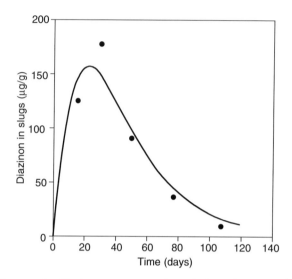

Figure 18.3. *Changes in diazinon concentration of slugs* (Agriolimax reticulatus), *sampled from a spring wheat plot to which the insecticide diazinon had been applied at a rate of 8 kg ha^{-1}. The solid line is a model curve fit using Eq. (18.5) with parameter values* $a_m = 19.6 \, \mu g \, g^{-1} d^{-1}$, $k = 0.039 \, d^{-1}$ *and* $k_0 = 0.055 \, d^{-1}$ *(Data are from Edwards, 1976).*

(*c.* 18 days versus *c.* 13 days), which, together with the bioaccumulation potential of the slugs explains the extended presence of diazinon in their bodies.

Most of the exposure regimes in a real environment will not resemble the input functions shown in Figure 18.2. In many cases, exposure concentrations will fluctuate erratically and cannot be captured in a simple model. Changes in internal concentrations can, however, still be described by numerical integration of Eq. (18.1), when the input is given as a time series. Deneer (1993) applied this approach in a study on the toxicokinetics of chlorpyrifos in guppy fish to estimate toxicokinetic parameters under time-variable exposure. Conversely, if the parameters are known, computer simulation can be used to generate predicted changes in body burden using any given exposure regime.

Knowledge of toxicokinetic parameters will help to predict how long an environmental contaminant will stay in the organism. Due to uptake and excretion processes lagging behind exposure changes, organisms have an "averaging" effect on the contaminant. Especially when the residence time in the organism is large compared to its residence in the environment (small *k*), the contaminant is more easily detected in the organism than in the environment. In the extreme case that an organism is not able to excrete at all, it carries the residue for the rest of its life. Some examples are cadmium in isopods and organochlorines in fat tissue of mammals. To illustrate this effect, the lower panels of Figure 18.2 give the increase of body burden for $k = 0$ in Eqs. (18.3) and (18.5); in this case the graph is a simple integral of the uptake curve. The time-integrating effect of internal body burdens is one of the important reasons why measurements of contaminants in biota are advocated.

The theory of compartment models can handle much more complicated situations than the simple one-compartment system considered above. However, analytical solutions for systems with more than one compartment quickly become very cumbersome and their application as indicators for bioavailability is therefore limited. Actual measurements of residues in organisms are rarely precise enough to allow a distinction between one or two compartments. In addition, the fact that organisms in reality do not consist of one anatomical compartment does not preclude their behaving as one compartment with respect to toxicokinetics. The use of two and three compartment systems is therefore restricted to studies of large animals for which data on different organs are available, and studies using radioactive tracers that allow very precise recordings to be made in time. The reader is referred to the specialist literature (Jacquez, 1972; Doucet and Sloep, 1992).

Toxicokinetic modelling can be extended to toxicodynamics by making assumptions on how the internal concentration affects the organism. For example, it may be assumed that the mortality rate increases linearly with the internal concentration (Bedaux and Kooijman, 1994; Widianarko and Van Straalen, 1996), or that there is a critical body concentration, which induces death when exceeded (McCarty et al., 1992). Depending on which assumptions are made, equations may be derived for the change of mortality with time, given a certain exposure regime.

A toxicokinetics-based interpretation of residues is important because it provides a system for estimating the risk associated with an observed body residue. Too many monitoring programmes just measure concentrations of contaminants in all kinds of biological tissues, without linking these data to the thresholds tolerated by the organisms involved. The ecotoxicological significance of such data is very limited.

McCarty and Mackay (1993) proposed the concept of *"critical body residue"* (CBR) to serve as a reference for evaluating measured concentrations in organisms. These authors reviewed the literature on toxicity of organic chemicals to fish and proposed a system of CBRs for eight categories of organic chemicals (Fig. 18.4).

Van Straalen (1996) proposed to link measured concentrations in organisms to critical body residues by expressing the concentration relative to the lethal body concentration (LBC) determined in toxicity experiments. The rationale is that external concentrations are often difficult to compare between laboratory

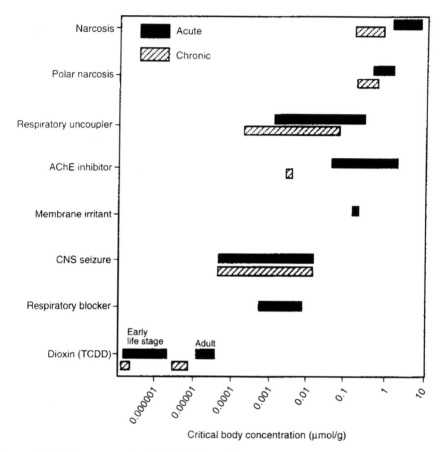

Figure 18.4. *Summary of critical body residues associated with acute and chronic toxicity endpoints for fish exposed to eight categories of organic chemicals. (Reprinted with permission from Environ. Sci. Technol. September 1993, 27(9), 1725–1730. Copyright 1993 American Chemical Society.)*

and field, due to the artificial media to which organisms are exposed in controlled laboratory experiments, but that internal critical concentrations are more comparable. It was also proposed to calculate the quotient of concentration and LBC for as many species as possible and to take the average for a certain site. This was expressed as a "bioindicator index for toxicant residues", as follows:

$$\beta = \frac{1}{s} \sum_{i=1}^{s} \frac{C_i}{LBC_i} \tag{18.6}$$

where C_i is the body concentration of species i, LBC_i the LBC of that species, and s the number of species assessed. Table 18.1 provides some data illustrating the

Table 18.1. *Risk assessment of two forest soils in The Netherlands, one affected by zinc smelter emissions, the other one a reference site, using the bioindicator index for toxicant residues, β, as defined in Eq. (18.6), estimated from cadmium concentrations in three species of soil invertebrates in the field (C_i) and the lethal body concentrations measured in laboratory experiments (LBC_i).*

Species		Smelter		Reference	
	LBC_i ($\mu g\,g^{-1}$)	C_i ($\mu g\,g^{-1}$)	C_i/LBC_i	C_i ($\mu g\,g^{-1}$)	LBC_i/C_i
Porcellio scaber	2117	65.1	0.026	12.0	0.006
Orchesella cincta	37	1.50	0.041	0.36	0.010
Platynothrus peltifer	234	15.0	0.064	1.43	0.006
Bioindicator index (β)			0.044		0.007

Source: Reproduced from Van Straalen (1996).

application of this index to body residues of cadmium in soil invertebrates at two different sites. Some interesting observations can be made on these data. First, the species containing the greatest amount of cadmium in its body (the isopod *Porcellio scaber*) is not the one most at risk. Second, species with a high internal concentration tend to have a high LBC as well, so that the quotient of C_i and LBC_i is much less variable among species than the concentration itself. Van Straalen (1996) tentatively proposed benchmark values of $\beta \geqslant 0.1$ to indicate a situation of "high risk", and $\beta < 0.01$ to indicate "negligible risk". The data in Table 18.1 suggest that, since residues are 4.4% of LBC at the smelter site, the presence of Cd in soil invertebrates poses some risk (but not a high risk) at this site; at the reference site the risk is to be considered negligible.

A similar argument may be developed in the case of vertebrates for which the critical levels in certain target organs are known from toxicological investigations. For example, Ma (1989) measured the concentrations of Pb in kidney of voles and shrews at a control and reference site and compared these to the critical levels derived from rodent toxicity studies. It appeared that shrews at the contaminated site accumulated Pb above the critical level, while voles at the same site were below this level.

Objections may be raised against the concept of critical body residue, specifically in relation to the question whether CBR or LBC can be considered a species-specific constant. Five issues deserve attention:

- CBR may depend on the physiological condition of the organism. An animal in bad condition (e.g. due to low energy reserves) may be affected by an internal concentration that is tolerated by the same animal in good condition.
- CBR for one contaminant may depend on the presence or absence of other contaminants in the body, e.g., Cd may be more toxic than expected when there is a shortage of Ca or Zn.

- CBR may be subject to selection. Tolerant genotypes can withstand a higher internal concentration than sensitive genotypes, so if the genetic background of the indicator organism is unknown, the use of a single CBR is doubtful.
- The CBR approach assumes that it is the concentration in the organism that is critical to a toxic effect. However, it may also be argued that it is not the concentration but the rate of entry into the body that is critical. This is especially important in the case of toxicants for which the body residue at any time does not correlate with the rate of uptake at that time (e.g. Zn in barnacles, see Rainbow and White (1989), and Zn in isopods, see Van Straalen et al., 2005).
- Further complications arise when the residue in the organism is distributed over two or more compartments that are not in equilibrium with each other and are not equally relevant to toxicity (Rainbow, 2002). In that case the concept of total body burden breaks down and measurements need to be focussed on certain target organs or on biochemically defined fractions of the total body burden (Wallace et al., 2003; Wallace and Luoma, 2003; Vijver et al., 2004).

The objections mentioned above do not take away the need for a consistent framework for interpreting internal concentrations. One may argue that the energy devoted to routine monitoring of residues in various biological matrices is better spent on toxicokinetic modelling, identification of target organs and new fractionation methods that may reveal the biological significance of these residues.

18.3 Target species for residue analysis

The choice of a species as a target for residue measurements is one of the most crucial steps in developing an indicator system for bioavailability. Several authors have developed lists of criteria which a species should meet in order to be considered (see e.g. Phillips, 1980). Among these criteria are obvious ones such as commonness and a sufficiently large body-size; however, the most critical issue is the existence of sufficient biological background information on reproductive cycles, physiology, and genetic constitution.

Species that are considered targets for residue analysis should not be too sensitive to the toxicant of interest. Sensitive species are unsuitable for indicating exposure concentrations in the high range, because toxicity will disturb the normal uptake and excretion processes. On the other hand, sensitive species, by their nature, are suitable for indicating effects. A distinction is often made between *"exposure oriented monitoring"* (using tolerant species) and *"effect oriented monitoring"* (using sensitive species). In a similar way, Hopkin (1993) separated *"type 2 monitoring"* (bioaccumulation) from *"type 1 monitoring"* (community effects) and *"type 3 monitoring"* (effects on single species).

The distinction is expressed in another way by the use of different terms for the species involved: *"biological indicators"* are said to indicate effects, while *"biological monitors"* indicate exposure concentrations (Martin and Coughtrey, 1982; Phillips and Rainbow, 1994). Although some of these terms have received acceptance and are used by many authors, in this chapter I use the term "bioindicator" for both effect and exposure, and the term "biomonitoring" for the repeated application of bioindicators.

Another aspect of terminology is the distinction between *"active biomonitoring"* (ABM) and *"passive biomonitoring"* (PBM). In the passive strategy, analysis is done on samples taken directly from the field. In the active strategy, organisms are transplanted from a clean site or cultured in the laboratory, and then exposed for a defined period at the site to be assessed. In this way many aspects of the age, size, and physiological condition of the biomonitor organisms can be controlled and the comparability of repeated measurements is greatly increased. ABM is the most promising approach for examining *in situ* bioavailability.

The level to which a bioindicator species concentrates toxicants relative to the environment varies greatly among species. The quotient of the internal concentration relative to the concentration in some compartment of the environment is often expressed as a *"concentration factor"*. Depending on which component of the environment is used as a reference, different terms are used. A summary of this terminology is provided in Table 18.2. Accordingly, species may be grouped on the basis of the value of the concentration factor. Dallinger (1993) made a distinction between "macroconcentrators" ($BSAF \geqslant 2$), "micro-concentrators" ($1 \leqslant BSAF < 2$), and "deconcentrators" ($BSAF < 1$).

The variability of concentration factors among species seems to be larger in the terrestrial than the aquatic environment. This may be due to the uniform exposure of aquatic organisms through the water phase, compared to the heterogeneity of exposure conditions in a terrestrial system, where organisms may take up contaminants through a variety of matrices. An example of inter-species variability of internal concentrations is given in a study by Janssen (1988). Eighteen species of invertebrate animals were collected from a single

Table 18.2. *Terminology used for concentration ratios.*

Bioconcentration factor (BCF)	The concentration in an aquatic organism divided by the concentration in the water. Also used for the concentration in a sediment- or soil-living animal divided by the concentration in the pore water
Biota-to-soil accumulation factor (BSAF)	The concentration in a soil- or sediment-living organism divided by the concentration in soil or sediment
Biomagnification factor (BMF)	The concentration in an organism divided by the concentration in its diet

contaminated pine forest and concentrations of Cd were measured in their bodies (Fig. 18.5). The data showed that, although all species lived close to each other in the same microenvironment, their Cd concentrations varied by a factor of 20. These differences were not related to body-size, nor to the trophic position of the species. In fact, animals feeding on the same food items accumulated Cd to different degrees, while animals feeding on different food items sometimes had a similar concentration. These differences apparently represent differences in the physiologies and metal accumulation strategies. Other authors have reached similar conclusions (Rainbow and White, 1989; Rabitsch, 1995).

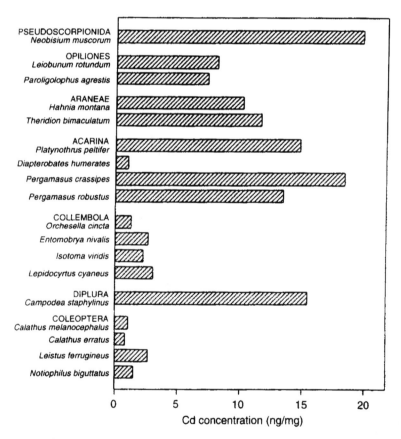

Figure 18.5. *Concentrations of cadmium measured by Janssen (1988) in different species of soil invertebrates in a contaminated pine forest in The Netherlands. The concentration in the forest floor litter was 7.7 µg g^{-1}. Cd was measured in individual specimens, all collected from the litter layer by hand sorting. Bars represent the mean of 5–56 individuals, depending on the species.*

A great variety of organisms have been proposed as targets for residue analysis in bioindicator systems. Some of the most important and generally accepted species or species groups are listed in Table 18.3. The reasons for selecting a particular species are rather variable. Commercial importance and the need to monitor residues in the context of human food safety is often a factor in the choice, although Phillips and Rainbow (1994) argued that the sampling plans designed to monitor human food safety need not be optimal in the context of environmental protection. Species that are most frequently used on a large scale across different countries are the blue mussel, *Mytilus edulis*, for the marine environment, various species of fish for the aquatic environment, mosses to monitor air pollutants, and higher plants to monitor soil in the terrestrial environment.

For the monitoring of residues in mussels an international "mussel watch" programme, including other bivalves such as oysters, has been set up. The measurement of residues in these organisms has yielded a wealth of information about temporal and geographical trends in contaminants. Mussels are particularly useful because a great deal of ecotoxicological information is available for these organisms, allowing the internal concentrations to be related to physiological responses and effect thresholds.

For the aquatic environment most attention has gone to fish. Due to the variety of species involved, the data are, however, not directly comparable between different countries. The mobility of fish is sometimes considered a drawback, because the geographical area over which they integrate pollutants may be too large to monitor discharges from point sources. Nevertheless, the good toxicological and physiological background knowledge available (cf. Fig. 18.4) is a strong point, since it allows residue measurements to be interpreted in terms of risks.

Mosses are usually seen as passive collectors of air pollutants. Not only naturally present mosses are used, but also transplanted mosses packed in bags (ABM). The moss bag technique has been developed in a greatly standardized fashion; the comparability of data obtained over large geographical areas makes it one of the favourite components of terrestrial monitoring programmes (Martin and Coughtrey, 1982).

Terrestrial plants are also frequently used as targets for residue analysis, especially for heavy metals. A large amount of information is available on heavy metals in plants and their use as bioindicator for available metal concentrations in soil (Markert, 1993; Ernst and Joosse, 1983). The use of heavy metal residues to monitor soil quality must be supported by a good background knowledge of metal physiology of the plant. Markert (1994) designed a "Biological System of the Elements" (BSE) on the basis of multiple correlations between metal concentrations in plant tissues. This system separates the biologically essential

Table 18.3. *Overview of organisms used as targets for residue analysis in indicator systems for bioavailability and the arguments mentioned in support of their use.*

Organisms	Arguments	References
Freshwater ecosystems		
Submerged mosses	Widespread Easily recognized Large surface area Concentrate metals	Say and Whitton (1983), Wehr et al. (1983)
Mussels	Important link in aquatic food chain Efficient accumulators through filter feeding Sedentary	Karbe et al. (1975), Foster and Bates (1978), Kraak et al. (1991)
Snails	Common Concentrate heavy metals	Elder and Collins (1991), Zadory and Müller (1981)
Chironomids	Abundant Reflect bioavailability of sediment contamination in addition to water	Krantzberg (1989), Hare and Campbell (1992)
Isopods	Common Easily sampled Accumulate metals	Van Hattum et al. (1996)
Eel	Commercially important Concentrate lipophilic contaminants to a high degree	Van der Oost et al. (1988), De Boer and Hagel (1994)
Pike	Ecologically and commercially important Integrates contamination as a top-predator	Olsson and Reutergårdh (1986)
Marine ecosystems		
Salt marsh plants	Residues reflect bioavailability in sediment Important as basal level of estuarine food chain	Gallagher and Kibby (1980), Otte et al. (1993)
Lugworms	Intimate contact with sediment Accumulate contaminants to a high degree Limited biotransformation capacity	Kaag et al. (1997)
Mussels	Commercially important Filter large amounts of water Efficient bioconcentrators of many contaminants Sedentary	Goldberg et al. (1978), Farrington et al. (1983), Bayne (1989), Sericano et al. (1993)
Crabs	Common Limited home range	Devescovi and Lucu (1995), John and Fernandez (1998)
Cod, herring, sea bass	Commercially important Indicate pollution of wide geographic range	Yasuno (1995), Kannan et al. (1992)
Bird feathers	Non-invasive Easy to collect Valued organisms for conservation and nature recreation	Goede and De Bruin (1984), Furness et al. (1986)

Table 18.3. (*Continued*)

Organisms	Arguments	References
Mammalian fat tissue	High bioaccumulation Important key organisms in nature conservation	Wells et al. (1994), Tanabe et al. (1994)
Terrestrial ecosystems		
Fungi	Fruit bodies easy to recognize and collect Specific capacity for accumulation of radionuclides and other metals	Ernst and Van Rooij (1987), Dietl et al. (1987), Schmitt (1989)
Lichens	Large surface area. Accumulate both from substrate and air Integrate over long time	Carlberg et al. (1983), Chettri et al. (1997)
Mosses	Intercept pollutants from air Close contact with air No uptake from soil	Thomas (1986), Glooschenko (1989), Tyler (1990)
Pine needles	Easy to collect Efficient accumulators for air pollutants	Calamari et al. (1994), Kratz (1996)
Herbal plants	Metal concentrations reflect bioavailability in soil and aerial deposition Connect below ground with terrestrial food chains	Posthumus (1983), Markert (1993)
Earthworms	Agricultural and ecological importance Intimate contact with soil Important link in terrestrial food chains	Morgan and Morgan (1988), Hendriks et al. (1995), Spurgeon et al. (1996)
Slugs, snails	Common in both natural and agricultural ecosystems Concentrate metals and pesticides	Popham and D'Auria (1980), Berger and Dallinger (1993), Gomot (1997)
Isopods	Common Accumulate extremely high amount of metals	Hopkin et al. (1986), Dallinger et al. (1992), Hopkin et al. (1993)
Ants	Common Integrate over defined area	Maavara et al. (1994), Rabitsch (1997)
Honey bees	Network of beekeepers Integrate contamination from different exposure routes (gas, liquid, particulates)	Bromenshenk et al. (1985), Morse et al. (1987)
Bird eggs	Correlation with reproductive success Valued organisms for conservation	Moriarty et al. (1986), Ratcliffe (1970), Peakall (1993)
Bird feathers	Non-invasive Easy to collect Valued organisms for conservation	Hahn et al. (1985), Solonen and Lodenius (1990), Denneman and Douben (1993)
Small mammal kidney and liver	Good connection with toxicological thresholds Reflect food-chain accumulation	Jefferies and French (1976), Ma (1994)

elements from the non-essential ones and subdivides the former into structural elements, electrolytical elements, and enzymatic elements. Ernst and Joosse (1983) emphasized the need to control for genetic variability within a species when using heavy metal residues in plants to indicate soil quality.

A basic requirement for the use of residue measurements in biomonitoring programmes is that the data represent true trends in time or space. Quality control of the analysis should be maintained by systematically analysing reference materials of which the composition is certified by the Community Bureau of Reference, the US National Institute of Standards and Technology, or comparable institutions. Several of these materials have been produced for example, oyster tissue, dogfish liver, and *Ulva lactuca* (Baudo et al., 1995). Round robin tests using these materials have sometimes revealed significant systematic differences between laboratories, while in other cases reported temporal trends appeared to be due to improvements in the analytical methodology. The need to adopt strict methodologies for quality control has now been recognized by all laboratories involved in routine biomonitoring programmes.

18.4 Confounding factors

The concentration of a contaminant in an organism will indicate the bioavailable concentration in the environment only if it is directly related to exposure. Several factors, intrinsic and extrinsic to the organism, may, however, interact with the uptake of the contaminant and the effects of these factors have to be investigated carefully before any organism is used as a biomonitor (Phillips and Segar, 1986). Four of these factors are discussed below.

18.4.1 Age

One of the most important variables that has to be taken into account when considering internal concentrations of pollutants is the age of the organism. In the case of organisms with a modular growth form (fungi, plants, and primitive invertebrates such as sponges) this not only refers to the age of the individual, but also to the age of the module that is sampled for analysis. For example, due to active reallocation processes occurring during budding, growth, and senescence of a plant leaf, the elemental composition is constantly changing and at any time a large variability may be expected among leaves and among individual plants (Ernst, 1990). In the case of animals, physiological changes associated with reaching sexual maturity may mobilize contaminants that were accumulated in the body during the juvenile stage. If there is a switch of feeding habits at some age, this may drastically alter the body burden of a pollutant, as in the case of blowflies that accumulate mercury during the larval stages, but in the pupal stage leave most of their body burden in the degenerated larval tissues, so that the emerging adult is clean once it has excreted the meconium (Nuorteva and Nuorteva, 1982).

The most common effect of age on contaminant concentrations in organisms is an increase due to gradual accumulation under continuous exposure (cf. Fig. 18.2). Long-lived species are expected to contain higher concentrations of slowly accumulating substances, simply because they have had more time to accumulate.

18.4.2 Body size

With increasing age most organisms attain a larger size and this may counteract the effects of age. The reason for this effect is that the rate of uptake of contaminants is largely proportional to the surface of the organism. Small individuals have a larger surface compared to their volume and on the basis of this fact alone are expected to have higher concentrations of contaminants than large individuals.

Van Straalen (1987) argued that the effects of weight and age on contaminant concentration call for a differential terminology. "*Accumulation sensu stricto*" was defined as an increase of concentration with age; in a growing organism, this will also imply an increase of concentration with weight. "*Accumulation sensu lato*" was defined as an increase of body burden with age. This broader term includes the situation where the concentration remains constant due to the fact that the rate of accumulation is equal to the increase in weight. Accumulation in the broad sense does not exclude a decrease of the concentration with age, if the rate of accumulation is slower than growth. This effect of "growth dilution" is not uncommon in plants during periods of rapid growth.

Effects of weight on pollutant concentrations are conveniently described as allometric equations of the form:

$$Q = aW^b \quad \text{or} \quad \log Q = \log a + b \log W \tag{18.7}$$

where Q is body burden, W weight, and a and b are parameters that may be estimated from sample data by means of a linear curve fit in a double logarithmic plot. Alternatively, the allometric model may be written in terms of the concentration, $C = Q/W$, as follows:

$$C = aW^{b-1} \quad \text{or} \quad \log C = \log a + (b-1) \log W \tag{18.8}$$

The value of the parameter b determines the type of accumulation. Accumation *sensu lato* implies $b > 0$; accumulation *sensu stricto* implies $b > 1$. If $b = 1$ accumulation is weight-proportional and there is no change of concentration.

Although Eqs. (18.6) and (18.7) are often used to describe weight-distributed residue data of contaminants, it should be borne in mind that the allometric model is purely descriptive and it is not based on any consideration about the toxicokinetic processes involved.

The effects of body weight on pollutant concentrations have been demonstrated in a variety of organisms. DDT and PCB concentrations decreased with body size in four species of marine fish (Ravid et al., 1985). Smock (1983) explained the higher concentrations of metals in small aquatic insects compared to large ones from the differences in surface/volume ratios. Williamson (1979) discussed the opposite effects of weight and age on cadmium residues in snails, while Janssen and Bedaux (1989) showed that cadmium concentrations in the pseudoscorpion *Neobisium muscorum* remained constant with weight within successive life-stages (protonymphs, deutonymphs, etc.), but decreased with weight when all stages were considered together.

Moriarty et al. (1984) showed the importance of adjusting for weight differences of fish sampled from populations with a different mean weight. This can be done using statistical methods such as analysis of covariance. Such an analysis assumes that the relationship between weight and pollutant body burden is linear and that the slope of the regression line is similar for all samples. Residue levels of different samples may then be compared after linear extrapolation of the residues to a common weight.

Griesbach et al. (1982) developed an allometric model for pesticide bioaccumulation consisting of five classes of organisms distinguished only by their body-size. The model predicted that following a pulse exposure, the pesticide would accumulate initially in the small organisms of a community, followed by a gradual transfer and final accumulation in the large organisms. Although the model lacked any trophic structure, it was in good agreement with data on DDT residues from Woodwell et al. (1967), although the latter authors explained their data in terms of food-chain effects. The computer simulations in the Griesbach et al. (1982) paper illustrate how strong the effect of allometric relationships can be.

Effects of body-size may interact with changes in growth rate to produce complicated seasonal changes in pollutant body burdens. This was the case in a study by Strong and Luoma (1981) on the marine bivalve *Macoma balthica*. Due to location-specific factors and seasonal variations in growth rate, the correlation between body-size and concentrations of Ag and Cu in the clams varied from strongly positive to non-significant and negative. In situations like this, the allometric descriptions in Eqs. (18.6) and (18.7) are useless.

18.4.3 Reproductive cycles

Another main factor influencing residues in biological tissues is reproduction. Phillips (1980) provided a very good documentation of cases in which seasonal changes in pollutant residues could be explained by the profound physiological changes occurring during a sexual cycle. For example, the whole body concentration of metals in bivalves often increases after spawning. This is due to the gonads containing low amounts of metals compared to other tissues, such that the release of gametes decreases body weight more than metal body burden. A similar mechanism was assumed in a study by Rabitsch (1997) on metal residues in worker ants of *Formica pratensis*. Like many insects, these animals accumulate lipids during hibernation, which increases their dry weight, but due to the low amount of metals in lipids this process decreases their metal concentration. When lipids were mobilized in spring, the metal concentration of the ants increased again.

For organochlorines and petroleum hydrocarbons the situation is different. Since these apolar chemicals are associated mainly with the lipid fraction of the organism, they follow the dynamics of lipids. In marine molluscs, gonadal tissues accumulate lipids during pre-spawning maturation and so the release of gametes may significantly decrease the body burden. These changes usually disappear when concentrations are expressed relative to lipid weight, rather than whole body weight. Another well-known case is the decrease of persistent organochlorine residues in lactating marine mammals. Tanabe et al. (1994) measured PCBs in the fat of seals and concluded that these animals accumulate PCBs until reaching sexual maturity at an age of 6 years, maintain a relatively low level while lactating and accumulate again when reproductive activities stop at age 20.

18.4.4 Temperature

A fourth factor affecting residues, especially in plants and ectothermic animals, is temperature. Both the rate of uptake and the rate of elimination will be affected by temperature, so body burdens will depend on the nett result of temperature effects on these toxicokinetic processes. Metabolic temperature responses in ectotherms can be described by means of the Arrhenius model (Sharpe and De Michele, 1977). An important parameter in this model is the activation energy, usually denoted H, which is a measure of the steepness of the temperature response. According to Eq. (18.1), the ultimate internal concentration under continuous exposure will be equal to the quotient of the uptake rate and the excretion rate constant: $Q_\infty = a/k$. If both a and k follow the Arrhenius model, the equilibrium body burden Q_∞ will depend on temperature, T, according to:

$$Q_\infty(T) = \frac{a_r}{k_r} \exp\left[\frac{H_a - H_k}{R} \left\{ \frac{1}{T_r} - \frac{1}{T} \right\} \right] \tag{18.9}$$

where a_r and k_r are the uptake and excretion rates at an arbitrary reference temperature T_r (in Kelvin), R the universal gas constant and H_a and H_k are the activation energies for the uptake process and the elimination process, respectively. Body burden will increase with temperature if the activation energy of the uptake process is larger than the activation energy of the elimination process and it will decrease if the converse is true. If the two processes have the same activation energies, then $H_a - H_k = 0$ and Eq. (18.8) predicts that body burden does not depend on temperature.

A study by Janssen and Bergema (1991) may serve as an illustration of the differential effects that temperature may have. Comparing cadmium kinetics in the springtail *Orchesella cincta* and the oribatid mite *Platynothrus peltifer*, these authors observed that increasing temperature affected uptake and excretion in a similar way in the former species, while uptake was affected more than excretion in the latter species. As a consequence, concentrations of Cd in springtails were invariant to temperature, while mites contained more Cd at high temperature.

18.5 Contaminants in food-webs

Since the early days of environmental toxicology biomagnification of persistent chemicals in successive steps of a food chain has been an issue of major concern. It is often argued that biomonitoring programmes should include determination of residues in top-predators because contaminant fluxes become more and more concentrated towards higher trophic levels. Biomonitoring high in the food chain is seen as a way of integrating contaminant exposure of the underlying trophic system. Is this argument really valid?

Despite the general acceptance of the idea, it must be pointed out that biomagnification is mainly limited to certain groups of contaminants, specifically organochlorines such as DDT and PCBs. In the terrestrial environment, food-chain biomagnification is not common among heavy metals, due to species-specific metal physiologies (Laskowski, 1991), and it is equally uncommon among polycyclic aromatic hydrocarbons, which are metabolized by many species at the base of the food chain (Van Brummelen et al., 1996; Faber and Heijmans, 1996). In the aquatic environment, uptake via the water usually dominates over food uptake and the latter becomes important only for very lipophilic chemicals ($\log K_{ow} > 5$, Thomann, 1989; Belfroid et al., 1996). In general, it seems that species-specific mechanisms for uptake, excretion or metabolism are more important than the position of a species in the food chain. This does not exclude the fact that some species at the top of a food chain may have physiologies that cause them to accumulate toxicants. For example, the low activities of certain biotransformation enzymes in fish-eating birds and birds of prey correlate with a high capacity to accumulate persistent organochlorines (Walker, 1980). In addition, some top-predators may be at risk because they happen to feed on species with a high bioconcentration potential. This is the case, for example, in gulls, woodcocks, shrews, and moles feeding on earthworms (Ma, 1987; Abdul Rida and Bouché, 1994).

The theoretical preconditions for biomagnification were discussed by Fagerström and Åsell (1974) and Fagerström (1991). These authors have pointed out that the levels in a trophic chain can be compared only if each level is in a steady state, that is, if the biomass and the total amount of contaminants in it are constant. The concentration in successive levels will increase if the contaminant is transferred more efficiently than the biomass itself. Biomass can thus be considered as a carrier for the contaminant and for biomagnification to occur, turnover of contaminant should be quicker than turnover of carrier. The ratio between these two turnovers was introduced as a critical parameter. Van Straalen (1987) called this ratio *"enrichment factor"* and derived mathematical expressions relating this factor to age-dependent growth, mortality, and accumulation in stationary populations with weight structure.

The development of various biomagnification models has not removed the great confusion in terminology. Specifically, the term "bioaccumulation" has been applied in so many different contexts that it may be better to avoid it altogether. A consistent set of definitions is provided in Table 18.4.

An example of a food-web distribution, heavy metals in freshwater invertebrates, is given in Figure 18.6. These authors did not find evidence for biomagnification processes within the littoral community studied. Factors that explained inter-species variation were species-specific physiologies for metal regulation and detoxification, body-size, and feeding habit. Filter feeders appeared to have a higher concentration than deposit feeders, while Zn, Cu, and Pb were higher in deposit feeders.

In congruence with the growing attention paid to food-web analysis in ecology it may be useful to approach the question of biomagnification in a more general way, and to analyse the distribution of pollutants in a complete

Table 18.4. *Terminology used to describe the behaviour of contaminants in food chains.*

Accumulation	The increase with time of body burden (accumulation *sensu lato*) or concentration (*sensu stricto*) in individual organisms
Biomagnification	The phenomenon that the concentration of a contaminant in a certain trophic level is higher than in the trophic level on which it feeds
Bioconcentration	The phenomenon that the concentration in a certain organism is higher than the concentration in the abiotic environmental compartment from which it absorbs the contaminant (specifically water)
Enrichment	A precondition for biomagnification saying that turnover of pollutant by a trophic level is higher than turnover of biomass

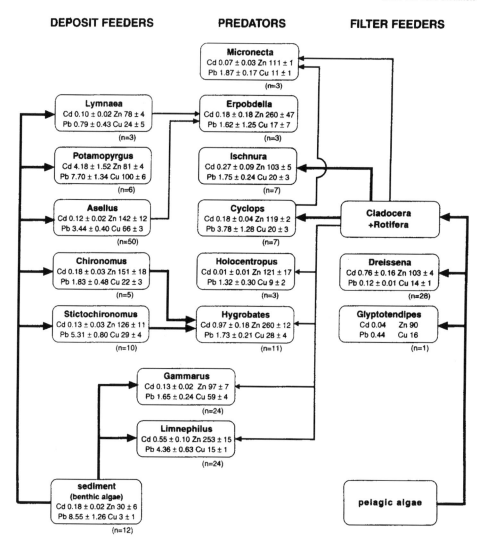

DEPOSIT FEEDERS   **PREDATORS**   **FILTER FEEDERS**

Micronecta
Cd 0.07 ± 0.03 Zn 111 ± 1
Pb 1.87 ± 0.17 Cu 11 ± 1
(n=3)

Lymnaea
Cd 0.10 ± 0.02 Zn 78 ± 4
Pb 0.79 ± 0.43 Cu 24 ± 5
(n=3)

Erpobdella
Cd 0.18 ± 0.18 Zn 260 ± 47
Pb 1.62 ± 1.25 Cu 17 ± 7
(n=3)

Potamopyrgus
Cd 4.18 ± 1.52 Zn 81 ± 4
Pb 7.70 ± 1.34 Cu 100 ± 6
(n=6)

Ischnura
Cd 0.27 ± 0.09 Zn 103 ± 5
Pb 1.75 ± 0.24 Cu 20 ± 3
(n=7)

Asellus
Cd 0.12 ± 0.02 Zn 142 ± 12
Pb 3.44 ± 0.40 Cu 66 ± 3
(n=50)

Cyclops
Cd 0.18 ± 0.04 Zn 119 ± 2
Pb 3.78 ± 1.28 Cu 20 ± 3
(n=7)

**Cladocera
+Rotifera**

Chironomus
Cd 0.18 ± 0.03 Zn 151 ± 18
Pb 1.83 ± 0.48 Cu 22 ± 3
(n=5)

Holocentropus
Cd 0.01 ± 0.01 Zn 121 ± 17
Pb 1.32 ± 0.30 Cu 9 ± 2
(n=3)

Dreissena
Cd 0.76 ± 0.16 Zn 103 ± 4
Pb 0.12 ± 0.01 Cu 14 ± 1
(n=28)

Stictochironomus
Cd 0.13 ± 0.03 Zn 126 ± 11
Pb 5.31 ± 0.80 Cu 29 ± 4
(n=10)

Hygrobates
Cd 0.97 ± 0.18 Zn 260 ± 12
Pb 1.73 ± 0.21 Cu 28 ± 4
(n=11)

Glyptotendipes
Cd 0.04   Zn 90
Pb 0.44   Cu 16
(n=1)

Gammarus
Cd 0.13 ± 0.02 Zn 97 ± 7
Pb 1.65 ± 0.24 Cu 59 ± 4
(n=24)

Limnephilus
Cd 0.55 ± 0.10 Zn 253 ± 15
Pb 4.36 ± 0.63 Cu 15 ± 1
(n=24)

sediment
(benthic algae)
Cd 0.18 ± 0.02 Zn 30 ± 6
Pb 8.55 ± 1.26 Cu 3 ± 1
(n=12)

pelagic algae

Figure 18.6. *Distribution of Cd, Zn, Pb, and Cu in the littoral food-web of a relatively unpolluted lake in the Netherlands. Concentrations are in $\mu g \, g^{-1}$. Major links are indicated by thick lines, minor ones by thin lines (Reprinted with permission from Sci. Total Environ., Elsevier Science.)*

food-web, rather than focussing on certain top-predators only (e.g. Traas et al., 1996). The food-web distribution of a contaminant may provide cues to critical interactions in ecosystems that confer stability. It would therefore be of considerable interest to analyse food-web distributions of contaminants along with food-web topologies and distributions of energy fluxes.

18.6 Conclusions

The theory and examples given above illustrate that there is a good scientific framework, developed over many years, for residue measurements in organisms as indicators of bioavailability. It is important to realize that the diversity of accumulation strategies among species is large and therefore only species for which kinetic patterns and possible confounding effects are known, should be used. There is also a need for standardization, especially in relation to ABM, the most logical approach for estimating bioavailability. In the same way as toxicity tests have become standardized in internationally accepted guidelines, the assessment of bioavailability needs to be based on a set of standardized biological targets.

References

Abdul Rida, A.M.M., Bouché, M.B., 1994. A method to assess chemical biorisks in terrestrial ecosystems. In: M.H. Donker, H. Eijsackers and F. Heimbach (Eds.), Ecotoxicology of Soil Organisms. Lewis Publishers, Boca Raton, pp. 383–394.

Banerjee, S., Sugatt, R.M., O'Grady, D.P., 1984. A simple method for determining bioconcentration parameters of hydrophobic compounds. Environ. Sci. Technol. 18, 79–81.

Baudo, R., Rossi, D., Quevauviller, Ph., 1995. Validation of the use of aquatic bioindicators by means of reference materials. In: M. Munawar, O. Hänninen, S. Roy, N. Munawar, L. Kärenlampi and D. Brown (Eds.), Bioindicators of Environmental Health. SPB Publishing, Amsterdam, pp. 211–225.

Bayne, B.L., 1989. Measuring the biological effects of pollution: the mussel watch approach. Water Sci. Technol. 21, 1089–1100.

Bedaux, J.J.M., Kooijman, S.A.L.M., 1994. Statistical analysis of bioassays, based on hazard modelling. Environ. Ecol. Stat. 1, 303–314.

Belfroid, A.C., Sijm, D.T.H.M., VanGestel, C.A.M., 1996. Bioavailability and toxicokinetics of hydrophobic aromatic compounds in benthic and terrestrial invertebrates. Environ. Rev. 4, 276–299.

Berger, B., Dallinger, R., 1993. Terrestrial snails as quantitative indicators of environmental pollution. Environ. Monit. Assess. 25, 65–84.

Bromenshenk, J.J., Carlson, S.R., Simpson, J.C., Thomas, J.M., 1985. Pollution monitoring of Puget Sound with honey bees. Science 227, 632–634.

Calamari, D., Tremeloda, P., Di Guardo, A., Vighi, M., 1994. Chlorinated hydrocarbons in pine needles in Europe: fingerprint for the past and recent use. Environ. Sci. Technol. 28, 429–434.

Carlberg, G.E., Ofstad, E.B., Drangsholt, H., Steinness, E., 1983. Atmospheric deposition of organic micropollutants in Norway studied by menas of moss and lichen analysis. Chemosphere 12, 341–356.

Chettri, M.K., Sawidis, T., Karataglis, S., 1997. Lichens as a tool for biogeochemical prospecting. Ecotox. Env. Saf. 38, 322–335.

Dallinger, R., 1993. Strategies of metal detoxification in terrestrial invertebrates. In: R. Dallinger, P.S. Rainbow (Eds.), Ecotoxicology of Metals in Invertebrates. Lewis Publishers, Boca Raton, pp. 245–289.

Dallinger, R., Berger, B., Birkel, S., 1992. Terrestrial isopods: useful biological indicators of urban metal pollution. Oecologia 89, 32–41.

De Boer, J., Hagel, P., 1994. Spatial differences and temporal trends of chlorobiphenyls in yellow eel (*Anguilla anguilla*) from inland water of the Netherlands. Sci. Total Environ. 141, 155–174.

Deneer, J.W., 1993. Uptake and elimination of chlorpyrifos in the guppy at sublethal and lethal aqueous concentrations. Chemosphere 26, 1607–1616.

Denneman, W.D., Douben, P.E.T., 1993. Trace metals in primary feathers of the barn owl (*Tyto alba guttatus*) in the Netherlands. Environ. Poll. 82, 301–310.

Devescovi, M., Lucu, C., 1995. Seasonal changes in the copper level in shore crabs *Carcinus mediterraneus*. Mar. Ecol. Prog. Ser. 120, 169–174.

Dietl, G., Muhle, H., Winkler, S., 1987. Höhere Pilze als Bioindikatoren für die Schwermetallbelastung von Böden. Verhandl. Gesell. Ökol. 16, 351–359.

Di Toro, O.M., Zarba, C.S., Hansen, D.J., Berry, W.J., Swartz, R.C., Cowan, C.E., Pavlou, S.P., Allen, H.E., Thomas, N.A., Paquin, P.R., 1991. Technical basis for establishing sediment quality criteria for nonionic organic chemicals using equilibrium partitioning. Environ. Toxicol. Chem. 10, 1541–1583.

Doucet, P.G., Sloep, P.B., 1992. Mathematical Modeling in the Life Sciences. Ellis Horwood, New York.

Edwards, C.A., 1976. The uptake of two organophosphorus insecticides by slugs. Bull. Environ. Contam. Toxicol. 16, 406–410.

Elder, J.F., Collins, J.J., 1991. Freshwater molluscs as indicator of bioavailability and toxicity of metals in surface water systems. Rev. Environ. Contam. Toxicol. 122, 37–79.

Ernst, W.H.O., 1990. Element allocation and (re)translocation in plants and its impact on representative sampling. In: H. Lieth, B. Markert (Eds.), Element Concentration Cadasters in Ecosystems. VHC, Weinheim, pp. 17–40.

Ernst, W.H.O., Joosse, E.N.G., 1983. Umweltbelastung durch Mineralstoffe – Biologische Effekte. VEB Gustav Fischer Verlag, Jena.

Ernst, W.H.O., Van Rooij, L.F., 1987. $^{134/137}$Cs fall out from Chernobyl in Dutch forest. Proceedings of International Conference Heavy Metals in the Environment, New Orleans. CEP Consultants, Edinburgh, pp. 248–286.

Faber, J.H., Heijmans, G.J.S.M., 1996. Polycyclic aromatic hydrocarbons in soil detritivores. In: N.M. van Straalen, D.A. Krivolutsky (Eds.), Bioindicator Systems for Soil Pollution. Kluwer Academic Publishers, Dordrecht, pp. 31–43.

Fagerström, T., 1991. Biomagnification in food chains and related concepts. Oikos 62, 257–260.

Fagerström, T., Åsell, B., 1974. On realism and generality of bio-accumulation models. Towards a general methodology. In: A.S.W. De Freitas, D.J. Kushner and S.U. Qadri (Eds.), Proceedings of International Conference on Transport of Persistent Chemicals in Aquatic Ecosystems. University of Ottawa Press, National Research Council of Canada, Ottawa, pp. IV-11–IV-16.

Farrington, J.W., Goldberg, E.D., Risebrough, R.W., Martin, J.H., Bowen, V.T., 1983. U.S. "Mussel Watch" 1976–1978: an overview of the trace-metal, DDE, PCB, hydrocarbon, and artificial radionuclide data. Environ. Sci. Technol. 17, 490–496.

Foster, R.B., Bates, J.M., 1978. Use of freshwater mussels to monitor point source industrial discharges. Environ. Sci. Technol. 12, 958–962.

Furness, R.W., Muirhead, S.J., Woodburn, M., 1986. Using bird feathers to measure mercury in the environment: relationships between mercury content and molt. Mar. Pollut. Bull. 17, 27–30.

Gallagher, J.L., Kibby, H.V., 1980. Marsh plants as vectors in trace metal transport in Oregon tidal marshes. Am. J. Bot. 67, 1069–1074.

Glooschenko, W.A., 1989. *Sphagnum fuscum* moss as an indicator of atmospheric cadmium deposition across Canada. Environ. Poll. 57, 27–33.

Gobas, F.A.P.C., McCorquodale, J.R., Haffner, G.D., 1993. Intestinal absorption and biomagnification of organochlorines. Environ. Toxicol. Chem. 12, 567–576.

Goede, A.A., De Bruin, M., 1984. The use of bird feather parts as a monitor for metal pollution. Environ. Poll. (Series B) 8, 281–298.

Goldberg, E.D., Bowen, V.T., Farrington, J.W., Harvey, G., Martin, J.H., Parker, P.L., Risebrough, R.W., Robertson, W., Schneider, E., Gamble, E., 1978. The mussel watch. Environ. Cons. 5, 101–126.

Gomot, A., 1997. Effets des métaux lourds sur le développement des escargots. Utilisation des escargots comme bio-indicateurs de pollution par les métaux lourds pour la préservation de la santé de l'homme. Bull. Acad. Nat. Méd. 181, 59–75.

Griesbach, S., Peters, R.H., Youakim, S., 1982. An allometric model for pesticide bioaccumulation. Can. J. Fish. Aquat. Sci. 39, 727–735.

Hahn, E., Ostapczuk, P., Ellenberg, H., Stoeppler, M., 1985. Environmental monitoring of heavy metals with birds as pollution integrating biomonitors. II. Cadmium, lead and copper in magpie (*Pica pica*) feathers from a heavily polluted and a control area. In: T.D. Lekkas (Ed.), Proceedings of International Conference on Heavy Metals in the Environment, Athens 1985. Vol. 1, CEP Consultants, Edinburgh, pp. 721–723.

Hare, L., 1992. Aquatic insects and trace metals: bioavailability, bioaccumulation, and toxicity. Crit. Rev. Toxicol. 22, 327–369.

Hare, L., Campbell, P.G.C., 1992. Temporal variations of trace metals in aquatic insects. Freshw. Biol. 27, 13–27.

Hendriks, A.J., Ma, W.-C., Brouns, J.J., De Ruiter-Dijkman, E.M., Gast, R., 1995. Modelling and monitoring organochlorine and heavy metal accumulation in soils, earthworms, and shrews in Rhine-delta floodplains. Arch. Environ. Contam. Toxicol. 29, 115–127.

Hopkin, S.P., 1993. *In situ* biological monitoring of pollution in terrestrial and aquatic ecosystems. In: P. Calow (Ed.), Handbook of Ecotoxicology. Blackwell Scientific Publications, Oxford, pp. 397–427.

Hopkin, S.P., Hardisty, G.N., Martin, M.H., 1986. The woodlouse *Porcellio scaber* as a "biological indicator" of zinc, cadmium, lead and copper pollution. Environ. Poll. Series B 11, 271–290.

Hopkin, S.P., Jones, D.T., Dietrich, D., 1993. The isopod *Porcellio scaber* as a monitor of the bioavailability of metals in terrestrial ecosystems: towards a global "woodlouse watch" scheme. Sci. Total Environ. Suppl., 357–365.

Jacquez, J.A., 1972. Compartmental Analysis in Biology and Medicine. Elsevier Publishing Company, Amsterdam.

Janssen, M.P.M., 1988. Species dependent cadmium accumulation by forest litter arthropods. Proceedings of International Conference on Environmental Contamination, Venice. CEP Consultants, Edinburgh, pp. 436–438.

Janssen, M.P.M., Bedaux, J.J.M., 1989. Importance of body-size for cadmium accumulation by forest litter arthropods. Neth. J. Zool. 39, 194–207.

Janssen, M.P.M., Bergema, W.F., 1991. The effect of temperature on cadmium kinetics and oxygen consumption in soil arthropods. Environ. Toxicol. Chem. 10, 1493–1501.

Janssen, M.P.M., Bruins, A., De Vries, T.H., Van Straalen, N.M., 1991. Comparison of cadmium kinetics in four soil arthropod species. Arch. Environ. Contam. Toxicol. 20, 305–312.

Jefferies, D.J., French, M.C., 1976. Mercury, cadmium, zinc, copper and organochlorine insecticide levels in small mammals trapped in a wheat field. Environ. Poll. 10, 175–182.

John, L., Fernandez, T.V., 1998. Incidence of trace metals in *Scylla serrata*, an edible crab from Ashtamudi estuary, India. J. Environ. Biol. 19, 99–106.

Kaag, N.H.B.M., Foekema, E.M., Scholten, M.C.Th., Van Straalen, N.M., 1997. Comparison of contaminant accumulation in three species of marine invertebrates with different feeding habits. Environ. Toxicol. Chem. 16, 837–842.

Kannan, K., Falandysz, J., Yamashita, N., Tanabe, S., Tatasukawa, R., 1992. Temporal trends of organochlorine concentrations in cod-liver oil from the southern Baltic proper, 1971–1989. Mar. Poll. Bull. 24, 358–363.

Karbe, L., Antonacopoulos, N., Schnier, C., 1975. The influence of water quality on accumulation of heavy metals in aquatic organisms. Verhandl. Internat. Verein. theor. angew. Limnol. 19, 2094–2101.

Kraak, M.H.S., Scholten, M.C.Th., Peeters, W.H.M., De Kock, W.C., 1991. Biomonitoring of heavy metals in the Western European rivers Rhine and Meuse using the freshwater mussel *Dreissena polymorpha*. Environ. Poll. 74, 101–114.

Krantzberg, G., 1989. Metal accumulation by chironomid larvae: the effects of age and body weight on metal body burden. Hydrobiologia 188/189, 497–506.

Kratz, W., 1996. Ecotoxicological bioindication: PAH, PCB and heavy metals studied in the natural resource monitoring programme of Berlin (Germany). In: N.M. van Straalen, A.D. Krivolutsky (Eds.), Bioindicator Systems for Soil Pollution. Kluwer Academic Publishers, Dordrecht, pp. 223–232.

Landrum, P.F., Eadie, B.J., Faust, W.R., 1992. Variation in the bioavailability of polycyclic aromatic hydrocarbons to the amphipod *Diporeia* (spp.) with sediment aging. Environ. Toxicol. Chem. 11, 1197–1208.

Laskowski, R., 1991. Are the top carnivores endangered by heavy metal biomagnification?. Oikos 60, 387–390.

Ma, W.-C., 1987. Heavy metal accumulation in the mole, *Talpa europea*, and earthworms as an indicator of metal bioavailability in terrestrial environments. Bull. Environ. Contam. Toxicol. 39, 933–938.

Ma, W.-C., 1989. Effect of soil pollution with metallic lead pellets on lead bioaccumulation and organ/body weight alterations in small mammals. Arch. Environ. Contam. Toxicol. 18, 617–622.

Ma, W.-C., 1994. Methodological principles of using small mammals for ecological hazard assessment of chemical soil pollution, with examples on cadmium and lead. In: M.H. Donker, H. Eijsackers and F. Heimbach (Eds.), Ecotoxicology of Soil Organisms. Lewis Publishers, Boca Raton, pp. 357–371.

Maavara, V., Martin, A.-J., Oja, A., Nuorteva, P., 1994. Sampling of different social categories of red wood ants (*Formica* s. str.) for biomonitoring. In: B. Markert (Ed.), Environmental Sampling for Trace Analysis. VCH, Weinheim, pp. 465–489.

Markert, B. (Ed.) 1993. Plants as Biomonitors. Indicators for Heavy Metals in the Terrestrial Environment. VCH, Weinheim.

Markert, B., 1994. The biological system of the elements (BSE) for terrestrial plants (glycophytes). Sci. Total Environ. 155, 221–228.

Martin, M.H., Coughtrey, P.J., 1982. Biological Monitoring of Heavy Metal Pollution. Applied Science Publishers, London.

McCarty, L.S., Mackay, D., 1993. Enhancing ecotoxicological modeling and assessment. Environ. Sci. Technol. 27, 1719–1727.

McCarty, L.S., Mackay, D., Smith, A.D., Ozburn, G.W., Dixon, D.G., 1992. Residue-based interpretation of toxicity and bioconcentration QSARs from aquatic bioassays: neutral narcotic organics. Environ. Toxicol. Chem. 11, 917–930.

Morel, J.-L., 1997. Bioavailability of trace elements to terrestrial plants. In: J. Tarradellas, G. Bitton and D. Rossel (Eds.), Soil Ecotoxicology. Lewis Publishers, Boca Raton, pp. 141–176.

Morgan, J.E., Morgan, A.J., 1988. Earthworms as biological monitors of cadmium, copper, lead and zinc in metalliferous soils. Environ. Poll. 54, 123–138.

Moriarty, F., 1984. Persistent contaminants, compartmental models and concentration along food-chains. Ecol. Bull. 36, 35–45.

Moriarty, F., Bell, A.A., Hanson, H., 1986. Does p,p'-DDE thin eggshells?. Environ. Poll. Series A 40, 257–286.

Moriarty, F., Hanson, H.M., Freestone, P., 1984. Limitations of body burden as an index of environmental contamination: heavy metals in fish *Cottus gobio* L. from the river Ecclesbourne, Derbyshire. Environ. Poll. Series A 34, 297–320.

Morse, R.A., Culliney, T.W., Gutenmann, W.H., Littman, C.B., Lisk, D.J., 1987. Polychlorinated biphenyls in honey bees. Bull. Environ. Contam. Toxicol. 38, 271–276.

Nuorteva, P., Nuorteva, S., 1982. The fate of mercury in sarcosaprophagous flies and in insects eating them. Ambio 11, 34–37.

Olsson, M., Reutergårdh, L., 1986. DDT and PCB pollution trends in the Swedish aquatic environment. Ambio 15, 103–109.

Otte, M.L., Haarsma, M.S., Broekman, R.A., Rozema, J., 1993. Relation between heavy metal concentrations in salt marsh plants and soil. Environ. Poll. 82, 13–22.

Peakall, D.B., 1993. DDE-induced eggshell thinning: an environmental detective story. Environ. Rev. 1, 13–20.

Phillips, D.J.H., 1980. Quantitative Aquatic Biological Indicators. Their Use to Monitor Trace Metal and Organochlorine Pollution. Applied Science Publishers, London.

Phillips, D.J.H., Rainbow, P.S., 1994. Biomonitoring of Trace Aquatic Contaminants. Chapman & Hall, London.

Phillips, D.J.H., Segar, D.A., 1986. Use of bio-indicators in monitoring conservative contaminants: programme design imperatives. Mar. Poll. Bull. 17, 10–17.

Popham, J.D., D'Auria, J.M., 1980. *Arion ater* (Mollusca: Pulmonata) as an indicator of terrestrial environmental pollution. Water Air Soil Poll. 14, 115–124.

Posthumus, A.C., 1983. Higher plants as indicators and accumulators of gaseous air pollution. Environ. Monitor. Assess. 3, 263–272.

Rabitsch, W.B., 1995. Metal accumulation in arthropods near a lead/zinc smelter in Arnoldstein, Austria. Environ. Poll. 90, 221–237.

Rabitsch, W.B., 1997. Seasonal metal accumulation patterns in the red wood ant *Formica pratensis* (Hymenoptera) at contaminated and reference sites. J. Appl. Ecol. 34, 1455–1461.

Rainbow, P.S., 2002. Trace metal concentrations in aquatic invertebrates: why and so what? Environ. Poll. 120, 497–507.

Rainbow, P.S., White, S.L., 1989. Comparative strategies of heavy metal accumulation by crustaceans: zinc, copper and cadmium in a decapod, an amphipod and a barnacle. Hydrobiologia 174, 245–262.

Ratcliffe, D.A., 1970. Changes attributable to pesticides in egg breakage frequency and eggshell thickness in some British birds. J. Appl. Ecol. 7, 67–115.

Ravid, R., Ben-Yosef, J., Hornung, H., 1985. PCBs, DDTs and other chlorinated hydrocarbons in marine organisms from the Mediterranean coast of Israel. Mar. Poll. Bull. 16, 35–38.

Salomons, W., Stigliani, W.M. (Eds.), 1995. Biogeodynamics of Pollutants in Soils and Sediments. Springer Verlag, Berlin.

Say, P.J., Whitton, B.A., 1983. Accumulation of heavy metals by aquatic mosses. 1: *Fontinalis antipyretica* Hedw. Hydrobiologia 100, 245–260.

Schmitt, J.A., 1989. Fungi and heavy metals. In: E. Borghi (Ed.), Fungi Atque Loci Natura. Borgo, Val di Taro, pp. 105–146.

Sericano, J.L., Wade, T.L., Brooks, J.M., Atlas, E.L., Fay, R.R., Wilkinson, D.L., 1993. National status and trends mussel watch program: chlordane-related compounds in Gulf of Mexico oysters, 1986–90. Environ. Poll. 82, 23–32.

Sharpe, P.J.H., De Michele, D.W., 1977. Reaction kinetics of poikilotherm development. J. Theor. Biol. 64, 649–670.

Smock, L.A., 1983. Relationships between metal concentrations and organism size in aquatic insects. Freshw. Biol. 13, 313–321.

Solonen, T., Lodenius, M., 1990. Feathers of birds of prey as indicators of mercury contamination in southern Finland. Hol. Ecol. 13, 229–237.

Southworth, G.R., Beauchamp, J.J., Schmieder, P.K., 1978. Bioaccumulation potential of polycyclic aromatic hydrocarbons in *Daphnia pulex*. Water Res. 12, 973–977.

Spurgeon, D.J., Sandifer, R.D., Hopkin, S.P., 1996. The use of macro-invertebrates for population and community monitoring of metal contamination – indicator taxa, effect parameters and the need for a soil invertebrate prediction and classification scheme (SIVPACS). In: N.M. van Straalen, D.A. Krivolutsky (Eds.), Bioindicator Systems for Soil Pollution. Kluwer Academic Publishers, Dordrecht, pp. 95–110.

Strong, C.R., Luoma, S.N., 1981. Variations in the correlation of body size with concentrations of Cu and Ag in the bivalve *Macoma balthica*. Can. J. Fish. Aquatic Sci. 38, 1059–1064.

Tanabe, S., Sung, J.-K., Choi, D.-Y., Baba, N., Kiyota, M., Yoshida, K., Tatsukawa, R., 1994. Persistent organochlorine residues in northern fur seal from the pacific coast of Japan since 1971. Environ. Poll. 85, 305–314.

Thomann, R.V., 1989. Bioaccumulation model of organic chemical distribution in aquatic food chains. Environ. Sci. Technol. 23, 699–707.

Thomas, W., 1986. Representativity of mosses as biomonitor organisms for the accumulation of environmental chemicals in plants and soils. Ecotox. Env. Saf. 11, 339–346.

Timmermans, K.R., Van Hattum, B., Kraak, M.H.S., Davids, C., 1989. Trace metals in a littoral foodweb: concentrations in organisms, sediment and water. Sci. Total Environ. 87/88, 477–494.

Traas, T.P., Stäb, J.A., Kramer, P.R.G., Cofino, W.P., Aldenberg, T., 1996. Modeling and risk assessment of tributyltin accumulation in the food web of a shallow freshwater lake. Environ. Sci. Technol. 30, 1227–1237.

Tyler, G., 1990. Bryophytes and heavy metals: a literature review. Bot. J. Linnean Soc. 104, 231–253.

Van Brummelen, T.C., Verweij, R.A., Wedzinga, S.A., Van Gestel, C.A.M., 1996. Polycyclic aromatic hydrocarbons in earthworms and isopods from contaminated forest soils. Chemosphere 32, 315–341.

Van der Oost, R., Heida, H., Opperhuizen, A., 1988. Polychlorinated biphenyl congeners in sediments, plankton, molluscs, crustaceans, and eel in a freshwater lake: implications of using reference chemicals and indicator organisms in bioaccumulation studies. Arch. Environ. Contam. Toxicol. 17, 721–729.

Van Gestel, C.A.M., Rademaker, M.C.J., Van Straalen, N.M., 1995. Capacity controlling parameters and their impact on metal toxicity in soil invertebrates. In: W. Salomons, W.M. Stigliani (Eds.), Biogeodynamics of Pollutants in Soils and Sediments. Springer Verlag, Berlin, pp. 171–192.

Van Hattum, B., Van Straalen, N.M., Govers, H.A.J., 1996. Trace metals in populations of freshwater isopods: influence of biotic and abiotic variables. Arch. Environ. Contam. Toxicol. 31, 303–318.

Van Straalen, N.M., 1987. Turnover of accumulating substances in populations with weight-structure. Ecol. Model. 36, 195–209.

Van Straalen, N.M., 1996. Critical body concentrations: their use in bioindication. In: N.M. van Straalen, D.A. Krivolutsky (Eds.), Bioindicator Systems for Soil Pollution. Kluwer Academic Publishers, Dordrecht, pp. 5–16.

Van Straalen, N.M., Donker, M.H., Vijver, M.G., Van Gestel, C.A.M. (2005). Bioavailability of contaminants estimated from uptake rates into soil invertebrates. Environ. Poll. 136, 409–417.

Vijver, M.G., Van Gestel, C.A.M., Lanno, R.P., Van Straalen, N.M., Peijnenburg, W.J.G.M., 2004. Internal metal sequestration and its ecotoxicological relevance: a review. Environ. Sci. Technol. 38, 4705–4712.

Walker, C.H., 1980. Species variations in some hepatic microsomal enzymes that metabolize xenobiotics. Progr. Drug Met. 5, 113–164.

Wallace, W.G., Lee, B.-G., Luoma, S.N., 2003. Subcellular compartmentalization of Cd and Zn in two bivalves. I. Significance of metal-sensitive fractions (MSF) and biologically detoxified metal (BDM). Mar. Ecol. Progr. Ser. 249, 183–197.

Wallace, W.G., Luoma, S.N., 2003. Subcellular compartmentalization of Cd and Zn in two bivalves. II. Significance of trophically available metal (TAM). Mar. Ecol. Progr. Ser. 257, 125–137.

Wehr, J.D., Empain, A., Mouvet, C., Say, P.J., Whitton, B.A., 1983. Methods for processing aquatic mosses used as monitors of heavy metals. Water Res. 17, 985–992.

Wells, D.E., Campbell, L.A., Ross, H.M., Thompson, P.M., Lockyer, C.H., 1994. Organochlorine residues in harbor porpoise and bottlenosed dolphins stranded on the coast of Scotland 1988–1991. Sci. Total Environ. 151, 77–99.

Widianarko, B., Van Straalen, N.M., 1996. Toxicokinetics-based survival analysis in bioassays using nonpersistent chemicals. Environ. Toxicol. Chem. 15, 402–406.

Williamson, P., 1979. Opposite effects of age and weight on cadmium concentrations of a gastropods mollusc. Ambio 8, 30–31.

Woodwell, G.M., Wurster, C.F. Jr., Isaacson, P.A., 1967. DDT residues in an east coast estuary: a case of biological concentration of a persistent insecticide. Science 156, 821–823.

Yasuno, M., 1995. Long-term biomonitoring of organochlorine and organotin compounds along the coast of Japan by the Japan Environment Agency. In: M. Munawar, O. Hänninen, S. Roy, N. Munawar, L. Kärenlampi and D. Brown (Eds.), Bioindicators of Environmental Health. SPB Academic Publishing, Amsterdam, pp. 179–193.

Zadory, L., Müller, P., 1981. Potential use of freshwater molluscs for monitoring river pollution. Geograph. J. 5, 433–445.

Developments in Soil Science, volume 32
Ravendra Naidu (Editor)

479

Chapter 19

BIOLOGICAL TOOLS TO ASSESS CONTAMINANT BIOAVAILABILITY IN SOILS

Z. Keatinge and I. Singleton

19.1 Introduction

The soil environment is a sink for many chemicals and it is inevitable that this leads to environmental contamination problems. Contaminated land may pose a threat to human health and the environment. Toxic effects may be acute (e.g. poisoning) or chronic (long-term damage, e.g. cancers or reproductive problems). The scale of the problem in the UK is such that the Environment Agency has estimated that over 20,000 sites or 300,000 hectares may be affected (Environment Agency, 2004a). The European Environment Agency has estimated 1.5 million sites in Europe may be contaminated although only 300,000 sites have been assessed and deemed contaminated so far (European Environment Agency, 2004). Similar contamination problems exist in other countries throughout the world.

Compounds which cause contamination include heavy metals, polyaromatic hydrocarbons (PAHs), pesticides and chlorinated solvents (Loomis and Hayes, 1996). For purposes of analysis the two major subgroups of contaminants are metals and organics. The division into these groups is slightly misleading as a large number of sites are contaminated with complex mixtures of chemicals.

Given the potential human and ecosystem problems associated with soil contamination, many countries have developed guidelines which set levels of contaminants, above which soils may be potentially considered to be a risk. Different countries have generic assessment criteria-set intervention values for specific contaminants. Variation in the intervention values given by different countries occur mainly from "uncertainty surrounding the toxicological database and differences in policy on exposure and averaging periods" (Environment Agency, 2002b).

Contaminated land regulations in England and Wales are set and administered by the Environment Agency. The primary aim of these regulations is to protect human health, with the environment as a secondary concern. Human health risk assessment is managed using the Contaminated Land Exposure Assessment (CLEA) regime (Environment Agency, 2004b). The principles behind it (and as such the stringent approach) have been developed using pharmaceuticals risk assessment as a model. CLEA gives soil guideline values (SGVs); a specific

Table 19.1. *Summary of UK Environment Agency Contaminated Land Exposure Assessment model soil guideline values given in mg/kg dry weight soil.*

SGV number	Compound	Residential with plant uptake and allotments	Residential without plant uptake	Commercial/ industrial
		SGV (mg/kg)	SGV (mg/kg)	SGV (mg/kg)
SGV1	Arsenic	20	20	500
SGV3	Cadmium	pH6 – 1, pH7 – 2, pH8 – 8	30	1400
SGV4	Chromium	130	200	5000
SGV5	Inorganic mercury	8	15	480
SGV7	Nickel	50	75	5000
SGV9	Selenium	35	260	8000
SGV10	Lead	450	450	750

Source: Environment Agency (2002b).

number in milligram of contaminant per kilogram of soil (Table 19.1), above which a site requires further assessment and/or remediation – usually at high cost.

Assessment of soil contamination for the sake of environmental protection is currently in draft form in the UK. It will use an Ecological Risk Assessment (ERA) scheme to protect ecosystems (Environment Agency, 2003). This is also regulated by the Environment Agency. In a scheme similar to CLEA, specific numbers – soil screening values (SSVs) – for priority contaminants will be given, above which intervention will be required (Table 19.2).

Site assessment is currently carried out by chemical analysis which determines "total" amounts of a contaminant in a soil (see Chapter 4 for an overview on risk based approach to assessing contaminated sites). Advantages of this approach are that it provides a clear picture of contaminants that may actually be present at a site. A major limitation of this approach is the fact that one may only find what is specifically being searched for (unknown or unexpected contaminants may be present), analysis takes a long time, exquisite sensitivity is needed to look for low amounts of chemicals and finally chemical analysis can be very costly. Indeed, CLEA only identify a small set of priority contaminants – ones known to be very toxic or mutagenic (Environment Agency, 2003, 2004b). These methods, although very convenient for regulation and analysis are unrealistic due to (historic) contamination being a highly complex "mixed bag" of compounds. Chemical analysis is unable to give any indication of potential synergistic/additive effects of contaminants or antagonistic effects (La Point and Waller, 2000). Contaminants may have the same or similar mode of action and as such magnify any toxic response or

Table 19.2. *List of priority contaminants likely to be found at potentially contaminated sites in the UK. Soil screening values will be developed for these contaminants in 2004/05.*

Metals	Arsenic, beryllium, cadmium, copper, lead, mercury, nickel, selenium, zinc
Organometals	Organolead compounds, organotin compounds, e.g. tributyltin
Inorganics	Cyanides
Aromatics	Total petroleum hydrocarbons, benzene, toluene, ethylbenzene, xylene(s), phenol
Polycyclic aromatic hydrocarbons (PAHs)	Benzo(a)pyrene, anthracene, naphthalene, 1,2,4-trichlorobenzene
Chlorinated hydrocarbons	Tetrachlorobenzene, pentachlorobenzene, 1,2-dichloroethane, 1,1,1,-trichloroethane, tricholorethene, tetrachloroethene, pentachlorophenol, chlorotoluenes, vinyl chloride, chloroform, hexachlorobuta-1,3-diene, polychlorinated biphenyls (total), dioxins and furans
Pesticides	Dieldrin, DDT (total), HCH (total)

Source: Environment Agency (2003).

alternatively contaminants may compete for a pathway, reducing each others capacity for toxicity. For example, An et al. (2004) investigated the toxicity of Cu, Cd and Pb to plant growth and found additive, synergistic and antagonistic effects. This means that chemical analysis alone is not able to accurately predict the biological response to mixtures of contaminants in soil.

Another weakness of chemical analysis approach is that it is unable to predict what percentage of any contaminant is biologically available (bioavailable) to cause harm. If a compound is not bioavailable it therefore cannot be toxic (see Chapter 4). Bioavailability therefore should be a major consideration when assessing potential soil toxicity. Bioavailability gives a more realistic view of the risk a site may pose to humans or the environment. For example, a recent British Geological Survey (BGS) survey of As in urban soils in Sheffield, UK found that over 60% of samples were contaminated to levels above SGVs set by CLEA (Environment Agency, 2002b) and it is likely that similar results will be found for all UK cities with an industrial history. The cost of remediating all such sites is prohibitive and clearly if bioavailability assays can indicate that contaminants in soils are not available to living organisms (therefore of low toxicity risk) and do not need to be remediated then this will lead to a significant cost saving (see Chapters 3 and 4). Another reason for examining bioavailability is for determining the potential success of any (bio)remediation techniques used. The complexity of soil makes defining bioavailability an even more complex question (see Chapter 3).

19.2 Bioavailability

As with many scientific terms, usage and definitions of bioavailability can vary (see Chapter 3 for definitions of bioavailability). The UK Environment Agency uses *bioavailable* as meaning "the fraction of the chemical that can be absorbed by the body, through the gastrointestinal system, the pulmonary system and the skin" and *bioaccessible* as meaning "the fraction of a substance that is available for absorption by an organism" (Environment Agency, 2004c). The US National Research Council report on bioavailability of contaminants in soils and sediments (Ehlers and Luthy, 2003; National Research Council, 2002) has no explicit definition of bioavailability. It defines *bioavailability processes* as the "individual physical, chemical, and biological interactions that determine the exposure of organisms to chemicals associated with soils and sediments". The major factor that these definitions do not include is the dynamic aspect of bioavailability over time. Schulin's summary of the 2003 bioavailability workshop (Schulin and Schroder, 2003) discusses the problems of defining bioavailability. It talks about how contaminant uptake depends on delivery rate from soil to an organism and how definitions based on momentary concentrations are inaccurate compared to ones based on fluxes and rates of re-supply – also see Peijnenburg and Jager (2003). It then goes on to say that for practical purposes the best approach is to describe bioavailability in terms of soluble or labile concentrations; that soil solutions concentrations of a contaminant are nearer to bioavailability than a total concentration in soil. For a detailed discussion on the definition of bioavailability readers are directed to Chapter 3.

19.3 Factors affecting bioavailability of contaminants in soil

Chapter 2 presents an overview of soil and environmental factors that underpin contaminant bioavailability in the soil environment. There are several well-documented factors which affect bioavailability including:

(a) Soil – basic soil properties which need to be considered that may influence bioavailability are: pH (Impellitteri et al., 2003; Lock and Janssen, 2003), redox potential (Rensing and Maier, 2003), ionic strength, organic matter (Pardue et al., 1996), type of soil (Lock et al., 2002), clay fraction (Babich and Statzky, 1977), water content, oxygen content, temperature and soil residents (organisms, plant roots, invertebrates, etc.).

(b) Contaminant physico-chemical properties that may affect bioavailability include molecular structure (see Chapter 10), polarity, aqueous solubility, hydrophobicity, lipophilicity (the major factor affecting bioaccumulation in food chains and secondary poisoning) and volatility (Reid et al., 2000a), speciation of metals (Arnold et al., 2003), mineral form (Davies et al., 2003), mobility and persistence. For more detailed information on the bioavailability of organics see Stokes et al. (2004); on pesticides see Gevao et al. (2003) and on metals see Rensing and Maier (2003). As touched upon in the introduction, co-contaminants may have an unpredictable effect and may act synergistically or competitively with each other.

(c)   The organism for which risk is being assessed – whether it be a microbe, invertebrate, plant, fish (sediment) or humans; different organisms all have completely different routes of exposure (for example microbes at soil pore microscopic level, humans at a macroscopic level. Different compounds will cause dissimilar toxic effects depending on species exposed (different reactions can and will be seen even at the individual level). For example, Lock et al. (2002) examined the toxicity of lindane to various soil invertebrates. They discovered that toxicity was species specific and not dependent (in this instance) on organic matter content of the soil. If organisms within the same feeding groups exhibit different sensitivities it may therefore be almost impossible to generalise or talk about the relevance of one organism to another.

(d)   Residence time of contaminants is one of the factors most talked about and difficult to define. Referred to as "ageing"; a time dependent interaction between the contaminant and the soil. Contaminants become sorbed to mineral and organic matter components of soil and trapped in micropores and become generally biologically inaccessible. The longer a contaminant is in contact with the soil, the more the soil and contaminant become associated, reducing bioavailability and consequent potential toxicity (Alexander, 2000; Hatzinger and Alexander, 1995; Chapter 2).

19.4 Assessment of contaminant bioavailability in soil

The bioavailability of contaminants tends to be mimicked using extraction methods (chemical methods of determining bioavailability). It is important to consider and understand the advantages and limitations of these extraction methods as many of them are used as the first step in assessing the bioavailability of pollutants to actual organisms (see below for details on biological tools). As mentioned previously, the soluble or labile fraction from a soil has acted as a guideline for bioavailability. The US NRC report (Ehlers and Luthy, 2003; National Research Council, 2002) gives a thorough review of the physical, chemical and biological tools used to measure bioavailability. Categories include physical and chemical characterisation of the soil (referred to as the solid phase as the report also covers sediment analysis), extracts for inorganic contaminants and extracts for organic contaminants.

Extracts for inorganic contaminants tend to fall into three groups; conventional or single, sequential and passive. Conventional tests simulate a simple leaching procedure or phytoavailability in a single extraction using water, acid, chelating agents or salt solutions (Rauret, 1998). Sequential methods use similar agents to single extractions but are designed to differentiate between elements associated with different physico-chemical soil phases (for example Ure et al., 1993). Passive techniques involve exchangeable resins or pore water measurements.

Extracts for organic contaminants tend to use solvents; "total" concentration of organics involves use of an organic solvent and either heating or shaking (Stokes et al., 2004). Table 19.3 from Stokes et al. (2004) gives a summary of non-exhaustive extraction techniques for a variety of target compounds. Readers are directed to Chapter 10 for a detailed consideration of organic contaminant speciation and bioavailability.

An area, mentioned briefly in the NRC report, of bioavailability research that has received increasing attention is that of *in vitro* tests to mimic human intake for both organics and inorganics. Such tests appear to be an interface between purely chemical extraction assays and those which use living organisms and so are briefly discussed here as being relevant to biological tools used to assess soil contaminant bioavailability.

Human exposure routes to contaminated soils are inhalation, ingestion and dermal contact. Ingestion of soils, or oral bioavailability, is considered the

Table 19.3. *Non-exhaustive extraction techniques of organic contaminants from soil.*

Method	Target compounds	References
Acetone shake extraction	Atrazine	Johnson et al. (1999)
Butanol shake extraction	PAHs	Kelsey et al. (1997), Liste and Alexander (2002), Reid et al. (2000b)
Methanol–water shake extraction	PAHs, PCBs, Atrazine	Kelsey et al. (1997), Krauss et al. (2000)
Artificial fresh water shake extraction	Pentachlorophenol and degradation products	Tuomela et al. (1999)
Calcium chloride shake extraction	Pesticides (Dicamba, Atrazine, Isoproturon)	Gevao et al. (2001), Loiseau and Barriuso (2002)
Hydroxypropyl-β-cyclodextrin (HPCD) shake extraction	PAHs	Cuypers et al. (2001), Reid et al. (2000b)
Surfactant (Triton X-100) shake extraction	PAHs, Pesticides (Trifluralin, Atrazine, Coumaphos)	Cuypers et al. (2001), Mata-Sandoval et al. (2002)
Potassium persulphate oxidation	PAHs	Cuypers et al. (2000, 2001)
CO_2 SFE	PAHs	Hawthorne and Grabanski (2000)
SPE using Tenax	PAHs, PCBs, Chlorobenzenes, Atrazine	Cornelissen et al. (1998, 1999), Cuypers et al. (2001), Xing et al. (1996)

Source: Adapted from Stokes et al. (2005).
Notes: SFE, Supercritical fluid extraction; SPE, solid phase extraction.

highest priority exposure route (Paustenbach, 2000). Different groups in different countries have designed their own version of an *in vitro* simulated gut extract, all based on the same principles but with significant variety in the experimental detail. Several of these are simultaneously assessed and reviewed in (Oomen et al., 2002). This topic is also expanded upon in the Environment Agency technical report of 2002 (Environment Agency, 2002a). The basics of the various techniques include processes that mimic the stomach (low pH and physical agitation) then the intestines (rise in pH and addition of digestive enzymes) all at physiological temperature (37°C). The features of each method that may lead to differences in contaminant bioavailability include the pH of the "stomach", time taken at different pH's and whether food is added as well as soil. Several methods have been connected to *in vivo* studies and validated for certain contaminants (for example, lead – Ruby et al., 1993).

In summary, these different *in vitro* extraction methods appear to give very different measures of bioavailability – also see Harkey and Young (2000) depending on the technique used and the contaminant being studied.

Advantages of chemical/*in vitro* extraction techniques to look at bioavailability of contaminants in soil are reproducibility of conditions, speed of analysis and potentially greater relevance to biological systems than measurement of "total" amounts of contaminants. The major limitation of the majority of methods described is that most of the extracts produced still have to be chemically analysed and are not suitable to then go on and use with biological tools to determine biological availability of contaminants/actual toxicity of a soil. For example, acid extracts and low pH simulated gut extracts would denature/kill pH sensitive molecules/organisms; organic solvents alone tend to be toxic to biological systems and *in vitro* intestinal enzymes will digest and break down biological matter. In addition the majority of extracts are designed to examine either metals or organics – which again is an unrealistic picture of mixture-contaminated soil. Extracts may over or under-estimate bioavailability to different species especially as different organisms are exposed to different soil fractions. Therefore the choice of extraction method used for bioassays is important and deserves critical attention.

19.5 Biological tools for assessment of contaminant bioavailability

The complexity of soil, contaminants and receptors means that it is impossible to have a "one size fits all" biological test for assessing pollutant bioavailability in soil. Things to be considered before choosing what biological test(s) to use are:

(a) Does the contaminant cause an acute, lethal response or is it a chronic, mutagenic, endocrine disruptive or reproductive failure effect? What is an important effect?

(b) Is an overall toxic response required or is the intention just to sense a specific compound?

(c) Is the test to be used a multicellular organism, a single-celled organism or a molecular (sub-cellular) test?

(d) Is the study to be at the site (*in situ*) or can samples be taken to the lab (*ex situ*)?

(e) What is the assay representing and trying to protect? Human risk? Ecosystem health?

(f) Is the test to be in direct contact with the soil (solid phase) or will an extract technique be used? If so, how relevant is it?

Of the tests discussed here, some are currently used with soil, and some are used for purposes such as sediment testing but could be adapted for use with soil extracts. Some are used for testing toxicity of pure compound or water samples, but again may be adapted for use with soil or soil extracts. The relevance of the biological assay will also depend on the relevance of any extract technique used.

Toxic effects can be categorised as systemic (organism biochemistry and histopathology changes), reproductive, genotoxic or population effects (e.g. abundance and diversity of arthropods, Borras and Nadal, 2004). As outlined above, biological tools to assess bioavailability can be subdivided in many different ways. This chapter uses a hierarchy of size in its outline.

19.5.1 Multicellular organism tests

Tests using whole animals/plants have been used for several decades and as such tend to have large data sets which lend validity to their usage. Initially designed to test factors such as toxicity of pesticides to plants or sediment toxicity, several methods have been enshrined as International Standards Organisation (ISO) Guidance notes (ISO, 1993a, 1993b, 1995, 1998, 1999a, 1999b, 2004) and as the Organisation for Economic Co-operation and Development (OECD) Chemicals Testing Guidelines (effects on biotic systems)(OECD, 1984a, 1984b, 2000a, 2000b, 2003). They include earthworm acute toxicity, earthworm reproduction, terrestrial plant growth (both monocotyledons and dicotyledons), inhibition of root growth, emergence and growth of higher plants and effects on invertebrate reproduction and survival. The advantage of these tests is that they are directly relevant to the specific species examined 'and represent *in situ* conditions?', i.e. they do not require a soil extract to be made and therefore represent actual pollutant bioavailability to a selected organism in soil over time. However the relevance of a particular test, e.g. on a plant species, is difficult to scale up or down to different parts of the ecosystem. Two reviews that go into greater detail are Hund-Rinke and Kordel (2003), which looks at bioavailability of metals, and the UK Environment Agency Review of ecotoxicological and biological test methods for the assessment of contaminated land, 2002 (Environment Agency, 2003). The UK Environment Agency report judges tests on the five "R" criteria: reproducible, representative, responsive, robust and relevant. A good example of bioavailability assessment of both organics and metals using multicellular organism toxicity tests is given by Cook et al. (2002). They found that soils with levels of contamination above intervention values, according to chemically based soil criteria, did not generate a toxic response to earthworms (mortality test) or seed germination and root elongation (algal growth inhibition and bacterial luminescence were also examined).

Mammalian tests (rodents, dogs, pigs, etc.), usually used as a surrogate for human risk assessment, are also reviewed by Hund-Rinke and Kordel (2003). They summarise that the tests are able to judge any toxic response that soil contaminants may induce, but due to complexity of diet, bioavailable concentrations of contaminants remain unknown. Various other mammalian *in vivo* studies of bioavailability of soil contaminants are summarised in National Research Council (2002) and Ehlers and Luthy (2003).

The overall advantages of using multicellular organisms to test for bio-availability are that the tests are directly relevant to the organism used, can show systemic changes, for example reactions to oestrogen-like substances, and can be reasonably cheap to carry out (e.g. plant assays). Limitations include no direct representation of other organisms, difficulties translating results (normalising) between different field sites and the time it takes to perform the tests – days/weeks rather than hours. Mammalian tests are further complicated by strict regulations governing animal welfare and expense.

19.5.2 Whole cell test organisms

The past two decades have seen an explosion of microbial/whole cell tests for use in estimating contaminant bioavailability. Genetic engineering and modification techniques have allowed indicator genes to be coupled to genes of specific interest to give a qualitative and quantitative response. The first use of a reporter gene to show a phenotypic response was the Ames test (Ames et al., 1973). This showed whether a compound caused a mutagenic response by causing a reversion from histidine dependence, i.e. if a mutagenic compound was present the *Salmonella typhimurium* would grow without additional histidine. Although not directly tested with contaminated soil, it is still suitable to use with soil extracts. The extract used (water/solvent/buffer, etc.) is as important as the biological test chosen when assessing the toxicity of a contaminated soil as the extract will be the factor that determines bioavailability. The Ames test and its successors are reviewed by Wegrzyn and Czyz (2003). These mutagenic biosensors tend to be sensitive, reasonably quick to perform (days or hours, not weeks) but are limited by the need for specialist (expensive) equipment and their relevance to other organisms.

The best established microbial test in environmental testing is the Microtox assay (http://www.azurenv.com). A bioluminescent marine bacterium, *Vibrio fischeri*, produces light as a by-product of normal cell functions. Any toxicity inhibits cell functions and proportionally, light emission, allowing toxicity to be quantitatively measured. Microtox itself is still widely used and has been shown to be highly sensitive, more so than various aquatic multicellular organism tests (Munkittrick et al., 1991). Advantages include sensitivity in its general toxic response, simplicity and rapidity, robustness and reproducibility and the sheer amount of data produced using it. Major limitations are that it relies on exposure to an extract, which may present difficulties in interpreting results and is a marine organism and therefore not strictly relevant in its response to soil contamination.

The principle behind Microtox, of light emission, has given rise to huge number of genetically engineered bacteria, yeast and mammalians cells which use light as an indication of bioavailability. "Lights off" systems, like Microtox, reduce light emission in response to general toxicity. "Lights on" systems, like Mutatox, have a luminescent gene coupled to a specific reporter gene (e.g. a stress response gene like rec A) or one that can detect specific classes of pollutants (e.g. metal sensitive biosensors (Rensing and Maier, 2003)) or nutrients (Joyner and Lindow, 2000) and emits light when suitably provoked. Microbial biosensors are eloquently reviewed in greater detail by Hansen and Sorensen (2001), Leveau and Lindow (2002) and Belkin (2003). Several UK biotechnology companies have emerged (for example http://www.remedios.uk.com and http://www.cysense. com) that use luminescent bacteria specifically to assess contaminated land.

The reviews by Leveau and Lindow (2002) and Belkin (2003) also discuss similar reporter gene-based systems that use β-galactosidase/lac Z and green fluorescent protein (GFP). The former was originally used in the SOS chromotest (Quillardet et al., 1982) an Ames-like test. The latter is now equally as popular as luminescent based tests, using fluorescence rather than luminescence. For example, Knight et al. (2004) presented a yeast/GFP genotoxicity assay used to test a range of environmentally relevant substances (pesticides, metals, solvents). Biosensors that utilise both luminescence and fluorescence, in order to test for acute and genotoxic threats simultaneously, are also available. The most high profile usage of these tools is as health monitors in the International Space Station (Rabbow et al., 2003).

Advantages of these light/colour-based toxicity indicator tests are ease of assay, speed of assay, versatility and sensitivity, up to a point – currently chemical analysis is more sensitive but that may change. Disadvantages are a lack relevance to other (multicellular) organisms, exposure to soil extracts rather than soil itself (soil extracts may not represent the actual *in situ* bioavailability of the pollutant) and, probably the factor most difficult to overcome, that these organisms have been genetically modified makes use beyond the lab (i.e. on site) awkward, if not impossible due to safety regulations (HMSO, 1990, 2002). Microbial and whole cell biosensors may be best employed as initial screening tools for environmental and soil contamination.

19.5.3 Sub-cellular or molecular tests

Rather than engineering reporter genes into cells it is possible to look directly at the molecular system of interest. Biosensors in common use for environmental and food contaminants include nucleic acid based (e.g. DNA hybridisation array (Fredrickson et al., 2001) and reverse transcriptase PCR for monitoring gene expression, looking for rises in specific stress genes (Environment Agency, 2003), enzyme based and antibody and receptor based. All these categories are reviewed in Baeumner (2003).

The UK Environment Agency has recommended the use of reverse transcriptase PCR to measure gene expression as a tool for ERA. It is a method that may be used to look at thousands of genes at once (and consequently compare species response) and their responses to vast arrays of contaminants. Sturzenbaum et al. (1998a, 1998b, 2001) examined changes in gene expression (in earthworms exposed to contaminated soil) in metallothionein, carboxypeptidase and other metal-sensitive genes and found transcription levels up to 100-fold greater in exposed organisms showing that the technique has a high degree of sensitivity. A great advantage of this technique is that organisms that have been directly in contact with contaminated soil can be analysed and therefore bioavailability is the parameter being assessed. A disadvantage is determining what "normal" levels of gene expression are in order to determine whether contamination has had an effect. It is also important to differentiate between 'normal stress responses', e.g. to drought and those actually related to pollution.

Antibody interactions are also highly sensitive and very specific. Immunoaffinity has been adapted from clinical research to quantify environmental pollutants like DDT and its breakdown products (Anfossi et al., 2004), metals (Chavez-Crooker et al., 2003), dioxins (Okuyama et al., 2004) and many others. Antibodies can be customised and raised against any contaminant and are supplied by various biotechnology companies (e.g. http://www.remedios.uk.com) The relevance of any antibody assay for bioavailability purposes will be dependant on what soil extraction method is used.

Assays that measure levels of DNA adducts (regarded as the first step in mutagenesis) have been used to determine exposure of earthworms to PAHs.

Short-term exposure gave levels of DNA adducts that were dose dependant (Saint-Denis et al., 2000). The potential use of this method to analyse soil DNA to assess *in situ* bioavailability is described in this publication (Singleton and Lyons, 2004).

These *in vitro* bioassays may be limited by their general applicability to organisms or ecosystems (and high cost of development) but in general due to their speed of assay (hours instead of days/weeks) they are very useful as initial bioavailability and toxicity screening tools.

19.6 Conclusion

There are a burgeoning number of bioavailability tests for seemingly huge numbers of scenarios (soil type, contaminant(s), receptor(s)). This review has compared the chemical and biological tools available for assessing bioavailability. The UK Environment Agency has acknowledged bioavailability as important for interpretation but considers the current body of research to be too disparate to make any concrete decisions to include the concept in regulation (Environment Agency, 2004c). Any bioavailability test has to be robust, clear in what it does and standardised before acceptance by any regulatory body. It is also important that the assay method is understood, e.g. are soil extracts exposed to selected organisms and if so, is the soil extract really representative of the bioavailable fraction. Does the assay determine the immediately available fraction or the fraction of pollutant available over time? The US NRC report argues that bioavailability is an integral part of risk assessment but acknowledges problems in communication of these concepts. It also notes that funding needs to be improved from a research perspective (as opposed to industry driven) in order that bioavailability is not seen merely as a tool for lowering clean-up levels (Ehlers and Luthy, 2003; National Research Council, 2002).

For every bioavailability test there are an even greater number of biological assays. Most tests cannot claim any widespread relevance and although a broad body of research in this area is healthy and required to further understand the concept of bioavailability, it makes decisions for regulators, as to what tests could be broadly accepted, incredibly difficult. The likelihood is that ecotoxicity and human health assessment, due to contaminated land, will continue to innovate as well as "borrowing" methods from pharmaceutical, food safety and water analysis procedures. The biological tools that will become most popular, and as a consequence more validated, will be those that can be taken into the field, are cheap, quick, reproducible, sensitive and physically robust. The weight of the literature would lead to a supposition that one single test will not predominate, using a battery of tests will remain usual, but that luminescent and fluorescent microbes created in the lab will eventually be used in the field.

More relevant tests need to be designed and more collaboration between soil scientists, plant experts and human toxicologists needs to be developed if biological tools to determine bioavailability are to be useful in the assessment of contaminated land.

References

Alexander, M., 2000. Aging, bioavailability, and overestimation of risk from environmental pollutants. Environ. Sci. Technol. 34, 4259–4265.

Ames, B.N., Lee, F.D., Durston, W.E., 1973. An improved bacterial test system for the detection and classification of mutagens. Proc. Natl. Acad. Sci. USA 70, 782–786.

An, Y.J., Kim, Y.M., Kwon, T.I., Jeong, S.W., 2004. Combined effect of copper, cadmium, and lead upon *Cucumis sativus* growth and bioaccumulation. Sci. Total Environ. 326, 85–93.

Anfossi, L., Giraudi, G., Tozzi, C., Giovannoli, C., Baggiani, C., Vanni, A., 2004. Development of a non-competitive immunoassay for monitoring DDT, its metabolites and analogues in water samples. Anal. Chim. Acta 506, 87–95.

Arnold, R.E., Langdon, C.J., Hodson, M.E., Black, S., 2003. Development of a methodology to investigate the importance of chemical speciation on the bioavailability of contaminants to *Eisenia andrei*. Pedobiologia 47, 633–639.

Babich, H., Stotzky, G., 1977. Reductions in toxicity of cadmium to microorganisms by clay minerals. Appl. Environ. Microbiol. 33, 696–705.

Baeumner, A.J., 2003. Biosensors for environmental pollutants and food contaminants. Anal. Bioanal. Chem. 377, 434–445.

Belkin, S., 2003. Microbial whole-cell sensing systems of environmental pollutants. Curr. Opin. Microbiol. 6, 206–212.

Borras, M., Nadal, J., 2004. Biomarkers of genotoxicity and other end-points in an integrated approach to environmental risk assessment. Mutagenesis 19, 165–168.

Chavez-Crooker, P., Pozo, P., Castro, H., Dice, M.S., Boutet, I., Tanguy, A., Moraga, D., Ahearn, G.A., 2003. Cellular localization of calcium, heavy metals, and metallothionein in lobster (*Homarus americanus*) hepatopancreas. Comp. Biochem. Physiol. C Toxicol. Pharmacol. 136, 213–224.

Cook, S.V., Chu, A., Goodman, R.H., 2002. Leachability and toxicity of hydrocarbons, metals and salt contamination from flare pit soil. Water Air Soil Pollut. 133, 297–314.

Cornelissen, G., Rigterink, H., Ferdinandy, M.M.A., Van Noort, P.C.M., 1998. Rapidly desorbing fractions of PAHs in contaminated sediments as a predictor of the extent of bioremediation. Environ. Sci. Technol. 32, 966–970.

Cornelissen, G., van der Pal, M., van Noort, P.C.M., Govers, H.A.J., 1999. Competitive effects on the slow desorption of organic compounds from sediments. Chemosphere 39, 1971–1981.

Cuypers, C., Grotenhuis, T., Joziasse, J., Rulkens, W., 2000. Rapid persulfate oxidation predicts PAH bioavailability in soils and sediments. Environ. Sci. Technol. 34, 2057–2063.

Cuypers, C., Clemens, R., Grotenhuis, T., Rulkens, W., 2001. Prediction of petroleum hydrocarbon bioavailability in contaminated soils and sediments. Soil Sediment Contam. 10, 459–482.

Davies, N.A., Hodson, M.E., Black, S., 2003. Is the OECD acute worm toxicity test environmentally relevant? The effect of mineral form on calculated lead toxicity. Environ. Pollut. 121, 49–54.

Ehlers, L.J., Luthy, R.G., 2003. Contaminant bioavailability in soil and sediment. Environ. Sci. Technol. 37, 295A–302A.

Environment Agency, 2002a. In-vitro methods for the measurement of the oral bioaccessibility of selected metals and metalloids: a critical review.

Environment Agency, 2002b. Soil Guideline Values Reports for individual soil contaminants, Report CLR 10 SGV 1–10.

Environment Agency, 2003. Ecological risk assessment – a public consultation on a framework and methods for assessing harm to ecosystems from contaminants in soil.

Environment Agency, 2004a. Land quality: contaminated land background.

Environment Agency, 2004b. Land quality: contaminated land exposure assessment (CLEA).

Environment Agency, 2004c. Land quality: frequently asked questions (FAQs)

European Environment Agency, 2004. Soil.

Fredrickson, H.L., Perkins, E.J., Bridges, T.S., Tonucci, R.J., Fleming, J.K., Nagel, A., Diedrich, K., Mendez-Tenorio, A., Doktycz, M.J., Beattie, K.L., 2001. Towards environmental toxicogenomics – development of a flow – through, high-density DNA hybridization array and its application to ecotoxicity assessment. Sci. Total Environ. 274, 137–149.

Gevao, B., Mordaunt, C., Semple, K.T., Piearce, T.G., Jones, K.C., 2001. Bioavailability of nonextractable (bound) pesticide residues to earthworms. Environ. Sci. Technol. 35, 501–507.

Gevao, B., Jones, K., Semple, K., Craven, A., Burauel, P., 2003. Nonextractable pesticide residues in soil. Environ. Sci. Technol. 37, 138A–144A.

Hansen, L.H., Sorensen, S.J., 2001. The use of whole-cell biosensors to detect and quantify compounds or conditions affecting biological systems. Microb. Ecol. 42, 483–494.

Harkey, G.A., Young, T.M., 2000. Effect of soil contaminant extraction method in determining toxicity using the Microtox (R) assay. Environ. Toxicol. Chem. 19, 276–282.

Hatzinger, P.B., Alexander, M., 1995. Effect of aging of chemicals in soil on their biodegradability and extractability. Environ. Sci. Technol. 29, 537–545.

Hawthorne, S.B., Grabanski, C.B., 2000. Correlating selective supercritical fluid extraction with bioremediation behavior of PAHs in a field treatment plot. Environ. Sci. Technol. 34, 4103–4110.

HMSO, 1990. Environmental Protection Act 1990 (c. 43), Part VI.

HMSO, 2002. Genetically Modified Organisms (Deliberate Release) Regulations 2002.

Hund-Rinke, K., Kordel, W., 2003. Underlying issues in bioaccessibility and bioavailability: experimental methods. Ecotoxicol. Environ. Saf. 56, 52–62.

Impellitteri, C.A., Saxe, J.K., Cochran, M., Janssen, G., Allen, H.E., 2003. Predicting the bioavailability of copper and zinc in soils: modeling the partitioning of potentially bioavailable copper and zinc from soil solid to soil solution. Environ. Toxicol. Chem. 22, 1380–1386.

ISO (1993a) Soil quality – Determination of the effects of pollutants on soil flora. Part 1: Method for the measurement of inhibition of root growth. International Standards Organisation, Geneva.

ISO (1993b) Soil quality – Effects of pollutants on earthworms (*Eisenia fetida*). Part 1: Determination of acute toxicity using artificial soil substrate. International Standards Organisation, Geneva.

ISO (1995) Soil quality – Determination of the effects of pollutants on soil flora. Part 2: Effects of chemicals on the emergence and growth of higher plants. International Standards Organisation, Geneva.

ISO (1998) Soil quality – Effects of pollutants on earthworms (*Eisenia fetida*). Part 2: Determination of effects on reproduction. International Standards Organisation, Geneva.

ISO (1999a) Soil quality – Effects of pollutants on earthworms. Part 3: Guidance on the determination of effects in field situations. International Standards Organisation, Geneva.

ISO (1999b) Soil quality – Inhibition of reproduction of Collembola (*Folsomia candida*) by soil pollutants. International Standards Organisation, Geneva.

ISO (2004) Soil quality – Effects of pollutants on Enchytraeidae (*Enchytraeus* sp.). Determination of effects on reproduction and survival. International Standards Organisation, Geneva.

Johnson, S.E., Herman, J.S., Mills, A.L., Hornberger, G.M., 1999. Bioavailability and desorption characteristics of aged, nonextractable atrazine in soil. Environ. Toxicol. Chem. 18, 1747–1754.

Joyner, D.C., Lindow, S.E., 2000. Heterogeneity of iron bioavailability on plants assessed with a whole-cell GFP-based bacterial biosensor. Microbiology 146, 2435–2445.

Kelsey, J.W., Kottler, B.D., Alexander, M., 1997. Selective chemical extractants to predict bioavailability of soil-aged organic chemicals. Environ. Sci. Technol. 31, 214–217.

Knight, A.W., Keenan, P.O., Goddard, N.J., Fielden, P.R., Walmsley, R.M., 2004. A yeast-based cytotoxicity and genotoxicity assay for environmental monitoring using novel portable instrumentation. J. Environ. Monit. 6, 71–79.

Krauss, M., Wilcke, W., Zech, W., 2000. Availability of polycyclic aromatic hydrocarbons (PAHs) and polychlorinated biphenyls (PCBs) to earthworms in urban soils. Environ. Sci. Technol. 34, 4335–4340.

La Point, T.W., Waller, W.T., 2000. Field assessments in conjunction with whole effluent toxicity testing. Environ. Toxicol. Chem. 19, 14–24.

Leveau, J.H.J., Lindow, S.E., 2002. Bioreporters in microbial ecology. Curr. Opin. Microbiol. 5, 259–265.

Liste, H.H., Alexander, M., 2002. Butanol extraction to predict bioavailability of PAHs in soil. Chemosphere 46, 1011–1017.

Lock, K., De Schamphelaere, K.A.C., Janssen, C.R., 2002. The effect of lindane on terrestrial invertebrates. Arch. Environ. Contam. Toxicol. 42, 217–221.

Lock, K., Janssen, C.R., 2003. Influence of ageing on zinc bioavallability in soils. Environ. Pollut. 126, 371–374.

Loiseau, L., Barriuso, E., 2002. Characterization of the atrazine's bound (nonextractable) residues using fractionation techniques for soil organic matter. Environ. Sci. Technol. 36, 683–689.

Loomis, T.A., Hayes, A.W., 1996. Loomis's Essentials of Toxicology. Academic Press, San Diego.

Mata-Sandoval, J.C., Karns, J., Torrents, A., 2002. Influence of rhamnolipids and Triton X-100 on the desorption of pesticides from soils. Environ. Sci. Technol. 36, 4669–4675.

Munkittrick, K.R., Power, E.A., Sergy, G.A., 1991. The relative sensitivity of Microtox Daphnid, Rainbow-Trout, and Fathead Minnow Acute Lethality Tests. Environ. Toxicol. Water Qual. 6, 35–62.

National Research Council (2002) Bioavailability of Contaminants in Soils and Sediments: Processes, Tools and Applications. US National Research Council, Washington, DC.

OECD (1984a) Chemicals Testing – Guidelines: 207 Earthworm, Acute Toxicity Tests. Organisation for Economic Co-operation and Development, Paris.

OECD (1984b) Chemicals Testing – Guidelines: 208 Terrestrial Plants, Growth Test. Organisation for Economic Co-operation and Development, Paris.

OECD (2000a) Chemicals Testing – Guidelines: 220 Enchytraedae Reproduction Test. Organisation for Economic Co-operation and Development, Paris.

OECD (2000b) Chemicals Testing – Guidelines: 222 Earthworm Reproduction Test. Organisation for Economic Co-operation and Development, Paris.

OECD (2003) Chemicals Testing – Guidelines: 208 Seedling Emergence and Seedling Growth Test. Organisation for Economic Co-operation and Development, Paris.

Okuyama, M., Kobayashi, N., Takeda, W., Anjo, T., Matsuki, Y., Goto, J., Kambegawa, A., Hod, S., 2004. Enzyme-linked immunosorbent assay for monitoring toxic dioxin congeners in milk based on a newly generated monoclonal anti-dioxin antibody. Anal. Chem. 76, 1948–1956.

Oomen, A.G., Hack, A., Minekus, M., Zeijdner, E., Cornelis, C., Schoeters, G., Verstraete, W., Van de Wiele, T., Wragg, J., Rompelberg, C.J.M., Sips, A., Van Wijnen, J.H., 2002. Comparison of five in vitro digestion models to study the bioaccessibility of soil contaminants. Environ. Sci. Technol. 36, 3326–3334.

Pardue, J.H., Kongara, S., Jones, W.J., 1996. Effect of cadmium on reductive dechlorination of trichloroaniline. Environ. Toxicol. Chem. 15, 1083–1088.

Paustenbach, D.J., 2000. The practice of exposure assessment: A state-of-the-art review. J. Toxicol. Environ. Health B Crit. Rev. 3, 79–291. Reprinted from Principles and Methods of Toxicology, 4th Ed., 2001.

Peijnenburg, W., Jager, T., 2003. Monitoring approaches to assess bioaccessibility and bioavailability of metals: matrix issues. Ecotoxicol. Environ. Saf. 56, 63–77.

Quillardet, P., Huisman, O., Dari, R., Hofnung, M., 1982. Sos Chromotest, a direct assay of induction of an Sos Function in *Escherichia Coli* K-12 to measure genotoxicity. Proc. Natl. Acad. Sci. USA 79, 5971–5975.

Rabbow, E., Rettberg, P., Baumstark-Khan, C., Horneck, G., 2003. The SOS-LUX-LAC-FLUORO-toxicity-test on the International Space Station (ISS). Adv. Space Res. 31, 1513–1524.

Rauret, G., 1998. Extraction procedures for the determination of heavy metals in contaminated soil and sediment. Talanta 46, 449–455.

Reid, B.J., Jones, K.C., Semple, K.T., 2000a. Bioavailability of persistent organic pollutants in soils and sediments – a perspective on mechanisms, consequences and assessment. Environ. Pollut. 108, 103–112.

Reid, B.J., Stokes, J.D., Jones, K.C., Semple, K.T., 2000b. Nonexhaustive cyclodextrin-based extraction technique for the evaluation of PAH bioavailability. Environ. Sci. Technol. 34, 3174–3179.

Rensing, C., Maier, R.M., 2003. Issues underlying use of biosensors to measure metal bioavailability. Ecotoxicol. Environ. Saf. 56, 140–147.

Ruby, M.V., Davis, A., Link, T.E., Schoof, R., Chaney, R.L., Freeman, G.B., Bergstrom, P., 1993. Development of an in-vitro screening-test to evaluate the in-vivo bioaccessibility of ingested mine-waste lead. Environ. Sci. Technol. 27, 2870–2877.

Saint-Denis, M., Pfohl-Leszkowicz, A., Narbonne, J.F., Ribera, D., 2000. Dose-response and kinetics of the formation of DNA adducts in the earthworm *Eisenia fetida andrei* exposed to B(a)P-contaminated artificial soil. Polycyclic Aromat. Compd. 18, 117–127.

Schulin, R., Schroder, P., 2003. Bioavailability of Soil Pollutants and Risk Assessment: Workshop summary/conclusion.

Singleton, I., Lyons, B., 2004. DNA-adduct analysis of soil DNA – a potential method to assess the in-situ bioavailability of polycyclic aromatic hydrocarbons.

Stokes, J.D., Paton, G.I., Semple, K.T., 2005. Behaviour and assessment of bioavailability of organic contaminants in soil: relevance for risk assessment and remediation. Soil Use Manage. 21(Suppl. 2), 475–486.

Sturzenbaum, S.R., Kille, P., Morgan, A.J., 1998a. Heavy metal-induced molecular responses in the earthworm, *Lumbricus rubellus* genetic fingerprinting by directed differential display. Appl. Soil Ecol. 9, 495–500.

Sturzenbaum, S.R., Kille, P., Morgan, A.J., 1998b. The identification, cloning and characterization of earthworm metallothionein. Febs Lett. 431, 437–442.

Sturzenbaum, S.R., Cater, S., Morgan, A.J., Kille, P., 2001. Earthworm pre-procarboxypeptidase: a copper responsive enzyme. Biometals 14, 85–94.

Tuomela, M., Lyytikainen, M., Oivanen, P., Hatakka, A., 1999. Mineralization and conversion of pentachlorophenol (PCP) in soil inoculated with the white-rot fungus *Trametes versicolor*. Soil Biol. Biochem. 31, 65–74.

Ure, A.M., Quevauviller, P., Muntau, H., Griepink, B., 1993. Speciation of heavy-metals in soils and sediments – an account of the improvement and harmonization of extraction techniques undertaken under the auspices of the Bcr of the Commission-of-the-European-Communities. Int. J. Environ. Anal. Chem. 51, 135–151.

Wegrzyn, G., Czyz, A., 2003. Detection of mutagenic pollution of natural environment using microbiological assays. J. Appl. Microbiol. 95, 1175–1181.

Xing, B.S., Pignatello, J.J., Gigliotti, B., 1996. Competitive sorption between atrazine and other organic compounds in soils and model sorbents. Environ. Sci. Technol. 30, 2432–2440.

http://www.azurenv.com, accessed on July 2004.

http://www.cysense.com, accessed on July 2004.

http://www.remedios.uk.com, accessed on July 2004.

Developments in Soil Science, volume 32
Ravendra Naidu (Editor)
© 2008 Elsevier B.V. All rights reserved

Chapter 20

CHEMICAL METHODS FOR ASSESSING CONTAMINANT BIOAVAILABILITY IN SOILS

G.S.R. Krishnamurti

20.1 Introduction

The ecotoxicological significance of the environmental impact of heavy metals in soils is determined more by the specific form and reactivity of their association with particulate forms (soil colloids) than by their accumulation rate by plants. Estimation of plant transfer and prediction of bioavailability, therefore, should be based on the proportion of the potentially 'active species' of the heavy metals. Quantitative assessment of these fractions involves speciation of both the soil solution (Krishnamurti and Naidu, 2008) and the active particulate fraction of the soils. The ecological consequences of heavy metal pollution of soils largely relate to the heavy metal solubility and mobility within the soil profile (see Chapter 7). These interrelated factors determine the leaching, availability to microbes and plants and ultimately to humans. Soil and environmental factors influencing contaminant bioavailability has been discussed in considerable detail elsewhere (see Chapter 2). The generalized schematic geochemical cycle of heavy metals in agro-ecosystems is shown in Fig. 20.1. Many studies have indicated the existence of different binding forms of heavy metals and a strongly pH-dependent solubility effect of trace elements. In addition, the fraction present in the associated soil solution can also be in such chemical forms which are not readily transported into the root systems and translocated into other parts of the plant (Pickering, 1986). The principal forms of mobile and mobilizable toxic elements in soil can be summarized as: (i) soil solution: ionic, molecular and chelated forms; (ii) exchange interface: readily exchangeable ions in inorganic and organic fractions; (iii) adsorption complex: more firmly bound ions; (iv) Fe, Mn oxides: incorporated in precipitated Fe, Mn and Al oxides; and (v) detrital (residual): fixed in crystal lattices of secondary minerals. The fate and dynamics of trace elements including heavy metals and nutrients in agro-ecosystems is illustrated in Fig. 20.2.

Metal speciation is essential in processes such as the toxicity and bioavailability of pollutants in natural systems. The speciation of metal cations governs their availability to plants and their potential to contaminate waterways (Bernhard et al., 1986). Available forms of metals are not necessarily associated

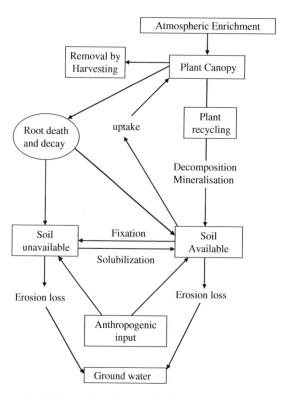

Figure 20.1. *Generalized biogeochemical cycle for trace elements in agro-ecosystems (from Krishnamurti, 2000).*

with one particular chemical species or a specific soil component. Hence to predict the availability of metal cations, we either have to establish the species involved and determine those forms only, or we have to establish an empirical relationship between an accepted measure of the metal and plant growth.

20.2 Solid-phase chemical fractionation

The mobility, transport and partitioning of trace metals in a natural terrestrial system is a function of the chemical form of the metal which, in turn, is controlled by the physicochemical and biological characteristics of that system. The tendency of the metal to accumulate in organisms in aquatic systems depends in particular upon the capacity of the system to retain this metal (see Chapter 18 of this book). Solid components in sediments govern the dissolved levels of the metal via sorption/desorption and dissolution/precipitation reactions coupled to complexation, acidification and redox

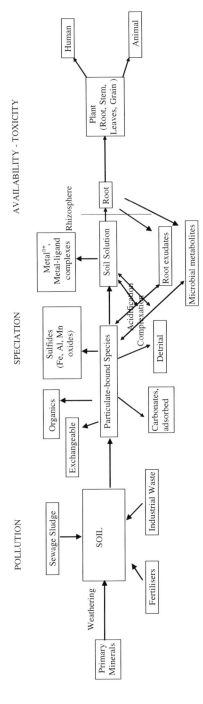

Figure 20.2. *Nutrient-trace element pollution-speciation-availability/toxicity cycle in terrestrial agro-ecosystems (from Krishnamurti, 2000).*

reactions (see Chapter 2 in this book; Violante et al., 2008). The distribution of trace elements among soil components is important for assessing the potential of soil to supply sufficient micronutrients to plant growth or to retain toxic quantities of heavy metals and for determining amelioration processes for soils at risk of causing trace element contamination. Trace elements in soils are normally found in many different physicochemical forms and may be associated with various soil components, such as silicate minerals, metal hydrous oxides and soil organic matter. The trace element species may be present as readily exchangeable, specifically adsorbed on to organic matter and hydrous metal oxides, substituted in stoichiometric compounds, or occluded in mineral structures. Therefore, identification of particulate trace metal species tends to be far more instructive and useful than any total metal determinations for understanding the mobility, bioavailability and toxicity of trace elements, and to develop useful environmental guidelines for potential toxic hazards (Davies, 1992).

20.2.1 Single chemical reagent extraction

Single extractants can be considered for their role in releasing elements bound or associated to specific soil phases. Extractants can be classified according to the different soil phases. More than one target soil phase may be attacked or the release from the target site may be less than complete. However, the concept of single extractant releasing elements associated with a specific soil phase is still useful and constitutes one of the methods for assessing the binding and mobilization of elements in soils. This aspect has been comprehensively discussed (Pickering, 1986; Beckett, 1989; Kersten and Foerstner, 1989; Ure et al., 1995; Ure, 1996).

Exchangeable/non-specifically adsorbed species: Use of electrolytes, such as $CaCl_2$, $MgCl_2$, KNO_3, $Mg(NO_3)_2$, $Ca(NO_3)_2$ as extractants promotes displacement of exchangeable metal ions which are held by electrostatic attraction to negative sites on soil surfaces. Usually 1 M solutions are employed as extractants, such as M $MgCl_2$ – Tessier et al. (1979), Hoffman and Fletcher (1979), Shuman (1979, 1983), Maher (1984), Hickey and Kittrick (1984), Nielsen et al. (1986), Elliot et al. (1990); M KNO_3 – Stover et al. (1976), Silviera and Sommers (1977), Schalscha et al. (1980, 1982), Miller and McFee (1983); M $Mg(NO_3)_2$ – Shuman (1983), Krishnamurti et al. (1995b), though more dilute solutions have also been preferred (0.5 M KNO_3 – Emmerich et al., 1982; Sposito et al., 1982; 0.5–0.05 M $CaCl_2$ – McLaren and Crawford, 1973; Shuman, 1979; Iyengar et al., 1981; McLaren et al., 1986; 0.25–0.5 M $Ca(NO_3)_2$ – Tiller et al., 1972; Miller et al., 1986a, 1986b; 0.5 M $MgCl_2$ – Gibbs, 1973, 1977), since they more closely resemble the electrolyte concentrations that can occur in natural systems. The use of nitrate salts is preferred to chloride salts on the grounds that chloride ion has a specific complexing effect (Tiller et al., 1972; Symeonides and McRae, 1977; Krishnamurti et al., 1995b).

Use of M NH₄OAc has also been advocated as an effective reagent for the extraction of exchangeable phase (Jackson, 1958; Gupta and Chen, 1975; Salomans and Foerstner, 1980, 1984; Schoer and Eggersgluess, 1982; Kersten and Foerstner, 1986; Rule and Alden, 1992; Castilho and Rix, 1993). However, the acetate ion has a special complexing effect, particularly with heavy metals.

Carbonates and specifically sorbed phases: Significant concentrations of trace elements may be associated with sediment carbonates. A mixture of M NaOAc with HOAc at pH 5 has been shown to extract >99% of total carbonate present in the soils (Jackson, 1958) and was used as a specific extractant for the determination of carbonate-bound heavy metals (Tessier et al., 1979; Foerstner et al., 1981; Harrison et al., 1981; Robbins et al., 1984; Hickey and Kittrick, 1984; Krishnamurti et al., 1995b). Less readily exchangeable phases, bound by covalent forces (i.e., specifically sorbed species) require a hydrogen ion or a soft cation such as Pb or Cu to displace from soil sites. 2.5% HOAc – Mclaren and Crawford (1973), Gupta and Chen (1975), Berrow and Mitchell (1980), Garcia-Mirgaya et al. (1981), Jarvis (1986); 0.05 M Na₂-EDTA (ethylene diamine tetraacetic acid) – Sposito et al. (1982), Ma and Uren (1995, 1998). 0.125 M Cu-acetate – Soon and Bates (1982); 0.05 M Pb nitrate – Miller et al. (1986a, 1986b).

Hydrous oxides of Fe and Mn: Fe and Mn oxides, present in soils as nodules, concretions, matrix component, cement between particles or as coatings on particles, are excellent scavengers of trace metals (Jenne, 1968). They sorb trace elements strongly, initially in exchange forms, but increasingly with time they are transformed to less mobile, specifically sorbed forms. A mixture of sodium dithionite, sodium citrate and sodium bicarbonate, buffered at pH 7.3 has been suggested as a suitable reagent for determining the total free iron contents (Mehra and Jackson, 1960) and has been used widely in soil/sediment studies (Tessier et al., 1979). This reagent dissolves both the crystalline and amorphous oxyhydroxides. However, the dithionite salt can be contaminated with metal impurities such as Zn (Shuman, 1982).

Acidified (pH 2) hydroxylamine hydrochloride (0.1 M) releases metals selectively from Mn oxyhydroxides and Mn oxides with little attack on Fe oxides (Chester and Hughes, 1967; Chao, 1972; Shuman, 1982; Chao and Zhou, 1983). Increasing the concentration to 0.5 M, or 1 M and decreasing pH to 1.5 provides affective attack on metals associated with both Fe and Mn oxides (Gupta and Chen, 1975; Stover et al., 1976; Tessier et al., 1979; Harrison et al., 1981; Miller and McFee, 1983; Miller et al., 1986a, 1986b; Rule and Alden, 1992; Krishnamurti et al., 1995b; Sahuquillo et al., 1999). For total recovery of amorphous Fe oxides, the treatment of samples with acidified ammonium oxalate (pH 3), in the absence of the catalyzing effect of light, has been proposed

(LeRiche and Weir, 1963; Schwertmann, 1964; McKeague and Day, 1966). This reagent has been extensively used for selectively extracting trace elements associated with amorphous Fe oxides (Shuman, 1979; Salomans and Foerstner, 1980, 1984; Kersten and Foerstner, 1986; Krishnamurti et al., 1995b). Schwertmann (1964) showed that in the dark the amorphous iron oxides were mainly attacked and under ultraviolet illumination the crystalline phases were dissolved as effectively as by the dithionite reagent. Heavy metals are released, with the exception of Pb and Cd, which co-precipitate with calcium oxalate. However, use of oxalic acid (pH 2.5) improved the performance relative to acid ammonium oxalate (Sahuquillo et al., 1999).

Organically complexed metal phases: Hydrogen peroxide is the commonly used extractant (Gupta and Chen, 1975; Tessier et al., 1979; Hickey and Kittrick, 1984; Gibson and Farmer, 1986; Rule and Alden, 1992; Krishnamurti et al., 1995b), even though it dissolves Mn oxides (Shuman, 1979, 1983; Keller and Vedy, 1994). However, Orsini and Bermond (1994) found the kinetics of the destruction of organic matter was slower and took almost 24 h.

Other reagents used include NaOCl buffered at pH 9.5 (Gibbs, 1973; Hoffmann and Fletcher, 1981; Shuman, 1983) and 0.1 M sodium or potassium pyrophosphate (Schalscha et al., 1982; Miller and McFee, 1983; McLaren et al., 1986; Krishnamurti et al., 1995b). Even though the C removal with NaOCl is said to be higher than that achieved by H_2O_2 (Lavkulich and Wiens, 1970), the associated precipitation of released metal ions in the alkaline medium and the possible alteration of mineral constituents are major disadvantages.

Extractants such as EDTA or DTPA (diethylene triamine pentaacetic acid) by virtue of their strong complexing ability displace metals from insoluble organic and organometallic complexes. EDTA, however, extracts more Cd and Zn than DTPA (Tlustos et al., 1994). The DTPA reagent was designed to minimize dissolution of carbonate-occluded Cu, Fe, Mn and Zn (Lindsay and Norvell, 1978).

Strong acid-extractable phases: Digestion of soils in strong acids, such as hydrochloric acid or nitric acid (Gupta and Chen, 1975; Stover et al., 1976; Foestner et al., 1981), which do not dissolve silicate minerals, can also give an estimate of the elements that are potentially mobilizable with changing environmental conditions. Aqua regia digestion has been used to dissolve metal pollutants, which largely enter the soil environment in non-silicate-bound forms and are also termed as pseudo-total pollutant metal content (Tiller et al., 1972; ISO, 1983).

Alumino-silicate lattice-bound phases: Trace elements that are occluded within silicate mineral lattice structures are usually determined with the $HF-HClO_4$ digestion technique (Jackson, 1958; Gupta and Chen, 1975; Tessier et al., 1979; Krishnamurti et al., 1995b).

20.2.1.1 Single chemical extraction and bioavailability

The use of chemical extraction for the assessment or prediction of crop plant metal content, plant growth and plant deficiency or toxicity has been the area of interest to scientists and environmental managers for the past five decades, although the term speciation was not been applied. Many soil extractants were used and for which plant contents and soil extractable contents are correlated and the methods validated over different crops and different soil types. The extractants developed remain soil and crop specific. A list of extractants commonly used by scientists over the years is presented in Table 20.1. A detailed review of trace element soil tests was recently published in *Methods of Analysis* edited by Bingham and Bartels (1996).

While many of the procedures listed in Table 20.1 are applied in polluted soils, they will not necessarily measure plant-available metal contents but rather the labile or mobile species present. They can at best indicate potentially

Table 20.1. *Examples of extractants used for assessing plant available trace element contents.*

Extractant	Elements	References
M NH$_4$OAc, pH 7	Mo, Ni, Pb, Zn	Johns (1972), John et al. (1972), Sedberry and Reddy (1976), Haq et al. (1980), Soon and Bates (1982), Sterckeman et al. (1996)
M NH$_4$OAc + 0.01 M EDTA, pH 7	Cu, Mn, Zn	Sterckeman et al. (1996)
0.5 M NH$_4$OAc + 0.02/0.05 M EDTA	Cu, Fe, Mn, Zn	Lakanen and Ervio (1971), Davis (1979), Sillenpaa (1982)
	Mo, Pb, Zn, Cd	Sauerbeck and Styperek (1985), Sillenpaa and Jansson (1992)
0.005 M DTPA + 0.01–0.1 M CaCl$_2$	Cd, Cu, Fe, Mn, Zn, Ni	Street et al. (1977), Lindsay and Norvell (1978), Soltanpour and Schwab (1977), Davis (1979), Soon and Bates (1982), Sillanpaa (1982), Houba et al. (1990), Whitten and Ritchie (1991)
0.01/0.05/0.1 M CaCl$_2$	Zn, Cd, Pb	Sauerbeck and Styperek (1985), Jackson and Alloway (1991), Smilde et al. (1992), Andrewes et al. (1996), Merkel (1996)
M NH$_4$NO$_3$	Cd, Ni	Symeonides and McRae (1977), DIN (1995), Merkel (1996)
0.1 M NaNO3	Cd, Zn, Cu, Pb	Hani and Gupta (1982)
M NH$_4$Cl	Cd	Krishnamurti et al. (1995a, 2000), Gray et al. (1999) Krishnamurti et al. (2007)
	Cu	Krishnamurti et al. (2007)

plant-available contents of the soil rather than actual contents. Comparison of many of the extractants in use indicated that 0.01 M or 0.05 M calcium chloride provides a measure of mobile heavy metal fraction in soils. This can be used to indicate potential availability of some heavy metals, particularly Cd, Zn, Mn and Pb, in contaminated soils (Hornburg et al., 1995; Andrewes et al., 1996).

Many extractants have been recommended for use as indicators of phytoavailability. Most methods are based on the establishment of a significant correlation between the amount of extractable heavy metal ions and the metal content of plants. Although such comparisons neither reflect soil processes controlling the phytoavailable metal fraction nor the mechanism of metal uptake by plants, strong correlations between the amount of extractable metals and metal content of plants have led to acceptance of such methods as indices of metal phytoavailability. The most commonly used extractants are diethylene-triamine penta aceticacid-triethanolamine (DTPA-TEA), buffered at pH 7.3 (Lindsay and Norvell, 1978) and 0.05 M $CaCl_2$ (Sauerbeck and Styperek, 1985). Some authors also reported non-significant relationships between extractable metals and test plant metal concentrations (e.g., Haq and Miller, 1972; Rappaport et al., 1988). O'Connor (1988) has subsequently identified a whole series of 'misuses' of the DTPA test which probably account for failure of this test. Two major constraints with the DTPA extractant include the high pH (that may not typify soil pH) and chelation effect of the ligand ion. The chelate-based extractants tend to extract significantly higher amounts of trace elements, and again may not necessarily reflect the plant-available content in the soils. To compensate for the high pH of the extractant, O'Connor (1988) suggested inclusion of pH as one of the variables in the correlation studies. There is nevertheless concern regarding the objective of such data processing, which is rather empirical and does not reflect processes that occur at the soil–plant–root interface.

Extraction of trace metals from soils by unbuffered salt solutions could be a rapid and simple way to evaluate their phytoavailability to plants (Jackson and Alloway, 1991; Lebourg et al., 1996). A number of extractants, such as 0.01 M $CaCl_2$ (Houba et al., 1990; Whitten and Ritchie, 1991; Andrewes et al., 1996), M NH_4NO_3 (Symeonides and McRae, 1977; DIN, 1995) and M NH_4Cl (Krishna-murti et al., 1995a, 2007) have been found to reflect plant available fraction of the metal content of soils. However, in some cases, salt solutions failed and the causes of failure could not be satisfactorily explained (Jackson and Alloway, 1991; Gupta and Aten, 1993; Singh et al., 1995). Examination of the data reported in those studies reveals a lack of appreciation of the mode of extraction of metals by these salt solutions. Reviewing bioavailability of heavy metals, Naidu et al. (1997) concluded that often soil extractants are used with little consideration of their mode of action and the chemistry of heavy metals. For instance, 0.01 M or

0.05 M CaCl$_2$ could never be expected to predict bioavailable Cu, given that Cu is largely present in soil associated with soluble organic matter. Where researchers have used this extractant, no correlation was found between extractable Cu and the Cu content of plants (e.g., Lee and Zheng, 1994).

The phytoavailability of a metal ion varies with the particular metal, soil properties and with plant species. Although there are many studies on plant metal availability as assessed by chemical extractants and plant uptake, comparisons between these studies can be virtually impossible because of the wide differences in the duration of treatments of soils and different extractant used for metal extraction. Moreover, few studies report metal availability tests, determined using chemical extractions, for native species under real field conditions (e.g., Gough et al., 1980; Krishnamurti et al., 1995a).

There are few guidelines for assessing the risk of contamination of the food chain with toxic metals, but the best extraction reagents for indicating soil-to-plant transfer appear to be weak, unbuffered, salt solutions (Lebourg et al., 1996).

Recent studies compared a number of commonly used extractants for assessing the plant-available Cd from soil extractions (Krishnamurti et al., 1995a, 2000; Gray et al., 1999). On the basis of the published data on native and uncontaminated soils, it appears that soil extraction with M NH$_4$Cl provides the best overall prediction of plant-available Cd (Table 20.2).

Use of ion-exchange membranes for the selective extraction of heavy metals from soil suspensions and soil solutions have been developed, especially for Cd, Cr, Ni and Pb, and were shown to correlate better with wheat seedling uptake than chemical extraction with 0.01 M CaCl$_2$ or 0.005 M DTPA (Lee and Zheng, 1994; Liang and Schoenau, 1996).

20.2.2 Sequential chemical extraction

Sequential extraction involves treating a sample of soil or sediment with a series of reagents to partition the trace element content. The reagents used are generally similar to those employed as single reagents to liberate metals bound to particular components of the sample matrix, such as exchangeable, bound to organics, and iron and manganese oxides. Conditional speciation assessment techniques such as sequential chemical extractions are not expected to yield information at the true molecular level in solid phases. The most recent definition of speciation has been given by IUPAC where chemical species are described as 'specific forms of chemical element' defined as to molecular, complex, electronic or nuclear structure (Templeton et al., 2000). This definition tends to restrict the term 'speciation' to well-defined chemical forms of elements. The term speciation should in principle not be applied to 'extractable trace elements', since these operationally defined determinations (obtained from

Table 20.2. *Concentrations of Cd extracted from the soils by various procedures and their correlation with the plant-available Cd.*

Extractant	Cd extracted (μg Cd/kg soil)		Simple correlation analysis[a]	
	Range	Mean	r^2	P
Soils of South Australia (Krishnamurti et al., 2000)				
0.01 M CaCl$_2$	10.0–45.0	23.7	0.051	5×10^{-1}
M NH$_4$NO$_3$	1.0–34.0	12.1	0.706	1×10^{-3}
0.05 M CaCl$_2$	21.0–103.0	46.3	0.581	6×10^{-3}
M NH$_4$Cl	22.0–162.0	75.8	0.862	9×10^{-5}
0.1 M EDTA	24.0–171.0	102.2	0.433	3×10^{-2}
Soils of Saskatchewan (Canada) (Krishnamurti et al., 1995a)				
M NH$_4$OAc	3.1–74.0	23.9	0.684	3×10^{-3}
ABDTPA	36.0–182.0	105.1	0.594	5×10^{-3}
M NH$_4$NO$_3$	0.2–15.9	3.4	0.320	2×10^{-2}
0.05 M CaCl$_2$	0.1–50.3	7.3	0.844	2×10^{-3}
M NH$_4$Cl	6.3–191.2	60.3	0.884	6×10^{-5}
Soils of New Zealand (Gray et al., 1999)				
M NH$_4$OAc		55.7	0.52	5×10^{-2}
AAAc-EDTA		174.4	0.52	5×10^{-2}
M NH$_4$NO$_3$		18.9	0.54	3×10^{-2}
0.05 M CaCl$_2$		58.2	0.51	5×10^{-2}
M NH$_4$Cl		121.9	0.59	1×10^{-2}

[a]Simple correlation analysis between the extractable Cd and Cd concentration in plant shoots and leaf.

single or sequential extractions) define groups of trace elements without clear identification. Applying strict definition, the determinations should clearly refer to the actual measurements, e.g., DTPA-extractable or NH$_4$Cl-extractable trace elements, and not to terms such as 'bioavailable' or 'mobile' which are rather interpretations of the measurements than the exact terminology of what is measured (Quevauviller, 2002). Hence, the more general terms 'forms' or 'fractions' are suggested for use rather than 'species' when referring to the results of such techniques (Kersten, 2002). Sequential chemical extraction experiments provide a convenient means for determining the metals associated with the principal accumulative phases in sedimentary deposits, to elucidate the mechanisms of their diagenetic transformation, and to elucidate their release processes upon changing environmental conditions (Kersten and Foerstner, 1989). A general goal of all studies involving sequential chemical extraction is the accurate determination of the partitioning of elements of environmental concern among different discrete phases of a sample. Mineralogical constituents of sediments considered important in controlling metal concentration in pore waters are hydrous oxides of iron and manganese, sulphides and organic matter (Horowitz, 1991).

Fractionation is usually performed by a sequence of 'selective' chemical extraction techniques, which include the successive removal of these phases and associated metals. In a typical procedure, the first species to be isolated are those that are loosely attached to the exchange sites, followed by a step-wise attack on the carbonate phase, iron and manganese oxyhydroxides, organic matter and finally the metals that are occluded within silicate mineral structures. Extended reviews are available on many of the techniques that are in use (Pickering, 1986; Ross, 1994; Kersten and Foerstner, 1995; Sheppard and Stephenson, 1997).

However, the schemes suggested by various environmental researchers for the speciation of particulate-bound heavy metals in soils and sediments during the past two decades show obvious similarities. Because, they are essentially modifications of the protocols introduced by Engler et al. (1977) and Tessier et al. (1979) which delineate the metal species sequentially as exchangeable, carbonate-bound, Fe and Mn oxide-bound, organically bound, and residual. With the use of additional reagents, the mineralogical bases may be further subdivided: for example, procedures involving separation of more labile, amorphous iron oxyhydroxides and the more refractory, crystalline forms. Shuman (1982, 1985), working on the speciation of Cu, Mn, Fe and Zn, modified the sequence of the extraction as: exchangeable, acid-soluble, easily reducible (Mn oxides-bound), moderately reducible (amorphous Fe oxides-bound), strongly reducible (crystalline Fe oxides-bound), oxidizable (organically bound) and residual. The low percentages of the oxidizable fractions obtained following the sequential extraction scheme of Tessier et al. (1979) as modified by Schuman (1985) were attributed to the lack of selectivity of the method used (Charlatchka et al., 1997).

20.2.3 Novel methods and applications of sequential extraction

Extraction with microwave heating: A major limitation to the widespread adoption of sequential extraction for trace element fractionation is the lengthy sample processing time (e.g., up to 16 h mechanical shaking). This has been reduced considerably in the recent developments by using microwave techniques (Real et al., 1994; Ginepro et al., 1996). The results obtained using a variation of Tessier's method involving microwave heating produced results (Gulmini et al., 1994) significantly different from those obtained using conventional procedures. However, broadly similar conclusions were drawn concerning the potential mobility and environmental impact of metal contaminants (Campos et al., 1998; Perez-Cid et al., 1998, 1999).

BCR sequential extraction procedure: To improve comparability between sequential extraction results obtained by different laboratories, the Community

Table 20.3. *The modified BCR three-step sequential extraction procedure.*

	Extractant	Fraction	Target phase
Step 1	0.11 M acetic acid	Exchangeable, acid- and water-soluble	Soil solution, exchangeable, carbonates
Step 2	475 ml of 0.5 M $NH_2OH \cdot HCl$+25 ml of 2 M HNO_3	Reducible	Iron and manganese oxyhydroxides
Step 3	H_2O_2 (30%) and M NH_4OAc at pH 2	Oxidizable	Organic matter and sulphides
Step 4[a]	$HF-HClO_4$	Residual	Silicate and non-silicate minerals

[a]Recommended to allow comparison for quality control purposes of sequential extraction ($\sum$ = step 1 + step 2 + step 3 + step 4) with results obtained from separate $HF-HClO_4$.

Bureau of Reference (BCR) proposed a simple, three-stage sequential extraction procedure for sediments (Ure et al., 1993; Rauret et al., 1999) (Table 20.3). The BCR sequential extraction has successfully been applied to a variety of matrices, including sediments from lacustrine (Fytianos et al., 1995), lagoon (Mester et al., 1998), riverine (Martin et al., 1998) and marine environments (Belazi et al., 1995; Marin et al., 1997; Userno et al., 1998), soil (Ho and Evans, 1997; Zhang et al., 1998), sewage sludge (Perez-Cid et al., 1996) and industrial dusts (Colandini et al., 1995; Davidson et al., 1998).

20.2.3.1 Sequential extraction and bioavailability

Metals in soil solutions are generally under-saturated with respect to the least soluble mineral species evaluated. However, the solid phase that controls the activity of metal in solution is unknown. Significant recent developments include attempts to bring together results of sequential extraction and bioavailability (Krishnamurti et al., 1995b; Qian et al., 1996; Wang et al., 1998; Krishnamurti and Naidu, 2000, 2002), to understand the specific fraction that actively participates in releasing the trace element to be termed as bioavailable.

The differentiation of the metal-organic complex-bound Cd species as distinct from the other organically bound species was the innovation in the selective sequential extraction scheme suggested by Krishnamurti et al. (1995a). This scheme proportioned the particulate-bound Cd species in soils as exchangeable, carbonate-bound, metal-organic complex-bound, easily reducible metal oxide-bound, organic-bound, amorphous mineral colloid-bound, crystalline Fe oxide-bound and residual. The metal-organic complex-bound Cd species were selectively extracted using 0.1 M pyrophosphate as the extractant in the sequential extraction scheme.

Krishnamurti et al. (1995a) determined the distribution of particulate-bound Cd species in a few typical soils of southern Saskatchewan (Canada) following the schemes of Tessier et al. (1979) and the modified scheme. They found that Cd in these soils was predominantly bound to the metal-organic complexes, accounting for on an average of 40% of the total Cd present in the soils (Fig. 20.3). Whereas on an average 36% of Cd in the soils was observed to be in the form Fe, Mn oxide-bound following the fractionation scheme of Tessier et al. (1979). The Cd in the surface horizons of the temperate soils (Krishnamurti et al., 1995a) is predominantly present in the form of metal-organic complex-bound, accounting for 31–55%, with an average of 40% of the total Cd present in the soils. The metal-organic complex-bound Cd is also generally the highest among the particulate-bound Cd species of the tropical soils (Onyatta and Huang, 1999), accounting for 25–46%, with an average of 37% of the total Cd in the soils. However, in certain tropical soils, residual or crystalline Fe oxide-bound Cd is predominant (Onyatta and Huang, 1999). An attempt to identify the specific species of Cd important to its bioavailability was made by Krishnamurti et al. (1995a), who performed multiple regression analysis between the Cadmium Available Index (CAI), as measured by ABDTPA-extractable Cd (Soltanpour and Schwab, 1977) and different forms of particulate-bound Cd. The importance of metal-organic complex-bound Cd species in the bioavailability of Cd and the nature of bonding of Cd sites was also worked out in detail using multiple regression analysis and differential FTIR (Fourier transform infrared) analysis (Krishnamurti et al., 1995a, 1997a). On the basis of the differential FTIR spectra of the metal-organic complexes, extracted by the $0.1\,M$ sodium pyrophosphate extractant used in the speciation scheme, Krishnamurti et al. (1997a) had shown that Cd in the soils was apparently bonded at the COO- of carboxyl and the OH of the phenolic groups and the OH of the Fe, Al and Mn in the metal-organic complexes.

Compared with bulk soils, solid-phase speciation of Cd differs substantially in phosphate fertilizer-treated rhizosphere soils (Krishnamurti et al., 1996). The amounts of carbonate-bound Cd and metal-organic complex-bound Cd species of the rhizosphere soils at two-week plant growth stage, particularly in the soils treated with Idaho phosphate fertilizer, are appreciably higher than those of the corresponding bulk soils. In comparison to the corresponding bulk soils, the amount of carbonate-bound Cd species of the rhizosphere soils increased by 15–18% in the control soils and by 79–92% in the soils treated with Idaho phosphate fertilizer. Whereas the metal-organic complex-bound Cd species increased by 4–7% in the control soils and by 2–3 times in the soils treated with Idaho phosphate fertilizer. The increase in the carbonate-bound Cd species in the rhizozphere soils is attributed to the increased amounts of carbonate, a product of plant respiration, present in soil–root interface. Xian and Shokohifard (1989) suggested that exudation of H_2CO_3 by roots may help to solubilize metal

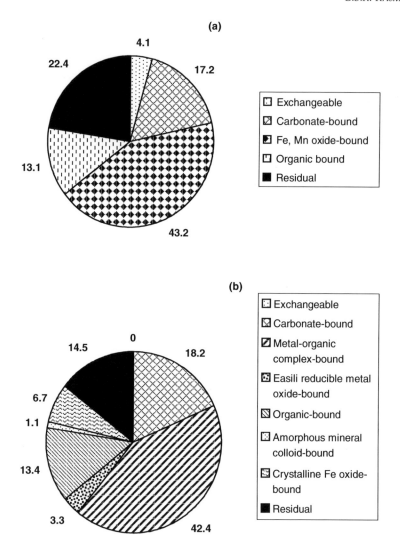

Figure 20.3. *Average percent distribution of solid-phase Cd fractions in selected soils of Saskatchewan (Canada) following the scheme of (a) Tessier et al. (1979) and (b) Krishnamurti et al., 1995a) (from Krishnamurti, 2000).*

carbonates and make them more bioavailable. The prolific root growth of the plants, particularly in the soils treated with Idaho phosphate fertilizer, may have resulted in the secretion of increased amounts of low-molecular-weight organic acids (LMWOAs) into the rhizosphere which resulted in high amounts of metal-organic complex-bound Cd species by chelation. Appreciable amounts of LMWOAs were detected in the root exudates of durum wheat, with the actual amount dependent on the cultivar (Cieslincki et al., 1998; Szmigielska et al., 1995). The sustained Cd release from the soils by LMWOAs was shown to have the same trend as that of Cd accumulated in the plant (Krishnamurti et al., 1997b).

Krishnamurti and Naidu (2000) modified the scheme of Krishnamurti et al. (1995b) sub-fractionating the metal-organic complexes, which may contain metal associated with both humic and fulvic acid fractions of soil organic matter, determined in the 0.1 M sodium pyrophosphate extracts. The sub-fractionation was performed following the method suggested by Schnitzer and Schuppli (1989) as metal–fulvic and metal–humic complexes. The metal solubilized by 6M HCl (at pH 1.0) in the pyrophosphate extracts is termed as metal–fulvic complexes and the metal bound strongly to humic acid and that resist solubilization at pH 1.0 is termed as metal–humic complexes. The details of the scheme are presented in Table 20.4. The distribution of the solid-phase

Table 20.4. *Multi-step selective sequential extraction scheme for fractionation of solid metal phases (from Krishnamurti and Naidu, 2000).*

	Target species	Reagent	Shaking time and temperature
Step 1	Exchangeable	10 mL of M NH_4NO_3 (pH 7)	4 h at 25°C
Step 2	Specifically adsorbed	25 mL of M CH_3COONa (pH 5)	6 h at 25°C
Step 3	Metal-organic complex-bound[a]	30 mL of 0.1 M $Na_4P_2O_7$ (pH 10)	20 h at 25°C
Step 4	Easily reducible metal oxide-bound	20 mL of 0.1 M $NH_2OH \cdot HCl$ in 0.01 M HNO_3	30 min at 25°C
Step 5	Organic-bound	5 mL of 30% H_2O_2 (pH 2), 3 mL of 0.02 M HNO_3	2 h at 85°C
		3 mL of 30% H_2O_2 (pH 2), 1 mL of 0.02 M HNO_3	2 h at 85°C
		cool, add 10 mL of 2 M NH_4NO_3 in 20% HNO_3	30 min at 25°C
Step 6	Amorphous mineral colloid-bound	10 mL of 0.2 M $(NH_4)_2C_2O_4$/0.2 M $H_2C_2O_4$ (pH 3)	4 h at 25°C (dark)
Step 7	Crystalline Fe oxide bound	25 mL of 0.2 M $(NH_4)_2C_2O_4$/0.2 M $H_2C_2O_4$ (pH 3) in 0.1 M ascorbic acid	30 min at 95°C
Step 8	Residual	Digestion with HF-$HClO_4$	

[a]30 mL of 0.1 M $Na_4P_2O_7$ extract was brought to pH 1.0 with the addition of 6M HCl and the suspension was left overnight for the coagulation of humic acid. The suspension was centrifuged at 12000 g for 10 min. Metal–fulvate complexes were determined in the supernatant. The residue was solubilized with 0.1 M $Na_4P_2O_7$ and the metal–humate complexes were determined in the solution.

fractions of Cd, Cu and Zn in a few typical soils of South Australia, performed following the fractionation scheme, is presented in Figs. 20.4 and 20.5.

The trace elements in these soils are predominantly (on an average 40% of Cu, 52.4% of Zn and 33.4% of Cd) associated with the mineral lattices, identified as a residual fraction in the scheme and considered not readily phytoavailable, followed by the fraction associated with organic sites (on an average 32.4% of Cu, 28.0% of Zn and 28.5% of Cd).

The mobility and phytoavailability of trace elements occur mainly through the solution phase. However, the plant uptake of an element depends not only on its activity in solution, but also on the relationship that exists between solution ions and solid-phase ions. The transfer of an element between soil and plant does not necessarily go through the solution phase (Kabata-Pendias, 2001). Trace elements might be phytoavailable directly from the solid phase, possibly through diffusion process. An attempt was made by Krishnamurti and Naidu

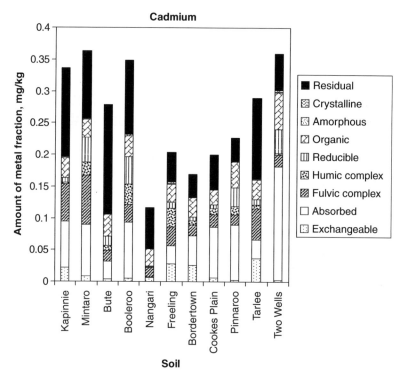

Figure 20.4. *Distribution of solid-phase Cd fractions in selected soils of South Australia using the sequential extraction scheme of Krishnamurti and Naidu (2000) (from Krishnamurti and Naidu, 2000).*

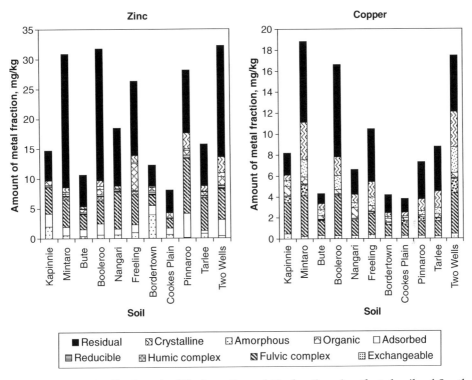

Figure 20.5. *Distribution of solid-phase Cu and Zn fractions in selected soils of South Australia using the sequential extraction scheme of Krishnamurti and Naidu (2000) (from Krishnamurti and Naidu, 2002).*

(2000, 2002) to understand the importance of solid-phase fractions in assessing phytoavailability of Cu, Zn and Cd using multiple regression analysis (Table 20.5). Phytoavailable Cu, Zn and Cd were found to correlate significantly with fulvic complex Cu ($r = 0.944$, $p < 0.0001$), exchangeable Zn ($r = 0.832$, $p = 0.002$) and fulvic complex Cd ($r = 0.824$, $p = 0.002$), respectively. It is observed that fulvic complex Cu could explain 89% of the variation in phytoavailable Cu, whereas fulvic complex element together with exchangeable element could explain 79 and 92% of the variation in phytoavailable Zn and Cd, respectively. The inclusion of solution element concentration in the regression analysis was not found to improve the predictability of the phytoavailable element. Regression analysis indicated that the phytoavailability of Cu, Zn and Cd in these soils is mainly through solid-phase fractions. However, more research is needed to understand the mechanisms involved in mobility and bioavailability and to develop rational predictive models.

Table 20.5. *Data on the multiple regression analysis between phytoavailable metal[a], soil solution metal concentration (M_s) and the solid-phase metal fractions (from Krishnamurti and Naidu, 2000, 2002).*

Simple correlation factors:

	Fulvic complex	Exchangeable	M_s
Phytoavailable Cu	0.944 (<0.0001)	0.424 (ns)	−0.029 (ns)
Phytoavailable Zn	−0.669 (0.02)	0.832 (0.002)	0.753 (0.008)
Phytoavailable Cd[a]	0.824 (0.002)	0.735 (0.01)	

Multiple regression analysis

Phytoavailable Cu $= \underset{(ns)}{1.79} \pm 1.48 + \underset{(<0.0001)}{5.11} \pm 0.59 \text{ Fulvic Cu}$ $R^2 = 0.89$ (< 0.0001)

$= \underset{(ns)}{1.02} \pm 2.23 + \underset{(<0.0001)}{5.13} \pm 0.62 \text{ Fulvic Cu} + \underset{(ns)}{3.89} \pm 8.14 \text{Cu}_s$ (1) $R^2 = 0.895$ (< 0.0001)

Phytoavailable Zn $= \underset{(0.003)}{20.62} \pm 3.54 + \underset{(0.0020)}{11.04} \pm 2.45 \text{ Exch Zn}$ $R^2 = 0.69$ (0.002)

$= \underset{(0.06)}{12.73} \pm 5.80 + \underset{(0.03)}{7.87} \pm 2.96 \text{ Exch Zn} + \underset{(ns)}{39.33} \pm 23.94 \text{ Zn}_s$ $R^2 = 0.77$ (0.003)

$= \underset{(0.002)}{33.58} \pm 7.43 + \underset{(0.007)}{8.81} \pm 2.45 \text{ Exch Zn} - \underset{(0.09)}{2.37} \pm 1.23 \text{ Fulvic Zn}$ $R^2 = 0.79$ (0.002)

$= \underset{(0.002)}{32.33} \pm 6.72 + \underset{(0.009)}{8.06} \pm 2.25 \text{ Exch Zn} - \underset{(0.06)}{2.43} \pm 1.11 \text{ Fulvic Zn} + \underset{(ns)}{26.19} \pm 15.52 \text{ Zn}_s$ (2) $R^2 = 0.85$ (0.003)

Phytoavailable Cd $= \underset{(ns)}{0.03} \pm 0.03 + \underset{(0.002)}{3.36} \pm 0.77 \text{ Fulvic Cd}$ $R^2 = 0.68$ (0.002)

$= \underset{(ns)}{0.00} \pm 0.02 + \underset{(0.0003)}{2.65} \pm 0.45 \text{ Fulvic Cd} + \underset{(0.002)}{3.57} \pm 0.76 \text{ Exch Cd}$ (3) $R^2 = 0.92$ (0.0001)

Note: The values in parentheses are the *P* (level of significance) values.
[a]Metal concentration in shoots and leaves of durum wheat plant harvested after 5-week growth.

20.3 Concluding remarks

The mobility and phytoavailability of trace elements occur mainly through the solution phase. However, trace elements may also be phytoavailable directly from the solid phase, possibly through diffusion process. Furthermore, heavy metal pollutants adsorbed onto the soil and bound to different active phases in the soil could be partly mobilized in the aquatic milieu by changes in pH and redox conditions, increased salinity, or concentration of organic chelators (Foerstner, 1984). Heavy metals in soils and sediments exist in eight distinctly different fractions (Krishnamurti et al., 1995 a). (1) Adsorptive and exchangeable, (2) bound to carbonate phases, (3) bound to metal-organic complexes, (4) bound to easily reducible metal (such as Mn) oxides, (5) bound to organic matter, (6) bound to amorphous mineral colloids, (7) bound to crystalline Fe oxides and (8) residual or detrital or lattice incorporated metals. Chemical species of toxic metals may behave differently with respect to remobilization under changing environmental conditions (Bernhard et al., 1986); knowledge of the total content of the toxic metals in a soil does not equate with a comprehensive understanding of the environmental situation. Therefore, it is necessary to carry out fractionation of these metals in soils and sediments to evaluate its role in bioavailability and ecotoxic potential.

The use of chemical extractants for assessing agricultural crop phytoavailability has been a major topic of study in agricultural laboratories for more than five decades (although the term fractionation or speciation was not used). There exists a voluminous literature in which plant and soil extractable contents are correlated and methods validated for different crop and soil types. The reagents commonly used for heavily contaminated soils are perhaps DTPA and calcium chloride. The extractant, M NH_4Cl, has been recently shown to be an effective reagent for assessing the phytoavailable Cd content of uncontaminated soils (Krishnamurti et al., 1995a, 2000; Gray et al., 1999).

The experimental determination of solid-phase species implies the use of fractionation with a sequential extraction procedure (e.g., Tessier et al., 1979; Shuman, 1985; Krishnamurti et al., 1995a; Krishnamurti and Naidu, 2000) consisting of a series of extractants chosen in order of increasing strength and decreasing availability. The pros and cons of sequential extraction have been described in detail by Kersten (2002). Although this approach is unlikely to provide precise information on the mineral phases to which the metals are bound, it does provide useful information on the potential mobility of heavy metal contaminants. This means the distribution of toxic heavy metals in different phases, their contamination risks to the terrestrial environment, and subsequently to plants and humans can be ascertained. Significant recent developments include the application of sequential extraction data for assessing phytoavailable heavy metals (Krishnamurti et al., 1995a, 2000, 2002). The exact mechanisms that enable solid-phase species to become phytoavailable remain to be ascertained.

References

Andrewes, P., Town, R.M., Hedley, M.J., Loganathan, P., 1996. Measurement of plant available cadmium in New Zealand soils. Aust. J. Soil Res. 34, 441–452.

Beckett, P.H.T., 1989. The use of extractions in studies on trace metals in soils, sewage sludges, and sludge-treated soils. Adv. Soil Sci. 9, 143–176.

Belazi, A.U., Davidson, C.M., Keating, G.E., Littlejohn, D., McCartney, M., 1995. Determination and speciation of heavy metals in sediments from the Cumbrian coast, NW England, UK. J. Anal. Atom. Spectrom. 10, 233–240.

Bernhardt, M., Brinkman, F.E. and Sadler, P.J. (Eds.), 1986. The Importance of Chemical Speciation in Environmental Processes. CRC Press, Boca Raton, FL.

Berrow, M.L., Mitchell, R.L., 1980. Location of trace elements in soil profiles: total and extractable contents of individual horizons. Trans. Royal Soc. Edinb. 71, 103–121.

Bingham, F.T., Bartels, J.M., 1996. Methods of Soil Analysis. Part 3. Chemical Methods. Soil Science Society of America, Madison, WI.

Campos, E., Barahona, E., Lachica, M., Mingorabce, M.D., 1998. A study of the analytical parameters important for the sequential extraction procedure using microwave heating for Pb, Zn and Cu in calcareous soils. Anal. Chim. Acta 369, 235–243.

Castilho, P, Del., Rix, I., 1993. Ammonium acetate extraction for soil heavy metal speciation, model aided soil test interpretation. Int. J. Environ. Anal. Chem. 51, 59–61.

Chao, T.T., 1972. Selective dissolution of manganese oxides from soils and sediments with acidified hydroxylamine. Soil Sci. Soc. Am. Proc. 36, 764–768.

Chao, T.T., Zhou, L., 1983. Extraction techniques for selective dissolution of amorphous iron oxides from soils and sediments. Soil Sci. Soc. Am. J. 47, 225–232.

Charlatchka, R., Cambier, P., Bourgeois, S., 1997. Mobilization of trace metals in contaminated soils under anaerobic conditions. In: R. Prost (Ed.), Contaminated Soils. (Proceedings of 3rd International Conference on Biogeochemistry of Trace Elements, Paris (France), May 15–19, 1995). INRA, Paris, pp. 159–174.

Chester, R., Hughes, M.J., 1967. A chemical technique for separation of ferro-manganese minerals, carbonate minerals and adsorbed trace elements from pelagic sediments. Chem. Geol. 2, 249–262.

Cieslincki, G., Van Rees, K.C.J., Szmigielska, A.M., Krishnamurti, G.S.R., Huang, P.M., 1998. Low-molecular-weight organic acids in rhizosphere soils of durum wheat and their effect on cadmium bioaccumulation. Plant Soil 203, 109–117.

Colandini, V., Legret, M., Brosseaud, Y., Balades, J.D., 1995. Metallic pollution in clogging materials of urban porous pavements. Water Sci. Technol. 32, 57–62.

Davidson, C.M., Duncan, A.L., Littlejohn, D., Ure, A.M., Garden, L.M., 1998. A critical evaluation of the three-stage BCR sequential extraction procedure to assess the potential mobility and toxicity of heavy metals in industrially-contaminated land. Anal. Chim. Acta 363, 45–55.

Davies, B.E., 1992. Trace elements in the environment: retrospect and prospect. In: D.C. Adriano (Ed.), Biogeochemistry of Trace Metals. Lewis Publishers, Boca Raton, FL, pp. 1–17.

Davis, R.D., 1979. Uptake of Cu, Ni and Zn by crops growing in contaminated soil. J. Sci. Food Agric. 30, 937–947.

DIN (Deutsches Instutut fur Normung)., 1995. Bodenbeschaffenheit. Extraektion von Spurenele-mente mit Ammonium-nitratlosung. Vornorm DIN V19730. Boden-Chemische Bodenunter-suchungsverfahren, Berlin.

Elliot, H.A., Dempsey, B.A., Maille, M.J., 1990. Content and fractionation of heavy metals in water treatment sludges. J. Environ. Qual. 19, 330–334.

Emmerich, W.E., Lund, L.J., Page, A.L., Chang, A.C., 1982. Solid phase forms of heavy metals in sewage sludge-treated soils. J. Environ. Qual. 11, 178–181.

Engler, R.M., Brannon, J.M., Rose, J., 1977. A practical selective extraction procedure for sediment characterization. In: T.F. Yen (Ed.), Chemistry of Marine Sediments. Ann Arbor Science Publishers, Ann Arbor, MI.

Foerstner, U., 1984. Metal pollution of terrestrial waters. In: J.O. Nriagu (Ed.), Changing Metal Cycles and Human Health. Springer-Verlag, Berlin, pp. 71–94.

Foerstner, U., Calmano, K., Conrad, H., Jaksch, H., Schimkus, C., Schoer, J., 1981. Chemical speciation of heavy metals in solid waste materials (sewage sludge, mining wastes, dredged materials, polluted sediments) by sequential extraction. Proceedings of International Conference on Heavy Metals in the Environment, WHO/EED, pp. 698–704.

Fytianos, K., Bovolernta, S., Muntau, H., 1995. Assessment of metal mobility from sediment of Lake Vegoritis. J. Environ. Sci. Health A 30, 1169–1190.

Garcia-Mirgaya, J., Castro, S., Paolini, J., 1981. Lead and zinc levels and chemical fractionation in roadside soils of Caracas, Venezuela. Water Air Soil Pollut. 15, 285–297.

Gibbs, R.J., 1973. Mechanisms of trace metals transport in rivers. Science 180, 171–173.

Gibbs, R.J., 1977. Transport phases of transition metals in Amazon and Yukon rivers. Bull. Geol. Soc. Am. 88, 829–843.

Gibson, J.J., Farmer, J.G., 1986. Multistep sequential chemical extraction of heavy metals from urban soils. Environ. Pollut. Series B11, 117–135.

Ginepro, M., Gulmini, M., Ostacoli, G., Zelano, V., 1996. Microwave desorption treatment after the oxidation step in Tessier sequential extraction scheme. Int. J. Environ. Anal. Chem. 63, 147–153.

Gough, L.P., McNeal, J.M., Severson, R.C., 1980. Predicting native plant copper, iron, manganese and zinc levels using DTPA and EDTA soil extractants, Northern Great Plains. Soil Sci. Soc. Am. J. 44, 1030–1036.

Gray, C.W., McLaren, R.G., Roberts, A.H.C., Condron, L.M., 1999. Cadmium phytoavailability in some New Zealand soils. Aust. J. Soil Res. 37, 461–477.

Gulmini, M., Ostacoli, G., Zelano, V., Torazzo, A., 1994. Comparison between microwave and conventional heating procedures in Tessier extraction of calcium, copper, iron and manganese in a lagoon sediment. Analyst (Lond.) 119, 2075–2080.

Gupta, S.K., Aten, C., 1993. Comparison and evaluation of extraction media and their suitability in a simple model to predict the biological relevance of heavy metal concentrations in contaminated soils. Int. J. Environ. Anal. Chem. 51, 25–46.

Gupta, S.K., Chen, K.Y., 1975. Partitioning of trace elements in selective chemical fractions of nearshore sediments. Environ. Lett. 10, 129–158.

Hani, H., Gupta, S.K., 1982. In: R.D. Davis, G. Hucker and P. L'Hermite (Eds.), Environmental Effects of Organic and Inorganic Contaminants in Sewage Sludge. Elsevier, Amsterdam, pp. 49–53.

Haq, A.U., Bates, T.E., Soon, Y.K., 1980. Comparison of extractants for plant-available zinc, cadmium, nickel, and copper in contaminated soils. J. Soil Sci. Soc. Am. 72, 772–777.

Haq, A.U., Miller, M.H., 1972. Prediction of available soil Zn, Cu and Mn using chemical extractants. Agron. J. 64, 779–782.

Harrison, R.M., Laxen, D.P.H., Wilson, S.J., 1981. Chemical associations of lead, cadmium, copper, and zinc in street dusts and roadside soils. Environ. Sci. Technol. 15, 1378–1383.

Hickey, M.G., Kittrick, J.A., 1984. Chemical partitioning of Cd, Cu, Ni and Zn in soils and sediments containing high levels of heavy metals. J. Environ. Qual. 13, 372–376.

Ho, M.D., Evans, G.J., 1997. Operational speciation of Cd, Cu, Pb and Zn in the NIST standard reference materials 2710 and 2711 (Montana soil) by the BCR sequential extraction procedure and FAAS. Anal. Commun. 34, 363–364.

Hoffman, S.J., Fletcher, W.K., 1979. Extraction of Cu, Zn, Mo, Fe and Mn from soils and sediments using a sequential procedure. In: J.R. Watterson, P.K. Theobald (Eds.), Geochemical Exploration, Rexdale, Association of Exploration Geochemists. Ontario, Canada, pp. 289–299.

Hoffman, S.J., Fletcher, W.K., 1981. Detailed lake sediment geochemistry of anomalous lakes on the Nechako Plateau, central British Colombia – comparison of trace metal distributions in Capoose and Fish Lakes. J. Geochem. Explor. 14, 221–244.

Hornburg, V., Welp, G., Brummer, G.W., 1995. Behaviour of heavy metals in soils. 2. Extraction of mobile heavy metals in soils with $CaCl_2$ and NH_4NO_3. Z. Pflanzener. Bodenkd. 158, 137–145.

Horowitz, A.J., 1991. Primer on Trace Metal-Sediment Chemistry. Lewis Publishers, Chelsia, p.136.

Houba, V.J.G., Novozamsky, I., Lexmond, T.M., van der Lee, J.J., 1990. Applicability of 0.01 M $CaCl_2$ as a single extraction solution for the assessment of nutrient status of soils and other diagnostic purposes. Commun. Soil Sci. Plant Anal. 21, 2281–2290.

ISO., 1983. Sludge and Sediment Digestion using Aqua Regia for Subsequent Determination of Acid Soluble Portion of Soil Zinc Fractions, DIN 38414, part 7, Boden-Chemische Bodenuntersuchungsverfahren, Berlin, p.4.

Iyengar, S.S., Martens, D.C., Miller, W.P., 1981. Distribution and plant availability of soil zinc fractions. Soil Sci. Soc. Am. J. 45, 735–739.

Jackson, A.P., Alloway, B.J., 1991. The bioavailability of cadmium to lettuce and cabbage in soils previously treated with sewage sludges. Plant Soil 132, 179–186.

Jackson, M.L., 1958. Soil chemical analysis. Elsevier, Englewood Cliffs, NJ.

Jarvis, S.C., 1986. Forms of Al in some acid permanent grassland soils. J. Soil Sci. 37, 211–222.

Jenne, E.A., 1968. Controls on Mn, Fe, Co, Ni, Cu and Zn concentrations in soils and waters: the dominant role of hydrous Mn and Fe oxides. In: R.A. Baker (Symp. Chairman), Trace Inorganics in Water. Advances in Chemistry Series 73. American Chemical Society, Washington, DC, pp. 337–387.

John, M.K., Van Laerhoven, J., Chauh, H.H., 1972. Factors affecting plant uptake and phytotoxicity of cadmium added to soils. Environ. Sci. Technol. 6, 1005–1009.

Johns, M.K., 1972. Lead availability related to soil properties and extractable lead. J. Environ. Qual. 1, 295–298.

Kabata-Pendias, A., 2001. Trace Elements in Soils and Plants. 3rd Ed., CRC Press, Boca Raton, FL.

Keller, C., Vedy, J.-C., 1994. Heavy metals in the environment: distribution of copper and cadmium fractions in two forest soils. J. Environ. Qual. 23, 987–999.

Kersten, M., 2002. Speciation of trace metals in sediments. In: A.M. Ure, C.M. Davidson (Eds.), Chemical Speciation in the Environment. 2nd Ed., Blackie, Glasgow, pp. 301–321.

Kersten, M., Foerstner, U., 1986. Chemical fractionation of heavy metals in anoxic estuarine coastal sediments. Water Sci. Technol. 18, 121–130.

Kersten, M., Foerstner, U., 1989. Speciation of trace elements in sediments. In: G. Batley (Ed.), Trace Element Speciation: Analytical Methods and Problems. CRC Press, Boca Raton, FL, pp. 245–317.

Kersten, M., Foerstner, U., 1995. Speciation of trace elements in sediments and combustion wastes. In: A.M. Ure, C.M. Davidson (Eds.), Chemical Speciation in the Environment. 2nd Ed., Blackie, Glasgow, pp. 234–275.

Krishnamurti, G.S.R., 2000. Speciation of heavy metals: An approach for remediation of contaminated soils. In: D.L. Wise, D.J. Trantolo, E.J. Cichori, H.I. Inyang and U. Stottmeister (Eds.), Remediation Engineering of Contaminated Soils. Marcel Dekker Inc., New York, pp. 693–714.

Krishnamurti, G.S.R., Cieslinski, G., Huang, P.M., Van Rees, K.C.J., 1997b. Kinetics of cadmium release from soils as influenced by organic acids: implication in cadmium availability. J. Environ. Qual. 26, 271–277.

Krishnamurti, G.S.R., Huang, P.M., Van Rees, K.C.J., 1996. Studies on soil rhizosphere: speciation and availability of cadmium. Chem. Spec. Bioavail. 8, 23–28.

Krishnamurti, G.S.R., Huang, P.M., Van Rees, K.C.J., Kozak, L.M., Rostad, H.P.W., 1995a. A new soil test method for the determination of plant available cadmium in soils. Commun. Soil Sci. Plant Anal. 26, 2857–2867.

Krishnamurti, G.S.R., Huang, P.M., Van Rees, K.C.J., Kozak, L.M., Rostad, H.P.W., 1995b. Speciation of particulate-bound cadmium in soils and its bioavailability. Analyst (Lond.) 120, 659–665.

Krishnamurti, G.S.R., Huang, P.M., Van Rees, K.C.J., Kozak, L.M., Rostad, H.P.W., 1997a. Differential FTIR study of pyrophosphate extractable material of soils: implication in cadmium-bonding sites and availability. In: R. Prost (Ed.), Contaminated Soils (Proceedings of the Third International Conference on the Biogeochemistry of Trace Elements, May 15–19, 1995, Paris, France). INRA, Paris, pp. 012pdf. 1–10

Krishnamurti, G.S.R., Naidu, R., 2000. Speciation and phytoavailability of cadmium in selected surface soils of south Australia. Aust. J. Soil Res. 38, 991–1004.

Krishnamurti, G.S.R., Naidu, R., 2002. Solid-solution speciation and phytoavailability of copper and zinc in soils. Environ. Sci. Technol. 36, 2645–2651.

Krishnamurti, G.S.R., Smith, L.H., Naidu, R., 2000. Method for assessing plant-available cadmium in soils. Aust. J. Soil Res. 38, 823–836.

Krishnamurti, G.S.R., Naidu, R., 2008. Chemical speciation and bioavailability of trace metals. In: A. Violante, P.M. Huang and G.M. Gadd (Eds.), Biophysico-Chemical Processes of Heavy Metals and Metalloids in Soil Environments. John Wiley & Sons Inc., New York, pp. 419–466.

Krishnamurti, G.S.R., Pigna, M., Arienzo, M., Violante, A., 2007. Solid-phase speciation and phytoavailability of copper in representative soils of Italy. Chem. Spec. Bioavail. 19, 57–67.

Lakanen, E., Ervio, R., 1971. A comparison of eight extractants for the determination of plant available micronutrients in soils. Acta Agral. Fenn. 123, 223–232.

Lavkulich, L.M., Wiens, J.H., 1970. Comparison of organic matter destruction by hydrogen peroxide and sodium hypochlorite and its effect on selected mineral constituents. Soil Sci. Soc. Am. Proc. 34, 755–758.

Lebourg, A., Sterckeman, T., Ciesielski, H., Proix, N., 1996. Suitability of chemical extraction to assess risks of toxicity induced by soil trace metal bioavailability. Agronomie 16, 201–215.

Lee, D.-Y., Zheng, H.-C., 1994. Simultaneous extraction of soil phytoavailable cadmium, copper, and lead by chelating resin membrane. Plant Soil 164, 19–23.

LeRiche, H.H., Weir, A.H., 1963. A method for studying trace elements in soil fractions. J. Soil Sci. 14, 225–235.

Liang, J., Schoenau, J.J., 1996. Speciation in metal contaminated soils as revealed by an ion exchange resin membrane fractionation procedure. Commun. Soil Sci. Plant Anal. 27, 3013–3026.

Lindsay, W.L., Norvell, W.A., 1978. Development of a DTPA soil test for zinc, iron, manganese, and copper. Soil Sci. Soc. Am. J. 42, 421–428.

Ma, Y.B., Uren, N.C., 1995. Application of new fractionation scheme for heavy metals in soils. Commun. Soil Sci. Plant Anal. 26, 3291–3303.

Ma, Y.B., Uren, N.C., 1998. Transformation of heavy metals added to soil-application of new sequential extraction procedure. Geoderma 84, 157–168.

Maher, W.A., 1984. Evaluation of a sequential extraction scheme to study associations of trace elements in estuarine and oceanic sediments. Bull. Environ. Contam. Toxicol. 32, 339–344.

Marin, B., Valladon, N., Polve, M., Monaco, A., 1997. Reproducibility testing of a sequential extraction scheme for the determination of trace metal speciation in a marine reference sediment by ICP-MS. Anal. Chim. Acta 342, 91–112.

Martin, R., Sanchez, D.M., Guttierrez, A.M., 1998. Sequential extraction of U, Th, Ce, La and some heavy metals in sediments from Ortigas river, Spain. Talanta 46, 1115–1121.

McKeague, J.A., Day, J.H., 1966. Dithionite and oxalate extractable Fe and Al as aids in differentiating various classes of soils. Can. J. Soil Sci. 46, 13–22.

McLaren, R.G., Crawford, D.V., 1973. Studies on soil copper. I. The fractionation of Cu in soils. J. Soil Sci. 24, 172–181.

McLaren, R.G., Lawson, D.M., Swift, R.S., 1986. The forms of cobalt in some Scottish soils as determined by extraction and isotope exchange. J. Soil Sci. 37, 223–234.

Mehra, O.P., Jackson, M.L., 1960. Iron oxide removal from soils and clays by a dithionite-citrate system buffered with sodium bicarbonate. Clays Clay Miner. 7, 317–327.

Merkel, D., 1996. Cadmium, copper, nickel, lead and zinc contents of wheat grain and soils extracted with $CaCl_2$/DTPA (CAD), $CaCl_2$ and NH_4NO_3. Agrobiol. Res. –Z. Agrobiol. Agrikulturchemie Okologie 49, 30–37.

Mester, Z., Cremisini, C., Ghiara, E., Morabito, R., 1998. Comparison of two sequential extraction procedures for metal fractionation in sediment samples. Anal. Chim. Acta 359, 133–142.

Miller, W.P., Martens, D.C., Zelazny, L.W., 1986a. Effect of sequence in extraction of trace metals from soils. Soil Sci. Soc. Am. J. 50, 558–560.

Miller, W.P., Martens, D.C., Zelazny, L.W., Kornegay, E.T., 1986b. Forms of solid phase copper in copper-enriched swine manure. J. Environ. Qual. 15, 69–72.

Miller, W.P., McFee, W.W., 1983. Distribution of Cd, Zn, Cu and Pd in soils of industrial north western Indiana. J. Environ. Qual. 12, 29–33.

Naidu, R., Kookana, R.S., Sumner, M.E., Harter, R.D., Tiller, K.G., 1997. Cadmium sorption and transport in variable charge soils – a review. J. Environ. Qual. 26, 602–617.

Nielsen, D., Hoyt, P.B., MacKenzie, A.F., 1986. Distribution of soil Zn fractions in British Colombia interior orchard soils. Can. J. Soil Sci. 66, 445–454.

O'Connor, G.A., 1988. Use and misuse of the DTPA soil test. J. Environ. Qual. 17, 715–718.

Onyatta, J.O., Huang, P.M., 1999. Cadmium speciation and bioavailability index for selected Kenyan soils. Geoderma 91, 87–101.

Orsini, L., Bermond, A.P., 1994. Copper biodisponibility in calcareous soil samples. Part 1. Chemical fractionation of copper. Environ. Technol. 15, 695–700.

Perez-Cid, B., Lavilla, I., Bendicho, C., 1996. Analytical assessment of two sequential extraction schemes for metal partitioning in sewage sludges. Analyst (Lond.) 121, 1479–1484.

Perez-Cid, B., Lavilla, I., Bendicho, C., 1998. Speeding up of a three-stage sequential extraction method for metal speciation using focused ultrasound. Anal. Chim. Acta 360, 35–41.

Perez-Cid, B., Lavilla, I., Bendicho, C., 1999. Application of microwave extraction for partitioning of heavy metals in sewage sludge. Anal. Chim. Acta 378, 201–210.

Pickering, W.F., 1986. Metal ion speciation – soils and sediments. Ore Geol. Rev. 1, 83–125.

Qian, J., Wang, Z.J., Shan, X.Q., Tu, Q., Wen, B., Chen, B., 1996. Evaluation of plant availability of soil trace metals by chemical fractionation and multiple regression analysis. Environ. Pollut. 91, 309–315.

Quevauviller, P.H., 2002. Quality control in speciation studies for environmental monitoring. In: A.M. Ure, C.M. Davidson (Eds.), Chemical Speciation in the Environment. 2nd Ed. Blackie Glasgow, pp. 132–158.

Rappaport, B.D., Martens, D.C., Reneau, R.B., Simpson, T.W., 1988. Metal availability in sludge amended soils with elevated metal levels. J. Environ. Qual. 17, 42–47.

Rauret, G., Lopez-Sanchez, J.F., Sahuquillo, A., Davidson, C., Ure, A., Quevauviller, P.H., 1999. Improvement of the BCR 3-step sequential extraction procedure prior to the certification of new sediment and soil reference materials. J. Environ. Monit. 1, 57–61.

Real, C., Barriero, R., Carballeira, A., 1994. The application of microwave heating in sequential extraction of heavy metals in estuarine sediments. Sci. Total Environ. 152, 135–146.

Robbins, L.M., Lyle, M., Heath, G.R., 1984. A sequential extraction procedure for partitioning elements among coexisting phases in marine sediments. College of Oceanography Report, 84-3. Oregon State University, Corvallis, Oregon, 55pp.

Ross, S.M., 1994. Sources and forms of potentially toxic metals in soil–plant systems. In: S.M. Ross (Ed.), Toxic Metals in Soil–Plant Systems. Wiley, New York.

Rule, J.H., Alden, R.W., 1992. Partitioning of cadmium in geochemical fractions of anaerobic estuarine sediments. Estuar. Coast. Shelf Sci. 34, 487–499.

Sahuquillo, A., Lopez-Sanchez, J.F., Rubio, R., Rauret, G., Thomas, R.P., Davidson, C.M., Ure, A.M., 1999. Use of certified reference material for extractable trace metals to assess sources of uncertainty in the BCR three-stage sequential extraction procedure. Anal. Chim. Acta 383, 317–327.

Salomans, W., Foerstner, U., 1980. Trace metal analysis of polluted sediments. II. Evaluation of environmental impact. Environ. Technol. Lett. 1, 506–517.

Salomans, W., Foerstner, U., 1984. Metals in the Hydrocycle. Springer Verlag, New York.

Sauerbeck, D., Styperek, P., 1985. Predicting the cadmium availability from different soils by CaCl$_2$ extraction. In: P. L'Hermite, H. Ott (Eds.), Processing and Use of Sewage Sludge. D Reidel Publishing Company, Dordrecht, The Netherlands, pp. 431–435.

Schalscha, E.B., Morales, M., Ahumada, I., Schirado, T., Pratt, P.F., 1980. Fractionation of Zn, Cu, Cr and Ni in waste water solids and in soil. Agrochimica 24, 361–368.

Schalscha, E.G., Marlaes, M., Vergara, I., Chang, A.C., 1982. Chemical fractionation of heavy metals in waste-water effected soils. J. Water Pollut. Control Fed. 54, 175–180.

Schoer, J., Eggersgluess, D., 1982. Chemical forms of heavy metals in sediments and suspended matter of Weser, Elbe and Ems Rivers. Mitt. Geol. Paleontol. Inst., Uni. Hamburg 52, 667–685.

Schnitzer, M., Schuppli, P., 1989. The extraction of organic matter from selected soils and particle size fractions with 0.5 M NaOH and 0.1 M Na$_4$P$_2$O$_7$ solutions. Can. J. Soil Sci. 69, 253–262.

Schwertmann, U., 1964. Differenzierung der Eisenoxide des Bodens durch Extraktion mit Ammoniumoxalat-losung. Z. Pflanzener. Bodenkd. 105, 194–202.

Sedberry, J.E., Reddy, C.N., 1976. The distribution of zinc in selected soils in Indiana. Commun. Soil Sci. Plant Anal. 7, 787–795.

Sheppard, M.I., Stephenson, M., 1997. Critical evaluation of selective extraction methods for soils and sediments. In: R. Prost (Ed.), Contaminated Soils. Proceedings of the third International Conference on the Biogeochemistry of Trace Elements, Paris, May 15–19, 1995. INRA, Paris, France, pp. 69–97.

Shuman, L.M., 1979. Zinc, manganese and copper in soil fractions. Soil Sci. 127, 10–17.

Shuman, L.M., 1982. Separating soil iron- and manganese oxide fractions for microelement analysis. Soil Sci. Soc. Am. J. 46, 1099–1102.

Shuman, L.M., 1983. Sodium hypochlorite methods for extracting microelements associated with soil organic matter. Soil Sci. Soc. Am. J. 47, 656–660.

Shuman, L.M., 1985. Fractionation method for soil microelements. Soil Sci. 140, 11–22.

Sillenpaa, M., 1982. Micronutrients and the nutrient status of soils: a global study. FAO Soils Bulletin 48, F.A.O., Rome, Italy, 444pp.

Silviera, D.J., Sommers, L.E., 1977. Extractability of Cu, Zn, Cd and Pb in soils incubated with sewage sludge. J. Environ. Qual. 6, 47–52.

Singh, B.R., Narwal, R.P., Jeng, A.S., Alamas, A., 1995. Crop uptake and extractability of cadmium in soils naturally high in metals at different pH levels. Commun. Soil Sci. Plant Anal. 26, 2123–2142.

Smilde, K.W., van Luit, B., van Driel, W., 1992. The extraction by soil and absorption by plants of applied zinc and cadmium. Plant Soil 143, 233–238.

Soltanpour, P.N., Schwab, A.P.B., 1977. A new soil test for simultaneous extraction of macro- and micro-nutrients in alkaline soils. Commun. Soil Sci. Plant Anal. 8, 195–207.

Soon, Y.K., Bates, T.E., 1982. Chemical pools of Cd, Ni and Zn in some polluted soils and some preliminary indications of their availability to plants. J. Soil Sci. 33, 477–488.

Sposito, G., Lund, L.J., Chang, A.C., 1982. Trace metal chemistry in arid-zone field soils amended with sewage sludge I: fractionation of Ni, Cu, Zn, Cd and Pb in solid phases. Soil Sci. Soc. Am. J. 46, 26–264.

Sterckeman, T., Gomez, A., Cielsielski, H., 1996. Soil and waste analysis for environmental risk assessment in France. Sci. Total Environ. 178, 63–69.

Stover, R.C., Sommers, L.E., Silviera, D.J., 1976. Evaluations of metals in waste water sludge. J. Water Pollut. Control Fed. 48, 2165–2175.

Street, J.J., Lindsay, W.L., Sabey, B.R., 1977. Solubility and plant uptake of cadmium in soils amended with cadmium and sewage sludge. J. Environ. Qual. 6, 72–77.

Symeonides, C., McRae, S.G., 1977. The assessment of plant available cadmium in soils. J. Environ. Qual. 6, 120–123.

Szmigielska, A.M., Van Rees, K.C.J., Cieslinski, G., Huang, P.M., Knott, D.R., 1995. Determination of low molecular weight dicarboxylic acids in root exudates by gas chromatography. J. Agric. Food Chem. 43, 956–959.

Templeton, D.M., Ariese, F., Cornelis, R., Danielson, L.G., Muntau, H., Van Leeuwen, V., Lobinski, R., 2000. IUPAC guidelines for terms related to chemical speciation and fractionation of trace elements: definitions, structural aspects and methodological approaches. Pure Appl. Chem. 72, 1453–1470.

Tessier, A., Campbell, P.G.C., Bissom, M., 1979. Sequential extraction procedure for the speciation of particulate trace metals. Anal. Chem. 51, 844–850.

Tiller, K.G., Honeysett, J.L., de Bries, M.P.C., 1972. Soil zinc and its uptake by plants. II. Soil chemistry in relation to prediction of availability. Aust. J. Soil Res. 10, 165–182.

Tlustos, P., van Dijk, D., Szakova, J., Parlikova, D., 1994. Cd and Zn release through the use of selected extractants. Rost. Vyroba 40, 1107–1121.

Ure, A.M., 1996. Single extraction schemes for soil analysis and related applications. Sci. Total Environ. 178, 3–10.

Ure, A.M., Davidson, C.M., Thomas, R.P., 1995. Single and sequential extraction schemes for trace metal speciation in soil and sediment. Tech. Instr. Anal. Chem. 17, 505–523.

Ure, A.M., Quevauviller, Ph., Muntau, H., Griepink, B., 1993. The speciation of heavy metals in soils and sediments: an account of the improvement and harmonization of extraction techniques undertaken under the auspices of Community Bureau of Reference (BCR). Int. J. Environ. Anal. Chem. 51, 135.

Userno, J., Gamero, M., Morillo, J., Gracia, I., 1998. Comparative study of three sequential extraction procedures for metals in marine sediments. Environ. Int. 24, 487–496.

Violante, A., Krishnamurti, G.S.R., Pigna, M., 2008. Factors affecting the sorption-desorption of trace elements in soil environments. In: A. Violante, P.M. Huang and G.M. Gadd (Eds.), Biophysico-Chemical Processes of Heavy Metals and Metalloids in Soil Environments. John Wiley & Sons Inc., New York, pp. 215–261.

Wang, W.H., Wong, M.H., Leharne, S., Fisher, B., 1998. Fractionation and biotoxicity of heavy metals in urban dusts collected from Hong Kong and London. Environ. Geochem. Health 20, 185–198.

Whitten, M.G., Ritchie, G.S.P., 1991. Calcium chloride extractable cadmium as an estimate of cadmium uptake by subterranean clover. Aust. J. Soil Res. 29, 215–221.

Xian, X., Shokohifard, G., 1989. Effect of pH on chemical forms and plant availability of cadmium, zinc and lead in polluted soils. Water Air Soil Pollut. 45, 265–273.

Zhang, T.H., Shan, H.Q., Li, F.L., 1998. Comparison of two sequential extraction procedures for speciation analysis of metals in soils and plant availability. Commun. Soil Sci. Plant Anal. 29, 1023–1034.

Developments in Soil Science, volume 32
Ravendra Naidu (Editor)

Chapter 21

MICROBIAL ACTIVITIES, MONITORING AND APPLICATION AS PART OF A MANAGEMENT STRATEGY FOR HEAVY METAL-CONTAMINATED SOIL AND GROUND WATER

J. Geets, J. Vangronsveld, L. Diels, S. Taghavi and D. van der Lelie

21.1 Introduction

Most heavy metals exist naturally in the earth's crust at trace concentrations, sufficient to provide living systems with essential nutrients but at levels too low to cause toxicity. Exceptions are soils that are naturally rich in heavy metals, such as the African copper belt, arsenopyrite minerals or the serpentine soils of New Caledonia, which are characterized by life forms (plants, microorganisms) that are well adapted to higher levels of bioavailable heavy metals. Since the industrial revolution, pollution by heavy metals has increased substantially through industrial effluent, emissions, landfill leaching, mining activities, use of fertilizers and agricultural pesticides, burning of waste and fossil fuels and municipal waste treatment. As heavy metals cannot be degraded they are persistent and will, over time, accumulate in the environment, including the food chain. Exposure to heavy metals through ingestion or uptake of drinking water and foods can lead to their accumulation in plants, animals and humans (Mulligan et al., 2001). There they exert inhibitory actions by blocking essential functional groups, displacing essential metal ions or modifying the active conformations of biological molecules (Hassen et al., 1998). These phenomena can result in acute or chronic metal toxicity, as well as mutagenesis or carcinogenesis. Of the different heavy metals, Cd, Cu, Pb, Hg, Ni, Zn and As are considered the most hazardous (Cameron, 1992).

Because of the threat they pose to human health and the extent of the problem of both natural and anthropogenic contamination by heavy metals, major efforts are being made to develop remediation technologies for the treatment of metal-contaminated soils, sediments and groundwater. These technologies are based on physical, chemical or on biological processes. The role of bioavailability in risk assessment and remediation is further considered in Chapters 25–30.

21.1.1 Physicochemical methods

Heavy metal pollution in soils and waters is conventionally treated using methods based on physical or chemical processes (Mulligan et al., 2001). Pump and treat, using precipitation or flocculation techniques followed by sedimentation and disposal of the resulting contaminated sludge is frequently used for treating heavy metal contamination in water. Other methods for heavy metal removal from water involve ion exchange, reverse osmosis and microfiltration. For *in situ* treatment of groundwater, a reactive barrier may be installed which removes the heavy metals either by chemical means, i.e. reduction by ferrous iron, or by biological means (Scherer et al., 2000; Benner et al., 2002; Nyman et al., 2002).

Polluted soil can be treated *ex situ* by physical separation (e.g. froth flotation, fluidized-bed separation, hydrocyclones), pyrometallurgical separation (i.e. volatilization in high temperature furnaces) or electrokinetic processes, involving passage of a low intensity electric current between a cathode and an anode embedded in the contaminated soil. The latter technique can also be used *in situ*. Soil washing ("chemical leaching") and *in situ* soil flushing involve the addition of water with or without additives including chelating agents such as EDTA, inorganic or organic acids, biosurfactants, etc., in order to desorb and leach the heavy metals. Soil washing can be followed by returning the clean soil to the original site. Chemical treatment of soils and wastewaters by reductive or oxidative reaction mechanisms are used to detoxify, precipitate or solubilize metals or to neutralize soil pH. However, these reactions are not specific and other metals can be converted into a more toxic or mobile form, resulting in novel contamination problems. Also, other essential trace elements and nutrients can be removed by leaving a treated soil that has no environmental value.

Other *in situ* methods for the treatment of contaminated soils include the isolation and containment of contaminants. These are used in order to prevent further movement, to reduce the permeability of the waste and to increase the strength or bearing capacity of the waste; these methods include vertical and horizontal containment with physical barriers made of steel, cement, bentonite and grout walls (Rumer and Ryan, 1995). Another method is solidification/stabilization, which contains the contaminants in an area by mixing or injecting agents. Solidification techniques such as encapsulation and vitrification encapsulate the contaminants in a solid matrix (Conner, 1990) while stabilization reduces the contaminant's mobility through the formation of chemical bonds. Vangronsveld and coworkers used berengite, a waste product from the burning of coal refuse, to immobilize heavy metals in a contaminated soil, thereby decreasing their phytotoxic effects (Vangronsveld et al., 1995, 1996) and allowing vegetation of the contaminated site. The combination of metal immobilization and phytostabilization resulted in reduced leaching of the metals to groundwater, as well as reduced wind erosion and spread of contaminated dust.

21.1.2 Biological treatment of heavy metals

Conventional methods of treatment are often expensive, lack specificity or even give rise to further environmental problems, so alternative cost-effective technologies generally based on biological processes are being developed to remediate heavy metal pollution. Bioremediation exploits microorganisms to deal with heavy metal pollution in a variety of methods such as bioleaching, biosorption, oxidation/reduction reactions, bioprecipitation and biomethylation. These techniques aim to change the speciation of the heavy metals, making them either more mobile in order to improve their removal, or decreasing their toxicity and mobility. Phytoremediation is a special situation in which plants and their associated microorganisms are used to assimilate and remove contaminants from the environment. Phytoremediation of heavy metals comprises several processes (Salt et al., 1995). Rhizofiltration is based on the ability of plant roots to adsorb, precipitate and concentrate toxic metals. Phytoextraction involves the uptake of metals by metal-accumulating plants such as *Thlaspi*, *Urtica*, *Chenopodium*, *Polygonum sachalase* and *Alyssum* into the harvestable parts of roots and shoots. Phytostabilization is a process of excreting components from the plants to decrease soil pH and form metal complexes. Finally, phytovolatilization involves the transformation of toxic elements into relatively harmless and gaseous forms which are volatilized by the plant's tissues (Rugh et al., 1998; Terry and Zayed, 1998). Although phytoremediation is a promising method, it is restricted to contamination at shallower depths and requires longer times compared to other methods (Cunningham et al., 1995). The general potential of phytoremediation was recently reviewed (van der Lelie et al., 2001; Adriano et al., 2004; Vassilev et al., 2004). Microorganisms can help plants to overcome heavy metal toxicity stress, either by decreasing metal toxicity or by counteracting the plant's stress response. In addition, they can assist the plants by rendering heavy metals more bioavailable, so improving their uptake.

In conclusion, all bioremediation technologies rely on the genetic and biochemical capacities of microorganisms to interact with and protect themselves against the toxic effects of heavy metals. An understanding of the ways how bacteria cope with toxic concentrations of heavy metals is therefore essential in order to exploit them for detoxification and removal of heavy metals.

21.2 Heavy metal resistance in bacteria

Microorganisms have co-existed with metals since the beginning of life. This is reflected by the presence of a wide range of heavy metals in the active sites of many enzymes, where the chemical properties of specific metal cations have been recruited for catalysing key metabolic reactions and for maintaining protein structures. These metals are considered as essential micronutrients because they are required in minute amounts for normal cell metabolism. However, other metals seem to serve no biologically relevant function. All metals, when present at high concentration, can damage cell membranes, block and inactivate enzymes, and damage DNA structure. From a physiological point of view, metals fall into three main categories: (i) essential and basically non-toxic (e.g. Ca and Mg), (ii) essential, but harmful at high concentrations (typically Fe, Mn, Zn, Cu, Co, Ni, Cr and Mo) and (iii) toxic and non-essential (e.g. Hg, Pb, As or Cd) (Valls and de Lorenzo, 2002).

Over geological timescales, bacteria have evolved in order to inhabit ecological niches containing high concentrations of heavy metals. In addition to natural heavy metal-rich biotopes, anthropogenic activities leading to increased atmospheric release as well as deposition into aquatic and terrestrial environments have created novel metal-loaded niches which exert a strong selective pressure for metal endurance (Gadd and White, 1993). For protection against the toxic effects of heavy metals, bacteria can adapt diverse chromosomal, transposon and plasmid-mediated resistance systems which confer upon them a certain range of metal tolerances. The physiological role of plasmid-encoded resistance determinants is generally to confer resistance, while the role of chromosomally encoded systems may also include metal ion homeostasis. For example, copper and zinc ions are required for growth, thus there must be a fine balance between uptake and efflux to provide metal ion homeostasis (Rensing et al., 1999). Overall, bacteria employ two major strategies for heavy metal tolerance. In some bacteria, metal tolerance is the outcome of their metabolism or is an intrinsic property related to their cell wall structure or the presence of extra-cellular polymeric substances. Other bacteria have developed specific resistance mechanisms to protect themselves against the toxic effects of heavy metals. These mechanisms include active transport efflux pumps, intra- and extra-cellular sequestration, enzymatic transformation to other, less toxic chemical species by redox reactions, methylation, alkylation/dealkylation and reduction in the sensitivity of cellular targets to metal ions (Gadd, 1992).

Almost all metal/microbe interactions have been examined in the context of environmental biotechnology as a possible means for the removal, recovery or detoxification of inorganic and organic metal or radionuclide pollutants. Advances in understanding of the roles of microorganisms in such processes, together with the ability to fine-tune their activities using the tools of molecular biology, has led to the development of novel or improved metal bioremediation processes (Lloyd and Lovley, 2001).

21.2.1 Efflux-based heavy metal resistance

Microorganisms use active transport or efflux systems to export toxic metals from their cytoplasm. These mechanisms can be chromosomal or plasmid-encoded. Non-essential metals normally enter the cell trough normal nutrient transport systems but are rapidly exported. These efflux mechanisms can be non-ATPase- or ATPase-linked and are highly specific for the cation or anion they export (Silver et al., 1989; Nies and Silver, 1995; Bruins et al., 2000). Several types of efflux-based mechanisms have been described. The most intensively studied mechanisms are (i) CBA efflux pumps, driven by protein complexes of the resistance-nodulation-cell-division (RND) protein complex family, which exports superfluous cations and (ii) P-type ATPases, which are the basic defence against heavy metal cations (Nies, 2003). Often, the efflux of heavy metals is combined with a form of heavy metal sequestration or precipitation, to avoid re-entry of the toxic metals.

Cupriavidus metallidurans and related species (formerly known as *Alcaligenes eutrophus*, *Ralstonia eutropha* and *thereafter as Ralstonia metallidurans*) are β-proteobacteria colonizing industrial sediments, soils and wastes with a high content of heavy metals (Mergeay et al., 1985). The *C. metallidurans*-type strain CH34 was found to be highly resistant to Zn(II), Cd(II) and Co(II) and later Ni(II), Cu(II), CrO_4^{2-}, Hg(II) and Pb(II) (Mergeay et al., 1985). *C. metallidurans* CH34 carries two endogenous megaplasmids, pMOL28 (171.5 kb) and pMOL30 (233.7 kb) bearing resistance determinants to combinations of cobalt and nickel (*cnr*), cadmium, zinc and cobalt (*czc*), mercury (*mer*), thallium (*tll*), copper (*cop*) and lead (*pbr*) (Diels et al., 1985; Taghavi et al., 1997; Borremans et al., 2001; Monchy et al., 2007). The genetic, physiological and biotechnological study of heavy metal resistance characteristics of *C. metallidurans* CH34 has contributed greatly to the knowledge of these mechanisms and is still increasing thanks to the progress made in the sequencing of its genome (see http://www.jgi.doe.gov). Efflux-mediated heavy metal resistance in prokaryotes has been recently reviewed in detail and will not be described here (Mukhopadhyay et al., 2002; Mergeay et al., 2003; Nies, 2003; Solioz and Stoyanov, 2003).

21.2.2 Enzymatic reduction

The toxicity of metals and metalloids such as Hg, As, U and Cr often depends on their oxidation state. As a resistance mechanism, several microorganisms are able to change this oxidation state from a more toxic to a less toxic form through enzymatic reduction or oxidation. These resistance systems, which do not support bacterial growth, have been studied in detail using molecular biological tools, and include resistance to Hg(II), As(V) and Ag(II). These systems were recently reviewed and will not be discussed here (Mukhopadhyay et al., 2002; Barkay et al., 2003; Silver, 2003). However, up to now, the molecular basis of respiratory metal-reduction processes is not understood in such fine detail, although rapid advances are expected in this area with the imminent availability of complete genome sequences for key metal-reducing bacteria, in combination with genomic and proteomic tools (Lloyd, 2003).

21.2.3 Dissimilatory metal-reduction

Dissimilatory metal-reducing bacteria couple the oxidation of H_2 or organic substrates to the reduction of ferric iron, manganese or other metals in order to conserve energy for growth (Lovley et al., 1993a). Dissimilatory Fe(III)-reducing communities can be responsible for the majority of organic matter oxidized in anoxic non-sulphidogenic environments (Nealson and Saffarini, 1994). These microorganisms can also play an important role in the fate of contaminant heavy metals through direct enzymatic reduction and also via indirect-reduction mechanisms, catalysed by biogenic Fe(II) (Lloyd, 2003). The bioreduction of Cr(VI), Co(III), U(VI) and Tc(VII) by Fe(III)-reducing microorganisms and metal-reducing enzymes have been proposed for use in remediating metal-contaminated and radionuclide-contaminated waters and sediments (Lloyd and Lovley, 2001).

Several Fe(III)-reducing bacteria have been isolated from various environments and it has been shown that they constitute a phylogenetically diverse group (Lonergan et al., 1996). *Geobacter metallireducens* (Lovley, 1993) and *Shewanella oneidensis* (formerly *Alteromonas putrefaciens*) (Myers and Nealson, 1988) were among the first bacteria studied in pure culture that were able to gain energy for growth by coupling the oxidation of organic matter and/or H_2 to the reduction of Fe and Mn. It is now clear that dissimilatory Fe and Mn reduction is not limited to these genera but is broadly distributed among the known bacterial taxa and include species like *Geobacter sulfurreducens*, *Desulfuromonas palmitatis*, *Desulfuromonas acetoxidans*, *Pelobacter carbinolicus* and members of the genus *Desulfuromusa*. Evidence suggest that members of *Geobacteraceae* are phylogenetically linked by a common origin and physiologically by the ability to use Fe(III) and/or S^0 as a terminal electron acceptor (for review, see Lonergan et al., 1996). With the exception of *Pelobacter* species, which are more restricted in the range of electron donors utilized, these organisms are able to completely oxidize a wide range of organic compounds, including acetate and aromatic compounds.

Recent reports demonstrate that the phylogenetic diversity of Fe(III)-reducing bacteria extends beyond the *Proteobacteria*. These include a.o. *Geovibrio ferrireducens* (Caccavo et al., 1996) and the related thermophile *Deferribacter thermophilus* (Green et al., 1997), *Geothrix fermentans* (Coates et al., 1999), *Ferribacter limneticum*, *Sulfurospirillum barnesii* (Oremland et al., 1994), the acidophilic sulphur-oxidizing bacterium *Thiobacillus ferrooxidans* (Pronk et al., 1992), and several hyperthermophilic archaea and bacteria (Vargas et al., 1998).

The mechanisms by which dissimilatory Fe(III)-reducing microorganisms transport electrons to Fe(III) and Mn(IV) have been studied in most detail in *S. oneidensis* and *G. sulfurreducens*. Research on these organisms has been strongly stimulated through the availability of their genome sequences (http://genome.jgi-psf.org/mic_home.html). Several studies demonstrated that electron transport to Fe(III) and Mn(IV) involved direct reduction by c-type cytochromes (Lloyd, 2003). At circum-neutral pH, Fe(III) oxides are highly insoluble, yet they are accessible to respiring metal-reducing bacteria. A direct transfer was proposed because reduction activities of cytochromes were localized to the outer membrane or surface of the cell (Lloyd, 2003). In addition, soluble extra-cellular quinones and humic acids function as "electron shuttles" between the metal-reducing prokaryotes and the mineral surface (Lovley et al., 1996). The reduced hydroquinone moieties abiotically transfer electrons to Fe(III) oxides. The oxidized humic acid is reduced by the microorganism, leading to further rounds of electron shuttling to the insoluble mineral (Nevin and Lovley, 2002). The release of a quinone-containing extra-cellular electron shuttle by *S. oneidensis*, which may promote electron transfer to Fe(III) and Mn(IV), was also suggested (Newman and Kolter, 2000).

An important new discovery was made recently when it was shown that *G. metallireducens* accesses Fe(III) oxide by chemotaxis (Childers et al., 2002). The strain synthesizes pili and flagella when grown on insoluble Fe(III) and Mn(IV) minerals, but not when grown on soluble forms of the metals. It was hypothesized that *Geobacter* species sense depletion of soluble electron acceptors and synthesize the appropriate appendages that allow movement and subsequent attachment to Fe(III) and Mn(IV) minerals.

In addition to the reduction of Fe(III) and Mn(IV), several Fe(III)-reducing bacteria have the capacity to reduce toxic metals and metalloids such as Cr, Se, As and U. However, the ability to enzymatically reduce these toxic metals is not restricted to Fe(III) reducers and in most cases, this metal-reduction metabolism does not support growth (Lovley et al., 1993b).

The reduction of highly toxic and mobile Cr(VI) to the less toxic, less mobile Cr(III) is likely to be a useful process for treatment of Cr(VI)-contaminated waters and soils (Palmer and Wittbrodt, 1991). Reduction of Cr(VI) is widespread and is found in species including *Escherichia coli*, *Pseudomonas* species, *S. oneidensis* and *Aeromonas* species, *Bacillus* species, *Streptomyces* spp., etc. (Wang et al., 2000). Cr(VI) reduction was demonstrated in membrane fractions of the dissimilatory Fe(III) reducer *S. oneidensis* (Myers et al., 2000). The reduction of Cr(VI) by the sulphate-reducing bacteria (SRB) *Desulfovibrio desulfuricans* and *Desulfovibrio vulgaris* was shown to be catalysed by cytochrome c_3 (Lovley and Phillips, 1994; Lloyd et al., 2001). The involvement of cytochromes in Cr(VI) reduction was also demonstrated in *Enterobacter cloacae* (cytochrome *c*) (Wang et al., 1989) and in *E. coli* (cytochrome b and d) (Shen and Wang, 1993). Cr(VI) reductase activity in *Pseudomonas putida* and *Pseudomonas ambigua* is located in the soluble fraction of the cell. Although anaerobic conditions are generally required to induce maximum activity against Cr(VI), analyses suggested that the enzyme systems of *P. putida* and *P. ambigua* operate under aerobic conditions, with NAD(P)H serving as the electron donor (Suzuki et al., 1992; Park et al., 2000). The Cr(VI) is initially reduced to Cr(V), which is subsequently reduced to Cr(III), the non-toxic form.

In the case of As(V), it should be noted that reduction results in the more mobile and toxic As(III) and as a consequence, this process has not proven useful for bioremediation. However, dissimilatory As reduction could be used to extract insoluble As(V) from contaminated soil (Lloyd and Lovley, 2001). Several organisms capable of growing through dissimilatory reduction of As(V) have been isolated, e.g. the dissimilatory Fe(III)-reducing bacterium *S. barnesii* (Stolz et al., 1997) and its close relative *Sulfurospirillum arsenophilum* (Stolz et al., 1999), and *Chrysiogenes arsenatis*, of which a periplasmic arsenate reductase could be purified and characterized (Macy et al., 1996; Krafft and Macy, 1998). The sulphate-reducing bacterium *Desulfosporosinus auripigmentum* reduces As(V) followed by sulphate, resulting in the formation of As_2S_3-precipitates (orpiment) as an indirect resistance mechanism (Newman et al., 1998).

In other microorganisms, the reduction of As(V) to As(III) is preceding the actual As-resistance mechanism. This is the case for the *ars*-operon encoded resistance mechanism, where the As(V) reduction step is catalysed by the ArsC reductase and precedes efflux of As(III) from the cell (Mukhopadhyay et al., 2002).

The dissimilatory Fe(III)-reducing bacteria *G. metallireducens* and *S. oneidensis* can conserve energy for anaerobic growth via uranium reduction. This reduction of the relatively mobile U(VI) to U(IV), which readily precipitates as the insoluble mineral uraninite, may be used to remove uranium from contaminated groundwater and soils (Anderson et al., 2002; Holmes et al., 2002). Other bacteria including *Clostridium* and *Desulfosporosinus* species (Suzuki et al., 2003), as well as for instance the SRB *D. desulfuricans* (Payne et al., 2002) and *D. vulgaris* (Spear et al., 2000) are also able to reduce U(VI) but no bacterial growth accompanies the reduction process. In *D. vulgaris*, hydrogenase transports electrons to cytochrome c_3, which functions as an U(VI) reductase (Lovley et al., 1993c). Cytochrome c_3 is also involved in hydrogen-dependent U(VI) reduction in *D. desulfuricans*, but in combination with organic electron donors, additional pathways were suggested that bypass the cytochrome (Payne et al., 2002).

In addition to the enzymatically direct-reduction pathway, SRB are also able to reduce metals indirectly. This process is an outcome of their metabolism: during sulphate respiration, hydrogen sulphide is produced which can reduce soluble toxic metals, often to less toxic or less soluble forms (Tebo, 1995).

21.2.4 Intra and extra-cellular complexation, sequestration and precipitation

The intrinsic properties of the bacterial cell, including those related to the cell wall structure, extra-cellular polymeric substances such as exopolysaccharides and binding or precipitation of metals inside or outside the cell (Gadd, 1992) provide alternative possibilities for excluding toxic heavy metals from the cellular metabolism. These mechanisms also provide means to avoid re-entry of toxic heavy metals into the cell after their export by resistance mechanisms.

Extra-cellular sequestration is an example of how bacteria protect metal-sensitive, essential cellular components and involves alterations in the cell wall, membrane or envelope of the microorganism (Bruins et al., 2000). Extra-cellular sequestration was observed in *C. metallidurans* CH34. In cultures of this strain, Zn and Cd removal resulted in the immobilization of the metal cations as complexes with carbonates, bicarbonates and hydroxides. These processes are induced by carbon dioxide emitted by the metabolic activity as well as the alkalization of the extra-cellular environment as a result of the CBA (proton/cation antiporter) transporter activity. These metal complexes are precipitated in the form of carbonate crystals at the cell surface (Diels et al., 1993a, 1993b; Collard et al., 1994). This extra-cellular sequestration of heavy metal cations at the cell surface is considered to be an important post-efflux mechanism which prevents re-entry of heavy metals into the cell, particularly when extra-cellular concentrations are high. It is estimated that extra-cellular polysaccharides and outer membrane proteins have important post-efflux functions as well.

Storage of excess Cu in the periplasmic space has also been reported. Some strains of *E. coli* can survive in Cu-rich environments that would normally overwhelm the chromosomally encoded Cu homeostatic systems. Such strains possess additional plasmid-encoded genes that confer Cu resistance (Rensing and Grass, 2003). Copper-resistance operons have been characterized from *Pseudomonas* sp. (Cooksey, 1994), *Xanthomonas* sp. (Lee et al., 1994) and *E. coli* (Brown et al., 1995), and *C. metallidurans* CH 34 (Monchy et al., 2006). The core genes from these Cu-resistant determinants are highly homologous and probably have similar functions, although many additional genes can be involved. Rensing and Grass (2003) proposed a mechanism for such plasmid-encoded Cu detoxification, based on the Pco-system in *E. coli*. Recently, a plasmid-encoded Cu-resistance operon from *Lactococcus lactis*, *lco*, was described and represents a new combination of structural genes for copper resistance which has not been identified in other bacteria (Liu et al., 2002).

Pb(II)-resistant strains of *Citrobacter freundii* and *Staphylococcus aureus* have also been isolated that accumulate Pb as an intracellular lead phosphate, though the molecular mechanism of detoxification remains to be elucidated (Levinson and Mahler, 1998). The Pb-resistance operon in *R. metallidurans*, in addition to a Pb efflux system, encodes for the PbrD-protein that may function as a chaperone for Pb(II) and might be involved in cytoplasmic Pb(II) sequestration (Borremans et al., 2001). In addition, the PbrB-protein may play a key role in precipitating Pb as Lead phosphate, avoiding re-entry of Pb into the cells.

Some bacteria possess metal-binding components that are involved in intracellular complexation processes. Such intracellular sequestration is defined as the accumulation of metals within the cytoplasm to prevent exposure to essential cellular components. Metals commonly sequestered are Cd^{2+}, Cu^{2+} and Zn^{2+}. Recently, a strategy for tolerance to metals in prokaryotes was described which involves the production of metallothioneins (MTs), a family of cysteine-rich proteins that bind and sequester metal ions (Cavet et al., 2003). MTs are thought to constitute the main mechanism by which eukaryotic cells regulate intracellular metal ion concentrations (Nordberg, 1998). Functionally homologous MTs are found in several bacteria (Robinson et al., 2001), but SmtA of the cyanobacterium *Synechococcus*, which confers resistance to Zn^{2+} and Cd^{2+}, is the best characterized (Blindauer et al., 2001). Expression of *smtA* is induced in response to elevated concentrations of zinc or cadmium via the action of SmtB, a zinc-responsive negative regulator of *smtA* transcription (Huckle et al., 1993). Recently, an MT (BmtA) was identified and characterized from the cyanobacterium *Oscillatoria brevis* (Liu et al., 2003). *P. putida* also produces a cysteine-rich protein, which may be related to MTs and confers to intracellular Cd^{2+} sequestration (Trevors et al., 1986). However, the production of MTs as the main mechanism of heavy metal tolerance is rather exceptional among bacteria.

Precipitation of metals as metal sulphides or phosphates is an alternative way that microorganisms increase their resistance. SRB couple the oxidation of organic compounds or molecular H_2 with the reduction of sulphate as an external electron acceptor under anaerobic conditions, a process known as dissimilatory sulphate reduction (Barton and Tomei, 1995). The end product of this reaction is hydrogen sulphide. When heavy metals are present above a certain threshold they can be toxic to SRB and so restrict their growth and sulphide production. However, once sulphide is formed, it reacts readily with the heavy metals to form metal sulphide (MeS) precipitates, resulting in a lower sulphide concentration and reduced metal toxicity. These MeS precipitates exhibit extremely low solubility and are relatively stable in environments under low redox conditions (Hao, 2000). Moreover, hydrogen sulphide can reduce soluble toxic metals, often to less toxic or less soluble forms (Tebo, 1995). However, Bridge and coworkers (Bridge et al., 1999) demonstrated that the sulphate-reducing bacterium *Desulfococcus multivorans* possesses extra-cellular metal-binding capacity that is unrelated to sulphide production.

Phosphate, an essential nutrient, is generally not released in quantity by living organisms (Lloyd and Lovley, 2001). However, *Citrobacter* sp. has been shown to couple biologically liberated phosphate to the formation of metal phosphate biominerals. In *Citrobacter* sp., a periplasmic acid-phosphatase (named PhoN) is associated with the outer membrane and exocellular lipopolysaccharide (LPS) and catalyses cleavage of phosphate from glycerol 2-phosphate, resulting in the accumulation of high concentrations of phosphate at the cell surface (Lovley, 1993). Mineral formation with the uranyl ion UO_2^{2+} was initiated by nucleation at the phosphate groups of the LPS, with crystal growth driven by enzymatically generated phosphate. As a result, cells precipitated large quantities of uranium and cadmium phosphates.

21.3 Methods for studying microbial community composition and activity

Remediation technologies using microorganisms are feasible alternatives to the physical cleansing of soils or concentrating heavy metals in polluted waters by physical or chemical means. Significant advances have been made in under-standing the roles of microorganisms in mineral cycling and in the application of these processes to the bioremediation of metals and radionuclides (Lloyd, 2003). Imminent developments in the field include the further genetic improvement of strains and the adaptation of existing methodologies to large-scale and *in situ* decontamination processes. Regarding the technical improvement of bacterial processes such as immobilization or biosorption of heavy metals, it is clear that the processes taking place in the cell microenvironment need to be better understood if mineralization is to be improved (Valls and de Lorenzo, 2002). For example, the development and optimization of an efficient immobilization process for heavy metal as metal sulphides, which is mediated by SRB, implies an insight in the SRB community composition and activity, and in the way this community is affected by different operational parameters such as supplementa-tion of nutrients, COD/SO_4^{2-} ratio and heavy metal concentration. Moreover, the presence of non-sulphate reducers should be taken into account (e.g. methano-gens, acetogens, fermenting bacteria) since these bacterial groups may influence the sulphate-reduction process by competition or synthrophy (Laanbroek and Veldkamp, 1982; Oude-Elferink et al., 1994, 1998a; Colleran et al., 1995). During treatment one should be able to demonstrate and follow the presence and activity of the metal transforming consortium in order to optimize process conditions. Microbial and molecular tools have been developed that aim to reveal and elucidate the relationships between structural and functional diversity in such microbial ecosystems, and to monitor bacterial activity during bioremedial treatment processes (Plaza et al., 2001). Such tools can be used for the follow-up of specific groups of bacteria, or to determine microbial community composition. Here, as an example, we will focus on techniques used for the study of SRB populations. Due to their great economic, environmental and biotechnological importance, populations of SRB have been intensively studied during the last decades. Several methods exist to characterize and quantify SRB populations.

21.3.1 Non-molecular identification and characterization

Initially, microbial identification methods are based on isolation, cultivation, metabolic characteristics and morphology of the strains. Culture methods for enumeration of SRB in the environment are generally based on the most-probable-number (MPN) technique (Beliaeff and Mary, 1993) with enumeration media that usually contain lactate as the main carbon and energy source and which indicate the presence of sulphate reducers by the formation of a black precipitate of ferrous sulphide (FeS) (Postgate, 1984; Tanner, 1989; Jain, 1995). Although important for ecological studies, these techniques are of limited usefulness for quantification and characterization of environmental populations, as it is now well recognized that most strains do not grow *in vitro*, either because cultivation media poorly resemble natural growth conditions or because different strains of microorganisms are interdependent (Gibson et al., 1987; Ward et al., 1990; Amann et al., 1995). Recently, a radiotracer MPN-technique (T-MPN) with natural media was developed which greatly improved the enumeration of sulphate reducers in environmental samples (Vester and Ingvorsen, 1998; Brandt et al., 2001). Microscope techniques such as scanning electron microscopy (SEM), phase contrast microscopy and confocal laser scanning microscopy (CLSM) can be useful for a rough characterization of biofilms and sludge (Labrenz and Banfield, 2000; Silva et al., 2002). Nevertheless, complementary characterization studies are essential because this way of identifying microbes is based on cell morphology only, which is for most SRB not very distinctive (Castro et al., 2000).

To avoid problems associated with traditional microbiological techniques, culture-independent methods for detection and enumeration of many bacterial species, including the sulphate reducers, have been developed. Immunodetection methods, such as the use of antibodies raised against SRB, successfully used for the detection of SRB in oil field waters and coastal sediments (Christensen et al., 1992; Lillebaek, 1995), and an immunoassay for the enzyme adenosine-5′-phosposulphate (APS) reductase (Odom et al., 1991) are promising but because the production of antibodies requires the availability of a pure culture of microorganisms, these methods cannot detect microorganisms that have not yet been obtained in pure culture (Raskin et al., 1995). Furthermore, antibodies can cross-react with other non-related strains (Smith, 1982). However, in the past few years, very efficient techniques have been developed to build and design large libraries of antibody fragments – such as phage display libraries – and ingenious selection procedures have been established to derive antibodies with the desired characteristics (Hoogenboom et al., 1998; Benhar, 2001).

Another culture-independent approach is the analysis of phospholipid ester-linked fatty acids, known as (PL) FAME-patterns or PLFA-profiles. A large database of PLFA-profiles was developed from pure culture studies (Taylor and Parkes, 1983; Dowling et al., 1986; Vainshtein et al., 1992; Kohring et al., 1994). Although this database was successfully used to identify and quantify SRB in sediments (Taylor and Parkes, 1985; Kleikemper et al., 2002) and granular sludge (Oude-Elferink et al., 1998b; Kaksonen et al., 2003a), it should be taken into account that yet undiscovered SRB-species and genera may inhabit anaerobic environments which might impede an accurate insight into community structure if PLFA alone are used.

21.3.2 Molecular-biological identification and characterization methods

Over the last decade, important advances in molecular biology led to the development of culture-independent approaches for describing bacterial communities, based on the analysis of DNA and RNA. Probe-hybridization techniques and polymerase chain reaction (PCR) fragment cloning, followed by restriction and/or sequencing analysis, enable assessment of the diversity of the microbial community in terms of the number of different species and, to a lesser extent, their relative abundance. Genetic fingerprinting techniques provide a global picture of the genetic structure of the bacterial community (Ranjard et al., 2000). The use of molecular ecology methods has provided an opportunity to study microbial communities without the need for cultivation. However, none of these techniques is all-encompassing or perfect (Muyzer, 1998). Problems and biases occur due to the PCR-amplification step (i.e. choice of primers, inhibition of the enzyme by humic compounds, formation of chimeric PCR products, preferential amplification) (Winzingerode et al., 1997). In addition, individual techniques have specific limitations. Despite these problems and limitations, a combination of these techniques can reveal a great deal about the microbial community (Torsvik and Ovreas, 2002). Nowadays, these molecular tools are routinely used to study the diversity and distribution of individual bacterial species, including SRB, in complex microbial communities such as sediments

(Di Giovanni et al., 1999; Joulian et al., 2001; Orphan et al., 2001; Kleikemper et al., 2002), granular sludge (Santegoedts et al., 1999; Kaksonen et al., 2003b), biofilms (Santegoedts et al., 1998) and microbial mats (Teske et al., 1998; Nakagawa et al., 2002). Molecular studies of bacterial populations using a single gene can be based on either core housekeeping genes or on functional genes. Analysing the microbial community genome – the "metagenome" – will result in huge amounts of genomic, evolutionary and functional information which is a stimulus to further biotechnological research (Lorenz and Schleper, 2002; Torsvik and Ovreas, 2002; Rodriguez-Valera, 2004). By applying molecular phylogenetic analysis, DNA microarrays, functional genomics and *in situ* activity measurements, the metabolic potential and activity of microbial communities can be revealed, as well as the interactions that occur within them.

21.3.2.1 Analyses based on a single gene

Molecular methods based on the sequence of PCR-amplified genes, especially 16S rRNA genes, have become very important for revealing the genetic and functional diversity of microbial communities. These methods include PCR fragment cloning, followed by restriction and/or sequence analysis, genetic fingerprinting techniques such as denaturing gradient gel electrophoresis (DGGE), single strand conformation polymorphism (SSCP), restriction fragment length polymorphism (RFLP), and hybridization techniques such as slot-blot hybridization and fluorescence *in situ* hybridization (FISH).

Cloning and sequencing of PCR-amplified fragments is frequently used to assess community diversity in terms of the number of different species and, to a lesser extent, their relative abundance. The cloning step allows separation of the amplified sequences so that they can be characterized individually using RFLP and/or sequencing analysis. Sequencing allows the identification of uncultured bacteria as well as an estimation of their relationship to known cultivable species (Ranjard et al., 2000) by comparing the clone sequences to sequences in the GenBank database using the BLAST program (Basic Local Alignment Search Tool, http://www.ncbi.nlm.nih.gov/BLAST/) (Altschul et al., 1990) of the Ribosomal Database Project (Maidak et al., 1994). Although successful in the exploitation of microbial diversity, the cloning approach is neither suitable for studying many different microbial communities simultaneously, nor for quantifying species abundance and spatial distribution or behaviour of microbial communities over time; the technique is time-consuming, labour-intensive and thus impractical for the high-throughput analysis of multiple samples (Muyzer and Smalla, 1998). Moreover, results are biased due to the PCR and the cloning strategy used, which leads to an inaccurate image of the diversity of the bacterial populations (Rayney et al., 1994; Winzingerode et al., 1997).

Genetic fingerprinting techniques are also based on PCR and suffer from the same biases, but because they are less time-consuming and relatively easier to perform they allow the simultaneous comparison of bacterial communities (Ranjard et al., 2000). Genetic fingerprinting is based on the principle of resolving the diversity of the amplified sequences simply by differential electrophoretic migration on agarose or polyacrylamide gels, which depend on their size or sequence.

DGGE is an example of a fingerprinting technique that is routinely used in many laboratories. Mixtures of PCR products are separated on polyacrylamide gels containing a linear gradient of DNA denaturants (urea and formamide). Sequence variation among the different DNA molecules influences their melting behaviour, so molecules with different sequences stop migrating at different positions on the gel (Muyzer et al., 1993). Another technique based on the same principle is temperature gradient gel electrophoresis (TGGE) (Riesner et al., 1991). The DGGE–TGGE approach can be used to study the diversity, dynamics and gene expression of microbial communities (Muyzer and Smalla, 1998). Furthermore, individual bands may be excised, reamplified and sequenced for the taxonomic identification of the bacterial species within a community (Ranjard et al., 2000). This is especially useful when analysing changes in microbial community composition, due to changes in environmental parameters for example. One of the limitations of the technique is that it separates only relatively small fragments, up to 500 base pairs, which restricts the information available for phylogenetic identification well as for probe or primer design. Another problem is the misinterpretation of fingerprints, e.g. when degenerate PCR-primers are used in the precedent PCR amplification step, double bands may appear in the DNA pattern. Furthermore, the formation of heteroduplex molecules during PCR might contribute to interpretation difficulties (Ferris and Ward, 1997). Also, some bacteria have multiple operons with sequence heterogeneity for genes such as the 16S rRNA gene, which may lead to overestimation of the number of bacteria (Nübel et al., 1996). Finally, some studies demonstrate that fragments of different sequences may migrate to the same position (Buchholz-Cleven et al., 1997; Vallaeys et al., 1997).

Another fingerprinting technique for studying microbial diversity is SSCP. Denatured PCR products are separated on a non-denaturing gel, based on differences in the folding conformation of single-stranded DNA. To simplify the SSCP pattern, one of the PCR primers is fluorescently labelled or biotinylated. SSCP electrophoresis can be used to reveal the presence of bacterial species in microbial populations and is also used to study population activity and dynamics (Delbès et al., 2001). A limitation of the method is that only very short fragments (ca. 150 bp) can be optimally separated.

RFLP, otherwise known as amplified ribosomal DNA restriction analysis (ARDRA), and terminal-RFLP (T-RFLP) are also frequently being used for the characterization of microbial communities. RFLP methods involve the digestion of PCR-amplified genes by restriction enzymes, followed by separation of the restricted fragments on agarose or polyacrylamide gels (Ranjard et al., 2000). A problem of estimating microbial diversity by ARDRA is that the number of fragments is not related to the number of different amplified DNA fragments, nor to the number of community members. T-RFLP solves this problem: the use of fluorescently labelled PCR primers makes it possible to detect only the terminal fragments and, as a consequence, the profiles obtained are simpler in terms of the number of bands. A databank has been produced of 16S rRNA T-RFLP patterns for a large set of bacterial species, which makes it possible to identify the bands present in a community by a profile-to-profile comparison (Marsh, 1999).

One of the techniques for quantifying the abundance of particular species in a bacterial population is slot-blot hybridization. In this approach, community DNA is spotted on nylon membranes and hybridized with labelled group- or species-specific probes. The relative abundance of the taxon can be estimated by comparing the signal obtained with the specific hybridization probe to that of a universal probe (Muyzer, 1998). A derived method is *in situ* hybridization or whole-cell hybridization, generally using fluorescent labelled probes (FISH). Because whole cells are hybridized, artefacts arising from biases in DNA-extraction, PCR amplification and cloning are avoided (Hill et al., 2000). To be detected, soil microbes must be metabolically active and possess cell walls sufficiently permeable to allow penetration of the probe. Progress is being made to overcome problems associated with penetration of probes in cells (Macnaughton et al., 1994). FISH provides a more accurate quantification of cells compared with slot-blot assays (Amann et al., 1995). However, they both have the limitation that only bacteria for which probes exist can be studied (Muyzer, 1998).

Nucleic acid microarrays – or DNA chips – are a recent advance in molecular technology. They provide a powerful tool for multiplexed detection of nucleic acids and offer tremendous potential for microbial community analysis and process monitoring in both basic and applied environmental science. DNA microarrays also demonstrate the potential to directly investigate microbial community dynamics and metabolic activity in environmental samples, even without using PCR (Small et al., 2001). Common approaches for microarray fabrication and analyses include cDNA and oligonucleotide probes affixed on planar, channel glass, gel element and microbial surfaces. The targets are fluorescent or radioactive-labelled total mRNA, DNA derived from mRNA by reverse transcription (RT)-PCR, or genomic DNA (Ball and Trevors, 2002). The most common method for introducing the microarray detection label includes PCR or RT-PCR amplification of a target sequence with a labelled primer or chemical labelling. After hybridization and washing, the microarray is scanned and the hybridization patterns can be qualitatively and quantitatively analysed.

The disadvantages to these techniques are that they perform poorly when comparing data from different experiments and when analysing novel sequences. Usually, individual fragments are extracted from the gels and the corresponding sequences are determined by direct DNA sequencing. This approach is labour-intensive and, in most cases, requires further PCR amplification or cloning of the eluted DNA. Recently, serial analysis of ribosomal sequence tags (SARST) was developed as a novel technique for characterizing microbial community composition (Neufeld et al., 2004). The SARST method captures sequence information from concatenates of short-PCR amplicons (tags) derived from the 16S rDNAs from complex bacterial populations. Depending on the protocol, tags are generated from either the V1 or V6 hyper-variable regions of bacterial 16S rDNA genes. The major advantage of the SARST method is the high-throughput generation of sequence data that can be directly used for species identification and comparisons between different experiments.

An emerging alternative approach to studying microbial communities is the use of microarrays designed to detect specific sequences from important lineages of microorganisms known or suspected to be present in the a particular population (Zhou, 2004). While this approach can provide a comprehensive quantitative survey for the presence or absence of a particular sequence, such as the presence of all recognized lineages of SRB (Loy et al., 2002, 2004), the technique has a closed architecture, i.e. it cannot identify novel sequences, nor can it easily distinguish between two or more closely related sequences in mixed populations. For microbial community analysis to be meaningful, the ability to identify previously uncharacterized members and discriminate between closely related sequences in a sample is essential for the comprehensive identification and assessment of organisms.

Phylogenetic identifying genes
The development of techniques to analyse nucleic acids made it possible to study the microbial diversity at the genetic level. Microorganisms are detected, identified and enumerated by the analysis of genes. The 16S rRNA gene is an excellent molecular marker for this purpose, because (i) it is present in all prokaryotes, (ii) the nucleotide sequence of 16S rRNA genes contain highly conserved as well as highly variable regions, which makes it possible to design general and specific primers and probes, (iii) it has sufficient sequence information for phylogenetic inference and (iv) it is an important cellular compound, which facilitates detection (Muyzer, 1998). The application of molecular biological techniques to detect and identify microorganisms by their 16S rRNA or its encoding gene is well established and is widely used to determine the microbial diversity and to analyse the structure of microbial communities. Depending on the objectives of the analysis, the molecular

methods can be applied to the community's DNA or its RNA. Whereas rDNA only reflects the presence of organisms, the amount of rRNA per cell is roughly proportional to metabolic activity (Wagner, 1994). Thus, to demonstrate the presence of specific bacterial groups and to analyse and follow the composition of bacterial populations, these molecular methods are applied to the community's DNA, whereas bacterial activity will be studied using the community's RNA.

To obtain insight into a specific group of bacteria such as SRB, the 16S rRNA gene-based molecular methods can be refined by using group-, genus- or strain-specific hybridization probes or PCR primers (Amann et al., 1992; Devereux et al., 1992; Daly et al., 2000; Stubner, 2002). Loy and coworkers successfully developed and used a microarray consisting of 132 16S rRNA gene-targeted oligonucleotide probes covering all recognized lineages of sulphate-reducing prokaryotes (SRP) for high-resolution screening of clinical and environmental samples (Loy et al., 2002). The microarray, named SRP-PhyloChip, has great potential for rapid screening of SRP diversity in complex samples and microarray SRP diversity fingerprints allow identification of relevant probes for further characterization of a sample by PCR or quantitative hybridization. This is a valuable option if large numbers of samples are to be analysed for temporal or spatial variations in SRP diversity.

Molecular identification based on the comparison of the variable regions in the 16S rRNA gene is considered one of the most accurate methods. However, in some cases, 16S rRNA gene sequences are not suitable for identification and differentiation between bacterial strains of the same genus because of high similarity percentages. For example, the 16S rRNA gene sequences of *Lactobacillus plantarum* and *Lactobacillus pentosus* share an identity value of >99% (Collins et al., 1991; Quere et al., 1997), and within the *Mycobacterium* genus, the interspecies percent similarity of the 16S rRNA gene sequences of *Mycobacterium kansasii* and *Mycobacterium gastri* is 100% (Rogall et al., 1990). To distinguish between closely related bacterial strains and allow for the identification of microbial community members, Single Point Genome Signature Tags (SP-GST) was developed as a generally applicable, high-throughput sequencing-based method that targets specific genes to generate identifier tags from well defined points in a genome (van der Lelie et al., 2006). Also, genetic relationships within a genus may be disturbed by recombination events, and in order to reflect the evolution of the genome as a whole, it is necessary to complete the interpretation of single-gene phylogenies, such as those based on 16S rRNA, by studying more loci (Gaunt et al., 2001). It has been proposed that the *recA* gene could be used as a phylogenetic marker. RecA is a small protein which takes part in homologous DNA recombination, SOS induction and DNA damage-induced mutagenesis. Due to its fundamental role, the *recA* gene is ubiquitous. Some regions of *recA* are conserved between species and other regions are highly variable, thus allowing comparison between both close and distant relatives, and large-scale phylogenies of *recA* in bacteria are consistent with the corresponding rRNA phylogenies. These criteria make it a useful marker for phylogenetic analyses (Lloyd and Sharp, 1993; Eisen, 1995). The *recA* gene was successfully used as an alternative sequencing target in several studies to complement identification methods based on 16S rRNA gene sequencing (Blackwood et al., 2000; Gaunt et al., 2001; Torriani et al., 2001).

Functional genes

Although the 16S rRNA approach provides the most general framework for studies of SRB-communities in the environment and has the potential to reveal interactions between sulphate reducers and other microorganisms such as methanogens or acetogens (Oude-Elferink et al., 1998b), it has its limitations. Retrieved sequences are frequently related to uncultivated organisms which makes it impossible to unambiguously predict the physiology or metabolic capabilities of the organism containing the gene. This is of concern in particular sulphate reducers. Indeed, SRP constitute a paraphyletic group with members among the *δ-proteobacteria*, gram-positive bacteria and even *Archaea* (Castro et al., 2000) and some lineages of sulphate reducers are closely related to organisms that are unable to carry out dissimilatory sulphate reduction. Thus, novel lineages of sulphate reducers cannot be identified by 16S rRNA gene sequence alone. For this reason the functional gene approach has been used to identify bacteria responsible for specific biogeochemical processes in the environment. In this, the organism's physiology is inferred from functional gene sequences coding for enzymes that are essential to it (Minz et al., 1999; Joulian et al., 2001; Dhillon et al., 2003).

First attempts to study SRB-populations via a functional gene approach were based on hydrogenase genes in *Desulfovibrio* species. Hydrogenases play an important role in the hydrogen metabolism of SRB (Voordouw et al., 1990). Distribution analysis of three types of hydrogenases among 22 members of the genus *Desulfovibrio* revealed that a [NiFe] hydrogenase was present in all *Desulfovibrio* species, while the genes for the [Fe] hydrogenase and [NiFeSe] hydrogenase had a more limited distribution (Voordouw et al., 1990). The [NiFe] hydrogenase is also thought to be essential in dissimilatory metal reduction by SRB. To understand the importance of metal reduction by SRB and to reveal the genetic diversity of *Desulfovibrio* species in environmental samples, PCR primers were designed for amplification of ∼0.45 kb fragment of the [NiFe] hydrogenase genes of *Desulfovibrio* species and used in combination with DGGE (Wawer and Muyzer, 1995). In this way, the number of different *Desulfovibrio* species could be determined. Later, these primers were used for direct assessment of *Desulfovibrio* species in groundwater of a uranium mill tailing site (Chang et al., 2001). Recently, both a T-RFLP method (Pérez-Jiménez and Kerkhof, 2005) and a dsrB-DGGE method (Geets et al., 2006) were developed for the diversity analysis of *dsr* genes in the environment.

The best-studied example is the *dsr* gene, which encodes for the dissimilatory sulphite reductase (DSR, EC 1.8.99.1). This enzyme catalyses the final step in sulphate respiration, the reduction of sulphite to sulphide, and hence is required by all sulphate reducers. Consequently, this enzyme has been found in all dissimilatory SRP investigated so far (Klein et al., 2001). Because of the remarkable conservation of the *dsr*-sequence, a general PCR primer set has been developed which allowed PCR-amplification of a 1,9 kb DNA fragment encoding most of the α and β subunits of the *dsrAB* genes from all recognized lineages of SRP (Karkhoff-Schweizer et al., 1995; Wagner et al., 1998). Recent studies of the environmental diversity of uncultured SRB populations indicate that sequence analysis of *dsr* genes is effective for the detection of SRB within a complicated microbial community structure such as those found in sediments (Joulian et al., 2001; Thomsen et al., 2001; Dhillon et al., 2003), microbial mats (Minz et al., 1999; Nakagawa et al., 2002), on the back of deep-sea worms (Cotrell and Cary, 1999), uranium tailing sites (Chang et al., 2001) and hydrocarbon-degrading consortia (Pérez-Jiménez et al., 2001). Moreover, sequence analysis of PCR-amplified environmental *dsr* genes often demonstrates the presence of unknown SRP that cannot be identified by 16S rRNA-sequencing analysis (Joulian et al., 2001; Thomsen et al., 2001; Nakagawa et al., 2002; Dhillon et al., 2003).

Although phylogenies of *dsr* sequences of reference strains are generally congruent with those inferred from the 16S rRNA, there are some contradictions caused by multiple lateral gene transfer events of the *dsrAB* genes (Wagner et al., 1998; Klein et al., 2001). It was discovered that *dsrAB* genes of several *Desulfotomaculum* spp., *Thermodesulfobacterium* spp. and *Desulfobacula toluolica* had been laterally transferred from unidentified ancestors of sulphate reducers within the δ-*proteobacteria*. This complicates the interpretation of phylogenetic *dsr* trees in environmental diversity studies of SRB.

In addition to the dissimilatory sulphite reductase, sulphate respiring prokaryotes possess APS reductase (EC 1.8.99.2) activity. This enzyme catalyses the reduction of APS to sulphite and AMP (adenosine monophosphate). Since the APS reductase is highly conserved (Fritz et al., 2000) and is generally found in SRP (Rabus et al., 2000), it appears to be an ideal candidate for phylogenetic analysis (Hipp et al., 1997). Recently, the diversity and distribution of SRB communities in gastrointestinal tracts were studied via PCR amplification and DGGE of a ~ 0.4 kb fragment of the APS reductase subunit A (*ApsA*) gene (Deplancke et al., 2000). To date, no environmental SRB population studies based on APS as a functional marker gene have been reported. This is probably due to the fact that a thorough phylogenetic framework of APS reductase from cultivated sulphate reducers has only been recently available. Nevertheless, a comparison of the *ApsA*- and 16S rRNA-based phylogenetic trees revealed major differences in the relative branching order for *Thermodesulfovibrio islandicus* and *Thermodesulfobacterium* spp., δ-*proteobacteria* belonging to the *"Syntrophobacteraceae"*, the *"Nitrospinaceae"* and the desulfobacteraceal *Desulfocella halophila*. These discrepancies in tree topologies can be explained by lateral transfer of *apsA*-genes across bacterial divisions. Hence, as is the case for the *dsr* gene as a genetic marker, environmental APS reductase gene sequences can now be linked to most of the recognized lineages of SRP. However, phylogenetic inferences in environmental diversity studies should be interpreted with caution (Friedrich, 2002).

21.3.2.2 Community genome analysis

Field data of bacterial population genetics and genomics render questionable the value of information about a single gene (Rodriguez-Valera, 2002). Recently, soil community genomes have been cloned using a bacterial artificial chromosome (BAC) vector. BACs are specialized cloning systems derived from an *E. coli* F plasmid and can maintain inserts exceeding 300 kb. Clones were found to express Dnase, antibacterial protein, lipase and amylase activity (Rondon et al., 2000). Metagenomic analysis provides functional information through genomic sequence and expression of traits, but other methods are required to link specific functions with the group responsible for them, and concomitant quantitative and comparative analysis of expressed rRNA genes and genes for key enzymes in relation to environmental factors are required to obtain information about the phylogeny and ecology of functional bacterial groups (Torsvik and Ovreas, 2002). However, metagenomic libraries are a powerful tool for exploring soil microbial diversity, providing access to the genetic information of uncultured soil microorganisms. Eventually, this will lead to a more realistic understanding of prokaryotic biodiversity and functionally complex microbial communities and will provide biotechnology with new tools (Rodriguez-Valera, 2002). The improvement of sequencing technologies has made metagenome shot-gun sequencing of an environmental sample feasible, however most environmental communities are far too complex to be fully sequenced in this manner. Reconstruction of community metagenomes has successfully been applied on samples from the Sargasso Sea (Venter et al., 2004) and an acid mine drainage (AMD) biofilm (Tyson et al., 2004).

21.4 Bioremediation processes based on microbial heavy metal detoxification mechanisms

Biological approaches for remediating heavy metal contamination offer potential for highly selective removal of toxic metals coupled with considerable operational flexibility, as they can be used both *in situ* or *ex situ*. Moreover, the availability of molecular monitoring tools for the detection and follow-up of the biological process – as discussed before – allows optimization of the treatment process at a lab scale, and adjustment of the on-site application. Biological approaches also have the potential for improvement through biostimulation, bioaugmentation and natural gene transfer (Dong et al., 1998).

21.4.1 Bioremediation of heavy metal-contaminated waters

Biological remediation processes for metal-polluted ground- or wastewater can be applied either *ex situ* or *in situ*. The *ex situ* strategies are based on the principle of pump-and-treat: water is pumped up and treated in an above-ground reactor by biosorption or bioprecipitation. There are various reactor designs such as the continuously stirred-tank reactor, the upflow anaerobic sludge blanket reactor (UASB), the fluidized-bed reactor and the packed-bed reactor. Although pump-and-treat processes are of demonstrated efficiency in heavy metal removal, installation costs and retention time are high, especially when large volumes of water have to be treated, which is often the case when treating polluted groundwater at industrial sites.

Lower in cost are *in situ* processes, such as the installation of a subsurface permeable barrier or the creation of a reactive zone or biobarrier. Permeable reactive barriers are receiving a great deal of attention as an innovative, cost-effective technology for *in situ* cleanup of contaminated groundwater. A wide variety of materials are proposed for use in subsurface barriers, including zero-valent metals (e.g. iron metal), humic materials, oxides, surfactant-modified zeolites, and oxygen- and nitrate-releasing compounds. These materials remove dissolved metals by immobilization within the barrier or their reduction to less harmful forms. The primary removal processes include sorption and precipitation, chemical reaction and biologically mediated reactions. These processes have been reviewed in detail by Scherer and coworkers (Scherer et al., 2000).

Another *in situ* approach is the creation of a reactive zone or biobarrier. This is a subsurface zone where the activity of metal transforming microorganisms is stimulated by the injection of nutrients. As groundwater passes through this zone, bacterial activity will result in metal retention resulting in clean groundwater, so avoiding further dispersion of the contaminating metals to the surrounding environment. The principle of a reactive zone is generally based on precipitation of metals as metal sulphides by SRB.

21.4.1.1 Biosorption

Biosorption is an emerging technology that uses biological materials to remove metals or radionuclides from solution through adsorption. Biosorption is the ability of biological materials to accumulate heavy metals from wastewater through metabolically mediated or physicochemical pathways of uptake or binding (Vieira and Volesky, 2000). Metal accumulative bioprocesses involve biosorptive (passive) uptake by dead biomass or bioaccumulation by living cells (Veglio and Beolcini, 1997). Many studies have demonstrated the efficiency of metal removal by microbial biomass in a range of reactor formats and under a range of physical and chemical conditions (Gadd, 2000). The main drawback in the use of biomaterials is that existing ion-exchange synthetic resins have a similar performances and their use is already well established. Nevertheless, biosorption methods seem to be more effective than their physicochemical counterparts in removing dissolved metals at low concentrations (below $2–10\,\text{mg}\,\text{l}^{-1}$) (Bunke et al., 1999) and demonstrate higher specificity, which avoids overloading of binding sites by alkaline-earth metals.

21.4.1.2 Bioprecipitation

Where reduction of a metal to a lower redox state occurs mobility and toxicity can be reduced, offering potential bioremediation applications for ground- and wastewater. Indirect precipitation by the formation of metal sulphides and phosphates is the strategy that has received most interest in the biotechnology of metal precipitation, although dissimilatory metal reduction can also be effectively utilized for decontamination (Valls and de Lorenzo, 2002). Remediation strategies which involve reducing metal toxicity and the mobility of contaminating metals in waters can be applied either *ex situ* in pump-and-treat reactor systems, or *in situ*, via permeable reactive barriers or by creating an *in situ* reactive zone or biobarrier.

Dissimilatory iron-reducing bacteria have an important role in oxidizing organic contaminants (e.g. aromatic hydrocarbons) in the subsurface but they are also capable of immobilizing contaminant metals by reduction to less soluble forms. For example, reduction of U(VI) to the less soluble U(IV) by *G. metallireducens* can be the basis of U(VI) removal from contaminated waters and leachates (Gorby and Lovley, 1992) and has been proposed as a remediation strategy for a uranium-contaminated site in New Mexico (Finneran et al., 2002; Holmes et al., 2002). The potential of microbiological reduction of Cr(VI) by *Shewanella alga* was studied in laboratory batch experiments and in field-scale by Friedly and coworkers (Friedly et al., 1995). They demonstrated that the dissimilatory metal-reducing bacterium *Shewanella algae* BrY is able to repeatedly generate surface reactive Fe(II) for the chemical reduction of Cr(VI) to the less toxic, less mobile Cr(III) in the presence of iron-chromium precipitates. Thus, besides its role in direct biological Cr(VI) reduction, the *Shewanella* strain also biologically re-reduces or regenerates surface-associated ferric iron (Fe(III)) to ferrous iron (Fe(II)) for the chemical reduction of Cr(VI) by surface-associated ferrous iron. Hence it follows that Dissimilatory Metal Reducing Bacteria (DMRB) can be used to establish permeable reactive barriers (Nevin and Lovley, 2002).

SRB have received much attention because of their reductive precipitation capacity of toxic metals such as Cr(VI) and U(VI), which is mediated by cytochrome c proteins. Biofilms of SRB enzymatically reduced and precipitated Cr(VI) (Smith and Gadd, 2000). Expression of the cytochrome c_7- gene from *D. acetoxidans* in *D. desulfuricans* led to enhanced expression of metal reductase activity in the recombinant strain. Such overproduction of active cytochrome c_7 could be important in fixed-enzyme reactors or in the production of organisms with enhanced metal reductase activity for bioremediation (Aubert et al., 1998).

Besides dissimilatory reduction, sulphate reducers are important for their indirect reductive metal precipitation mechanisms, for example, reduction of Cr(VI) can be a result of bacterial respiration or indirect reduction by Fe^{2+} and the resulting sulphide, or reduction by the resulting H_2S (Lloyd et al., 2001). SRB are anaerobic heterotrophs utilizing a wide range of organic substrates and SO_4^{2-} as electron acceptor. The immobilization of metals as metal sulphide complexes due to the sulphide production activities of SRB is one of the best-known natural precipitation systems. Bioprecipitation of metals by SRB requires an anaerobic environment, low redox conditions and an electron donor such as lactate or methanol (Mulligan et al., 2001). Mixed sulphate-reducing bacterial consortia are more effective than pure cultures in the removal of heavy metals from solution (White and Gadd, 1998, 2000).

SRB have been successfully used in the treatment of waters and leachates in large-scale bioreactors and in pilot laboratory surveys (Gadd and White, 1993; Gadd, 2000). Sulphate-reducing bacterial biofilm reactors may offer a means to intensify the process entrap or precipitate metals such as Cd or Cu at the biofilm surface (White and Gadd, 1998, 2000). For the treatment of contaminated waters such as AMD (acid mine drainage), biotechnological applications of SRB might be a promising alternative for chemical treatment, which generally involves the addition of lime (Dvorak et al., 1992; Drury, 1999; Chang et al., 2000; Jong and Parry, 2003). SRB activity in porous and permeable reactive walls installed in the path of contaminated waste may provide a promising alternative for remediating metal-contaminated wastes (Waybrant et al., 1998; Benner et al., 2002). Several patents have been granted for *in situ* systems that remove dissolved metals (Waybrant et al., 1998; Benner et al., 2002). These inventions provide an *in situ* anaerobic reactive zone in the path of migrating groundwater to precipitate dissolved metals as metal sulphides. The *in situ* bioprecipitation process (ISBP) by SRB was recently investigated for metals such as Cu, Zn, Cd, Ni, Co, Fe, Cr and As. The first field tests showed that ISBP is feasible as a strategy for sustaining groundwater quality (Geets et al., 2003). However, the effect of some factors like the choice of electron donor, the frequency of the electron donor injection, the soil type, the pH increase due to molasses fermentation, etc. on ISBP still have to be elucidated.

High concentrations of different heavy metals inhibit SRB (Hao, 2000). However, some strains of SRB can produce exopolymers that complex metals which results in a decrease of their bioavailability and toxicity. Microbial organisms can also adapt to adverse environments including metal-contaminated systems. Thus, metals may inhibit SRB in batch systems, but not in continuous flow systems (Beech and Cheung, 1995).

Metal precipitation can also be mediated by the liberation of inorganic phosphate from organic donor molecules. A *Citrobacter* sp. which was isolated from metal-polluted soil was found to be capable of accumulating high levels of uranium, zirconium and nickel through the formation of highly insoluble metal phosphates (Macaskie et al., 1994, 1997). Hence, the phosphate precipitation process by *Citrobacter* sp. is clearly promising for bioremediation.

21.4.2 Treatment of heavy metal-contaminated soil

Metals in soil, sludge or solid waste often need to be removed from the matrix by solubilization in a liquid phase. Afterwards, these water-solubilized metals can be concentrated in a second step – the desolubilization phase. The solubilization of heavy metals can be achieved via autotrophic or heterotrophic leaching, the use of metallophores or by chemical leaching followed by microbial treatment. Metals displaced in this way into the water phase can be desolublized via biologically induced adsorption, precipitation and transformation or complexation processes (Diels et al., 1999).

Bioleaching is a simple and effective technology for metal extraction of mineral ores, based on the activity of acidophilic iron- and sulphur-oxidizing bacteria, mainly *T. ferrooxidans* and *Thiobacillus thiooxidans* (Olson et al., 2003). Bacterial leaching is carried out in an aerobic and acid environment (pH 4) and can be performed by direct or indirect means. In direct leaching, the bacteria need to make physical contact with the mineral sulphide surface. Metal sulphides, such as CuS, ZnS and NiS, are oxidized to produce sulphuric acid, which can desorb the metals from the soil by substitution of protons (Mulligan et al., 2001). In indirect bioleaching, the bacteria generate a lixiviant which chemically oxidizes the sulphide mineral. In acid solution, this lixiviant is Fe(III): the bacteria convert Fe(II) to Fe(III) which in turn oxidizes sulphur minerals (Bosecker, 1997). Although most interest arises from a hydrometallurgical perspective, leaching of contaminating metals from soils and other matrices is also possible (White and Gadd, 1998). In fact, bioleaching of heavy metals from contaminated aquatic sediments using indigenous sulphur-oxidizing bacteria can be better than sulphuric acid treatment for metal solubilization (Seidel et al., 1998). Bioleaching has also been shown to be efficient and cost-effective means of removing heavy metals from sewage sludge without seriously affecting its soil conditioning and fertilizing properties (Wong and Henry, 1984; Couillard and Mercier, 1991; Chan et al., 2003). There is growing interest in microbial production of organic acids or heterotrophic leaching as an alternative in pollution treatment and metal recovery. Heterotrophic leaching may be particularly appropriate for wastes of high pH, as most *Thiobacilli* cannot solubilize effectively above pH 5.5 (Gadd, 2000). Members of the genus *Bacillus* are most effective in heterotrophic metal solubilization (Bosecker, 1997).

SRB reactors have been used in the removal of heavy metals from soil in an integrated decontamination process: sulphur-oxidizing bacteria were used to leach metals from the contaminated soils and the released metals were subsequently removed using bacterial sulphate reduction in an internal sedimentation reactor (White and Gadd, 1998). This combination of bioleaching followed by separate bioprecipitation of leached metals by SRB has proved effective in removing and concentrating a range of metals, including Zn, Cu and Cd, from contaminated soil. The bioleaching process can be applied *ex situ*, but where soil and water conditions are appropriate, the soil leaching component may be carried out *in situ*.

The patented Bio Metal Sludge Reactor (BMSR) (Diels et al., 1992) relies on siderophore-mediated metal solubilization and biocrystallization by *R. metallidurans* CH34 to treat sandy soils contaminated with heavy metals. This bacterium solubilizes the metals via the production of metal complexing siderophores and adsorbs the metals in its biomass on metal-induced outer membrane proteins and by bioprecipitation. The BMSR system consists of a continuous stirred-tank reactor that is fed with contaminated soil to which water and nutrients are added. The soil treated with CH34 will sedimentate while the bacteria stay in suspension, which allows separation of the metal-loaded bacteria and soil to take place in a settling device. The bacteria can be recovered by either a flotation or a flocculation process. This soft treatment system has attained a large decrease in the bioavailability of Cd, Zn and Pb (Diels et al., 1999). This is in contrast to chemical leaching by acids or complexing agents, which removes all ions.

Recently, a novel approach for removal of heavy metals from polluted soils was described. A mouse MT was expressed on the surface of *C. metallidurans* to promote biosorption. The recombinant strain was found to have an enhanced ability for immobilizing Cd^{2+} ions. Furthermore, inoculation of the recombinant strain in Cd^{2+}-polluted soil resulted in a significant decrease of the toxic effects of the heavy metal on the growth of tobacco plants. Thus, microbial metal immobilization and the resulting decrease in metal bioavailability can be used in moderately polluted fields, allowing their use in agriculture (Valls et al., 2000). It has also been shown that inoculation of metal-resistant bacteria into soils protected the indigenous bacterial community from the effects of heavy metals (Stephen et al., 1999). Such bioaugmentation strategies can be applied for the *in situ* protection of micro- and macrobiota from metal toxicity.

21.5 Conclusions

During recent decades, major advances have been made in understanding of the interaction mechanisms between microorganisms and heavy metals, and in the application of specialized microorganisms for the *in situ* and *ex situ* treatment of heavy metal and radionuclide-contaminated soils, wastes and groundwater. The efficiency of heavy metal bioremediation depends on the presence and activities of the microorganisms involved which is, in turn, affected by environmental conditions, operational parameters and the local composition of the overall microbial community. When opting for a biological remediation strategy, important questions to be answered include: (i) Are microorganisms with the desired characteristics and activities present at the contaminated site? (ii) What is their activity? and (iii) How is the microbial community composition and function influenced by environmental parameters and process conditions?

Recently, molecular and non-molecular methods for the identification and characterization of bacteria and their specific properties have been used to assess the composition and activity of microbial communities found at heavy metal-contaminated sites. These techniques promise to become complementary tools to classical chemical and physiological analysis (heavy metal concentrations and speciation, redox potential, etc.) for monitoring spatial and temporal changes in microbial community composition and function.

Molecular and microbial monitoring tools, when combined with batch, column and pilot studies, enable the follow-up and optimization of the biological processes during the developmental phase, as well as during the bioremediation treatment itself. At the lab scale, they will demonstrate the presence of active microbial communities and identify the "key players" whose activities are crucial to successful remediation. They will also become important management tools for following up the efficiency of bioremediation processes, especially when applying *in situ* remediation. When combined with batch, column and pilot studies, they can be used as part of the remediation startup phase to define optimal process conditions: comparison of the bacterial community composition and activity for different process conditions and their respective metal removal efficiency will predict the success or failure of the final remediation, and will help in the selection of optimal process conditions. Once the bioremediation strategy is applied, these monitoring methods can be used for follow-up, and as a decision tool for necessary process adjustments.

The only limitations for the large-scale, high-throughput application of microbial monitoring as part of a management strategy for the remediation of contaminated sites are the cost of analysis and the need for specialized labour and equipment. However, recent developments in tests for the detection of pathogens indicate that cheap tests for the detection of specific groups of environmentally important microorganisms or their functions, such as SRB, will become available in future. As a result, monitoring tools will become an integrated part of the management decision system for the remediation of contaminated sites when efficient, cost-effective and reliable bioremediation technologies are applied.

References

Adriano, D.C., Wenzel, W.W., Vangronsveld, J., Bolan, N.S., 2004. Role of assisted natural remediation in environmental cleanup. Geoderma. 122, 121–142.

Altschul, S.F., Gish, W., Miller, W., Myers, E.W., Lipman, D.J., 1990. Basic local alignment search tool. J. Mol. Biol. 215, 403–410.

Amann, R., Ludwig, W., Schleifer, K., 1995. Phylogenetic identification and in situ detection of individual microbial cells without cultivation. Microbiol. Rev. 59, 143–169.

Amann, R., Stromley, J., Devereux, R., Key, R., Stahl, D.A., 1992. Molecular and microscopic identification of sulfate-reducing bacteria in multi-species biofilms. Appl. Environ. Microbiol. 58, 614–623.

Anderson, R.T., Vrionis, H.A., Ortiz-Bernad, I., Resch, C.T., Long, P.E., Dayvault, R., Karp, K., Marutzky, S., Metzler, D.R., Peacock, A., White, D.C., Lowe, M., Lovley, D.R., 2002. Stimulating the in situ activity of Geobacter species to remove uranium from the groundwater of a uranium-contaminated aquifer. Appl. Environ. Microbiol. 69, 5884–5891.

Aubert, C., Lojou, E., Bianco, P., Rousset, M., Durand, M.-C., Bruschi, M., Dolla, A., 1998. The *Desulfuromonas acetoxidans* triheme cytochrome c_7 produced in *Desulfovibrio desulfuricans*: biocatalyst characterization and use in a flow-through bioreactor. Appl. Environ. Microbiol. 64, 1308–1312.

Ball, K.D., Trevors, J.T., 2002. Bacterial genomics: the use of DNA microarrays and bacterial artificial chromosomes. J. Microbiol. Methods 49, 275–284.

Barkay, T., Miller, S.M., Summers, A.O., 2003. Bacterial mercury resistance from atoms to ecosystems. FEMS Microbiol. Rev. 27, 355–384.

Barton, L.L., Tomei, F.A., 1995. Characteristics and Activities of Sulfate-Reducing Bacteria. Plenum Press, New York.

Beech, I.B., Cheung, C.W.S., 1995. Interactions of exopolymers produced by sulfate-reducing bacteria with metal-ions. Int. Biodeterior. Biodegrad. 35, 59–72.

Beliaeff, B., Mary, J.Y., 1993. The 'most probable number' estimate and its confidence limits. Water Res. 5, 799–805.

Benhar, I., 2001. Biotechnological advances of phage and cell display. Biotechnol. Adv. 19, 1–33.

Benner, S.G., Blowes, D.W., Ptacek, C.J., Mayer, K.U., 2002. Rates of sulfate reduction and metal sulfide precipitation in a permeable reactive barrier. Appl. Geochem. 17, 301–320.

Blackwood, K.S., He, C., Gunton, J., Turenne, C., Wolfe, J., Kabani, A.M., 2000. Evaluation of recA sequences for identification of *Mycobacterium* species. Appl. Environ. Microbiol. 38, 2846–2852.

Blindauer, C.A., Harrison, M.D., Parkinson, J.A., Robinson, A.K., Cavet, J.S., Robinson, N.J., Sadler, P.J., 2001. A metallothionein containing a zinc finger within a four-metal cluster protects a bacterium from zinc toxicity. Proc. Natl. Acad. Sci.USA. 98, 9593–9598.

Borremans, B., Hobman, J.L., Provoost, A., Brown, N.L., van der Lelie, D., 2001. Cloning and functional analysis of the pbr lead resistance determinant of *Ralstonia metallidurans* CH34. J. Bacteriol. 183, 551–568.

Bosecker, K., 1997. Bioleaching: metal solubilization by microorganisms. FEMS Microbiol. Rev. 20, 591–604.

Brandt, K.K., Vester, F., Jensen, A.N., Ingvorsen, K.K., 2001. Sulfate reduction dynamics and enumeration of sulfate-reducing bacteria in hypersaline sediments of the Great Salt Lake (Utah, USA). Microbiol. Ecol. 41, 1–11.

Bridge, T.A.M., White, C., Gadd, G.M., 1999. Extracellular binding activity of the sulfate-reducing bacterium *Desulfococcus multivorans*. Microbiol. 145, 2987–2995.

Brown, N.L., Barrett, S.R., Camakaris, J., Lee, B.T.O., Rouch, D.A., 1995. Molecular genetics and transport analysis of the copper resistance determinant (*pco*) from *Escherichia coli* plasmid pRJ1004. Mol. Microbiol. 17, 1153–1166.

Bruins, M.R., Kapil, S., Oehme, F.W., 2000. Microbial resistance to metals in the environment. Ecotox. Environ. Saf. 45, 198–207.

Buchholz-Cleven, B.E.E., Rattunde, B., Straub, K.L., 1997. Screening for genetic diversity of isolates of anaerobic Fe(II)-oxidizing bacteria using DGGE and whole-cell hybridization. Syst. Appl. Microbiol. 20, 301–309.

Bunke, G., Götz, P., Buchholz, R., 1999. Metal Removal by Biomass: Physico-Chemical Elimination Methods. Wiley-VCH Verlag, Weinheim.

Caccavo, F. Jr., Coates, J.D., Rossello-Mora, R.A., Ludwig, W., Schleifer, K.H., Lovley, D.R., McInerney, M.J., 1996. *Geovibrio ferrireducens*, a phylogenetically distinct dissimilatory Fe(III)-reducing bacterium. Arch. Microbiol. 165, 370–376.

Cameron, R.E., 1992. Guide to site and soil description for hazardous waste site characterization. Metals, Environmental Protection Agency (EPA)/600/4-91/029 1.

Castro, H.F., Williams, N.H., Ogram, A., 2000. Phylogeny of sulfate-reducing bacteria. FEMS Microbiol. Ecol. 31, 1–19.

Cavet, J.S., Borrelly, G.M., Robinson, N.J., 2003. Zn, Cu and Co in cyanobacteria: selective control of metal availability. FEMS Microbiol. Rev. 27, 165–181.

Chan, L.C., Gu, X.Y., Wong, J.W.C., 2003. Comparison of bioleaching of heavy metals from sewage sludge using iron- and sulfur-oxidizing bacteria. Adv. Environ. Res. 7, 603–607.

Chang, Y.J., Peacock, A.D., Long, P.E., Stephen, J.R., McKinley, J.P., Macnaughton, S.J., Hussain, A.K., Saxton, A.M., White, D.C., 2001. Diversity and characterization of sulfate-reducing bacteria in groundwater at a uranium mill tailings site. Appl. Environ. Microbiol. 67, 3149–3160.

Chang, I.S., Shin, P.K., Kim, B.H., 2000. Biological treatment of acid mine drainage under sulfate-reducing conditions with solid waste materials as substrate. Water Res. 34, 1269–1277.

Childers, S.E., Ciufo, S., Lovley, D.R., 2002. *Geobacter metallireducens* access Fe(III) oxide by chemotaxis. Nature 416, 767–769.

Christensen, B., Torsvik, T., Lien, T., 1992. Immunomagnetically captured thermophilic sulfate-reducing bacteria from North Sea oil field waters. Appl. Environ. Microbiol. 58, 1244–1248.

Coates, J.D., Ellis, D.J., Gaw, C.V., Lovley, D.R., 1999. *Geothrix fermentans* gen. nov., sp. nov., a novel Fe(III)-reducing bacterium from a hydrocarbon-contaminated aquifer. Int. J. Syst. Bacteriol. 49, 1615–1622.

Collard, J., Corbisier, P., Diels, L., Dong, Q., Jeanthon, C., Mergeay, M., Taghavi, S., van der Lelie, D., Wilmotte, A., Wuertz, S., 1994. Plasmids for heavy metal resistance in *Alcaligenes eutrophus* CH34: mechanisms and applications. FEMS Microbiol. Rev. 14, 405–414.

Colleran, E., Finnegan, S., Lens, P., 1995. Anaerobic treatment of sulfate-containing waste streams. Antonie van Leeuwenhoek 67, 29–46.

Collins, M.D., Rodrigues, U.M., Ash, C., Aguirre, M., Farrow, J.A.E., Martinez-Murcia, A., 1991. Phylogenetic analysis of the genus Lactobacillus and related lactic acid bacteria as determined by reverse transcriptase sequencing of 16S rRNA. FEMS Microbiol. Lett. 77, 5–12.

Conner, J.R., 1990. Chemical Fixation and Solidification of Hazardous Wastes. Van Nostrand Reinhold, New York.

Cooksey, D.A., 1994. Molecular mechanisms of copper resistance and accumulation in bacteria. FEMS Microbiol. Rev. 14, 381–386.

Cotrell, M.T., Cary, S.C., 1999. Diversity of dissimilatory sulfite reductase genes of bacteria associated with the deep-sea hydrothermal vent polychaete annelid *Alvinella pompejana*. Appl. Environ. Microbiol. 65, 1127–1132.

Couillard, D., Mercier, G., 1991. An economic evaluation of biological removal of heavy metals from wastewater sludge. Water Environ. Res. 66, 32–39.

Cunningham, S.D., Berti, W.R., Huang, J.W., 1995. Phytoremediation of contaminated soils. TIBTech. 13, 393–397.

Daly, K., Sharp, R.J., McCarthy, A.J., 2000. Development of oligonucleotide probes and PCR primers for detecting phylogenetic subgroups of sulfate-reducing bacteria. Microbiol. Ecol. 146, 1693–1705.

Delbès, C., Leclerc, M., Zumstein, E., Godon, J.-J., Moletta, R., 2001. A molecular method to study population diversity and activity dynamics in anaerobic digestor. Water Sci. Technol. 43, 51–57.

Deplancke, B., Hristova, K.R., Oakley, H.A., McCracken, V.J., Aminov, R., Mackie, R.I., Gaskins, H.R., 2000. Molecular ecological analysis of the succession and diversity of sulfate-reducing bacteria in the mouse gastrointestinal tract. Appl. Environ. Microbiol. 66, 2166–2174.

Devereux, R., Kane, M.D., Wilfrey, J., Stahl, D.A., 1992. Genus- and group-specific hybridization probes for determinative and environmental studies of sulfate-reducing bacteria. Syst. Appl. Microbiol. 15, 601–609.

Dhillon, A., Teske, A., Dillon, J., Stahl, D.A., Sogin, M.L., 2003. Molecular characterization of sulfate-reducing bacteria in the Guaymas basin. Appl. Environ. Microbiol. 69, 2765–2772.

Diels, L., Carpels, M., Geuzens, P., Mergeay, M., Rymen, T., 1992. Method and device for cleaning soil polluted by at least one heavy metal. European patent: 92203049.9.

Diels, L., De Smet, M., Hooyberghs, S., Corbisier, P., 1999. Heavy metals bioremediation of soil. Mol. Biotechnol. 12, 149–158.

Diels, L., Faelen, M., Mergeay, M., Nies, D., 1985. Mercury transposons from plasmids governing multiple resistance to heavy metals in *Alcaligenes eutrophus* CH34. Arch. Int. Physiol. Biochem. 93, 27–28.

Diels, L., Van Roy, S., Mergeay, M., Doyen, W., Taghavi, S., Leysen, R., 1993a. Immobilization of Bacteria in Composite Membranes and Development of Tubular Membrane Reactors for Heavy Metal Recuperation. Mechanical Engineering Publications Limited, London, UK.

Diels, L., Van Roy, S., Taghavi, S., Doyen, W., Leysen, R., Mergeay, M., 1993b. The use of Alcaligenes eutrophus immobilized in a tubular membrane reactor for heavy metal recuperation. In Biohydrometallurgical Technologies. M.L.A.a)C.L.B. A.E. Torma, T.m., metals & materials society (ed). USA: Warrendale, PA, pp. 133–144.

Di Giovanni, G.D., Watrud, L.S., Seidler, R.J., Widmer, F., 1999. Comparison of parental and transgenic alfalfa rhizosphere bacterial communities using BIOLOG GN metabolic finger-printing and enterobacterial repetitive intergenic consensus sequence-PCR (ERIC-PCR). Microbiol. Ecol. 37, 129–139.

Dong, Q., Springael, D., Schoeters, J., Nuyts, G., Mergeay, M., Diels, L., 1998. Horizontal gene transfer of bacterial heavy metal resistance genes and its applications in activated sludge systems. Water Sci. Technol. 37, 465–468.

Dowling, N.J.E., Widdel, F., White, D.C., 1986. Phospholipid ester-linked fatty acid biomarkers of acetate-oxidizing sulphate-reducers and other sulphide-forming bacteria. J. Gen. Microbiol. 132, 1815–1825.

Drury, W.J., 1999. Treatment of acid mine drainage with anaerobic solid-substrate reactors. Water Environ. Res. 71, 1244–1250.

Dvorak, D.H., Hedin, R.S., Edenborn, H.M., McIntire, P.E., 1992. Treatment of metal-contaminated water using bacterial sulfate-reduction: results from pilot-scale reactors. Biotechnol. Bioeng. 40, 609–616.

Eisen, J.A., 1995. The RecA protein as a model molecule for molecular systematic studies of bacteria: comparison of trees of RecA and 16S rRNAs from the same species. J. Mol. Evol. 41, 1105–1123.

Ferris, M.J., Ward, D.M., 1997. Seasonal distributions of dominant 16S rRNA defined populations in a hot spring microbial mat examined by detaturing gradient gel electrophoresis. Appl. Environ. Microbiol. 63, 1375–1381.

Finneran, K.T., Housewright, M.E., Lovley, D.R., 2002. Multiple influences of nitrate on uranium solubility during bioremediation of uranium-contaminated subsurface sediments. Environ. Microbiol. 4, 510–516.

Friedly, J.C., Davis, J.A., Kent, D.B., 1995. Modeling of hexavalent chromium reduction in field-scale transport and laboratory batch-experiments. Water Resour. Res. 31, 2783–2794.

Friedrich, M.W., 2002. Phylogenetic analysis reveals multiple lateral transfers of adenosine-5'-phosphosulfate reductase genes among sulfate-reducing microorganisms. J. Bacteriol. 184, 278–289.

Fritz, G., Buchert, T., Huber, H., Stetter, K.O., Kroneck, P.M.H., 2000. Adenylylsulfate reductases from archaea and bacteria are 1:1 alpha beta-heterodimeric iron-sulfur flavoenzymes-high similarity of molecular properties emphasizes their central role in sulfur metabolism. FEBS Lett. 473, 63–66.

Gadd, M.G., 1992. Metals and microorganisms: a problem of definition. FEMS Microbiol. Lett. 100, 197–204.

Gadd, G.M., 2000. Bioremedial potential of microbial mechanisms of metal mobilization and immobilization. Curr. Opin. Biotechnol. 11, 271–279.

Gadd, M.G., White, C., 1993. Microbial treatment of metal pollution – a working biotechnology? Trends Biotechnol. 11, 353–359.

Gaunt, M.W., Turner, S.L., Rigottier-Gois, L., Lloyd-Macgilp, S.A., Young, J.P.W., 2001. Phylogenies of atpD and recA support the small subunit rRNA-based classification of rhizobia. Int. J. Syst. Evol. Microbiol. 51, 2037–2048.

Geets, J., Diels, L., Van Geert, K., 2003. In situ metal bioprecipitation from lab scale to pilot tests. Proceedings of the 8th International FKZ/TNO conference on contaminated soil (Consoil). Ghent, Belgium, pp. 1641–1648.

Gibson, G.R., Parkers, R.J., Herbert, R.A., 1987. Evaluation of viable counting procedures for the enumeration of sulfate-reducing bacteria in estuarine sediments. J. Microbiol. Methods 7, 201–210.

Gorby, Y.A., Lovley, D.R., 1992. Enzymatic uranium reduction. Environ. Sci. Technol. 26, 205–207.

Green, A.C., Patel, B.K.C., Sheehy, A.J., 1997. *Deferribacter thermophilus* gen. sp. nov., a novel thermophilic manganese- and iron-reducing bacterium isolated from a petroleum reservoir. Int. J. Syst. Bacteriol. 47, 505–509.

Hao, O.L., 2000. Metal Effects on Sulfur Cycle Bacteria and Metal Removal by Sulfate-Reducing Bacteria. IWA Publishing, London.

Hassen, A., Saidi, N., Cherif, M., Boudabous, A., 1998. Resistance of environmental bacteria to heavy metals. Biores. Technol. 64, 7–15.

Hill, G.T., Mitkowski, N.A., Aldrich-Wolfe, L., Emele, L.R., Jurkonie, D.D., Ficke, A.etal., 2000. Methods for assessing the composition and diversity of soil microbial communities. Appl. Soil Ecol. 15, 25–36.

Hipp, W.M., Pott, A.S., Thum-Schmitz, N., Faath, I., Dahl, C., Trüper, H.G., 1997. Towards the phylogeny of APS reductases and siroheam sulfite reductases in sulfate-reducing and sulfur-oxidizing prokaryotes. Microbiology 143, 2891–2902.

Holmes, D.E., Finneran, K.T., O'Neil, R.A., Lovley, D.R., 2002. Enrichment of members of the family Geobacteriaceae associated with stimulation of dissimilatory metal reduction in uranium contaminated aquifer sediments. Appl. Environ. Microbiol. 68, 2300–2306.

Hoogenboom, H.R., de Bruine, A.P., Hufton, S.E., Hoet, R.M., Arends, J.-W., Roovers, R.C., 1998. Antibody phage display technology and its applications. Immunotechnology 4, 1–20.

Huckle, J.W., Morby, A.P., Turner, J.S., Robinson, N.J., 1993. Isolation of prokaryotic metallothionein locus and analysis of transcriptional control by trace metal ions. Mol. Microbiol. 7, 177–187.

Jain, D.K., 1995. Evaluation of the semisolid Postgate's B medium for enumerating sulfate-reducing bacteria. J. Microbiol. Methods 22, 27–38.

Jong, T., Parry, D.L., 2003. Removal of sulfate and heavy metals by sulfate reducing bacteria in a short term bench scale upflow anaerobic packed bed reactor runs. Water Res. 37, 3379–3389.

Joulian, C., Ramsing, N.B., Ingvorsen, K., 2001. Congruent phylogenies of most common small-subunit rRNA and dissimiliatory sulfite reductase gene sequences retrieved from estuarine sediments. Appl. Environ. Microbiol. 67, 3314–3318.

Kaksonen, A.H., Plumb, J.J., Franzmann, P.D., Puhakka, J.A., 2003a. Simple organic electron donors support diverse sulfate-reducing communities in fluidized-bed reactors treating acidic metal- and sulfate-containing wastewater. FEMS Microbiol. Ecol. 47, 279–289.

Kaksonen, A.H., Riekkola-Vanhanen, M.L., Puhakka, J.A., 2003b. Optimization of metal sulphide precipitation in fluidized-bed treatment of acidic wastewater. Water Res. 37, 255–266.

Karkhoff-Schweizer, R.R., Huber, D.P.W., Voordouw, G., 1995. Conservation of the genes for dissimilatory sulfite reductase from *Desulfovibrio vulgaris* and *Archaeoglobus fulgidus* allows their detection by PCR. Appl. Environ. Microbiol. 61, 290–296.

Kleikemper, J., Pelz, O., Schroth, M.H., Zeyer, J., 2002. Sulfate-reducing bacterial community response to carbon source amendments in contaminated aquifer microcosms. FEMS Microbiol. Ecol. 42, 109–118.

Klein, M., Friedrich, M., Roger, A.J., Hugenholtz, P., Fishbain, S., Abicht, H., Blackall, L.L., Stahl, D.A., Wagner, M., 2001. Multiple lateral transfers of dissimilatory sulfite reductase genes between major lineages of sulfate-reducing prokaryotes. J. Bacteriol. 183, 6028–6035.

Kohring, L.L., Ringelberg, D.B., Devereux, R., Stahl, D.A., Mittelman, M.W., White, D.C., 1994. Comparison of phylogenetic relationships based on phospholipid fatty acid profiles and ribosomal RNA sequence similarities among dissimilatory sulfate-reducing bacteria. FEMS Microbiol. Lett. 119, 303–308.

Krafft, T., Macy, J.M., 1998. Purification and characterization of the respiratory arsenate reductase of *Chrysiogenes arsenatis*. Eur. J. Biochem. 255, 647–653.

Laanbroek, H.J., Veldkamp, H., 1982. Microbial interactions in sediment communities. Phil. Trans. R. Soc. Lond. 297, 533–550.

Labrenz, M., Banfield, J.F., 2000. Microbial diversity in a natural ZnS-producing biofilm of sulfate-reducing bacteria in a subsurface acid mine drainage system. Unpublished data.

Lee, Y.A., Hendson, M., Panopoulus, N.J., Schroth, M.N., 1994. Molecular cloning, chromosomal mapping, and sequence analysis of copper resistance in genes from *Xanthomonas campestris pv. fuglandis*: homology with blue copper proteins and multicopper oxidase. J. Bacteriol. 176, 173–188.

Levinson, H.S., Mahler, I., 1998. Phosphatase activity and lead resistance in *Citrobacter freundii* and *Staphylococcus aureus*. FEMS Microbiol. Lett. 161, 135–138.

Lillebaek, R., 1995. Application of antisera raised against sulfate-reducing bacteria for indirect immunofluorescent detection of immunoreactive bacteria in sediment from the German Baltic Sea. Appl. Environ. Microbiol. 61, 3436–3442.

Liu, C.-Q., Charoechai, P., Khunajakr, N., Deng, Y.-M., Dunn, N.W., 2002. Genetic and transcriptional analysis of a novel plasmid-encoded copper resistance operon from *Lactococcus lactis*. Gene 297, 241–247.

Liu, T., Nakashima, S., Hirose, K., Uemura, Y., Shibasaka, M., Katsuhara, M., Kasamo, K., 2003. A methallothionein and Cpx-ATPase handle metal tolerance in the filamentous cyanobacterium *Oscillatora brevis*. FEBS Lett. 542, 159–163.

Lloyd, J.R., 2003. Microbial reduction of metals and radionuclides. FEMS Microbiol. Rev. 27, 411–425.

Lloyd, J.R., Lovley, D.R., 2001. Microbial detoxification of metals and radionuclides. Curr. Opin. Biotechnol. 12, 248–253.

Lloyd, J.R., Mabbaett, A., Williams, D.R., Macaskie, L.E., 2001. Metal reduction by sulfate-reducing bacteria: physiological diversity and metal specificity. Hydrometallurgy 59, 327–337.

Lloyd, A.T., Sharp, P.M., 1993. Evolution of the recA gene and the molecular phylogeny of bacteria. J. Mol. Evol. 37, 399–407.

Lonergan, D.J., Jenter, H.L., Coates, J.D., Phillips, E.J.P., Schmidt, T.M., Lovley, D.R., 1996. Phylogenetic analysis of dissimilatory Fe(III)-reducing bacteria. J. Bacteriol. 178, 2402–2408.

Lorenz, P., Schleper, C., 2002. Metagenome-a challenging source of enzyme discovery. J. Mol. Catal. B: Enzymatic 19–20, 13–19.

Lovley, D.R., 1993. Dissimilatory metal reduction. Annu. Rev. Microbiol. 47, 263–290.

Lovley, D.R., Coates, J.D., Blunt-Harris, E.L., Phillips, E.J.P., Woodward, J.C., 1996. Humic substances as electron acceptors for microbial respiration. Nature 382, 445–447.

Lovley, D.R., Giovannoni, S.J., White, D.C., Champine, J.E., Phillips, E.J.P., Gorby, Y.A., Goodwin, S., 1993a. *Geobacter metallireducens* gen. nov. sp. nov., a microorganism capable of coupling the complete oxidation of organic matter to the reduction of iron and other metals. Arch. Microbiol. 159, 336–344.

Lovley, D.R., Roden, E.E., Phillips, E.J.P., Woodward, J.C., 1993b. Enzymatic iron and uranium reduction by sulfate-reducing bacteria. Mar. Geol. 113, 41–53.

Lovley, D.R., Phillips, E.J.P., 1994. Reduction of chromate by *Desulfovibrio vulgaris* and its c_3 cytochrome. Appl. Environ. Microbiol. 60, 726–728.

Lovley, D.R., Widman, P.K., Woodward, J.C., Phillips, E.J., 1993c. Reduction of uranium by cytochrome c_3 of *Desulfovibrio vulgaris*. Appl. Environ. Microbiol. 59, 3572–3576.

Loy, A., Kusel, K., Lehner, A., Drake, H.L., Wagner, M., 2004. Microarray and functional gene analyses of sulfate-reducing prokaryotes in low-sulfate, acidic fens reveal cooccurrence of recognized genera and novel lineages. Appl. Environ. Microbiol. 70, 6998–7009.

Loy, A., Lehner, A., Lee, N., Adamczyk, J., Meier, H., Ernst, J., Schleifer, K.H., Wagner, M., 2002. Oligonucleotide microarray for 16S rRNA gene-based detection of all recognized lineages of sulfate-reducing prokaryotes in the environment. Appl. Environ. Microbiol. 68, 5064–5081.

Macaskie, L.E., Jeong, B.C., Tolley, M.R., 1994. Enzymatically-accelerated biomineralization of heavy metals: application to the removal of americum and plutonium from aqueous flows. FEMS Microbiol. Rev. 14, 351–368.

Macaskie, L.E., Yong, P., Doyle, T.C., Roig, M.G., Diaz, M., Manzano, T., 1997. Bioremediation of uranium-bearing wastewater: biochemical and chemical factors affecting bioprocess application. Biotechnol. Bioeng. 53, 100–109.

Macnaughton, S.J., O'Donnell, A.G., Embley, T.M., 1994. Permeabibilzation of mycolic acid-containing actinomycetes for in situ hybridization with fluorescently labeled oligonucleotide probes. J. Microbiol. Methods 26, 279–285.

Macy, J.M., Nunan, K., Hagen, K.D., Dixon, D.R., Harbour, P.J., Cahill, M., Sly, L.I., 1996. *Chrysiogenes arsenatis*, gen. sp. nov., a new arsenate respiring bacterium isolated from gold mine wastewater. Int. J. Syst. Bacteriol. 46, 1153–1157.

Maidak, B.L., Larsen, N., McCaughey, M.J., Overbeek, R., Olsen, G.J., Fogel, K., Blandy, J., Woese, C.R., 1994. The Ribosomal Database Project. Nucleic Acid Res. 22, 3485–3487.

Marsh, T.L., 1999. Terminal restriction fragment length polymorphism (T-RFLP): an emerging method for characterizing diversity among homologous populations of amplification products. Curr. Opin. Microbiol. 2, 323–327.

Mergeay, M., Monchy, S., Vallaeys, T., Auquier, V., Benotmane, A., Bertin, P., Taghavi, S., Dunn, J., van der Lelie, D., Wattiez, R., 2003. *Ralstonia metallidurans*, a bacterium specifically adapted to toxic metals: towards a catalogue of metal-responsive genes. FEMS Microbiol. Rev. 27, 385–410.

Mergeay, M., Nies, D., Schlegel, H.G., Gerits, J., Charles, P., Van Gijsegem, F., 1985. *Alcaligenes eutrophus* CH34 is a facultative chemolitotroph with plasmid-bound resistance to heavy metals. J Bacteriol. 162, 328–334.

Minz, D., Flax, J.L., Green, S.J., Muyzer, G., Cohen, Y., Wagner, M., Rittman, B.E., Stahl, D.A., 1999. Diversity of sulfate-reducing bacteria in oxic and anoxic regions of a microbial mat characterized by comparative analysis of dissimilatory sulfite reductase genes. Appl. Environ. Microbiol. 65, 4666–4671.

Monchy, S., Benotmane, M.A., Janssen, P., Vallaeys, T., Taghavi, S., van der Lelie, D., Mergeay, M., 2007. Plasmids pMOL28 and pMOL30 of *Cupriavidus metallidurans* are specialised in the maximal viable response to heavy metals. J. Bacteriol. 189, 7417–7425.

Monchy, S., Benotmane, M.A., Wattiez, R., van Aelst, S., Auquier, V., Borremans, B., Mergeay, M., Taghavi, S., van der Lelie, D., Vallaeys, T., 2006. Proteomic analyses of the pMOL30 encoded copper resistance in *Cupriavidus metallidurans* strain CH34. Microbiology 152, 1765–1776.

Mukhopadhyay, R., Rosen, B., Phung, L., Silver, S., 2002. Microbial arsenic: from geocycles to genes and enzymes. FEMS Microbiol. Rev. 26, 311–325.

Mulligan, C.N., Yong, R.N., Gibbs, B.F., 2001. Remediation technologies for metal-contaminated soils and groundwater: an evaluation. Eng. Geol. 60, 193–207.

Muyzer, G., 1998. Structure, Function and Dynamics of Microbial Communities: The Molecular Biological Approach. IOS Press, Amsterdam.

Muyzer, G., de Waal, E.C., Uitterlinden, A.G., 1993. Profiling of complex microbial populations by denaturing gradient gel electrophoresis analysis of polymerase chain reaction-amplified genes encoding for 16S rRNA. Appl. Environ. Microbiol. 59, 695–700.

Muyzer, G., Smalla, K., 1998. Application of denaturing gradient gel electrophoresis (DGGE) and temperature gradient gel electrophoresis (TGGE) in microbial ecology. Antonie van Leeuwenhoek 73, 127–141.

Myers, C.R., Carstens, B.P., Antholine, W.E., Myers, J.M., 2000. Chromium (VI) reductase activity is associated with the cytoplasmic membrane of anaerobically grown Shewanella putrefaciens MR-1. J. Appl. Microbiol. 88, 98–106.

Myers, C.R., Nealson, K.H., 1988. Bacterial manganese reduction and growth with manganese oxide as the sole electron acceptor. Science 240, 1319–1321.

Nakagawa, T., Hanada, S., Maruyama, A., Marumo, K., Urabe, T., Fukui, M., 2002. Distribution and diversity of thermophilic sulfate-reducing bacteria within a Cu-Pb-Zn mine (Toyoha, Japan). FEMS Microbiol. Ecol. 138.

Nealson, K.H., Saffarini, D., 1994. Iron and manganese in anaerobic respiration: environmental significance, physiology and regulation. Annu. Rev. Microbiol. 48, 311–343.

Neufeld, J.D., Yu, Z., Lam, W., Mohn, W.W., 2004. Serial analysis of ribosomal sequence tags (SARST): a high-throughput method for profiling complex microbial communities. Environ. Microbiol. 6, 131–144.

Nevin, K.P., Lovley, D.R., 2002. Mechanisms for Fe(III) oxide reduction in sedimentary environments. Geomicrobiol. J. 19, 141–159.

Newman, D.K., Kennedy, E.K., Coates, J.D., Ahmann, D., Ellis, D.J., Lovley, D.R., Morel, F.F.M., 1998. Dissimilatory arsenate and sulfate reduction in *Desulfotomaculum auripigmentum* sp. nov. Arch. Microbiol. 165, 380–388.

Newman, D.K., Kolter, R., 2000. A role for excreted quinones in extracellular electron transfer. Nature 405, 94–97.

Nies, D.H., 2003. Efflux mediated heavy metal resistance in prokaryotes. FEMS Microbiol. Rev. 27, 313–319.

Nies, D.H., Silver, S., 1995. Ion efflux systems involved in bacterial metal resistances. J. Ind. Microbiol. 14, 189–199.

Nordberg, M.A., 1998. Metallothioneins: historical review and state of knowledge. Talanta 46, 243–254.

Nübel, U., Engelen, B., Felske, A., 1996. Sequence heterogeneities of genes encoding 16S rRNAs in *Paenibacillus polymyxa* detected by temperature gradient gel electrophoresis. J Bacteriol. 178, 5363–5643.

Nyman, J.L., Caccavo, J.F., Cunningham, A.B., Gerlach, R., 2002. Biogeochemical elimination of chromium(VI) from contaminated water. Biorem. J. 6, 39–55.

Odom, J.M., Jessie, K., Knodel, E., Emptage, M., 1991. Immunological cross-reactivities of adenosine-5′-phosphosulfate reductases from sulfate-reducing and sulfur-oxidizing bacteria. Appl. Environ. Microbiol. 57, 727–733.

Olson, G.J., Brierley, J.A., Brierley, C.L., 2003. Bioleaching review, part B: Progress in bioleaching: applications of microbial processes by the mineral industries. Appl. Microbiol. Biotechnol. 63, 249–257.

Oremland, R.S., Switzer Blum, J., Culbertson, C.W., Visscher, P.T., Miller, L.G., Dowdle, P., Strohmaier, R.E., 1994. Isolation, growth and metabolism of an obligately anaerobic, selenate-respiring bacterium, strain SES-3. Appl. Environ. Microbiol. 60, 3011–3019.

Orphan, V.J., Hinrichs, K.U., Ussler, W., Paull, C.K., Taylor, L.T., Sylva, S.P., Hayes, J.M., 2001. Comparative analysis of methane-oxidizing Archaea and sulfate-reducing bacteria in anoxic marine sediments. Appl. Environ. Microbiol. 67, 1922–1934.

Oude-Elferink, S.J.W.H., Boschker, H.T.S., Stams, A.J.M., 1998a. Identification of sulfate-reducers and *Syntrophobacter* sp. in anaerobic granular sludge by fatty-acid biomarkers and 16S rRNA probing. Geomicrobiol. J. 15, 3–17.

Oude-Elferink, S.J.W.H, Visser, A., Hulshoff Pol, L.W., Stams, A.J.M., 1994. Sulfate reduction in methanogenic bioreactors. FEMS Microbiol. Rev. 15, 119–136.

Oude-Elferink, S.J.W.H., Vorstman, W.J.C., Sopjes, A., Stams, A.J.M., 1998b. Characterization of the sulfate-reducing and syntrophic population in granular sludge from a full-scale anaerobic reactor treating papermill wastewater. FEMS Microbiol. Ecol. 27, 185–194.

Palmer, C.D., Wittbrodt, P.R., 1991. Processes affecting the remediation of chromium-contaminated sites. Environ. Health Perspect. 92, 25–40.

Park, C.H., Keyhan, M., Wielinga, B., Fendorf, S., Matin, A., 2000. Purification to homogeneity and characterization of a novel *Pseudomonas putida* chromate reductase. Appl. Environ. Microbiol. 66,

Payne, R.B., Gentry, D.M., Rapp-Giles, B.J., Casalot, L., Wall, J.D., 2002. Uranium reduction by *Desulfovibrio desulfuricans* strain G20 and a cytochrome c3 mutant. Appl. Environ. Microbiol. 68, 3129–3132.

Pérez-Jiménez, J.R., Kerkhof, L.J., 2005. Phylogeography of sulfate-reducing bacteria among disturbed sediments, disclosed by the analysis of the dissimilatory sulfite reductase genes. Appl. Environ. Microbiol. 71, 1004–1011.

Pérez-Jiménez, J.R., Young, L.Y., Kerkhof, L.J., 2001. Molecular characterization of sulfate-reducing bacteria in anaerobic hydrocarbon-degrading consortia and pure cultures using the dissimilatory sulfite reductase (*dsrAB*) genes. FEMS Microbiol. Ecol. 35, 145–150.

Plaza, G., Ulfig, K., Hazen, T.C., Brigmon, R.L., 2001. Use of molecular techniques in bioremediation. Acta Microbiol. Pol. 50, 205–218.

Postgate, 1984. The Sulfute-Reducing Bacteria. Cambridge University Press, Cambridge, Great Britain.

Pronk, J.T., De Bruyn, J.C., Bos, P., Keunen, J.G., 1992. Anaerobic growth of *Thiobacillus ferrooxidans*. Appl. Environ. Microbiol. 58, 2227–2230.

Quere, F., Deschamps, A., Urdaci, M.C., 1997. DNA probe and PCR-specific reaction for *Lactobacillus plantarum*. J. Appl. Microbiol. 82, 783–790.

Rabus, R., Hansen, T. Widdel, F., (2000). Dissimilatory sulfate- and sulfur-reducing prokaryotes. In: M. Dworkin, S. Falkow, E. Rosenberg (Eds.), The Prokaryotes, release 3.3, Springer, New York, available at: http://www.prokaryotes.com.

Ranjard, L., Poly, F., Nazaret, S., 2000. Monitoring complex bacterial communities using culture-independent molecular techniques: application to soil environment. Res. Microbiol. 151, 167–177.

Raskin, L., Amann, R.I., Poulsen, L.K., Rittmann, B.E., Stahl, D.A., 1995. Use of ribosomal RNA-based molecular probes for characterization of complex microbial communities in anaerobic biofilms. Water Sci. Technol. 31, 261–272.

Rayney, F.A., Ward, N., Sly, L.I., Stackebrandt, E., 1994. Dependence on the taxon composition of clone libraries for PCR amplified, naturally occuring 16S rDNA, on the primer pair and the cloning system used. Experientia 50, 796–797.

Rensing, C., Ghosh, M., Rosen, B., 1999. Families of soft-metal-ion transporting ATPases. J. Bacteriol. 18, 5891–5897.

Rensing, C., Grass, G., 2003. *Escherichia coli* mechanisms of copper homeostasis in a changing environment. FEMS Microbiol. Rev. 27, 197–213.

Riesner, D., Henco, K., Steger, G., 1991. Temperature gradient gel electrophoresis: a method for the analysis of conformational transitions and mutations in nucleic acids and proteins. Adv. Electrophor. 4, 169–250.

Robinson, N.J., Whitehall, S.K., Cavet, J.S., 2001. Microbial metallothioneins. Adv. Microb. Physiol. 44, 183–213.

Rodriguez-Valera, F., 2002. Approaches to prokaryotic biodiversity: a population genetic perspective. Environ. Microbiol. 4, 628–633.

Rodriguez-Valera, F., 2004. Environmental genomics, the big picture?. FEMS Microbiol. Lett. 231, 153–158.

Rogall, T., Wolters, J., Flohr, T., Bottger, C., 1990. Towards a phylogeny and definition of species at the molecular level within the genus *Mycobacterium*. Int. J. Syst. Bacteriol. 40, 323–330.

Rondon, M.R., August, P.R., Bettermann, A.D., Brady, S.F., Grossman, T.H., Liles, M.R., Loiacono, K.A., Lynch, B.A., MacNeil, I.A., Minor, C., Tiong, C.L., Gilman, M., Osburne, M.S., Clardy, J., Handelsman, J., Goodman, R.M., 2000. Cloning the soil metagenome: a strategy for accessing the gene and functional diversity of uncultured microorganisms. Appl. Environ. Microbiol. 66, 2541–2547.

Rugh, C.L., Senecoff, J.F., Meager, R.B., Merkle, S.A., 1998. Development of transgenic yellow poplar for mercury phytoremediation. Nat. Biotechnol. 16, 925–928.

Rumer, R.R., Ryan, M.E., 1995. Barrier Containment Technologies for Environmental Remediation Applications. Wiley, New York.

Salt, D.E., Blaylock, M., Kumar, N.P.B.A., Dushenkov, V., Ensley, B.D., Chet, I., Raskin, I., 1995. Phytoremediation: a novel strategy for the removal of toxic metals from the environment using plants. Biotechnology 13, 468–474.

Santegoedts, C.M., Damgaard, L.R., Hesselink, G., Zopfi, J., Lens, P., Muyzer, G., de Beer, D., 1999. Distribution of sulfate-reducing and methanogenic bacteria in anaerobic aggregates determined by microsensor and molecular analysis. Appl. Environ. Microbiol. 65, 4618–4629.

Santegoedts, C.M., Ferdelman, T.G., Muyzer, G., de Beer, D., 1998. Structural and functional dynamics of sulfate-reducing populations in bacterial biofilms. Appl. Environ. Microbiol. 64, 3731–3739.

Scherer, M.M., Richter, S., Valentine, R.L., Alvarez, P.J.J., 2000. Chemistry and microbiology of permeable reactive barriers for in situ groundwater clean up. Crit. Rev. Microbiol. 26, 221–264.

Seidel, H., Ondruschka, J., Morgenstern, P., Stottmeister, U., 1998. Bioleaching of heavy metals from contaminated aqautic sediments using indigenous sulfur-oxidizing bacteria: a feasibility study. Water Sci. Technol. 37, 387–394.

Shen, H., Wang, Y.-T., 1993. Characterization of enzymatic reduction of hexavalent chromium by *Escherichia coli* ATCC 3346. Appl. Environ. Microbiol. 59, 3771–3777.

Silva, A.J., Varesche, M.B., Foresti, E., Zaiat, M., 2002. Sulphate removal from industrial wastewater using packed-bed anaerobic reactor. Process Biochem., 927–935.

Silver, S., 2003. Bacterial silver resistance: molecular biology and uses and misuses of silver compounds. FEMS Microbiol. Rev. 27, 341–354.

Silver, S., Nucifors, G., Chu, L., Misra, T.K., 1989. Bacterial resistance ATPases: primary pumps for exporting toxic cations and anions. Trends Biochem. Sci. 14, 76–80.

Small, J., Call, D.R., Brockman, F., Straub, T.M., Chandler, D.P., 2001. Direct detection of 16S rRNA in soil extracts by using oligonucleotide microarrays. Appl. Environ. Microbiol. 67, 4708–4716.

Smith, A.D., 1982. Immunofluorescence of sulphate-reducing bacteria. Arch. Microbiol. 133, 118–121.

Smith, W.L., Gadd, G.M., 2000. Reduction and precipitation of chromate by mixed culture sulphate-reducing bacterial biofilms. J. Appl. Microbiol. 88, 983–991.

Solioz, M., Stoyanov, J.V., 2003. Copper homeostasis in *Enterococcus hirae*. FEMS Microbiol. Rev. 27, 183–195.

Spear, J.R., Figueroa, L.A., Honeyman, B.D., 2000. Modeling reduction of uranium U(VI) under variable sulfate concentrations by sulfate-reducing bacteria. Appl. Environ. Microbiol. 66, 3711–3721.

Stephen, J.R., Chang, Y.J., Macnaughton, S.J., Kowalchuk, G.A., Leung, K.T., Flemming, C.A., White, D.C., 1999. Effect of toxic metals on indigenous soil beta-subgroup proteobacterium ammonia oxidizer community structure and protection against toxicity by inoculated metal-resistant bacteria. Appl. Environ. Microbiol. 65, 95–101.

Stolz, J.F., Ellis, D.J., Blum, J.S., Ahmann, D., Lovley, D.R., Oremland, R.S., 1999. *Sulfurospirillum barnesii* sp. nov. and *Sulfurospirillum arsenophilum* sp. nov., new members of the *Sulfurospirillum* clade of the epsilon Proteobacteria. Int. J. Syst. Bacteriol. 49, 1177–1180.

Stolz, J.F., Gugliuzza, T., Switzer Blum, J., Oremland, R.S., Murillo, F.M., 1997. Differential cytochrome content and reductase activity in *Geospirillum barnesii* strain SES-3. Arch. Microbiol. 167, 1–5.

Stubner, S., 2002. Enumeration of 16S rDNA of *Desulfotomaculum* lineage 1 in rice field soil by real-time PCR with SybrGreen^{TM} detection. J. Microbiol. Methods 50, 155–164.

Suzuki, Y., Kelly, S.D., Kemner, K.M., Banfield, J.F., 2003. Microbial populations stimulated for hexavalent uranium reduction in uranium mine sediment. Appl. Environ. Microbiol. 69, 1337–1346.

Suzuki, T., Miyata, H., Hoitsu, K., Kawai, K., Takamizawa, Y., Tai, Y., Okazaki, M., 1992. NAD(P)H-dependent chromium (VI) reductase of *Pseudomonas ambigua* G-1: a Cr(V) intermediate is formed during the reduction of Cr(VI) to Cr(III). J. Bacteriol. 174, 5340–5345.

Taghavi, S., Mergeay, M., van der Lelie, D., 1997. Genetic and physical maps of the *Alcaligenes eutrophus* CH34 megaplasmid pMOL28 and its derivative pMOL50 obtained after temperature-induced mutagenesis and mortality. Plasmid 27, 22–34.

Tanner, R.S., 1989. Monitoring sulfate-reducing bacteria: comparison of enumeration media. J. Microbiol. Methods 10, 83–90.

Taylor, J., Parkes, J., 1983. The cellular fatty acids of the sulfate-reducing bacteria *Desulfobacter* sp., *Desulfobulbus* sp. and *Desulfovibrio desulfuricans*. J. Gen. Microbiol. 129, 3303–3309.

Taylor, J., Parkes, J., 1985. Identifying different populations of sulfate-reducing bacteria within marine sediment systems, using fatty acid biomarkers. J. Gen. Microbiol. 131, 631–642.

Tebo, B., 1995. Metal Precipitation by Marine Bacteria: Potential for Biotechnological Applications. Plenum Press, New York.

Terry, N., Zayed, A.M., 1998. Phytoremediation of selenium. Environmental Chemistry of Selenium. In: W.T. Frankenberger Jr. and R.A. Engberg. (Eds.), Marcel Dekker, Inc., New York, NY, pp. 633–656.

Teske, A., Ramsing, N., Habicht, K., Fukui, M., Küver, J., Jørgensen, B., Cohen, Y., 1998. Sulfate-reducing bacteria and their activities in cyanobacterial mats of Solar Lake (Sinai, Egypt). Appl. Environ. Microbiol. 64, 2943–2951.

Thomsen, T.R., Finster, K., Ramsing, N.B., 2001. Biogeochemical and molecular signatures of anaerobic methane oxidation in a marine sediment. Appl. Environ. Microbiol. 67, 1646–1656.

Torriani, S., Felis, G.E., Dellaglio, F., 2001. Differentiation of *Lactobacillus plantarum, L. pentosus,* and *L. paraplantarum* by recA gene sequence analysis and multiplex PCR assay with recA gene-derived primers. Appl. Environ. Microbiol. 67, 3450–3454.

Torsvik, V., Ovreas, L., 2002. Microbial diversity and function in soil: from genes to ecosystems. Curr. Opin. Microbiol. 5, 240–245.

Trevors, J.T., Stratton, G.W., Gadd, G.M., 1986. Cadmium transport, resistance and toxicity in bacteria, algae and fungi. Can. J. Microbiol. 32, 447–464.

Tyson, G.W., Chapman, J., Hugenholtz, P., Allen, E.E., Ram, R.J., Richardson, P.M., Solovyev, V.V., Rubin, E.M., Rokhsar, D.S., Banfield, J.F., 2004. Community structure and metabolism through reconstruction of microbial genomes from the environment. Nature 428, 37–43.

Vainshtein, M., Hippe, H., Kroppenstedt, R.M., 1992. Cellular fatty acid composition of *Desulfovibrio* species and its use in classification of sulfate-reducing bacteria. Syst. Appl. Microbiol. 15, 554–566.

Vallaeys, T., Topp, E., Muyzer, G., 1997. Evaluation of denaturing gradient gel electrophoresis in the detection of 16S rDNA sequence variation in rhizobia and methanotrophs. FEMS Microbiol. Ecol. 24, 279–285.

Valls, M., Atrian, S., de Lorenzo, V., Fernandez, L.A., 2000. Engineering a mouse metallothionein on the cell surface of *Ralstonia eutropha* CH34 for immobilization of heavy metals in soil. Nat. Biotechnol. 18, 661–665.

Valls, M., de Lorenzo, V., 2002. Exploiting the genetic and biochemical capacities of bacteria for the remediation of heavy metal pollution. FEMS Microbiol. Rev. 26, 327–338.

van der Lelie, D., Lesaulnier, C., McCorkle, S., Geets, J., Taghavi, S., Dunn, J., 2006. Use of single-point genome signature tags as a universal tagging method for microbial genome surveys. Appl. Environ. Microbiol. 72, 2092–2101.

van der Lelie, D., Schwitzguebel, J.P., Glass, D.J., Vangronsveld, J., Baker, A., 2001. Assessing phytoremediation's progress in the United States and Europe. Environ. Sci. Technol. 35, 446A–452A.

Vangronsveld, J., Colpaert, J.V., Van Tichelen, K.K., 1996. Reclamation of a bare industrial area contaminated by non-ferrous metals: physico-chemical and biological evaluation of the durability of soil treatment and revegetation. Environ. Pollut. 94, 131–140.

Vangronsveld, J., van Assche, F., Clijsters, H., 1995. Reclamation of a bare industrial area contaminated by non-ferrous metals: in situ immobilization and revegetation. Environ. Pollut. 87, 51–59.

Vargas, M., Kashefi, K., Blunt-Harris, E.L., Lovley, D.R., 1998. Microbial evidence for Fe(III) reduction on early Earth. Nature 395, 65–67.

Vassilev, A., Schwitzguebel, J.P., Thewys, T., van der Lelie, D., Vangronsveld, J., 2004. The use of plants for remediation of metal-contaminated soils. ScientificWorldJournal 16, 9–34.

Veglio, F., Beolcini, F., 1997. Removal of metals by biosorption: a review. Hydrometallurgy 44, 301–316.

Venter, J.C., Remington, K., Heidelberg, J.F., Halpern, A.L., Rusch, D., Eisen, J.A., Wu, D., Paulsen, I., Nelson, K.E., Nelson, W., Fouts, D.E., Levy, S., Knap, A.H., Lomas, M.W., Nealson, K., White, O., Peterson, J., Hoffman, J., Parsons, R., Baden-Tillson, H., Pfannkoch, C., Rogers, Y.H., Smith, H.O., 2004. Environmental genome shotgun sequencing of the Sargasso Sea. Science 304, 66–74.

Vester, F., Ingvorsen, K.K., 1998. Improved most-probable-number method to detect sulfate-reducing bacteria with natural media and a radiotracer. Appl. Environ. Microbiol. 64, 1700–1707.

Vieira, R.H., Volesky, B., 2000. Biosorption: a solution to pollution?. Int. Microbiol. 3, 17–24.

Voordouw, G., Niviere, V., Ferris, F.G., Fedorak, P.M., Westlake, D.W.S., 1990. Distribution of hydrogenase genes in *Desulfovibrio* spp. and their use in identification of species from the oil field environment. Appl. Environ. Microbiol. 56, 3748–3754.

Wagner, R., 1994. The regulation of ribosomal RNA synthesis and bacterial cell growth. Arch. Microbiol. 161, 100–109.

Wagner, M., Roger, A.J., Flax, J.L., Brusseau, G.A., Stahl, D.A., 1998. Phylogeny of dissimilatory sulfite reductases supports an early origin of sulfate respiration. J. Bacteriol. 180, 2975–2982.

Wang, C.L., Maratukulam, P.D.L., Clark, D.S., Keasling, J.D., 2000. Metabolic engineering of an aerobic sulfate reduction pathway and its application to precipitation of cadmium to the cell surface. Appl. Environ. Microbiol. 66, 4497–4502.

Wang, P.C., Mori, T., Komori, K., Sasatsu, M., Toda, K., Ohtake, H., 1989. Isolation and characterization of an *Enterobacter cloacae* strain that reduces hexavalent chromium under anaerobic conditions. Appl. Environ. Microbiol. 55, 1665–1669.

Ward, D.W., Weller, R., Bateson, M.M., 1990. 16S rRNA sequences reveal numerous uncultured microorganisms in a natural community. Nature 345, 63–65.

Wawer, C., Muyzer, G., 1995. Genetic diversity of *Desulfovibrio* spp. in environmental samples analyzed by denaturing gradient gel elctrophoresis of [NiFe] hydrogenase gene fragments. Appl. Environ. Microbiol. 61, 2203–2210.

Waybrant, K.R., Blowes, D.W., Ptacek, C.J., 1998. Selection of reactive mixtures for use in permeable reactive walls for treatment of mine drainage. Environ. Sci. Technol. 32, 1972–1979.

White, C., Gadd, G.M., 1998. Accumulation and effects of cadmium on sulfate-reducing bacterial biofilms. Microbiol. 144, 1407–1415.

White, C., Gadd, G.M., 2000. Copper accumulation by sulfate-reducing bacterial biofilms. FEMS Microbiol. Lett. 183, 313–318.

Winzingerode, F., Göbel, U.B., Stackebrandt, E., 1997. Determination of microbial diversity in environmental samples: pitfalls of PCR-based rRNA analysis. FEMS Microbiol. Rev. 21, 213–229.

Wong, L., Henry, J.G., 1984. Decontaminating biological sludge for agricultural use. Water Sci. Technol. 17, 575–586.

Zhou, J., 2004. Microarrays for bacterial detection and microbial community analysis. In: W. Verstraete (Ed.), European Symposium on Environmental Biotechnolgy (ESEB). Ostend, Belgium.

Developments in Soil Science, volume 32
Ravendra Naidu (Editor)

Chapter 22

DNA ADDUCT ANALYSIS OF ENVIRONMENTAL DNA: A POTENTIAL METHOD TO ASSESS THE *IN SITU* BIOAVAILABILITY OF POLYCYCLIC AROMATIC HYDROCARBONS

I. Singleton and B. Lyons

22.1 Introduction

Environmental pollution by mutagenic compounds, such as polycyclic aromatic hydrocarbons (PAHs), is a significant problem and there is a need to assess ecosystem and human risk associated with such contamination. Generally risk analysis is based on the total level of pollutants present but it is being recognised that measuring the biological availability of pollutants may be a more accurate method (see Chapters 2, 3, 4 and 23). Toxicity tests for mutagenic compounds in soil have been developed and are extremely useful. However, such tests generally involve soil extraction/leaching and subsequent exposure of extracts/leachates to single species (Backhaus et al., 2000; Bekaert et al., 1999; Bispo et al., 1999; see Chapters 18 and 19). The use of extracts (often solvents) does not truly demonstrate the environmental bioavailability of pollutants and assays with single species are not representative of complex ecosystems. This is particularly relevant for the known diversity of bacteria and fungi. To overcome these limitations the ideal method would be economic, examine multiple species and determine real compound bioavailability in complex environmental samples.

This work presents initial findings on the potential of using DNA isolated from environmental samples which represent a wide range of species (in particular microorganisms) to determine *in situ* bioavailability of mutagenic compounds. Microbial species have intimate pollutant contact and differential abilities to access pollutants and these factors, together with microbial involvement in significant environmental processes, means that they are very useful indicators of compound bioavailability. It is hypothesised that an *in situ* measure of PAH bioavailability can be made by determining the levels of DNA adducts present on isolated environmental DNA. DNA adducts are widely regarded as the first step in mutagenesis and involve the binding of mutagenic pollutants (e.g. PAH metabolites) to DNA. This means that DNA adducts will only be formed if pollutants are bioavailable in soil to organisms thus giving a true measure of the bioavailability of pollutants associated with a particular sample. The method has potential for measuring the mutagenic potential of contaminated soil but is more likely to be of use for assessing the success of remediation strategies. For example bioremediation of PAH-contaminated soil is used frequently to reduce contamination but it is known that many microbial transformation products of PAHs are mutagenic and can cause DNA adduct formation (Cerniglia and Sutherland, 2001). The presence of DNA adducts caused by PAH metabolites indicates that PAHs are bioavailable to microbes in soils as the parent PAHs have been metabolised. By measuring DNA adduct formation during PAH bioremediation in soil we could potentially infer that certain methods, e.g. surfactant use, increase PAH bioavailability. Other possibilities include using bioremediation methods that avoid DNA adduct formation in soil by production of microbial PAH metabolites that are not carcinogenic, i.e. the DNA adduct method will allow us to follow the success of bioremediation and determine particular environmental conditions that avoid production of mutagenic metabolites. This could be achieved through use of specific microbial inocula known not to produce such metabolites.

First of all, it is necessary to demonstrate the production of DNA adducts in microbes in pure culture after exposure to PAHs as very little work has been done in this area although mutations have been found in bacteria (*Salmonella* reversion assay) exposed to PAHs (Sasaki et al., 1997). The second step is to look for presence of DNA adducts in DNA isolated from contaminated soil. The only previous work that the authors are aware of in this regard involves determination of DNA adducts in earthworms exposed to contaminated soil (Walsh et al., 1997; Saint-Denis et al., 2000).

The work below describes initial investigations on selected microbial species exposed to benzo[a]pyrene (BaP), analysis of DNA adducts in contaminated soil, presents results and discusses findings.

22.2 Materials and methods

22.2.1 Microbial culture conditions

Escherichia coli, Streptomyces and an undefined mixed culture obtained from soil were incubated in nutrient broth (1/10 dilution) for 48 h at 150 rpm and 30°C before addition of BaP solution in acetone to a final concentration of 100 mg BaP per litre.

Bjerkandera species (white rot fungus) was incubated statically in malt extract broth (1/10 dilution) for 4 days at 22°C before addition of BaP as above. Subsequently all cultures were incubated for 5 weeks and samples for measurement of DNA adduct analysis removed after 31 days incubation. Controls consisted of microbial cultures plus acetone without addition of BaP.

22.2.2 DNA isolation and DNA adduct analysis

DNA was isolated from both pure microbial cultures and from all soil samples using a commercial BIO101 kit according to manufacturers instructions (BIO101 Inc., Carlsbad, California, USA). Amounts of DNA isolated from cultures were determined by UV spectroscopy. DNA was stored at $-20°C$ until analysed for adducts. DNA adduct levels were determined by the ^{32}P post-labelling assay which is non-specific and capable of detecting a wide range of genotoxin-DNA adducts.

More specifically DNA adducts were determined using the butanol-enhanced version of the ^{32}P post-labelling assay (for details see Harvey et al., 1997). Briefly, 10 μg samples of DNA were digested to deoxyribonucleoside 3'-monophosphates in a total volume of 30 μl of digestion mix. This mix (29 μl) was then subjected to butanol enhancement, evaporated to dryness and then redissolved in 9.5 μl of water. The remaining 1 μl of the DNA digest was diluted and held for the labelling of the normal undamaged nucleotides for subsequent quantification. Adducted and normal nucleotides were labelled separately, but simultaneously, for 2 h using 2 μl of labelling buffer (100 mM bicine NaOH, pH 9.5, 100 mM $MgCl_2$, 100 mM dithiothreitol, 10 mM spermidine), 0.5 μl T4 polynucleotide kinase (30 U/μl; Amersham) and 5 and 2 μl of $[\gamma\text{-}^{32}P]$ ATP (>3000 Ci/mol, 10 μCi/μl; Amersham), respectively.

The adducted deoxyribonucleoside-3'-5'-biphosphates present in the sample were then purified and separated from their normal undamaged counterparts using multidimensional anion exchange thin layer chromatography (TLC), on 10×10 cm polyethyleneimine (PEI)-cellulose plates (Camlab, Cambridge, UK). The levels of DNA adduct radioactivity were determined using an AMBIS radioanalytical scanning system (LabLogic, Sheffield, UK). Upon the quantification of both the adducted nucleotides and the normal nucleotides, the relative adduct labelling values were calculated and converted to the number of adducted nucleotides per 10^8 undamaged nucleotides. Appropriate negative and positive DNA controls were analysed throughout the studies. Negative controls consisted of calf thymus DNA, while positive controls were generated *in vitro* by the addition of 1.5 mM BaP diol epoxide (NCI Chemical Carcinogens Repository, Kansas City, Missouri) to 1 mg of flounder (*Platichthys flesus*) DNA, followed by phenol-chloroform-isoamyl alcohol (24:24:1) extraction and isopropanol precipitation.

22.2.3 DNA adduct analysis: PAH-contaminated soil

Soil (air-dried, sieved to 2 mm) from Cockle Park Farm (University of Newcastle upon Tyne) was purposely spiked with 250 mg BaP, 250 mg napthalene and 250 mg pyrene all dissolved in acetone (Soil pH 6.2, Soil texture: Clay loam). Excess acetone was evaporated from the soil in a fume hood. After this soils were adjusted to 60% moisture holding capacity and incubated at 20°C for 3 months. Control uncontaminated soils were held under identical conditions. After 3 months soil samples (0.5 g) were taken and DNA isolated as described above except that DNA quantification was carried out after gel electrophoresis.

Isolated soil DNA from uncontaminated and contaminated soil was assayed for presence of DNA adducts as described above.

22.3 Results and discussion

22.3.1 Microbial culture studies

DNA adduct levels found in pure microbial cultures exposed to BaP are displayed in Table 22.1. *E.coli* control and test cultures along with the mixed test and control samples all contained quantifiable DNA adducts. These adducts appeared to be related to the test regime and therefore regarded as background adducts. DNA extracted from *Streptomyces* control, *Bjerkandera* control and *Bjerkandera* test cultures did not contain any detectable adducts under current assay conditions. In contrast, DNA samples from *Streptomyces* exposed to BaP for 31 days contained quantifiable DNA adducts with chromatographic characteristics similar to that of BaP diol epoxide treated DNA (Fig. 22.1). The lack of identical adducts in the *Streptomyces*

Table 22.1. *Numbers of DNA adducts detected in microbial cultures exposed to benzo[a]pyrene.*

Sample	DNA adducts per 10^8 undamaged nucleotides	Notes
E. coli control	8.3	Likely background adduct unrelated to PAH exposure regime
E. coli test	7.8	Likely background adduct unrelated to PAH exposure regime
Mixed culture control	2.7	Likely background adduct unrelated to PAH exposure regime
Mixed culture test	3.4	Likely background adduct unrelated to PAH exposure regime
Streptomyces control	N.D.	
Streptomyces test	113.0	DNA adducts detected with chromatographic characteristics similar to BaP-diol epoxide positive control
Bjerkandera control	N.D.	
Bjerkandera test	N.D.	

Note: N.D., no DNA adducts detectable under current assay conditions.

A : Control. *Streptomyces* incubated without benzo(a)pyrene
B : Test. *Streptomyces* incubated with benzo(a)pyrene

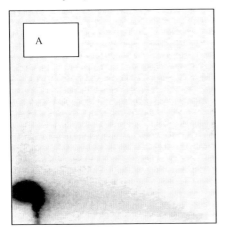
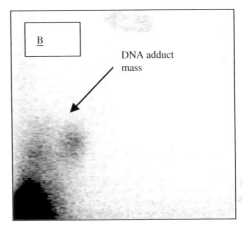

Figure 22.1. *Autoradiography of DNA adducts found in* Streptomyces *species incubated with benzo[a]pyrene. The quantification of DNA adducts was determined as described in the materials and methods section. A: Control,* Streptomyces *incubated without benzo[a]pyrene; B: Test,* Streptomyces *incubated with benzo(a)pyrene.*

control sample would suggest that the DNA adducts detected in *Streptomyces* test culture were related to the PAH exposure conditions.

The presence of adducts in the *Streptomyces* culture can be explained as the cytochrome P450 enzyme system is widespread in actinomycetes and is involved in pollutant transformation (Lamb et al., 2000; De Schrijver and De Mot, 1999). Specifically cytochrome P450 enzymes are known to be involved in PAH transformation with concomitant production of diol epoxides which cause DNA adducts (Cerniglia and Sutherland, 2001). The other organisms used in the experiment appear unable to access and transform BaP but this could be related to culture conditions used and needs further validation. Alternative explanations for the lack of DNA adducts include:

(a) The metabolic transformation pathway utilised by the organisms did not result in the production of mutagenic metabolites.
(b) Subsequent transformation of mutagenic metabolites into non-mutagens.
(c) Certain DNA adducts formed may not have been quantifiable under present DNA adduct labelling and/or chromatography procedures.
(d) DNA repair systems of the microbial species used were able to excise bases with adducts during the incubation period meaning that few adducts were detected upon analysis.

The above explanations require further experimental verification.

22.3.2 PAH-contaminated soil study

A significant amount of DNA adducts were found in soil exposed to PAHs whereas no adducts were detected in control uncontaminated soils (Table 22.2 and Fig. 22.2). This indicates that PAHs were being transformed by soil dwelling microbes and therefore available to the soil microbes. There could also be potential chemical transformation of PAHs occurring resulting in the formation of mutagenic compounds. Even so this means that chemically transformed PAHs are bioavailable to microbes in soil and does not invalidate the method for assessing the success of remediation methods. BaP was the only potentially mutagenic PAH in the soil and therefore it could be assumed that the compound

Table 22.2. *Numbers of DNA adducts found in uncontaminated and PAH-contaminated soil.*

Sample	DNA adducts per 10^8 undamaged nucleotides
Uncontaminated soil	No adducts detected
PAH-contaminated soil	39.6+/−17.2

A: Control. Soil without benzo(a)pyrene addition
B: Test : Soil with benzo(a)pyrene added

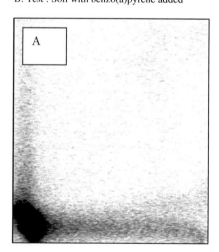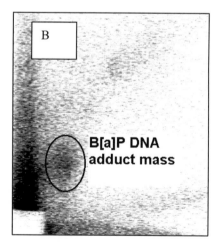

Figure 22.2. *Autoradiography of DNA adducts from PAH-contaminated soil and uncontaminated control soil. DNA adducts were determined as described in materials and methods section. A: Control, soil without benzo[a]pyrene addition; B: Test, soil with benzo[a]pyrene added.*

was bioavailable to soil microbes. This does not take into account potential synergistic interactions between different PAHs and non-specific chemical reactions occurring in the soil.

22.4 Conclusions and future work

DNA adducts were found in *Streptomyces* indicating that the method has potential for assessing the bioavailability of PAHs in the environment. Subsequently DNA adducts were detected in PAH-contaminated soils indicating the bioavailability of PAHs in the soil and that the PAHs were transformed into mutagenic compounds. We believe that this method has potential for indicating the bioavailability of PAHs in the environment and could be used for assessing the success of bioremediation of PAH-contaminated soils. However further work is necessary to validate and standardise the technique.

Acknowledgements

The authors gratefully acknowledge the support of the NERC and University of Newcastle upon Tyne, UK for funding this work. The authors also acknowledge supply of *Bjerkandera* from the Department of Industrial Microbiology, Wageningen University and *Streptomyces* from Dr. Gayle Payne, University of Newcastle upon Tyne.

References

Backhaus, T., Scholze, M., Grimme, L.H., 2000. The single substance and mixture toxicity of quinolones to the bioluminescent bacterium *Vibrio fischeri*. Aquat. Toxicol. 49, 49–61.

Bekaert, C., Rast, C., Ferrier, V., Bispo, A., Jourdain, M.J., Vasseur, P., 1999. Use of in vitro (Ames and Mutatox tests) and in vivo (Amphibian Micronucleus test) assays to assess the genotoxicity of leachates from a contaminated soil. Org. Geochem. 30(8B), 953–962.

Bispo, A., Jourdain, M.J., Jauzein, M., 1999. Toxicity and genotoxicity of industrial soils polluted by polycyclic aromatic hydrocarbons (PAHs). Org. Geochem. 30(8B), 947–952.

Cerniglia, C.E., Sutherland, J.B., 2001. Bioremediation of polycyclic aromatic hydrocarbons by lignolytic and non-lignolytic fungi. In: G.M. Gadd (Ed.), Fungi in Bioremediation. Cambridge University Press, Cambridge, UK, pp. 136–187.

De Schrijver, A., De Mot, R., 1999. Degradation of pesticides by actinomycetes. Crit. Rev. Microbiol. 25, 85–119.

Harvey, J.S., Lyons, B.P., Waldock, M., Parry, J.M., 1997. The application of the ^{32}P-postlabelling assay to aquatic biomonitoring. Mutat. Res. 378, 77–88.

Lamb, D.C., Kelly, D.E., Masaphy, S., Jones, G.L., Kelly, S.L., 2000. Engineering of heterologous Cytochrome P450 in *Acinetobacter* sp.: application for pollutant degradation. Biochem. Biophys. Res. Comm. 276, 797–802.

Saint-Denis, M., Pfohl-Leszkowicz, A., Narbonne, J.F., Ribera, D., 2000. Dose-response and kinetics of the formation of DNA adducts in the earthworm *Eisenia fetida andrei* exposed to B(a)P-contaminated artificial soil. Polcyclic Arom. Compd. 18, 117–127.

Sasaki, J.C., Arey, J., Eastmond, D.A., Parks, K.K., Grosovsky, AJ., 1997. Genotoxicity induced in human lymphoblasts by atmospheric reaction products of naphthalene and phenanthrene. Mutat. Res. 393, 23–35.

Walsh, P., ElAdlouni, C., Nadeau, D., Fournier, M., Coderre, D., Poirier, G.G., 1997. DNA adducts in earthworms exposed to a contaminated soil. Soil Biol. Biochem. 29, 721–724.

Developments in Soil Science, volume 32
Ravendra Naidu (Editor)
© 2008 Elsevier B.V. All rights reserved

Chapter 23

CAN BIOAVAILABILITY ASSAYS PREDICT THE EFFICACY OF PAH BIOREMEDIATION?

A.L. Juhasz

23.1 Introduction

Numerous genera of microorganisms have been observed to oxidise PAHs (see reviews by Cerniglia (1992), Juhasz and Naidu (2000), Kanaly and Harayama (2000)). In laboratory-based studies, the rate and extent of PAH degradation can be quantified following simple liquid culture or soil microcosm degradation experiments where the 'disappearance' of the compound of interest is monitored over time. Degradation experiments using ^{14}C-labelled compounds provide a more sophisticated methodology for determining the fate of PAHs (i.e. whether they are mineralised or transformed) during biodegradation.

The microbial degradation of PAHs, like other organic contaminants, is dependent on the rate of uptake and metabolism of PAHs by the cells and by the rate of transfer of PAHs to the cell. When comparing PAH degradation rates from liquid to soil matrices, it is evident that solid matrices have significant effects on degradation processes. Contaminant and microbial interactions with organic carbon, clay and mineral fractions influence sorption/desorption processes which inturn inhibit degradation processes. Such mechanisms leading to contaminant sequestration will be discussed later in the chapter.

A degradation curve similar to that depicted in Figure 23.1 is typically observed during bioremediation of PAH-contaminated soil or sediment – this 'hockey stick' curve represents biphasic PAH degradation kinetics (Cornelissen et al., 1998; Cuypers et al., 2002; Semple et al., 2003; Thiele-Bruhn and Brümmer, 2004; Huesemann et al., 2004). The initial phase of PAH removal, depicted as 'A' in Figure 23.1, represents the rapid degradation of PAHs following stimulation of PAH-degrading microorganisms. During phase A, removal of PAHs is limited by the microbial degradation kinetics and not the desorption of PAHs from the soil matrix, i.e. the rate of PAH mass transfer is greater relative to the intrinsic activity of the catabolic microorganisms (Bosma et al., 1997). In contrast, section B of Figure 23.1 represents the slow degradation kinetics phase. Prior to this phase, the readily available PAHs have been depleted and the rate of PAH degradation is limited by the bioavailability or slow desorption of PAHs

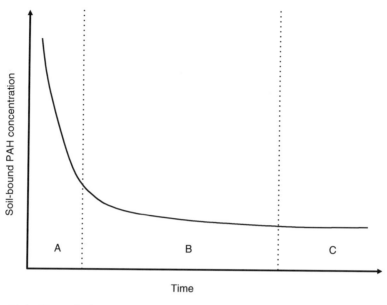

Figure 23.1. *Degradation curve typically observed during bioremediation of PAH-contaminated soil or sediment. The initial phase of PAH removal (A) represents the rapid degradation of PAHs where PAH removal is limited by microbial degradation kinetics and not the desorption of PAHs from the soil matrix. Section B represents the slow degradation kinetics phase where the rate of PAH degradation is limited by the slow desorption of PAHs from the soil matrix to soil solution. Section C represents the sequestered or unavailable PAH fraction.*

from the soil matrix to soil solution (Juhasz et al., 2000). Finally, section C of Figure 23.1 represents the sequestered or unavailable PAH fraction.

Numerous researchers have demonstrated that the bioavailability of organic compounds can appreciably affect the bioremediation of the contaminated matrix. Weissenfels et al. (1992) showed that PAH contaminants in coking plant soil were unable to be degraded even after inoculation of the soil with bacteria known to be effective in degrading PAHs. However, rapid degradation of PAHs was observed after the PAHs were extracted from the coking plant soil and reapplied into the extracted soil material. Erickson et al. (1993) observed similar results when investigating the bioremediation of PAH-contaminated soil from a manufacturing gas plant. Incubation of soil with the indigenous microflora (for 3 months), at different temperatures, soil moistures or nutrient conditions, did not result in a reduction in PAH concentration. Augmentation of soils with organisms known to be capable of degrading PAHs did not stimulate PAHs loss in the manufacturing gas plant soil. Erickson et al. (1993) concluded that the failure to observe PAH loss in any of the soils may be due to the toxicity of the soil or the poor availability of the PAHs to the soil microorganisms. However, when naphthalene or phenanthrene was spiked into augmented or non-augmented soils, rapid degradation of these compounds occurred. It also appeared that background naphthalene and phenanthrene from the contaminated soil remained undegraded at the conclusion of the incubation period. These results led to the conclusion that the soils were not toxic to the indigenous or augmented microorganisms, however, the PAHs within the soil were bound in such a way that made them unavailable for degradation.

The decrease in bioavailability, as a function of time, is often referred to ageing or weathering which may occur due to the compounds incorporation into organic matter by chemical oxidation, diffusion into small soil pores, absorption onto organic matter or by sequestering into non-aqueous phase liquids (NAPL) with high resistance towards NAPL–water mass transfer. The term 'ageing' does not include reactions that alter the structure of the molecule such as polymerisation or covalent binding to humic substances (Hatzinger and Alexander, 1995). The effect of ageing on contaminant bioavailability and risk has been reviewed in Chapters 2, 3, 4 and 19.

23.2 PAH sequestration and ageing

Following entry into the soil environment, a number of physical and chemical processes can influence the fate and dynamics of PAHs. PAH sequestration via association, retention or sorption mechanisms with solid matrices in soils and sediments are important processes that affect bioavailability. PAHs may bind to soil or sediment components through hydrophobic partitioning or via chemical or physical bond formation. Non-polar, hydrophobic compounds, such as PAHs, are usually retained on the organic components of the soil including condensed humic material or soot particles (see Chapters 2 and 10).

With increasing contaminant–soil residence time, diffusional and diagenetic processes affect the retention of organic compounds in a process termed ageing. PAHs may be incorporated into natural organic matter, diffuse into nanopores or absorb into organic matter (Fig. 23.2). The concentration and composition of soil organic mater will significantly influence the sequestration of hydrophobic organic contaminants such as PAHs. Bogan and Trbovic (2003) suggested that the susceptibility of PAHs to chemical oxidation was a function of total organic carbon (TOC) in soils containing greater than 5% TOC whereas in low TOC soils, susceptibility to oxidation was dependent on soil porosity. In addition, the importance of these two sequestration mechanisms changed with increasing contaminant–soil residence time: porosity-mediated sequestration of PAHs gained greater importance with increasing time (Bogan and Trbovic, 2003). Nam and Alexander (1998) investigated the importance of nanopore

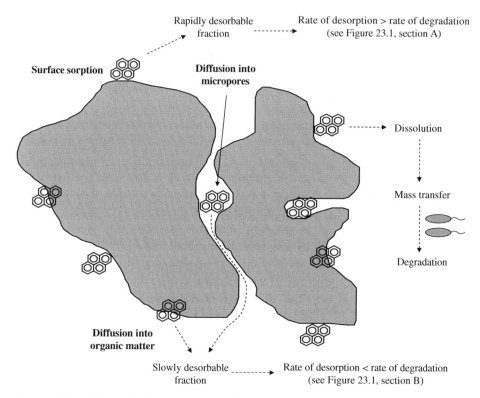

Figure 23.2. *Schematic diagram representing the sorption–desorption of PAHs in soil/ sediment matrices.*

hydrophobicity on the bioavailability and subsequent degradation of phenanthrene. During experiments with silica, 3-aminopropyl-bonded or diatomite beads with varying pore sizes (2.4 nm–5.4 µm), between 10 and 99% of phenanthrene could be sorbed after 15 h; however, 48–100% of the sorbed PAH could be desorbed after 120 h and rapidly biodegraded. Conversely, limited phenanthrene degradation (<7%) was observed following sorption of phenanthrene on polystyrene beads with a pore size of <400 nm. Nam and Alexander (1998) concluded that contaminant bioavailability can be significantly reduced by sequestration of PAHs into hydrophobic nanopores.

In addition, physical, chemical and biological processes may cause changes in the soil sorptive matrix which influences the sequestration of PAHs. For example, the change in aromaticity of soil organic matter over time results in greater sorption capacity for hydrophobic organic contaminants. Gauthier et al. (1987) correlated the sorptive uptake of PAHs with the degree of aromaticity in soil humic material at normalised soil organic carbon levels. In these experiments, partitioning coefficients for the binding of pyrene to 14 different humic and fulvic acids varied by up to a factor of 10. From these findings, Gauthier et al. (1987) suggested that when modelling the fate and transport of hydrophobic organic contaminants in the environment, the structure and composition of the humic material should be taken into consideration.

The ability of soils to release (desorb) contaminants controls its susceptibility to microbial degradation thereby influencing the effectiveness of the bioremediation process. Therefore desorption is considered to be an important process controlling contaminant persistence in soil. Similar to the degradation kinetics outlined in Figure 23.1, kinetics for the desorption of PAHs from contaminated matrices is biphasic: an initial rapidly desorbable phase is followed by a slowly desorbable phase. Microbial activity is an important parameter controlling desorption kinetics during the initial desorption phase. Removal of PAHs via microbial catabolic activity will steepen concentration gradients between soil/sediment matrices and soil solution/water phases thereby accelerating PAH desorption (Cornelissen et al., 1998). Slow desorption may occur via diffusion through organic matter or diffusion through and along micropore walls (Pignatello and Xing, 1996; Wu and Gschwend, 1986) although the mechanisms controlling diffusion have not been fully elucidated. An unavailable PAH fraction may also exist in contaminated soils and sediments: this residual fraction may occur as a result of incorporation of sorbed contaminants into the soil/sediment organic matrix.

23.3 Determination of contaminant bioavailability

Numerous researchers have demonstrated a close link between bioavailability and the rate of biodegradation of organic contaminants by microorganisms. This suggests that all factors that control the binding and release of contaminants are important for enhancing the effectiveness of the bioremediation process. Assuming that microorganisms are present in contaminated soil with the capacity to degrade the target contaminants, the major limiting factor for the successful application of bioremediation is contaminant bioavailability. In order to ensure the successful application of bioremediation strategies, a contaminant bioavailability characterisation test is required. Historically, only total contaminant concentrations were determined, based on exhaustive extraction methods, prior to the selection of remediation strategy which is often not relevant to processes involving living organisms. Alternatively, bench scale treatability trials were used to determine the ability of microorganisms to degrade target contaminants, however, these methodologies are time consuming and laborious. Recently, a number of methodologies have been developed for

Table 23.1. *Methodologies used for assessing PAH bioavailability for bioremediation.*

Bioavailability assay	Method	Comments	References
Non-exhaustive extraction			
Cosolvents	Soils extracted with low molecular weight primary alcohols (methanol, ethanol, propanol, butanol), acetone or toluene or dilute solutions of the aforementioned solvents. Soil/solution ratios of up to 1:20 may be used with extraction times of up to 24 h	Good correlation between biodegradable and bioavailable PAHs in spiked soils for low molecular weight PAHs ($\leq$ three-ring PAHs). Over- and under-estimation of PAH bioavailability in contaminated soils	Thiele-Bruhn and Brümmer (2004), Macleod and Semple (2003), Northcott and Jones (2001), Chung and Alexander (1998), Chung and Alexander (1999), Liste and Alexander (2002), Breedveld and Karlsen (2000), Kelsey et al. (1997), Reid et al. (2000), Reid et al. (2004), Juhasz et al. (2005), Macleod and Semple (2000), Oleszczuk and Baran (2004)
Surfactants	Triton X-100, Genapol UDD88 and Synperonic LF/RA30 have been used at soil/solution ratios of up to 1:20 with extraction times of up to 172 h with surfactant concentrations of up to 240 × CMC	Triton X-100 overestimated the PAH bioavailable fraction. Good correlation between biodegradable and bioavailable PAHs for Genapol UDD88 and Synperonic LF/RA30	Cuypers et al. (2002), Thiele-Bruhn and Brümmer (2004)
Hydroxypropyl-β-cyclodextrin	Soils extracted with 40 mM HPCD solution for 20 h with a soil/solution ratio of 1:20	Good correlation between biodegradable and bioavailable PAHs for three- and four-ring compounds. Overestimates five- and six-ring PAH bioavailability	Cuypers et al. (2002), Reid et al. (2000), Reid et al. (2004), Juhasz et al. (2005)
Supercritical fluid extraction	CO_2 used at 200 bar at 50°C for up to 200 min. Up to 4 g of soil is used with a CO_2 flow rate of 1 ml min^{-1}	Good correlation between biodegradable and bioavailable PAHs for low molecular weight compounds ($\leq$ three-ring PAHs). Overestimates high molecular weight PAH bioavailability ($\geq$ four-ring compounds)	Hawthorn and Grabanski (2000), Hawthorn et al. (2001), Hawthorn et al. (2002), Szolar et al. (2004)
Persulphate oxidation	Potassium persulphate added to achieve a persulphate to organic matter ratio of 12 g g^{-1} equivalent to an aqueous persulphate concentration of 35.7 mg l^{-1}. Slurries are incubated at 70°C for 3 h	Good correlation between biodegradable and bioavailable PAHs for three- and four-ring compounds. Overestimates five- and six-ring PAH bioavailability	Cuypers et al. (2000), Cuypers et al. (2001), Juhasz et al. (2005)
Aqueous desorption	Desorption experiments conducted with Tenax TA, Amberlite XAD-2, Amberlite carboxylic acid cation exchange resin, biobeads polyacrylic resin (SM7) or biobeads divinyl benzene resin (SM2). Resin/soil ratios of up to 1:8 have been used with extraction times of up to 16 weeks.	Good correlation between biodegradable and bioavailable PAHs for three- and four-ring compounds. Overestimates five- and six-ring PAH bioavailability	Grosser et al. (2000), Lei et al. (2004), Carmichael et al. (1997), Huesemann et al. (2004), Cornelissen et al. (1998), Cuypers et al. (2001), Braida et al. (2004)

assessing the bioavailability of soil-bound contaminants for bioremediation potential (Table 23.1). The majority of work has focussed on the assessment of non-exhaustive extraction methods using mild extractants such as low molecular weight primary alcohols, hydroxypropyl-β-cyclodextrin (HPCD), surfactants and supercritical fluid extraction. In addition, aqueous desorption tests utilising porous polymers (e.g. XAD resin, Tenax TA) and persulphate oxidation are alternative methods that have been developed for the assessment of contaminant bioavailability for bioremediation. These methodologies have been conducted in conjunction with biodegradation assays in order to correlate PAH bioavailability with PAH biodegradability. The following sections of this chapter will discuss the development and application of these methodologies for the assessment of PAH bioavailability.

23.4 Non-exhaustive extractants

23.4.1 Extraction with low molecular weight primary alcohols

Non-exhaustive extraction methods using low molecular weight primary alcohols have been utilised for estimating the fraction of soil-bound PAHs available for microbial degradation. These methodologies are simplistic and rely on the cosolvence power of the selected alcohol. Contaminated soil is shaken with methanol, ethanol, propanol or butanol (at soil solution ratios of up to 1:20) for a predetermined period of time. Following centrifugation, the concentration of PAHs in the alcohol extract is determined (the desorbable or bioavailable fraction) or the soil is recovered, dried and residual soil-bound PAHs determined (the sequestered or non-bioavailable fraction).

A number of researchers have suggested that mild butanol extraction may be an appropriate means for predicting PAH bioavailability (Kelsey et al., 1997; Breedveld and Karlsen, 2000; Liste and Alexander, 2002). Liste and Alexander (2002) and Kelsey et al. (1997) found good relationships between PAH desorption using butanol and PAH bioavailability (using microbial degradation and earthworm uptake assays) in aged spiked soil containing single PAH compounds. Conversely, in a recent study by Macleod and Semple (2003) using pyrene-spiked soils, butanol extraction overestimated the bioavailability of pyrene compared to bacterial mineralisation measured over a 24-week period. The above studies exemplify the variable results present in the literature of bioavailability research using spiked soils and low molecular weight primary alcohols. Spiking and ageing of soil under laboratory conditions and time frames may not be long enough to truly reflect conditions in the field. In addition, contaminant bioavailability may be influenced by the presence of co-contaminants which would not be reflected in single compound spiked soil studies. As a result, spiked soil studies (determining contaminant availability and sequestration) often fail to mimic conditions found in 'real' contaminated soils (Hawthorn et al., 2002).

Using aged creosote-contaminated soil, Breedveld and Karlsen (2000) estimated the bioavailability of PAHs using biodegradability and desorption methods. While butanol extraction was able to provide a good qualitative estimate of the potentially biodegradable PAH fraction, quantitatively it under- or over-estimated PAH bioavailability depending on soil type. In soil with a high organic carbon content (16.6%), PAH bioavailability underestimated three- and five/six-ring PAH biodegradability by 52 and 36%, respectively, while in soil containing 4.2% total organic carbon, PAH biodegradability was over-estimated by up to 300%.

Juhasz et al. (2005) found a poor relationship between propanol or butanol extractable PAHs and the biodegradable PAH fraction in an aged creosote-contaminated soil. Increasing the cosolvent carbon chain length and percentage composition resulted in increased amounts of PAHs desorbed from the soil. This was expected as increasing the cosolvent percentage composition increases the cosolvency power while increasing the carbon chain length decreases the influence of the polar group on the partitioning of PAHs (Rao et al., 1985, 1990). However, even when using longer carbon chain length solvents, a poor correlation between bioavailable and biodegradable PAHs was observed in this study. Low molecular weight PAHs bioavailability was underestimated whereas high molecular weight PAH bioavailability was overestimated (Juhasz et al., 2005).

23.4.2 Extraction with hydroxypropyl-β-cyclodextrin

Reid et al. (2000) proposed the use of an aqueous-based extraction technique, utilising hydroxypropyl-β-cyclodextrin (HPCD), for estimating the PAH bioavailable fraction in soil. HPCD was chosen because of its high solubility, the prevalence of hydroxyl functional groups on the exterior of the torus and a hydrophobic organic cavity, making it possible to form an inclusion complex with PAHs (Reid et al., 2000). An optimised HPCD extraction method for assessing PAH bioavailability in spiked soil involves extracting contaminated soil (1.25 g) with HPCD solution (40 mM) for a 20 h-period (at room temperature) (Reid et al., 2000). Following extraction, samples are centrifuged, the supernatants discarded and the soil recovered for PAH determination (the sequestered or non-bioavailable fraction).

This non-exhaustive extraction method was tested against the PAH catabolic activity of a *Pseudomonas* sp. in phenanthrene-spiked soils. Labelled and unlabelled phenanthrene was spiked into dystrochrept soil and the bioavailability assessed using HPCD extraction, butanol extraction and phenanthrene mineralisation.

HPCD extraction was in good agreement with phenanthrene mineralisation (slope 0.977; intercept 0.162; R^2 = 0.964), however, butanol extraction overestimated phenanthrene bioavailability by >60% (Reid et al., 2000).

Cuypers et al. (2002) and Juhasz et al. (2005) used the HPCD extraction method for estimating the bioavailable PAH fraction in contaminated sediments and creosote-contaminated soil. In Assendelft and Petroleum Harbour sediments, biodegradation was greatest for the three-ring compounds and progressively decreased as the molecular weight of the PAHs increased (Cuypers et al., 2002). Comparison of the residual PAH concentration after biodegradation (21 days) and HPCD extraction (172 h) showed that the extent of three- and four-ring biodegradation could be predicted using this method, however, five- and six-ring PAH biodegradation was overestimated. Similar results were obtained by Juhasz et al. (2005) when bioavailability assessment was run in parallel with biodegradation experiments using creosote-contaminated soil. HPCD extraction was able to predict three- and four-ring PAH bioavailability, however, the bioavailability of five-ring PAHs was overestimated.

23.4.3 Extraction with surfactants

Non-exhaustive extraction utilising surfactants (e.g. Triton X-100) have also been trialed for their ability to predict PAH bioavailability. Using surfactant concentrations of up to 240 times their critical micelle concentration (CMC), soils are shaken for a predetermined period of time at soil/solution ratios of up to 1:20. Following extraction, the soil is recovered, dried and residual soil-bound PAHs are determined (the sequestered or non-bioavailable fraction). In a recent study by Cuypers et al. (2002), PAH bioavailability in contaminated sediment was estimated following extraction with Triton X-100 with results compared to HPCD extraction, aqueous desorption tests using Tenax TA and biodegradation studies. Initial studies demonstrated that increasing the Triton X-100 dose (up to 240 times the CMC) resulted in enhanced extraction of PAHs from contaminated sediments. At high Triton X-100 doses, however, the amount of PAH extracted plateaued, resulting in optimum Triton X-100 extraction concentrations of 155–190 times CMC depending on the sediment used.

Extraction of PAHs with Triton X-100 was rapid and residual PAH concentrations were low compared to HPCD extraction and aqueous desorption tests. It was observed that Triton X-100 extraction of PAHs from contaminated sediments was biphasic – an initial rapid extraction phase was followed by a relatively slow extraction phase. When compared to biodegradation assays, extraction with Triton X-100 overestimated the PAH biodegradable fraction. Cuypers et al. (2002) proposed that the overestimation of PAH bioavailability resulted from the ability of Triton X-100 to extract both readily and poorly available PAHs. This resulted from interactions of Triton X-100 with the solid matrix which enhanced the desorption behaviour of PAHs. The slowly desorbable PAH fraction may be sorbed in flexible pores in natural organic matter (Van Noort et al., 2000) and these pores may plasticise as a result of interactions with Triton X-100. The decreased affinity of PAHs for the organic matter following plasticisation leads to enhanced desorption of poorly bioavailable PAHs (Cuypers et al., 2002). While extraction of PAH-contaminated sediments with Triton X-100 was a poor indicator of PAH bioavailability, HPCD extraction was in good agreement with biodegradation results.

In another study investigating the applicability of surfactant extraction for estimating PAH biodegradability, Thiele-Bruhn and Brümmer (2004) trialed two non-ionic surfactants, Genapol UDD 88 and Synperonic LF/RA 30, on contaminated soils collected from former gasworks and coking plants. Results were compared to biodegradation data obtained from pot experiments conducted under outdoor conditions over a 74- or 168-week treatment period. PAH degradation followed coupled first-order exponential functions representing a readily degradable mobile phase and a back diffusion-controlled mobilisable fraction (Thiele-Bruhn and Brümmer, 2004). Extraction kinetics for PAHs from contaminated soils by Genapol UDD 88 and Synperonic LF/RA 30 resembled the biphasic PAH biodegradation kinetics, however, the time frames required to exhaust the back diffusion-controlled mobilisable fraction were considerably shorter (80 h). When compared to biodegradation data for 30 soils, extraction with Genapol UDD 88 or Synperonic LF/RA 30 accounted for 106–147% of the PAH fraction which was removed through biodegradation. Although a good correlation was found between total surfactant extractable and PAH degradable concentrations, correlation coefficients varied for low and high molecular weight PAHs. For example, using Genapol UDD 88, correlation coefficients for log–log transformed data ranged from 0.62 for six-ring PAHs to 0.80 for three-ring compounds (Thiele-Bruhn and Brümmer, 2004). Consequently, it was proposed that PAHs should be separated into groups according to their degree of condensation prior to the determination of relatedness between surfactant extractability (bioavailability) and PAH biodegradability. Thiele-Bruhn and Brümmer (2004) concluded from this research that the fraction of PAHs extracted by surfactants characterised the mobile and mobilisable PAH fractions in soils and non-exhaustive extraction using surfactants is a suitable methodology for estimating the PAH fraction degradable through bioremediation.

23.4.4 Supercritical fluid extraction

Supercritical fluid extraction utilising CO_2 has also been applied for estimating PAH bioavailability for biodegradation (Hawthorn and Grabanski, 2000;

Hawthorn et al., 2001, 2002). CO_2 was used in these studies since its polarity is similar to biological lipids and the solubility of PAHs in water and supercritical CO_2 are similar (Hawthorn et al., 2001). In studies by Hawthorn et al. (2001), short-term mild supercritical fluid extraction was proposed to estimate the readily or fast-desorbing PAH fraction. PAH extraction experiments were conducted using CO_2 at 200 b/50°C for up to 200 min and the results compared to long-term non-equilibrium studies using XAD-2 (for up to 120 days) and land treatment bioremediation studies (for up to 343 days).

Although the time frames for estimating PAH bioavailable fractions were significantly different (200 min compared to 120 days), PAH desorption profiles were similar when using CO_2 supercritical fluid extraction and XAD-2 methodologies. Both methods showed the decrease in PAH bioavailability with increasing molecular weight and decreasing PAH bioavailability with increasing land treatment time. When compared to biodegradation data, both methodologies gave good predictions for PAH removal, in particular for the lower molecular weight PAH fraction. For the four-ring compounds (fluoranthene, pyrene, benz[a]anthracene, chrysene), both methods overestimated PAH release with correlation coefficients of 0.67 and 0.80 for XAD-2 and CO_2 supercritical fluid extraction methods for PAH biodegradation at 343 days. Whilst good correlations were determined for bioavailable and biodegradable PAH fractions, Hawthorn et al. (2001) found that varying the treatment time used for PAH desorption significantly influence these correlations. For example, longer XAD-2 desorption times overestimated PAH removal while a-40 min supercritical fluid extraction was optimal for correlation with 343 day biodegradation data. This indicates that method optimisation may be required in order to accurately estimate bioavailable fractions. In further studies, Hawthorn et al. (2002) determined that PAH release rates using CO_2 supercritical fluid extraction did not correlate with soil characteristics such as PAH concentration, C, N or S composition or organic carbon fractions indicating that PAH bioavailability cannot be estimated on the basis of sample matrix composition.

23.4.5 Persulphate oxidation

Persulphate oxidation has been employed for assessing PAH bioavailability in contaminated soils and sediments. When heated, persulphate decomposes to form sulphate radicals which can react with organic substances. As PAH sequestration involves adsorption onto, and diffusion into organic matter, reactions with persulphate removes expanded organic matter and readily available PAHs sorbed in it. This available PAH fraction could be removed using persulphate oxidation which selectively oxidises soft expanded organic matter (Weber et al., 1992; Young and Weber, 1995). As a result, the amount of PAHs that remain in the soil after persulphate oxidation may be a measure of the non-bioavailable or non-biodegradable fraction (Cuypers et al., 2000). In this method, contaminated soil (5 g) is mixed with potassium persulphate and deionised water to achieve a persulphate to organic matter ratio of $12 \, g \, g^{-1}$, equivalent to an aqueous persulphate concentration of $35.7 \, mg \, l^{-1}$. The soil-persulphate slurry is incubated at 70°C, shaking end to end for 3 h after which slurries are centrifuged, supernatants discarded and soils recovered for PAH extraction (the sequestered or non-bioavailable fraction).

Cuypers et al. (2000) utilised the persulphate oxidation method for estimating the PAH bioavailable fraction in 14 contaminated soils and sediments. The extent of PAH removal by persulphate oxidation was in good agreement with the extent of two-, three- and four-ring PAH removal by biodegradation. However, persulphate oxidation of five- and six-ring PAHs significantly overestimated the biodegradation potential of these compounds. Regression lines for residual PAHs did not differ significantly (two-sided t-test, 95%) from the 1:1 correlation line, however, the good correlation for five- and six-ring PAHs was influenced by the low PAH removal by both persulphate oxidation and biodegradation. Variability in the low concentration range of these compounds only had a minor effect on the overall fit of residual PAHs (Cuypers et al., 2000).

In experiments conducted by Juhasz et al. (2005) using creosote-contaminated soil, persulphate oxidation was able to predict three- and four-ring PAH biodegradability, however, the biodegradability of five-ring PAHs was overestimated. In biopile experiments where indigenous PAH degrading microorganisms were stimulated through the addition of nutrients, no significant decrease in benzo[*a*]pyrene or benzo[*b+k*]fluoranthene concentration was observed. However, approximately 36% ($36 \, mg \, kg^{-1}$) and 22% ($78 \, mg \, kg^{-1}$) of benzo[*a*]pyrene and benzo[*b+k*]fluoranthene were estimated to be bioavailable after persulphate treatment, respectively. Juhasz et al. (2005) suggested that the inability of five-ring PAH compounds to be removed during bioremediation may stem from biological factors, such as high activation energies, unfavourable Gibbs free energy or slow transport across the cell membrane rather than bioavailability issues. In addition, bioavailability estimates for high molecular weight PAHs, such as benzo[*a*]pyrene, using persulphate oxidation may not be reflective of the biodegradable fraction but may represent the fraction of PAHs potentially available to impact on ecological receptors.

In an attempt to determine whether the biodegradable and bioavailable PAH fractions in the creosote-contaminated soil were similar, Juhasz et al. (2005) performed persulphate oxidation on treated soil following bioremediation. Only a small amount of 'biodegradable' PAHs (four-ring compounds) could be further removed following persulphate oxidation. Juhasz et al. (2005) concluded that this may represent the release of a slowly desorbing PAH fraction from the soil matrix. Presumably, degradation of the slowly desorbing fluoranthene, pyrene, benz[*a*]anthracene and chrysene fraction would occur over time provided viable PAH-degrading microorganisms were present, environmental conditions were suitable for bioremediation and the concentration of bioavailable PAHs was above threshold values required to synthesise enzymes for PAH degradation.

23.4.6 Aqueous desorption tests

Another methodology receiving increased attention for the assessment of PAH bioavailability involves aqueous desorption tests (Carmichael et al., 1997; Cornelissen et al., 1998; Huesemann et al., 2004). These tests utilise porous polymers (e.g. XAD resin, Tenax TA) which act as a contaminant sink for the determination of contaminant desorption rates in soil slurries under abiotic conditions. Amberlite XAD resins have been widely used in desorption experiments due to their ability to sorb organic compounds through hydrophobic bonding (Lei et al., 2004). Tenax TA, a porous polymer based on 2,6-diphenyl-*p*-phenylene oxide, has been applied successfully for the extraction of PAHs from contaminated sediments (Cuypers et al., 2002). Desorption rates can then be compared to biodegradation rates to determine whether mass transfer limitations (bioavailability) or microbial factors are responsible for contaminant biodegradation kinetics. Desorption experiments are prepared by mixing the porous polymer with contaminated soil/ a slurry is then formed through the addition of milli-Q-water. During the mixing process, the porous polymer absorbs hydrophobic compounds that are released into the aqueous phase from the soil matrix by either desorption or dissolution (Huesemann et al., 2004). At the completion of desorption experiments, the polymer is recovered and extracted to determine the released or bioavailable fraction.

In a study by Lei et al. (2004) the bioavailability of PAHs in contaminated sediment was assessed using XAD-2 assisted desorption in conjunction with biodegradation experiment. Contaminated sediment was obtained from the East River, New York, and contained approximately $750 \, mg \, kg^{-1}$ of the 16 USEPA priority PAHs. Biodegradation experiments were conducted using the sediment as the inoculum and desorption experiments were prepared using XAD/ sediment ratios of 1, 2, 4 and 8. After 24 weeks, constant residual PAH levels were reached in biodegradation experiments. Degradation of 2–4 ring PAHs was observed, however, only limited degradation of five-ring compounds (benzo-[*b*+*k*]fluoranthene, benzo[*a*]pyrene, benzo[*e*]pyrene) occurred. No significant decrease in the concentration of six-ring PAHs (indeno[1,2,3-*c*,*d*]pyrene, benzo-[*g*,*h*,*i*]perylene) was observed over the 24-week incubation period. In contrast, desorption rates for two-, three- and most four-ring compounds approached zero after 2 weeks whereas five- and six-ring compounds desorbed from the sediment for up to 8 weeks. When residual PAH concentrations following biodegradation and XAD-2 desorption were compared, similar sediment-bound levels were observed for two-, three- and four-ring PAHs. Lei et al. (2004) suggested that the XAD-2 assisted desorption methods was a promising assays for determining the bioavailability of these compounds. However, for the higher molecular weight PAH fraction ($\geq$ five-rings), considerable more PAH was desorbed from the sediment compared to concentrations that were degraded. It was proposed that these compounds may be bioavailable in terms of the accessibility to the aqueous phase but not susceptible to biodegradation under the environmental conditions provided (Lei et al., 2004).

In a similar study, Cornelissen et al. (1998) investigated the relationship between PAH biodegradation and PAH bioavailability using Tenax TA as the matrix for aqueous desorption tests. Remediation of PAH-contaminated sediments was undertaken in bioreactors (4 and $30\,m^3$) or in a landfarm until PAH degradation plateaued, presumably due to the depletion of bioavailable contaminants. Desorption studies were preformed with contaminated sediment slurries and Tenax TA (60–80 mesh) at a Tenax TA/sediment ratio of 0.6:1.0. The Tenax TA was refreshed periodically and PAH concentration determined following extraction with hexane.

The extent of PAH bioremediation varied depending on the sediment treated. In Petrol Harbour sediment A, all PAHs (two- to six-ring compounds) were significantly degraded over the 4-month treatment period, while only the lower molecular weight PAHs (naphthalene, acenaphthene, fluorene, phenanthrene, anthracene, fluoranthene) were degraded in Petrol Harbour sediment B. In Wemeldinge sediment following landfarming treatment, significant degradation was observed for all PAHs with the exception of some five- (benzo[k]fluoranthene, benzo[a]pyrene, dibenz[a,h]anthracene) and six-ring compounds (benzo[g,h,i]-perylene, indeno[1,2,3-c,d]pyrene).

Aqueous desorption tests using Tenax TA identified a rapidly desorbing and slow desorbing PAH fractions in the contaminated sediments. Rate constants for the rapidly desorbing fractions were 100–3,000 times greater than rate constants for slowly desorbing fractions. Following biodegradation, however, the rapidly desorbing fractions for biodegradable PAHs were significantly smaller while for compounds such as benzo[g,h,i]perylene (a recalcitrant PAH), the rapidly desorbing fraction remained essentially the same. Cornelissen et al. (1998) suggested that aqueous desorption tests utilising Tenax TA was a suitable method for predicting the extent of PAH biodegradation without desorptional limitations provided microorganisms with the necessary catabolic potential were present. For high molecular weight PAHs ($\geq$ five-ring compounds), the magnitude of the rapidly desorbing fraction meant that microbial factors and not bioavailability issues are probably responsible for their persistence.

Furthermore, for biodegradable PAHs ($\leq$four-ring compounds), the persistence of a residual fraction following bioremediation is probable due to small rate constants for the slowly desorbing fractions (Cornelissen et al., 1998; Huesemann et al., 2004).

23.5 Predicting PAH biodegradation using bioavailability assays

The above examples demonstrate that a wide variety of methodologies have been employed, to varying degrees, for predicting PAH bioavailability. A common thread amongst this research is the ability to predict three- and four-ring PAH bioavailability but the overestimation of high molecular weight ($\geq$ five-ring compounds) PAH bioavailability. In a number of studies (Cuypers et al., 2000; Juhasz et al., 2005; Cornelissen et al., 1998; Huesemann et al., 2002), microbial degradation of high molecular weight PAHs ($\geq$ five-ring compounds) was not observed over extensive incubation periods, although bioavailability assessment suggested that these compounds were available for degradation. In the study of Juhasz et al. (2005), an attempt to stimulate the degradation of five-ring PAHs was initiated via bioaugmentation of a PAH-degrading *Mycobacterium* sp. Although *Mycobacterium* sp. strain 1B had the metabolic capacity to degrade high molecular weight PAHs such as benzo[*a*]pyrene in liquid culture and in spiked soils, no significant degradation of these compounds was observed in creosote-contaminated soil. This suggests that the inability of $\geq$ five-ring PAH compounds to be removed during bioremediation may stem from biological factors (e.g. high activation energies, unfavourable Gibbs free energy) (Bonten, 2001), rather than bioavailability issues (Cuypers et al., 2000; Juhasz et al., 2005; Cornelissen et al., 1998; Huesemann et al., 2002).

To gain a better understanding of the capacity of chemical bioavailability methodologies to predict individual PAH biodegradability, the ratio of bioavailable to biodegradable PAH concentrations were plotted against PAH octanol–water partitioning coefficients (K_{ow}) (Fig. 23.3). A ratio approaching 1 indicates the ability to predict PAH biodegradation based on bioavailability assessment. A ratio <1 indicates that bioavailability assessment under-predicts PAH biodegradability while a ratio >1 indicates that bioavailability assessment over-predicts PAH biodegradability. The data presented in Figure 23.3 from Juhasz et al. (2005) and Hawthorn et al. (2001) for four different bioavailability methodologies demonstrates that persulphate oxidation, HPCD extraction, XAD-2 desorption and mild CO_2 supercritical fluid extraction are good predictors of PAH bioavailability for compounds with K_{ow} values <6.0. For PAHs with K_{ow} values ≥ 6.0, bioavailability estimates using these methods may not reflect the biodegradable fraction but may represent the fraction of PAHs potentially available to impact on ecological receptors.

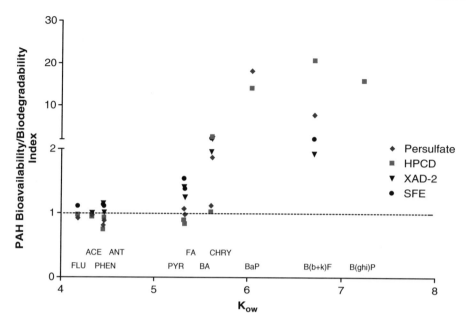

Figure 23.3. *Relationship between PAH octanol–water partitioning coefficient and the ability to predict PAH biodegradability using bioavailability (♦, persulphate oxidation; ■, HPCD extraction (Juhasz et al., 2005); ▼, XAD-2; ●, mild CO₂ supercritical fluid extraction (Hawthorn et al., 2001)) methodologies. The PAH bioavailability/biodegradability ratio was calculated by dividing the bioavailable PAH concentration by the biodegradable PAH concentration. The PAH bioavailability/ biodegradability ratio was then plotted against PAH octanol–water partitioning coefficient. A ratio of 1 indicates the ability to predict PAH biodegradation based on bioavailability assessment. A ratio <1 indicates that bioavailability assessment under-predicts PAH biodegradability while a ratio >1 indicates that bioavailability assessment over-predicts PAH biodegradability. Some PAH bioavailability/biodegradability ratios were unable to be calculated due to zero biodegradation or bioavailability determination. PAH nomenclature: FLU, fluorene; ACE, acenaphthene; PHEN, phenanthrene; ANT, anthracene; PYR, pyrene; FA, fluoranthene; BA, benz[a]anthracene; CHRY, chrysene; BaP, benzo[a]pyrene; B[b+k]F, benzo[b+k]fluoranthene; B[ghi]P, benzo[g,h,i]perylene.*

23.6 Conclusion

A variety of methodologies utilising non-exhaustive extraction techniques, chemical oxidation and aqueous desorption tests have been developed for estimating the bioavailability of environmental contaminants for bioremediation purposes. Whilst the biodegradability of three- and four-ring compounds can be predicted (based on bioavailability) using these methodologies, generally the bioremediation potential of high molecular weight PAHs ($\geq$ five-ring compounds) is overestimated. In order to predict the behaviour of PAHs in soil, the mechanisms associated with soil–PAH interactions and PAH–microbial interactions need to be understood. Our knowledge of PAH sorption–desorption mechanisms and processes leading to PAH ageing and sequestration is ever increasing, however, a dearth of information still exists on mechanisms involved in high molecular weight PAH degradation. Conceivably, the overestimation of high molecular weight PAH bioavailability using the aforementioned methodologies may stem from biological issues associated with the inability of microorganisms to degrade these compounds when present in a bioavailable form. In order to address this shortfall, a greater understanding of high molecular weight PAH biodegradation is required including PAH transport mechanisms across the cell wall, inducers for transport or degradative enzymes and metabolic pathways of high molecular weight PAHs.

Acknowledgement

This research was supported by the Centre for Environmental Risk Assessment and Remediation, University of South Australia.

References

Bogan, B.W., Trbovic, V., 2003. Effect of sequestration on PAH degradability with Fenton's reagent: roles of total organic carbon, humin and soil porosity. J. Hazard. Mater. 100, 285–300.

Bonten, L.T.C., 2001. Improving bioremediation of PAH contaminated soil by thermal pre-treatment. Ph.D. Thesis, Wangeningen University, The Netherlands.

Bosma, T.N.P., Middeldorp, P.J.M., Schraa, G., Zehnder, A.J.B., 1997. Mass transfer limitation of biotransformations: quantifying bioavailability. Environ. Sci. Technol. 31, 248–252.

Braida, W.J., White, J.C., Pignatello, J.J., 2004. Indices for bioavailability and biotransformation potential of contaminants in soil. Environ. Toxicol. Chem. 23, 1585–1591.

Breedveld, G.D., Karlsen, D.A., 2000. Estimating the availability of polycyclic aromatic hydrocarbons for bioremediation of creosote-contaminated soils. Appl. Microbiol. Biotechnol. 54, 255–261.

Carmichael, L.M., Christman, R.F., Pfaender, F.K., 1997. Desorption and mineralisation kinetics of phenanthrene and chrysene in contaminated soils. Environ. Sci. Technol. 31, 126–132.

Cerniglia, C.E., 1992. Biodegradation of polycyclic aromatic hydrocarbons. Biodegradation 3, 351–368.

Chung, N., Alexander, M., 1998. Differences in sequestration and bioavailability of organic compounds in aged and dissimilar soils. Environ. Sci. Technol. 32, 855–860.

Chung, N., Alexander, M., 1999. Effect of concentration on sequestration and bioavailability of two polycyclic aromatic hydrocarbons. Environ. Sci. Technol. 33, 3605–3608.

Cornelissen, G., Rjgterink, H., Ferdinandy, M.M.A., Van Noort, P.C.M., 1998. Rapidly desorbing fractions of PAHs in contaminated sediments as a predictor of the extent of bioremediation. Environ. Sci. Technol. 32, 966–970.

Cuypers, C., Grotenhuis, T., Joziasse, J., Rulkens, W., 2000. Rapid persulfate oxidation predicts PAH bioavailability in soils and sediments. Environ. Sci. Technol. 34, 2057–2063.

Cuypers, C., Clemens, R., Grotenhuis, T., Rulkens, W., 2001. Prediction of petroleum hydrocarbon bioavailability in contaminated soils and sediments. Soil Sediment Contam. 10, 459–482.

Cuypers, C., Pancras, T., Grotenhuis, T., Rulkens, W., 2002. The estimation of PAH bioavailability in contaminated sediments using hydroxypropyl-β-cyclodextrin and Triton X-100 extraction techniques. Chemosphere 46, 1235–1245.

Erickson, D.C., Loehr, R.C., Neuhauser, E.F., 1993. PAH loss during bioremediation of manufactured gas plant site soil. Water Res. 27, 911–919.

Gauthier, T.D., Seitz, W.R., Grant, C.L., 1987. Effects of structural and compositional variations of dissolved humic materials on pyrene K_{oc} values. Environ. Sci. Technol. 21, 243–248.

Grosser, R.J., Friedrich, M., Ward, D.W., Inskeep, W.M., 2000. Effect of model sorptive phases on phenanthrene biodegradation: different enrichment conditions influence bioavailability and selection of phenanthrene degrading isolates. Appl. Environ. Microbiol. 66, 2695–2702.

Hatzinger, P.B., Alexander, M., 1995. Effect of aging of chemicals in soil on their biodegradability and extractability. Environ. Sci. Technol. 29, 537–545.

Hawthorn, S.B., Grabanski, C.B., 2000. Correlating selective supercritical fluid extraction with bioremediation behaviour of PAHs in a field treatment plot. Environ. Sci. Technol. 34, 4103–4110.

Hawthorn, S.B., Poppendieck, D.G., Grabanski, C.B., Loehr, R.C., 2001. PAH release during water desorption, supercritical carbon dioxide extraction and field bioremediation. Environ. Sci. Technol. 35, 4577–4583.

Hawthorn, S.B., Poppendieck, D.G., Grabanski, C.B., Loehr, R.C., 2002. Comparing PAH bioavailability from manufactured gas plant soils and sediments with chemical and biological tests. 1. PAH release during water desorption and super critical carbon dioxide extraction. Environ. Sci. Technol. 36, 4795–4803.

Huesemann, M.H., Hausmann, T.S., Fortman, T.J., 2002. Microbial factors rather than bioavailability limit the rate and extent of PAH biodegradation in aged crude oil contaminated model soils. Bioremediat. J. 6, 321–336.

Huesemann, M.H., Hausmann, T.S., Fortman, T.J., 2004. Does bioavailability limit biodegradation? A comparison of hydrocarbon biodegradation and desorption rates in aged soil. Biodegradation 15, 261–274.

Juhasz, A.L., Megharaj, M., Naidu, R., 2000. Bioavailability: The major challenge (constraint) to bioremediation of organically contaminated soils. In: D.L. Wise, D.J. Trantolo, E.J. Cichon, H.I. Inyang and U. Stottmeister (Eds.), Remediation of Hazardous Waste Contaminated Soils, 2nd Ed., Vol. 1: Engineering Considerations and Remediation Strategies, Section 1–1: Engineering Issues in Waste Remediation. Marcel Dekker, New York, NY, pp. 217–241.

Juhasz, A.L., Waller, N., Stewart, R., 2005. Predicting the efficacy of PAH bioremediation in creosote-contaminated soil using bioavailability assays. Bioremediat. J. 9, 1–16.

Juhasz, A.L., Naidu, R., 2000. Bioremediation of high molecular weight polycyclic aromatic hydrocarbons: a review of the microbial degradation of benzo[a]pyrene. Int. Biodeterior. Biodegradation 45, 57–88.

Kanaly, R.A., Harayama, S., 2000. Biodegradation of high molecular weight polycyclic aromatic hydrocarbons by bacteria. J. Bacteriol. 182, 2059–2067.

Kelsey, J.W., Kottler, B.D., Alexander, M., 1997. Selective chemical extractants to predict bioavailability of soil-aged organic chemicals. Environ. Sci. Technol. 31, 214–217.

Lei, L., Suidan, M.T., Khodadoust, A.P., Tabak, H.H., 2004. Assessing the bioavailability of PAHs in field-contaminated sediment using XAD-2 assisted desorption. Environ. Sci. Technol. 38, 1786–1793.

Liste, H-H., Alexander, M., 2002. Butanol extraction to predict bioavailability of PAHs in soil. Chemosphere 46, 1011–1017.

Macleod, C.J.A., Semple, K.T., 2000. Influence of contact time on extractability and degradation of pyrene in soil. Environ. Sci. Technol. 34, 4952–4957.

Macleod, C.J.A., Semple, K.T., 2003. Sequential extraction of low concentrations of pyrene and formation of non-extractable residues in sterile and non-sterile soil. Soil Biol. Biochem. 35, 1443–1450.

Nam, K., Alexander, M., 1998. Role of nanoporosity and hydrophobicity in sequestration and bioavailability: tests with model solids. Environ. Sci. Technol. 32, 71–74.

Northcott, G.L., Jones, K.C., 2001. Partitioning, extractability and formation of nonextractable PAH residues in soil. 1. Compound differences in aging and sequestration. Environ. Sci. Technol. 35, 1103–1110.

Oleszczuk, P., Baran, S., 2004. The concentration of mild-extracted polycyclic aromatic hydrocarbons in sewage sludges. J. Environ. Sci. Health A Tox. Hazard. Subst. Environ. Eng. 39, 2799–2815.

Pignatello, J.J., Xing, B., 1996. Mechanism of slow sorption or organic chemicals to natural particles. Environ. Sci. Technol. 30, 1–11.

Rao, P.S.C., Hornsby, A.G., Kilcrease, D.P., Nkedi-Kizza, P., 1985. Sorption and transport of hydrophobic organic chemicals in aqueous and mixed solvent systems: model development and preliminary evaluation. J. Environ. Qual. 14, 376–383.

Rao, P.S.C., Lee, L.S., Pinal, R., 1990. Cosolvency and sorption of hydrophobic organic compounds. Environ. Sci. Technol. 24, 647–654.

Reid, B.J., Stokes, J.D., Jones, K.C., Semple, K.T., 2000. Nonexhaustive cyclodextrin-based extraction technique for the evaluation of PAH bioavailability. Environ. Sci. Technol. 34, 3174–3179.

Reid, B.J., Stokes, J.D., Jones, K.C., Semple, K.T., 2004. Influence of hydroxypropyl-beta-cyclodextrin on the extraction and biodegradation of phenanthrene in soil. Environ. Toxicol. Chem. 23, 550–556.

Semple, K.T., Morriss, A.W.J., Paton, G.I., 2003. Bioavailability of hydrophobic organic contaminants in soil: fundamental concepts and techniques for analysis. Eur. J. Soil Sci. 54, 809–818.

Szolar, O.H.J., Rost, H., Hirmann, D., Hasinger, M., Braun, R., Loibner, A.P., 2004. Sequential supercritical fluid extraction (SSFE) for estimating the availability of high molecular weight polycyclic aromatic hydrocarbons in historically polluted soils. J. Environ. Qual. 33, 80–88.

Thiele-Bruhn, S., Brümmer, G.W., 2004. Fractionated extraction of polycyclic aromatic hydrocarbons (PAHs) from polluted soil: estimation of the PAH fraction degradable through bioremediation. Eur. J. Soil Sci. 55, 567–578.

Van Noort, P.C.M., Belfroid, A., Cornelissen, G., ten Hulscher, D.E.M., Rigterink, H., Vrind, B., 2000. Organic sorbate planarity in relation to desorption kinetics and extent of non-linear sorption in sediments. 220th ACS National Meeting, Washington, DC, August 20–24, Vol. 40(2), p. 181.

Weber, W.J. Jr., McGinley, P.M., Katz, L.E., 1992. A distributed reactivity model for sorption by soils and sediments. 1. Conceptual basis and equilibrium assessments. Environ. Sci. Technol. 26, 1955–1962.

Weissenfels, W.D., Klewer, H.-J., Langhoff, J., 1992. Adsorption of polycyclic aromatic hydrocarbons (PAHs) by soil particles: influence on biodegradability and biotoxicity. Appl. Microbiol. Biotechnol. 36, 689–698.

Wu, S.C., Gschwend, P.M., 1986. Sorption kinetics of hydrophobic organic compounds to natural sediments and soils. Environ. Sci. Technol. 20, 717–725.

Young, T.M., Weber, W.J. Jr., 1995. A distributed reactivity model for sorption by soils and sediments. 3. Effects of diagenetic processes on sorption energetics. Environ. Sci. Technol. 29, 92–97.

Developments in Soil Science, volume 32
Ravendra Naidu (Editor)

Chapter 24

THE APPLICATION OF FIBRE OPTIC CHEMICAL SENSORS FOR HEAVY METAL MONITORING IN CONTAMINATED ENVIRONMENTS

P. Gangaiya and N. Mahendra

24.1 Introduction

Environmental monitoring can be conducted for many different reasons but one of the most common, as far as contaminants such as heavy metals are concerned, is to ensure that concentrations in the environment do not exceed established risk levels (see Chapters 4, 19 by Singleton). Heavy metals are present in the environment in very low concentrations except where geochemical mineralisation processes have resulted in a build up of levels such as in areas close to mineral veins (Knezek and Ellis, 1980). Some of the metals, such as Cu, Zn, and Mn, are also micronutrients and required by all living organisms in small amounts whereas others such as Cd, Hg and Pb are not required at all and toxic from very low concentrations. For all heavy metals increases above background can result in significant harm to the living environment (Richardson, 1997; see chapter 18 by van Straalen). In seawater for instance releases from copper-based paints used as an anti-fouling agent on the hulls of ships are known to have caused serious toxicity problems for marine life, and the use of such paints is now severely restricted. Significant releases can also occur from metal-based industries with no rigorous controls on the management of their wastes and discharges. Even in the relatively less industrialised island nations of the Pacific, increased metal concentrations have been detected in areas close to industrial areas (Gangaiya et al., 2001). From an environmental standpoint therefore, there is much interest in monitoring and controlling the levels and releases of heavy metals into the environment and for this suitable techniques are required.

The use of measurement devices such as sensors would present greater advantage in any environmental monitoring situation because sensors by their very nature are meant to provide rapid, real-time information. Should there be an undesirable increase in contaminant levels, quick detection using a sensor-based approach will allow the situation to be controlled before significant damage to the environment is caused. Sensors for various purposes and based on different operating principles and designs are available in the market but in the last twenty years or so a new development has emerged in the form of sensors based on fibre optic technology (Wolfbeis, 1991; Narayanaswamy, 1993). These are generally referred to as fibre optic chemical sensors (FOCS), although optical fibre chemical sensors (OFCS) and optrodes or optodes have also been used. This chapter provides a general introduction to fibre optic chemical sensing with particular application for heavy metal monitoring. It looks at the general features of these devices, their analytical applications and limitations, and developments and future research needs.

24.2 Fibre optic chemical sensors: general characteristics

Optical fibres have created a revolution in the communications industry because of their ability to transmit optical signals over vast distances without significant loss of the signal intensity. This is because of the way they are designed with a central core surrounded by a cladding which has a higher refractive index than the core. As a result of the difference in the refractive indices of the core and cladding, light is propagated along the optical fibre via total internal reflection (Fig. 24.1).

The propagation of light along the optical fibre allows the design of fibre optic-based sensors in which the sensing region can be separated from the other components of the measurement device by significant distances, making remote sensing a real possibility. Optical fibres transmit a source of incident light to the sensing region which contains the analyte and following interaction between the analyte and the light, the modulated optical signal is transmitted back to a suitable detector (Fig. 24.2). The instrumentation associated with fibre optic chemical sensing requires both optical and electronic components and typically involves a light source, optical fibres, filter or monochromator, a photodetector, signal processing unit and an output display (Fig. 24.3). Various radiation sources such as light-emitting diodes (LEDs) (Hauser and Tan, 1993), incandescent lamps (Narayanaswamy et al., 1988) and lasers (Crane et al.,

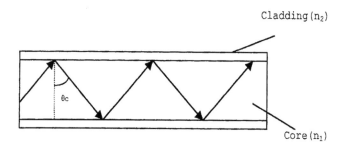

Figure 24.1. *Total internal reflection in an optical fibre θ_c – critical angle for incident light; n_1 and n_2 are the refractive indices of core and cladding, respectively.*

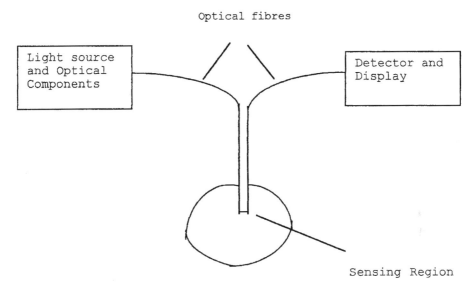

Figure 24.2. *Schematic representation of fibre optic chemical sensing.*

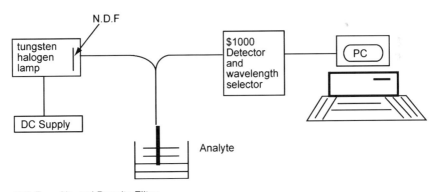

N.D.F. = Neutral Density Filter

Figure 24.3. *Example of an instrument set up for fibre optic sensing.*

1995; Lu and Zhang, 1995) have been used. Amongst detectors, photomultiplier tubes (PMT) and photodiodes including charge-coupled devices (CCDs) have been used.

Very few analytes have analytically useful direct interaction with optical light and for most a transducer-based system has to be used in which a reagent whose optical properties are modified by contact with the analyte is incorporated into the system. Any of the already known optical interactions such as absorption, fluorescence, reflectance can be made the basis for measurement (Seitz, 1984; Wolfbeis, 1987; Narayanaswamy, 1993). The optically active reagent can be immobilised directly on the optical fibre where it contacts the measurement sample or it can be immobilised on another substrate and then a layer of the substrate containing the immobilised reagent attached to the fibre. Various substrates have been investigated and found to be useful. For example, Saari and Seitz (1982) and Pulido et al. (1993) tried sintered and porous glass, and a number of others (e.g. Kirkbright et al., 1984; Ervin et al., 1993; Mahendra et al., 2002) have used hydrophobic polymeric materials. Hydrophilic supports of various kinds such as cellulose (Jones and Porter, 1988), polyacrylamide (Peterson et al., 1980), poly vinyl chloride (Bakker and Simon, 1992) and poly vinyl alcohol (Zhang et al., 1989) have also been used.

Immobilisation of the reagent on the substrate can be carried out either through physical or chemical means (Seitz, 1988). Physical methods include gel entrapment (Baldini et al., 1995), encapsulation within a selectively permeable membrane (Bright et al., 1988), adsorption and electrostatic attraction onto supports such as ion-exchange resins (Qin et al., 1998) and organic polymers of Amberlite XAD series (Alder, 1987; Mahendra et al., 2002). Chemical methods involve the formation of a covalent bond between the reagent molecule and the substrate surface (e.g. Saari and Seitz, 1984a, 1984b; Jones and Porter, 1988) and could include complicated chemical reactions in which structures of either or both are modified to achieve the immobilisation.

As already mentioned, optical fibres connecting the sensing region to the other optical components of the measurement system can be of considerable length which makes it possible to locate the sensor directly in the environment that is to be monitored, particularly that of a hazardous nature where other methods may be restrictive. Sensors based on fibre optics thus have a number of inherent advantages which apart from rapid, real-time monitoring also include remote sensing, electronic passivity, flexibility in design and low cost. As with any other analytical technique however there are disadvantages as well (Table 24.1). The nature of the advantages and disadvantages are explained in detail by Narayanaswamy (1993) but the disadvantage relating to limited dynamic range

Table 24.1. *Advantages and disadvantages of FOCS as an analytical technique.*

Advantages	Disadvantages
Electronically passive	Interference by ambient light
Flexible and easily miniaturised	Long-term stability limited
Inexpensive	Long response times
Capable of real-time monitoring	Limited dynamic range of measurement
Useful in remote sensing	

needs further discussion here, as it may not be a total disadvantage where environmental monitoring is concerned. The limited dynamic range can be explained in terms of the chemistry of the sensor response. The reaction between any analyte (A) that is to be detected and the optical reagent (R) utilised for its monitoring can be represented as in Eq. (24.1), and for which the equilibrium constant is written as Eq. (24.2).

$$A_{(aq)} + R_{(s)} = AR_{(s)} \tag{24.1}$$

$$K = \frac{[AR]}{[A][R]} \tag{24.2}$$

The total amount of reagent in the system, C_R, can be represented in terms of Eq. (24.3) and by rearrangement, Eq. (24.4) can be obtained.

$$C_R = [AR] + [R] \tag{24.3}$$

$$[A] = \frac{1}{K} \times \frac{C_R - [R]}{[R]} \tag{24.4}$$

According to Eq. (24.4), the analyte concentrations that can be detected depends on the equilibrium constant of the reaction and the proportion of reacted reagent against the unreacted reagent. Since this parameter is detected optically, this in turn depends on the optical properties of the reagent. Where absorption is used, this might be the extinction coefficient or the molar absorptivity. The range of concentrations around $1/K$ that can be detected will then be determined by how much of a change in R will result in a suitable measurable optical signal. If this change is of the order of 9–10% of the reagent, then this will mean that an approximate 10-fold change on either side of $1/K$ can in principle be detected. With smaller detectable changes, the range can be expanded. The limited dynamic range however is not altogether a serious disadvantage as in some environmental situations concentrations within a certain narrow range may be of interest. By suitable choice of reagent, sensors can be designed for this appropriate range.

24.3 Development of heavy metal sensors

The literature on fibre optic sensors reveals that many different reagent types, immobilisation techniques and substrates have been utilised for developing sensors for heavy metal monitoring. Many different heavy metals have been the focus of attention including the more common ones such as Cd, Cr, Cu, Pb and Zn. Table 24.2 gives a summary of the specifications for a selection of these

Table 24.2. *A selection of sensor configurations for heavy metal monitoring.*

Analyte	Reagent, analytical wavelength (nm)	Immobilisation method, support matrix	Working range, limit of detection (LOD)	Other details	References
Cd(II)	8-hydroxy-7-iodoquinoline-5-sulphonic acid, 505 nm	Adsorption, Sephadex	0.2–70 μM, 0.08 μM	Fluorescence, kinetic equilibrium methods, regeneration with K(I), Al(III) and Fe(II) interfere	Lu and Zhang (1995)
Cd(II)	5,10,15,20-tetra (sulphonato-phenyl) porphyrin (TPPS), 442 nm	Entrapment, PVC	LOD 20 μM	Absorbance, response time 20 min, leaching evident	Czolk et al. (1991)
Cd(II) Pb(II)	PAR, 500 nm	Reagent solution, Nafion tube	LOD 3 nM for both	Absorbance, flow mode, Co(II), Mn(II), Ni(II) and Zn(II) do not interfere	Booksh et al. (1994); Lin and Burgess (1994); Lin et al. (1994)
Cd(II) Zn(II)	8-(benzenesulfonaamido)-quinoline, 370 nm (492 nm)	Adsorption, XAD-7	5–500 ppb LOD (Zn) 1.6 ppb LOD Cd 1.9 ppb	Fluorescence, FIA system, kinetic method, regeneration with 0.1 M HCl	Compano et al. (1994)
Cd(II) Pb(II) Cu(II)	PAR disulfide derivative, IR region	Anchored to silver substrate	LOD (Cd) 503 ppb LOD (Pb) 522 ppb LOD (Cu) 1.49 ppb	Surface-enhanced Resonance Raman Spectroscopy used	Crane et al. (1995)
Co(II)	Rhodamine 6G, 500 nm (550 nm)	Electrostatic attraction Nafion	LOD 1 μM	Fluorescence quenching and enhancement, also responds to Cr(III), Fe(II) and (III), Cu(II), Ni(II)	Bright et al. (1988)
Cr(VI)	1,5-diphenylcarbohydeazide, 540 nm	Absorption, C$_{18}$-styrene divinyl-benzene beads	LOD 0.03 ppm	Reflectance, flow injection	Egorov and Ruzicka (1995)

Cu(II)	Fast Sulphon Black F, 612.9 nm	Adsorption, XAD-7	11–27 ppm, 10 ppm	Reflectance, dip probe design, kinetic assay, regenerated with EDTA	Mahendra et al. (2002)
Cu(II)	α-benzoinoxime, 560 nm	Adsorption, XAD-2	5–127 ppm, 5 ppm	Reflectance, dip probe design, kinetic assay, regenerated with HCl	Mahendra et al. (2003)
Cu(I)	2,2'-biquinoline, 610 nm	Adsorption, XAD-2	0.8–20 ppm, 0.8 ppm	Absorbance, flow cell design, kinetic assay, regenerated with HNO_3	Mahendra (1999)
Hg(II) Ag(I)	Dithiocarbamate, 665 nm	Entrapment, PVC	LOD 2.5 nM	Absorbance, pH 4.7, regenerated with 0.01 M HCl	Lerchi et al. (1994)
Hg(II) Ag(I) Cu(II) Mn(II)	Urease, thymol blue, 600 nm N,N'-diethylaniline, 250 nm	Covalent, amino propyl glass Absorption, fibrous fluorocarbon	LOD (Hg) 2 ppb LOD 0.1 nM	Reflectance, inhibition of urease measured Fluorescence	Andres and Narayanaswamy (1995) Klinkhammer (1994)
Ni(II)	PAN, 555 nm	Electrostatic attraction, Nafion	0.02 0.12–mM	Absorbance, regenerated with 0.5 M HNO_3; also detects Cd(II), Pb(II), Zn(II), Hg(II)	Madden et al. (1996)
Pb(II)	Dithizone, 650 nm	Adsorption, XAD-4	0.3–10 μM	Reflectance, kinetic assay, flow mode, regenerated with 0.01 M HCl	Oliveira and Narayanaswamy (1992)
Zn(II)	Carbonic anhydrase and dansylamide, 362 nm (460/560 nm)	Retention, dialysis membrane	40–1000 nM	Recognition of Zn by enzyme transduced by dansylamide fluorescent probe	Thompson and Jones (1993)
Zn(II)	Lipophilised PAR, 523 nm	Entrapment, PVC	1 μM–3 mM	Absorbance, pH 4.8, response time 5 min, Co and Cu interfere	Wang et al. (1992)

sensors. In some cases, fluorometric reagents such as morin, rhodamine and calcein have been used which either lose or gain fluorescence in the presence of heavy metals and in others, photometric chelating agents are used which change colour in the metal's presence. Examples are various derivatives of phenanthro-line: 4-(2-pyridylazo)resorcinol (PAR) and 1-(2-pyridylazo-)-2-naphthol (PAN). Many of the reagent systems also require careful pH control, as reactions with metals will only occur at certain pH conditions. The choice of reagent is primarily dictated by the need to have maximum sensitivity and selectivity for the particular heavy metal. Sometimes in the interest of sensitivity, reagents are chosen which have a very high binding constant for the metal, that is the reaction is strongly favoured in the forward direction and has a high equilibrium constant (K). A high K value is desirable for achieving better sensitivity but this is usually at the expense of reversibility because the reagent does not easily release the analyte when exposed to an analyte-free environment. In this situation, regeneration is required with the use of another reagent which strips the sensing layer of the analyte before it can be used for a subsequent measurement. Table 24.2 also includes information on the kind of regenerants that have been used.

Another noteworthy feature of some of the sensors being proposed for heavy metal monitoring is that the response time can be of the order of several minutes if not more (Czolk et al., 1991). This is because the sensing reagent is often immobilised on a solid phase and the analyte which is in the aqueous phase has to diffuse across the solid phase for equilibrium to be established and this process is time dependent. If reactions are not able to reach equilibrium within a reasonable time period then rather than using a steady state signal, a slightly different approach is used in which the rate of change of signal is used as an indication of the analyte concentration. This is referred to as the kinetic approach and has been used in several of the examples cited in Table 24.2. This approach leads to a reduction in response time but it can be at the expense of not achieving detection limits that are theoretically possible. But this may not be of much concern if reasonably contaminated environments are being investigated as concentrations in these environments can be significantly higher than background levels.

Interference from other metals is also possible for many of the sensors although it has been shown that immobilisation of a chelating photometric reagent can have the effect of increasing their selectivity towards metals (Ditzler et al., 1986). This is due to the fact that the immobilisation process restricts the formation of complexes with a high reagent-to-analyte ratio thereby making the reagent selective to complexes with low stoichiometric ratios. Interference from other metals could however be used to advantage if their respective optical signals can be sufficiently differentiated from each other in which case multi-element sensing could be possible. This could again be very useful for monitoring contaminated environments where more than one metal might be present and be a cause for concern. The metals can be detected and quantified in one single run.

24.4 Challenges for application to real-life monitoring

Currently the main applications of fibre optic sensing lie in the monitoring of heavy metals in aqueous samples, such as contaminated water and wastewater. Solids containing heavy metals will require extraction into the aqueous phase before they can be subjected to this method of measurement and therefore their application as a true sensor in these environments is limited, although other sensors that will work in such environments are also not easily available. Despite the many advantages they can provide, fibre optic-based sensors are yet to be utilised as a routine method of measurement for environmental monitoring. There could be many reasons for this including the lack of general awareness of its potential applications but there are issues relating to technical performance as well. One of these could be the lack of true reversibility for many of the developed sensors. Full reversibility would mean that signal change with both increasing and decreasing concentrations of analyte is possible without the need for any external interference. This is usually not the case for sensors where there is chemical binding of the analyte with immobilised reagent and the reaction is strongly favoured in the forward direction, as explained above. Regeneration is usually required but where continuous in-situ monitoring is required, the regeneration step could be difficult to incorporate in the measurement set-up. Also many of the reagents investigated so far also require careful pH control and this introduces another complication if continuous *in-situ* measurement is desired. Obviously, new reagents need to be identified or synthesised which, while showing good sensitivity and selectivity, also react reversibly with heavy metals.

Response times and detection limits also have to be lowered considerably if there is to be widespread use for heavy metal monitoring as contamination could be a concern at very low concentrations for some metals. Almost instantaneous response would be desirable if on-line monitoring is to be conducted and for this sensor designs have to investigated which allow significant reduction in the equilibration time. Thin films on which a layer of the active reagent is deposited on the external surface may result in better response times but mechanisms on how this could be achieved need to be investigated. Shorter response times can also be expected to result in lower detection limits but another way of lowering detection limits is through the selection of more sensitive reagents.

Another reason for the lack of widespread adoption for environmental monitoring could be that so far most development work has been conducted under carefully controlled laboratory conditions and the sensors have not been rigorously tested in real environmental situations. In the real environment there could be other variables affecting sensor response which cannot be easily controlled. What this means is that for fibre optic sensing to work in any given environmental situation, a sensing system would have to be custom-built and this obviously detracts from the possibility of a system being commercially promoted. Custom-building for any given situation will also require a high level of technical expertise which may not be easily available.

24.5 Conclusions

FOCS have great potential as an analytical tool in environmental monitoring, particularly in situations where capacity for remote sensing is desirable. This might be so in environments of a hazardous nature such as radioactive sites where direct human access might not be advisable. Despite many advantages, fibre optic sensors have not been used much as an analytical tool for the purpose of heavy metal monitoring. This is not because of the lack of research effort in this area as the literature is profuse with examples of development work for many different heavy metals. However, issues relating to performance such as sensitivity, specificity or selectivity, reversibility, reproducibility and stability have to be fully resolved and shown to be superior to other analytical techniques before this new technique will enjoy widespread acceptance. Considerable progress has been made on all these fronts and with continuing research it will not be long before such sensors are accepted as a very viable alternative to other methods of analysis.

References

Alder, J.F., 1987. New sensor developments. Anal. Proc. 24, 110–112.

Andres, R.T., Narayanaswamy, R., 1995. Effect of coupling reagent on the metal inhibition of immobilised urease in an optical biosensor. Analyst 120, 1549–1554.

Bakker, E., Simon, W., 1992. Selectivity of ion-sensitive bulk optodes. Anal. Chem. 64, 1805–1812.

Baldini, G.E., Grattan, K.T.V., Tseung, A.C.C., 1995. Impregnation of a pH-sensitive dye into sol-gels for fibre optic chemical sensors. Analyst 120, 1025–1028.

Booksh, K.S., Lin, Z., Wang, Z., Kowalski, B.R., 1994. Extension of trilinear decomposition method with an application to the flow probe sensor. Anal. Chem. 66, 2561–2569.

Bright, F.V., Poirier, G.E., Hieftje, G.M., 1988. A new ion sensor based on fibre optics. Talanta 35, 113–118.

Compano, R., Ferrer, R., Guiteras, J., Dolors Prat, M., 1994. Spectrofluorimetric detection of zinc and cadmium with 8-(Benzenesulfonamido)-quinoline immobilised on a polymeric matrix. Analyst 119, 1225–1228.

Crane, L.G., Wang, D., Sears, M., Heyns, B., Carron, K., 1995. SERS surfaces modified with a 4-(2-pyridylazo) resorcinol disulphide derivative: detection of copper, lead and cadmium. Anal. Chem. 67, 360–364.

Czolk, R., Reichert, J., Ache, H.J., 1991. An optical sensor for the determination of cadmium (II) ion. Sens. Actuators A 25-27, 439–441.

Ditzler, M.A., Pierre-Jacques, H., Harrington, S.A., 1986. Immobilisation as a mechanism for improving the inherent selectivity of photometric reagents. Anal. Chem. 58, 195–200.

Egorov, O., Ruzicka, J., 1995. Flow injection renewable fibre optic sensor system: principle and validation on spectrophotometry of chromium (VI). Analyst 120, 1959–1962.

Ervin, A.M., Ewing, K.J., Lamontagne, R.A., Aggarwal, I.D., Rowley, D.A., 1993. Development of a fibre optic sensor for trace metal detection in aqueous environments. Appl. Opt. 32, 4287–4290.

Gangaiya, P., Tabudravu, J., South, R., Sotheeswaran, S., 2001. Heavy metal contamination of the Lami coastal environment, Fiji. S. Pac. J. Nat. Sci. 19, 24–29.

Hauser, P.C., Tan, S.S.S., 1993. All solid state instrument for fluorescence based fibre optic chemical sensors. Analyst 118, 991–995.

Jones, T.P., Porter, M.D., 1988. Optical pH sensor based on the chemical modification of a porous polymer film. Anal. Chem. 60, 404–406.

Kirkbright, G.F., Narayanaswamy, R., Wethi, N.A., 1984. (The Late) Fibre optic pH probe based on the use of an immobilised colorimetric indicator. Analyst 109, 1025–1028.

Klinkhammer, G.P., 1994. Fibre optic spectrometers for in-situ measurements in the oceans, the ZAPS probe. Mar. Chem. 47, 13–20.

Knezek, B.D., Ellis, B.G., 1980. Essential Micronutrients IV: Copper, Iron, Manganese, and Zinc. In: B.E. Davies (Ed.), Applied Soil Trace Elements. Wiley, pp. 259–286.

Lerchi, M., Reitter, E., Simon, W., Pretsch, E., Chowdhury, D.A., Kamata, S., 1994. Bulk optodes based on neutral dithiocarbamate ionophores with high selectivity and sensitivity for silver and mercury cations. Anal. Chem. 66, 1713–1717.

Lin, Z., Booksh, K.S., Burgess, L.W., Kowalski, B.R., 1994. Second-order fibre optic heavy metal sensor employing second order tensorial calibration. Anal. Chem. 66, 2552–2560.

Lin, Z., Burgess, L.W., 1994. Chemically facilitated Donnan dialysis and its application in a fibre optic heavy metal sensor. Anal. Chem. 66, 2544–2551.

Lu, J., Zhang, Z., 1995. Optical fibre fluorosensor for cadmium with diethylaminoethyl-sephadex as a substrate. Analyst 120, 453–455.

Madden, J.E., Cardwell, T.J., Cattrall, R.W., Deady, L.W., 1996. Nafion-based optode for the detection of metal ions in flow analysis. Anal. Chim. Acta 319, 129–134.

Mahendra, N., 1999. Development of a fibre optic chemical sensor for determination of copper. MSc Thesis, The University of the South Pacific, Suva, Fiji, 161 p.

Mahendra, N., Gangaiya, P., Sotheeswaran, S., Narayanaswamy, R., 2002. Investigation of an optical fibre Cu(II) sensor using Fast Sulphon Black F (FSBF) immobilised onto XAD-7. Sens. Actuators B 81, 196–201.

Mahendra, N., Gangaiya, P., Sotheeswaran, S., Narayanaswamy, R., 2003. Investigation of an optical fibre copper based on immobilised α-benzoinoxime (cupron). Sens. Actuators B 90, 118–123.

Narayanaswamy, R., 1993. Optical chemical sensors: transduction and signal processing. Analyst 118, 317–322.

Narayanaswamy, R., Russell, D.A., Sevilla III, F., 1988. Optical fibre sensing of fluoride ions in a flow stream. Talanta 35, 83–88.

Oliveira, W., Narayanaswamy, R., 1992. A flow cell optosensor for lead based on immobilised dithiozone. Talanta 39, 1499–1503.

Peterson, J.I., Golstein, S.R., Fitzgerald, R.V., 1980. Fibre optic pH probe for physiological use. Anal. Chem. 52, 864–869.

Pulido, P., Barrero, J.M., Perez-conde, M.C., Camara, C., 1993. Evaluation of three supports for an optical fibre ferric ion sensor. Quim. Anal. 12, 49–52.

Qin, W., Zhujun, Z., Huajun, L., 1998. Chemiluminescence flow through sensor for copper based on an anodic stripping voltammetric flow cell and an ion exchange column with immobilised reagents. Anal. Chem. 70, 3579–3584.

Richardson, H.W., 1997. Handbook of Copper Compounds and Application. Marcel Dekker, New York.

Saari, L.A., Seitz, W.R., 1982. PH sensor based on immobilised fluoresceinamine. Anal. Chem. 54, 821–823.

Saari, L.A., Seitz, W.R., 1984a. Immobilised calcein for metal ion preconcentration. Anal. Chem. 56, 810–813.

Saari, L.A., Seitz, W.R., 1984b. Optical sensor for beryllium based on immobilised morin fluorescence. Analyst 109, 655–657.

Seitz, W.R., 1984. Chemical sensors based on fibre optics. Anal. Chem. 56(1), 16A–34A.

Seitz, W.R., 1988. Chemical sensors based on immobilised indicators and fibre optics. CRC Crit. Rev. Anal. Chem. 19, 135–173.

Thompson, R.B., Jones, E.R., 1993. Enzyme based fibre optic zinc biosensor. Anal. Chem. 65, 730–734.

Wang, K., Seiler, K., Rusterholz, B., Simon, W., 1992. Characterisation of an optode membrane for Zn(II) incorporating a lipophilized analogue of the dye 4-(2-pyridylazo) resorcinol. Analyst 117, 57–60.

Wolfbeis, O.S., 1987. Fibre-optic probes for chemical sensing. Anal. Proc. 24, 14–15.

Wolfbeis, O.S. (Ed.) 1991. Chemical Sensors and Biosensors. Vols. I and II, CRC Press, Boca Raton.

Zhang, Z., Yunke, Z., Russell, R., Shanksher, Z.M., Grant, C.L., Seitz, W.R., 1989. Poly(vinyl alcohol) as a substrate for indicator immobilisation for fibre optic chemical sensors. Anal. Chem. 61, 202–205.

F: The role of bioavailability in risk assessment and remediation

Remediation of contaminated sites using traditional technology is estimated to cost in excess of US $10 billion. These technologies operate within legislation that considers total contaminant content rather than the fraction that is free or bioavailable. Such an approach is costly and unattractive to end users of contaminated land. Where contaminants are tightly bound to the soil and not bioavailable, as evidenced by bioavailability bioassays and other toxicity assays, exhaustive clean up of soils is not necessary as the contaminants may not pose risk to the environment or humans. This approach to remediation is classed as "Risk Based Land Management" (RBLM). Technology associated with RBLM is considered attractive as it can save millions of dollars in remediation costs. Not surprisingly, RBLM is now considered as the most appropriate method for managing contaminated land. One key step in RBLM technology is establishing contaminant bioavailability. This section builds on Sections A–E by relating bioavailability to risk assessment and showing how this could be used for managing the risks associated with contaminated land.

Developments in Soil Science, volume 32
Ravendra Naidu (Editor)

Chapter 25

CONCEPT FOR RISK ASSESSMENT OF SOIL CONTAMINANTS BASED ON TOTAL AND BIOAVAILABLE CONCENTRATION – IMPLEMENTATION IN SWITZERLAND

S.K. Gupta and M. Haemmann

25.1 Introduction

Sustainable ecosystems depend upon a healthy aquatic, atmospheric and terrestrial environment. In order to achieve this it is necessary to ensure a reasonable balance in the loss and production of renewable and non-renewable resources. Food quality, potable water quality, sustainable use of agricultural land and environmental quality are intimately linked to one another.

The production of sufficient quality food depends on the availability of fertile, uncontaminated soil, an adequate supply of moisture and nutrients and on the biological functioning of the agro-ecosystem. In addition to inputs of heavy metal(loid)s and xenobiotics (which may originate from the use of fertilizers, pesticides, etc.), a growing global concern is the increasing pollution of agricultural soils through inavertent use of waste residues containing both nutrients and trace but significant levels of toxic substances.

This raises the question of how agriculture can define and warrant food quality standards which are trusted by consumers. At the same time the public expects agricultural land use to be sustainable, to maintain the recreational and aesthetic values of the landscape and to ensure its life-supporting ecological functions. Last but not the least, economic aspects must be considered because food production is not viable if farmers cannot make a living.

Soil quality plays a particularly crucial and complex role in the quality and health of agricultural food products. Intact soil performs vital ecological functions such as filtering and inactivating pollutants, decomposing organic waste substances and recycling nutrients. Soil quality is directly affected by agricultural land management. Once a soil is degraded it is very difficult or even impossible to repair. For example, if its retention and filtering capacities are exceeded, it also can be a source for undesirable substances to enter the food chain. On the other hand, the Maximum Contaminant Level Goals (MCLG) and Optimum Levels of Soil fertility (OLSF) are recorded and documented worldwide for many agricultural sites, groundwater, surface water and farm products.

This discussion makes clear the necessity of developing standard values which establish the degree of contamination, provide a basis for comparison and allow food quality to be assessed across international borders. These values should be based on risks to the soil system itself or else risks to other receptors such as animals, plants and humans. These values will allow the selection of appropriate precautionary, protective and remedial measures for affected sites. This concept is based on a Swiss ordinance relating to the impact of contaminants on soils (VBBo, 1998). It takes into account both the degree of soil contamination and the concentration of bioavailable pollutants. Methods used to measure bioavailable concentrations in soil should be well tested, especially under field conditions. The three-steps risk assessment concept used in Switzerland (Vollmer and Gupta, 1995) is compiled, accepted by the Ministry of Agriculture, Swiss Government and implemented in the modified VSBO (Ordinance relating to pollutants, 1986) which is named as VBBo (Ordinance relating to impacts on soils, 1996). The basic principles in preparation of concept were preventive, predictive and curative soil protection measures. In this concept three values were suggested for example guide, trigger and clean-up values. Later in another document (Haemmann and Gupta, 1997) limit of three values were derived and basics were presented in this report. These principles can be used worldwide. Swiss Government (1998) specially Ministry of Agriculture and The Swiss Agency for the Environment, Forests and Landscape (SAEFL) accepted the concept in Law on Environment Protect (USB$_{rev}$, 1997). As consequence, three-step evaluation concept for example guide, trigger and clean-up values were included in the Ordinance on Soil Soil Impacts (VBBo, 1998). Further these values are not only total concentration of contaminants in soils but also easily soluble concentrations of metals in soils. The easily soluble metal concentration is measured in neutral salt extract of soils and in most of the cases reflects the concentration of metals which are available to roots and soil organism (see Chapter 3 for definition of bioavailability). This chapter presents an overview of concepts for risk assessment currently being used in Switzerland. Readers interested in risk based approach to managing contaminated land are directed to Chapters 4 and 19.

25.2 Soil, soil fertility and soil quality

Soil fertility is a central concept of qualitative soil protection both under natural and managed ecosystems (VSBo, 1986; VBBo, 1998). A soil is deemed fertile if it:

(a) Is characterized by a diverse and biologically active animal and plant communities, possessing a site-specific structure and an undisturbed capacity for decomposition.

(b)   Makes possible undisturbed growth of plants and plant community (both and influenced by man) and does not affect their characteristic properties.

(c)   Assures that plant products are of good quality and wholesome for man and animals.

(d)   Produces food and fodder of good quality which do not present a hazard to the health of humans or animals.

Soils may become contaminated chemically, biologically or physically, through natural or anthropogenic pollution, pathogenic organisms or changes in soil structure. Trigger values are used to indicate whether certain impacts on the soil may pose a hazard to human, plants or animal life, depending on land use practices. They also indicate whether soil usage should be reduced or an alternative land use system can be adopted.

Verifying these criteria is a difficult task, especially in intensively farmed soils. The Swiss *Guidelines on Soil Fertility Assessment* (Gysi et al., 1991) contains a wide range of assessment criteria which are of assistance.

25.3 Characterisation of soil contamination

25.3.1 Definition

A distinction should be drawn between *soil contaminations* and *abandoned contaminated sites.* The proposed assessment plan applies only to instances of soil contamination and excludes abandoned contaminated sites. A separate plan already exists for the assessment and treatment of abandoned contaminated sites (BUWAL, 1994a, b).

Soil contamination is distinguished by a number of criteria (Table 25.1) from abandoned contaminated sites. The principal criteria concern the *extent* of the pollution and its *impact*. Soil contamination in the rural environment usually affects mainly the top layer of the soil, and impacts on soil fertility. In abandoned industrial contaminated sites, however, the contamination can extend into

Table 25.1. *Criteria to establish instances of soil contamination and abandoned sites.*

Parameters	Polluted soils	Abandoned contaminated sites
Extent	Large-area "two-dimensional" Topmost soil layer	Localised "three-dimensional" Topsoil and subsoil
Genesis	Diffuse, of atmospheric origin or from agriculture, traffic, etc.	Local, due to deliberate tipping or leaching
Effects	Impairment of soil fertility	Effects on aquatic environment, air, soil fertility, direct impact on man and animals
Measures	Primary: on emission sources	Primary: on the contamination itself

Source: Vollmer and Gupta (1995).

the subsoil. Such sites can potentially impact on the aquatic environment, the air and soil fertility. There are many examples of subsoil contamination including the presence of excessive levels of contaminants in groundwater at historically contaminated sites. Readers interested in contaminant behaviour including migration into groundwater are directed to Chapter 2 and Naidu et al. (1996).

25.3.2 Selection of pollutants

The criteria used for pollutant selection in Switzerland are necessity, enforce-ability and the availability of adequate information on individual pollutants, exposure pathways and the effects of the pollutants to receptors.

The selection criteria are evaluated in Table 25.2 on a scale extending from the absence of legally binding criteria for a pollutant (indicated by negative sign) to the existence of several such criteria (indicated by plus sign).

On the basis of this evaluation, trigger values for Cd, Cu, Pb and Zn are suggested by the authors which may be different from that reported in other countries. No trigger and clean-up values are proposed for Ni and F. The basic geogenic burden of both pollutants is relatively high (Tuchschmid, 1995) and accounts for most instances where the guide values are exceeded. Again, hardly any evaluation criteria along the exposure pathway exist and the information base is also inadequate.

25.3.2.1 Necessity

Where pollutants may pose risk to the environment and human health, it is necessary to determine trigger and clean-up values. Trigger values are generally established following extensive toxicological research. In Switzerland, trigger values for heavy metals were established for a number of plant species including human health. For example, trigger values for vegetable crops

Table 25.2. *Evaluation criteria for the selection of pollutants for this study.*

Pollutants	Legally binding criteria	Other evaluation criteria	Information on exposure/dose–effect ratio	Suitability for trigger and clean-up values
Cadmium	+++	+++	++	++
Copper	+++	+++	++	++
Fluorine	+	+	−	−
Lead	+	++	++	+
Nickel	−	+++	+	0
Zinc	+	+++	++	+

Source: Hämmann and Gupta (1997).

were established in Switzerland for Cd and Pb because these two heavy metals are toxic to humans. However, in the case of fodder plant cultivation, trigger values were set for Cd, Pb and Cu. Although Zn is toxic to animals in relatively high doses, no cases of poisoning attributable to soil contamination are known. For this reason, no trigger value was proposed for Zn. Trigger values for direct exposure were set for Pb and Cd for the same reasons as with food plant cultivation.

Clean-up values were set in for agricultural and horticultural land uses for lead and cadmium, as these elements are toxic to humans and for Cu and Zn because of their phytotoxicity. Clean-up values for cadmium, copper, lead and zinc were also set, for the same reasons, for domestic gardens and allotments. Clean-up values were likewise proposed only for lead and cadmium in the case of children's playgrounds.

Table 25.3 shows that most of the inorganic pollutants are phototoxic. If metals such as Cd, Cu, F, Pb, Ni and Zn exceed soil guide values, plants growing on these areas usually exceed the zootoxic concentrations.

Table 25.3. *Areas in which guide values are exceeded and toxicity characteristics.*

Pollutant	Areas	Human toxicity	Zoo- toxicity	Phyto- toxicity
Cadmium	Special crops and contaminated areas	x	x	
Chromium	In subsoils only in exceptional cases	x	x	x
Cobalt	In subsoils only in exceptional cases			
Copper	Agricultural areas, special crops, residential areas, and contaminated areas		x	x
Lead	Special crops, residential areas and contaminated areas	x	x	
Mercury	In residential and contaminated areas in exceptional cases only	x	x	x
Molybdenum	In contaminated areas in exceptional cases only		x	
Nickel	Agricultural, forest, residential and contaminated areas The basic geogenic burden is 6–124% of the guide value in subsoils and 28–112% of the guide value in topsoils (80% of all values according to Tuchschmid, 1995)			x
Thallium	In contaminated areas only in exceptional cases	x	x	
Zinc	Residential, contaminated and forest areas (mainly soluble contents)			x

Source: Meyer (1991), Tuchschmid (1995), Lindt et al. (1990), Merian (1984), Haemmann and Gupta (1997).

25.3.2.2 Enforceability (legal imposition)

Enforceability is achieved through legal ordinances. It relates to the requirements by the regulatory authorities in order to minimise risk to the human, animal and environmental health. Enforceability is conditional upon the existence of evaluation criteria along the exposure pathway. If, for example, the authorities order a plant analysis on farmland because a trigger value has been exceeded, there must be an evaluation criterion for the concentration of the relevant pollutant in the plant in order to determine whether a hazard to the risk receptor does or does not exist. In the absence of such a criterion, no evaluation is possible.

25.3.2.3 Availability of adequate information needed for risk assessment

No trigger or clean-up values can be derived unless sufficient information is available on the behaviour of the pollutant in the environment. This depends on the extent of exposure to a risk-creating pollutant, the dose–effect ratio of the pollutant in the risk receptor and whether the pollutant is currently causing major problems. The resulting general picture then guides the selection of pollutants for triggers and clean-up values.

25.4 Concept for setting up three levels of standard values

There are three main components of the concept for risk assessment presented here: guide, trigger and clean-up values (Fig. 25.1).

This concept can be used to establish values irrespective of type of soil degradation, the land use or the nation affected. Drawbacks and suggestions for further development are also given in this paper.

Figure 25.2 shows the relationship between three standard values, the degree of contamination and type of soil protection for example precautionary, preventive, etc.

25.4.1 Deriving guide values

These values serve to prevent soil pollution and provide an indicator for assessing the long-term fertility of soil. They indicate the level of contamination above which the long-term fertility of the soil is at risk and the soil cannot completely perform its many functions.

Guide values are not the only way to protect soil fertility. Where guide values do not exist, case-specific assessments must be carried out to determine whether or not soil fertility is assured.

Guide values have been set for the total content (HNO_3 extract) and in some cases for the soluble content ($NaNO_3$ extract) of certain heavy metals and fluorine. They are independent of land use and are applicable to all soils with

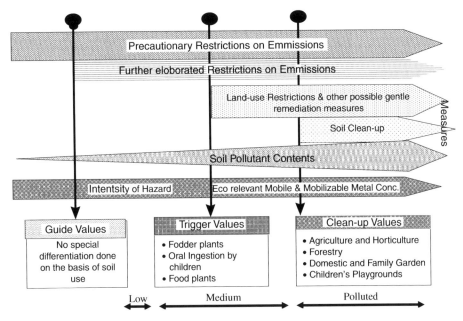

Figure 25.1. *Soil and site protection concept.*

humus content up to 15%. In Switzerland, a guide value of an element is deemed to have been exceeded if at least one of the two related values, for example HNO_3 extract or $NaNO_3$ extract guide value prescribed in Ordinance on Soil Impacts (VBBo, 1998) is exceeded.

According to the preventive principle, the derivation of guide values is based on susceptible land uses and soil properties. Their derivation has taken into consideration the impacts on plant growth and the growth and activities of soil organisms. In many cases this implies that although long-term soil fertility is no longer assured, a real hazard to humans, animals and plants is unlikely to arise given the current land use (Gupta, 1984; Gupta, 1989; Gupta and Aten, 1993; Gupta et al, 1996).

The Swiss Ordinance relating to soil impacts (VBBo, 1998) stipulates that if pollutant levels increase or guide values are exceeded, the cantons must investigate the source of pollution. If necessary they take field measurements of air pollution (LRV, 1985), hazardous chemicals (StoV, 1986) or waste (TVA, 1990) and enforce the relevant controls.

25.4.2 Deriving trigger values

Under certain circumstances it is possible that even where the clean-up values are not exceeded, a danger to health may exist. When trigger values are

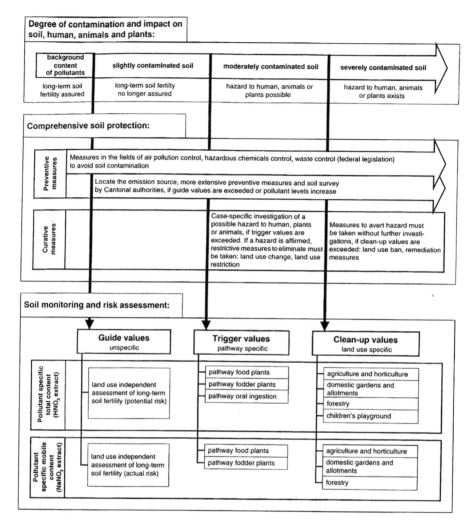

Figure 25.2. *Interrelationship between degree of contamination and impacts on human, animals and plants (Hämmann and Gupta, 1997).*

exceeded, the soil protection authorities are obliged to assess whether a concrete hazard exists. This does not, however, lead to any direct protection measures (e.g. restrictions on use). Direct protection measures only become necessary if the investigation confirms the suspected hazard in particular case. The existence of trigger values in no way compromises the effectiveness of precautionary soil fertility protection.

The revision of the Swiss Federal Environmental Protection Law (USG, 1983; USG, 1997) laid the legal foundations for the assessment of soil-related hazards to man, animals and plants and for the adoption of appropriate measures. For this purpose trigger and clean-up values were introduced in the Ordinance for Soil Quality (VBBo, 1998).

Trigger values indicate the threshold of a possible hazard. If they are exceeded, a more detailed investigation must be carried out to determine whether or not a hazard exists. Clean-up values reveal a level of contamination of the soil at which a hazard exists. The VBBo lays down the measures that must be adopted if trigger and clean-up values are exceeded. In order to evaluate risk of soil pollutants in other countries, the concept after considering national factors can be applied to other countries.

Hazard assessment for exposure pathways:

Pollutants reach man, plants and animals by a variety of pathways, particularly the following:

- Food plants: Soil→food plant→man
- Direct exposure: Soil→man (e.g. children playing outside home)
- Fodder plants: Soil→fodder plant→animal→man and/or
                    Soil→animal→man (oral ingestion of contaminated soil)
- Plant growth: Soil→plant

The hazard for each of these exposure pathways can be assessed in different ways. A distinction is made here between hazard analysis, in which pollutant transfer from soil to risk receptor is examined, and hazard evaluation by a criterion which indicates the level of exposure above which a hazard exists.

In the food plant exposure pathway, the hazard is analysed by quantifying pollutant transfer from soil to plant. This transfer is determined by qualitative data analysis. (Regression analyses are dispensed with because the requirements for their application are not met and soil–plant transfer factors cannot be applied because they are too inexact.) The evaluation criteria adopted are the maximum concentrations in foodstuffs set by the Ordinance for Food Quality (FIV, 1995) in Switzerland. However other countries may apply their own regulations.

The fodder plant exposure pathway combines three separate pathways: the animal's direct ingestion of contaminants in the fodder plant, soil contamination of the plant and direct oral ingestion of soil during grazing. These individual pathways are aggregated by mathematical relations between them. The evaluation criteria used are the maximum concentrations applicable to feedstuffs in accordance with the Swiss Ordinance for Feeding stuff Quality (FMBV, 1995) and limiting values.

In the direct exposure pathway, hazard analysis is carried out in two ways. First, a quantitative exposure analysis can be performed inductively by calculating exposure from the soil ingestion rate, soil pollutant concentration and body weight. This evaluation is based on protection for external exposure such as Provisional Tolerable Weekly Intake (PTWI). The results are inexact because the parameters can vary over two orders of magnitude. Second, it is possible to evaluate epidemiological studies in which concentrations in the soil can be correlated deductively with parameters such as blood or urine concentrations in man. The risk is then evaluated on the basis of protective values for internal exposure (e.g. the US-EPA (1996) Level of Concern for lead in the blood). Depending on the pollutant and the available data, the inductive method, the deductive method or a combination of both will be necessary.

Hazard analysis for the plant growth exposure pathway is based on yield or on observable qualitative growth disorders.

These values serve hazard protection purposes and are a measure of an existing risk to humans, farm animals and crops. A Realistic Worst Case (RWC) scenario derived from them is defined for each exposure pathway. With the aid of suitable data, the maximum tolerable concentration in the soil is then determined and can be listed as a trigger value.

The following approach is used to derive trigger value for food plant and fodder cultivation (combine) in Switzerland.

• Choice of relevant pollutants based on human toxicology and/or animal toxicology.
• Choice of a food plant with a high accumulation potential for each pollutant (RWC scenario).
• Choice of the Ordinance for Food Quality (FIV, 1995)-based tolerance value as an evaluation criterion in the RWC scenario.
• Derivation of the maximum permissible concentration in the soil by qualitative data analysis of soil–plant relations.
• Choice of a fodder plant that makes up a high proportion of the total rations of the relevant livestock animal (RWC scenario). This will frequently be grass consumed by grazing animals.
• Assumption of a maximum direct soil ingestion rate or maximum soil contamination.
• Choice of maximum concentration, using either the Swiss Ordinance for Feed Quality (FMBV).
• Ordinance for Feedstuff Quality or the handicap? Threshold as the evaluation criterion.
• Derivation of maximum permissible concentration for other fodder/livestock animal combinations and proportions of total rations, so as to test the calculated value.
• Literature-related critical appraisal and setting of trigger value on the basis of the real worst case scenario.
• Comparison of values with those applying in other countries.

25.4.3 Deriving clean-up values

Clean-up values serve to protect consumers from contamination hazards and are a measure of an existing risk to humans, animals and crops. Different forms of land use have different exposure pathways. Clean-up values are determined in accordance with a best case (BC) scenario. Under Swiss Environmental Protection Law (USB$_{rev}$, 1997), clean-up values indicate a level of contamination at which the site can no longer be used without presenting a hazard to man, plants or animals. This means that even in the BC, a hazard persists when all possible limitations on use have already been imposed.

Table 25.4 shows the individual exposure pathways with the evaluation criteria and assumptions for the BC scenario. Unlike the trigger-value situation, here the plant growth pathway is considered, because a minimum level of plant growth is necessary for agricultural or horticultural use of the land or for its use as a domestic garden or allotment.

The pathway-specific maximum permissible concentrations have to be calculated. These must be aggregated into a clean-up value depending on the main land use-agriculture and horticulture, domestic gardens and allotments, or children's playgrounds). Table 25.5 shows how this was done.

The presently used values are given in Table 25.6. These are in the Swiss Ordinance relating to soil impacts (VBBo, 1998).

Table 25.4. *Assumptions to establish BC-Scenario.*

Exposure pathway	Evaluation criterion	BC-scenario
Food plants	Three times the tolerance value or limit value (FIV)	The evaluation criterion is exceeded at minimum soil–plant pollutant transfer
Fodder plants	Lead and cadmium maximum: concentration in green forage; copper: handicap threshold	Grazing by an insensitive livestock animal, assumption of minimum proportion of direct oral soil ingestion and minimum soil–plant pollutant transfer
Direct exposure	Lead: blood lead level; cadmium: PTWI	Lead: significant observable increase in blood level; cadmium: residual utilisation of PTWI compared with other exposure pathways (Swiss average), body weight 15 kg, soil in question 0.1 per day
Plant growth	See BC-scenario	25% reduction in yield or description of manifest qualitative growth disorders

Source: Haemmann and Gupta (1997).

Table 25.5. *The pathway specific maximum permissible concentration.*

Type of exposure pathways	Agriculture and horticulture	Domestic gardens and allotments	Children's playground
Relevant exposure	Food plant FoP Fodder plants FoP Plant growth	Food plant FoP Fodder plants FoP Direct exposure (DE)	Direct exposure (DE) Direct exposure (DE)
Combinations	Max (min (FoP); min (FdP; PG)) either fodder plant or food plant cultivation should be possible	Use as garden and allotments and as playgrounds should be possible	Use as playgrounds should be possible

25.5 Methodological foundations

25.5.1 Relevant exposure pathways

Relevant exposure pathways have already been discussed in Section 25.3. The choice of guide, trigger and clean-up values are fixed by the Swiss Environment Protection Law (USG, 1997), which permits only protection of man and of livestock and plants In addition, some individual pathways are not covered by the Ordinance relating to impacts on soils (VBBo, 1998) but are regulated by the Ordinance on Abandoned Contaminated Sites (AltlV, 1996).

25.5.2 Hazard assessment

The system for assessing a hazard is illustrated in Figure 25.3.

Hazard analysis is used to determine the exposure of a risk receptor. Hazard evaluation decides whether the exposure of the risk receptor does or does not constitute a hazard. The definitions of different terms used in this chapter are given in Table 25.7.

25.5.3 Methods of hazard analysis

Hazard analysis is conducted in different ways depending on the exposure pathway (food plants, fodder plants, plant growth and direct exposure). However the same methods can be used in some cases. These methods are presented in the following sections.

Availability of pollutants in the soil

The central issue in hazard assessment is the fraction of the soil pollutant content available to the risk receptor and how this fraction is measured. In principle it is possible to distinguish between the mobile, mobilisable and total metal which is extracted by boiling HNO_3 is not the real total (mineral bound is

Table 25.6. *Guide, trigger and clean-up values in existing Swiss Ordinance relating to Soil Impacts.*

A. Guide values

Pollutants	Contents in soils mg per kg dry soil up to 15% humus mg per l greater than 15%	
	t = Total (HNO$_3$)	s = Soluble (NaNO$_3$)
Chromium	50	–
Nickel	50	0.2
Copper	40	0.7
Zinc	150	0.5
Molybdenum	5	–
Cadmium	0.8	0.02
Mercury	0.5	–
Lead	50	–
Fluorine	700	20

t = total content (2 M HNO$_3$ ratio soil to extractant 1:10)
s = soluble (NaNO$_3$ ratio soil to extractant 1:2.5)
Total fluorine (NaOH smelt 0.5–200)
Soluble fluorine (water extract 1:509)

B. Trigger values

Use	Contents in soils mg per kg dry soil up to 15% humus mg per l greater than 15%						Sampling depth (cm)
	Lead (Pb)		Cadmium (Cd)		Copper (Cu)		
	t	s	t	s	t	s	
Plants for human consumption	200	–	2	0.02	–	–	0–20
Plants for animal consumption	200	–	2	0.02	150	0.7	0.20
Use of soils with possible direct[a] uptake	300	–	10	–	–	–	0–5

Notes: t; total content (HNO$_3$), S; soluble in NaNO$_3$.
[a]oral, inhaled, dermal.

C. Clean-up values

Use	Contents in soils mg per kg dry soil up to 15% humus mg per l greater than 15%								Sampling depth (cm)
	Lead (Pb)		Cadmium		Copper (Cu)		Zinc (Zn)		
	t	s	t	s	t	s	t	s	
Agriculture and horticulture	2000	–	30	0.1	1000	4	2000	5	0–20
Home and family gardening	1000	–	20	0.1	1000	4	2000	5	0.20
Children's play grounds	1000	–	20	–	–	–	–	–	0–5

Source: VBBo (1998).

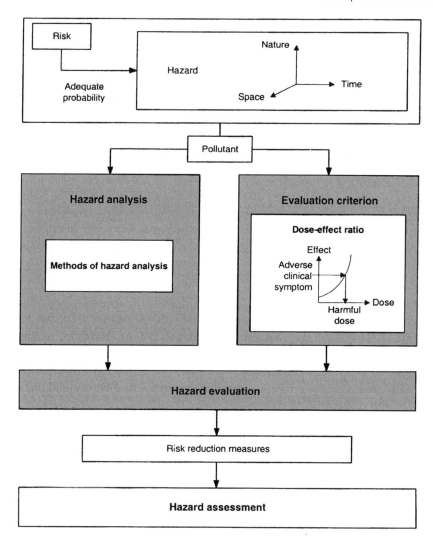

Figure 25.3. *Systematisation of hazard assessment in contaminated soils (Hämmann and Gupta, 1997).*

not extracted) so it is pseudo-total in the soil (Gupta, 1984; Gupta and Aten, 1993; Gupta et al, 1996). In order to evaluate metal concentration in agricultural soils, the mobile and pseudo-total contents are measured. The mobile concentration is determined as the soluble content by 0.1 M $NaNO_3$ extraction and the pseudo-total content as the total content by 1 M HNO_3 extraction. These two methods are specified for the assessment of soil contamination and have been in use since 1986 (VSBo, 1986; VBBo, 1998) (Fig. 25.4).

Table 25.7. *Definitions of terms used in the field of hazard assessment.*

Term	Definition
Concrete	A concrete hazard exists if there is already a situation of impairment or if one is likely to arise in the normal course of events.
Dose–effect ratio	Ratio of the dose of a pollutant to its clinical effect on a risk receptor.
Evaluation criterion	Criterion used to evaluate whether or not a hazard exists at a given exposure level. The evaluation criterion distinguishes between "risk present" and "risk not present". It is derived from the dose–effect ratio (e.g. level of concern according to US EPA for lead in the blood: $10\,\mu g\,dl^{-1}$).
Exposure analysis hazard	Quantification of exposure of a risk receptor to a pollutant. A hazard specified in terms of nature, direction and time.
Hazard analysis	Qualitative and/or quantitative determination of the actual or possible effects of a contaminated soil on the environment.
Hazard assessment	A comprehensive procedure for hazard analysis and hazard evaluation, which, in particular, also includes recommendations on the avoidance, reduction or elimination of the hazard.
Hazard evaluation	Evaluation on the basis of the legislation in force of whether a contaminated soil presents a hazard to the environment and is therefore legally relevant. This evaluation is based on the soil protection requirements and on requirements applicable to the relevant risk receptors (e.g. maximum concentrations in foods in the case of man). In the absence of such requirements, the evaluation must be based on established toxicological knowledge, social values and the state of the art.
Risk	Statement, made in context, on the probability of the release of pollutants from the soil into the environment and the degree of the harmful or undesirable effects of these pollutants on the risk receptors.
Risk receptor	The entity to be protected. The entities concerned are man, animals (mainly livestock animals) and plants (mainly agricultural plants).
Risk reduction measures	All measures likely to reduce exposure of the risk receptor and hence to avoid the hazard (e.g. a ban on use).

Source: Haemmann and Gupta (1997).

Determination of the mobil stable fraction is desirable in principle (Gupta, 1984). This involves identifying the pollutant fraction in the soil that might become a problem should the characteristics of the soil change over time (for example through soil acidification, climate change, humus decay, etc.). The proportion of contaminant in the soil which is not yet mobile may then be mobilised and may threaten certain receptors. Hence the mobil sable fraction is a measure of the potential hazard to a risk receptor, whereas the mobile fraction defines the actual hazard. First, however, a standard extraction method must be developed and specified (Fig. 25.5).

For the sake of continuity, trigger and clean-up values are suggested for the two extraction methods used hitherto. Since no trigger or clean-up values are

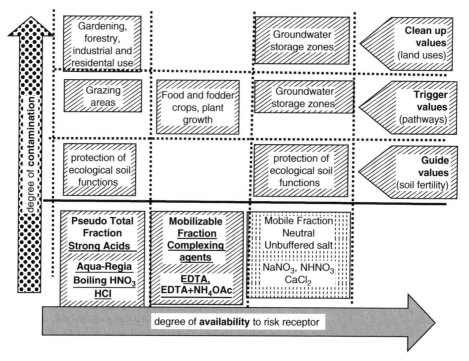

Figure 25.4. *The relationship among degree of availability to risk receptors (chemical forms of pollutants, chemical extractants) risk receptors and degree of contamination.*

derived for the mobil sable fraction, it is assumed as a first-order approximation that the total content is also a measure of maximum mobil sable metal concentration and hence is of potential hazard.

Whether the soluble or the total content is significant depends on the risk receptor, the exposure pathway and whether the hazard is current or potential. Table 25.8 gives a general view of the risk receptors, exposure pathways, extraction methods and relevant types of hazard.

Only the total content is relevant to the exposure pathways of direct human or animal soil ingestion. The total content is assumed to be a measure of availability because the pH of the human or livestock animal gastro-intestinal tract is very low at which most of the metals are soluble.

Soil–plant pollutant transfer is of primary importance in all other exposure pathways. The hazard analysis may be based either on total content or on soluble content. The scatter of the plant concentration values is usually less where the soluble content is used because soluble content is, as a rule, a measure of availability. Figure 25.6 illustrates this clearly for the Dornach and La Chaux-de-Fonds sites; these sites are in Switzerland. The total content of Cd is

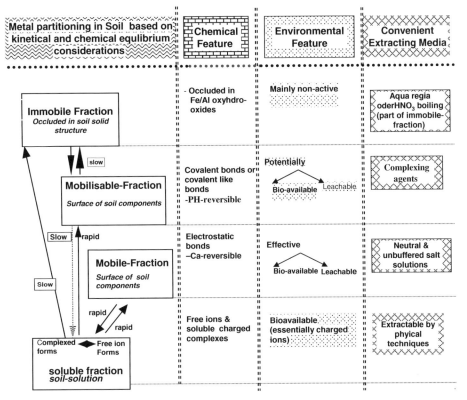

Figure 25.5. *Relationship of chemical and kinetically based metal partitioning and their importance (Gupta, 2004).*

very high in Dornach, but the pollutant concentration in the plants is low. If soluble content is considered, it is found to be very low. The Cd is not available to the plant. Conversely, plant concentrations of Cd at La Chaux-de-Fonds are very high although the total content is lower than at Dornach. Because the soluble content is very high, the Cd is readily available, so that transfer from soil to plant is high.

Analysis of pollutant transfer from soil to risk receptor

There are several methods for quantifying the transfer of a pollutant from soil to risk receptor. They are discussed below.

Qualitative data analysis

In this method, a graphic presentation is used to separate and quantify the areas of maximum and minimum transfer from soil to risk receptor. An example of

Table 25.8. *Relationship among type of metal concentrations in soils, risk receptors, exposure pathways and type of hazard indicated.*

Risk receptor	Exposure pathway	Extraction	Type of hazard indicated
Man	Direct exposure	Total content	Effective
	Food plants	Total content	Potential
	Food plants	Soluble content	Effective
Animal	Fodder plants contamination of harvested crop	Total content	Effective/potential
	Fodder plants (direct exposure)	Total content	Effective/potential
	Fodder plants (soil–pollutant transfer)	Soluble content	Effective
	Fodder plants I (soil–plant–pollutant transfer)	Total content	Potential
Plant	Plant growth	Soluble content	Effective
	Plant growth	Total content	Potential

Source: Haemmann and Gupta (1997).

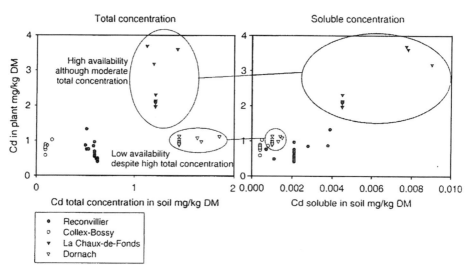

Figure 25.6. *Relationship among plant concentration, total and soluble concentrations of pollutants in soils (Hämmann and Gupta, 1997).*

qualitative data analysis is given in Figure 25.7. This offers a qualitative statistical statement on the relationship between soil concentration and the relevant variable in the risk receptor. Qualitative data analysis is applicable if the relationship is one-dimensional that is, if only the soil concentration and another variable need to be correlated.

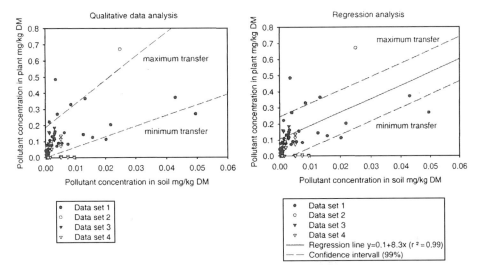

Figure 25.7. *Pollutant concentrations in soils and its influence on concentration in plants: analysis of data by qualitative and regression approaches (Hämmann and Gupta, 1997).*

The main advantages of this method are its simplicity and reproducibility. However, as soon as other variables are introduced, it can no longer be used. Another disadvantage is that the method is subjective however even with these draw backs, the method is used in the derivation of trigger and clean-up values.

Regression analyses

In this case the data are not considered qualitatively but are analysed by mathematical regression. The advantage of this approach is that it yields a quantitative-statistical statement of the expected value of the relevant variable (e.g. plant concentration or reduction in yield) for a given soil concentration. The statistical evaluation is usually based on simple or multiple linear regressions derived by the least squares method. The explanatory variable in the case of a simple linear regression is soil concentration. Additional magnitudes are used as explanatory variables in multidimensional linear regressions.

Where a linear regression is used, it is tacitly assumed that certain conditions are satisfied.

- The model must be linear. As a rule, this is not the case.
- The explanatory variable must be fixed that is, the error in its determination is negligible. This is not the case in the measurement of soil concentrations, mainly because of substantial heterogeneity in the soil.

- The standard deviation of the explanatory variable must be constant over the entire range. This too is seldom the case in the measurement of soil concentrations.

Figure 25.6 shows an example of a regression analysis. The maximum and minimum transfers can be determined, for example, by the 99% confidence intervals.

While assessing the coefficient of determination (r^2), account must be taken of the fact that there are no leverage points, as is the case, for example, in Figure 25.6. This means that a point in the data set must not have excessive influence. Such points are immediately visible in simple regressions. More sophisticated statistical methods must be used to detect them in multivariate problems.

The advantage of regression analysis is that the position of the line in the graph never depends on the observer. The disadvantage is that the conditions set out above are not satisfied in most cases. This creates a spurious impression of scientific validity. For this reason, regression analyses are not used for the derivation of trigger and clean-up values.

Transfer factors in soil–plant transfer

Soil–plant transfer factors are excellent indicator to assess accumulative capacity of plants. This is the ratio between soil and plant concentrations as defined by the following equation.

$$t = a(b/c_{tot/sol})$$

$$t = \text{Transfer factor}$$

$$c_{tot} = \text{Total content or soluble content } of \text{ pollutant in the soil}(mg\,kg^{-1}\,DM)$$

$$c_{sol} = \text{Concentration } of \text{ pollutant in the plant}(mg\,kg^{-1}\,DM)$$

The parameter a in this equation indicates the dependence of the plant concentration on the soil concentration, while b represents a basic concentration in the plant that is present even at very low soil concentrations.

This function is a hyperbola and can be interpreted as follows: if the basic concentration b in the plant is low, the second term is virtually negligible and the transfer factor is more or less constant over the entire range of soil concentrations. If, however, the basic concentration b is high, the transfer factor is much higher at low soil concentrations than at high concentrations. An example of this situation is shown in Figure 25.8.

A large number of studies quote transfer factors for specific pollutants and plants. Due to its simplicity it is often used parameter. At first sight, their application to a hazard analysis is temptingly simple.

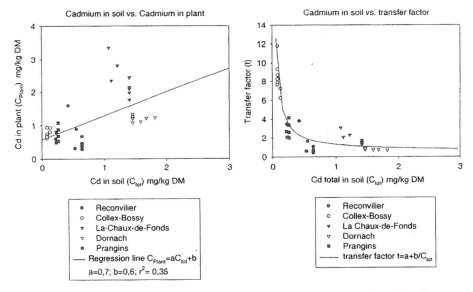

Figure 25.8. *Inter-relationship among pollutants concentrations in the soils, plants and transfer factors (Hämmann and Gupta, 1997).*

Table 25.9. *Input variables for qualitative exposure analysis.*

Exposure pathways	Input variables
Food plants	Food basket, pollutant concentrations of the various foods, daily food consumption
Fodder plants	Fodder mixture, quantity of soil ingested daily (earthy contamination and direct oral soil ingestion), pollutant concentrations in fodder and soil, daily fodder consumption
Plant growth	–
Direct exposure	Quantity of soil consumed daily, pollutant concentration in soil

Source: Haemmann and Gupta (1997).

Quantitative exposure analysis

Quantitative exposure analysis is used for inductive calculation of the exposure of a risk receptor by means of a formula. The input variables are the pollutant content of the ingested product (food, fodder or soil) and the daily consumption of food, fodder or soil.

The method shown in Table 25.9 has the advantage of simple and quick calculation of exposure, and is comparable to a maximum tolerable daily exposure for the purpose of deriving a trigger or clean-up value. The range for the calculated exposure is very wide.

Epidemiological studies

Epidemiological studies can be used to confirm or disprove a correlation between soil contamination and exposure of the organism (e.g. lead in the blood or cadmium in the urine). The precise mechanism of action in this case is of secondary importance. The prime object of consideration is whether the pollutant concentration in the soil correlates statistically with that found in the receptor (e.g. human).

An advantage of this method is its quantitative-statistical result. This can be used to check the quantitative exposure analysis described above. A disadvantage is that little is known about the exposure pathways from the soil to the risk receptor.

25.5.4 Criteria for hazard evaluation

Human health

There are several ways to evaluate the hazard to human health. These can be either by way of maximum concentrations in foodstuffs or by protective values for external or internal exposure. Maximum concentrations in foodstuffs are relevant only in the case of the food plant exposure pathway, because the hazard is evaluated on the food plant. Protective values can be used with all exposure pathways.

Maximum concentrations in foods

Tolerance values are maximum concentrations of substances above which the foodstuff is deemed to be contaminated or otherwise reduced in value (FIV, 1995). Limit values are maximum concentrations above which the foodstuff is deemed unfit for human consumption (FIV, 1995). These values are set not only on the basis of a dose–effect ratio – that is toxicologically – but also take account of, for example, the technically unavoidable concentration of a substance that may be introduced to the product by food processing. The origin of many values is not exactly known. However, they are used unconditionally by the Cantonal laboratories and by the Federal Office of Public Health for foodstuff quality appraisal (Zimmerli, 1997). If no values are mentioned in the FIV for a given heavy metal and individual food plants, international literature and international ordinances may be used (BGA, 1993).

Protective values for external exposure

The PTWI (FAO and WHO, 1996) and Provisional Maximum Daily Intake (PMTDI) (FAO and WHO, 1996) most commonly used for evaluation of chronic exposure due to the consumption of foodstuffs.

However, other protective values for evaluation of hazards also exist. Hassauer et al. (1992) determined reference values, distinguishing between acute and chronic and between carcinogenic and other chronic-toxic effects. These values are based on the idea of the Reference Dose (EPA, 1995) and on the PMTDI or, as applicable, the PTWI. They indicate a No Observed Adverse Effect Level (NOAEL) from a regulatory rather than a scientific viewpoint (Konietzka and Dieter, 1994). Reference values are usually one order of magnitude lower than the PTWI and PMTDI and are thus on a similar level to the RF values of the EPA (1995).

Protective values for internal exposure

Protective values for internal exposure are medically based threshold values for pollutant concentrations in the blood, urine, etc. Unlike the external exposure values, the concentrations evaluated here are those already resorbed by the human body. An example is the level of concern for blood-lead content in children (Table 25.10).

Protective values for internal exposure may be used where epidemiological studies have demonstrated a correlation between pollutant concentration in the soil and the concentration in organs or in the blood.

25.5.5 Maximum concentrations in feedstuffs

Animal health and hygiene of animal products

As in the case of human health evaluation, two different approaches are possible: maximum concentrations in livestock feedstuffs and protective values for external and internal exposure. Evaluation in terms of food hygiene is also relevant, because animal products ultimately reach man.

Maximum concentrations in feed stuffs

In the case of maximum concentrations in feedstuffs, a distinction is made between (a) unwanted substances in feedstuffs, and (b) permitted animal feed additives. The latter are essential elements mixed with the feed. There is little point in the level added through feed additives for evaluation because they are usually added to the feed in the form of salt, so that they are much more readily available to the animal organism than if they were ingested via fodder plants or soil particles. Legally binding maximum concentrations exist, as in the case of human foodstuffs, for commercial stock feeds. It is also necessary to distinguish between single and composite feeds and between complete feeds and supplementary feeds. Detailed information on this point may be found in the Swiss Ordinance for Feeding Stuffs Quality (FMBV, 1995).

Table 25.10. *Definition of some protective values.*

Protective values	Definition
PTWI	Provisional Tolerable Weekly Intake (FAO and WHO, 1996). The endpoint used for food contaminants such as heavy metals with cumulative properties. Its value represents permissible human weekly exposure to those contaminants unavoidably associated with the consumption of otherwise wholesome and nutritious foods.
PMTDI	Provisional Maximum Tolerable Daily Intake (FAO and WHO, 1996) The endpoint used for contaminants with no cumulative properties. Its value represents permissible human exposure as a result of the natural occurrence of the substance in food and in drinking-water. In the case of trace elements that are both essential nutrients and unavoidable constituents of food, a range is expressed, the lower value representing the essential level and the upper value the PMTDI.
Reference values	Reference values are whole-body doses of a hazardous substance at which there is sufficient probability of no adverse health effects where individual substances are considered in accordance with the present state of our knowledge or in the case of which only a slight health risk is assumed to exist (Hassauer et al., 1992).
Red	Reference dose (TERA, 1997). An estimate (with uncertainty spanning perhaps an order of magnitude) of a daily exposure to the human population (including sensitive subgroups) that is likely to be without an appreciable risk of deleterious effects during a lifetime. RfDs are based on non-carcinogenic effects and are usually calculated by applying uncertainty factors to a NOAEL or LOAEL. Used by US EPA (United States Environmental Protection Agency).
MRL	Minimum risk levels (TERA, 1997). An estimate of daily human exposure to a dose of a chemical that is likely to be without an appreciable risk of adverse non-calcareous effects over a lifetime of exposure. Expressed in mg per kg per day. Used by US ATSDR (Agency for toxic Substances and Disease Registry of the Centres for Disease Control of the United States).

Source: Haemmann and Gupta (1997).

Protective values for external exposure

As in the case for humans, protective values can be defined for livestock animals. Handicap thresholds are an example. These are defined as the value of a relevant trace element that must not be exceeded otherwise deviations from the "normal" metabolism may occur (Kessler, 1993). In the case of cadmium, the handicap threshold is based on considerations of food quality rather than the toxicology of livestock animals. This means that the exposure must not exceed the handicap threshold, or else certain organs of the animal cannot be used because they will not satisfy food quality standards.

Maximum Tolerable Levels have been set by the U.S. National Academy of Sciences (NRC, 1980). They are defined as "the dietary level that, when fed for a limited period will not impair animal performance and should not produce unsafe residues in human food derived from the animal". The same values are also found in Hapke (1988).

Toxicity thresholds are mentioned in Anonymous (1976). They indicate a level of pollutants administered experimentally (usually in the form of salt) at which toxic manifestations were observed. Kirchgessner (1992) also gives tolerance and toxicity thresholds, which, however, do not quite coincide with the above data.

Protective values for internal exposure

Protective values for internal exposure are found in the veterinary literature. They are not discussed further here because they are not needed for the derivation of regulatory standards.

Plant growth

The principal criteria for hazard evaluation are maximum permissible yield reduction, phytotoxic limit concentrations and qualitative evaluation of growth disorders.

25.6 Accuracy and limitations in the derivation of standard values

25.6.1 Complexity

Complexity has to be reduced in ecotoxicological studies. If regulatory standards are to be science-based, it is inevitable that limits to testing will be found. The necessary reduction in complexity may entail oversimplification owing to the selection of parameters (Schroeder et al., 1993). Scharamm (1989) also shows that scientific methods for the determination of limit values in the soil environmental system are insufficient. In his view, political as well as scientific considerations must be taken into account here.

25.6.2 Accuracy

Accuracy limitations arising in deriving regulatory standards are attributable mainly to the following factors:

Evaluation criteria: In principle it is impossible to lay down precise evaluation criteria because the dose–effect ratio always lies in a wide range. For this reason, uncertainty factors are often used.

Data recording: Measurement of the concentration of a pollutant is always subject to error. This effect is relatively well known and can be minimised by suitable methods such as interlaboratory comparisons (Desaules et al., 1996).

The RFD values of the USEPA are an order of magnitude lower than those of the FAO/WHO. There is no scientific justification for this, because it may be assumed that both the EPA and the FAO/WHO have the same knowledge base. This problem will not therefore be eliminated by the accumulation of new knowledge. The fixing of precise evaluation criteria is ultimately a political act.

25.6.3 Cumulation in exposure analysis

A pollutant is transferred from soil to risk receptor through different environmental compartments, in which the pollutant concentration must be determined. The accuracy of measurement varies at each environmental compartment and other natural ranges of variation are also likely to occur. Individual measurement errors and possible ranges of variation are thus cumulative.

25.6.4 Comparability of different studies

Studies of the transfer of pollutants from soil to risk receptor are used for the derivation of regulatory standards. The comparability of the various studies is limited because different procedures are used. The extraction methods in particular, as well as experimental design, in principle allow comparison only if they are similar. Outside Switzerland, total content is often determined by extraction with aqua regia, HCI or other acids. Because not enough data sets for the derivation of regulatory standards exist in Switzerland, international literature can be considered for this reason only studies in which $NaNO_3$ was used in accordance with the VSBo (1986) are considered here. However, comparison between different extractants were made and presented elsewhere (Gupta, 1984; Gupta and Aten, 1993; Gupta et al, 1996).

25.6.5 Limited number of studies

The number of sites in Switzerland with very high levels of soil contamination (approaching to clean-up values) that have been studied systematically is very small. Although some studies do exist, most of them were conducted on land contaminated between five and ten times more than the guide value for a pollutant. They thus fall within the range of trigger values. There are even fewer studies at the level of clean-up values. International studies can be resorted to for the derivation of total content limits. This is not possible in the case of soluble content, for the reasons stated above. This greatly limits the number of studies available. The precision of the derived limits therefore decreases.

25.7 Conclusions

The three-steps concept for risk assessment of soil contamination (1985) based on total and bioavailable concentration was implemented in Switzerland through Ordinance on Soil Impacts (VBBo, 1998). This legal document is based on Swiss Federal Law on Environment Protection (USG$_{rev}$, 1993). This law laid foundation for the evaluation of soil-related hazards to man, animals and plants and for the adoption of appropriate measures.

In the Ordinance on Soil Impacts (VBBo, 1998) besides guide values, trigger and clean-up values are added. The trigger values indicate the threshold values of possible hazard. If the trigger value either total or soluble concentration of element exceeded the standard prescribed values, a more detailed investigation must be carried out to determine whether or not a hazard exists. Clean-up values reveal a level of contamination of soil at which a hazard exists.

This chapter shows how these trigger and clean-up values are derived and what errors are associated with this process. For this purpose the authors show how the hazard analysis can be carried out for the relevant exposure pathways soil→food plant→man, soil→fodder plant→animal→man, soil→plant, soil→man and what evaluation criteria can be applied. This information was used as basis to establish derivation of trigger and clean-up values independently of the relevant pollutants. Trigger and clean-up values for the inorganic pollutants Pb, Cd, Cu and Zn were derived, justified and compared with similar existing values in literature.

References

AltIV Verordnung über die Sanierung von belasteten Standorten vom 26. August 1998 (1996) (Altlasten-Verordnung), SR 814.680, Schweizerischer Bundesrat.

Anonymous: Daten und Dokumente zum Umweltschutz: 11. (1976) Spurenelemente mit toxischer Wirkung-Tabellen über Gehalte in Futtermitteln, Dokumentationsstelle der Universität Hohenheim, Stuttgart.

BGA (1993). Richtwerte für Schadstoffe in Lebensmitteln., Bundesgesundheitsblatt 5 (Bekanntmachung des BGA), 210–211.

BUWAL (1994a). Kreisschreiben Nr. 20. Bundesamt für Umwelt, Wald und Landschaft, Eidgenössische Forstdirektion, 4.

BUWAL (1994b). Anleitungsentwurf zur Durchführung einer Risikoanalyse. Bundesamt für Umwelt, Wald und Landschaft, Eidgenössische Forstdirektion, 7.

Desaules A., Dahinden, R., Lischer, P. (1996). VSBo-Ringversuch-Bericht (1995), Eidgenössische Forschungsanstalt für Agrarökologie und Landbau, Institut für Umweltschutz und Landwirtschaft (IUL), Liebefeld-Bern, 1–47.

FAO & WHO (1996) Summary of evaluations performed by the joint FAO/WHO expert committee on food additives. ILSI Press, Rome.

FIV (1995). Verordnung über Fremd- und Inhaltsstoffe in Lebensmitteln (Fremd- und Inhaltsstoffverordnung), SR 817.021.23, Eidgenössisches Departement des Innern, Bern.

FMBV (1995). Futtermittel: Futtermittelverordnung und Futtermittelbuch-Verordnung mit Anhängen, SR 916.307/916.307.1, Schweizerischer Bundesrat und Eidgenössisches Volkswirtschaftsdepartement, Bern.

Gupta, S.K., 1984. Importance of soil solution composition in deciding the best suitable analytical cirteria for guidelines on maximum tolarable metal load an in assessing bio-significance of metals in soil. Schweiz. Landw. Fo. 23, 209–225.

Gupta, S.K., 1989. Metallverteilung zwischen fester Phase und löslicher Phase des Bodens und ihre Bedeutung zur Beurteilung oekologischer Problems. Bull. BGS 13, 69–74.

Gupta, S.K. (2004). Personal communication.

Gupta, S.K., Aten, C., 1993. Comparision and evaluation of extraction media and their suitability in a simple model to predict the biological relevance of heavy metal concentrations in contaminated soils. Int. J. Environ. Anal. Chem. 51, 25–46.

Gupta, S.K., Vollmer, M.K., Krebs, R., 1996. The importance of mobile, mobilisable and pseudo total heavy metal fractions in soil for three level risk assessment and risk management. Sci. Total Environ. 178, 11–20.

Gysi, C., Gupta, S., Jaggi, W., Neyroud, J.-A., 1991. Wegleitung zur Beurteilung der Bodenfruchtbarkeit, Bundesamt für Umwelt, Wald und Landschaft und Eidgenössische Forschungsanstalt für Agrikulturchemie und Umwelthygiene, Liebefeld-Bern. BUWAL, FAC.

Hämmann M., Gupta, S.K. (1997). Derivation of Trigger and Clean-up Values for Inorganic Pollutants in the Soil. Swiss Federal Agency for the Environment, Forests and Landscape (SAFEL), Bern Environmental Document Nr. 83 Soil, pp. 1–100.

Hapke, H.-J., 1988. Toxikologie für Veterinärmediziner. Enke, Stuttgart.

Hassauer M., Kalberlah, F., Oltmanns, J., Schneider, K. (1992). Basisdaten Toxikologie für umweltrelevante Stoffe zur Gefahrenbeurteilung bei Altlasten, Nr. 102 03 443/01, Forschungs- und Beratungsinstitut Gefahrstoffe GmbH, Freiburg i.Br., 213 S.

Kessler, J., 1993. Schwermetalle in der Tierproduktion. Landw. Schweiz, 6, 273–277.

Kirchgessner, M., 1992. Tierernährung. DLG-Verlag, Frankfurt (Main).

Konietzka, R., Dieter, H.H., 1994. Kriterien für die Ermittlung gefahrenverknüpfter chronischer Schadstoffzufuhren per Bodenaufnahme. In: D. Rosenkranz, G. Einsele, H.-M. Harress (Eds.), Bodenschutz: Ergänzbares Handbuch der Massnahmen und Empfehlungen für Schutz, Pflege und Sanierung von Böden, Landschaft und Grundwasser, Erich Schmidt Verlag, Berlin, 3530.

Lindt, T.J., Fuhrer, J., Stadelmann, F.X., 1990. Kriterien zur Beurteilung einiger Schadstoffgehalte von Nahrungs – und Futterpflanzen (Schriftenreihe der FAC Liebefeld Nr. 8) Eidgenösische Forschungsanstalt für Agrikulturchemie und Umwelthygiene, Liebefeld-Bern, 156 p.

LRV 1985. Luftreinhalte-Verordnung vom 16. Dezember 1985 (LRV) SR 814.318.142.1 BAFU Switzerland, Bern.

Merian, E. (Ed.) 1984. Metalle in der umwelt, Verlag chemie, Basel, 722p.

Meyer, K. 1991. Bodenverschmutzung in der Schweiz: Untersuchungen und ausgewählte Ergebnisse über die Stoffbelastung von Böden, Stand 1990, Nationales Forschungsprogramm "Boden" (NFP22), Liebefeld-Bern.

Naidu, R., Kookana, R.S., Oliver, D., Rogers, S., McLaughlin, M.J., 1996. Contaminants and the Soil Environment in the Australasia-Pacific Region. Kluwer Academic Publishers Group, Hingham, MA, 717 pages.

NRC (1980). Mineral Tolerance of Domestic Animals. National Research Council, Washington, DC, 577 S.

Scharamm, E., 1989. Lobbies zur Festlegung von Grenzwerten im Bodenschutz. Forum Wissenschaft (studienheft), 7, 115–117.

Schroeder, W., Frönzle, O., Daschkeit, A., 1993. Festsetzung, bodensützender Normwerte. Pratische Notwendigkeit im Spannungsfeld von Erkentntnismöglichkeit der Oekosystemforschungund Anforderung des Umweltschutzes, Deutsche MAB-Mitteilung, 37, 79–90.

StoV (1986). Verordnung über umweltgefährdende Stoffe vom 9. Juni 1986, SR 814.013.

TERA (1997). International Toxicity Estimates for Risk (ITER), Toxicology Excellence for Risk Assessment, http://www.tera.org/iter/glossary.html.

Tuchschmid, M.P., 1995. Qunatifizierung und regionalisierung von Schwermetall- und Flourgehalten bodenbildender Gesteine in der Schweiz (Umweltmaterialien Nr. 32 –Boden), Bunddesamt für Umwelt, Wald und Landschaft (BUWAL now BAFU), Bern.

TVA 1990. Technische Verordnung über Abfälle vom 10. Dezember 1990 (TVA) 10. Dezember 1990 *Inkrafttreten*1. 814.600 Februar 1991 AS 1991 169.

US-EPA (1996) Soil Screeening Guidance: User's Guide, Nr. 9355.4-23. Office of Emergency and Remediation Response. Environmental Protection Agency, Washington.

USG (1983). Bundesgesetz über den Umweltschutz (Umweltschutzgesetz) vom 7. Oktober 1983, SR 814.01, Bundesversammlung der Schweizerischen Eidgenossen-schaft, Bern.

USG (1997), Bundesgesetz über den Umweltschutz (Umweltschutzgesetz9 vom 7. Oktober 1983, Stand 1. Juli 1997, SR 814.01, Bundesversammelung der Schweizerischen Eidgenössenschft, Bern.

USG$_{rev,}$ 1993. Botschaft zu einer Aenderung des Bundesgesetzes über den Umweltschutz, SR93.053, Sschweizerischer Bundesrat, Switzerland, Bern.

VBBo. (1998). Verordnung über Belastungen des Bodens SR 814.12, Schweizerischer Bundesrat, Switzerland, Bern.

Vollmer, M., Gupta, S.K. (1995). Risk Assessment Plan for Contaminated Soils in Switzerland. FAC and now Agroscope FAL, Reckenholz, Swiss Federal Research Station of Agroecology and Agriculture (FAL), Reckenholzstrasse 191, 8046 Zürich (www. Reckenholz.ch).

VSBo. (1986). Verordnung über Schadstoffe im Boden (aufgehoben)SR 814.12, Schweizerischer Bundesrat, Bern.

Zimmerli, B. (1997). persönliche Mitteilung.

Developments in Soil Science, volume 32
Ravendra Naidu (Editor)

Chapter 26

CONTAMINANTS IN THE ROOTZONE: BIOAVAILABILITY, UPTAKE AND TRANSPORT, AND THEIR IMPLICATIONS FOR REMEDIATION

B.E. Clothier, S.R. Green, B.H. Robinson, T. Thayalakumaran, D.R. Scotter, I. Vogeler, T.M. Mills, M. Deurer, M. van der Velde and Th. Granel

26.1 Introduction

26.1.1 Roots are the big movers of water and chemicals in soil

As soil solution moves through the porous medium of the rootzone, its chemical concentration changes in response to exchange processes with the reactive surfaces of the soil's matrix (see Chapters 2, 6, 7 and 11). The chemical makeup of the soil solution also changes as a result of the distributed and active uptake of water and solutes by plant roots. Chemicals that are taken up by roots and incorporated into the plant, whether they are nutrients or contaminants, are, not surprisingly, said to be bioavailable (see Chapter 3 for definition). Bioavailability is intimately linked to soil–chemical exchange processes, and it is also integrally dependent upon uptake mechanisms at the root–soil interface, as summed across the root system that is distributed throughout the soil.

Bioavailability of chemicals in the rootzone is a phenomenon that is not only critical for the productive uses of fertile soil, but it is also important for protecting surface and groundwaters from degradation by leachates (Clothier, 1997). Additionally, bioavailability is a key mechanism in the burgeoning use of plants, and other soil biota, for remediating contaminated sites.

The dual dimensions of bioavailability have, it seems, led to a polarisation in scientific emphasis. Both have, nonetheless, been productive research directions that have broadened and deepened our understanding of bioavailability. On the one hand, soil scientists appear to have concentrated on bioavailability as being a process mediated by chemical exchange mechanisms. Their focus has been on understanding the nature and 'strength' of the binding, or sequestration, of chemicals by the soil's reactive exchange sites. Plant scientists, on the other hand, have sought to understand bioavailability from the perspective of the roots themselves, and whole-plant physiology. Their goal has been to determine the 'strength' of the roots to suck up, or reject, chemicals, and to understand the cellular processes whereby this exchange can either be enhanced, or mitigated. Beneficially, there has been a convergence of these soil-oriented and plant-based approaches, which has resulted in an improved understanding and management of bioavailability. The reason for this alignment has, it is considered, been a result of improved measurement devices and monitoring techniques for observing transport, exchange and uptake processes in the rootzone, complemented by the rapid development of quantitative modelling tools that can describe the complex of biophysical mechanisms that operate in the rootzone.

Here we employ this measurement-modelling dualism, in a review of our team's recent studies. We detail our attempts to decipher the mechanisms that control the bioavailability, transport and plant uptake of both metals and inorganic contaminants in soil, and we discuss their implications for phytoremediation of contaminated sites using poplars and willows. Nyer and Gatliff (1998) "... predict that phytoremediation will be the next 'hot' technology for the environmental remediation field", yet they caution "... that this technology is not simply the buying of plants from the local K-mart and placing them in the soil near a contaminated site"!

We begin with a description of the results from screening experiments that have identified a huge range in the ability of our willow clones to extract cadmium from contaminated soil. Beyond these pot experiments with our willow clones, our emphasis has been on the entire soil–plant–atmosphere continuum so that we can detail the dynamics and ultimate fate of water and contaminants in the rootzone. Pot trials fail to mimic field conditions where the soil is heterogeneous, and where the root system possesses a complex topology. We will show how new devices and better measurement techniques can provide data to improve our understanding of bioavailability under field conditions. These data serve to highlight remaining nescience concerning the functioning, and leakiness, of phytoremediation systems (Schnoor et al., 1995; Salt et al., 1998; Pivetz, 2001). We describe a field project, and large-scale lysimeter experiments that are exploring the ability of poplars both to dewater and remove boron from contaminated sawdust piles. Next we detail our experiments and analyses that have determined the fate of Cu that is mobilised from soil by ethylenediamine tetra-acetic acid (EDTA) in an effort to enhance phytoextraction. Again these results highlight the leakiness of the soil of the rootzone, and this identifies the need for still greater understanding of the role of dynamic transport processes in soil, and active uptake mechanisms of chemicals in the rootzone.

We then demonstrate the power of quantitative modelling for providing predictions of the fate of rootzone water and contaminants. The predictions made by comprehensive mechanistic models of transport and uptake in the rootzone can be seductive, but we stress that they must be interpreted cautiously. To illustrate this, we conclude with a warning about the difficulty in separating out, and modelling the consequences of the soil-determined and plant-controlled mechanisms that operate contemporaneously in the rootzone to control plant uptake, and thereby determine bioavailability.

26.2 Clonal variation in the bioavailability of cadmium

Cadmium (Cd) can be a soil contaminant as a result of biosolids application, road run-off, mining or industrial activities, as well as through the extended use of marine phosphatic fertilizers. Willows have been found effective at extracting Cd from soil (Östman, 1994; Robinson et al., 2000). Mills et al. (2000) found that Tangoio willow (*Salix matsudana* × *S. alba* NZ 1040), growing in soil with 20.6 mg-Cd kg^{-1}, could accumulate up to 250 mg-Cd kg^{-1} (dry weight) in its leaves. Thus, willows would seem to offer good prospects for phytoremediation of Cd-contaminated soils, especially since Mills et al. (2000) noted the higher rate of water usage by willows in comparison to poplars. Hence we began a search of the willow clones bred by HortResearch for their ability to take-up, or exclude Cd. This would, we felt, allow selection of high Cd-accumulating clones for phytoremediation, and tailored choice of the low Cd-accumulators for stock feed.

Poles of 15 clones were inserted into replicated 20 L containers of Manawatu fine sandy loam with a naturally resident Cd concentration of 0.3 mg/kg. The cadmium in this soil reflects its history of fertilisation with marine superphosphate. Eight clones were shrub willows (sub-genus *Caprisalix*) and seven were tree willows (sub-genus *Salix*). Further details can be found in Granel et al. (2001). After some 3 months growth in a glasshouse over summer, the trees were harvested, and the concentration of metals was measured in the leaves. The results for Cd are shown in Figure 26.1.

The mean bioaccumulation coefficient, namely the plant–soil concentration ratio of Cd is 12, but the distinguishing characteristic of the Cd-concentration data is the great variation, from a high of 33, down to 6. Whereas these simple pot-experiments do not allow for an in-depth exploration of the mechanisms controlling the bioavailability of Cd, a piecewise application of the non-parametric Spearman rank correlation test, r_s, (Zar, 1972) does permit some interesting speculation. A comparison of the clones when ranked in terms of Cd and water use per unit leaf area was not significant ($r_s = -0.023$, ns $n = 15$), whereas there was a significant Spearman rank correlation between the Cd and manganese content of the leaves ($r_s = 0.786$, $P = 0.05\%$, $n = 15$). Yet there was only a weak correlation between the Cd and zinc concentrations ($r_s = 0.489$, $P = 5\%$, $n = 15$). Whereas there was no Cd-water use linkage, there was a weak correlation in the ranking of the clones with respect to leaf Zn and water use per unit leaf area ($r_s = 0.414$, $P = 10\%$, $n = 15$). We did not measure levels of metal accumulation in the roots, although Negri et al. (1996) found high levels of Zn sequestration in the dry tissue of the root system of poplars.

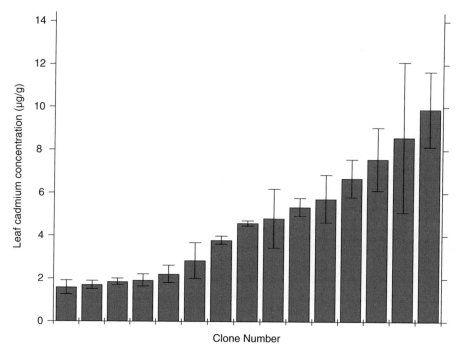

Figure 26.1. *The variation in cadmium uptake in the leaves and stems of replicated treatments of 15 clones of willow growing in 20 L of soil with a cadmium concentration of 0.3 ppm Cd (following Granel et al., 2001).*

This rank analysis and non-parametric testing demonstrates that, under these controlled conditions of a pot-trial, the phytoextraction capacity, or bioavailability of Cd is clonally dependent on rhizosphere factors unrelated to water-use, and the level of Cd in the soil. It would seem that there have been genetically controlled, plant-mediated increases in Cd bioavailability. An increase in the bioavailability of contaminants has been attributed to root exudates either increasing the solubility of the contaminant itself, or by competing with the contaminant for binding sites on the soil's matrix (Siciliano and Germida, 1998). Exploring these mechanisms, and their genetic control, would appear to be an exciting and profitable area of phytoremediation research.

26.3 Boron uptake from contaminated sawdust by a poplar

Borated compounds such as boric trioxide, along with chromated copper arsenate, are used commonly as timber preservatives (CCA-B). The post-treatment dressing of lumber produces sawdust contaminated with heavy metals and boron. Poor past practices of sawdust disposal have created numerous point sources of contamination, as the CCA-Bs in the bare-sawdust pile can leach into both surface and groundwaters. Concern has also focused on the environmental risks associated with the use of treated timber in playgrounds and marinas (Kluger, 2001).

Near Kopu, in the Coromandel Peninsula of New Zealand, there is a 5 ha pile of contaminated sawdust, which has been formed over 30 years or so. It is up to 15 m deep in places. As well as organic leachates, and traces of heavy metals, the pile is leaking boron at approximately 6–7 ppm, which is well above the New Zealand Drinking Water Standard (NZDWS) of 1.4 ppm. In conjunction with Fletcher Challenge Forests we initiated, in 2000, a project to phytoremediate the pile using HortResearch's poplar and willow hybrid clones. Willows, and especially poplars are known to accumulate high levels of boron, and other contaminants, in their foliage (Schnoor et al., 1995; Bañuelos et al., 1999). A trial planting with our 1 m poplar and willow poles in July 2000 permitted the identification of superior clones, and allowed us to develop a site management strategy to enhance the growth of the trees on this high C/N medium. Full planting of the site was performed in July 2001, and by early spring in October 2001, the poplars and willows had developed extensive canopies (Fig. 26.2).

Figure 26.2. *A contaminated sawdust pile undergoing active phytoremediation with a mixture of eucalypts, willows and poplars. Leachate collected from the pile is being understorey irrigated onto the 1- and 2-year-old trees.*

Figure 26.3. *Large weighing lysimeters containing 750 L of contaminated sawdust growing one poplar clone, with multiplexed TDR measurements of water content and electrical conductivity. Leachate volume was also monitored continuously, and the chemical concentration assessed regularly. Treatments included varying levels of irrigation and fertilizer application.*

Meanwhile, we initiated a large lysimeter project to determine, in greater detail, the mechanisms and controls on water transmission and leachate purification. The 750 L lysimeters, growing one clone of poplar (Kawa, *Populus deltiodes × P. yunnanensis* NZ 5006), were weighed continuously, and their drainage monitored (Fig. 26.3). The sawdust water content was measured frequently using our multiplexed Time Domain Reflectometry (TDR) system (Green and Clothier, 1999). Leachate samples were collected regularly, and the metal and boron accumulations in the leaves of the poplars were frequently measured throughout the growing season. Eight lysimeters were used, and there was a range of three irrigation treatments, and three fertilizer regimes.

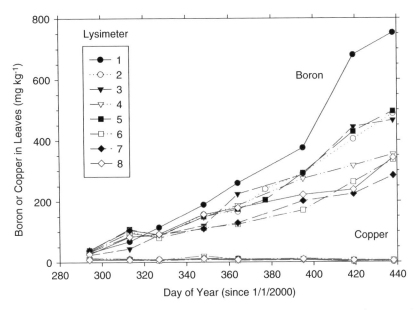

Figure 26.4. *The seasonal growth in the concentration of both boron and copper in the leaves of first-year poplars growing in the eight lysimeters shown in Figure 26.3. The average level of boron in the contaminated sawdust was just 19 ppm.*

The seasonal development in the boron concentration in the leaves of the eight poplars is shown in Figure 26.4. The average concentration of B in the sawdust used to fill the eight lysimeters was $19.4\,\mathrm{mg\,kg^{-1}}$, so that the maximum bioaccumulation coefficient for B was 39. The level of B in the wood of the tree was more that an order of magnitude lower than that in the leaves. Mills et al. (2000) found a two-fold ratio between cadmium in the leaves and stems for poplars and willows. Since the sawdust has a bulk density of just $0.2\,\mathrm{T\,m^{-3}}$, the top metre of sawdust would have a load of $38\,\mathrm{kg\text{-}B\,ha^{-1}}$. Each poplar in its first year of growth produced approximately 275 g of leaves. In subsequent years, if they were not annually coppiced, they would produce much more. If we take an average leaf B concentration of $500\,\mathrm{mg\,kg^{-1}}$, then if the trees were planted at a $2 \times 2\,\mathrm{m}$ spacing (2500 trees $\mathrm{ha^{-1}}$), there would be an annual phytoextraction of $0.34\,\mathrm{kg\text{-}B\,ha^{-1}}$. This simple calculation would suggest that a maximum of less than 100 years would be needed to remove all the B from the top metre of the pile. Not coppicing every year would increase the amount of B extracted, for the second, and later years' leaf growth will exceed 250 g, and reduce the remediation timescale. Means of collecting the annual leaf fall would need to be developed though. A longer coppicing cycle would probably be acceptable, as long as offsite leachate impacts were eliminated. The optimum frequency of

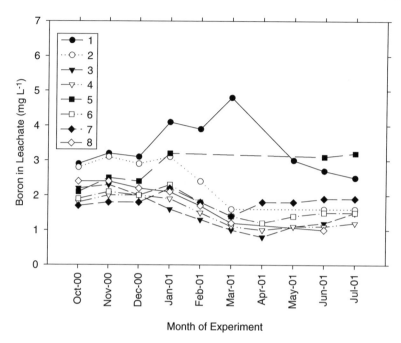

Figure 26.5. *The level of boron measured in the leachate from the eight lysimeters of contaminated sawdust shown in Figures 26.3 and 26.4.*

coppicing will thus need to be determined from the balance between greater phytoextraction of the contaminant, and leachate minimisation.

The seasonal trend in the levels of B in the leachate from the eight lysimeters is shown in Figure 26.5. The NZDWS for B is $1.4\,\mathrm{mg\,L^{-1}}$, so the level in the leachate from the lysimeters exceeds drinking water standards. The leachate from the lysimeters measured here is of the same order as that observed in the leachate from the field site. If we focus on the lysimeters with the medium and high irrigation treatments, that is we ignore lysimeters 1 and 5 because of their low drainage amounts, it can be seen that the trees have had a positive effect in reducing the B concentration in the leachate. In October, the medium and high water treatments had an average leachate concentration of $2.1\,\mathrm{mg\,L^{-1}}$ B ($\pm$ 0.4, $n = 6$), which dropped to $1.5\,\mathrm{mg\,L^{-1}}$ B ($\pm$ 0.4, $n = 6$) in July. The extraction of B from the sawdust by the trees is seen to be having a positive effect on the B concentration in the leachate of these lysimeters. Indeed, here, with the higher rates of drainage where dilution offers no solution, the level of B has almost reached the NZDWS.

Bañuelos et al. (1999) found that when they irrigated poplars with solutions up to $5\,\mathrm{mg\,L^{-1}}$ B, the trees could accumulate up to $750\,\mathrm{mg\,Kg^{-1}}$ B in their lower, and oldest, leaves. Our results are similar. However, in our case, we have a good measure of the concentration of B in the mobile soil solution (Fig. 26.5), and furthermore we have very good measurements of the trees' cumulative transpiration (T; L) over the duration of the experiment. As well, we have good measures of the leaf yield of the trees (Y; kg), and the B concentration in the leaves (C_B; mg kg^{-1}). From these data we can then back-calculate what must have been the effective concentration of B entering the root system (C_{eff}; mg L^{-1}) of the trees, since over the growth season, mass balance demands

$$C_{eff}T = C_B Y \qquad (26.1)$$

as long as we ignore B accumulation in other organs of the trees. The values we calculated for C_{eff} for each of the lysimeters are compared in Figure 26.6 with the seasonally averaged value of the B concentration in the leachate (from Fig. 26.5).

Unlike the data of Hinkley (1979), here the effective concentration of boron taken up by the trees is only approximately 10% of that in the leachate stream. This has significant phytoremediation implications for the hydrology of a polluted site, should there be any drainage emanating from the base of the contamination. This stresses the need for phytoremediating trees to be such

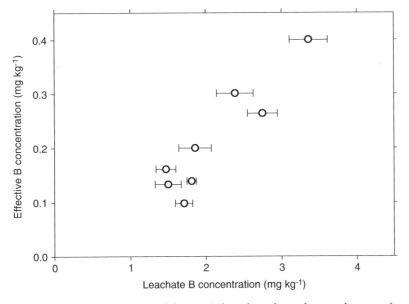

Figure 26.6. *The effective level of boron inferred to have been taken up from the observed concentration in leaves, given the transpiration, relative to that measured in the leachate. The scales on the axes are arranged such that the unit slope would correspond to a 1:10 ratio of uptake efficiency of the leachate loading in boron.*

effective dewaterers as to minimise drainage. Leaf area is critical for transpiration, and regular coppicing would, through the frequent removal of leaves and stems, result in lower transpiration, and thereby a greater likelihood of drainage. For our contaminated sawdust pile, leachate was collected in a bunded dam, and the boron-laden leachate irrigated back on to the pile (Fig. 26.2), so that the trees could have another chance to extract the contaminants. As the trees grow bigger, there will be a diminution of drainage. So, selectively staged coppicing is considered the appropriate form of contaminant removal from the site, as there is a continuous need to ensure that offsite hydrologic effects are eliminated. Hydraulic control of leaching is critical, and this demands good understanding of root system functioning and tree transpiration. As concluded by Pivetz (2001), "field studies will be necessary [and] successful phytoremediation is likely to be achieved only through the combining of expertise from numerous scientific disciplines". Off-site hydrologic impacts will, we suggest, be a critical part of phytoremediation risk assessments.

Although our sawdust pile contains a significant amount of Cu (approximately $13 \, mg \, Kg^{-1}$), Cu was neither found in the leachate at high levels (approximately $0.1 \, mg \, L^{-1}$), nor did it accumulate significantly in the leaves of the trees ($< 5 \, ppm$). This recalcitrance has led to the investigation of techniques that seek to enhance metal ion bioavailability by the addition of metal chelates, the so-called method of chelate-assisted phytoextraction (Salt et al., 1998).

26.4 Consequences of chelation for enhancing copper extraction by plants

Certainly pot studies have demonstrated the potential for chelates such as EDTA, *inter alia*, to enhance the bioavailability and plant uptake of Cu, and other metals (Deram et al., 2000). What is less clear, however, is how chelation can affect the transport of the now-mobilised contaminants through the rootzone. Nonetheless, there have been significant studies concerning the transport of organically chelated radionuclides resulting from efforts to decontaminate sites polluted by nuclear weapons production (e.g. Jardine and Taylor, 1995). In an earlier study, we added a Cu solution to artificially contaminate the surface soil of a large lysimeter growing a poplar tree (Vogeler et al., 2001a). Later, a pulse of EDTA was added to the soil. A breakthrough of Cu was observed in the leachate some 17 days after the addition of the chelate. By destructive sampling at the end of the experiment, the solubilisation of the Cu as a result of chelation was found to have smeared the profile of resident Cu down through the soil to the base of the lysimeter at 1.2 m.

Greater understanding is thus required of the hydraulics of chelation therapy for phytoremediation, and this has been the focus of the doctoral studies of Thabonithy Thayalakumaran. Here, we review salient findings of her research, with a particular emphasis on determination of the partitioning between the greater plant uptake of Cu through chelate-enhanced bioavailability, and the greater risk of its leaching loss beyond the rootzone to receiving waters.

First, we need to consider a theoretical description for the coupled transport phenomena. During steady water flow, and at this stage ignoring plant uptake, the transport of the three forms of EDTA, and that of a halide tracer, can be described using the convection–dispersion equation in the form

$$\theta \frac{\partial C}{\partial t} = \theta D \frac{\partial^2 C}{\partial x^2} - q \frac{\partial C}{\partial x} + S \qquad (26.2)$$

where θ is the volumetric water content ($m^3\,m^{-3}$), C is either the concentration of the halide, or of the free-, Cu- or Fe-EDTA in the soil solution ($mol\,m^{-3}$), t the time (s), D the dispersion coefficient ($m^2\,s^{-1}$), x the distance in the direction of flow (m), q the steady Darcy flux density ($m\,s^{-1}$), and S a term accounting for any chemical reactions bringing that chemical species into, or out of solution ($mol\,m^{-3}$ of soil s^{-1}).

For the halide tracers $S = 0$. For EDTA, we consider three major reactions. The first two relate to the added EDTA ($EDTA_0$) reacting with Cu and Fe in the soil to form complexes that we will denote as $CuEDTA^{2-}$, and $Fe(III)EDTA^-$. The third reaction is $CuEDTA^{2-}$ reacting with Fe in the soil to form $Fe(III)EDTA^-$ and Cu ions, which are then adsorbed by the soil. So we assume

$$S_{Cu} = K_1(C_0 M_{Cu})^n - K_3(C_{Cu} M_{Fe})^n \qquad (26.3)$$

$$S_{Fe} = K_2(C_0 M_{Fe})^n + K_3(C_{Cu} M_{Fe})^n \qquad (26.4)$$

$$S_0 = -S_{Cu} - S_{Fe} \qquad (26.5)$$

Here, S_0, S_{Cu} and S_{Fe} are the source/sink terms for $EDTA_0$, $CuEDTA^{2-}$ and $Fe(III)EDTA^-$, respectively, and n is a dimensionless constant indicating the order of the reaction. K_1 and K_2 are the rate constants for the reactions between $EDTA_0$ and the soil copper and the iron, respectively. K_3 is the rate constant the reaction between $CuEDTA^{2-}$ and the soil's iron. C_0, C_{Cu} and C_{Fe} denote the soil solution concentrations of $EDTA_0$, $CuEDTA^{2-}$ and $Fe(III)EDTA^-$. M_{Cu} is the EDTA-extractable copper concentration in unit soil volume ($mol\,m^{-3}$), and M_{Fe} is the concentration in unit soil volume of readily EDTA-extractable iron in the soil ($mol\,m^{-3}$).

Initial experiments were conducted to examine this theory, and to determine the impact of the addition of chelate on the mobility of the resident copper and iron. A small vertical column was packed with Manawatu fine sandy loam that had Cu added to it some weeks prior to the experiments. Water was then added to the column via a peristaltic pump until a steady-state saturated flow rate of 24 mm h^{-1} was achieved, and aliquots of the distal efflux were collected. Then, a 13 mm pulse of 0.001 M EDTA, along with a Br tracer, was infiltrated. Next, after some 900 mm of infiltration had occurred, another 13 mm pulse of EDTA and Br was infiltrated, but this time at 0.01 M. The concentrations of Br, Cu and Fe are shown in Figure 26.7, along with the predictions made using Eqs. (26.2)–(26.5). There it can be seen that the EDTA was effective at mobilising the Cu, as well as the resident iron, and transporting it as rapidly as the inert solute Br. Also, as can be seen by the relative areas under the curves in Figure 26.7, the stronger the EDTA solution, the greater the yield of mobilised Cu, and Fe. So not only is EDTA effective in rendering Cu more bioavailable, it can create a mobility for the resident contaminant that could, under certain circumstances, lead to it being leached into receiving water bodies. Thus, we were keen to explore the partitioning between the chelate-enhanced plant uptake, and leaching following the EDTA-mobilisation treatment.

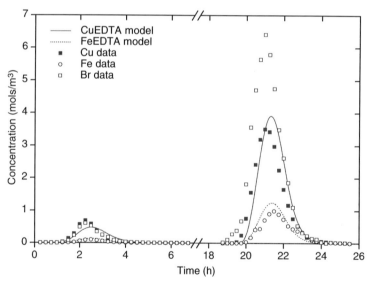

Figure 26.7. *Measured effluent concentrations during leaching through repacked Manawatu fine sandy loam of bromide tracer, plus that of mobilised iron and copper following the application of an EDTA solution. At first, a 13 mm pulse of 0.001 M EDTA was applied, and then after 900 mm of infiltration another 13 mm EDTA pulse, but this time at 0.01 M, was infiltrated. Predictions, using Eqs. (26.2)–(26.5), are shown for both copper (solid line) and iron (dashed line).*

Both undisturbed cores, and repacked columns, of Opotiki sandy loam previously contaminated with Cu were planted with the copper-tolerant pasture grass *Agrostis tenuis*. The undisturbed cores came from an orchard where long-term use of Cu sprays had resulted in copper levels of $290\,\mu g\,g^{-1}$ in the top 20 mm of soil, ranging down to $20\,\mu g\,g^{-1}$ at 100–150 mm. The material for the repacked column was taken from a surface sample of this orchard's soil, and when repacked into the columns the Cu was found to be uniformly at $265\,\mu g\text{-Cu}\,g^{-1}$. The pots were 100 mm in diameter, and 150 mm deep, so that the total amount of Cu in the columns was 3500 μmol, and 1750 μmol in the cores. The plants achieved a yield of 380 (± 8) $g\,m^{-2}$, or $3.8\,T\,ha^{-1}$ equivalent. Two months before harvest, the plants were irrigated with 23 mm of 0.01 M EDTA. Then on every day, the cores and columns were wetted to bring them back to pot capacity, while ensuring that there was no leachate loss from the base. Then following harvest, disc permeameters containing water and set to a tension head of −10 mm, were used to leach seven pore volumes through the cores and columns. The Cu concentrations in the leachate, and that resident in the cores were then determined.

Figure 26.8a shows the amount of Cu found in the plants at harvest. Certainly in the repacked column, the level of Cu in the plant had become elevated, although the Cu concentration in the pasture growing on the undisturbed core was little different to that of the control. It would seem that the heterogeneous distribution of both the roots, and the Cu contaminant in the undisturbed soil, along with the likely preferential invasion of the EDTA solution, has conspired to render the mobilisation treatment a lot less effective for enhancing plant uptake.

The fate of the total Cu in the columns and cores is shown in Figure 26.8b. A little of the Cu in the control columns was either taken up by the plant, or leached. Note the broken scale in Figure 26.8b. For the repacked column, even though the plant concentration of Cu became elevated after the EDTA treatment (Fig. 26.8a), in terms of the total Cu in the system, less than 0.5% actually ended up in the plant, whereas some 10% of the copper was leached out the base of the column. This uptake–leaching disparity was even greater for the undisturbed core, with 0.2% taken up by the plant, and nearly 20% leached.

These results identify the need for caution with respect to leaching losses if enhanced phytoremediation strategies seek to use chelation as a means of encouraging phytoextraction of contaminants. As with the work of Grčman et al. (2001), here, the rootzone has again been shown to be 'leaky', and the interaction between distributed uptake by plant root systems, and the topology of preferential flow paths, the bane of soil physics, is highlighted. Nonetheless, better understanding of the mechanisms and patterns of distributed uptake of water and chemical across the root system of plants may itself lead to improved tactics for phytoremediation.

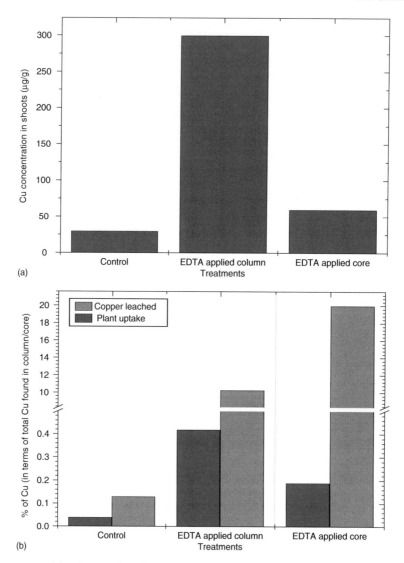

Figure 26.8. *(a) The uptake of copper by pasture from copper-contaminated soil at 280 µg g⁻¹, under natural conditions (left), and in a repacked column following the application of 23 mm of 0.01 M EDTA (middle), and for pasture growing on an undisturbed core following the same EDTA treatment (right). (b) The proportion of copper taken up by the plant, and that leached, as a proportion of the total copper found in the soil. The total amount of leaching following harvest was seven pore volumes.*

26.5 Distributed uptake across the root system

Roots are distributed laterally and vertically throughout surface soil, and the manner in which plants extract water, and chemicals across their root system is not well understood. Even if bioavailable chemicals were uniformly distributed throughout the rootzone, the pattern of active root uptake will determine the sources and quantities of chemical extracted by the plant. Soil physicists and physiologists have long tried to unravel this distributed uptake pattern simply for water (Gardner, 1960; Feddes et al., 1978; Clothier and Green, 1997; Green and Clothier 1999). The uptake of chemicals will be even more complicated, depending on whether active or passive uptake mechanisms are employed by the plant. There is often an even further dimension to this complexity, in that the bioavailable chemical is generally not uniformly distributed depthwise throughout the profile. So, the amount of bioavailable chemical that is actually taken up by the plant will be critically dependent upon the depthwise pattern of water-uptake activity by the roots. Here, we show how new measurement technologies can be employed to determine the details of the distributed pattern of water uptake by roots. These results permit development of irrigation tactics that could be used to enhance the efficacy of phytoremediation systems, especially where the contaminants are located close to the soil surface.

The results we present here come from an intensive study of the pattern of water uptake by the roots of irrigated apple trees. The soil surrounding two 12-year-old apple trees was instrumented with a multiplexed array of 150 mm long, TDR probes. The three-wire probes were attached to the end of tubes of various lengths so that they could be used to measure the depthwise profile of changing water content. The depthwise suites of probes were grouped into nine compact sites at various distances away from the trees. Here, we present the average results from the TDR probes that measured the soil's volumetric water content at the centred depths of 100 and 500 mm. The trees were enrobed in an understorey rainout shelter below which were located microsprinkler irrigation emitters. During mid- to late-summer, daily measurements of the soil's water content were made. Under one tree (Fig. 26.9) the schedule of irrigation was set to apply approximately 30 mm of water every 10, or so, days. The amounts of irrigation were recorded by an array of small rain gauges. For the other tree (Fig. 26.10), a similar irrigation schedule was initiated, however after three wettings, the regime was changed to smaller aliquots of 5 mm applied more frequently at 1–2 day frequencies. This change in irrigation tactics can be seen to have dramatically altered the depthwise pattern of the root extraction of water.

When the irrigation of the soil was by large, infrequent amounts (Fig. 26.9), the roots at both 0.1 and 0.5 m were active in extracting the water from the soil,

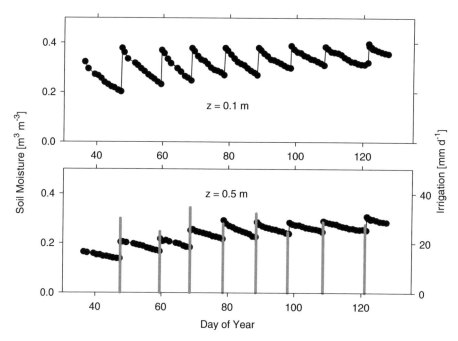

Figure 26.9. *Measurements using 150 mm long TDR waveguides of the average soil water content around an apple tree (n = 9) with the probes centred at depths of 0.1 m (top) and 0.5 m (bottom). The soil is a Manawatu fine sandy loam. Also shown are the eight microsprinkler irrigation events applied during the first 4 months (1996).*

as evidenced by the changing pattern of soil water content. However, when the regime was changed to small, frequent wettings, the tree elected to preferentially extract the surface water at 0.1 m, while 'switching' off those roots at the depth of 0.5 m (Fig. 26.10). This ability to control, by irrigation tactics, the depthwise pattern of root activity, should allow enhancement of phytoremediation of surface-bound pollutants by repeatedly focussing root activity in the zone of contamination. Conversely, in semi-arid or arid environments, even if there were bioavailable chemicals located near the surface, the plant is less likely to extract these compounds by virtue of its transference of root-uptake activity to the nether regions of its root zone. This again reinforces Pivetz's (2001) call for a multidisciplinary approach to understand the bioavailability of contaminants and the prospects for phytoremediation.

Thus, there is a need to combine our understanding of the physical controls on the transport and uptake of chemicals, with our knowledge of the physiological functioning of whole-plant systems. Measurements are providing us with vision of greater acuity of the biophysical processes operating in the

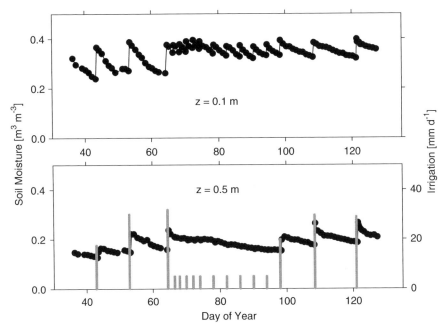

Figure 26.10. *Measurements of the soil's water content, as in Figure 26.9, but on a neighbouring tree. In this case, however, after the third wetting, the frequency of irrigation was increased and the aliquot reduced from 30 to 5 mm.*

rootzone, while the burgeoning development of quantitative modelling schema offer great potential for understanding and predicting bioavailability, transport and uptake across heterogeneous root systems.

26.6 Modelling and parameterising biophysical transport mechanisms

Whereas the theory describing water flow and chemical transport through unsaturated soil, in the presence of distributed uptake of both by roots, has long been defined (Richards 1931; Gardner, 1960), it has been the phenomenal growth in computing power for numerically solving these coupled, non-linear, partial differential equations that has led to readily available packages for quantitatively predicting of solute movement and uptake in the rootzone. Comprehensive mechanistic models can now provide the biophysics researcher with detailed predictions of chemical status, solute mobility and plant uptake under realistic boundary conditions. This surge in model availability has been complemented by the development of new measurement devices that are capable of measuring rootzone processes on spatial and temporal scales unthought of just a decade ago. Nonetheless, problems of model verification and process parameterisation remain. We conclude with a review of some of our recent work (Vogeler et al., 2001b) that not only highlights the predictive power of mechanistic models, but it serves to identify remaining difficulties in unravelling the plant-controlled and soil-based connections that determine the bioavailability of chemicals in the rootzone.

The transient transport and exchange of reactive solutes moving through soil during transient water flow can be described by the one-dimensional convection–dispersion equation, written as:

$$\frac{\partial(\rho S)}{\partial t} + \frac{\partial(\theta C)}{\partial t} = \frac{\partial}{\partial z}\left(\theta D_s \frac{\partial C}{\partial z}\right) - \frac{\partial(q_w C)}{\partial z} - U_s(z) \tag{26.6}$$

where C is the concentration of chemical in solution (mol m^{-3}) and S is that chemical's concentration adsorbed on the soil's matrix (mol kg^{-1}), and θ the soil's volumetric water content ($\text{m}^3 \text{m}^{-3}$) and ρ the bulk density of the soil (kg m^{-3}).

The solute diffusion–dispersion coefficient is D_s ($\text{m}^2 \text{s}^{-1}$), and q_w is the flux density of water flow (m s^{-1}) and U_s ($\text{mol m}^{-3}\text{s}^{-1}$) is the distributed uptake of solute by the root system of the plant. The solute diffusion–dispersion coefficient we take as being

$$D_s = \tau D_0 + \frac{\lambda q_w}{\theta} \tag{26.7}$$

where τ is the tortuosity of the soil, D_o the free-water diffusion coefficient of the solute ($\text{m}^2 \text{s}^{-1}$), and λ the dispersivity of the soil (m). The parameter λ describes the degree of smearing a solute pulse undergoes as it moves through soil. This dispersion is a result of both diffusive processes and hydromechanical mechanisms, and of the interactions between them.

The local uptake of solute U_s by roots we can describe as either being passive,

$$U_s = U_w C \quad \forall C \tag{26.8}$$

or it can be represented as being passive just up to a certain critical threshold concentration, C_1, after which there is active exclusion of solutes, viz.

$$\begin{aligned} U_s = U_w C \quad & C \le C_1 \\ U_w C_1 \quad & C > C_1 \end{aligned} \tag{26.9}$$

Experiments were performed in lysimeters of repacked Manawatu fine sandy loam that were 900 deep, and 305 mm in diameter. The set-up is described in detail in Vogeler et al. (2001b), so only salient features are repeated here. Tomato plants were grown in the lysimeters, and a modified Hoagland solution

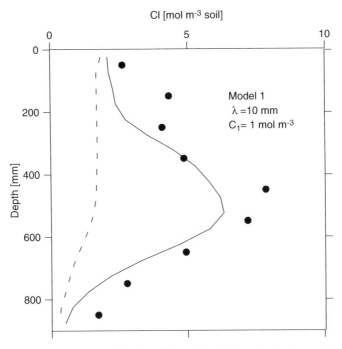

Figure 26.11. *The measured profile of resident chloride in a lysimeter of Manawatu fine sandy loam at the end of a tomato growth cycle. The dashed line is the prediction using the assumption that Cl uptake is passive (Eq. 26.8), whereas the solid line assumes that the maximum uptake of Cl is 1 mmol L^{-1} as in Eq. (26.9) (from Vogeler et al., 2001b).*

regularly irrigated on to the soil. This solution had a Cl concentration of 5.3 mol m^{-3}. The top 100 mm of the soil in the lysimeter described here, lysimeter A, contained ^{35}S-labelled superphosphate. Thus, there were two tracers available for tracking of water movement and plant uptake. TDR rods were inserted through the walls of the lysimeter at increments of 100 mm to monitor the depthwise pattern of root uptake of water. At the end of the experiment the soil was destructively sampled and the depthwise pattern in Cl determined (Fig. 26.11).

Given the dispersivity assumption of $\lambda = 10$ mm, the measured profile of Cl was not well described when we considered that uptake of Cl could be described by Eq. (26.8). Rather, we found that a better description resulted if we took into account root exclusion of Cl with the critical threshold C_1 being 1 mol m^{-3} (Eq. 26.9). So by first assuming the soil's hydrodynamic smearing to be given by $\lambda = 10$ mm, we could obtain good agreement by fitting to the plant's exclusion threshold C_1. This inverse modelling to invoke a separation of the soil-based and plant-controlled mechanisms of bioavailability seems logical and reasonable.

However, when we looked at the ^{35}S data, a sensitivity analysis revealed problems with this approach, and raised the spectre of 'equifinality'. Beven (1993) describes equifinality as being the achievement of an equally good description of observed dynamics by a number of models, or various parameter sets within a given model, all of which may be biophysically reasonable. Equifinality thus cautions us against the use of inverse procedures to parameterise the complex of dynamic and exchange processes that we have presumed to be in operation in the rootzone (Fig. 26.12).

In looking at the depthwise distribution of ^{32}S in the soil of lysimeter A (Fig. 26.12), Vogeler et al. (2001b) concluded that the best fit to the data could be achieved with the hydromechanical dispersivity λ being 10 mm, and the exclusion threshold by roots for sulphate being $C_1 = 6\,mol_c\,m^{-3}$. Comparison with published data would suggest that these values for the soil-based and plant-controlled processes are reasonable.

This 'reasonableness' could, however, be challenged when a sensitivity analysis was carried out. Accepting the exclusion threshold value, and varying the dispersivity, produces the three depthwise patterns shown in Figure 26.12a; whereas holding the dispersivity constant, and varying the exclusion limit is shown to produce the indistinguishable curves in Figure 26.12b. This identifies the bogey of equifinality. Despite representing two disparate mechanisms; one hydromechanical dispersion in the soil, the other root exclusion of solutes, we might not be able to identify a single, clear, optimum-parameter set. In the absence of other independent evidence, it may be difficult to identify a unique set of parameters that will produce accurate rootzone modelling across a range of different initial and boundary conditions.

26.7 Prognosis

Equifinality places considerable constraints on the value and veracity of comprehensive models. Despite these limitations, there has been mounting enthusiasm for the development, sale and application of comprehensive mechanistic models. Yet, on the 75th anniversary of the journal *Soil Science*, Philip (1991) lamented that "... open slather to play around with glamorous space-age toys [means that] it may well be that by 2066 we shall be deep into the electronic Dark Ages". Passioura (1996) pondered whether simulation models were science, snake oil, education or engineering. Whither modelling?

The enthusiasm and ability to predict, using comprehensive mechanistic models, the transport, exchange and fate of rootzone solutes will not abate, for these schemes seductively offer great potential. Thus, we must hasten the

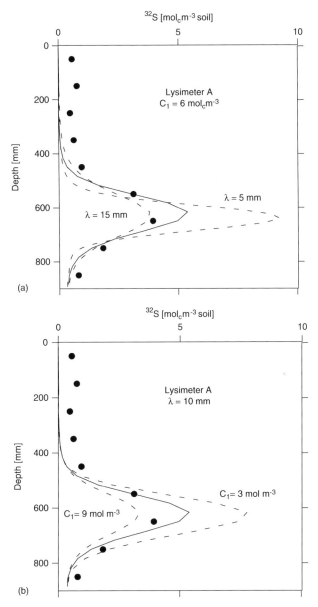

Figure 26.12. *(a) The measured concentration of sulphur in the soil at the end of the tomato growth cycle, and three predictions achieved assuming a fixed C_1, and varying the dispersivity $\pm 50\%$ approximately 10 mm. (b) As for (a), but now the predictions assume a fixed dispersivity, and the maximum uptake concentration C_1 is varied $\pm 50\%$ approximately $6\ mol_c\ m^{-3}$.*

development of tools to overcome the constraints posed by equifinality. On the one hand, statistical procedures such as generalised likelihood uncertainty estimation (GLUE), are seen as a way of dealing with multiple, acceptable parameter sets (Schulz et al., 1999). However, we consider that it is the development of new and exciting measurement devices, along with powerful datalogging technologies, that will help to overcome equifinality through the provision of better and less equivocal observations of rootzone processes. In the future, modelling developments will be enhanced by better measurements, complemented with clever thinking.

Nuttle (2000) posed the question "... can ecosystem managers rely on mechanistic simulation models to guide their decisions?" He warned that they "... are vulnerable as long as managers regard the mechanistic numerical models as the primary tool for synthesising scientific information [but they] can reduce this vulnerability by relying on other components of the whole-system approach. Observation, experiment and modelling together are the essential components of the whole system approach".

Perspicacious experimentation using novel devices, in tandem with comprehensive modelling will lead to new understanding of what controls the bioavailability, uptake, transport and fate of chemicals in the rootzone.

Acknowledgement

Many parts of this research were carried out under FRST Contract C06X0004 "Knowledge Tools for Environmental Action".

References

Bañuelos, G.S., Shannnon, M.C., Ajwa, H., Draper, J.H., Jordhal, J., Licht, L., 1999. Phytoextraction and accumulation of boron and selenium by polar (*Populus*) hybrid clones. Int. J. Phytorem. 1, 81–96.

Beven, K.J., 1993. Prophecy, reality and uncertainty in distributed hydrological modelling. Adv. Water Resour. 16, 41–51.

Clothier, B.E., 1997. Can soil management be regulated to protect water quality? Land Contam. Reclam. 5(4), 337–342.

Clothier, B.E., Green, S.R., 1997. Roots: the big movers of water and chemical in soil. Soil Sci. 162, 534–543.

Deram, A., Petit, D., Robinson, B.H., Brooks, R.R., Gregg, P.E.H., van Halluwyn, C., 2000. Natural and induced heavy-metal accumulation by *Arrhenatherum elatius*: implications for phytoremediation. Commun. Soil Sci. Plant Anal. 31, 413–421.

Feddes, R.A., Kowalik, P., Zaradny, H., 1978. Simulation of field water use uptake by plants using a soil water dependent root extraction function. J. Hydrol. 31, 13–26.

Gardner, W.R., 1960. Dynamic aspects of water availability to plants. Soil Sci. 89, 63–73.

Granel, Th., Robinson, B.H., Mills, T.M., Clothier, B.E., Green, S.R., Fung, L.E., 2001. Cadmium accumulation by willow clones used for conservation, stock fodder and phytoremediation. Aust. J. Soil Res. 40, 1331–1337.

Grčman, H., Velikonja-Bolta, Š., Vodnik, D., Kos, B., Leštan, D., 2001. EDTA enhanced heavy metal phytoextraction: metal accumulation, leaching and toxicity. Plant Soil 235, 105–114.

Green, S.R., Clothier, B.E., 1999. The rootzone dynamics of water uptake by a mature apple tree. Plant Soil 206, 61–77.

Hinkley, T., 1979. Concentrations of metals in very small volumes of soil solution. Nature 277, 444–446.

Jardine, P.M., Taylor, D.L., 1995. Fate and transport of ethylenediaminetetraacetate chelated contaminants in subsurface environments. Geoderma 67, 125–140.

Kluger, J., 2001. Toxic playgrounds. Time July 16 (2001), 38–39.

Mills, T.M., Robinson, B.H., Green, S.R., Clothier, B.E., Fung, L.E, Hurst, S., 2000. Difference in Cd uptake and distribution within polar and willow species. Proceedings of the 42nd Annual Conference New Zealand Water and Waste Association, Rotorua, New Zealand, 27–29 September.

Negri, M.N., Hicnchman, R.R., Gatliff, E.G., 1996. Phytoremediation: using green plants to clean up contaminated soil, groundwater, and wastewater. Proceedings of the International Topical Meeting on Nuclear and Hazardous Waste Management, Spectrum 96, Seattle, WA, August, 1996, American Nuclear Society.

Nuttle, W.K., 2000. Ecosystem managers can learn from past successes. Eos (American Geophysical Union, Washington, DC) 81(25), 278, 284.

Nyer, E.K., Gatliff, E.G., 1998. Phytoremediation. In: E.K. Nyer (Ed.), Groundwater and Soil Remediation: Practical Methods and Strategies, Ann Arbor Press, Chelsea, Michigan, USA, 226pp.

Östman, G., 1994. Cd in *Salix* – a study of the capacity of *Salix* to remove Cd from arable soils. In: P. Aronsson and K. Perttu (Eds.), Willow Vegetation Filters for Municipal Wastewaters and Sludges A Biological Purification System. Proceedings of a Study Tour, Conference and Workshop in Sweden, 5–10 June, 1994, Sveriges Lantbruksuniversitet, Sweden, pp. 153–155.

Passioura, J.B., 1996. Simulation models: science, snake oil, education or engineering. Agron. J. 88, 690–694.

Philip, J.R., 1991. Soils, natural science, and models. Soil Sci. 151, 91–98.

Pivetz, B.E., 2001. Phytoremediation of contaminated soil and groundwater at hazardous waste sites. Ground Water Issue, US Environmental Protection Agency EPA/540/S-01/500, February, 2001, Washington DC, 36pp.

Richards, L.A., 1931. Capillary conduction of liquid through porous mediums. Physics 1, 318–333.

Robinson, B.H., Mills, T.M., Petit, D., Fung, L.E., Green, S.R., Clothier, B.E., 2000. Natural and induced cadmium-accumulation in poplar and willow: implications for phytoremediation. Plant Soil 227, 301–306.

Salt, D.E., Smith, R.D., Raskin, I., 1998. Phytoremediation. Ann. Rev. Plant Physiol. Plant Mol. Biol. 49, 643–668.

Schnoor, J.L., Licht, L.A., McCutcheon, S.C., Wolfe, N.L., Carreira, L.H., 1995. Phytoremediation of organic and nutrient contaminants. Environ. Sci. Technol. 29, 318–323.

Schulz, K., Beven, K.J., Huwe, B., 1999. Equifinality and the problem of robust calibration in nitrogen budget simulations. Soil Sci. Soc. Am. J. 63, 1934–1941.

Siciliano, S.D., Germida, J.J., 1998. Mechanisms of phytoremediation: biochemical and ecological interactions between plants and bacteria. Environ. Rev. 6, 65–79.

Vogeler, I., Green, S.R., Clothier, B.E., Kirkham, M.B., Robinson, B.H., 2001a. Contaminant transport in the root zone. In: I.K. Iskandar, M.B. Kirkham (Eds.), Bioavailability, Fluxes and Transfer of Trace Elements in Soils and Soil Components. Lewis Publishers, CRC Press, Chapter 9 pp. 175-198.

Vogeler, I., Green, S.R., Scotter, D.R., Clothier, B.E., 2001b. Measuring and modelling the transport and root uptake of chemicals in the unsaturated zone. Plant Soil 231, 161–174.

Zar, J.H., 1972. Significance testing of the Spearman rank correlation coefficient. J. Am. Stat Assoc. 67(339), 578–580.

Developments in Soil Science, volume 32
Ravendra Naidu (Editor)

Chapter 27

MANIPULATING BIOAVAILABILITY TO MANAGE REMEDIATION OF METAL-CONTAMINATED SOILS

N.S. Bolan, B.G. Ko, C.W.N. Anderson, I. Vogeler, S. Mahimairaja and R. Naidu

27.1 Introduction

Unlike organic contaminants, most metals do not undergo microbial or chemical degradation and the total concentration of these metals in soils persists for a long time after their introduction (see Chapter 2). With greater public awareness of the implications of contaminated soils on human and animal health there has been increasing interest amongst the scientific community in the development of technologies to remediate contaminated sites. For diffuse distribution of metals (e.g., fertilizer-derived Cd input in pasture soils), remediation options generally include amelioration of soils to minimize the metal bioavailability. Bioavailability can be minimized through chemical and biological immobilization of metals using a range of inorganic compounds, such as lime and phosphate (P) compounds (e.g., apatite rocks), and organic compounds, such as 'exceptional quality' biosolid (Fig. 27.1; Bolan and Duraisamy, 2003). Reducing metal availability and maximizing plant growth through inactivation may also prove to be an effective method of *in situ* soil remediation on industrial, urban, smelting and mining sites. The more localized metal contamination found in urban environments (e.g., Cr contamination in timber-treatment plants) is remediated by metal mobilization processes that include bioremediation (including phytoremediation) and chemical washing. Removal of metals through phytoremediation techniques and the subsequent recovery of the metals (i.e., phytomining) or their safe disposal are attracting research and commercial interests (Fig. 27.2). However, when it is not possible to remove the metals from the contaminated sites by phytoremediation, other viable options, such as *in situ* immobilization should be considered as an integral part of risk management.

Since bioavailability is a key factor for remediation technologies (see Chapters 3 and 4), *in situ* (im)mobilization using inorganic and organic compounds that are low in heavy metal content may offer a promising option. In this chapter, after a brief introduction of sources of metal inputs and their dynamics in soils, the role of various inorganic and organic soil amendments in the (im)mobilization of metals in soils in relation to managing their remediation,

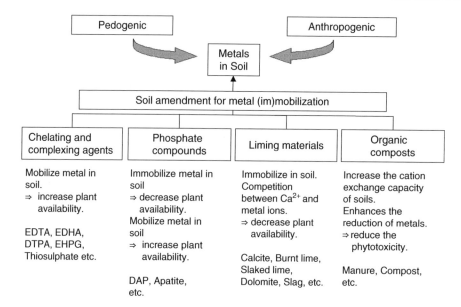

Figure 27.1. *Role of soil amendments in the im(mobilization) and bioavailability of metals in soils.*

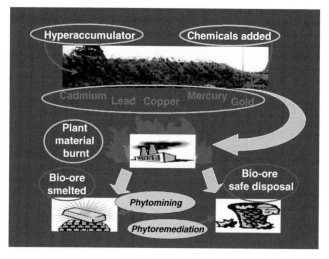

Figure 27.2. *Conceptual diagram illustrating the process of mobilization of metals and the subsequent removal using plants.*

will be reviewed. Those interested in the application of bioavailability to risk-based approach to managing soil and groundwater contamination are directed to Chapter 4.

27.2 Sources of heavy metals

Trace elements include both biologically essential (e.g., Cu, Cr and Zn) and non-essential (e.g., Cd, Pb and Hg) elements. The essential elements (for plant, animal or human nutrition) are required in low concentrations and hence are known as 'micronutrients'. The non-essential trace elements are phytotoxic and/or zootoxic even at low concentration and are widely known as 'toxic elements' or 'heavy metals'. Both groups are toxic to plants, animals and humans above certain concentrations specific to each element (Alloway, 1990; Adriano, 2001).

Trace elements reach the soil environment through both pedogenic and anthropogenic processes. Most trace elements occur naturally in soil parent materials, chiefly in forms that are not readily bioavailable for plant uptake. Often the concentrations of trace elements released by natural pedogenic processes are largely related to the origin and nature of the parent material. Unlike pedogenic inputs, trace elements added through anthropogenic activities typically have a high bioavailability (Naidu et al., 1996a). Anthropogenic activities, primarily associated with industrial processes, manufacturing, the disposal of domestic and industrial waste materials and the application of phosphate fertilizers are the major source of trace element enrichment in soils (Adriano, 2001; Bolan et al., 2003a). Furthermore, in industrialized regions worldwide, the urban-based industrial processes contribute to metal addition through atmospheric deposition (Galloway et al., 1982; Gray et al., 2003).

27.3 Dynamics of heavy metals in soils

Metal ions can be retained in the soil by sorption, precipitation and complexation reactions (see Chapter 2) and are removed from soil through plant uptake and leaching (see Chapter 26). Although most metals are not subject to volatilization losses, some metals such as arsenic, mercury and selenium tend to form gaseous compounds (Bolan et al., 2005; Mahimairaja et al., 2005). The lower the metal solution concentration and the more sites available for sorption, the more likely that sorption/desorption processes will determine the soil solution concentration (Tiller, 1988). However, the fate of metals in the soil environment is dependent on both soil properties and environmental factors.

27.3.1 Sorption process

Charged solute species (ions) are attracted to the charged soil surface by electrostatic attraction and/or through the formation of specific bonds (Mott, 1981). Retention of charged solutes by charged surfaces is broadly grouped into specific and non-specific retention (Bolan et al., 1999a). In general terms, non-specific adsorption is a process in which the charge on the ions balances the charge on the soil particles through electrostatic attraction; whereas specific adsorption involves chemical bond formation between the ions in the solution and those in the soil surface (Sposito, 1984).

Both soil properties and soil solution composition determine the dynamic equilibrium between metals in solution and the soil solid phase. The concentration of metals in soil solution is influenced by the nature of both organic (citrate, oxalate, fulvic, dissolved organic carbon) and inorganic (HPO_4^{2-}, NO_3^-, Cl^- and SO_4^{2-}) ligand ions, and soil pH through their influence on metal sorption processes (Harter and Naidu, 1995; Bolan et al., 2003b). Two reasons have been given for the effect of inorganic anions on the sorption of heavy metal cations (Naidu et al., 1994). Firstly, inorganic anions form ion pair complexes with heavy metals, thereby reducing the sorption of heavy metals. Secondly, the specific sorption of ligand anions is likely to increase the negative charge on soil particles, thereby increasing the sorption of heavy metals.

The effect of pH values >6 in lowering free metal ion activities in soils has been attributed to the increase in pH-dependent surface charge on oxides of Fe, Al and Mn, chelation by organic matter or precipitation of metal hydroxides (Stahl and James, 1991). The activity of heavy metals in solution in naturally acidic soils is found to decrease with increasing pH. The gradual decrease in heavy metal activity with increasing pH is attributed to increasing cation exchange capacity (CEC) (Shuman, 1986). Similarly, Stahl and James (1991) observed that with an increase in surface charge there was an increase in Zn^{2+} retention but specific sorption was favoured over non-specific adsorption. In general, both the CEC and the total amount of heavy metal removed from soil solution increase with an increase in soil pH.

27.3.2 Complex formation

Other chemical interactions that contribute to metal retention by colloid particles include complexation reaction between metals and the inorganic and organic ligand ions. As might be expected, the organic component of soil constituents has a high affinity for heavy metal cations because of the presence of ligands or groups that can form chelates with metals (Harter and Naidu, 1995). With increasing pH, the carboxyl, phenolic, alcoholic and carbonyl functional groups in soil organic matter dissociate, thereby increasing the affinity of ligand ions for metal cations.

The extent of metal-organic complex formation, however, varies with a number of factors including temperature, steric factors and concentration. All these interactions are controlled by solution pH and ionic strength, the nature of the metal species, dominant cation and inorganic and organic ligands present in the soil solution.

A number of studies have indicated that Cl^- tends to form a complex with Cd^{2+} as $CdCl^+$, thereby decreasing the sorption of Cd^{2+} onto soil particles (Naidu et al., 1994). In contrast to inorganic ligand ions, Haas and Horowitz (1986) found that Cd^{2+} adsorption by kaolinite, a variable charge mineral, was enhanced by the presence of organic matter which was attributed to the formation of an adsorbed organic layer on the clay surface.

27.3.3 Precipitation

Precipitation appears to be the predominant process in high pH soils and in the presence of anions such as SO_4^{2-}, CO_3^{2-}, OH^- and HPO_4^{2-} and when the concentration of the heavy metal ion is high (Naidu et al., 1996b). Increasingly precipitation of metals using phosphate compounds is becoming a routine technique to immobilize metals such as lead.

27.4 Definition and indicators of bioavailability

27.4.1 Definition

The bioavailability of a chemical in the soil environment has often been defined as the fraction of the total contaminant in the interstitial porewater (i.e., soil solution) and soil particles that is available to the receptor organism. Considerable controversy exists in the literature relating to 'what constitutes the bioavailable fraction' including the definition itself and the methods used for its measurements. Some of these controversies and the various definitions are summarized in Chapter 3. For instance, microbiologists often regard the concentration that can induce a change either in morphology or physiology of the organism as the bioavailable fraction, whereas plant scientists regard the plant-available pool as the bioavailable fraction. Moreover, chemists and plant scientists have often used a single chemical extraction as an index of bioavailability assuming that bioavailability is a static phenomenon. Recent studies have, however, indicated that the transformation of contaminants in soils is a dynamic process, which means bioavailability changes with time.

A more generic definition of *bioavailability* is the potential for living organisms to take up chemicals from food (i.e., oral) or from the abiotic environment (i.e., external) to the extent that the chemicals may become involved in the metabolism of the organism. More specifically, it refers to the biologically available chemical fraction (or pool) that can be taken up by an organism and can react with its metabolic machinery (Campbell, 1995); or it refers to the fraction of the total chemical that can interact with a biological target (Vangronsveld and Cunningham, 1998). In order to be bioavailable, the contaminants (e.g., metals) have to come in contact with the organism (i.e., physical accessibility). Moreover, metals need to be in a particular form (i.e., chemical accessibility) to be able to enter a plant root. In essence, for a metal to be bioavailable, it will have to be mobile and transportable and be in an accessible form to the plant.

27.5 Indicators of bioavailability

Bioavailability of metals in soils can be examined using chemical extraction and bioassay tests (see Chapters 18–21). Chemical extraction tests include single extraction and sequential fractionation (Ruby et al., 1996; Basta and Gradwohl, 2000). Bioassay involves plants, animals and microorganisms (Yang et al., 1991).

27.5.1 Chemical

Many chemical techniques have been used to assess the bioavailable fraction of metals in soils, but these have serious limitations in providing a good estimate of the pool of toxic metals ions that are accessible to plants and microbes (see Chapter 20). A range of chemical extractants including mineral acids (e.g., 1 N HCl), salt solutions (e.g., 0.1 M $CaCl_2$), buffer solutions (e.g., 1 M NH_4OAc) and chelating agents (e.g., DTPA) have been used to predict the bioavailability of metals in soils (Sutton et al., 1984). Chelating agents, such as EDTA and DTPA have often been found to be more reliable in predicting the plant availability of metals (Sims and Johnson, 1991), since they are more effective in removing soluble metal-organic complexes that are potentially bioavailable.

Sequential fractionation schemes are often used to examine the redistribution or partitioning of metals in various chemical forms that include soluble, adsorbed (exchangeable), precipitated, organic and occluded (see Chapter 20). Although the extraction procedures vary among chemical fractionation schemes, generally the solubility and bioavailability of metals in soils decrease with each successive step of the scheme (Basta and Gradwohl, 2000). Specific chemical pools measured by chemical fractionation schemes have been correlated with plant uptake of metals and have been successful in predicting the plant availability of metals in soils (Shuman, 1986). However, the ability of chemical extractants to dissolve metals is matrix dependent. Therefore, chemical extraction should be validated for different metals sources, such as inorganic fertilizer, and organic sewage sludge and manure byproducts.

The diversity of reagents used to extract specific metal forms from contaminated soils makes comparison of results cumbersome. Even when the same reagent is employed, the efficiency of extraction depends on the nature of sample, its particle size distribution, duration of extraction, pH, temperature, strength of extractant and solid:solution ratio (Miller et al., 1986). Chemical reagents used for the extraction may themselves alter the indigenous speciation of metals, and in general less vigorous extractants will probably be more selective for specific fractions than more severe reagents, which may extract other forms, although the overall efficiency of extraction may be lower (Lake et al., 1984; Ross, 1994).

Plants take up most of the nutrients from soluble and exchangeable fractions (Adriano, 2001). It is important to emphasize that there is a dynamic equilibrium between these fractions, and any depletion of the available pool (soluble and exchangeable fractions) due to plant uptake or leaching losses will result in the continuous release from other fractions to replenish the available 'pool' (Adriano, 2001).

The bioavailability of metals in soils has recently been examined using the physiologically based *in vitro* chemical fractionation schemes that include physiologically based extraction test (PBET), the potentially bioavailable sequential extraction (PBASE) and the gastrointestinal (GI) test. These innovative tests predict the bioavailability of metals in soil and sediments when ingested by animals and humans. As in the case of traditional sequential extraction scheme, the ability of the PBET extractant to solubilize metals increases with each successive extraction step. Metals extracted earlier in the PBET sequence are more soluble and, therefore more potentially bioavailable than metals extracted later by the more aggressive extractants used (Basta and Gradwohl, 2000). Despite the recognized non-specific (i.e., operational) nature of chemical extraction methods, their analytical simplicity and rapidity renders them most suitable for routine identification of metal form and estimation of the bioavailability of metals under field conditions. However, the distribution of a given metal among the various fractions can only be considered as an estimate at best due to the subjectivity of the steps involved.

27.5.2 Biological

In situ techniques involving growing of the organisms of interest in the contaminated material and quantifying the uptake of metal into the organism or assessing the toxicological response are being used by many researchers as bioindicators of contamination (see Chapter 18). Measurements of metal bioavailability and toxicity in soils using soil microorganisms are receiving increasing attention, as microorganisms are more sensitive to heavy metal stress than plants or soil macrofauna. The methods using microflora and protozoa have the potential to provide a measure of bioavailability of heavy metals in the short-term and even facilitate the measurement of temporal changes. In contrast, responses by mesofauna (microarthropods) and macrofauna (enchytridae, invertebrates and earthworms) are cumulative effects. These methods, however, are time consuming and can only provide an overall effect of heavy metal bioavailability to the species tested. The rapid development of molecular techniques and their continued successful application to the study of soil microbial ecology and function provides significant future potential for the monitoring of soil pollution impacts. However, molecular tools are comparatively expensive, and thus the level of information these techniques provide, over and above more traditional microbial indicator techniques, needs to be clearly demonstrated. The application of molecular tools towards the identification and characterization of microbes has been discussed in detail by Geet et al. in Chapter 21.

27.6 Soil amendments for metal (im)mobilization

A number of amendments are used either to mobilize or immobilize heavy metals in soils. The basic principle involved in the mobilization technique is to release the metals into soil solution, which is subsequently removed using higher plants. In contrast, in the case of the immobilization technique the metal concerned is removed from soil solution either through adsorption, complexation and precipitation reactions, thereby rendering the metals unavailable for plant uptake and leaching to groundwater (Table 27.1). In this section, the potential value of these soil amendments in the (im)mobilization of metals in relation to remediation will be discussed.

27.6.1 Chelating and complexing agents

When a metal ion combines with an electron donor, the resulting substance is called a complex or coordination compound. If the substance which combines with the metal contains two or more donor groups so that one or more rings are formed, the resulting structure is called a metal chelate, and the donor is called chelating agent. The term 'chelate' is derived from the Greek word 'chela' which means 'claw' and it is so named because these species can coordinate at several or all positions around a central metal ion by literally wrapping themselves around the metal ion. Chelating agents which have high affinity for metal ions can be used to enhance the solubilization of metals in soils through the

Table 27.1. *Selected references on the potential value of soil amendments in the (im)mobilization of metals in soils.*

Amendments	Metal	Observations	References
Chelating/complexing agents			
Thiosulphate	Hg	Shoot Hg accumulation in the presence of thiosulphate salt was dependent upon plant species characteristic and humic acid content	Moreno et al. (2005b)
EDTA	Cu	Mobilized and increased plant uptake	Thayalakumaran et al. (2003b)
EDTA	Cu, Fe, Mn, Pb, Zn	EDTA had only a limited effect on metal uptake by plants	Walker et al. (2003)
EDTA	Pb	Plant uptake in roots and shoots increased	Chen et al. (2004)
Humic acid	Cd	Plant uptake enhanced significantly	Evangelou et al. (2004)
EDTA	Cd	Mobilized Cd in soil, but did not increase plant total uptake	Jiang et al. (2003)
CCA EDTA	Pb, Zn	CCA improved bioavailability in soil, enhanced shoot content of Pb but not of Zn	Li et al. (2005)
Phosphate compounds			
KH_2PO_4	Cd	Enhanced immobilization, decreased plant availability	Bolan et al. (2003c)
Apatite, zeolite, Fe oxide	Cd, Pb	Reduced the mobility and uptake by plant	Chlopecka and Adriano (1997)
KH_2PO_2	Pb, Zn, Cd	Reduced Pb level but not Zn and Cd in earthworms	Pearson et al. (2000)
Phosphate	As	Increasing phosphate supply decreased As uptake. Increasing As supply decreased the P concentration in the root	Wang et al. (2002)
Phosphate	Pb, As	Increased plant uptake of soil As. No effect on soil Pb phyto-availability	Creger and Peryea (1994)
Liming materials			
$Ca(OH)_2$	Cd	Transformed to less mobile fractions,	Bolan et al. (2003d)

Table 27.1 (*Continued*)

Amendments	Metal	Observations	References
Lime, FBA, bark	Cr	reduced phytoavailability Reduced the availability for plant uptake and leaching groundwater	Bolan and Thiyagarajan (2001)
Lime	Cd	Did not reduce uptake and transfer to the kernels of sunflower	Li et al. (1996)
Lime	Cd	Reduced uptake by lettuce	Lehoczky et al. (2000)
Lime	Pb, Cd	Decreased Cd uptake, but little influence Pb uptake by radish	Han and Lee (1996)
Ca(OH)$_2$, CaCO$_3$	Cd	Ca(OH)$_2$ prevented phytotoxicity, but CaCO$_3$ was no effective in reducing phytotoxicity	Chaney et al. (1977)
Lime	Cd, Ni, Zn, Cu	Reduced uptake of Cd, Ni, Zn, but did not reduce Cu	Brallier et al. (1996)
Organic matter			
Biosolid	Cd	Reduced the bioavailability	Bolan et al. (2003b)
Biosolid, manure	Cr	Reduced the phytotoxic effect	Bolan et al. (2003e)
PS, SS	Cd, Zn	Cd uptake increased, but Zn unaffected	Merrington and Madden (2000)
Biosolid	Cd	Reduced plant availability	Brown et al. (1998)
Manure	Zn, Mn	Corn and leaf concentrations of Zn and Mn were enhanced	Wallingford et al. (1975)
Biosoilid	Zn	Decreased plant availability	Shuman et al. (2000)
Biosolid	Cd	Cl complexation of Cd increased the phytoavailability of Cd in biosolid-amended soil	Weggler-Beaton et al. (2000)
Organic matter	Ni, Cd, Zn	Reduced Ni uptake but not Cd, Zn in rice	Kashem and Singh (2001)
Compost, manure	Cr	Reduced the phytotoxicity	Bolan et al. (2003g)

Note: EDTA, ethylene diaminetetraacetic acid; CCA, coated chelating agent; FBA, fluidized bed boiler ash; PS, papermill sludge; and SS, sewage sludge.

formation of soluble metal chelates. A number of synthetic chelating agents are available including EDTA, EDHA, DTPA and EHPG.

The value of chelating agents in enhancing the availability of trace elements such as Fe, Cu and Zn has been well recognized in plant nutrition and various chelated compounds are available as nutrient sources. Recently, the potential value of chelating agents in the remediation of contaminated soils through mobilization of metals has been explored (Thayalakumaran et al., 2003a; Chen et al., 2004). These compounds have been found to be very effective in the solubilization of metals such as Cu and Pb, thereby enhancing their subsequent uptake by plants. The effectiveness of any chelate in mobilizing soil metals depends on several factors, including metal species, metal:chelate ratio, thermodynamic stability constants, presence of competing cations, soil pH, stability of the metal–chelate complex, extent of metal retention onto soil constituents and the ageing of the contaminating metal. It has been suggested that synthetic chelates mobilize metals from the exchangeable fraction, the organic matter and the carbonate-bound fraction, but not from the metals in oxides (Elliott and Shastri, 1999). In contrast, Barona et al. (2001) found that EDTA extraction of Pb resulted in a substantial decrease in oxides and organic-bound fractions.

A number of issues need to be considered when using chelates to accelerate the remediation of metal-contaminated sites. The addition of these chelating agents is likely to induce the solubilization of other than the target metals which may be phytotoxic (e.g., Al and Mn). The increased solubilization of metals can result in their increased leaching to groundwater, especially in the absence of active plant growth. Several column and lysimeter studies have shown enhanced concentrations of heavy metals in the leachates following addition of synthetic chelates (Sun et al., 2001; Wenzel et al., 2003; Wu et al., 2004). For example, Grčman et al. (2001) found that while EDTA resulted in an enhanced heavy metal uptake by *Brassica rapa*, it also resulted in leaching of Pb, Zn and Cd.

An important factor in chelate-induced phytoremediation is the stability of the metal–chelate complex with time. As shown by Thayalakumaran et al. (2003a), EDTA application to a Cu-contaminated soil resulted in an initial formation of the $CuEDTA^-$ complex. Over time $CuEDTA^-$ reacted with soil iron leading to the formation of the more stable $Fe(III)EDTA^-$ complex in the soil solution, resulting in the release of Cu to soil solution and its readsorption by the soil. Thayalakumaran et al. (2003b) demonstrated the importance of the time on the stability of the metal–chelate complex in a pot experiment growing *Agrostis tenuis*. Numerous small applications of EDTA resulted in a higher copper accumulation in the herbage compared with EDTA applied in one or two larger amounts. They also found that EDTA-enhanced copper uptake was much smaller in plants growing in intact cores compared to those growing in repacked pots, which they attributed to the spatial separation between the Cu adsorption sites and the roots in the intact cores resulting in a lower Cu uptake.

Alkaline hydrolysis of metals with the chelating compounds induces the formation of metal hydroxy precipitates. Some of the chelating compounds are subjected to microbial degradation, thereby affecting their long-term effectiveness. However, the persistence of soluble EDTA–metal complexes in soils can cause negative effects on soil microfauna.

Recently, there has been interest in the use of complexing agents such as thiosulphate in mobilizing Hg in soils. For example, Moreno et al. (2005a) demonstrated that the addition of sulphur-containing ligand compounds such as sodium or ammonium thiosulphate to mine tailings contaminated with Hg can induce the solubilization of Hg and its subsequent uptake and accumulation in both roots and shoots of plants. The concentration of mercury in plants after treatment is a function of the extractable concentration of Hg in the soil, hence the chemistry of metal in the soil is of paramount importance. Moreno et al. (2005a) noticed that the concentration of Hg in *Brassica juncea* plants reached a maximum of 85 mg/kg when the plants were grown in a soil with a Hg content of 3.4 mg/kg, thereby reaching a maximum transfer factor (i.e., the ratio between plant and soil concentration) of 25. Moreno et al. (2005b) described a field demonstration for Hg phytoremediation conducted at a base-metal tailings site in the Coromandel district of New Zealand. These authors showed that Hg uptake could be induced in the field, although the results were generally variable.

Mercury is a metal often targeted for remediation through volatilization. Transgenic plants encoding genes from Hg-detoxifying bacteria that have increased resistance and enhanced volatilization capacity are able to extract Hg(II) and methylmercury from contaminated soil and convert these forms to Hg(0). This less toxic form of the metal is then released as Hg vapour into the atmosphere (Rugh et al., 1996; Bizily et al., 1999). However, Moreno et al. (2005b) observed that application of thiosulphate to soil reduced the amount of Hg volatilized by *Brassica juncea* plants as well as increasing the Hg concentration stored in the plant tissues. During an experiment conducted in sealed volatilization chambers, the daily amount of Hg volatilized from planted mine tailings irrigated with water was recorded as 23 times greater than the amount released from non-planted mine tailings, but the daily amount of Hg volatilized from planted tailings irrigated with thiosulphate solution was only 6 times greater than the amount released from the non-planted control.

Clearly, then there is potential to increase the uptake of a wide range of metals into plants through the judicious application of chemicals to soil. But as discussed above, for Hg the effect of this increased uptake on other physiological functions, such as volatilization, must also be considered.

27.6.2 Phosphate compounds

A large number of studies have provided conclusive evidence for the mitigative value of both water-soluble (e.g., diammonium phosphate, DAP) and water-insoluble (e.g., apatite, also known as PR) P compounds to immobilize metals in soils, thereby reducing their bioavailability for plant uptake and mobility for transport (Bolan et al., 2003c, 2003f). Phosphate compounds enhance the immobilization of metals in soils through various processes including direct metal adsorption by P compounds, phosphate anion-induced metal adsorption and precipitation of metals with solution P as metal phosphates.

Depending on the source, soil application of P compounds can cause direct adsorption of metals onto these compounds through increased surface charge and enhanced anion-induced metal adsorption. Metal adsorption onto apatite is facilitated through the exchange of Ca^{2+} from the apatite particle with the metal cations in soil solution (Eq. (27.1)).

Precipitation as metal phosphates has been proved one of the main mechanisms for the immobilization of metals, such as Pb and Zn in soils. These fairly stable metal-phosphate compounds have extremely low solubility over a wide pH range, which makes P application an attractive technology for managing metal-contaminated soils. In typical arable soils, precipitation of metals is unlikely, but in highly metal-contaminated soils, this process can play a major role in the immobilization of such metals.

The ability of apatite to immobilize Pb in solution or Pb in contaminated soils through precipitation as Pb phosphates has been well documented. Such precipitates are more commonly manifested as hydroxypyromorphite or as chloropyromorphite. Two processes for the reaction of dissolved Pb with apatite have been proposed. Firstly, Pb^{2+} can substitute for Ca^{2+} in apatite (Eq. (27.1)). Thus (Ca, Pb) apatite could be potentially formed by adsorption of Pb or by dissolution of hydroxyapatite (HA, $Ca_{10}(PO_4)_6(OH)_2$), followed by coprecipitation of mixed apatites. Secondly, Pb^{2+} can react with apatite through HA dissolution (Eq. (27.2)), followed by precipitation (Eq. (27.3)) of pure hydroxypyromorphite ($Pb_{10}(PO_4)_6(OH)_2$).

$$Ca_{10}(PO_4)_6(OH)_2(s) + xPb^{2+} \rightarrow (Ca_{10-x}Pb_x)(PO_4)_6(OH)_2(s) + xCa^{2+} \quad (27.1)$$

$$Ca_{10}(PO_4)_6(OH)_2(s) + 14H^+(aq) \rightarrow 10Ca^{2+}(aq) + 6H_2PO_4^-(aq) + 2H_2O \quad (27.2)$$

$$10Pb^{2+}(aq) + 6H_2PO_4^-(aq) + 2H_2O \rightarrow Pb_{10}(PO_4)_6(OH)_2(s) \quad (27.3)$$

Soils which have a high affinity for ligand anions may enhance metal ion adsorption through modifications in the soil surface charge density. In strongly weathered soils, ligand adsorption on clay surfaces enhances metal adsorption. It has often been shown that the adsorption of specifically sorbing ligands such as HPO_4^{2-} and SO_4^{2-} in strongly weathered and variable charge soils can induce Cd^{2+} adsorption through increased negative surface charge (Naidu et al., 1994; Bolan et al., 1999b). An alternate mechanism, which appears to be important in temperate soils, involves metal–ligand complexation in solution and subsequent reduction in cation charge, which probably reduces adsorption (Harter and Naidu, 1995). The decrease in Cd^{2+} sorption on to soils in the presence of phosphate (Krishnamurti et al., 1996) was attributed to the formation of soluble Cd-phosphate complexes. This is consistent with the concept that free Cd^{2+} activity, rather than total dissolved Cd^{2+} concentration, is a controlling factor in Cd^{2+} sorption (O'Connor et al., 1984). This mechanism may also be operative in high pH (limed) tropical soils in which anion adsorption will be low and any effect on adsorption will be through ligand–metal complexation process.

However, phosphate ions induce the mobilizations of As in soil, thereby increasing its bioavailability. Phosphorus and arsenic belong to the same chemical group and both have comparable dissociation constants for their acids and solubility products for their salts. Therefore, $H_2AsO_4^-$ and $H_2PO_4^-$ ions compete for the same sorption sites. So, phosphate addition is likely to enhance phytoremediation of arsenic-contaminated soil.

For example, Woolson et al. (1973) observed that phosphate addition to an As-contaminated soil displaced about 77% of the total As in the soil. Although phosphate addition increases As solubility, Peryea (1991) reported that desorption of As was dependent on the soil type since no increase in As concentration in soil solution from a volcanic soil (with high anion-fixing and pH buffering capacity) was observed. This suggests that only large additions of P (>400 mg/kg) would affect the As solubility in these soils (Smith et al., 1998; Chen et al., 2002). In a long-term poultry litter-amended agricultural soils, Arai et al. (2003) observed that the extent of As desorption from the litter increased with increasing pH from 4.5 to 7, but only 15% of the total As was released at pH 7, indicating the presence of insoluble phases and/or strongly retained soluble compounds. Elkhatib et al. (1984) have suggested that the sorption of As(III) is not reversible in soil.

Again the role of $H_2PO_4^-$ ions in enhancing the mobility of As, especially AsO_4^{2-} ions should be noted. For example, Qafoku et al. (1999) noticed that the leaching of As in a column containing mineral soil incorporated with As-rich poultry manure increased with the addition of phosphate compound. The Arsenic concentration in the leachate was approximately ten times higher when $Ca(H_2PO_4)_2$ was used to leach the soil column as compared to $CaSO_4$ solution. In the presence of $Ca(H_2PO_4)_2$ solution, a maximum As concentration of $800\,\mu g/L$

was found in the leachate, much higher than the WHO maximum permissible limit of $10\,\mu g/L$ for drinking water.

Arsenic uptake by plants is associated with the $H_2PO_4^-$ uptake mechanism, where presumably As(V) is taken up as a $H_2PO_4^-$ analogue (Pickering et al., 2000). Therefore, there is a growing interest in using P fertilizer to enhance As uptake by plants. Tu and Ma (2003) suggested that phosphate application may be an important strategy for efficient use of Chinese brake (*Pteris vittata* L.) to phytoremediate As-contaminated soils. The addition of P fertilizer to As-contaminated soil was found to increase As solubility and mobility and thus increase plant uptake of soil As (Creger and Peryea, 1994).

In an hydroponic experiment, Wang et al. (2002) investigated the interactions of As(V) and $H_2PO_4^-$ on the uptake and distribution of As and P, and As speciation in *P. vittata*. They found that the plants accumulated As in the fronds up to 27,000 mg/kg dry weight, and the frond As to root As concentration ratio varied between 1.3 and 6.7. Increasing phosphate supply decreased the As uptake markedly, with the effect being greater on root As concentration than on shoot concentration. They concluded that As(V) is taken up by *P. vittata* via the $H_2PO_4^-$ transporters, reduced to As(III) and sequestered in the fronds primarily as As(III). Whereas, in a fly ash-amended soil, Qafoku et al. (1999) observed that $H_2PO_4^-$ displaced both As(III) and As(V), thereby increasing the mobility of As in soils. Thus, $H_2PO_4^-$-induced plant uptake of As could be employed in phytoremediation of As-contaminated sites.

Davenport and Peryea (1991) observed that high rates of monoammonium phosphate (MAP) or monocalcium phosphate (MCP) fertilizers significantly increased the amount of As leached from the soil. Mixing high rates of MAP or MCP fertilizers with orchard soil, Peryea (1991) reported that As release from lead-arsenate-contaminated soil was positively related to the level of P input but was not significantly influenced by P source. Arsenic solubility was regulated by specific $H_2PO_4^-–AsO_4$ exchange, while $H_2PO_4^-$ solubility was controlled by the equilibria of metastable P minerals. The results indicate that the use of P fertilizers on such soils has the potential to greatly enhance downward movement of As (Peryea and Kammereck, 1997). Thus the increased mobilization of As resulting from phosphate input can result in its increased leaching to groundwater, especially in the absence of active plant growth. Hence attempts to use plants to remove As from soils need to take the multiple effects of phosphate into consideration.

27.6.3 Liming materials

Although liming is primarily aimed at ameliorating soil acidity, it is increasingly being accepted as an important management tool in reducing the toxicity of heavy metals in soils (Bolan et al., 2003a, 2003d). A range of liming materials is available, which vary in their ability to neutralize the acidity. These include calcite (CaCO₃), burnt lime (CaO), slaked lime (Ca(OH)₂), dolomite (CaMg(CO₃)₂) and slag (CaSiO₃). The acid neutralizing value of liming materials is expressed in terms of calcium carbonate equivalent (CCE), defined as the acid neutralizing capacity of a liming material expressed as a weight percentage of pure CaCO₃.

Liming, as part of the normal cultural practices, has often been shown to reduce the concentration of Cd, Pb and other metals in edible parts of crops. Similarly, liming serpentine soils containing toxic levels of Ni has shown to alleviate the phytotoxic effects of Ni. In these cases, the effect of liming materials in decreasing metal uptake by plants has been attributed to both decreased mobility in soils (through adsorption/precipitation) and to the competition between Ca^{2+} and metals ions on the root surface.

Removal of Cr(III) from industrial effluent is achieved using lime or magnesium oxide to precipitate as chromic hydroxide. Precipitation is reported to be most effective at pH 8.5–9.5 due to the low solubility of chromic hydroxide in that range. This method can decrease Cr concentrations to very low levels and hence precipitation systems are extensively used by major tanneries (Bolan and Thiyagarajan, 2001).

27.6.4 Organic composts

The major sources of organic composts include biosolid and animal manures (Bolan et al., 2004). Traditionally biosolid is viewed as one of the major sources of metal accumulation in soils. Advances in the treatment of sewage water and isolation of industrial wastewater in the sewage treatment plants have resulted in a steady decline in the metal content of biosolid. Furthermore, stabilization using alkaline materials has resulted in the immobilization of metals in biosolid.

Most manure products contain low levels of heavy metals (except Cu and Zn in swine manure and As in poultry manure) (Bolan et al., 2004). Furthermore, recent advances in the treatment of manure byproducts have resulted in reduced bioavailability of metals. Similarly, treatment of poultry manure with alum [Al₂(SO₄)₃] decreased the concentration of water-soluble Zn, Cu and Cd. Hence unlike sewage sludge application, where land application is limited based on allowable trace element loadings, regulations governing livestock and poultry manure byproducts are generally based on total N and P loading. Manure byproducts that are low in metal content can be used to immobilize metal contaminants in soils.

Although a number of studies have examined the role of biosolid as a source of metal contamination in soil, only limited work has been reported on the beneficial effect of organic amendments as a sink for the immobilization of metals in soils. Recent studies have shown that alkaline-stabilized biosolid that are low in total and/or bioavailable metal content (known as 'exception quality' biosolid or 'designer sludge') can be used as an effective sink for reducing the bioavailability of metals in contaminated soils and sediments. Immobilization of metals by such amendments is achieved through adsorption, complexation and redox reactions.

Addition of organic amendments has often been shown to increase the CEC of soils, thereby resulting in increased metal adsorption (Bolan et al., 2003b). The presence of phosphates, aluminium compounds and other inorganic minerals in some organic amendments, such as typical municipal sewage sludge, is also believed to be responsible for the retention of metals, thereby inducing the 'plateau effect' in metal uptake by crops and preventing the increased metal availability suggested in the 'time bomb' hypothesis.

Metals form both soluble and insoluble complexes with organic constituents in soils, a process which apparently depends on the nature of the organic matter (Bolan et al., 2003e). As might be expected, the organic component of soil constituents has a high affinity for metal cations because of the presence of ligands or functional groups that can form chelates with metals. With increasing pH, the carboxyl, phenolic, alcoholic and carbonyl functional groups in soil organic matter dissociate, thereby increasing the affinity of ligand ions for metal cations.

A number of studies have shown that addition of organic soil amendments enhances the reduction of metals/metalloids such as Cr and Se (Bolan et al., 2005). For example, the addition of cattle manure has resulted in the reduction of Cr(VI) to less toxic and less mobile Cr(III) (Bolan et al., 2003g). Various reasons could be attributed to the increase in the reduction of Cr(VI) in the presence of organic manure compost including the supply of carbon and protons, and the stimulation of microbial activity thereby facilitating the relevant redox reactions. Although Cr(VI) reduction can occur through both chemical and biological processes, the biological reduction is considered to be the dominant process in most agricultural soils which are low in ferrous [Fe(II)] ions, and the addition of manure compost is likely to increase the biological reduction of Cr(VI).

27.7 Conclusions

Since one of the primary objectives of remediating contaminated sites is to reduce the bioavailability of metals, *in situ* immobilization using soil amendments that are low in heavy metal content may offer a promising option. However, a major inherent problem associated with immobilization techniques is that although the heavy metals become less bioavailable, their total concentration in soils remains unchanged. The immobilized heavy metal may become plant available with time through natural weathering process or through breakdown of high-molecular-weight organic-metal complexes. Although the formation of soluble metal-organic complexes reduces the phytoavailability of metals, the mobility of the metal may be facilitated greatly in soils receiving alkaline-stabilized biosolid because of the reduction of metal adsorption and increased concentration of soluble metal-organic complex in solution.

A large number of studies have provided conclusive evidence for the potential value of both water-soluble and water-insoluble P compounds to immobilize metals in soils, thereby reducing their bioavailability for plant uptake. However, it should be recognized that, depending on the nature of P compounds and the heavy metal species, application of these materials could cause either mobilization (e.g., As) or immobilization (e.g., Pb) of the metals. While mobilization by certain P compounds enhances the bioavailability of metals, immobilization inhibits their plant uptake and reduces their transport in soils and subsequent groundwater contamination. Furthermore, some of these materials contain high levels of metals (e.g., Cd) and can act as an agent of metal introduction to soils. Accordingly, these materials should be scrutinized prior to their large-scale use as immobilizing agent in contaminated sites.

References

Adriano, D.C., 2001. Trace Elements in Terrestrial Environments; Biogeochemistry, Bioavailability and Risks of Metals., 2nd Ed., Springer, New York, NY.

Alloway, B.J., 1990. Cadmium. In: B.J. Alloway (Ed.), Heavy Metals in Soil. Wiley, New York, pp. 100–124.

Arai, Y., Lanzirotti, A., Sutton, S., Davis, J.A., Sparks, D.L., 2003. Arsenic speciation and reactivity in poultry litter. Environ. Sci. Technol. 37, 4083–4090.

Barona, A., Aranguiz, I., Elias, A., 2001. Metal associations in soils before and after EDTA extractive decontamination: implications for the effectiveness of further cleanup procedures. Environ. Pollut. 113, 79–85.

Basta, N.T., Gradwohl, R., 2000. Estimation of Cd, Pb, and Zn bioavailability in smelter-contaminated soils by a sequential extraction procedure. J. Soil Contam. 9, 149–164.

Bizily, S.P., Rugh, C.L., Summers, A.O., Meagher, R.B., 1999. Phytoremediation of methylmercury pollution: MerB expression in *Arabdopsis thaliana* confers resistance to organomercurials. Proc. Natl. Acad. Sci.USA 96, 6808–6813.

Bolan, N.S., Adriano, D.C., Curtin, D., 2003a. Soil acidification and liming interactions with nutrient and heavy metal transformation and bioavailability. Adv. Agron. 78, 216–272.

Bolan, N.S., Adriano, D.C., Duraisamy, A., Mani, P., 2003b. Immobilization and phytoavailability of cadmium in variable charge soils. III. Effect of biosolid compost addition. Plant Soil 256, 231–241.

Bolan, N.S., Adriano, D.C., Mahimairaja, S., 2004. Distribution and bioavailability of trace elements in livestock and poultry manure by-products. Crit. Rev. Environ. Sci. Technol. 34, 291–338.

Bolan, N.S., Adriano, D.C., Mani, P., Duraisamy, A., Arulmozhiselvan, S., 2003c. Immobilization and phytoavailability of cadmium in variable charge soils: I. Effect of phosphate addition. Plant Soil 250, 83–94.

Bolan, N.S., Adriano, D.C., Mani, P., Duraisamy, A., Arulmozhiselvan, S., 2003d. Immobilization and phytoavailability of cadmium in variable charge soils: II. Effect of lime addition. Plant Soil 250, 187–198.

Bolan, N.S., Adriano, D.C., Mani, S., Khan, A.R., 2003e. Adsorption, complexation and phytoavailability of copper as influenced by organic manure. Environ. Toxicol. Chem. 22, 450–456.

Bolan, N.S., Adriano, D.C., Naidu, R., 2003f. Role of phosphorus in (im)mobilization and bioavailability of heavy metals in the soil-plant system. Rev. Environ. Contam. Toxicol. 177, 1–44.

Bolan, N.S., Adriano, D.C., Natesan, R., Koo, B.J., 2003g. Effects of organic amendments on the phytoavailability of chromate in mineral soil. J. Environ. Qual. 32, 120–128.

Bolan, N.S., Duraisamy, D., 2003. Role of inorganic and organic soil amendments on immobilisation and phytoavailability of heavy metals: a review involving specific case studies. Aust. J. Soil Res. 41, 533–555.

Bolan, N.S., Mahimairaja, S., Megharaj, M., Naidu, R., Adriano, D.C., 2005. Biotransformation of arsenic in soil and aquatic environments in relation to its bioavailability and bioremediation. , Chapter 25. In: R. Naidu, R. Smith, G. Owens, P. Bhattacharya, P. Nadebaum (Eds.), Arsenic in Environment. CSIRO Publication, VIC 3066, Australia.

Bolan, N.S., Naidu, R., Syers, J.K., Tillman, R.W., 1999a. Surface charge and solute interactions in soils. Adv. Agron. 67, 88–141.

Bolan, N.S., Naidu, R., Syers, J.K., Tillman, R.W., 1999b. Effect of anion sorption on cadmium sorption by soils. Aust. J. Soil Res. 37, 445–460.

Bolan, N.S., Thiyagarajan, S., 2001. Retention and plant availability of chromium in soils as affected by lime and organic amendments. Aust. J. Soil Res. 39, 1091–1103.

Brallier, S., Harrison, R.B., Henry, C.L., Dongsen, X., 1996. Liming effects on availability of Cd, Cu, Ni and Zn in a soil amended with sewage sludge 16 years previously. Water Air Soil Pollut. 86, 195–206.

Brown, S.L., Chaney, R.L., Angle Scott, J., Ryan, J.A., 1998. The phytoavailability of cadmium to lettuces in long-term biosolids-amended soil. J. Environ. Qual. 27, 1071–1078.

Campbell, P.G.C., 1995. Interactions between trace metals and aquatic organisms: a critic of the free ion model. In: A. Tessier, D.R. Turner (Eds.), Metal Speciation and Bioavailability in Aquatic Systems. John Wiley & Sons, Ltd., pp. 45–102.

Chaney, W.R., Strickland, R.C., Lamoreaux, R.J., 1977. Phytotoxicity of cadmium inhibited by lime. Plant Soil 47, 275–278.

Chen, T.B., Fan, Z.L., Lei, M., Huang, Z.C., Wei, C.Y., 2002. Effect of phosphorus on arsenic accumulation in As-hyperaccumulator *Pteris vittata* L. & its implication. Chinese Sci. Bull. 47, 1876–1879.

Chen, Y., Shen, Z., Li, X., 2004. The use of vetiver grass (*Vetiveria zizanioides*) in the phytoremediation of soils contaminated with heavy metal. Appl. Geochem. 19, 1553–1565.

Chlopecka, A., Adriano, D.C., 1997. Influence of zeolite, apatite and Fe-oxide on Cd and Pb uptake by crops. Sci. Total Environ. 207, 195–206.

Creger, T.L., Peryea, F.J., 1994. Phosphate fertilizer enhances arsenic uptake by apricot liners grown in lead-arsenate-enriched soil. HortScience 29, 88–92.

Davenport, J.R., Peryea, F.J., 1991. Phosphate fertilizers influence leaching of lead and arsenic in soil contaminated with lead arsenate. Water Air Soil Pollut. 57–58, 101–110.

Elkhatib, E.A., Bennett, O.L., Wright, R.J., 1984. Arsenite sorption and desorption in soils. Soil Sci. Soc. Am. J. 48, 1025–1030.

Elliott, H.A., Shastri, N.L., 1999. Extractive decontamination of metal-polluted soils using oxalate. Water Air Soil Pollut. 110, 335–346.

Evangelou, M.W.H., Daghan, H., Schaeffer, A., 2004. The influence of humic acids on the phytoextraction of cadmium from soil. Chemosphere 57, 207–213.

Galloway, J.N., Thorntond, J.D., Norton, S.A., Volchok, H.I., McLean, R.A.N., 1982. Trace-metals in atmospheric deposition – a review and assessment. Atmos. Environ. 16, 1677–1700.

Gray, C.W., McLaren, R.G., Roberts, A.H.C., 2003. Atmospheric accessions of heavy metals to some New Zealand pastoral soils. Sci. Total Environ. 305, 105–115.

Grčman, H., Velikonja-Bolta, D., Vodnik, D., Kos, B., Leštan, D., 2001. EDTA enhanced heavy metal phytoextraction: metal accumulation, leaching, and toxicity. Plant Soil 235, 105–114.

Haas, C.I., Horowitz, N.D., 1986. Adsorption of cadmium to kaolinite in the presence of organic material. Water Air Soil Pollut. 27, 131–140.

Han, D.H., Lee, J.H., 1996. Effect of liming on uptake of lead and cadmium by *Raphanus sativa*. Arch. Environ. Contam. Toxicol. 31, 488–493.

Harter, R.D.R., Naidu, R., 1995. Role of metal-organic complexation in metal sorption by soils. Adv. Agron. 55, 219–264.

Jiang, X.J., Luo, Y.M., Zhao, Q.G., A. Baker, J.M., Christie, P., Wong, M.H., 2003. Soil Cd availability to Indian mustard and environmental risk following EDTA addition to Cd-contaminated soil. Chemosphere 50, 813–818.

Kashem, M.A., Singh, B.R., 2001. Metal availability in contaminated soil: II. Uptake of Cd, Ni and Zn in rice plants grown under flooded culture with organic matter addition. Neth. Nutr. Cycl. Agroecosyst. 61, 257–266.

Krishnamurti, G.S.R., Huang, P.M., Van Rees, K.C.J., 1996. Studies on soil rhizosphere: speciation and availability of cadmium. Chem. Spec. Bioavailab. 8, 23–28.

Lake, D.L., Kirk, P.W.W., Lester, J.N., 1984. Fractionation, characterization, and speciation of heavy metals in sewage sludge and sludge-amended soils. J. Environ. Qual. 13, 175–183.

Lehoczky, E., Marth, P., Szabados, I., Szomolanyi, A., 2000. The cadmium uptake by lettuce on contaminated soil as influenced by liming. Commun. Soil Sci. Plant Anal. 31, 2433–2438.

Li, H., Wang, Q., Cui, Y., Dong, Y., Christie, P., 2005. Slow release chelate enhancement of lead phytoextraction by corn (*Zea mays* L.) from contaminated soil – a preliminary study. Sci. Total Environ. 339, 179–187.

Li, Y.M., Chaney, R.L., Schneiter, A.A., Johnson, B.L., 1996. Effect of limestone applications on cadmium content of sunflower (*Helianthus annuus* L.) leaves and kernels. Plant Soil 180, 297–302.

Mahimairaja, S., Bolan, N.S., Adriano, D.C., Robinson, B., 2005. Arsenic contamination and its risk management in complex environmental settings. Adv. Agron. 86, 1–82.

Merrington, G., Madden, C., 2000. Changes in the cadmium and zinc phytoavailability in agricultural soil after amendment with papermill sludge and biosolid. Commun. Soil Sci. Plant Anal. 31, 759–776.

Miller, W.P., Martens, D.C., Zelazny, L.W., Kornegay, E.T., 1986. Forms of solid-phase copper in copper-enriched swine manure. J. Environ. Qual. 15, 69–72.

Moreno, F.N., Anderson, C.W.N., Stewart, R.B., Robinson, B.H., 2005b. Mercury volatilisation and phytoextraction from base-metal mine tailings. Environ. Pollut. 136, 341–352.

Moreno, F.N., Anderson, C.W.N., Stewart, R.B., Robinson, B.H., Ghomshei, M., Meech, J.A., 2005a. Induced plant uptake and transport of mercury in the presence of sulphur-containing ligands and humic acids. New Phytol. 166, 445–454.

Mott, C.J.B., 1981. Anion and ligand exchange. In: D.J. Greenland, M.H.B. Haynes (Eds.), The Chemistry of Soil Processes. Wiley, New York, pp. 179–219.

Naidu, R., Bolan, N.S., Kookana, R.S., Tiller, K.G., 1994. Ionic-strength and pH effects on the adsorption of cadmium and the surface charge of soils. Eur. J. Soil Sci. 45, 419–429.

Naidu, R., Kookana, R.S., Oliver, D.P., Rogers, S., McLaughlin, M.J., 1996a. Contaminants and the Soil Environment in the Australasia-Pacific Region. Kluwer Academic Publishers, London.

Naidu, R., Kookana, R.S., Sumner, M.E., Harter, R.D., Tiller, K.G., 1996b. Cadmium adsorption and transport in variable charge soils: a review. J. Environ. Qual. 26, 602–617.

O'Connor, G.A., O'Connor, C., Cline, G.R., 1984. Sorption of cadmium by calcareous soils: influence of solution composition. Soil Sci. Soc. Am. J. 48, 1244–1247.

Pearson, M.S., Maenpaa, K., Pierzynski, G.M., Lydy, M.J., 2000. Effects of soil amendment on the bioavailability of lead, zink and cadmium to earthworms. J. Environ. Qual. 29, 1611–1617.

Peryea, F.J., 1991. Phosphate-induced release of arsenic from soils contaminated with lead arsenate. Soil Sci. Soc. Am. J. 55, 1301–1306.

Peryea, F.J., Kammereck, R., 1997. Phosphate-enhanced movement of arsenic out of lead arsenate-contaminated topsoil and through uncontaminated subsoil. Water Air Soil Pollut. 93, 243–254.

Pickering, I.J., Prince, R.C., George, M.J., Smith, R.D., George, G.N., Salt, D.E., 2000. Reduction and coordination of arsenic in Indian mustard. Plant Physiol. 122, 1171–1177.

Qafoku, N.P., Kulier, U., Summer, M.E., Radcliffe, D.E., 1999. Arsenate displacement from fly ash in amended soils. Water Air Soil Pollut. 114, 185–198.

Ross, S.M., 1994. Retention, transformation and mobility of toxic metals in soils. In: S.M. Ross (Ed.), Toxic Metals in Soil-Plant Systems. Wiley, New York, pp. 63–152.

Ruby, M.V., Davis, A., Schoof, R., Eberle, S., Sellstone, C.M., 1996. Estimation of bioavailability using a physiologically based extraction test. Environ. Sci. Technol. 30, 420–430.

Rugh, C.L., Wilde, H.D., Stacks, N.M., Thompson, D.M., Summers, A.O., Meagher, R.B., 1996. Mercury ion reduction and resistance in transgenic *Arabdopsis thaliana* plants expressing a modified bacterial merA gene. Proc. Natl. Acad. Sci. USA 93, 3182–3187.

Shuman, L.M., 1986. Effect of ionic strength and anions on zinc adsorption by two soils. Soil Sci. Soc. Am. J. 50, 1438–1442.

Shuman, L.M., Dudka, S., Das, K., 2000. Zink forms and plant availability in a compost amended soil. Water Air Soil Pollut. 128, 1–11.

Sims, J.T., Johnson, G.V., 1991. Micronutrient soil test. In: J.J. Mortvedt (Ed.), Micronutrients in Agriculture. 2nd Ed. Soil Science Society of America, Madison, WI, pp. 427–476.

Smith, E., Naidu, R., Alston, A.M., 1998. Arsenic in the soil environment. A review. Adv. Agron. 64, 149–195.

Sposito, G., 1984. The Surface Chemistry of Soils. Oxford University Press, New York.

Stahl, R.S., James, B.R., 1991. Zinc sorption by B horizon soils as a function of pH. Soil Sci. Soc. Am. J. 55, 1592–1597.

Sun, B., Zhao, F.J., Lombi, E., McGrath, S.P., 2001. Leaching of heavy metals from contaminated soils using EDTA. Environ. Pollut. 113, 111–120.

Sutton, A.L., Nelson, D.W., Mayrose, V.B., Kelly, D.T., Nye, J.C., 1984. Effect of copper levels in swine manure on corn and soil. J. Environ. Qual. 13, 198–203.

Thayalakumaran, T., Robinson, B.H., Vogeler, I., Scotter, D.R., Clothier, B.E., Percival, H.J., 2003b. Plant uptake and leaching of copper during EDTA-enhanced phytoremediation of repacked and undisturbed soil. Plant Soil 254, 415–423.

Thayalakumaran, T., Vogeler, I., Scotter, D.R., Percival, H.J., Robinson, B.H., Clothier, B.E., 2003a. Leaching of copper from contaminated soil following the application of EDTA. II. Intact core experiments and model testing. Aust. J. Soil Res. 41, 335–350.

Tiller, K.G., 1988. Heavy metals in soils and their environmental significance. Adv. Soil Sci. 9, 13–142.

Tu, C., Ma, L.Q., 2003. Effects of arsenate and phosphate on their accumulation by an arsenic-hyperaccumulator. *Pteris vittata* L. Plant Soil 249, 373–382.

Vangronsveld, J., Cunningham, S.D., 1998. Introduction to the concepts. In: J. Vangronsveld, S.D. Cunningham (Eds.), Metal Contaminated Soils: *In situ* Inactivation and Phytorestoration. Springer, Berlin, pp. 219–225.

Walker, D.J., Clemente, R., Roig, A., Bernal, P.M., 2003. The effects of soil amendments on heavy metal bioavailability in two contaminated Mediterranean soils. Environ. Pollut. 122, 303–312.

Wallingford, G.W., Murphy, V.S., Powers, W.L., Manges, H.L., 1975. Effect of beef-feedlot manure and lagoon water on iron, zink, manganese and copper content in corn and in DTPA soil extracts. Soil Sci. Soc. Am. J. 39, 482–487.

Wang, J., Zhao, F.J., Meharg, A.A., Raab, A., Feildmann, J., McGrath, S.P., 2002. Mechanisms of arsenic hyperaccumulation in *Pteris vittata*. Uptake kinetics, interactions with phosphate and arsenic speciation. Plant Physiol. 130, 1552–1561.

Weggler-Beaton, K., McLaughlin, M.J., Graham, R.D., 2000. Salinity increases cadmium uptake by wheat and Swiss chard from soil amended with biosolids. CSIRO Aust., 37–45.

Wenzel, W.W., Unterbrunner, R., Sommer, P., Sacco, P., 2003. Chelate-assisted phytoextraction using canola (*Brassica napus* L.) in outdoors pot and lysimeter experiments. Plant Soil 249, 83–96.

Woolson, E.A., Axely, J.H., Kearney, P.C., 1973. The chemistry and phytotoxicity of arsenic in soil: II. Effect of time and phosphorus. Soil Sci. Soc. Am. J. 37, 254–259.

Wu, L.H., Luo, Y.M., Xing, X.R., Christie, P., 2004. EDTA-enhanced phytoremediation of heavy metal contaminated soil with Indian mustard and associated potential leaching risk. Agric. Ecosyst. Environ. 102, 307–318.

Yang, J.E., Skogley, E.O., Georgitis, S.J., Ferguson, A.H., 1991. Phytoavailability soil test: development and verification of theory. Soil Sci. Soc. Am. J. 55, 1358–1365.

Developments in Soil Science, volume 32
Ravendra Naidu (Editor)

Chapter 28

THE VALUE OF NITRILOTRIACETATE IN CHELATE-ASSISTED PHYTOREMEDIATION

K. Wenger, S.K. Gupta and R. Schulin

28.1 Introduction

Although phytoremediation of heavy metals contaminated soils has been proposed as a low-cost and "green" and aesthetically pleasant remediation technique (e.g. Jørgensen, 1993; Cunningham et al., 1995; McGrath et al., 1998), its success has not been demonstrated in field applications so far. A limiting factor often appears to be the low phytoavailability of the metals, in particular in neutral to alkaline soils. One approach to overcome this limitation is to increase dissolved metal concentrations by adding complex-forming water-soluble organic ligands (see Chapters 26 and 29).

A suitable ligand for chelate-assisted phytoextraction of metals should increase both the solubility of the target metals in soil and also their uptake by plants. Increased solubility of a metal will not necessarily enhance its uptake and transfer into harvestable plant organs. Considerable debate exists on whether metal complexes are taken up into the stele of roots as such or only after first being split from the chelate. In general, the free metal ion is considered to be taken up most readily across membranes through carriers and ion channels, while stable complexes with large organic ligands may rather be excluded than taken up. Chaney et al. (1972) found that Fe was taken up by plants after first being split from the Fe-chelate complex. Chelator-buffered nutrient solutions are now routinely used for studying plant-micronutrient interactions (e.g. Fe, Cu, Zn or Mn) (Parker et al., 1950), while the uptake of the chelated metals is normally considered to play no role (Halvorson and Lindsay, 1977). However, there are numerous observations that chelating agents are taken up by plants. Radiotracer studies showed that Fe–EDTA actually enters the plant, at least partially, as an intact complex (Wallace et al., 1955; Stewart, 1963). It was only in 1998 that a direct analytical determination of an intact metal–EDTA complex in the xylem sap of *Brassica juncea* proved that the metal and the chelating agent form a complex inside the plant (Vassil et al., 1998). Ethylendiaminetetraacetate (EDTA) complexes of Mn, Zn, Al, Cd and Cu have been detected inside the plant using chromatography/masspectrometry (Collins et al., 2001, 2002). Collins et al. (2002) suggested a passive uptake of Zn–EDTA via an apoplastic pathway in barley (*Hordeum vulgare* L.) and potato (*Solanum tuberosum* L.). Wenger et al. (2005) proposed in their conceptual model that besides an active high-affinity transport system for micronutrients required by the plant a less specific additional pathway for micronutrients and other solutes may also exist. However, a lower rate of uptake of metal cations is reported from metal–organic complexes than of free cations (Jarvis, 1987).

While the formation of soluble chelates may enhance plant accumulation of metals under certain conditions, it will generally also increase the mobility of metals in soil and thus the risk of leaching and pollution of ground and surface water (see Chapter 29 by Kirkham and Chapter 26 by Clothier). This risk is particularly high in the case of ligands that are not only strong chelators, but also persistent in soil due to low biodegradability, as in the case of EDTA (Kari and Giger, 1996). The leaching risk can be avoided if the percolating soil solution is completely captured. This usually requires a fully contained system, which is only achievable at the expense of many advantages that phytoremediation has been advocated for. In particular, containment, water collection and water treatment will considerably raise costs. If chelate-assisted phytoextraction is applied *in situ* without such measures, it is only possible to reduce leaching risks, but not to avoid them completely. In order to minimize these risks it is necessary to limit the solubilizing effect in time, space and degree in such a way that the mobilized metals are maximally available for root uptake and do not exceed the capacity of the plants used to accumulate them. This means that the ligands should be applied as close to the roots as possible, e.g. by directly injecting them into the root zone rather than spraying them onto the soil surface. Additionally, the ligands applied should be biodegraded in due time and not persist longer than the period during which the roots are active. Even if the polluting metals are efficiently accumulated and immobilized within the roots of remediation plants, little is gained in the long term, if they are not translocated into plant parts which can be easily harvested. In general these are the aerial parts of plants. In order to be useful in phytoremediation, a ligand should therefore not only enhance metal uptake by the roots but also facilitate metal translocation into the harvestable parts of the accumulator plant (Fig. 28.1).

In this chapter, we evaluate the potential use of nitrilotriacetate (NTA) as an alternative to EDTA and other poorly degradable aminocarboxylates that have been proposed as ligands in chelate-assisted phytoextraction. NTA is known to be fairly biodegradable (Bolton et al., 1996; Tiedje and Mason, 1974). For this reason it has replaced EDTA in detergents, although its metal complexes are not as strong. However, NTA was shown to solubilize soil-polluting metals much more efficiently than low-molecular weight organic acids such as citrate and oxalate (Wenger et al., 1998), and it also has the advantage of being commercially available in large quantities at reasonable costs. In the following

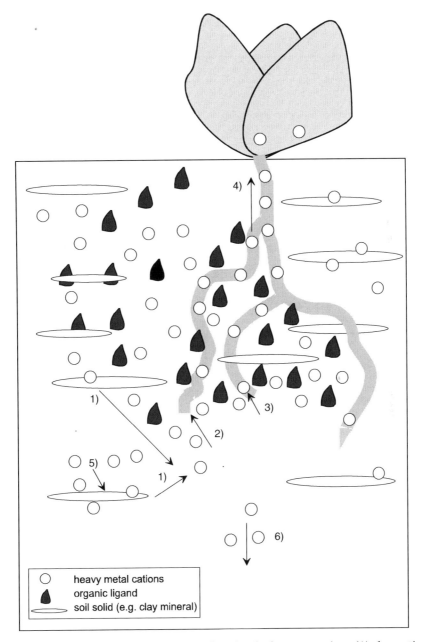

Figure 28.1. *Processes involved in ligand-assisted phytoextraction: (1) desorption or dissolution from the soil solids, (2) transport to the roots by diffusion and mass flux, (3) absorption and uptake by the roots, (4) transfer to the xylem and translocation into the leaves, (5) re-adsorption or precipitation and (6) leaching.*

sections we present outcomes of our study that focused on batch experiments, hydroponic cultures and finally pot and field studies assessing the effect of cheating agents on the bioavailability and uptake of metals by accumulator plants.

28.2 Effect of NTA on metal solubility in clay suspension solutions and in soils

To investigate the processes governing heavy metal mobilization and uptake in natural soils we used well-defined model systems. Figure 28.2 shows the influence of NTA in comparison to citrate on Cu and Zn sorption to the clay minerals montmorillonite and kaolinite as found by Neubauer et al. (2000) in batch suspension experiments. NTA prevented sorption of Cu and Zn to Na–montmorillonite and kaolinite (Fig. 28.2) over the pH range from 4 to 10,

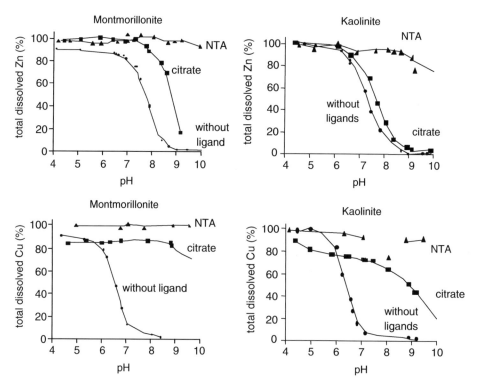

Figure 28.2. *Effect of NTA and citrate on heavy metal sorption to montmorillonite and kaolinite (Neubauer et al., 2000). Experimental conditions: $1 \, g \, l^{-1}$ clay mineral, 0.1 M $NaClO_4$ as background electrolyte, metal concentration of approximately $10^{-5} M$ and ligand concentration of about $3 \times 10^{-4} M$.*

while citrate due to its lower complex stability constant only shifted the Cu and Zn sorption edges towards higher pH. According to speciation calculations, metal–NTA$^-$ and metal–NTA$_2^{4-}$ complexes are the dominating species in this pH range for both metals. Due to their negative charge they are not bound electrostatically by Na–montmorillonite and kaolinite to any significant degree.

Formation of negatively charged metal complexes also explains the influence of citrate on heavy metal sorption by montmorillonite and kaolinite (Fig. 28.2), but due to the lower stability of metal–citrate complexes compared with metal–NTA complexes the effect was limited to a lower range of pH. Apart from shifting Cu and Zn sorption edges to higher pH, citrate also increased sorption of Cu, but not of Zn, at low pH. Neubauer et al. (2000) explained this effect with the formation of ternary Cu–citrate surface complexes at positively charged edge sites of the clay minerals. Because Cu exhibits a higher affinity than Zn to the oxygen donor atoms of citrate, only Cu interacted with the surface-bound citrate. Overall, the batch experiments of Neubauer et al. (2000) suggested that the ligand effects on Cu and Zn sorption to clay minerals can be explained by the competition of ligands and clay mineral surfaces for metal complexation, electrostatic interactions between metal–ligand complexes and surface charge of clay minerals and the formation of ternary surface complexes.

In addition to the above experiments with well-defined mineral suspensions, we studied the efficiency of organic ligands to extract Zn and Cu from real soil samples with aged contamination. One soil was calcareous and had been contaminated with Zn and Cu by dust emission from a nearby brass smelter. The other soil was non-calcareous and was polluted in particular with Zn and Pb due to former sewage sludge applications. Besides citrate and NTA, two other synthetic chelating agents, EDTA and diethylenetraminepentaacetate (DTPA), were used. Parameters and metal concentrations of the two soils are given in Table 28.1.

Consistent with the effects of metal sorption in clay mineral suspension, NTA mobilized Cu and Zn to a much higher extent than citrate (Fig. 28.3). Differences between the three synthetic ligands in extracting Zn and Cu from the two different soils were not very large, but NTA was generally the least effective. In total, about 60–80% of the HNO$_3$-extractable metals could be solubilized by the synthetic organic ligands in the acidic soil, while in the calcareous soil only 20–40% of the HNO$_3$-extractable Zn and Cu were solubilized. The lower efficiency of the synthetic organic ligands in the calcareous soil compared to the acidic soil may be due to differences in the metal binding fractions of the two soils. Sequential extractions showed that in the calcareous soil about 76% of the total Zn was in the crystalline Fe-oxide occluded and residual metal form, while in the acidic soil only about 44% was found in the strongly bound fractions and a higher portion was in the more readily extractable fractions (data not shown).

Table 28.1. *Soil parameters and metal concentrations of the calcareous soil, Dornach, and the acidic soil, Rafz.*

Soil characteristics		Dornach	Rafz
Soil type		Calcaric Regosol	Haplic Luvisol
pH (H$_2$O)		7.6	6.8
CaCO$_3$ (%)		13.4	0.6
C$_{org}$ (%)		4.2	1.6
Clay (%)		32	18
Silt (%)		50	28
Sand (%)		18	54
CEC (meq 100 g^{-1})		25.9	16.9
Metals	Zn	695	813
(mg kg^{-1} dry matter	Cd	2.8	0.9
soil) extractable with	Cu	524	58
2 M HNO$_3$	Pb	54	492
Metals	Zn	0.08	4.0
(mg kg^{-1} dry matter	Cd	0.003	0.007
soil) extractable with	Cu	0.79	0.18
0.1 M NaNO$_3$	Pb	<0.04	0.04

The higher portion in the strongly bound fractions in the calcareous soil may have been due to the origin of the pollution, which were ZnO and CuO containing dust emission from a non-ferrous metal smelter and the accumulation of the metals in the surface horizon of the soil.

28.3 Influence of NTA on metal uptake by tobacco in hydroponic culture

Figures 28.4 and 28.5 show the influence of NTA on Cu, Zn and Pb uptake by tobacco (*Nicotiana tabacum* L.) in nutrient solutions. Treatments included the addition of 100 µm Cu, 100 µm Zn and 100 µm Pb as single elements and in combination with 500 µm NTA, as well as control treatments. One series of experiments was carried out without addition of P to the nutrient solution in order to prevent metal phosphate precipitations. The plants were harvested 6 days after the start of the experiment.

NTA reduced metal accumulation in the roots by a factor of about 70 for Cu and Pb and about 4 and 20 for Zn without P and with P, respectively (Fig. 28.4). On the other hand, NTA increased Cu and Pb uptake into the shoots (Fig. 28.5). Cu and Pb concentrations in the shoots were 2–3 times greater than in the respective treatments without NTA and without P. In the presence of P the relative increase in the uptake of these metals caused by NTA was even larger. In contrast, zinc uptake into the shoots was much higher without than with NTA

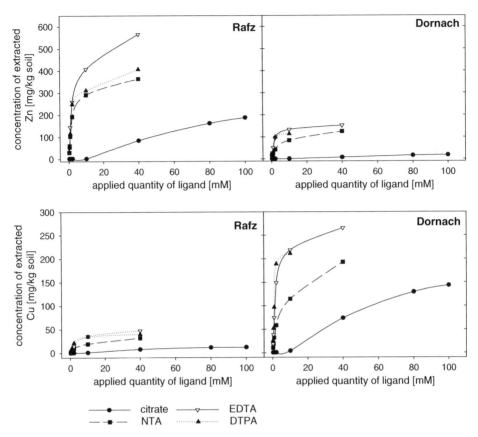

Figure 28.3. *Effect of various organic ligands on Zn and Cu extraction from two different soils (see Table 28.1).*

(Fig. 28.5). This might have been due to an efficient active Zn uptake and translocation into the tobacco shoots that could not proceed as effectively in the presence of NTA.

According to speciation calculations the free metal cations were the dominant species of all three metals in the treatments without NTA (Table 28.2). In the absence of NTA, similar amounts of Cu and Pb were found in the roots, but less Zn, probably due to the higher translocation of Zn into the shoots. The addition of P did not significantly affect the metal accumulation in the roots, but tended to decrease Cu and Pb uptake by the shoots. Whether this was due to precipitation of Cu and Pb phosphates, as predicted by speciation calculations and confirmed for Pb by measurements, remained an open question. For Zn, the addition of P led to the opposite tendency. Whether this was due to a promotion of Zn uptake and translocation by P was not clear. In no case a significant

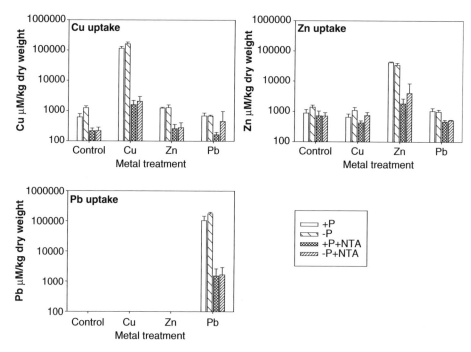

Figure 28.4. *Uptake of Cu, Zn and Pb by tobacco roots under hydroponic conditions in presence or absence of 500 μm NTA and phosphorus (added as 100 μm KH₂PO₄) for various metal treatments ("Cu": addition of 100 μm Cu; "Zn": addition of 100 μm Zn; "Pb": addition of 100 μm Pb).*

growth reduction or any other visible sign of metal toxicity was found in the presence of NTA in these experiments. Indeed, NTA was found in the shoots in NTA and metal and NTA treatments, suggesting that the toxicity of the accumulated metals was mitigated by the formation of metal–NTA complexes.

There are two parallel transport pathways of ions (solutes) and water across the root cortex towards the stele: a pathway of passive transport through the apoplasm (cell walls and intercellular spaces), and a pathway of active transport from cell to cell in the symplasm (Fig. 28.6). In general, the apoplasm of the root cortex is directly accessible to solutes from the external solution, although it does not provide a free space for movement of charged solutes. Within the cell wall, carboxylic groups and other moities act as cation exchangers. Cations are preferentially bound to these sites, whereas anions are repelled. Cation binding in the apoplasm can significantly contribute to the total cation accumulation of roots (Marschner, 1995). The high metal concentrations found in the roots in the treatments without NTA (Fig. 28.4) may have resulted primarily from binding in the apoplasm of the root cortex, rather than from "true" uptake into the root

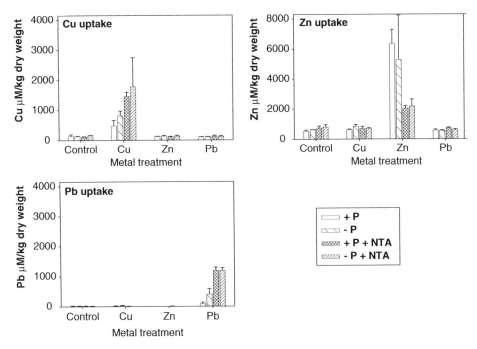

Figure 28.5. *Uptake of Cu, Zn and Pb by tobacco shoots under hydroponic conditions in presence or absence of 500 μm NTA and phosphorus (added as 100 μm KH₂PO₄) for various metal treatments ("Cu": addition of 100 μm Cu; "Zn": addition of 100 μm Zn; "Pb": addition of 100 μm Pb).*

cells. The reduction of metal concentrations in the roots by addition of NTA may then simply be explained by the prevention of metal sorption to these cation exchange sites in the apoplast due to formation of negatively charged metal–NTA complexes. Indeed, the retention of unchelated Pb and unchelated Cu in cell walls of roots, particularly around intercellular spaces, was observed in an ultra-structural study using transmission electron microscopy for Pb (Jarvis and Leung, 2001) and for Cu by means of EDX microanalysis (energy dispersive X-ray microanalysis) (Jung et al., 2003). The main barrier against passive movement into the stele via the apoplastic pathway is the endodermis, the innermost layer of cells of the cortex. In the radial and transverse walls of the endodermis, hydrophobic incrustations (suberin), i.e. the Casparian band, constitute an effective barrier against passive solute transfer into the stele. The endodermis is not a perfect barrier for apoplastic transport, however. In addition to passage cells, there appear to be sites distributed along the root axis where this barrier is "leaky". At the root apex for instance the Casparian stripe is not yet fully developed and, thus, allows apoplastic transport to reach the stele

Table 28.2. *Concentrations (μm) and aqueous speciation (%) of Cu, Zn and Pb in hydroponic experiments with tobacco seedlings as affected by metal and NTA treatments.*

Treatments	Metal input (mM)	Measured total metal concentration in solution (mM)			Speciation % (calculated)	
		$P-$	$P+$		$P-$	$P+$
Cu	100	79	77	Cu^{2+}	92	93
				$CuSO_4$	6	4
				$CuOH^+$	1	2
Cu&NTA	100	86	89	$CuNTA^-$	100	100
Zn	100	94	93	Zn^{2+}	94	10
				$ZnSO_4$	6	–
				$Zn_3(PO_4)_{2aq}$	–	90
Zn&NTA	100	97	96	$ZnNTA^-$	100	100
Pb	100	89	6	Pb^{2+}	89	88
				$PbSO_4$	9	9
				$PbOH^+$	1	2
				$PbNO_{3+}$	1	–
Pb&NTA	100	96	97	$PbNTA^-$	100	100

Note: Species distribution calculated with ChemEQL (Müller, 1996). Species were considered relevant if their concentration exceeded 1% of the measured total dissolved concentration.

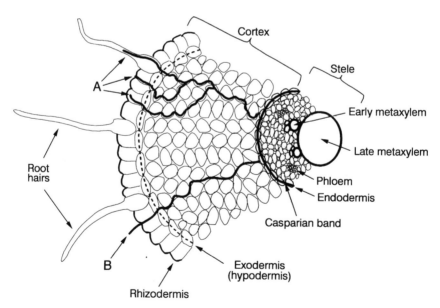

Figure 28.6. *Partial cross-section of a maize root showing the symplastic (A) and the apoplastic (B) pathway of ion (solute) transport across the root (Marschner, 1995).*

(Huang and van Steveninck, 1989). Crowdy and Tanton (1970) showed that the uptake of Pb–EDTA was restricted to a region between 3 and 140 mm behind the root tip where the suberization of the cell walls had not yet occurred. This pathway would allow metal–chelates to bypass the impermeable Casparian stripe through the apoplast.

Based on the literature cited, we hypothesize that in addition to the free metal ions also metal–NTA complexes are taken up by the tobacco seedlings and translocated with the stream of transpiration water into the shoots. Because these complexes were biochemically inactive, they did not cause any toxicity for the 6 days duration of the experiment, although much higher concentrations of Cu and Pb were accumulated in the shoots than without NTA.

28.4 Influence of NTA on metal uptake from nutrient solution with montmorillonite

Figure 28.7 shows the effect of NTA on Cu uptake by tobacco seedlings in the same type of hydroponic experiments as described before, but with the difference that $10 \, g \, L^{-1}$ montmorillonite was added to the system. Four treatments (control, NTA, Cu and Cu+NTA) were applied (Wenger et al., 2003). In the Cu-enriched suspension, NTA increased Cu uptake and translocation into shoots by a factor of 26, while it decreased Cu concentrations in the roots by a factor of 3.6 (Fig. 28.7). Thus, NTA had a much stronger effect on Cu uptake and translocation by tobacco seedlings than in the absence of the clay mineral. Inferring from the results of Neubauer et al. (2000) presented above that NTA

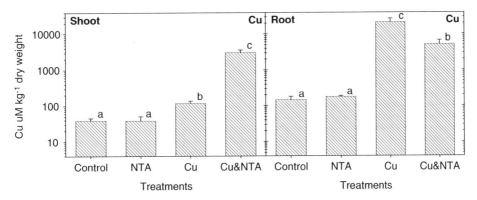

Figure 28.7. *Copper uptake ($\mu m \, kg^{-1}$) in shoots and roots of tobacco in different treatments for the nutrient solution experiment with montmorillonite. Columns with common index are not significantly different (p > 0.05) (Wenger et al., 2003).*

prevented Cu sorption to montmorillonite, more Cu was available in the solution for plant uptake than in the absence of NTA.

28.5 Pot and field Experiments

Encouraged by the promising influence of NTA on the phytoextraction of metals under hydroponic conditions, we studied NTA-assisted phytoextraction from real soil with aged contamination. A pot experiment was performed with tobacco (*Nicotiana tabacum* L.) and maize (*Zea mays* L.) using the same topsoil (0–20 cm) material as in the extraction experiments described before (Table 28.1). Compared to the marked increase of soluble Zn, Cu, Cd and Pb concentrations in the two soils, metal accumulation in the shoots of the two plants was much less enhanced by NTA (Wenger et al., 2002a). For example, soluble Pb was increased by a factor 50 in the non-calcareous soil, while Pb concentration in tobacco shoots increased only fivefold.

The potential of NTA to enhance phytoextraction of Zn, Cd and Cu from the calcareous soil was also examined in a field experiment using various plant species, including tobacco and maize (Kayser et al., 2000). The addition of NTA increased the solubility ($NaNO_3$-extraction) of Zn, Cd and Cu in the soil by factors of around 80, 30 and 8, respectively, but accumulation of these metals in average increased only by factors in the range 2–3 for the tested plant species (Table 28.3). This increase was much less than achieved by addition of NTA in the montmorillonite suspensions. Comparing the results of these experiments, it must be considered, however, that the experimental conditions were very different. In the hydroponic system roots have unlimited access to nutrients and water, while in real soil nutrient supply is limited by desorption and transfer processes (mass transfer by convection and diffusion) from the soil to the root surfaces. Also competition with other plants and soil microorganism for water and nutrients can limit metal uptake. Another reason for the higher efficiency of NTA in the hydroponic experiments probably lies in the very short duration of the experiments and the early growth stage of the plants. Pot and field experiments were carried out over 90 days until full maturity of the plants. According to Marschner (1995), there is a fairly clear decline in mineral nutrient content of dry matter with the growth and ageing of plants. Furthermore, in our experiments the solubilizing effect of NTA was only of rather short duration in soil after each application. Soluble metal concentrations strongly increased and remained high for about 7 days after the application of NTA, but then dropped again and reached almost the initial values within 20 days (Kayser et al., 2000; Wenger et al., 2002a). For this reason, NTA was applied repeatedly, but there were still periods of reduced metal solubility due to the decrease of NTA activity in between. A further difference between hydroponic and soil

Table 28.3. *Relative Treatment Effects (RTE =$T_{1998}/T_{1997} * C_{1997}/C_{1998}$) (mean and SD) and results of analysis of variance (ANOVA) of log-transformed plant tissue metal concentrations (Kayser et al., 2000).*

	Cd		Zn		Cu	
	Control	NTA	Control	NTA	Control	NTA
Brassica juncea	1	2.1±0.9	1	2.0±0.5	1	1.8±0.3
	a	a	a	b	a	b
Nicotiana tabacum	1	2.0±0.8	1	2.0±0.3	1	1.9±0.2
	a	b	a	b	a	b
Salix viminalis	1	1.7±0.2	1	1.8±0.3	1	4.2±3.8
	a	b	a	a	a	a
Helianthus annuus	1	1.6±0.8	1	1.9±0.4	1	1.8±0.4
	a	a	a	b	a	c
Zea mays	1	2.1±0.7	1	1.7±0.3	1	2.2±1.2
	a	a	a	a	a	b
Average RTE	1	1.9±0.2	1	1.9±0.1	1	2.4±0.9

Notes: The RTE relates the heavy metal concentration of a plant grown on a particular subplot in the year of treatment (T_{1998}) to that of the first year (T_{1997}), in which no treatments were applied. The ratio C_{1997}: C_{1998} between the heavy metal concentrations of controls is a correction term to account for differences in heavy metal uptake between the two experimental years due to other factors, e.g. climate. RTE values for control equal 1 by definition. Treatments with the same letters as in the respective control had no significant effect ($p > 0.05$).

conditions is that NTA and metal distribution in the hydroponic system is homogenous, whereas in soil such homogeneity is impossible to achieve. NTA was applied by a handheld fertilizer injector. With this method it was possible to inject NTA close to the roots at depths of about 20 cm below soil surface, but unavoidably the distribution of NTA in the root zone was not uniform. Therefore, it is likely that solubilization was not optimal everywhere in the root zone. The plants may have actually accentuated the inhomogeneity effect by actively avoiding the growth of roots into local zones with high soluble metal contents.

28.6 Conclusion

The results clearly show that NTA does not only have a strong potential to solubilize heavy metals in soil, but also enhance the uptake and translocation of heavy metals into the shoots of remediation plants. For Zn this effect was only found under real soil conditions, indicating that for this element not the uptake by the plants, but the supply from the bulk soil to the root surfaces was the limiting step. Overall, phytoextraction efficiencies in real soil were rather modest and did not by far match the increase in metal solubilization in the soil.

In our experiments with real soil we did not find a similar chelate-induced surge of Pb translocation and accumulation as reported by Blaylock et al. (1997) and Huang et al. (1997), who applied EDTA at very high dosages. Vassil et al. (1998) speculated that the high uptake of Pb–EDTA observed by the latter authors may have been due to destruction of the physiological barrier(s) in the roots. Such a harsh solubilization is quite problematic under *in situ* conditions. It bears a high risk of metal leaching into deeper soil layers and into the groundwater.

The observed enhancement of phytoextraction is insufficient to make this technique suitable for practical applications in the clean-up of soil contaminations. There is still much potential for further improvement though. Alternative chelating agents may be found or developed, which are more efficient than NTA in enhancing metal uptake by accumulator plants without having the problem of low biodegradability such as EDTA. The application technique could be improved in such a way that roots are more homogenously exposed to solubilized metals. For example, the chelating agents could be added to the soil in combination with fertilizer in form of tablets, which slowly release chelating agents together with the fertilizer as roots grow towards them. But even if a 20-fold increase in metal accumulation by the employed plants could be achieved in the field in comparison to our results, a 50% reduction of Pb in the non-calcareous soil of this study would still require around 250 years for Pb and 100 years for Cu 100 in the case of the calcareous soil, assuming an annual yield of $10\,t\,ha^{-1}$ dry matter (Fig. 28.8).

On the other hand, also the removal and off-site treatment of contaminated soil is hardly a desirable solution. It becomes prohibitively expensive for large areas, and abandoned sites without vegetation cover bear a high risk that the contamination spreads by erosion and leaching. Therefore, phytoremediation should still be considered as a potential option for the long-term treatment of contaminated land where stabilization is the main goal. If phytostabilization is the treatment of choice, it might as well be combined with phytoextraction in order to ameliorate the soil gradually with time. In certain cases, e.g. soil contamination with Pb or with radionuclides, which are not readily taken up and translocated into the above ground parts of remediation plants, also the use of chelating agents such as NTA may be considered to accelerate the process, which may be termed phytomelioration. Chelate-assisted phytomelioration of course requires that it is possible to prevent leaching of the solubilized pollutants as in any other remediation scheme. Phytostabilization and phytomelioration will be particularly attractive also from an economic perspective, if they can be combined with commercial production of fibre, non-food oil, biofuel and even feedstuff. Many plant parts used for commercial purposes do not accumulate significantly higher concentrations of heavy metals on contaminated

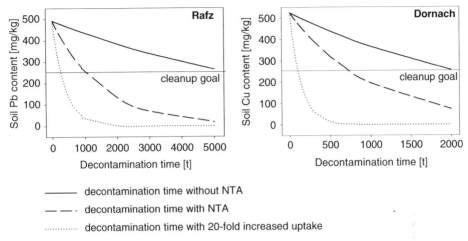

Figure 28.8. *Calculated reduction of Pb in Rafz soil and of Cu in Dornach soil by tobacco plants, assuming an annual harvest of 10 t ha^{-1} dry matter: (i) with observed metal accumulation rate in absence of NTA, (ii) with observed metal accumulation rate in presence of NTA and (iii) with a hypothetical 20-fold increase of metal accumulation rate. Calculations are based on first-order kinetics: t = (k^{-1}) * ln (C$_o$*C$_z^{-1}$) (C$_o$, C$_z$ = soil metal concentration at beginning and the end of phytoextraction period; k = specific constant).*

than on uncontaminated soils. If these plant parts are separated from those parts accumulating high contaminant concentrations, their utilization should be no problem. In a field study in which we investigated the efficiency of Zn phytoextraction by maize plants (*Zea mays* L.) we found for example that Zn accumulation in maize cobs remained far below the tolerable maximum Zn concentration of 250 mg kg^{-1} for fodder plants according to Swiss regulations (FMBV, 1995), while Zn concentrations in the leaf reached up to about 1500 mg kg^{-1} (Wenger et al., 2002b). For these reasons, further research should also focus on the allocation of heavy metals in remediation plants and on alternative commercial uses of those plant parts showing lower metal accumulation.

Acknowledgements

We are thankful to Dr. Uta Neubauer und Dr. Achim Kayser for contributing their results and we are gratefully indebted to Dr. Susan Tandy for proofreading the English.

References

Blaylock, M.J., Salt, D.E., Dushenkov, S., Zakharova, O., Gussman, C., Kapulnik, Y., Ensley, B.D., Raskin, I., 1997. Enhanced accumulation of Pb in Indian mustard by soil-applied chelating agents. Environ. Sci. Technol. 31, 860–865.

Bolton, H. Jr., Girvin, C.C., Plymale, A.E., Harvey, S.D., Workman, D.J., 1996. Degradation of metal-nitrilotriacetate complexes by *Chelatobacter heintzii*. Environ. Sci. Technol. 30, 931–938.

Chaney, R.L., Brown, J.C., Tiffin, L.O., 1972. Obligatory reduction of ferric chelates in iron uptake by soybeans. Plant Physiol. 50, 208–213.

Collins, R.N., Merrington, G., McLaughlin, M.J., Knudsen, C., 2002. Uptake of intact zinc-ethylenediaminetetraacetic acid from soil is dependent on plant species and complex concentration. Environ. Toxicol. Chem. 21, 1940–1945.

Collins, R.N., Onisko, B.C., McLaughlin, M.J., Merrington, G., 2001. Determination of metal-EDTA complexes in soil solution and plant xylem by ion chromatography-electrospray mass spectrometry. Environ. Sci. Technol. 35, 2589–2593.

Crowdy, S.H., Tanton, T.W., 1970. Water pathways in wheat leaves. I. Free space in wheat leaves. J. Exp. Bot. 21, 102–111.

Cunningham, S.C., Berti, W.R., Huang, J.W., 1995. Phytoremediation of contaminated soils. Trends Biotechnol. 13, 393–397.

FMBV, 1995. Futtermittel: Futtermittelverordnung und Futtermittelbuch-Verordnung mit Anhängen. SR 916.307/916.307.1. Schweizerischer Bundesrat und Eidgenössisches Volkswirtschaftsdepartement, Bern.

Halvorson, A.D., Lindsay, W.L., 1977. The critical zinc^{2+} concentration for corn and the nonabsorption of chelated zinc. Soil Sci. Soc. Am. J. 41, 531–534.

Huang, C.X., van Steveninck, R.F.M., 1989. The role of particular pericycle cells in the apoplastic transport in root meristems of barley. J. Plant Physiol. 135, 554–558.

Huang, J.W., Chen, J., Berti, W.R., Cunningham, S.D., 1997. Phytoremediation of lead-contaminated soils: role of synthetic chelates in lead phytoextraction. Environ. Sci. Technol. 31, 800–805.

Jarvis, S.C., 1987. The uptake and transport of silicon by perennial ryegrass and wheat. Plant Soil 97, 429–437.

Jarvis, S.C., Leung, D.W.M., 2001. Chelated lead transport in *Chamaecytisus proliferus* (L.f.) link ssp. *proliferus* var. palmensis (H. Christ): an ultrastructural study. Plant Sci. 161, 433–441.

Jørgensen, S.E., 1993. Removal of heavy metals from compost and soil by ecotechnological methods. Ecol. Eng. 2, 89–100.

Jung, C., Maeder, V., Funk, F., Frey, B., Sticher, H., Frossard, E., 2003. Release of phenols from *Lupinus albus* L. roots exposed to Cu and their possible role in Cu detoxification. Plant Soil 252, 301–312.

Kari, F.G., Giger, W., 1996. Speciation and fate of ethylenediaminetetraacetate (EDTA) in municipal wastewater treatment. Water Res. 30, 122–134.

Kayser, A., Wenger, K., Keller, A., Attinger, W., Felix, H.R., Gupta, S.K., Schulin, R., 2000. Enhancement of phytoextraction of Zn, Cd and Cu from calcareous soil: the use of NTA and sulphur amendments. Environ. Sci. Technol 34, 1778–1783.

Marschner, H., 1995. Mineral Nutrition of Higher Plants. Academic Press, London.

McGrath, S.P., Sidoli, C.M.C., Baker, A.J.M., Reeves, R.D., 1998. Phytoextraction for soil remediation. In: R.R. Brooks (Ed.), Plants that Hyperaccumulate Heavy Metals: Their Role in Phytoremediation, Microbiology, Archaeology, Mineral Exploration and Phytomining. CABI, Wallingford, UK, pp. 261–287.

Müller, B., 1996. ChemEQL V.2.0. A program to calculate chemical speciation and chemical equilibria. Eidgenössische Anstalt für Wasserversorgung, Dübendorf, Switzerland.

Neubauer, U., Furrer, G., Kayser, A., Schulin, R., 2000. Siderophores, NTA and citrate: potential soil amdendments to enhance heavy metal mobility in phytoremediation. Int. J. Phytoremediation 2, 353–368.

Parker, D.R., Chaney, R.L., Norvell, W.A., 1950. Chemical equilibrium models: applications to plant nutrition research. In: S. Goldberg (Ed.), Chemical Equilibrium and Reaction Models. Soil Science Society of America, Madison, Vol. 42, SSSA, pp. 163–200.

Stewart, I., 1963. Chelation in the absorption and translocation of mineral elements. Ann. Rev. Plant Physiol. 14, 295–310.

Tiedje, J.M., Mason, B.B., 1974. Biodegradation of nitrilotriacetate (NTA) in soils. Soil Sci. Soc. Am. Proc. 38, 278–283.

Vassil, A.D., Kapulnik, Y., Raskin, I., Salt, D.E., 1998. The role of EDTA in lead transport and accumulation by Indian mustard. Plant Physiol. 117, 447–453.

Wallace, A., North, C.P., Mueller, R.T., Hemaidan, N., 1955. Behavior of chelating agents in plants. Proc. Am. Soc. Hort. Sci. 65, 9–16.

Wenger, K., Hari, T., Gupta, S.K., Krebs, R., Rammelt, R., Leumann, C.D., 1998. Possible approaches for in-situ restoration of soils contaminated by zinc. Adv. Geoecol. 31, 745–753.

Wenger, K., Kayser, A., Gupta, S.K., Furrer, G., Schulin, R., 2002a. Comparison of NTA and elemental sulfur as potential soil amendments in phytoremediation. Soil Sediment Contam. 11, 655–672.

Wenger, K., Kayser, A., Gupta, S.K., Furrer, G., Schulin, R., 2002b. Zinc extraction potential of two common crop plants, Nicotiana tabacum and Zea mays. Plant Soil 242, 217–225.

Wenger, K., Kayser, A., Gupta, S.K., Furrer, G., Schulin, R., 2003. The role of nitrilotriacetate in copper uptake by tobacco. J. Environ. Qual. 32, 1669–1676.

Wenger, K., Tandy, S., Nowack, B., 2005. Effects of chelating agents on trace metal speciation and bioavailability. In: B. Nowack and J.M. van Briesen (Eds.), Biogeochemistry of Chelating Agents, ACS Symposium Series 910, pp. 204–224.

Developments in Soil Science, volume 32
Ravendra Naidu (Editor)

Chapter 29

EDTA-ASSISTED PHYTOSTABILIZATION BY BARLEY ROOTS CONTAMINATED WITH HEAVY METALS

F. Madrid, M.S. Liphadzi and M.B. Kirkham

29.1 Introduction

Phytoremediation is defined as the use of green plants to remove pollutants from the environment or to render them harmless. It is divided into the following six areas (Salt et al., 1998):

(a) Phytoextraction: The use of pollutant-accumulating plants to remove metals or organics from the soil by concentrating them in the harvestable parts.
(b) Phytodegradation: The use of plants and associated microorganisms to degrade organic pollutants.
(c) Rhizofiltration: The use of plant roots to absorb and adsorb pollutants, mainly metals, from water and aqueous waste streams.
(d) Phytostabilization: The use of plants to reduce the bioavailability of pollutants in the environment.
(e) Phytovolatilization: The use of plants to volatilize pollutants; and the use of plants to remove pollutants from air.

Phytoextraction of metallic contaminants in soil is an area of recent, major scientific, and technological progress. Two basic strategies are used in phytoextraction: chelate-assisted phytoextraction and long-term continuous phytoextraction. Of the two processes, chelate-assisted phytoextraction is the more developed and is presently being used commercially (Salt et al., 1998). Many studies show that chelate-facilitated phytoremediation can increase bioavailability of heavy metals for clean-up of contaminated sites (Blaylock et al., 1997; Huang et al., 1997; Brooks, 1998; Cooper et al., 1999; Barona et al., 2001; Vogeler et al., 2001; McGrath and Zhao, 2003).

Phytostabilization is a component of phytoremediation and involves absorption and precipitation of contaminants, principally metals, by plants, reducing their mobility and preventing their migration to ground water (leaching) or to air (wind transport) or entry into the food chain (Miller, 1996). While there has been much work studying chelate-facilitated phytoextraction, limited work has been conducted to study the use of chelates to phytostabilize metals. In addition, most research with chelates has focussed on metal content of plant shoots and not roots. The ability of roots to minimize contaminant bioavailability is not well documented, and essentially very few studies report the effect of chelating agents on metal uptake by roots (Jiang et al., 2003).

Under non-polluted, field conditions, chelates are used to make essential micronutrients available to crops. Among the major chelating agents that are used in agriculture are EDTA (ethylenediaminetetraacetic acid), HEDTA (*N*-(2-hydroxyethyl) ethylenediaminetetraacetic acid), DTPA (diethylenetrinitrilopentaacetic acid), CDTA (*trans*-1,2-cyclohexylenedinitrilotetraacetic acid), and EDDHA (ethylenediamine di(o-hydroxyphenyl) acetate) (Leymonie, 2001). The essential micronutrients for plants are B, Cu, Fe, Mn, Mo, and Zn. Only cations can be chelated, and chelates of B and Mo do not exist (Leymonie, 2001). The synthetic chelates are crucial micronutrient sources in many of the leading brands of high technology soluble fertilizers used in hydroponics and fertigation. They can be applied to leaves directly or to the soil. Soil addition is preferable to build up soil fertility (Leymonie, 2001). Because the chelates are fast acting, they can cure a micronutrient deficiency immediately if they are applied to leaves. Soil applied chelates can also cure rapidly a deficiency (e.g., FeEDDHA). Due to their high stability, some chelates do not easily release their cationic micronutrient and this makes them slow-acting when applied to the leaf, but ideal for the soil where root exudates can break them down into a simpler form for uptake and translocation (Leymonie, 2001).

In agriculture, chelates are used rather than the inorganic salt of the micronutrient, because of their ability to protect a micronutrient from becoming unavailable for plant uptake. Chelates used for crop growth must be stable. The most stable chelates are the Fe chelates, followed by Zn and Cu chelates. Manganese chelates have a very low stability. For example, with EDTA as the chelate, after 30 days, 55% of Fe remains chelated, while only 22% of Zn and 8% of Cu remain chelated. For Mn, it is 0% after 1 day. The micronutrients revert to the inorganic form and are unavailable in the soil (Leymonie, 2001).

We have recently reviewed phytoremediation (Liphadzi and Kirkham, 2005), heavy-metal toxicity to plants (Liphadzi and Kirkham, 2006), and methods to test soil availability of heavy metals (Madrid and Kirkham, 2007). We also studied heavy-metal displacement in chelate-irrigated soil using barley (*Hordeum vulgare* L.) (Madrid et al., 2003). Barley was selected for this study given that it is the most salt tolerant of any commercially important plant (Richards, 1954, p. 67), and numerous researchers (e.g., Ebbs and Kochian, 1998) have suggested that barley has phytoremediation potential at least equal to Indian mustard (*Brassica juncea* (L.) Czern; also called leaf mustard or Chinese mustard), a plant widely used in phytoremediation research. We found that under irrigated conditions, the chelate mobilized the metals and they accumulated in drainage water (Madrid et al., 2003). However, the focus of that work was not uptake by the roots and shoots. Also, in most semi-arid areas, little drainage occurs. It is important to know how barley roots take up metals when water is not moving by the roots. Limited work has been conducted to assess the tolerance of barley to a variety of heavy metals. Ebbs and Kochian (1998) conducted a pot experiment using a Zn-contaminated soil and showed that the addition of EDTA to the soil did not increase Zn accumulation by barley shoots. They did not measure root concentrations. The ability of barley to reduce the bioavailability of heavy metals in the environment, when water is not leaching through the soil, is unknown. Therefore, the objective of this work was to study the uptake by barley roots and shoots of Cd, Cu, Ni, Pb, and Zn, when salts of these heavy metals were added to an agricultural soil (Haynie very fine sandy loam) with or without the chelating agent, EDTA, and no drainage was allowed to occur.

29.2 Materials and methods

The experiment was carried out in a greenhouse located at Kansas State University in Manhattan, KS (39° 08′ N, 96° 37′ W, 314 m ASL). On 26 January 2001, 104 square plastic pots (8 cm wide; 8 cm length; 7.3 cm tall) each were filled with 350 g of a Haynie very fine sandy loam soil (coarse-silty, mixed, superactive, calcareous, mesic Mollic Udifluvents). The field capacity of the soil is $0.33 \, m^3/m^3$ and its bulk density is $1.52 \, Mg/m^3$ (Song et al., 1999). It has a pH of 7.6 (Zhu, 2001). Drainage holes in the pots were sealed off with masking tape. Table 29.1 gives a summary of treatments. Salts of five different heavy metals (Cd, Cu, Ni, Pb, and Zn) were dissolved in water and poured on to the surface of the dry soil to provide two concentrations of the metals, a low and high one. The salts were: $CdSO_4$; $CuSO_4 \cdot 5H_2O$; $NiSO_4 \cdot 6H_2O$; $Pb(NO_3)_2$; and $ZnSO_4 \cdot 7H_2O$. Two treatments consisted of adding all metals together in the soil at the two concentrations. These treatments were designated M1 an M2 for all salts at the low and high concentrations, respectively. Metals were added as the corresponding sulfate salt, except Pb, which was added as the nitrate salt due to the low solubility of its sulphate salt. A higher concentration of the essential heavy metals, Cu and Zn, was used than for the non-essential heavy metals (Cd, Ni, and Pb). The low and high concentrations of the Cu and Zn salts provided 100 mg/kg and 500 mg/kg of Cu and 100 mg/kg and 500 mg/kg Zn. The low and high concentrations of the Cd, Ni, and Pb salts provided 50 mg/kg and 250 mg/kg of each heavy metal.

The procedure adopted to add salts to the pots is summarized in the following sections. Salt solutions were prepared as follows:

Cu: 13.700 g/L $CuSO_4 \cdot 5H_2O$
Zn: 15.400 g/L $ZnSO_4 \cdot 7H_2O$
Cd: 3.220 g/L $CdSO_4$

Table 29.1. *Summary of treatments.*

Treatment designation	Metal salt added	Metal conc. mg/kg	EDTA
CA0 (control)	None	0	No
Cu1A0	$CuSO_4 \cdot 5H_2O$	100	No
Cu2A0	$CuSO_4 \cdot 5H_2O$	500	No
Zn1A0	$ZnSO_4 \cdot 7H_2O$	100	No
Zn2A0	$ZnSO_4 \cdot 7H_2O$	500	No
Cd1A0	$CdSO_4$	50	No
Cd2A0	$CdSO_4$	250	No
Pb1A0	$Pb(NO_3)_2$	50	No
Pb2A0	$Pb(NO_3)_2$	250	No
Ni1A0	$NiSO_4 \cdot 6H_2O$	50	No
Ni2A0	$NiSO_4 \cdot 6H_2O$	250	No
M1A0	[a]	[b]	No
M2A0	[a]	[c]	No
CA1 (control)	None	0	Yes
Cu1A1	$CuSO_4 \cdot 5H_2O$	100	Yes
Cu2A1	$CuSO_4 \cdot 5H_2O$	500	Yes
Zn1A1	$ZnSO_4 \cdot 7H_2O$	100	Yes
Zn2A1	$ZnSO_4 \cdot 7H_2O$	500	Yes
Cd1A1	$CdSO_4$	50	Yes
Cd2A1	$CdSO_4$	250	Yes
Pb1A1	$Pb(NO_3)_2$	50	Yes
Pb2A1	$Pb(NO_3)_2$	250	Yes
Ni1A1	$NiSO_4 \cdot 6H_2O$	50	Yes
Ni2A1	$NiSO_4 \cdot 6H_2O$	250	Yes
M1A1	[a]	[b]	Yes
M2A1	[a]	[c]	Yes

[a] $CuSO_4 \cdot 5H_2O$, $ZnSO_4 \cdot 7H_2O$, $CdSO_4$, $Pb(NO_3)_2$, $NiSO_4 \cdot 6H_2O$.
[b] Cu, Zn: 100 mg/kg; Cd, Pb, Ni: 50 mg/kg.
[c] Cu, Zn: 500 mg/kg; Cd, Pb, Ni: 250 mg/kg.

Ni: 7.800 g/L $NiSO_4 \cdot 6H_2O$
Pb: 2.800 g/L $Pb(NO_3)_2$.

With these solutions, 10 mL per pot was added for the low concentration and 50 mL per pot was added for the high concentration. For example, 10 mL of the $Cd(SO_4)$ solution was needed for each pot of treatments Cd1A0 (low Cd conc.; no EDTA; see Table 29.1 for designation of treatments) and Cd1A1 (low Cd conc.; with EDTA) (eight pots), and 50 mL was needed for the treatments Cd2A0 (high Cd conc.; no EDTA) and Cd2A1 (high Cd conc.; with EDTA) (eight pots). Ten milliliter of the $Cd(SO_4)$ solution was needed for the treatments M1A0 (mixture of all salts; low conc.; no EDTA) and M1A1 (mixture of all salts; low conc.; with EDTA) (eight pots) and 50 mL was needed for the treatments M2A0 (mixture of all salts; high conc.; no EDTA) and M2A1 (mixture of all salts; high conc.; with EDTA) (eight pots). In this way, each high-metal pot received

50 mL per pot. M1A0 and M1A1 received 5 times 10 mL (one dose per metal; 50 mL). Pots with the low concentration of metal addition received 10 mL of metal solution plus 40 mL of tap water to get the 50 mL that the pots with the high concentration of metal addition received. Control pots received 50 mL of tap water. The only exceptions to this volume of solution or water added (50 mL) were M2A0 and M2A1, which received 250 mL (5 times 50 mL). This was added during 1.5 days to avoid leaching.

KNO$_3$ (17.100 g/L) was added to pots, so each treatment received the same amount of nitrogen. To provide the KNO$_3$, 3.0 mL of solution were added to the Cu, Zn, Cd, and Ni treatments (sulphate salts) and 2.5 mL of KNO$_3$ were added to the Pb1 and M1 treatments. The Pb2 and M2 treatments did not receive KNO$_3$, as they had received the highest amount of NO$_3$ with the metal addition.

Four days after the addition of the metal solutions or tap water (control pots), pots were rewatered by adding 70 mL to each pot and 10 winter barley (*Hordeum vulgare* L. 'Weskan') seeds were planted in each pot. Plants were thinned to seven plants per pot 17 days after planting (DAP). Plants were randomly chosen to be cut, when they were thinned. Pots were weighed each day and about 70 mL of water were added daily to each pot to bring the weight to 440 g (the pot weighed 20 g). During the last two weeks of the experiment pots received about 80 mL of water daily.

After 31, 35, and 38 DAP, the disodium salt of EDTA was added at a rate of 0.5 g/kg soil (0.02 mmol/cm^2 surface) each day. Until EDTA was added to the soil, each of the 13 treatments (5 salts at low and high concentrations; mixture of salts at low and high concentrations; control) was replicated eight times. After the addition of EDTA, there were 26 treatments (Table 29.1) and each treatment was replicated four times.

Germination was recorded on 14 DAP, and height was measured on 14, 21, 28, and 37 DAP. The three tallest plants in each pot were measured with a ruler and then the values were averaged together and recorded as height for that pot. The eight replications (up to the first addition of EDTA, 31 DAP) or four replications (after EDTA addition) were then averaged together to give values for mean and standard deviation.

Shoots were harvested 41 DAP, and fresh and dry weights were measured. Shoot water content was determined by using the following equation:

$$\% \text{ Plant water content} = [(FW - DW)/FW] \times 100 \qquad (29.1)$$

where FW is fresh weight and DW is dry weight.

After the soil was air dried, roots were extracted from the soil by dry sieving (1-mm mesh size), a method that can be used with sandy soil (Böhm, 1979,

p. 115). Fine roots were not captured by this method. The main roots radiated out from the crown and remained structurally intact during the extraction.

Plant parts were digested by using a nitric-perchloric acid digest (Kirkham, 2000), and they were analysed for Cd, Cu, Ni, Pb, Zn, Fe, and Mn with an atomic absorption spectrophotometer (Perkin Elmer Model No. 3110). Soil was analysed for total amount of the same heavy metals using a method similar to that of Sposito et al. (1982). In their method, total content of the heavy metals in the soil is determined on filtered extracts obtained from 2-g samples, which are digested overnight by 12.5 mL 4 M HNO_3 at 80°C. We used 2-g samples, but added 20 mL 4 M HNO_3 and heated the mixture for 18 h at 85°C in a water bath. The extract was analysed with ICP-ES (inductively coupled plasma-atomic emission spectroscopy) (Fison Instruments, now owned by Thermo Optek Corp., Franklin, Massachusetts, USA).

The experimental design was a completely random one. Data were analysed using software from the SAS Institute (1998). Tukey's procedure (Steel and Torrie, 1980, pp. 185–186) was used to compare means.

29.3 Results and discussion

29.3.1 Growth and plant water content

All seeds germinated except those in the pots with the mixture of metals at the high concentration (M2 treatment) (Table 29.2). Germination was similar among all treatments (88.8% or greater), except for the pots with Ni at the high salt concentration, which had a low germination percent (65.0%).

No differences in growth occurred due to EDTA treatment (Table 29.3), probably because EDTA was added prior to harvest (3, 6, and 10 days before harvest). Plants grown with Ni at the high concentration and without EDTA had the lowest shoot and root dry weight. Plants grown with Cu at the high concentration and with EDTA had the highest shoot dry weight, and plants grown with Cu at the low concentration with no EDTA had the highest root weight.

Plants treated with the high concentration of metals grew taller than the other plants (Fig. 29.1). This is probably because the plants were etiolated. The metal-treated plants were chlorotic, even before EDTA was added. Ebbs and Kochian (1998) also noted chlorosis in grasses grown in nutrient solution with high levels of Cd, Cu, and Zn. They added foliar micronutrient sprays containing Mn-EDTA and Fe-EDHHA to overcome the chlorosis. We added no remedial foliar chelates. The chlorosis was caused probably by the heavy metals replacing Fe or Mg. In plants, the cyclic tetrapyrrole structure called protoporphyrin IX (Conn and Stumpf, 1963, p. 264) is the precursor both to chlorophyll, which has the metallic element Mg at its centre, and to heme, which has Fe at its centre (Buchanan et al., 2000, pp. 576–577). Heme or ferroprotoporphyrin, binds with

Table 29.2. *Germination of barley in Haynie very fine sandy loam soil treated with two different concentrations of five metal salts. Table 29.1 gives the salts used. No EDTA had been added to the soil when germination was determined.*

	Germination (%)
Control	88.8a[a]
Cadmium	
50 mg/kg	97.5a
250 mg/kg	90.0a
Copper	
100 mg/kg	96.3a
500 mg/kg	98.8a
Lead	
50 mg/kg	95.0a
250 mg/kg	93.8a
Nickel	
50 mg/kg	91.3a
250 mg/kg	65.0b
Zinc	
100 mg/kg	97.5a
500 mg/kg	96.3a
Mixture of salts[b]	
Low conc. (M1)	91.3a
High conc. (M2)	0c

[a]Means followed by the same letter do not differ significantly at the 0.05 level, according to Tukey's procedure ($n = 8$).
[b]See Table 29.1 for the description of the treatments.

globins or basic proteins, to form haemoglobins (White et al., 1964, p. 193). (The function of haemoglobin in plants is not definitely known, but it is the red pigment that occurs in root nodules. The haemoglobin of nodules is spectro-scopically indistinguishable from mammalian haemoglobin, but differs some-what in its amino acid composition (Burris, 1976, p. 891). It may function as an oxygen-transport system that supports respiration in nodules.) Thus, the chlorosis was due, most likely, to a disruption by the toxic heavy metals of the biochemical pathway for chlorophyll synthesis. Future studies should measure chlorophyll content in heavy-metal-treated plants. This can be done simply and cheaply by using a SPAD meter (Minolta Camera Company, Ltd., Osaka, Japan), which is a light-weight, hand-held meter for measuring the chlorophyll content of leaves without causing damage to the plants.

The chlorosis and tallness of the metal-treated plants were probably related. Plants that are etiolated (blanched because of little chlorophyll) become slender and tall. Height is hormonally determined. Gibberellin makes dwarf plants grow tall and its action can be likened to etiolation (Thimann, 1977, p. 61). Auxin tends to inhibit elongation (Jacobs, 1979, p. 63). Etiolation is usually

Table 29.3. *Shoot and root dry weight of barley grown in Haynie very fine sandy loam soil treated with two different concentrations of five metal salts. Table 29.1 gives the salts used. EDTA was added 3, 6, and 10 days before harvest, which was 41 days after planting.*

	Shoots		Roots	
	No EDTA	With EDTA	No EDTA	With EDTA
	(g/pot)			
Control	0.50ab[a]	0.51ab	0.68ab	0.51abc
Cadmium				
50 mg/kg	0.49ab	0.49ab	0.65abc	0.52abc
250 mg/kg	0.54ab	0.52ab	0.62abc	0.48abc
Copper				
100 mg/kg	0.52ab	0.55ab	0.71a	0.59abc
500 mg/kg	0.55ab	0.58a	0.53abc	0.47bc
Lead				
50 mg/kg	0.49ab	0.48ab	0.60abc	0.54abc
250 mg/kg	0.49ab	0.50ab	0.63abc	0.60abc
Nickel				
50 mg/kg	0.52ab	0.47ab	0.67abc	0.55abc
250 mg/kg	0.41b	0.56ab	0.44c	0.52abc
Zinc				
100 mg/kg	0.52ab	0.51ab	0.64abc	0.63abc
500 mg/kg	0.53ab	0.54ab	0.62abc	0.66abc
Mixture of salts[b]				
Low conc. (M1)	0.57a	0.51ab	0.59abc	0.52abc
High conc. (M2)	0c	0c	0d	0d

[a]Means followed by the same letter do not differ significantly at the 0.05 level, according to Tukey's procedure ($n = 4$).
[b]See Table 29.1 for the description of the treatments.

caused by darkness. It makes internodes colourless and leaf blades small and pale. Gibberellin also tends to make the leaves both small and pale and the internodes pale. Cobalt can mimic the effect of gibberellin (Thimann, 1977, p. 62). The effects on growth exerted by Co may be due to its inhibiting hormonal action, although it is possible that it has a specific direct effect on growth (Thimann, 1977, p. 62). Although unknown, this effect may be one that changes plant–water relations. Turgor pressure and growth are directly related (Gardner, 1973). If a heavy metal like Co, or the ones used in this study, reduces turgor pressure, then it could reduce growth directly.

The water content of the shoots was not affected by metal- or EDTA-treatment, except that of shoots grown with Ni at the high salt concentration without EDTA (Table 29.4). These plants had a higher water content (81.9%) than the others, which had water contents less than 80%. It is not known why the plants grown with Ni had a higher water content. EDTA was not present, so it

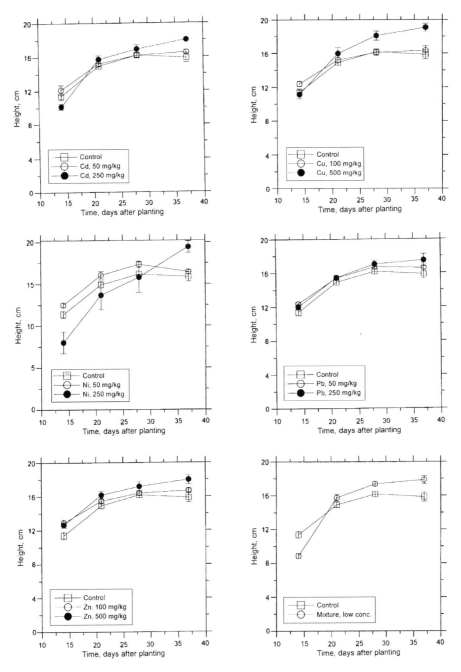

Figure 29.1. *Height of barley grown in Haynie very fine sandy loam soil treated with two different concentrations of five metal salts and a mixture of them. Table 29.1 gives the treatments and salts used. EDTA was added 3, 6, and 10 days before harvest, which was 41 days after planting (DAP). The control treatment is shown in each part of the figure for comparison. Vertical bars show the standard error (n = 8 for DAP 14, 21, and 28, and n = 4 for DAP 37).*

Table 29.4. *Water content at harvest of shoots of barley grown in a Haynie very fine sandy loam soil treated with two different concentrations of five metal salts. Table 29.1 gives the salts used. EDTA was added 3, 6, and 10 days before harvest, which was 41 days after planting.*

	Plant water content	
	No EDTA (%)	With EDTA (%)
Control	77.0b[a]	76.3b
Cadmium		
50 mg/kg	77.5b	76.3b
250 mg/kg	76.3b	77.4b
Copper		
100 mg/kg	76.9b	76.6b
500 mg/kg	77.3b	78.8ab
Lead		
50 mg/kg	76.6b	77.1b
250 mg/kg	77.4b	76.5b
Nickel		
50 mg/kg	76.7b	77.6b
250 mg/kg	81.9a	78.8ab
Zinc		
100 mg/kg	76.6b	76.7b
500 mg/kg	75.5b	77.0b
Mixture of salts[b]		
Low conc. (M1)	77.6b	77.6b
High conc. (M2)	No plants	No plants

[a]Means followed by the same letter do not differ significantly at the 0.05 level, according to Tukey's procedure ($n = 4$).
[b]See Table 29.1 for the description of the treatments.

was not affecting the plant water content. These plants also had the lowest germination percentage (Table 29.2) and grew the poorest (Table 29.3).

No systematic study concerning the effects of metals on plant–water relations has been carried out. An experiment such as the current one needs to be repeated and the water potential, osmotic potential, turgor potential (also called turgor pressure), stomatal resistance, and transpiration rate should be measured. Kirkham (1977) reviewed early literature concerning physiological effects of metals on transpiration. Wilting and loss of turgor pressure often are symptoms of heavy metal toxicity. In chrysanthemum (*Chrysanthemum morifolium* Ramat. 'Indianapolis White') leaves, concentrations in nutrient solution of Cd as low as 0.01 μg/mL reduced stomatal resistance and increased turgor pressure, compared to controls grown with no Cd. But higher concentrations of Cd (0.1 and 1.0 μg/mL) increased stomatal resistance and reduced turgor pressure. Effects on plant–water relations were evident within 24 h after application of the Cd (Kirkham, 1978). In winter wheat (*Triticum aestivum* L.), a cultivar with a higher stomatal resistance ('KanKing') accumulated less Cd than a cultivar with a lower stomatal resistance ('Ponca') (Pearson and Kirkham, 1981). These results indicated that Cd moved through the plants in the transpiration stream. Cadmium apparently increased the permeability of membranes to ions and water, because osmotic potentials usually were lower, and turgor potentials higher, with Cd than without. The results suggested that measurements of stomatal resistance might be a simple method to screen cultivars for Cd-resistance. Cultivars with a higher stomatal resistance might accumulate less Cd than those with a lower stomatal resistance.

Salt et al. (1995) noted that movement of Cd in xylem vessels appears to be mainly dependent on transpiration-driven mass flow. However, Salt et al. (1998, p. 655) said that because xylem cell walls have a high cation exchange capacity, they are expected to retard the upward movement of metal cations. Therefore, noncationic metal-chelate complexes, such as Cd-citrate, should be transported efficiently in the transpiration stream. In the current experiment, even though the metal-treated barley was chlorotic, it did not wilt. Its tolerance to high salt concentrations in the root zone might make it less susceptible to wilting compared to salt-sensitive plants, like field beans (*Phaseolus vulgaris* L.), which are the most salt-sensitive of field crops (Richards, 1954, p. 67).

Little work has been done to determine the effect of chelating agents on plant–water relations. Kirkham (1979) grew two cultivars of winter wheat (*T. aestivum* L. 'KanKing' and 'Ponca') in nutrient culture media with 0.001 or 0.01 M FeEDDHA. (The recommended level for use in Hoagland's nutrient solution is 0.001 M FeEDDHA). For both cultivars, the high level of FeEDDHA increased water potentials, turgor potentials, and stomatal resistances, and decreased osmotic potentials. The results showed that the chelate affected the water balance of the plants. The results of Kirkham (1979) agreed with information from Kozlowski (1968, p. 17), who said that chelating compounds are antitranspirants. Vassil et al. (1998) found that exposure of Indian mustard (*Brassica juncea* (L.) Czern.) to high concentrations of Pb and EDTA caused reductions in both the transpiration rate and the shoot water content. Epstein et al. (1999) found that transpiration of Indian mustard was not affected at EDTA concentrations less than 5 mmol/kg, but at 10 mmol/kg EDTA, transpiration decreased by 80%. In contrast, Wu et al. (1999), who used sap-flow gauges to monitor transpiration, found that EDTA increased transpiration of corn (*Zea mays* L.).

Based on the literature, it appears that heavy metals and chelates may have opposing effects on transpiration: heavy metals tend to increase transpiration rate (at least at low concentrations, in the case of Cd; Kirkham, 1978) and chelates generally tend to decrease it. The comparative effect of different chelating agents (e.g., those mentioned previously: EDTA, HEDTA, DTPA, CDTA, EDDHA) on transpiration has not been studied. It might be beneficial to find a chelate used in phytoremediation that increases transpiration rate, because if a toxic heavy metal is solubilized by the chelate, it should then be taken up by the plant and moved in the transpiration stream to the shoot, where it can be immobilized. Rate of ion translocation is often directly related to rate of transpiration, because many ions are carried up in the transpiration stream (Hylmö, 1953).

29.3.2 Metals in shoots, roots, and soil

For each heavy metal at each level of salt added, concentrations of heavy metals in shoots and roots of barley treated with EDTA were higher than those in shoots and roots of barley with no EDTA (Table 29.5). For example, at the high level of Cd addition, shoots with and without EDTA had 95.0 and 14.9 mg/kg Cd, respectively; roots had 742 and 137 mg/kg Cd, respectively. The only exception was barley grown with Ni at the low level of Ni addition. Its shoots had about the same amount of Ni with EDTA (12.8 mg/kg) as without EDTA (12.5 mg/kg). The maximum amount of an added heavy metal found in a tissue was 1099 mg/kg Cu in roots grown with EDTA at the high level of Cu addition.

According to the literature (Beeson, 1941; Chapman, 1973; Kirkham, 1975, 1977), the normal concentration ranges of the five added heavy metals in plants are (in mg/kg): Cd, 0.1–5; Cu, 4–40; Ni, 0.1–5; Pb, 0.1–12; and Zn, 10–100. (The literature usually does not distinguish between concentrations in roots and shoots; the values are probably more representative of what is found in shoots rather than roots, because roots are rarely analysed). In the presence of the added metals in this study, roots and shoots were taking up abnormal quantities of metals and the amount was increased with EDTA.

Without EDTA, the soil always had a higher concentration of an added metal than the roots. But with EDTA, the roots always had a higher concentration of an added metal than the soil. For example, again taking Cd at the high level of Cd salt addition, without EDTA the roots and soil had 137 mg/kg and 181 mg/kg Cd, respectively. With EDTA, the roots and soil had 742 mg/kg and 186 mg/kg Cd, respectively. With EDTA, the concentration factors varied from 5.5 (Cu, low level; 473 mg/kg in root and 89 mg/kg in soil) to 1.3 (Ni, high level; 256 mg/kg in root and 202 mg/kg in soil).

When the soil was not contaminated, the roots did not consistently have a higher concentration of a metal than the soil. This was true both for plants grown in soil with and without EDTA. For example in the control treatment, Pb was higher in the soil than in the roots both with and without EDTA (Table 29.5). For the control treatment, the amount of Ni in the roots and soil was about the same (c. 8 mg/kg), and EDTA had little effect on root uptake. However, both with and without EDTA, more Cd was in the root than in the soil. In New Zealand, Robinson et al. (2000) found that hybrid poplars (*Populus deltoides* × *P. yunnanensis* NZ 5006 and *P. deltoides* × *P. nigra* NZ 5015) and willow (*Salix*

Table 29.5. *Concentrations of cadmium, copper, nickel, lead, zinc, iron, and manganese in roots, shoots, and soil at harvest of barley grown in a Haynie very fine sandy loam soil treated with two different concentrations of five metal salts. Table 29.1 gives the salts used. EDTA was added 3, 6, and 10 days before harvest, which was 41 days after planting. Mean and standard deviations are given (n = 4).*

	Cadmium	Copper	Nickel	Lead	Zinc	Iron	Manganese
				mg/kg			
Control							
No EDTA							
Shoot	0.9 ± 0.4	4.2 ± 0.6	3.2 ± 1.2	3.4 ± 0.9	12.4 ± 2.8	64.4 ± 6.1	69.8 ± 4.2
Root	2.0 ± 0.7	19.0 ± 1.6	8.2 ± 1.9	4.5 ± 1.2	16.5 ± 0.7	2402 ± 209	55.8 ± 6.0
Soil	0.78 ± 0.12	7.29 ± 0.57	8.04 ± 0.05	17.6 ± 1.7	14.6 ± 0.4	7961 ± 158	130 ± 3
With EDTA							
Shoot	0.9 ± 0.6	5.4 ± 1.6	2.4 ± 1.7	2.7 ± 3.4	17.2 ± 2.3	66.6 ± 4.6	65.5 ± 6.9
Root	2.8 ± 1.0	42.9 ± 18.9	7.7 ± 0.9	9.8 ± 2.1	25.0 ± 7.1	2139 ± 200	45.0 ± 1.6
Soil	0.73 ± 0.07	8.06 ± 1.71	8.32 ± 0.15	17.2 ± 0.4	15.0 ± 0.3	8139 ± 248	138 ± 1.2
Cadmium							
50 mg/kg							
No EDTA							
Shoot	4.0 ± 0.7	6.6 ± 0.7	6.2 ± 1.8	15.3 ± 4.5	10.8 ± 1.8	59.7 ± 7.0	54.4 ± 4.4
Root	39.9 ± 11.1	18.9 ± 3.5	7.6 ± 1.6	6.0 ± 2.2	16.4 ± 2.5	2283 ± 393	53.2 ± 7.5
Soil	40.0 ± 1.4	7.32 ± 1.03	8.57 ± 0.5	17.0 ± 0.1	15.1 ± 0.3	7794 ± 182	129 ± 2
With EDTA							
Shoot	30.5 ± 4.1	11.5 ± 1.3	6.5 ± 0.8	17.4 ± 2.6	15.4 ± 2.3	64.4 ± 5.0	47.0 ± 1.7
Root	170 ± 32	38.4 ± 7.4	7.9 ± 0.7	10.5 ± 0.8	23.3 ± 0.7	2447 ± 142	51.8 ± 4.6
Soil	37.7 ± 1.3	7.40 ± 0.26	8.46 ± 0.23	17.1 ± 0.3	15.4 ± 0.4	7826 ± 176	133 ± 3
Cadmium							
250 mg/kg							
No EDTA							
Shoot	14.9 ± 2.9	7.2 ± 0.2	5.7 ± 0.7	14.7 ± 1.8	11.3 ± 1.7	56.0 ± 7.7	47.4 ± 7.7
Root	137 ± 49	20.2 ± 4.0	8.1 ± 0.8	7.0 ± 0.9	14.5 ± 0.9	2288 ± 666	58.7 ± 10.6
Soil	181 ± 8	7.43 ± 0.75	8.15 ± 0.12	16.8 ± 0.4	14.8 ± 0.3	7743 ± 107	131 ± 6
With EDTA							
Shoot	95.0 ± 8.2	10.1 ± 2.8	7.7 ± 0.4	19.5 ± 2.9	16.6 ± 4.2	90.2 ± 56.9	38.9 ± 2.3
Root	742 ± 159	40.7 ± 5.0	8.5 ± 1.4	7.8 ± 1.3	29.1 ± 18.3	2446 ± 159	45.4 ± 3.3
Soil	186 ± 5	8.60 ± 2.35	8.31 ± 0.2	16.9 ± 0.1	14.6 ± 0.4	7910 ± 224	131 ± 3
Copper							
100 mg/kg							
No EDTA							
Shoot	1.5 ± 0.6	7.5 ± 1.0	6.3 ± 2.3	7.7 ± 3.4	20.7 ± 10.4	63.5 ± 9.3	57.5 ± 3.6
Root	2.0 ± 0.5	40.2 ± 5.3	7.9 ± 1.3	6.6 ± 1.2	17.5 ± 2.4	2258 ± 81	52.1 ± 2.6
Soil	0.81 ± 0.06	95.8 ± 8.4	8.08 ± 0.11	16.3 ± 0.2	14.8 ± 0.35	7770 ± 117	130 ± 4
With EDTA							
Shoot	3.6 ± 0.9	35.1 ± 5.3	10.6 ± 2.2	21.3 ± 5.4	17.0 ± 1.8	71.7 ± 5.5	72.7 ± 9.1
Root	1.9 ± 0.3	473 ± 45	9.5 ± 2.3	9.8 ± 1.3	22.3 ± 2.5	2403 ± 249	53.8 ± 6.2
Soil	0.67 ± 0.02	89 ± 1.2	8.13 ± 0.05	16.6 ± 0.2	15.0 ± 0.2	7901 ± 128	133 ± 2
Copper							
500 mg/kg							
No EDTA							
Shoot	5.1 ± 0.2	11.4 ± 0.6	18.0 ± 4.8	31.7 ± 1.4	17.8 ± 3.1	109 ± 52	61.0 ± 17.2
Root	2.2 ± 0.4	111 ± 75	9.6 ± 1.7	7.2 ± 1.5	16.8 ± 3.4	2470 ± 270	59.6 ± 6.9

Table 29.5. (*Continued*)

	Cadmium	Copper	Nickel	Lead	Zinc	Iron	Manganese
				mg/kg			
Soil	0.63±0.01	415±77	8.16±0.15	16.7±0.5	15.2±0.4	7691±241	135±3
With EDTA							
Shoot	0.2±0.3	77.0±24.9	1.0±1.1	2.8±3.3	11.2±2.2	56.3±4.5	48.1±5.7
Root	1.9±0.2	1099±167	8.7±0.8	8.1±1.2	22.1±1.2	2670±415	54.8±4.2
Soil	0.66±0.02	408±15	8.30±0.32	16.5±0.3	15.0±0.2	7831±42	135±1
Nickel							
50 mg/kg							
No EDTA							
Shoot	3.5±2.3	7.9±2.7	12.5±8.1	30.5±20.5	8.8±1.6	48.3±5.8	46.7±8.9
Root	2.2±0.4	20.0±2.4	50.1±13.5	7.5±1.2	16.3±2.5	2482±212	59.8±3.6
Soil	0.64±0.05	7.16±1.22	52.1±1.5	16.1±0.3	14.3±0.1	7569±23	129±1
With EDTA							
Shoot	1.5±1.1	6.3±1.0	12.8±5.1	12.6±9.8	12.7±1.4	50.6±5.1	50.0±4.3
Root	2.2±0.2	41.4±8.4	72.4±6.9	12.0±1.4	22.9±3.3	2483±136	50.3±3.1
Soil	0.72±0.04	7.59±1.58	45.9±2.9	16.8±0.8	14.3±0.4	7631±160	127±3
Nickel							
250 mg/kg							
No EDTA							
Shoot	3.8±0.5	10.1±0.9	16.0±1.7	32.8±4.6	12.1±3.3	48.4±4.5	30.1±4.1
Root	6.7±4.6	31.5±17.0	150±58	11.8±3.9	23.3±8.6	3185±1018	74.1±21.6
Soil	0.71±0.04	6.82±0.07	218±9	16.7±0.3	15.5±0.7	7961±147	134±3
With EDTA							
Shoot	4.4±0.8	10.8±1.8	34.7±6.8	38.1±6.3	13.1±2.7	46.9±6.6	39.0±4.0
Root	2.4±0.6	41.5±4.7	256±50	11.0±1.7	56.3±63.5	2338±102	48.9±3.5
Soil	0.73±0	7.79±0.92	202±11	16.7±0.2	15.0±0.1	7952±80	133±2
Lead							
50 mg/kg							
No EDTA							
Shoot	2.8±0.2	8.0±0.5	11.2±0.6	28.7±2.5	12.4±2.0	55.2±6.0	58.4±8.5
Root	2.5±1.1	17.3±3.4	9.1±2.4	31.8±11.8	15.3±2.7	2592±362	58.4±6.7
Soil	0.70±0.02	7.10±0.88	8.34±0.32	60.2±7.5	14.8±0.04	7888±99	133±1
With EDTA							
Shoot	0.9±0.8	8.0±1.0	3.1±2.9	45.5±13.8	14.2±1.9	54.0±8.8	52.3±4.6
Root	2.9±0.3	45.6±7.2	9.4±1.2	252±114	24.6±2.2	2197±175	46.8±6.1
Soil	0.74±0.03	7.25±0.21	8.34±0.33	62.1±2.2	14.9±0.2	8031±46	136±2
Lead							
250 mg/kg							
No EDTA							
Shoot	3.0±0.2	7.5±0.4	11.1±1.0	28.3±1.5	9.7±1.9	49.7±5.2	53.6±5.5
Root	2.6±0.8	22.1±6.5	8.7±1.7	170±76	16.2±1.4	2441±371	61.4±6.3
Soil	0.81±0.09	6.95±0.19	8.53±0.12	244±22	15.2±0.1	7990±137	138±1
With EDTA							
Shoot	4.0±0.3	10.4±1.3	13.8±1.0	149.5±8.9	11.8±0.4	47.9±1.0	46.4±2.0
Root	2.6±0.4	40.4±3.5	8.9±0.9	898±153	24.1±1.3	2179±144	44.3±1.9
Soil	0.81±0.07	7.72±0.81	8.64±0.28	220±6	14.9±0.2	7942±123	135±1
Zinc							
100 mg/kg							
No EDTA							
Shoot	0.8±0.8	5.5±3.1	7.3±8.2	21.3±31.4	23.4±6.8	91.0±37.5	61.6±10.1

Table 29.5. (*Continued*)

	Cadmium	Copper	Nickel	Lead	Zinc	Iron	Manganese
				mg/kg			
Root	1.8±0.2	18.0±5.3	7.0±0.9	6.1±0.9	63.0±13.4	2281±321	56.5±5.8
Soil	0.74±0.04	6.98±0.78	8.75±0.17	17.1±0.5	77.0±7.2	7863±102	135±2
With EDTA							
Shoot	2.1±0.4	10.5±3.8	7.7±1.6	18.1±3.4	76.8±8.6	68.0±9.5	55.3±2.2
Root	2.4±0.8	34.8±2.4	9.6±1.5	10.4±1.5	279±34	2281±274	48.5±3.8
Soil	0.65±0.02	7.27±0.97	8.61±0.18	16.5±0.2	69.9±5.0	7818±111	137±2
Zinc							
500 mg/kg							
No EDTA							
Shoot	1.9±1.4	7.3±4.4	6.8±5.0	16.6±11.8	94.2±18.8	74.2±7.8	50.4±8.7
Root	2.4±0.7	18.2±5.8	7.7±1.1	5.8±1.1	217±110	2153±253	49.8±4.7
Soil	0.76±0.10	7.48±0.99	8.65±0.52	17.2±0.3	322±17	8031±229	134±3
With EDTA							
Shoot	0.1±0.3	5.0±1.9	0.8±1.5	1.9±3.7	142±22	49.2±6.7	36.3±5.4
Root	2.3±0.3	26.5±2.1	8.6±1.0	8.6±2.3	629±99	2633±400	55.9±7.4
Soil	0.78±0.10	6.89±0.43	9.13±2.10	16.8±0.4	305±21	8090±91	131±2
Mixture[a]							
Low level							
No EDTA							
Shoot	11.1±1.6	11.3±0.9	22.2±1.4	45.6±2.8	16.6±1.7	44.6±3.8	26.1±1.7
Root	69.5±12.9	88.2±17.9	74.6±12.9	30.8±7.0	66.3±13.4	2666±227	59.9±3.4
Soil	41.6±1.0	100.7±5.3	50.7±1.1	63.4±4.6	74.2±1.0	8010±139	130±3
With EDTA							
Shoot	26.2±6.0	33.7±7.4	31.6±5.4	53.2±5.8	47.4±11.3	49.1±5.1	30.7±2.2
Root	115±6	528±14	108±6	62.9±7.7	139±15	2577±272	50.8±5.2
Soil	40.6±1.5	94.5±7.0	49.8±2.6	60.3±2.9	72.5±4.0	8100±85	134±2
Mixture							
High level[b]							
No EDTA							
Shoot	...[b]	...	...	...	...	...	...
Root	...	...	...	...	...	...	...
Soil	187±6	430±21	271±3	230±17	291±7	8022±102	131±1
With EDTA							
Shoot	...	...	...	...	...	...	...
Root	...	...	...	...	...	...	...
Soil	170±12	371±23	163±20	192±7	265±17	7904±88	131±3

[a]See Table 29.1 for the description of the mixture of salts.
[b]No plants germinated.

matsudana × *S. alba* NZ 1040), grown with the chelating agents NTA (nitrilo-triacetic acid) and EDTA in soil (Manawatu silt loam), had a higher concentration of Cd in the above ground-parts than in the soil. On average, root Cd concentrations were 0.9 times the concentrations in the shoots. That is, shoots of poplar and willow were concentrating Cd.

The EDTA increased uptake of metals into the roots, but relatively small quantities were transported from the roots to the shoots. In botany, the point at or just below the surface of the ground where the shoot and root join is called the crown. The connection between the shoot and the root at the crown is complex (Esau, 1965, pp. 523–531). The vascular systems of the shoot and the root involve spatial adjustments between systems with differently oriented parts. It is a transition region between an organ with an axial vascular system (the root) and one whose vascular system develops in relation to leaves. This experiment suggested that, because even with EDTA relatively small quantities of the heavy elements were transported to the shoots, this junction must form a barrier for the transport of the metals higher up the plant. Studies need to be done to determine how this junction controls the movement of the heavy metals solubilized by a chelate. Such studies would need to show the location of the metal-chelate compound in the vascular system of the root and the traces of the first foliar organs of the plant.

The water potential of the soil at the crown and in different parts of the root zone will also affect the direction of flow of water in the plant (Kirkham, 1983). If part of a root system is in dry soil and part in wet soil, heavy metals solubilized by a chelate can move from the wet region through the crown to the dry region. This movement could be through hydraulic lift, which occurs when plant roots extract water from a moist subsoil and release it into a dry topsoil (Song et al., 2000). In the field, barley has a deep root system (1.5 m; Schwab et al., 1996, p. 258), and it might be effective in moving metals solubilized at depth with a chelate to the surface of the soil through hydraulic lift.

Even though small quantities of the heavy metals were transported to the shoots, higher concentrations of the metals still were in the shoots with EDTA than without. While phytoremediation has the goal of removing contaminants from the soil and sequestering them in plant tissue which then can be ashed, the increased uptake of heavy metals in the presence of EDTA may be advantageous for the production of phytopharmaceuticals (Bisset, 2000). The chelates might increase concentrations of essential heavy metals in herbal drugs. Currently, rice (*Oryza sativa* L.) is being genetically engineered for higher concentrations of Fe to prevent anaemia in the millions of people worldwide who depend upon rice for food (Pinstrup-Andersen and Schiøler, 2001, pp. 98–99). Mutants of *Arabidopsis*, the weedy plant used by plant geneticists to understand basic molecular biology, have been developed to increase Fe uptake. Plants (like *Arabidopsis*) must first reduce Fe(III) to Fe(II) before transporting it into the cell (Guerinot and Salt, 2001). The mutants lacking this activity have allowed cloning of the gene encoding the enzyme responsible for catalysing this rate-limiting step in Fe acquisition (Robinson et al., 1999). However, not all countries accept genetically engineered food. Chelates could provide a non-controversial method to produce high-iron rice and increase concentrations of other essential metals in crops.

29.3.3 Recovery of metals from the soil

The method used to determine total concentration of heavy metals (Sposito et al., 1982) extracted almost all the Pb in the soil (as much as 92.9%, Table 29.6), but recovered lesser amounts of the other heavy metals. Zinc had the lowest percentage of recovery. Stronger acids than the one used in this study (4.0 M HNO_3) have been used to determine total concentrations of heavy metals in soils. Isaac and Kerber (1971) describe a technique in which the dried soil sample is heated to 900°C in a crucible and then the ignited sample is treated with hydrochloric acid (HCl) and 48% hydrofluoric acid (HF). However, if high heat and strong acids are used, some metals may volatilize and not be recovered (Dr. Gary R. Griffith, Kansas State Univ., Soil Testing Lab., personal communication). Consequently, the less harsh treatment of heating to 80°C and using 4 M HNO_3 was used. The fact that recovery of the heavy metals was not 100% indicates that some of them were binding to soil and not being removed by the 4 M HNO_3.

Table 29.6. *Recovery of heavy metals from a Haynie very fine sandy loam soil treated with two different concentrations of five metal salts. Table 29.1 gives the salts used. EDTA was added 3, 6, and 10 days before harvest of barley, which grew in the pots for 41 days.*

	Recovery	
	No EDTA (%)	With EDTA (%)
Cadmium		
50 mg/kg	78.8	74.3
250 mg/kg	72.0	74.2
Copper		
100 mg/kg	89.2	82.9
500 mg/kg	81.7	80.5
Lead		
50 mg/kg	91.0	92.9
250 mg/kg	91.3	82.6
Nickel		
50 mg/kg	89.3	78.5
250 mg/kg	84.2	78.3
Zinc		
100 mg/kg	67.0	60.8
500 mg/kg	62.6	59.3

29.4 Future research dealing with chelate-assisted phytoremediation

The research suggested the following questions that might be answered in the future.

What is the maximum amount of a heavy metal bioavailable with a chelate?

We can trick nature by using the chelate, so abnormal amounts of a heavy metal are taken up by roots. But what is the maximum? What is the maximum concentration of a heavy metal in a root that permits survival? How much of a heavy metal would we have had to have added to the soil so that the barley would have died after the first addition of the chelate 31 DAP? We added the chelate three different times, 31, 35, and 38 DAP, but yet the barley lived.

Why does a plant take up more of one toxic metal than another?

Roots of barley grown with 250 mg/kg Cd and EDTA had 742 mg/kg Cd and roots grown with 250 mg/kg Ni and EDTA had 256 mg/kg Ni. Why did the roots accumulate more Cd than Ni, when treated with the same concentration of a heavy metal?

Were the chelate-solubilized metals adsorbed on to the surface of the roots?

Large quantities of heavy metals were in the roots, but not transported to the shoots. What is the 'isotherm' for roots with and without the chelate? Vogeler et al. (2001) determined the adsorption isotherms for copper with and without EDTA. The isotherms were determined in batch experiments by shaking Manawatu fine sandy loam with copper solutions that either contained EDTA or did not contain EDTA. With EDTA, more copper was in solution than without. Instead of shaking the Manawatu fine sandy loam with different copper-EDTA solutions, could we shake a mat of roots with copper solutions with and without EDTA? What would the adsorption isotherms be for the mat of roots?

Can we use a tension infiltrometer filled with a metal-chelate solution to determine the depth of penetration of the metal into the soil?

The chelate helps mobilize heavy metals in the soil solution. A tension infiltrometer (Clothier and White, 1981) allows one to determine the depth of penetration into soil of a solute. With the tension infiltrometer, the amount of macropore flow is controlled by applying water to soil at water potentials, ψ_o, less than zero. We can use the tension infiltrometer to determine the depth of penetration of a solute into soil by using the following equation:

$$Z = I^* / \theta_m \tag{29.2}$$

where Z is the depth of front (m), I^* the depth of water infiltrated (m), and θ_m = mobile water content (m^3/m^3).

Clothier et al. (1992) developed a method to measure mobile water content in the field. The equation is:

$$\theta_{\mathrm{m}} = \theta(c^*/c_{\mathrm{m}}) \tag{29.3}$$

where c^* is the concentration of a tracer that one finds in the soil after infiltrating it into the soil using a tension infiltrometer; c_{m} the concentration of the tracer in the original solution that one puts into the sealed reservoir of the tension infiltrometer; θ the soil water content measured after infiltration.

We could use the tension infiltrometer to get θ_{m} and then use this value in Eq. (29.2). Could we then infiltrate a metal-chelate solution into the soil using a tension infiltrometer to get Z, the depth of the front?

Can we use chelates to remove deleterious heavy metals from lunar soil?

Lunar soils might be used for rooting media in future expeditions to the moon. However, they contain deleterious phases such as free Fe, Ni, Cr, and Co (Helmke and Corey, 1989). Could we use chelate-facilitated phytoremediation to remove these metals? After clean-up, they could be used for agricultural production.

Can we use the dual-pipe, subirrigation-drainage system to keep heavy metals solubilized by chelates in the root zone?

A heavy metal, solubilized by chelates and not taken up by plant roots, escapes below the root zone and pollutes ground water. One way to prevent this pollution might be the use of the dual-pipe subirrigation-drainage system developed in Iowa (Kirkham and Horton, 1993). The system consists of two series of pipes – one nearer to the soil surface than the other. The higher level of pipes serves as the subirrigation tubes, and the pipes below them serve as the drainage tubes. The dual system provides an irrigation system that irrigates below the soil surface and thus avoids the evaporative losses of aboveground irrigation. In this type of subirrigation system, water that is intercepted by the drain tubes is recycled by pumping it back into the irrigation supply to conserve nutrients and reduce the potential for ground water pollution. The key to the system is a barrier at depth. If there is no barrier, the solutes move to ground water. The barrier can be natural, like a clay layer, or artificial. When dealing with phytoremediation, it might be worth the expense to install an artificial barrier, so that the dual-pipe subirrigation-drainage system could be used to keep the solubilized metal in the root zone. Installing the barrier might be cheaper than excavation of the polluted soil.

In summary, our work showed that the chelate concentrated the metals in the barley roots. The next step is to take the experiment to the field, and using our new devices, like the tension infiltrometer and the dual-pipe subirrigation-drainage system, determine the effect of chelates on uptake of heavy metals.

29.5 Conclusion

The chelate EDTA increased uptake of metals into barley roots. Without EDTA, soil contaminated with one of the five heavy metals (Cd, Cu, Ni, Pb, and Zn) always had a higher concentration of the added metal than the roots. But with EDTA, the roots always had a higher concentration of the added metal than the soil. The results suggested that EDTA can concentrate metals in roots and stabilize them in the soil. Metals in a mat of roots at the soil surface would be less prone to erode by water or wind or to move to ground water through preferential flow paths.

Acknowledgement

This is contribution no. 02-177-B from the Kansas Agricultural Experiment Station, Manhattan, Kansas 66506, USA.

References

Barona, A., Aranguiz, I., Elías, A., 2001. Metal associations in soils before and after EDTA extractive decontamination: implications for the effectiveness of further clean-up procedures. Environ. Pollut. 113, 79–85.

Beeson, K.C. (1941). The mineral composition of crops with particular reference to the soils in which they were grown. USDA Misc. Pub. 369. U.S. Dep. Agr., Washington, DC, 164 pp.

Bisset, N.G. (Ed.), 2000. Herbal Drugs and Phytopharmaceuticals. CRC Press, Boca Raton, FL, p. 565.

Blaylock, M.J., Salt, D.E., Dushenkov, S., Zakharova, O., Gussman, C., Kapulnik, Y., Ensley, B.D., Raskin, I., 1997. Enhanced accumulation of Pb in Indian mustard by soil-applied chelating agents. Environ. Sci. Technol. 31, 860–865.

Böhm, W., 1979. Methods of Studying Root Systems. Springer-Verlag, Berlin, 188 pp.

Brooks, R.R., 1998. General introduction. In: R.R. Brooks (Ed.), Plants that Hyperaccumulate Heavy Metals. CAB International, Wallingford, Oxon, United Kingdom, pp. 1–14.

Buchanan, B.B., Gruissem, W., Jones, R.L. (Ed.) 2000. Biochemistry and molecular biology of plants. American Society of Plant Physiologists. Rockville, MD, 1367 pp.

Burris, R.H., 1976. Nitrogen fixation. In: J. Bonner, J.E. Varner (Eds.), Plant Biochemistry. 3rd Ed. Academic Press, NY, pp. 887–908.

Chapman, H.D. (Ed.), 1973. Diagnostic Criteria for Plants and Soils. Second printing. Quality Printing Co, Abilene, TX, 793 pp.

Clothier, B.E., Kirkham, M.B., McLean, J.E., 1992. In situ measurement of the effective transport volume for solute moving through soil. Soil Sci. Soc. Am. J. 56, 733–736.

Clothier, B.E., White, I., 1981. Measurement of sorptivity and soil water diffusivity in the field. Soil Sci. Soc. Am. J. 45, 241–245.

Conn, E.E., Stumpf, P.K., 1963. Outlines of Biochemistry. Wiley, NY, 391 pp.

Cooper, E.M., Sims, J.T., Cunningham, S.D., Huang, J.W., Berti, W.R., 1999. Chelate-assisted phytoextraction of lead from contaminated soils. J. Environ. Qual. 28, 1709–1719.

Ebbs, S.D., Kochian, L.V., 1998. Phytoextraction of zinc by oat (Avena sativa), barley (Hordeum vulgare), and Indian mustard (Brassica juncea). Environ. Sci. Technol. 32, 802–806.

Epstein, A.L., Gussman, C.D., Blaylock, M.J., Yermiyahu, U., Huang, J.W., Kapulnik, Y., Orser, C.S., 1999. EDTA and Pb-EDTA accumulation in Brassica juncea grown in Pb-amended soil. Plant Soil 208, 87–94.

Esau, K., 1965. Plant Anatomy. , 2nd Ed., Wiley, NY, 767 pp.

Gardner, W.R. (1973). Internal water status and plant response in relation to the external water régime. In: R.O. Slatyer (Ed.), Plant Response to Climatic Factors. U.N. Educ., Sci., Cult. Organ., Paris, pp. 221–225.

Guerinot, M.L., Salt, D.E., 2001. Fortified foods and phytoremediation. Two sides of the same coin. Plant Physiol. 125, 164–167.

Helmke, P.A., Corey, R.B. 1989. Physical and chemical considerations for the development of lunar-derived soils. In: D.W. Ming and D.L. Hennniger (Eds.), Lunar Base Agriculture: Soils for Plant Growth. American Society of Agronomy, Crop Science Society of America, Soil Science Society of America, Madison, WI, pp. 193–212.

Huang, J.W., Chen, J., Berti, W.R., Cunningham, S.D., 1997. Phytoremediation of lead-contaminated soils: role of synthetic chelates in lead phytoextraction. Environ. Sci. Technol. 31, 800–805.

Hylmö, B., 1953. Transpiration and ion absorption. Physiol. Plant 6, 333–405.

Isaac, R.A., Kerber, J.D. 1971. Atomic absorption and flame photometry: techniques and uses in soil, plant, and water analysis. In: L.M. Walsh (Ed.), Instrumental Methods for Analysis of Soils and Plant Tissue. Soil Science Society of America, Madison, WI, pp. 17–37.

Jacobs, W.P., 1979. Plant Hormones and Plant Development. Cambridge University Press, Cambridge, 339 pp.

Jiang, X.J., Luo, Y.M., Zhao, Q.G., Baker, A.J.M., Christie, P., Wong, M.H., 2003. Soil Cd availability to Indian mustard and environmental risk following EDTA addition to Cd-contaminated soil. Chemosphere 50, 813–818.

Kirkham, D., Horton, R., 1993. Modeling water flow from subirrigation with drainage. Soil Sci. Soc. Am. J. 57, 1451–1457.

Kirkham, M.B., 1975. Trace elements in corn grown on long-term sludge disposal site. Environ. Sci. Technol. 9, 765–768.

Kirkham, M.B. 1977. Trace elements in sludge on land: effect on plants, soils, and ground water. In: R.C. Loehr (Ed.), Land as a Waste Management Alternative. Ann Arbor Science Publisher, Ann Arbor, MI, pp. 209–247.

Kirkham, M.B., 1978. Water relations of cadmium-treated plants. J. Environ. Qual. 7, 334–336.

Kirkham, M.B., 1979. Effect of FeEDDHA on the water relations of wheat. J. Plant Nutr. 1, 417–424.

Kirkham, M.B., 1983. Physical model of water in a split-root system. Plant Soil 75, 153–168.

Kirkham, M.B., 2000. EDTA-facilitated phytoremediation of soil with heavy metals from sewage sludge. Int. J. Phytoremediation 2, 159–172.

Kozlowski, T.T. 1968. Introduction. In: T.T. Kozlowski (Ed.), Water Deficits and Plant Growth, Vol. 1. Development, Control, and Measurement. Academic Press, NY, pp. 1–21.

Leymonie, J.-P. (Ed.) 2001. Lost in the Jungle of Micronutrient Products? New Ag International. The World's Leading Publication on High Tech Agriculture. Sept. issue. NAI, Ltd, Teddington, Middlesex TW11 8GF, United Kingdom, pp. 40–48.

Liphadzi, M.S., Kirkham, M.B., 2005. Phytoremediation of soil contaminated with heavy metals: A technology for rehabilitation of the environment. S. Afr. J. Bot. 71, 24–37.

Liphadzi, M.S., Kirkham, M.B. 2006. Physiological effects of heavy metals on plant growth and function. In: B. Huang (Ed.), Plant-Environment Interactions. CRC, Taylor & Francis, Boca Raton, Florida, pp. 243–269.

Madrid, F., Liphadzi, M.S., Kirkham, M.B., 2003. Heavy metal displacement in chelate-irrigated soil during phytoremediation. J. Hydrol. 272, 107–119.

Madrid, F., Kirkham, M.B. 2007. Testing the manipulation of soil availability of metals. In: N. Willey (Ed.), Phytoremediation. Humana Press, Totowa, New Jersey, pp. 121–129.

McGrath, S.P., Zhao, F.-J., 2003. Phytoextraction of metals and metalloids from contaminated soils. Curr. Opin. Biotechnol. 14, 277–282.

Miller, R.R. 1996. Phytoremediation. Technology Overview Report TO-96-03. Ground-Water Remediation Technologies Analysis Center, Pittsburgh, PA. iii+17 pages. (Homepage: http://www.gwrtac.org).

Pearson, C.H., Kirkham, M.B., 1981. Water relations of wheat cultivars grown with cadmium. J. Plant Nutr. 3, 309–318.

Pinstrup-Andersen, P., Schiøler, E. 2001. Seeds of Contention. World Hunger and the Global Controversy over GM Crops. English language translation. The Johns Hopkins University Press, Baltimore, MD, 164 pp.

Richards, L.A. (Ed.) 1954. Diagnosis and Improvement of Saline and Alkali Soils. Agr. Handbook No. 60. U.S. Dep. Agr., Washington, DC, 160 pp.

Robinson, N.J., Procter, C.M., Connolly, E.L., Guerinot, M.L., 1999. A ferric-chelate reductase for iron uptake from soils. Nature 397, 694–697.

Robinson, B.H., Mills, T.M., Petit, D., Fung, L.E., Green, S.R., Clothier, B.E., 2000. Natural and induced cadmium-accumulation in poplar and willow: Implications for phytoremediation. Plant Soil 227, 301–306.

Salt, D.E., Prince, R.C., Pickering, I.J., Raskin, I., 1995. Mechanisms of cadmium mobility and accumulation in Indian mustard. Plant Physiol. 109, 1427–1433.

Salt, D.E., Smith, R.E., Raskin, I., 1998. Phytoremediation. Annu. Rev. Plant Physiol. Plant Mol. Biol. 49, 643–668.

SAS Institute 1998. SAS/STAT, release 6.12 edition. Statistical Analysis Systems (SAS) Institute, Cary, NC.

Schwab, G.O., Fangmeier, D.D., Elliot, W.J., 1996. Soil and Water Management Systems. , 4th Ed., Wiley, NY, 371 pp.

Song, Y., Kirkham, M.B., Ham, J.M., Kluitenberg, G.J., 1999. Dual probe heat pulse technique for measuring soil water content and sunflower water uptake. Soil Tillage Res. 50, 345–348.

Song, Y., Kirkham, M.B., Ham, J.M., Kluitenberg, G.J., 2000. Root-zone hydraulic lift evaluated with the dual-probe heat-pulse technique. Aust. J. Soil Res. 38, 927–935.

Sposito, G., Lund, L.J., Chang, A.C., 1982. Trace metal chemistry in arid-zone field soils amended with sewage sludge: I. Fractionation of Ni, Cu, Zn, Cd, and Pb in solid phases. Soil Sci. Soc. Am. 46, 260–264.

Steel, R.G.D., Torrie, J.H., 1980. Principles and Procedures of Statistics. McGraw-Hill Book Co., NY, 633 pp.

Thimann, K.V., 1977. Hormone Action in the Whole Life of Plants. University of Massachusetts Press, Amherst, MA, 448 pp.

Vassil, A.D., Kapulnik, Y., Raskin, I., Salt, D.E., 1998. The role of EDTA in lead transport and accumulation by Indian mustard. Plant Physiol. 117, 447–453.

Vogeler, I., Green, S.R., Clothier, B.E., Kirkham, M.B., Robinson, B.H., 2001. Contaminant transport in the root zone. In: I.K. Iskandar, M.B. Kirkham (Eds.), Trace Elements in Soils: Bioavailability, Fluxes and Transfer. Lewis Publishers, Boca Raton, FL, pp. 175–197.

White, A., Handler, P., Smith, E.L., 1964. Principles of Biochemistry. , 3rd Ed., McGraw-Hill Book Co., NY, 1106 pp.

Wu, J., Hsu, F.C., Cunningham, S.D., 1999. Chelate-assisted Pb phytoextraction: Pb availability, uptake, and translocation constraints. Environ. Sci. Technol. 33, 1898–1904.

Zhu, L. 2001. Crop growth in soil from beneath closed animal waste lagoons. M.S. Thesis, Kansas State University, Manhattan, KS, 128 pp.

Developments in Soil Science, volume 32
Ravendra Naidu (Editor)
© 2008 Elsevier B.V. All rights reserved

Chapter 30

LAND RECLAMATION USING EARTHWORMS IN METAL CONTAMINATED SOILS

M.H. Wong and Y. Ma

30.1 Introduction

Earthworms are suitable bioindicators of chemical contamination of the soil, and are easy and economical to handle (Bustos-Obregón and Goicochea, 2002). Analysis of earthworm tissue may also provide an excellent index of bioavailability of heavy metals in soils (Helmke et al., 1979; Pearson et al., 2000, see Chapter 18). The acute earthworm toxicity test was first developed by Edwards (1984) and is the internationally agreed standard protocol for assessing the chemical toxicity of contaminants in soils (OECD, 1984; USEPA, 1988). However, the standardization of a chronic toxicity test to detect subtle effects of contaminants on earthworms – such as disturbances in behaviour, retarded development, lowered fertility, teratogenic effects, etc. – over a longer exposure period has not yet been fully achieved (Reinecke and Reinecke, 1998). Venabes et al. (1992) described a suite of biomarkers in earthworms that could be used as potential indicators of ecotoxicity, including bioavailability of the chemicals, reproductive/developmental, neurological and immunological markers.

In a laboratory experiment, earthworms were used to enhance the dispersal of the bioaugmented polychlorinated biphenyls (PCB)-degrading microorganisms, and simultaneously improve soil aeration, increasing soil C and N content, and modifying the soil microbial community. Inoculating with the earthworm *Pheretima hawayana* achieved 55% removal of total soil PCB, compared with only 39% in identically treated soils without earthworms (Singer et al., 2001). A plant-based remediation technique is designed to remove petroleum from contaminated soils, and it was discovered that earthworm addition both improved plant growth and reduced the time needed to restore soil ecosystem function (Callaham et al., 2002).

Early attempts at land reclamation involved the establishment of vegetation on an overburden of innocuous amendments (Wong, 1987, 2002). As the desired end-point has moved from reclamation for agriculture to restoration for amenity and wildlife conservation (Dickinson, 2000), reduced engineering and lower cost methods of treating reclaimed land have been a more recent goal (Vangronsveld and Cunningham, 1998). Heavy metal contamination of soils poses a major environmental and human health problem all over the world, making them difficult and expensive to reclaim (Nriaga, 1984). The success of reclamation may be limited by elevated toxic metal concentrations, poor soil structure and low soil fertility (Satchell, 1983). To maintain sustainable vegetation on metal contaminated soils, a thorough understanding of metal mobility as influenced by plant–soil–animal interactions is necessary.

Often, a vital component of a healthy, sustainable soil is the earthworm community that develops and maintains the structure of the soil, promoting the breakdown of organic matter and other processes that provide the integrity of soil (Edwards and Bohlen, 1996). There have been numerous studies on heavy metal uptake, toxicity and accumulation by earthworms. Earthworm activities influence soil physico-chemical and biological processes, as well as the growth of plant roots (Springett and Gray, 1997), which may alter metal mobility and bioavailability in soils and are likely to play a significant role in the process of soil reclamation. This chapter reviews past experience in the use of earthworms for land reclamation, with special emphasis on metal contaminated soils.

30.2 The use of earthworms for land reclamation

Many attempts have been made to inoculate earthworms into poor soils (in particular soils with coarse texture and the lack of organic matter and nutrients) that have no earthworms or to encourage the build-up of earthworm populations by adding organic matter or fertilizers (Edwards and Bohlen, 1996). Earthworm inoculation has been used to a limited extent in land improvement and reclamation and has been particularly successful where there is poor soil structure and low inherent fertility (Butt et al., 1995). A number of studies indicate that earthworms play an important role in improving reclaimed soils.

30.2.1 Earthworms in poor non-toxic soils

Earthworm population densities correlated significantly with improvements in soil structure, decline of soil bulk density and the breakdown of undecomposed litter in an irrigated sown pasture on sandy loam soil in Australia (Noble et al., 1970). The successful introduction of an alien species, *Aporrectodea caliginosa*, into limed coniferous forest plots in Finland has been achieved, and the indigenous species *Lumbricus rubellus* and *Eisenia rubida* also benefited from liming and addition of deciduous litter (Huhta, 1979). Earthworm populations in virgin peat are limited by high moisture, low pH and the poor quality of organic litter. However, once peat soils are drained and reclaimed for agriculture they can support moderate densities of earthworm populations (Curry and Cotton, 1983).

In soils of drained peat meadows, it has been observed that high densities of earthworms (*L. rubellus*, above 200 individuals m^{-2}) can increase the rate of organic matter mineralization, intensify competition for nutrients and indirectly contribute to the biological decomposition of organic matter (Makulec, 2002). In a degraded Alfisol in southwestern Nigeria, an increase in earthworm population through fallows resulted in higher leaf-litter decomposition, availability of soil organic matter, availability of P, and extractable cations and pH; it also decreased soil bulk density and penetrometer resistance in the fallow plots (Tian et al., 2000). From these and other examples we can generally conclude that earthworms can ameliorate disturbed soils by improving aggregation, reducing bulk density and increasing nutrient availability (Ponder et al., 2000). Nevertheless, different plantation species differ in their influence on earthworms (Warren and Zou, 2002), and the biomass and abundance of earthworms can be regulated through careful selection of plantation species in degraded land.

30.2.2 Earthworms in mine waste sites

Most studies using earthworms to reclaim mine wastes are related to coal mine wastes. *Lumbricus terrestris* has been successfully introduced into reclaimed coal mine spoil sites, ranging in pH from 3.5 to over 7 in the USA (Vimmerstedt and Finney, 1973). It has been noted that once earthworm populations were established in coal mine spoils under *Alnus glutinosa*, plant litter incorporation was greatly accelerated (Dunger, 1991). Earthworm activity increased stable aggregate and resulted in a higher proportion of organic matter, when introduced into physically degraded soils restored after opencast coal mining (Scullion and Malik, 2000). In opencast mine sites, grass-cutting systems of management inhibited the recovery of earthworm populations but sheep-grazing increased populations of *A. caliginosa*, *Allolobophora chlorotica* and *L. rubellus* (Stewart and Scullion, 1988) possibly due to the continuous supply of manure. The application of various organic waste materials significantly improved population density, fresh biomass, food consumption and the burrowing activity of earthworms in mine soils from opencast coal mining (Emmerling and Paulsch, 2001). By increasing food supply and stabilizing temperature and moisture conditions in the surface soil, the organic amendment of newly reclaimed sites of a 14-year-old colliery spoil heap was found to favour earthworm establishment (Luff and Hutson, 1977).

Research into earthworm succession in afforested colliery spoil heap in the Sokolov region (Czech Republic) indicated that the density and biomass of worms increased from youngest alder stand to the older ones; earthworm populations were often higher in reclaimed sites than in the control alder stand (Piael, 2001). Normally, 10–20 years, or even longer, is required for earthworm establishment in opencast coal mine sites (Armstrong and Bragg, 1984). Regular sub-soiling could significantly increase earthworm population densities, the increase in numbers of *A. caliginosa* being greatest (Scullion and Malik, 2000).

30.3 The role of earthworms in reclaiming heavy metal contaminated soils

Soils are a major sink for heavy metals released into the environment both naturally and anthropogenically. Heavy metal contamination is among the most widespread of land contamination, and is the most difficult and expensive to remove. The use of vegetation is regarded as one of the most realistic and effective methods for stabilizing and reclaiming metalliferous lands (Bradshaw and Johnson, 1992). The chemistry, mobility, availability and toxicity of heavy metals in soils are functions of numerous factors including pH, organic matter, nutrient status and microbial activity within the soil environment (Luo and Christie, 1998). Acidity, metal toxicity, unfavourable moisture conditions and the lack of suitable food are among the factors likely to inhibit earthworm establishment in recently reclaimed mining soils (Edwards and Bohlen, 1996). Besides having fundamental effects on soil structure, soil fertility and plant growth, stimulating numerous biological processes and interactions; earthworms also influence metal mobility in soil through their feeding and burrowing activities. The mixing and comminution of soils, humic materials and detritus in the earthworm gut may also increase the immobilization of heavy metals.

30.3.1 Metal tolerance in earthworms

In general, earthworms appear to be more sensitive to heavy metals than other soil invertebrates (Bengtsson et al., 1992) and they accumulate much higher levels of metals from polluted soils than most other soil animals (Beyer et al., 1982). Different earthworm species may respond differently to heavy metals, and juvenile earthworms are more sensitive than adults (Spurgeon and Hopkin, 1999). The toxicity of heavy metals in soil to earthworms depends heavily on soil type and metal availability as well as earthworm tolerance (Ma, 1982; Marinussen, et al., 1997). Table 30.1 shows a comparison of metal toxicity on selected earthworm species based on LC_{50} values. Some earthworm populations were found to tolerate heavy metal concentrations well above the concentration known to induce lethal effects in non-tolerant populations (Lagisz et al., 2002).

It has been observed that earthworms can survive metal-contaminated soils and accumulate heavy metals in their tissues (Helmke et al., 1979; Hartenstein et al., 1980). Several species heave been reported to be tolerant to heavy metals, including Pb (*L. rubellus* from disused Pb mines; Morgan and Morgan, 1988), As and Cu (*L. rubellus* and *Dendrodrilus rubidus* from mine spoils; Langdon et al.,

Table 30.1. *Comparison of metal toxicity (LC$_{50}$ values) of earthworms from the present study and other published literature.*

Earthworm species	Tested substrates[a]	LC$_{50}$ (mg kg^{-1})			References
		Pb	Zn	Cu	
Pheretima sp.	Pb/Zn mine tailings	1382 (14d), 1216 (28d)	1876 (14d), 1651 (28d)	146 (14d), 129 (28d)	Ma et al. (2002) unpublished data
P. guillelmi	Pb/Zn mine tailings	1567 (14d), 1297 (28d)	2127 (14d), 1761 (28d)	166 (14d), 137 (28d)	Ma et al. (2002) unpublished data
P. guillelmi	1, 2, 3	3705 (14d)	767 (14d)	311 (7d)	Ma et al. (2002) unpublished data
E. fetida	2	–	1131 (14d)	–	Spurgeon et al. (1997)
E. fetida	1, 2, 3	4480 (14d), 3760 (56d)	1010 (14d), 745 (56d)	683 (14d), 555 (56d)	Spurgeon et al. (1994)
E. fetida	1, 3	5941 (14d)	–	643 (14d)	Neuhauser et al. (1984)
E. fetida	4	–	–	1105 (14d)	Edwards and Bater (1992)
E. fetida	Brass powder	190 mg/kg for a mixture of 20% Cu and 30% Zn (14)	Wentsel and Gueltar (1987)		

[a]Tested substrates: 1, Pb(NO$_3$)$_2$; 2, Zn(NO$_3$)$_2$ · 6H$_2$O; 3, Cu(NO$_3$)$_2$ · 3H$_2$O; 4, Cu(SO$_4$).

1999, 2001) and Pb and Zn (*Pheretima guillelmi* from Pb/Zn mine tailings; Ma et al., 2002).

Eisenia fetida has developed resistance to Cd following a 2-week period of exposure (Reinecke et al., 1999). This species also has the ability to recover within weeks from periods of sub-lethal but toxic exposure to metals (Neuhauser et al., 1984). It appears that Zn and Fe can be regulated by earthworms (Ireland, 1975; Ash and Lee, 1980), and Ca secretion may play a role in detoxification of Pb and Zn (Ireland, 1975; Morgan and Morgan, 1998). Arsenic speciation has also been taken into account when studying the resistance of *L. rubellus* to arsenate. This is due to the fact that the earthworm altered As speciation after accumulating As in their tissues, which was assumed to have a role in the resistance of *L. rubellus* to arsenate (Langdon et al., 2002).

A genetically based tolerance mechanism to Pb in *L. rubellus* naturally exposed to Pb has been suggested (Morgan and Morgan, 1988). Spurgeon and Hopkin (2000) investigated the development of genetically inherited resistance to heavy metals in laboratory-selected generations of the earthworm *E. fetida*, and found that physiological responses, other than differences in kinetic parameters, are responsible for the increased resistance to Zn and Cu in the selected earthworms. The earthworm's response to metal contamination has been linked to the induction and expression of metallothionein (MT) proteins, a detoxification strategy analogous to that found in other biological systems (Gruber et al., 2000), while the earthworm Enchytraeus can express a gene encoding the non-MT CRP protein to detoxify Cd in Cd-polluted environments (Tschuschke et al., 2002).

On the basis of the above information, it seems that the capacity of earthworm species to adapt both physiologically and genetically to contaminants may allow populations to persist in polluted environments. Earthworms may rapidly invade reclaimed soils, they can survive quite severe contamination, including that found in some mining spoils, and there may even be selection of metal tolerance in some species (Spurgeon and Hopkin, 1999; Langdon et al., 2001). Metal-resistant earthworms that will readily enter contaminated soils may play a valuable role in remediation of mine spoils and industrial contaminated sites, as do metal-tolerant plants (Smith and Bradshaw, 1979).

30.3.2 Effects of earthworms on metal speciation and bioavailability

It is commonly recognized that soil pH, salinity, redox potential, organic matter, the solid and solution components and their relative concentrations and affinities for an element, and time will all contribute to soil metal speciation and availability (Ure and Davidson, 1995). In general, total metal concentrations in soils are less meaningful than the fractions that are available for plant and animal uptake. The specific bioavailability, mobility and reactivity of metals in soil are directly related to the real risk that metal cations impose on the global ecosystem (McBride, 1995). The chemical forms of metals determine their fate in the soil (Ure, 1996). The physical nutrient enrichment as well as the gut-related processes of earthworms imply a potential increase of plant nutrient and metal availability. Earthworms are known to have a significant effect on metal distribution in soil and they are also likely to alter the chemical form of metals that are egested. Earthworm activity modifies soil pH, which is a key chemical factor affecting the bioavailability of nutrient elements and heavy metals in the soil (Edwards and Bohlen, 1996).

Several studies revealed that earthworms increase metal availability, e.g. an increase in Cr and Co availability from 20 to 39%, and in Cu by 6% (by *L. terrestris* using a soil core experiment; Devliegher and Verstraete, 1996, 1997); an increase of 50% in availability of water-soluble Pb (in the faeces of *Dendrobaena rubida* from a heavy metal contaminated soil; Ireland, 1975); and an increase of 48 and 25% in diethylenetriamene pentaacetate (DTPA)-extractable Pb and Zn (*Pheretima* in Pb/Zn mine tailings amended with mineral soil; Ma et al., 2002).

Our earlier study indicated that *P. guillelmi* activity significantly increased DTPA-Zn in red soil, and organic-Zn in paddy soil, but had little effect on neither DTPA-Zn nor organic-Zn in alluvial soil (Cheng and Wong, 2002). The increase in DTPA-Zn by earthworms was consistent with a decrease in pH in red soil (Fig. 30.1).

On the basis of a modified three-step sequential extraction, it was found that *P. guillelmi* significantly increased HOAc-Zn and oxidisable-Zn, but decreased reducible-Pb in soil (Ure et al., 1993). One of our recent experiments investigated Pb and Zn speciation and bioavailability in a plant-metal spiked soil system using both single and sequential chemical extraction procedures. *P. guillelmi* significantly increased DTPA- and NH_4NO_3-Zn, but had little effect on either DTPA-, NH_4OAc-, or NH_4NO_3-Pb in soil. The increase in Zn bioavailability in this study can be attributed to the decreased pH and increased dissolved organic matter in the soil, which was affected by earthworm activity. Pyromorphites that formed with phosphates reduced the solubility and bioavailability of Pb within the soil (Pearson et al., 2000).

When comparing earthworm casts with surrounding soils, a significant increase of DTPA-Pb and -Zn in earthworm casts was revealed (Table 30.2), indicating the importance of the gut-related processes on the increase of metal bioavailability in metal contaminated soil to which earthworms were added (Ma et al., 2003). These included (1) the consequence of absorption of organic materials during gut passage; (2) the posterior alimentary canal may excrete

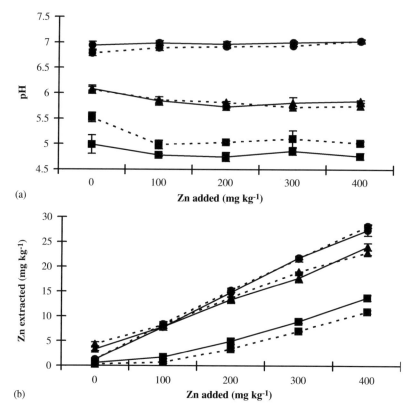

Figure 30.1. *Effect of* P. guillelmi *and Zn on (a) pH and (b) DTPA-extractable Zn in soils. Solid line, with worms; dashed line, without worms.* ■, *red soil;* ●, *paddy soil;* ▲, *alluvial soil (Cheng and Wong, 2002).*

Table 30.2. *Comparison of DTPA-extractable Pb and Zn in soils and the casts of* P. guillelmi *after 3 months' incubation in soils mixed with Pb/Zn mine tailings (Ma et al., 2002). Parenthesis indicates the percentage difference in casts relative to soil. T25: 25% tailings+75% soil; T50: 50% tailings+50% soil.*

		DTPA-Pb (mg kg^{-1})		DTPA-Zn (mg kg^{-1})	
		Soil	Cast	Soil	Cast
6-month	T25	135	208 (+54.1%)	69.8	114 (+63.3%)
Leucaena planted	T50	267	378 (+41.6%)	98.1	178 (+81.4%)
10-month	T25	142	297 (+109%)	70.8	129 (+82.2%)
Leucaena planted	T50	279	372 (+33.3%)	106	193 (+82.1%)

metals, such as Pb and Zn; and (3) metals may preferentially accumulate through metal-tolerant strains of gut microflora resident in the earthworm gut (Ireland, 1976; Morgan and Morgan, 1992).

30.3.3 Earthworms and soil microorganisms

Microbial respiration, biomass and fungal volume in fresh earthworm casts strongly depend on the type of soil and litter material consumed by earthworms (Tiunov and Scheu, 2000). Earthworms and arbuscular mycorrhizal fungi (AMF) are ubiquitous and abundant co-inhabitants of the soil environment (Ingham, 1992). AMF form an important mutualistic association with plant roots to benefit plant growth. Earthworms enhance the dispersal of microorganisms by ingesting them at one location from a particular food source and egesting them elsewhere, or by transporting microorganisms that adhere to their body surface (Edwards and Bohlen, 1996). Rabatin and Stinner (1989) reported that most earthworms found in no-till corn and pasture contained AMF in their gut. Moreover, the influence of earthworms on infective AMF propagules was even greater, and the earthworm casts contained up to 10 times as many infective AMF propagules as did the surrounding soil (Gange, 1993). In greenhouse experiments, earthworms have been shown to enhance the spread of AMF on the roots of soybean (McIlveen and Cole, 1976) and the seedlings of tropical fruit trees (Ydrogo, 1994). Although grazing by earthworms on the soil mycelium of AMF may also limit its development or disconnect it from the root mycelium, this negative effect may be more than offset by the beneficial effects of stimulation of hyphae and spore production to plant growth (Fitter and Sanders, 1992). The removal of senescing soil mycelia by earthworms may also result in the elimination of growth-inhibitory secondary metabolites (Moore, 1988).

Dispersal of nitrogen-fixing bacteria can also be enhanced by earthworm activity. Doube et al. (1994) demonstrated the ability of earthworms to disperse *Rhizobia* through soil, to increase levels of colonization of legume roots by *Rhizobia*, and to increase levels of root nodulation. Nodulation and arbuscular mycorrhiza formation appear as physiological and biochemical processes involved in plant–AMF–*Rhizobium* interactions (Kucey and Paul, 1982). It is well established that AMF play a positive role in plant nodulation. *Rhizobia* and arbuscular mycorrhiza often interact synergistically and result in better root nodulation, nutrient uptake and plant yield (Krishna and Bagyaraj, 1984). Nodulated legumes with arbuscular mycorrhizae are well-adapted to cope with nutrient-deficient situations (Herrera et al., 1993). Management of appropriate microsymbionts can help legumes to promote the stabilization of a self-sustaining ecosystem. The identification, screening and cultivation of metal-tolerant legume plants, accompanied by inoculation of tolerant earthworm species and suitable AMF, as well as their associated N-fixing bacteria, may present a new opportunity in ecological reclamation.

30.4 Trials of earthworm inoculation for the reclamation of Pb/Zn mine tailings from Lechang (China)

30.4.1 The role of earthworms

A fast-growing leguminous shrub *Leucaena leucocephala* was used to study the feasibility of reclaiming Pb/Zn mine tailings in a greenhouse trial. *P. guillelmi* significantly stimulated the growth of *Leucaena* in the contaminated soils (Table 30.3). Earthworms increased N, P uptake by plants remarkably, especially at 50% tailings treatment, and significantly enhanced total Pb and Zn uptake by *Leucaena*. There were strong positive correlations between the soil DTPA-extractable Pb, Zn concentrations and Pb, Zn uptake by plants. Pb was less mobile in plant tissues and almost exclusively retained in the roots. This study showed that earthworms increased the rate of metal uptake into plants from spoils by at least 16% and as much as 53%; most metals were taken up and stabilized in the root of *Leucaena*. Metal-tolerant plants that exclude metals or at least restrict the translocation of metals to shoot would be essential for stabilization of mine spoils, and the context of ecological reclamation should be broadened, by considering plant–soil–animal interactions.

30.4.2 Earthworms, mycorrhizal fungi and nitrogen-fixation

Our recent study involving dual inoculation of earthworms (*P. guillelmi*) and AMF indicated that earthworm activity plays a positive role in mycorrhizal root colonization in *Leucaena* (Fig. 30.2, from Ma et al., 2002). There was a synergistic effect between earthworms and AMF inoculation on plant growth especially under higher proportions of tailings treatment. Total nutrient and metal uptake by the plant showed that earthworms significantly enhanced N, P uptake by

Table 30.3. *Relative growth of* L. leucocephala *in the soils mixed with Pb/Zn mine tailings, compared to growth in clean soil without earthworms (expressed as a percentage), after 3 months in the presence and absence of earthworms (Ma et al., 2002). T25: 25% tailings+75% soil; T50: 50% tailings+50% soil; T75: 75% tailings+25% soil.*

	T25		T50		T75
	With worms	No worms	With worms	No worms	No worms
6-month plants	114 ± 15.7	96.3 ± 13.3	110.6 ± 10.9	78.1 ± 11.5	56.0 ± 8.53
10-month plants	85.3 ± 10.7	75.7 ± 7.73	82.2 ± 4.03	55.8 ± 3.55	69.0 ± 5.19

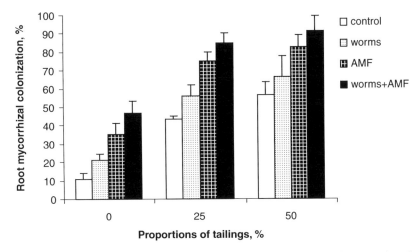

Figure 30.2. *Root mycorrhizal colonization of* L. leucocephala *inoculated with earthworms (worms), arbuscular mycorrhizal fungi (AMF), both earthworms and arbuscular mycorrhizal fungi (worms+AMF) and non-inoculated (control), as influenced by the amount of Pb/Zn mine tailing added (from Ma et al., unpublished data).*

Leucaena but had no effect on K uptake, while AMF inoculation significantly increased N, P, K uptake. In terms of metal uptake by the plant, earthworms tended to increase metal uptake, while arbuscular mycorrhizae decreased metal uptake especially under conditions of higher metal contamination. The results also showed that earthworms considerably improved root mycorrhizal infection and somewhat stimulated plant root nodulation, while arbuscular mycorrhizae exerted positive effects on plant nodulation and nodule efficiency. Another study noted that addition of the same species of earthworms to Zn contaminated soils (0, 100, 200, 300 and 400 mg kg^{-1}) resulted in significant increases in N mineralization of three soils (red soil, alluvial soil and paddy soil) and the amount of N mineralized among different Zn treatments did not differ significantly (Table 30.4, from Cheng and Wong, 2002). This indicated that earthworm activity activated the bacteria responsible for N mineralization.

30.5 Conclusion

The rapid industrial development of modern society has drastically altered the soil environment and negatively affected earthworm populations. However, earthworms may be used to reclaim lands made derelict through human

Table 30.4. *Effect of earthworm* P. guillelmi *on N mineralization (mg N/kg) after 40 days' incubation (mean ± SD) (Cheng and Wong, 2002). The values between treatments (with and without worms) of the same soil type followed by the same letter are not significantly different at* $p < 0.05$.

Treatment		Zn added (mg/kg)				
		0	100	200	300	400
Red soil	− worms	24.6b	26.4b	29.3b	26.4b	28.4b
	+ worms	87.0a	76.5a	83.4a	74.2a	67.4a
Paddy soil	− worms	80.9a	71.6a	73.1b	73.3b	76.5b
	+ worms	91.6a	88.3a	91.7a	83.8a	87.1a
Alluvial soil	− worms	88.3b	98.5b	95.2b	86.6b	93.8b
	+ worms	139a	160a	146a	148a	131a

activity. Besides serving as valuable indicator in soil ecotoxicological evaluation and risk assessment, earthworms also play an important role in terrestrial ecosystems by improving soil physical structure, soil fertility and plant growth.

For successful reclamation of metal contaminated soils, it is necessary to improve unfavourable physical and chemical conditions of the substrates to allow plant establishment. In addition, it is also important to address the issues of (i) sustainability of vegetation and (ii) mobility of toxic metals. Inoculation of earthworms into metal contaminated soils can accelerate plant growth resulting in higher yields, increase nutrient and metal bioavailability and enhance metal uptake by plants. Addition of earthworms to reclaimed soils is also essential for improving soil biodiversity and ecosystem sustainability. The development of healthy, sustainable soil and vegetation in the longer term needs a thorough understanding of plant–soil–animal interactions which affect metal mobility. Characterization of metal compartmentation and distribution in earthworm tissues, as well as the application of gene technology may provide valuable insights into the mechanisms of metal tolerance, detoxification, and accumulation by earthworms.

Soil pH appears to be a key factor affecting metal mobility in soils and evaluation of the effects of earthworms on metal mobility and bioavailability at a wider range of pH conditions should be explored to broaden the use of earthworms for reclaiming different metal contaminated soils.

Acknowledgements

This study is supported by a research grant (No. HKBU 2049/00M) of the Research Grants Council of the University Grants Committee, Hong Kong.

References

Armstrong, M.J., Bragg, N.C., 1984. Soil physical parameters and earthworm populations associated with opencast coal working and land restoration. Agric. Ecosyst. Environ. 11, 131–143.

Ash, C.P.J., Lee, D.K., 1980. Lead, cadmium, copper, and iron in earthworms from roadside sites. Environ. Pollut. (Series A) 22, 59–67.

Bengtsson, G., Ek, H., Rundgren, S., 1992. Evolutionary response of earthworms to long-term metal exposure. Oikos 63, 289–297.

Beyer, W.N., Patte, O.H., Sileo, L., Hoffman, D.J., Mulhern, B.M., 1982. Metal contamination in wildlife living near two zinc smelters. Environ. Pollut. A38, 63–86.

Bradshaw, A.D., Johnson, M., 1992. Revegetation of Metalliferous Mine Waste: The Range of Practical Techniques Used in Western Europe. Elsevier, Manchester.

Bustos-Obregón, E., Goicochea, R., 2002. Pesticide soil contamination mainly affects earthworm male reproductive parameters. Asian J. Androl. 4, 195–199.

Butt, K.R., Frederickson, J., Morris, R.M., 1995. An earthworm cultivation and soil inoculation technique for land restoration. Ecol. Eng. 4, 1–9.

Callaham, M.A. Jr., Stewart, A.J., Alarcon, C., McMillen, S.J., 2002. Effects of earthworm (*Eisenia fetida*) and wheat (*Triticum aestivum*) straw additions on selected properties of petroleum-contaminated soils. Environ. Toxicol. Chem. 21, 1658–1663.

Cheng, J.M., Wong, M.H., 2002. Effects of earthworms on Zn fractionation in soils. Biol. Fertil. Soils 36, 79–86.

Curry, J.P., Cotton, D.C.F., 1983. Earthworms and land reclamation. In: J.E. Satchell (Ed.), Earthworms and Land Reclamation: From Darwin to Vermiculture. Chapman and Hall, London, pp. 215–228.

Devliegher, W., Verstraete, W., 1996. *Lumbricus terrestris* in a soil core experiment: effects of nutrient-enrichment processes (NEP) and gut-associated processes (GAP) on the availability of plant nutrients and heavy metals. Soil Biol. Biochem. 28, 489–496.

Dickinson, N.M., 2000. Strategies for sustainable woodlands on contaminated soils. Chemosphere 41, 259–263.

Doube, B.M., Ryder, M.H., Davoren, C.W., Stephens, P.M., 1994. Enhanced root nodulation of subterranean clover *Trifolium subterraneum* by *Rhizobium trifolii* in the presence of the earthworm *Aporrectodea trapezoids*. Biol. Fertil. Soils 6, 237–251.

Dunger, W., 1991. Primary succession of humiphagous soil animals on coal mined areas. Zool. Jb. Abt. Fuer Syst. Oekol. Geog. Tiere. 118, 423–447.

Edwards, C.A., 1984. Report on the second stage in development of a standardized laboratory method for assessing the toxicity of chemical substances to earthworms. Commission of the European Communities, Brussels, Luxembourg, EUR 9360 EN.

Edwards, C.A., Bater, J.E., 1992. The use of earthworms in environmental management. Soil Biol. Biochem. 24, 1683–1689.

Edwards, C.A., Bohlen, P.J., 1996. Biology and Ecology of Earthworms. Chapman and Hall, London.

Emmerling, C., Paulsch, D., 2001. Improvement of earthworm (Lumbricidae) community and activity in mine soils from open-cast coal mining by the application of different organic waste materials. Pedobiologia 45, 396–407.

Fitter, A.H., Sanders, I.R., 1992. Interactions with the soil fauna. In: M.F. Allen (Ed.), Mycorrhizal Functioning. Chapman & Hall, New York, pp. 333–354.

Gange, A., 1993. Translocation of mycorrhizal fungi by earthworms during early succession. Soil Biol. Biochem. 25, 1021–1026.

Gruber, C., Stuerzenbaum, S., Gehrig, P., Sack, R., Hunziker, P., Berger, B., Dallinger, R., 2000. Isolation and characterization of a self-sufficient one-domain protein (Cd)-metallothionein from *Eisenia foetida*. Eur. J. Biochem. 267, 573–582.

Hartenstein, R., Niuhauser, E.F., Collier, J., 1980. Accumulation of heavy metals in the earthworm *Eisenia fetida*. J. Environ. Qual. 8, 322–327.

Helmke, P.A., Robarge, W.P., Korotev, R.L., Schomberg, P.J., 1979. Effects of soil-applied sewage sludge on concentrations of elements in earthworms. J. Environ. Qual. 8, 322–327.

Herrera, M.A., Salamanca, C.P., Barea, J.M., 1993. Inoculation of woody legumes with selected arbuscular mycorrhizal fungi and rhizobia to recover desertified Mediterranean ecosystems. Appl. Environ. Microbiol. 59, 129–133.

Huhta, V., 1979. Effects of liming and deciduous litter on earthworm (Lumbricidae) populations of a spruce forest, with an inoculata experiment on *Allolobophora caliginosa*. Pedobiologia 19, 340–345.

Ingham, R.E., 1992. Interactions between invertebrates and fungi: effects on nutrient availability. In: G.C. Carrol, D.T. Wicklow (Eds.), The Fungal Community. Marcel Dekker, New York, pp. 669–690.

Ireland, M.P., 1975. The effect of earthworm *Dendrobaena rubiada* on the solubility of lead, zinc, and calcium in heavy metal contaminated soil in Wales. J. Soil Sci. 26, 313–318.

Ireland, M.P., 1976. Excretion of lead, zinc and calcium by the earthworm *Dendrobaena rubida* living in soil contaminated with zinc and lead. Soil Biol. Biochem. 8, 347–350.

Krishna, K.R., Bagyaraj, D.J., 1984. Growth and nutrient uptake of peanut inoculated with the mycorrhizal fungus *Glomus fasciculatum* compared with non-inoculated ones. Plant Soil 77, 405–408.

Kucey, R.M.N., Paul, E.A., 1982. Carbon flow photosynthesis, and N_2 fixation in mycorrhizal nodulated faba beans (*Vicia faba* L). Soil Biol. Biochem. 14, 407–412.

Lagisz, M., Kramarz, P., Laskowski, R., Tobor, M., 2002. Population parameters of the beetle *Pterostichus oblongopunctatus* F. from metal contaminated and reference areas. Bull. Environ. Contam. Toxicol. 69, 243–249.

Langdon, C.J., Meharg, A.A., Feldmann, J., Balgar, T., Charnock, J., Farquhar, M., Piearce, T.G., Semple, K.T., Cotter-Howells, J., 2002. Arsenic-speciation in arsenate-resistant populations of the earthworm, *Lumbricus rubellus*. J. Environ. Monit. 4, 603–608.

Langdon, C.J., Piearce, T.G., Black, S., Semple, K.T., 1999. Resistance to arsenic-toxicity in a population of the earthworm *Lumbricus rubellus*. Soil Biol. Biochem. 31, 1963–1967.

Langdon, C.J., Piearce, T.G., Meharg, A.A., Semple, K.T., 2001. Resistance to copper toxicity in populations of the earthworms *Lumbricus rubellus* and *Dendrodrilus rubidus* from contaminated mine wastes. Environ. Toxicol. Chem. 20, 2336–2341.

Luff, M.L., Hutson, B.R., 1977. Soil faunal populations. In: B. Hackett (Ed.), Landscape Reclamation Practice. IPC science and Technology Press, Guildford, UK, pp. 125–147.

Luo, Y.M., Christie, P., 1998. Bioavailability of copper and zinc in soils treated with alkaline stabilized sewage sludges. J. Environ. Qual. 27, 335–342.

Ma, W., 1982. The influence of soil properties and worm related factors on the concentrations of heavy metals in earthworms. Pedobiologia 24, 109–119.

Ma, Y., Dickinson, N.M., Wong, M.H., 2002. Toxicity of Pb/Zn mine tailings to the earthworms *Pheretima guillelmi* and the effects of burrowing on metal availability. Biol. Fertil. Soils 36, 79–86.

Ma, Y., Dickinson, N.M., Wong, M.H., 2003. Remediation of Pb/Zn mine tailings from Guandong, China: earthworms (*Pheretima guillelmi*), trees (*Leucaena leucocephala*), soil nutrition and metal mobility. Soil Biol. Biochem. 35, 1369–1379.

Makulec, G., 2002. The role of *Lumbricus rubellus* Hoff in determining biotic and abiotic properties of peat soils. Pol. J. Ecol. 50, 301–339.

Marinussen, M.P.J.C., Van der Zee, S.E.A.T.M., de Haan, F.A.M., Bouwman, L.M., Hefting, M.M., 1997. Heavy metal (copper, lead, and zinc) accumulation and excretion by the earthworm, Dendrobaena veneta. J. Environ. Qual. 26, 278–284.

McBride, M.B., 1995. Toxic metal accumulation from agricultural use of sludge: are USEPA regulations protective? J. Environ. Qual. 24, 5–18.

McIlveen, W.D., Cole, H. Jr., 1976. Spore dispersal of Endogonadaceae by worms, ants, wasps, and birds. Can. J. Bot. 54, 1486–1489.

Moore, J.C., 1988. The influence of microarthropods on symbiotic and non-symbiotic mutualisms in detrital-based below-ground food webs. Agric. Ecosyst. Environ. 24, 147–159.

Morgan, J.E., Morgan, A.J., 1988. Cadmium-lead interactions involving earthworms. Part2: the effect of accumulated lead on endogenous calcium in *Lumbricus rubellus*. Environ. Pollut. 55, 41–54.

Morgan, J.E., Morgan, A.J., 1992. Heavy metal concentrations in the tissue ingesta and faeces of ecophysiologically different earthworm species. Soil Biol. Biochem. 12, 1691–1697.

Morgan, J.E., Morgan, A.J., 1998. The distribution and intracellular compartmentation of metals in the endogeic earthworm *Aporrectodea caliginosa* sampled from an unpolluted and a metal-contaminated site. Environ. Pollut. 99, 167–175.

Neuhauser, E.G., Loehr, R.C., Milligan, D.J., Malecki, M.R., 1984. Toxicity of metals to the earthworm *Eisenia foetida*. Biol. Fertil. Soils 1, 149–152.

Noble, J.C., Gordon, W.T., Kleinig, C.R., 1970. The influence of earthworms on the development of mats of organic matter under irrigated pasture in Southern Australia. Proceedings of the 11th International Grassland Conference, Brisbane, pp. 465–468.

Nriaga, J.O., 1984. Changing Metal Cycles and Human Health (Dahlem Konferenaen). Springer-Verlag, Berlin, 445pp.

OECD (Organization for Economic Cooperation and Development), 1984. Guideline for Testing Chemicals No. 207. Earthworm Acute Toxicity Tests, Paris, France.

Pearson, M.S., Maenpaa, K., Pierzynski, G.M., Lydy, M.J., 2000. Effects of soil amendments on the bioavailability of lead, zinc and cadmium to earthworms. J. Environ. Qual. 29, 1611–1617.

Piael, V., 2001. Earthworm succession in afforested Colliery spoil heaps in the Sokolov Region, Czech Republic. Restor. Ecol. 9, 359–364.

Ponder, F. Jr., Li, F., Jordan, D., Berry, E.C., 2000. Assessing the impact of *Diplocardia ornate* on physical and chemical properties of compacted forest soil in microcosms. Biol. Fertil. Soils 32, 166–172.

Rabatin, S.A., Stinner, B.R., 1989. The significance of vesicular-arbuscular mycorrhizal faunal-soil-macroinvertebrate interactions in agroecosystems. Agric. Ecosyst. Environ. 27, 195–204.

Reinecke, A.J., Reinecke, S.A., 1998. The use of earthworms in ecotoxicological evaluation and risk assessment: new approaches. In: C.A. Edwards (Ed.), Earthworm Ecology. St. Lucie Press, London, pp. 273–295.

Reinecke, S.A., Prinsloo, M.W., Reinecke, A.J., 1999. Resistance of *Eisenia fetida* (Oligochaeta) to cadmium after long-term exposure. Ecotoxicol. Environ. Saf. 42, 75–80.

Satchell, J.E., 1983. Earthworm Ecology, From Darwin to Vermiculture. Chapman and Hall Ltd, London.

Scullion, J., Malik, A., 2000. Earthworm activity affecting organic matter, aggregation and microbial activity in soils restored after opencast mining for coal. Soil Biol. Biochem. 32, 119–126.

Singer, A.C., Jury, W., Luepromchai, E., Yahng, C.-S., Crowley, D.E., 2001. Contribution of earthworms to PCB bioremediation. Soil Biol. Biochem. 33, 765–776.

Smith, R.A.H., Bradshaw, A.D., 1979. The use of metal tolerant plant populations for the reclamation of metalliferous wastes. J. Appl. Ecol. 16, 595–612.

Springett, J., Gray, R., 1997. The interaction between plant roots and earthworm burrows in pasture. Soil Biol. Biochem. 24, 1739–1744.

Spurgeon, D.J., Hopkin, S.P., 1999. Seasonal variation in the abundance, biomass and biodiversity of earthworms in soils contaminated with metal emissions from a primary smelting works. J. Appl. Ecol. 36, 173–183.

Spurgeon, D.J., Hopkin, S.P., 2000. The development of genetically inherited resistance to zinc in laboratory-selected generations of the earthworm *Eisenia fetida*. Environ. Pollut. 109, 193–201.

Spurgeon, D.J., Hopkin, S.P., Jones, D.T., 1994. Effects of cadmium, copper, lead and zinc on growth, reproduction and survival of the earthworm *Eisenia fetida*: assessing the environmental impact of point-source metal contamination in terrestrial ecosystems. Environ. Pollut. 84, 123–130.

Spurgeon, D.J., Tomlin, M.A., Hopkin, S.P., 1997. Influence of temperature on the toxicity of zinc to the earthworm *Eisenia fetida*. Bull. Environ. Contam. Toxicol. 58, 283–290.

Stewart, V.I., Scullion, J., 1988. Earthworms, soil structure and the rehabilitaion of former open-cast coal-mining land. In: C.A. Edwards, E.F. Neuhauser (Eds.), Earthworms in Waste and Environmental Management. SPB Academic Press, The Hague, The Netherlands, pp. 263–272.

Tian, G., Olimah, J.A., Adeoye, O., Kang, B.T., 2000. Regeneration of earthworm populations in a degraded soil by natural and planted fallows under humid tropical conditions. Soil Sci. Soc. Am. J. 64, 222–228.

Tiunov, A.V., Scheu, S., 2000. Microbial biomass, biovolume and respiration in *Lumbricus terrestris* L. cast material of different age. Soil Biol. Biochem. 32, 265–275.

Tschuschke, S., Schmitt-Wrede, H., Greven, H., Wunderlich, F., 2002. Cadmium resistance conferred to yeast by a non-metallothionein-encoding gene of the earthworm Enchytraeus. J. Biol. Chem. 277, 5120–5125.

Ure, A.M., 1996. Single extraction schemes for soil analysis and related applications. Sci. Total Environ. 178, 3–10.

Ure, A.M., Davidson, C.M., 1995. Chemical Speciation in the Environment. Blackie Academic & Professional, London.

Ure, A.M., Quevauviller, Ph., Muntau, H., Griepink, B., 1993. Speciation of heavy metals in soils and sediments. An account of the improvement and harmonization of extraction techniques undertaken under the auspices of the BCR of the CEC. Int. J. Environ. Anal. Chem. 51, 135–151.

USEPA, 1988. Analytical methods for the national sewage sludge survey. USEPA, Washington, DC.

Vangronsveld, J., Cunningham, S.D., 1998. Metal-contaminated Soils: in situ Inactivation and Phytorestoration. RG Landes, Georgetown, TX.

Venabes, B.J., Fitzpatrick, L.C., Goven, A.J., 1992. Earthworms as indicators of toxicity. In: P.W. Greig-Smith, H. Becjer, P.J. Edwards and F. Heimback (Eds.), Ecotoxicology of Earthworms. Intercept Ltd, Hants, UK.

Vimmerstedt, J.P., Finney, J.H., 1973. Impact of earthworm introduction on litter burial and nutrient distribution in Ohio strip mine spoil banks. Soil Sci. Soc. Am. Proc. 37, 388–391.

Warren, M.W., Zou, X., 2002. Soil macrofauna and litter nutrients in three tropical tree plantations on a disturbed site in Puerto Rico. For. Ecol. Manage. 170, 161–171.

Wentsel, R.S., Gueltar, M.A., 1987. Toxicity of brass powder in soil to the earthworm *Lumbricus terrestris*. Environ. Toxicol. Chem. 6, 741–745.

Wong, M.H., 1987. Reclamation procedures of wastes contaminated by Cu, Pb, and Zn. Environ. Manage. 10, 707–713.

Wong, M.H., 2002. Ecological restoration of mine degraded soils. With emphasis on metal contaminated soils. Chemosphere 50, 775–780.

Ydrogo, H.F.B., 1994. Effecto de las lombrices de tierra (*Pontoscolex corethrurus*) en las micorrizas vesiculo arbusculares (M.V.A.) en la etapa de crecimiento de araza (*Eugenia stipitata*), achiote (*Bixa orellana*), pijuayo (*Bactris gasipaes*), en suelos ultisoles de Yurimaguas. Thesis, University Nac. San Martin, Peru.

AUTHOR INDEX

SUBJECT INDEX